Spring

March 1, 12:30 a.m. Local Standard Time
April 1, 10:30 p.m. Local Standard Time
May   1,   9:30 p.m. Local Daylight Time

NORTH

W9-ATB-989

δ Cephei

CASSIOPEIA   Double Cluster

CEPHEUS

Algol

PERSEUS

Vega

Eltanin

Polaris

CAMELOPARDALIS

DRACO   URSA MINOR

Capella

Kochab

HERCULES   M 13

AURIGA

Aldebaran

CORONA BOREALIS   Alcor   Dubhe

TAURUS

Mizar   Merak

M 35

BOOTES   URSA MAJOR

Castor

CANES VENATICI   LYNX

SERPENS CAPUT   GEMINI

Pollux

Arcturus   LEO MINOR

Betelgeuse   ORION

COMA BERENICES   M 44   CANIS MINOR

Ecliptic

M 5   CANCER   M 42

Denebola   LEO

Regulus   Procyon

MONOCEROS

VIRGO   Sirius

LIBRA   Equator

SEXTANS   CANIS MAJOR

Spica   Alphard

CRATER   HYDRA

CORVUS

PUPPIS

WEST

SOUTH
SPRING

Contemporary Astronomy

second edition

Jay M. Pasachoff

Williams College-Hopkins Observatory
Williamstown, Massachusetts

SAUNDERS GOLDEN SUNBURST SERIES

SAUNDERS COLLEGE PUBLISHING
Philadelphia New York Chicago
San Francisco Montreal Toronto
London Sydney Tokyo Mexico City
Rio de Janeiro Madrid

Address orders to:
383 Madison Avenue
New York, NY 10017
Address editorial correspondence to:
West Washington Square
Philadelphia, PA 19105

This book was set in Caledonia by Hampton Graphics.
The editors were John Vondeling, Lloyd Black and Cathryn Baskin.
The art director, and text and cover designer was Nancy E. J. Grossman.
The production manager was Tom O'Connor.
New artwork was drawn by Dimitri Karetnikov.
The printer was Hampshire Press.

Front cover: Saturn and its satellites Tethys *(outer left)*, Enceladus *(inner left)*, and Mimas *(right of rings)* are seen in this mosaic of images taken by Voyager 1 on October 30, 1980, from a distance of 18 million km (11 million miles). The projected width of the rings at the center of the disk is 10,000 km (6000 miles). (JPL/NASA photo)

Back cover: A mosaic of Jupiter and its four Galilean satellites made from the Voyager 1 spacecraft. Callisto with its impact bull's-eye is at lower right, Ganymede is at lower left, relatively bright Europa is shown against Jupiter, and Io appears in the background at left. (JPL/NASA photo)

LIBRARY OF CONGRESS

CATALOG CARD NO.: 80-53923

    Pasachoff, Jay M.
      Contemporary astronomy.

    Philadelphia, Pa.: Saunders College

608 p.

8101   801010

CONTEMPORARY ASTRONOMY   ISBN 0-03-057861-2

123   146   10 98765432

CBS COLLEGE PUBLISHING
Saunders College Publishing
Holt, Rinehart and Winston
The Dryden Press

PREFACE

Astronomy, as a science, combines the best of new and old traditions. Every year we find something new or unexpected; a profusion of x-ray quasars, molecules in interstellar space, pulsars, coronal holes, x-ray and gamma-ray bursts, black holes, Jupiter's ring, and volcanoes on Io are but a few examples from recent times. At the same time, astronomy has a rich and fascinating history.

I have attempted, in this book, to describe the state of astronomy as it is now. I have tried to fit the latest discoveries in their places alongside established results in order to give a contemporary picture of the state of our science.

Contemporary Astronomy is written for students with no background in mathematics or physics. It discusses astronomy in non-mathematical terms. In an attempt to make the material easier to understand, I have followed educational theory and paid special attention to developing material from the concrete to the abstract. The many examples of celestial objects demonstrating different phenomena should also help in comprehension.

I have found in my own classes that most students taking astronomy do so because they have heard a bit about black holes or some other fascinating new phenomenon and want to know more. I have thus organized this book to plunge right into real astronomy from the beginning. This differs from the traditional approach of presenting many chapters of history of astronomy, philosophy of science, and basic physics right away. These topics are treated where appropriate throughout the text.

The book begins with the stars, since so much of astronomy deals with them, and since the methods we use in studying the stars are generally applicable throughout much of astronomy. Also, students want to get to topics like pulsars and black holes as soon as possible. Planets, as companions of our sun and perhaps other stars as well, come next. (These chapters, however, do not depend on any of the material after the beginning of Chapter 3, and so can be taken up earlier or later in the course if desired.) Then we consider our own galaxy and continue outward through other galaxies, quasars, and cosmological consideration of the universe as a whole. A final chapter summarizes current hopes for astronomy in the next few years, and offers some descriptions of how research astronomers function, something that students have found very interesting and enlightening.

My text *Astronomy: From the Earth to the Universe* treats many of the same topics on a similar level. The principal difference is that *Astronomy: From the Earth to the Universe* treats the history of astronomy and the planets first (and has expanded historical sections), while *Contemporary Astronomy* treats the stars first. Much can be said in favor of the logic of each of these choices.

Marc Kutner and I have prepared an expanded version of this book, *University Astronomy*, for students with backgrounds in mathematics or physics. *University Astronomy* uses algebra and trigonometry but not calculus.

All the versions emphasize topics that are of current interest to astronomers and in which current research is very active. My point of view is that astronomers find astronomy fascinating, and that students will find it so too if the reasons for our excitement are explained. I have attempted to cover all fields of astronomy, without major gaps. In preparing this second edition, I am again excited by all the new results in so many fields of astronomy, and by how much has changed in even the last year.

A major theme of the books has been the expansion of our senses to all parts of the spectrum. We see time and again throughout the entire book how observations of gamma rays, x-rays, ultraviolet light, infrared light, and radio waves are fit together with observations of visible light and with theory to improve our knowledge of the universe and the objects in it. Such new results appear throughout the entire book, rather than being segregated in a few chapters.

A major goal of the books has been to describe astronomy in clear, thorough, understandable, and colloquial terms. All the steps in chains of reasoning have been included in order to make certain that the major points are clear.

In order to depict the widespread activity in astronomy, and to demonstrate the human aspect of astronomical research, I have sometimes named individual astronomers and institutions. Further, thorough explanations are sometimes used to embellish a central point. The ability to expand in this way has enabled me to set many matters in especially interesting contexts. But at no time should students find themselves memorizing material instead of understanding concepts. As an aid in determining just what is most important to remember, I have provided summaries of each chapter in this book. A glossary and a detailed index should also help students in finding references and explanations.

As part of my attempt to show students why astronomy is exciting instead of just a collection of facts, I have described the recent space exploration of each planet in somewhat chronological terms. Thus the students not only can see what we know about the planets at present but also how modern research has changed our views. This should give a base for keeping up with still newer research in years to come.

My stress has continually been on how to understand what is going on, and why astronomers think or act as we do. My criterion has been to include mostly material that can be remembered for years and that provides a basic understanding of the topic, rather than facts that are merely learned for examinations and soon forgotten. In order to give a feel of what it is like to be an astronomer, I have at several places in the text described just what an astronomer does in different circumstances, such as observing at a large telescope, observing with a satellite, or working at a solar eclipse. I have myself been fortunate to be able to observe with a variety of optical, radio and space telescopes in this country and abroad, and I am glad to have the opportunity to describe the excitement an observer feels.

It also seems to me that after completing a course in astronomy a student should be able to carry on a reasonable conversation with an astronomer—and this includes an amateur astronomer as well as a professional one. This has led me to continue the usage of the somewhat strange units that are retained by astronomers for historical reasons rather than for reasons of logic. For example, magnitudes are used to describe the brightness of stars, though I try to make clear the relation of the magnitude scale to a direct scale of brightness. I also draw the Hertzsprung-Russell graphs with the y-axis labeled in both luminosity and magnitude, so that professors can choose to skip the section on working with magnitudes. Anyway, I find it easier to remember magnitudes, which are simple numbers like 2 or 16, rather than luminosities like 2×10^4. If these units are madness, there is method in it.

In view of the importance to scientific research of governmental support, and the need for everyone to be informed about the value of such research, I have described from time to time the benefits that might accrue to society from certain types of astronomical research.

After the text, there are a series of appendices that reflect the latest research from space vehicles and from the ground. The glossary and detailed index follow.

ACKNOWLEDGMENTS

The publishers and I have placed a heavy premium on accuracy, and have made certain that the manuscript and proof have been read not only by students for clarity and style but also by several astronomers for their professional comments. I would like to thank, in particular, David S. Evans (University of Texas, Austin), John E. Gaustad (University of California, Berkeley), Edward L. Robinson (University of Texas, Austin), Eugene Tademaru (University of Massachusetts, Amherst), and David Theison (University of Maryland, College Park) for reading and commenting upon the entire manuscript of the second edition. I am also grateful to James B. Pollack (NASA/Ames Research Center) and Joseph Veverka (Cornell University) for their comments on the chapters about the solar system.

The contribution made by readers of the prior edition and its alternative versions is a lasting one. I thank Thomas T. Arny (University of Massachusetts, Amherst), Marc L. Kutner (Rensselaer Polytechnic Institute), Richard L. Sears (University of Michigan), and Joseph F. Veverka (Cornell University) for reading the entire manuscript of the first edition. I am also grateful to David E. Koltenbah (Ball State University), Karl F. Kuhn (Eastern Kentucky University), and Stephen T. Gottesman (University of Florida) for consulting on the preparation of *Astronomy Now*. I thank Owen Gingerich (Harvard-Smithsonian Center for Astrophysics), Joe S. Tenn (Sonoma State College), Richard L. Sears (University of Michigan), LeRoy A. Woodward (Georgia Institute of Technology), Dennis de Cicco (*Sky and Telescope*), and John A. Eddy (High Altitude Observatory) for reviewing the additional material for *Astronomy: From the Earth to the Universe*.

I would also like to thank other astronomers and physicists who have commented on parts of the manuscript in their particular areas of expertise. They include (in the approximate order in which topics are treated in the book):

Jean Pierre Swings (Institut d'Astrophysique, Liège, Belgium), stars and quasars; Jeffrey L. Linsky (Joint Institute for Laboratory Astrophysics, Boulder), radiative transfer; David Park (Williams College), atomic physics; James G. Baker (Harvard College Observatory), optics; Jonathan E. Grindlay (Harvard-Smithsonian Center for Astrophysics), black holes and x-ray bursts; Peter V. Foukal (Atmospheric and Environmental Research, Inc.), the sun; Robert F. Howard (Hale Observatories), the sun; Lawrence Cram (Sacramento Peak Observatory), opacity effects; John A. Eddy (High Altitude Observatory), the sun and climate; J. Craig Wheeler (University of Texas), stellar evolution; Raymond Davis, Jr. (Brookhaven National Laboratory), solar neutrinos; David N. Schramm (University of Chicago), stellar evolution and nucleosynthesis; Joseph H. Taylor, Jr. (University of Massachusetts, Amherst), pulsars; Bruce Margon (UCLA), SS 433; John A. Wheeler (University of Texas), black holes; Kenneth Brecher (Boston University), black holes; Peter Conti (University of Colorado), stars; Bruce E. Bohannan (University of Colorado), stars; Maurice Shapiro (Naval Research Laboratory), cosmic rays;

Owen Gingerich (Harvard-Smithsonian Center for Astrophysics), history of astronomy; Lawrence S. Lerner (California State University, Long Beach), history of astronomy; David D. Morrison (University of Hawaii), the solar system; Farouk El-Baz (National Air and Space Museum, Smithsonian Institution), the moon; Ewen A. Whitaker (Lunar and Planetary Laboratory, University of Arizona), the moon; Gerald J. Wasserburg (California Institute of Technology), lunar chronology; Robert G. Strom (Lunar and Planetary Laboratory, University of Arizona), Mercury; Reinhard A. Wobus (Williams College), geology; William R. Moomaw (Williams College), ozone; Ian Halliday (Herzberg Institute of Astrophysics, Ottawa), Neptune and Pluto; J. Donald Fernie (David Dunlap Observatory, Toronto), the Adams-Leverrier affair; James W. Christy (U.S. Naval Observatory), Pluto; Dale Cruikshank (University of Hawaii), Pluto; P. Kenneth Seidelmann (U.S. Naval Observatory), ephemerides and planetary parameters; Brian Marsden (Harvard-Smithsonian Center for Astrophysics), comets, asteroids, and meteoroids; George D. Gatewood (Allegheny Observatory), Barnard's star; Everett Mendelsohn (Harvard University), superstition; Agris Kalnajs (Mt. Stromlo Observatory, Australia), spiral structure; Barry Turner (National Radio Astronomy Observatory), interstellar molecules; Lewis E. Snyder (University of Illinois), interstellar molecules; Juri Toomre (Joint Institute for Laboratory Astrophysics), gravitational effects on galaxies; Leonid Weliachew (Observatoire de Paris, France), galaxies; George K. Miley (Leiden Observatory, the Netherlands), interferometry; Bernard J. T. Jones (Institute for Astronomy, Cambridge, England), galaxy formation; Jerome Kristian (Hale Observatories), quasars, Seyfert and N galaxies, and cosmology; Bernard Burke (M.I.T.), the double quasar; Ray Weymann (University of Arizona), the double quasar; John Lathrop (Williams College), cosmology; and Alan T. Moffet (California Institute of Technology), radio astronomy; Laird Thompson (University of Hawaii), cosmology; Herbert Rood (Michigan State University and Institute for Advanced Studies, Princeton), cosmology; and Brent Tully (University of Hawaii), clusters of galaxies.

Many others have read shorter bits of manuscript, cleared up particular points, or provided special illustrations, and I thank them as well. In particular, let me acknowledge the assistance of Riccardo Giacconi (Harvard-Smithsonian Center for

Astrophysics), Maarten Schmidt (Hale Observatories), Hyron Spinrad (University of California, Berkeley), Richard B. Dunn (Sacramento Peak Observatory), Freeman D. Miller (University of Michigan), Peter van de Kamp (Sproul Observatory), Kenneth D. Tucker (Fordham University), Edwin C. Krupp (Griffith Observatory), Sarah Lee Lippincott (Sproul Observatory), Dennis di Cicco *(Sky and Telescope),* and R. Newton Mayall (American Association of Variable Star Observers). The many institutions and individuals who provided photographs or other illustrations are acknowledged separately; I am grateful to all of them.

Marc L. Kutner, Nancy P. Kutner, and Naomi Pasachoff have worked closely with me on the questions, the index, and the glossary, respectively.

I thank many people at Saunders College Publishing for their efforts on my books. John J. Vondeling merits special thanks for his continued support and his interest in this project since its inception. Project and developmental editor Lloyd Black has worked long and hard on the manuscript and figures. Lorraine Battista designed the first edition books and covers, and has my eternal gratitude for this major contribution. New design elements and the revised color section present in the second edition are the work of art director Nancy E. J. Grossman. The detailed layout is by Joan de Lucia. The support and efforts of George Laurie of the Art Department (first edition) and of production manager Tom O'Connor (both editions) helped the book immeasurably.

Many of the drawings were executed by George Kelvin of Science Graphics from my sketches. Other drawings were executed by Grant Lashbrook and John Hackmaster at Saunders. New drawings for the second edition were executed by Dimitri Karetnikov.

I remember fondly the influence that Donald H. Menzel of the Harvard College Observatory had on my career over a period of many years.

Cathryn Baskin has expertly contributed to many aspects of the manuscript, illustrations, and proof, and has been of tremendous assistance. I thank her.

I thank John T. Jefferies and Donald A. Landman for their hospitality at the Institute for Astronomy of the University of Hawaii during the final proof stages of this book.

Various members of my family have provided vital and valuable editorial services. In particular, my wife, Naomi Pasachoff, my father, Dr. Samuel S. Pasachoff, and my mother, Anne T. Pasachoff, have worked many hours on the manuscript and proof. Now that my daughter Eloise is five, she is starting on her own writing projects. And my daughter Deborah, who is three, comes in regularly to supervise my work and to ask me to help her with her own projects. Now the alphabet all the way up to Z; what next?

The index was originally put together in collaboration with Nancy P. Kutner.

NOTES TO TEACHERS

A *Teacher's Guide to Contemporary Astronomy*, 2nd edition, is available. It contains possible syllabi, labs, tests, and answers to questions, and lists films, tapes, and other audio and visual aids for use as supplementary materials. The publishers and I are making available artwork and photographs in formats that can be projected in class. If you are interested in such material, please contact Textbook Marketing, Saunders College Publishing % Holt, Rinehart & Winston, Inc., 383 Madison Avenue, New York, NY 10017. A set of over 250 black-and-white photographs and drawings on 100 overhead transparency pages is available.

NOTES TO READERS

A number of sections and boxes have been marked with asterisks (*). This indicates that they are not in the main line of discussion, and can be omitted without loss of understanding of later chapters.

I am extremely grateful to all of the individuals named above for their assistance. Of course, it is I who have put this all together, and I alone am responsible for any errors that have crept through our sieve. I would appreciate learning from readers, not just about typographical or other errors, but also with suggestions for presentation of topics or even with comments about specific points that need clarifying. I invite readers to write me % Williams College, Hopkins Observatory, Williamstown, Mass. 01267. I promise a personal response to each writer.

JAY M. PASACHOFF

Williamstown, Massachusetts

TABLE OF CONTENTS

Part V   THE MILKY WAY GALAXY 414

Part VI   GALAXIES AND BEYOND 466

LIST OF TABLES

Contemporary Astronomy

PART I

THE UNIVERSE: AN OVERVIEW

CHAPTER 1

AIMS:
To get a feeling for the variety of objects in the Universe and a sense of scale for size and time

The universe is a place of great variety—after all, it has everything in it! Some of the things astronomers study are of a size and scale that we humans can easily comprehend: the planets, for instance. Most astronomical objects, however, are so large and so far away that our minds have trouble grasping the sizes and distances involved.

Moreover, astronomers study the very small in addition to the very large. The radiation we receive from distant bodies is emitted by atoms, which are much too small to see with the unaided eye. Also, the properties of the large astronomical objects are often determined by changes that take place on a minuscule scale—that of atoms. Thus the astronomer must be an expert in the study of objects the size of atoms as well as in the study of objects the size of galaxies.

It should come as no surprise to hear that the variety of objects at very different distances from us or with very different properties often must be studied with widely differing techniques. Clearly, different tests are required to analyze the properties of solid particles like Martian soil with a spacecraft sitting on the Martian surface than are required to interpret the composition of a gaseous body like a quasar deep in space. There is one unifying method, however, that links much of astronomy. This method is *spectroscopy,* which involves, as we shall see, analyzing components of the light or other radiation that we receive from distant objects and studying the components in detail. Throughout this book, we shall return to spectroscopic methods time and again, to study not only visible light but also other types of radiation. Not all of astronomy is spectroscopy, of course; much important information is gained by making images—pictures—of planets, the sun, and galaxies, for example, by studying the variation of the amount of light coming from a star or a galaxy over time, or by visiting the moon and planets.

The explosion of astronomical research in the last few decades has been fueled by our new ability to study radiation other than light—gamma rays, x-rays, ultraviolet radiation, infrared radiation, and radio waves. Astronomers' use of their new abilities to study such radiation is a major theme of this book. All the kinds of radiation together make up the *electromagnetic spectrum,* which will be discussed in Chapter 2. As we shall see there, we

The Whirlpool Galaxy, M51, in the constellation Canes Venatici.

can think of radiation as waves, and all the types of radiation have similar properties except for the length of the waves. Still, although x-rays and visible light may be similar to a scientist, our normal experiences tell us that very different techniques are necessary to study them.

The earth's atmosphere shields us from most kinds of radiation. Not only light waves but also radio waves penetrate the atmosphere, though most other types of radiation are blocked. Over the last 50 years, radio astronomy has become a major foundation of our astronomical knowledge. For the last 25 years, we have been able to send satellites into orbit outside the earth's atmosphere, and we are no longer limited to the study of radio and visible (light) radiation. Many of the fascinating discoveries of this decade—the probable observation of a black hole, for example—were made because of our newly extended senses. We shall be discussing how all parts of the spectrum are used to help us understand the universe. And besides the information that the spectrum provides, we can also get information from direct sampling of bodies in our solar system (as we shall discuss in Part IV) and from cosmic rays (particles whizzing through space, which we shall discuss in Sections 10.3 and 22.3). Perhaps one day we will also be able to observe gravitational waves (as we shall discuss in Section 10.12), whose existence is predicted by Einstein's general theory of relativity (Section 7.11).

1.1 A SENSE OF SCALE

Let us try to get a sense of scale of the universe, starting with sizes that are part of our experience and then expanding toward the infinitely large. In the margin, we can keep track of the size of our field of view as we expand in powers of 100: each diagram will show a square 100 times greater on a side.

We shall use the metric system, which is in common use by scientists. The basic unit of length is the meter, which is equivalent to 39.37 inches, slightly more than a yard. Prefixes are commonly used (see Appendix 2) in conjunction with the word "meter," abbreviated "m," to define new units. The most commonly used are "milli-," meaning 1/1000, "centi-," meaning 1/100, and "kilo-," meaning 1000 times. Thus 1 millimeter is 1/1000 of a meter, or about .04 inch, and a kilometer is 1000 meters, or about 5/8 mile. We will keep track of the powers of 10 by which we multiply 1 m by writing the number of tens we multiply together as an exponent; 1000 m, for example, is 10^3 m.

We can also keep track of distance in units that are based on the length of time that it takes light to travel. The speed of light is, according to Einstein's 1905 special theory of relativity (Section 7.11), the greatest speed that is physically attainable. Light travels at 300,000 km s^{-1}* (186,000 m s^{-1}), fast enough to circle the earth 7 times in a single second. Even at that fantastic speed, we shall see that it would take years for us to reach the stars. Similarly, it has taken years for the light we see from stars to reach us, and we are thus really seeing the stars as they were years ago. In a sense, we are looking backward in time. The distance that light travels in a year is called a *light year;* note that the light year is a unit of length rather than a unit of time even though the term "year" appears in it.

Literary scholars could say that a "milli-Helen" is a unit of beauty sufficient to launch one ship.

The distance travelled by an object is equal to the rate at which the object is travelling (its velocity) times the time spent travelling (d = vt).

---

*In SI units, which we will use in this book, km s^{-1} = km/s (kilometers per second).
SI, Système International, is the international form of the metric system now in use.

1.1a Survey of the Universe

Let us begin our journey through space with a view of a sphere reproduced here one millimeter in diameter (Fig. 1–1).

1 mm = 0.1 cm

A square 100 times larger on each side is 10 centimeters × 10 centimeters. (Since the area of a square is the length of a side squared, the area of a 10 cm square is 10,000 times the area of a 1 mm square). The area encloses a flower (Fig. 1–2).

10 cm = 100 mm

As we move far enough away to see an area 10 meters on a side, we are seeing an area approximately that taken up by half a tennis court (Fig. 1–3).

10 m = 1000 cm

A square 100 times larger on each side is now 1 kilometer square, about 250 acres. An aerial view of several square blocks in New York City shows how big an area this is (Fig. 1–4).

1 km = 10³ m

The next square, 100 km on a side, encloses a major city, New York. Note that though we are still bound to the limited area of the earth, the area we can see is increasing rapidly (Fig. 1–5).

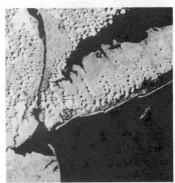

100 km = 10⁵ m

10,000 km = 10⁷ m

A square 10,000 km on a side covers nearly the entire earth (Fig. 1–6).

When we have receded 100 times farther, we see a square 100 times larger in diameter: 1 million kilometers across. It encloses the orbit of the moon around the earth (Fig. 1–7). We can measure with our wristwatches the amount of time that it takes light to travel this distance. If we were carrying on a conversation by radio with someone at this distance, there would be pauses of noticeable length after we finished speaking before we heard an answer. This is because radio waves, even at the speed of light, take that amount of time to travel. Astronauts on the moon have to get used to these pauses when speaking to earth. Laser pulses from earth bounced off the moon to find the distance to the moon take a noticeable time to return. The photograph was taken by the Voyager I spacecraft en route to Jupiter and Saturn. The earth (bottom) is in the foreground and the moon (top) is in the background. Eastern Asia, the Western Pacific Ocean, and part of the Arctic are on the illuminated portion of the earth. Because the moon is many times fainter than the earth, it was artificially brightened in the computer by a factor of three relative to the earth so that it would show up better in this print.

1,000,000 km = 10⁹ m = 3 lt sec

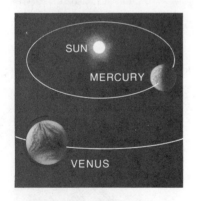

10¹¹ m = 5 lt min

When we look on from 100 times farther away still, we see area 100 million kilometers across, 2/3 the distance from the earth to the sun. We can now see the sun and the inner planets in our field of view (Fig. 1–8). This is the scale representing approximately the limit of our current ability to send spacecraft to sample the planets directly. We have now sent spacecraft to the moon, Mercury, Venus, Mars, Jupiter, Saturn, and Uranus; the results will be discussed at length in Part IV.

An area 10 billion kilometers across shows us the entire solar system in good perspective. It now takes light 8 hours to travel this distance. The outer planets have become visible and are receding into the distance as our journey outward continues (Fig. 1–9). This artist's conception shows a Voyager spacecraft near Saturn.

10¹³ m = 8 lt hrs

From 100 times farther away, we see little that is new. The solar system seems smaller and we see how vast is the empty space around us. We have not yet reached the scale at which another star besides the sun is in our field of view, although, of course, many stars are visible in the background (Fig. 1–10).

ORBIT OF PLUTO
5.9 BILLION KM

10^{15} m = 38 lt days

• BARNARD'S STAR   • SUN

α CENTAURI
PROXIMA CENTAURI

LALANDE 21185
WOLF 359 •

10^{17} m = 10 ly

As we continue to recede from the solar system, the nearest stars finally come into view. We are seeing an area 10 light years across, which contains only a few stars (Fig. 1–11). Part II of this book discusses the properties of stars.

By the time we are 100 times farther away, we can see a fragment of our galaxy, the Milky Way Galaxy (Fig. 1–12). We see not only many individual stars but also many clusters of stars and many areas of glowing, reflecting, or opaque gas or dust called nebulae. There is even a lot of material between the stars that is mostly invisible to our eyes but that can be studied with radio or space telescopes. Part V of this book is devoted to the study of our galaxy and its contents.

10^{19} m = 10^3 ly

In a field of view 100 times larger in diameter, we can now see an entire galaxy. The photograph (Fig. 1–13) shows a galaxy called M74, located in the direction of the constellation Pisces, though it is far beyond the stars in that constellation. This galaxy shows arms wound in spiral form. Our galaxy also has spiral arms, though they are wound more tightly.

10^{21} m = 10^5 ly

Next we move sufficiently far away so that we can see an area 10 million light years across at a glance (Fig. 1–14). There are 10^{25} centimeters in 10 million light years, about as many centimeters as there are grains of sand in all the beaches of the earth. Our galaxy is in a cluster of galaxies, called the Local Group, that would take up only 1/3 of our angle of vision. In this group are all types of galaxies, which we will discuss in Chapter 25. The photograph shows part of a cluster of galaxies in the constellation Leo.

10^{23} m = 10^7 ly

Box 1.1 Scientific Notation

In astronomy we often find ourselves writing numbers that have strings of zeros attached, so we use what is called *scientific notation*. In scientific notation, which we have already employed in the previous section under Figures 1–4 to 1–14, we merely count the number of zeros, and write it as a superscript to the number 10. Thus the number 10,000,000,000, a 1 followed by 10 zeros, is written 10^{10}. If the number had been 30,000,000,000, we would have written 3×10^{10}. This method of writing numbers simplifies our writing chores considerably, and helps prevent making mistakes in copying long strings of numbers.

Scientific notation also allows simplification in calculation. The superscript 5 in 10^5 is called the *exponent*. When a number with an exponent is multiplied by the same number with a different exponent, to find the result we simply add the exponents and use the sum as the new exponent. Thus $10^5 \times 10^7 = 10^{5+7} = 10^{12}$. When a number with an exponent is itself raised to an exponent, as we say, then one simply multiplies the exponents. Thus, since $100 = 10^2$, we see that $(100)^5 = (10^2)^5$ and by merely multiplying 2×5 to get 10, we have 10^{10}.

We can handle numbers less than one with negative exponents. A minus sign in the exponent of a number means that the number is actually one divided by what the quantity would be if the exponent were positive. Thus $10^{-2} = 1/10^2$. One can compute the exponent that follows the minus sign by counting the number of places by which the decimal point has to be moved to the right until it is at the right of the first non-zero digit. Thus, for example, $0.000001435 = 1.435 \times 10^{-6}$, since the decimal point on the left side has to be moved six places to the right to come after the digit 1. Positive exponents, similarly, are the number of places that the decimal point has to be moved to the left to be on the right side of the first digit.

Some of the units used in astronomy and prefixes used with them are listed in Appendix 2.

$$10^0 = 1$$
$$10^1 = 10$$
$$10^2 = 100$$
$$10^3 = 1000$$

Note that anything to the zero power is 1 and anything to the first power is itself.

If we could see a field of view 1 billion light years across, our Local Group of galaxies would appear as but one of many clusters. It is difficult to observe on this scale.

Before we could enlarge our field of view another 100 times we might see a supercluster—a cluster of clusters of galaxies. We would be seeing almost to the distance of the quasars, which are the topic of Chapter 26. Quasars, the most distant objects known, may be explosive events in the cores of galaxies. Light from the most distant quasars observed may have taken 10 billion years to reach us on earth, and if this is so we are thus looking back to times billions of years ago. Since we think that the big bang that began the universe took place approximately 13 billion years ago, we are looking back almost to the beginning of time.

We even think that we have detected the radiation given off at the time of the big bang itself. A combination of radio, ultraviolet, x-ray, and optical studies is allowing us to explore the past and predict the future of the universe. All this is discussed in Chapter 27.

1.2 A SENSE OF TIME

A day can seem to us to pass so slowly that it is difficult for us to realize how fleetingly we humans appear on the time line of history (Fig. 1–15). Let us consider that the time between the origin of the universe—13 to 20 billion

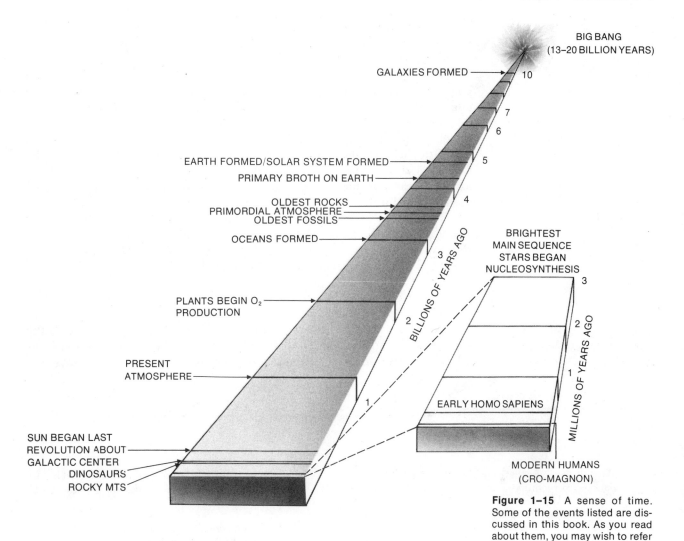

Figure 1–15 A sense of time. Some of the events listed are discussed in this book. As you read about them, you may wish to refer back to this figure.

years ago—and the year 2000 is one day. Then it wasn't until late afternoon, about 4 P.M., that the earth formed. Plants didn't begin to produce oxygen until about 8 P.M., and we have no fossils from before 10 P.M. The first human beings appeared only about 2 seconds ago, and it is only three thousandths of a second ago that Columbus discovered America. It is only about a ten-thousandth of a second before the new millennium arrives, January 1, 2000. And the sun should live, as a normal star, for another 8 hours.

1.3 THE VALUE OF ASTRONOMY

Throughout history, observations of the heavens have led to discoveries that have had major impact on people. Even the dawn of mathematics may have followed ancient observations of the sky, made in order to keep track of seasons and seasonal floods in the fertile areas of the earth. Observations of the motions of the moon and the planets, which are free of such complicating terrestrial forces as friction and which are massive enough so that gravity dominates their motions, led to an understanding of gravity and of the forces that govern all motion.

We can consider the regions of space studied by astronomers as a cosmic laboratory for the study of matter or radiation under a variety of conditions, often under conditions that we cannot duplicate on earth.

Many of the discoveries of tomorrow—perhaps the control of nuclear fusion or the discovery of new sources of energy, or perhaps something so revolutionary that it cannot now be predicted—will undoubtedly also be based on discoveries made through such basic research as the study of astronomical systems. Considered in this sense, astronomy is an investment in our future.

The impact of astronomy on our conception of the universe has been strong through the years. Discoveries that the earth is not in the center of the universe, or that the universe has been expanding for billions of years, affect our philosophical conceptions of ourselves and our relations to space and time.

Yet most of us study astronomy not for its technological and philosophical benefits but for its grandeur and inherent interest. We must stretch our minds to understand the strange objects and events that take place in the far reaches of space. The effort broadens us and continually fascinates us all. Ultimately, we study astronomy because of its fascination and mystery.

QUESTIONS

1. Why do we say that our senses have been expanded in recent years?

†2. The speed of light is 3×10^5 km s^{-1}. Express this number in m s^{-1} and in cm s^{-1}.

†3. During the Apollo explorations of the moon, we had a direct demonstration of the finite speed of light when we heard ground controllers speak to the astronauts. The sound from the astronauts' earpieces was sometimes picked up by the astronauts' microphones and retransmitted to earth as radio signals. We then heard our words repeated. What is the time delay between the original and the "echo," assuming that no other delays were introduced in the signal?

†4. The distance to the Andromeda galaxy is 2×10^6 light years. If we could travel at one-tenth the speed of light, how long would a round trip take? (Hint: The distance to something is equal to the velocity at which you are going multiplied by the time it takes to get there; "distance = rate times time.")

†5. How long would it take to travel to Andromeda at 1000 km hr^{-1}, the speed of a jet plane?

6. List the following in order of increasing size: (a) light year, (b) distance from earth to sun, (c) size of Local Group, (d) size of a football stadium, (e) size of our galaxy, (f) distance to a quasar.

7. Of the examples of scale in this chapter, which would you characterize as part of "everyday" experience? What range of scale does this encompass? How does this range compare with the total range covered in the chapter?

8. What is the largest of the scales discussed in this chapter that could reasonably be explored in person by humans with current technology?

†9. (a) Write the following in scientific notation: 4642; 70,000; 34.7. (b) Write the following in scientific notation: 0.254; 0.0046; 0.10243. (c) Write out the following in an ordinary string of digits: 2.54×10^6; 2.0004×10^2.

†10. What percentage of the age of the universe has elapsed since the appearance of homo sapiens on the earth?

---

†This indicates a question requiring a numerical solution.

TOPIC FOR DISCUSSION

What is the value of astronomy to you? Save your answer for comparison with your answer to a similar question in Chapter 28, or with your answer when you have completed this course.

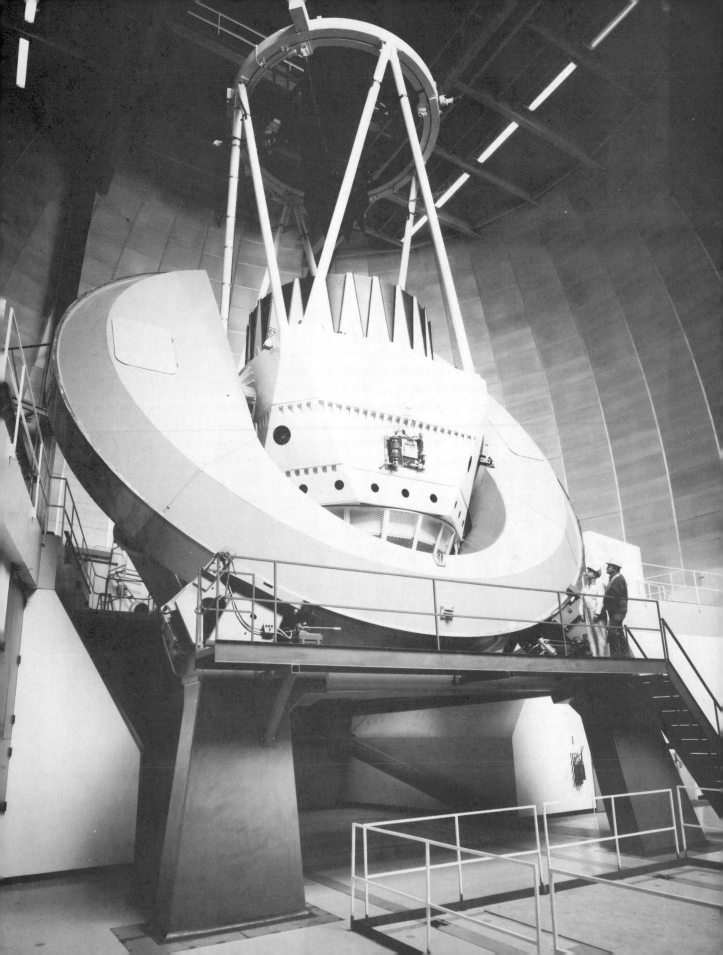

CHAPTER 2

LIGHT AND
TELESCOPES

AIMS:
To understand the electromagnetic
spectrum and the different types of
observing techniques and instruments
used in astronomy

If you forced a group of astronomers to choose one and only one instrument to be marooned with on a desert island, they would probably choose—a computer. The notion that astronomers spend most of their time at telescopes is far from the case in modern times. Still, telescopes have been and continue to be very important to the development of astronomy, and in this chapter we will discuss how we study the stars with telescopes and other observational devices.

Although the word "telescope" makes most of us think of objects with lenses and long tubes, that type of telescope is only used to observe "light." Ordinary "visible" light, the kind we see with our eyes, is not the only way we can study the universe. We shall see in this chapter that light is only one type of **radiation**—a certain way in which energy moves through space. Other types of radiation are gamma rays, x-rays, ultraviolet light, infrared light, and radio waves. They are all fundamentally identical to ordinary light, though in practice we must usually use different methods to observe them.

First, however, we discuss the properties of radiation that enable scientists to study the universe; after all, we cannot touch a star, and though we have brought bits of the moon back to earth for study, we are not able to do the same for even the nearest planets.

2.1 THE SPECTRUM

It was discovered over 300 years ago that when ordinary light is passed through a prism, a band of color like the rainbow comes out the other side.

The 4-meter Mayall telescope at the Kitt Peak National Observatory.

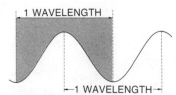

Figure 2–1 The definition of the wavelength of electromagnetic radiation.

Thus "white light" is composed of all the colors of the rainbow (see Color Plate 3).

These colors are always spread out in a specific order, which has traditionally been remembered by the initials of the friendly fellow **ROY G. BIV,** which stand for **R**ed, **O**range, **Y**ellow, **G**reen, **B**lue, **I**ndigo, **V**iolet. No matter what rainbow you watch, or what prism you use, the order of the colors never changes.

We can understand why light contains the different colors if we think of light as waves of radiation. These waves are all travelling at the same speed, 3×10^8 m s^{-1} (186,000 miles s^{-1}). This speed is normally called the *speed of light* (even though light normally travels at a slightly lower speed because the light is not usually in a perfect vacuum; "the speed of light" usually really means "the speed of light in a vacuum").

The distance between one crest of the wave to the next or one trough to the next, or in fact between any point on a wave and the similar point on the next wave, is called the *wavelength* (Fig. 2–1). Light of different wavelengths appears as different colors. It is as simple as that. Red light has approximately 1½ times the wavelength of blue light. Yellow light has a wavelength in between the two. Actually there is a continuous distribution of wavelengths, and one color blends subtly into the next.

The wavelengths of light are very short: just a few ten-millionths of a meter. Astronomers use a unit of length called an angstrom, named after the Swedish physicist A. J. Ångstrom. One angstrom (1 Å) is 10^{-10} m. (Remember that this is 1 divided by 10^{10}, which is .000 000 000 1 m.) The wavelength of violet light is approximately 4000 angstrom units, usually written 4000 angstroms (4000 Å). Yellow light is approximately 6000 Å, and red light is approximately 6500 Å in wavelength.

The human eye is not sensitive to radiation whose wavelength is much shorter than 4000 Å or much longer than 6600 Å, but other devices exist that can measure light at shorter and longer wavelengths. At wavelengths shorter than violet, the radiation is called *ultraviolet;* at wavelengths longer than red, the radiation is called *infrared.* (Note that it is not "infared," a common misspelling; "infra" is a Latin prefix that means "below.")

All these types of radiation—visible or invisible—have much in common. Many of us are familiar with the field of force produced by a magnet—a magnetic field—and the field of force produced by a charged object like a hair comb on a dry day—an electric field. It has been known for over a century that fields that vary rapidly behave in a way that no one would guess from the study of magnets and combs. In fact, light, x-rays, and radio waves are all examples of rapidly varying electric fields and magnetic fields that have become detached from their sources and move rapidly through space. For this reason, these radiations are referred to as *electromagnetic radiation.*

We can draw the entire *electromagnetic spectrum,* often simply called the *spectrum* (plural: *spectra*), ranging from radiation of wavelength shorter than 1 Å to radiation of wavelength many meters long and longer (Fig. 2–2).

Note that from a scientific point of view there is no real qualitative difference between types of radiation at different wavelengths. They can all be thought of as *electromagnetic waves,* waves of varying electric and magnetic fields. Light waves comprise but one limited range of wavelength. When an electromagnetic wave has a wavelength of 1 Å, we call it an x-ray.

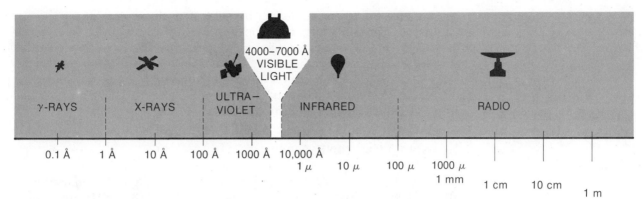

Figure 2–2 The electromagnetic spectrum.

When it has a wavelength of 5000 Å, we call it light. When it has a wavelength of 1 cm (which is 10^8 Å), we call it a radio wave. Of course, there are obvious practical differences in the methods by which we detect x-rays, light, and radio waves, but the principles that govern their existence are the same.

Note also that light occupies only a very small portion of the entire electromagnetic spectrum. It is obvious that the new ability that astronomers have to study parts of the electromagnetic spectrum other than light waves enables us to increase our knowledge of celestial objects manyfold.

(Only certain parts of the electromagnetic spectrum can penetrate the earth's atmosphere. We say that the earth's atmosphere has "windows" for the parts of the spectrum that can pass through it. The atmosphere is transparent at these windows and opaque at other parts of the spectrum. One window passes what we call "light," and what astronomers technically call *visible light,* or *the visible,* or *the optical part of the spectrum.* Another window falls in the radio part of the spectrum, and modern astronomy uses "radio telescopes" to detect that radiation)(Fig. 2–3).

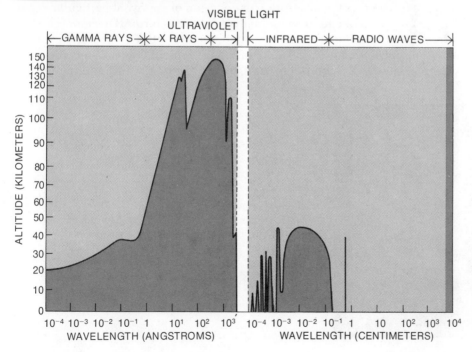

Figure 2–3 Windows of transparency in the terrestrial atmosphere allow only part of the solar spectrum to penetrate to the earth's surface. The curve specifies the altitude where the intensity of arriving radiation is reduced to half its original value. When this happens high in the atmosphere, little or no radiation of that wavelength reaches the ground. From "Ultraviolet Astronomy" by Leo Goldberg. Copyright ©1969 by Scientific American. All rights reserved.

But we of earth are no longer bound to our planet's surface; balloons, rockets, and satellites carry telescopes above the atmosphere to observe in parts of the electromagnetic spectrum that do not pass through the earth's atmosphere. By now, we have made at least some observations in each of the named parts of the spectrum. It seems strange, in view of the long-time identification of astronomy with visible observations, to realize that optical studies no longer dominate astronomy.

2.2 SPECTRAL LINES

As early as 1666, Isaac Newton showed that sunlight was composed of all the colors of the rainbow. William Wollaston in 1804 and Joseph Fraunhofer in 1811 also studied sunlight as it was dispersed (spread out) into its rainbow of component colors (Fig. 2–4). Wollaston and Fraunhofer were able to see that at certain colors there were gaps that looked like dark lines across the spectrum at those colors (Color Plate 5). These gaps are thus called *spectral lines;* the continuous radiation in which the gaps appear is called the *continuum* (pl: *continua*). The dark lines in the spectrum of the sun and in the spectra of stars are gaps that represent the diminution of electromagnetic radiation at those particular wavelengths. These dark lines are called *absorption lines;* they are also known (for the sun, in particular) as *Fraunhofer lines.*

It is also possible to have wavelengths at which there is somewhat more radiation than at neighboring wavelengths; these are called *emission lines.* We shall see that the nature of a spectrum, and whether we see emission or absorption lines, can provide considerable information about the nature of the body that was the source of the light. We say that the lines are "in emission" or "in absorption."

It was discovered in laboratories on earth that patterns of spectral lines can be explained as the absorption or emission of energy at particular wavelengths by atoms of chemical elements in gaseous form. If a vapor (gaseous form) of any specific element is heated, it gives off a characteristic set of emission lines. That element, and only that element, has that specific set of spectral lines. If, on the other hand, a continuous spectrum radiated by a

Note that "absorption" is spelled with a "p," not a second "b."

On our home radios, the band over which we can tune is in a range of the radio spectrum, and the stations, places in the spectrum where there is energy, represent spectral lines. In particular, they are emission lines.

We are not here giving examples of how continuous radiation is generated in a gas; different mechanisms apply at different temperatures and spectral ranges.

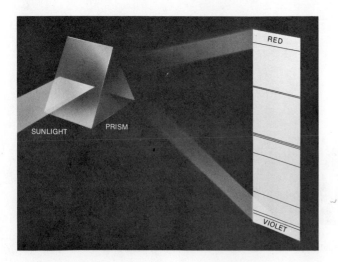

Figure 2–4 When a narrow beam of sunlight is dispersed by a prism, we see not only a continuous spectrum but also dark Fraunhofer lines.

Color Plate 1 (top, left): The 4-m Mayall optical telescope on Kitt Peak, near Tucson, Arizona. (Kitt Peak National Observatory photo)

Color Plate 2 (top, right): The Soviet 6-m telescope, the largest in the world. Note the figures standing on top of the dome. (Courtesy of J.M. Kopylov, Special Astrophysical Observatory)

Color Plate 3 (bottom, left): The Greenbelt, Maryland, control room of the International Explorer Spacecraft. The image displayed on the screen shows a stellar spectrum in the ultraviolet. (Photo by the author)

Color Plate 4 (bottom, right): The 67-m radio telescope at the Australian National Radio Observatory at Parkes, N.S.W. (Photo by the author)

Color Plate 5 (bottom): Fraunhofer's original spectrum from 1814. Only the D and H lines retain their notation from that time. The C line is Hα.

𝒜 ℬ 𝒞   𝒟   ℰ   ℱ   𝒢   ℋ

Color Plate 6 (top): HEAO-1 (High-Energy Astronomy Observatory 1), launched in 1977 to study x-rays and gamma rays. (NASA and TRW Systems Group photo)

Color Plate 7 (bottom): The VLA (Very Large Array), the aperture synthesis radio telescope near Socorro, New Mexico. It is composed of 27 dishes, each 26 m in diameter, arranged in the shape of a "Y" over a flat area 27 km in diameter. The third arm of the "Y" extends to the right. (National Radio Astronomy Observatory photo)

Color Plate 8 (top): Multiple exposures at hourly intervals taken in Alaska. The upper section shows the phenomenon of the midnight sun (which never dips below the horizon) on the day of the summer solstice, and the lower section shows the sun on the shortest day of the year. (Photo by Mario Grassi)

Color Plate 9 (bottom): Lightning flashes illuminate the Kitt Peak National Observatory in this 45-second exposure. (Photo © 1972 Gary Ladd)

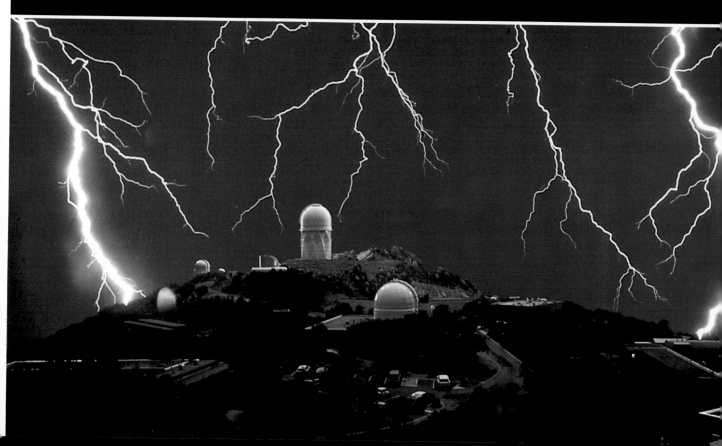

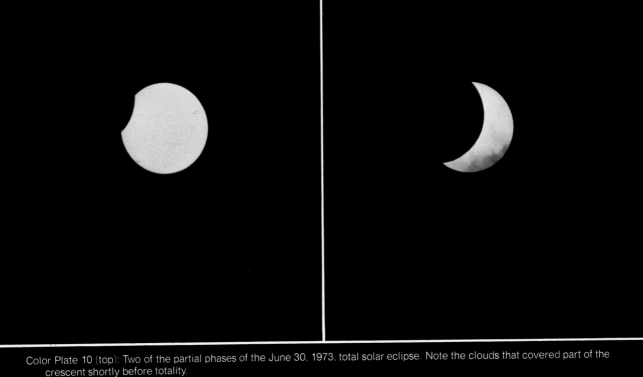

Color Plate 10 (top): Two of the partial phases of the June 30, 1973, total solar eclipse. Note the clouds that covered part of the crescent shortly before totality.

Color Plate 11 (bottom): The diamond ring effect at the beginning of totality at the June 30, 1973, eclipse. These photographs were taken from Loiengalani, Kenya, by the author's expedition.

Color Plate 12 (top): The solar corona at the 1980 total eclipse, photographed by the author's expedition to India. Note the many streamers typical of solar maximum.

Color Plate 13: (top) The flash spectrum of the solar corona at the 1977 total solar eclipse. The crescents are chromospheric spectral lines. The horizontal streak is an overexposed spectrum of a Baily's bead. (bottom) The spectrum during totality shows the green and red emission lines from the corona and the emission lines of prominences dotted around the solar limb. (Dennis di Cicco photos; Williams College expedition)

Color Plate 14 (top): An erupting prominence on the sun, photographed on August 9, 1973, in the ultraviolet radiation of ionized helium at 304 Å with the Naval Research Laboratory's slitless spectroheliograph aboard Skylab. Because of the technique used, radiation at nearby wavelengths appears as a background. A black dot visible on the solar surface slightly to the left of the base of the prominence may be the site of the flare that caused the eruption. Supergranulation is visible on the solar disk in the light of He II, and a macrospicule shows on the extreme right. (NRL/NASA photo)

Color Plate 15 (bottom): Contours of equal intensity computed for another eruptive prominence observed in the radiation of ionized helium with the NRL instrument aboard Skylab. Each color represents a different strength of emission. This view shows the prominence 90 minutes after its eruption, when the gas had moved 500,000 km from the solar surface. (NRL/NASA photo)

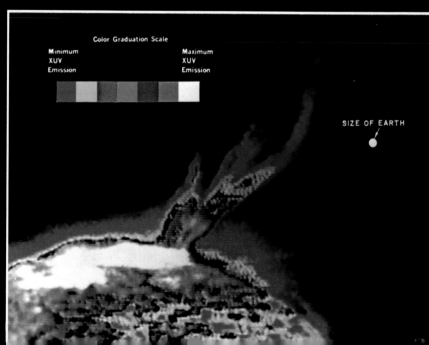

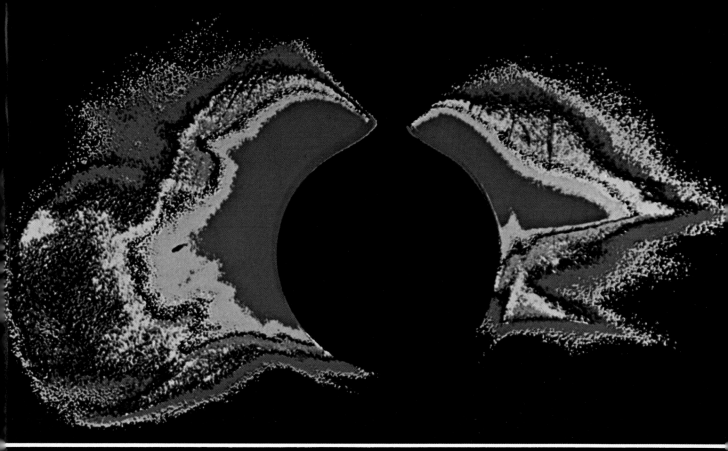

Color Plate 16 (top): Isophotes showing the intensity of the white light solar corona, reduced from observations made by the High Altitude Observatory's coronagraph aboard Skylab. The solar photosphere has been artificially occulted. (HAO/NASA photo)

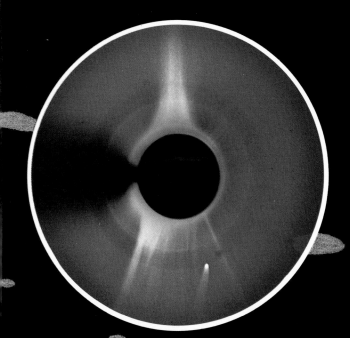

Color Plate 17: On the few days when Comet Kohoutek was closest to its perihelion, it was invisible to observers on Earth. The Skylab astronauts, fortunately, were able to sketch and photograph it. Here we see the image of Comet Kohoutek in the High Altitude Observatory's coronagraph on December 27, 1973. (HAO/NASA photo)

Color Plate 18: An eruption of the sun observed with the coronagraph aboard Solar Maximum Mission, launched in 1980. The image is 6 solar radii across. (HAO/NCAR/NSF and NASA)

HAO SMM CORONAGRAPH/POLARIMETER
DOY 103 UT= 1837

HAO SMM CORONAGRAPH/POLARIMETER
DOY 103 UT= 1416 POL=0

Color Plate 19: Two false-color views of the corona in visible light with the coronagraph aboard Solar Maximum Mission. The different colors show the intensities, which correspond roughly to the density of the gas present. Blue is the densest and yellow is the least dense. The occulting disk, which is the quarter-circle in the corner of each image, is 1.75 times the diameter of the sun. It is immediately surrounded by a series of colored rings that are an optical effect in the system. The image on the right shows a sharp-edged narrow feature known as a coronal spike. (Courtesy of Lewis House, Ernest Hildner, William Wagner, and Constance Sawyer—HAO/NCAR/NSF and NASA)

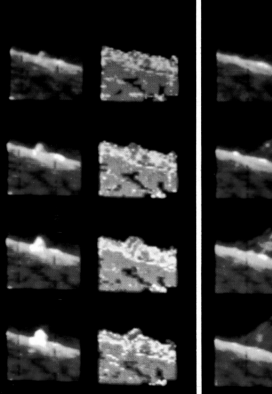

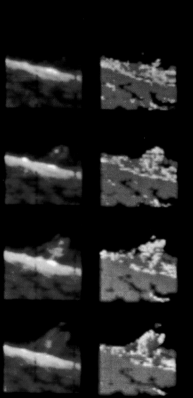

Color Plate 20 (left): A solar flare observed on April 30, 1980 from the Solar Maximum Mission. The sequence on the left, at 2.5-minute intervals shows the radiation of three-times ionized carbon at 1548 A The sequence on the right shows line-of-sight velocities found from the Doppler shift. Red represents material moving away and blue represents approaching material.

We see a jet of hot gas rising from the base of a magnetic loop and filling the top of the loop. Some of the gas may even go over the top of the loop. The velocity picture supports this last idea since it shows material coming toward us at the top of the loop. A bright flare was detected in x-rays 3 minutes after the last frame shown.

Color Plage 21 (right): A similar sequence of events observed from the Solar Maximum Mission on its next orbit 90 minutes later. Each picture here is separated by 7.5 minutes. Although at first glance the material appears to be rising, detailed measurement gives a different picture. Actually, features are not moving but new ones are appearing successively higher up. This may occur because hotter gas is cooling to the 200,000 degrees necessary to be visible at this temperature. (Color Plates 20 and 21 are courtesy of Bruce E. Woodgate, Einar Tandberg-Hanssen and colleagues at NASA's Marshall Space Flight Center)

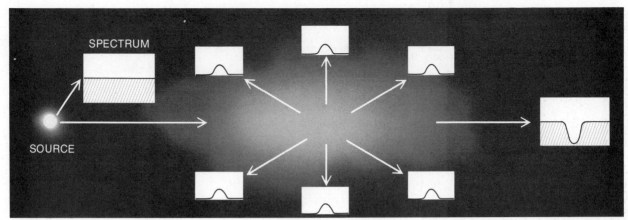

source of energy at a high temperature is permitted to pass through cooler vapor of any specific element, a set of absorption lines (Fig. 2–5) appears in the continuous spectrum at the same characteristic wavelengths as those of the emission lines of that element. Thus the vapor of an element through which light has passed has subtracted energy from the continuous spectrum at the set of wavelengths that is characteristic of that element (Fig. 2–6 and Color Plate 5). This was discovered by the German chemist Gustav Kirchhoff in 1859. If a continuous spectrum is directed first through the vapor of one absorbing element and then through the vapor of a second absorbing element, or through a mixture of the two gases, then the absorption spectrum that results will show the characteristic spectral absorption lines of both elements.

The same characteristic patterns of spectral lines that we detect on earth are observed in the spectra of stars, and we conclude basically that the same chemical elements are in the outermost layers of the stars. They absorb radiation from a continuous spectrum generated below them in the star, and thus cause the formation of absorption lines. Evidently, since each element

Figure 2–5 When we view a source that emits a continuum through a vapor that emits emission lines, we may see absorption lines at wavelengths that correspond to the emission lines. Each of the inset boxes shows a graph of intensity (in the vertical direction) against wavelength (in the horizontal direction). Each graph of the spectrum shown is the one you would measure if you were at the tip of the arrow looking back along the arrow. Note that the view from the right shows an absorption line, while from any other angle the vapor in the center appears to be giving off an emission line. Only when you look through one source silhouetted against a hotter source do you see any absorption lines.

Though the exact German pronunciation is difficult for native speakers of English, the name Kirchhoff is normally pronounced in English with a "hard" ch (that is, like k): Kirk'-hoff.

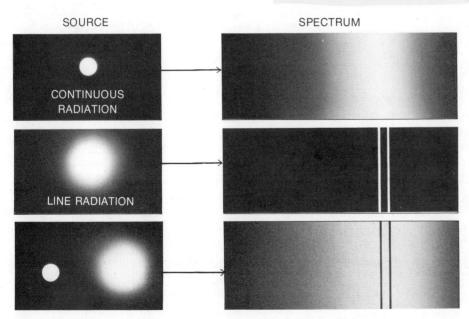

SOURCE   SPECTRUM

CONTINUOUS RADIATION

LINE RADIATION

Figure 2–6 In the left column, we see sources of continuous or spectral line radiation. When we look at these sources from the right side of the diagram (that is, looking through the spectral line source at the continuous source in the bottom case), we see the spectra displayed in the right column. When continuous radiation from a source *(top)* shines through a gas that by itself has an emission line spectrum *(center)* then an absorption line spectrum results *(bottom)* whenever the gas in the middle is cooler than the source of the continuous radiation. In this figure we are looking at the appearance of a spectrum to the eye, rather than at its graph.

We shall say more about the formation of spectral lines in Section 3.3.

has its own characteristic pattern of lines when it is in a certain range of conditions of temperature and density (a measure of how closely the star's matter is packed), the absorption spectrum of a star can be used (1) to identify the chemical constituents of the star's atmosphere, in other words, the types of atoms that make up the gaseous outer layers of the star, (2) to find the temperature of the surface of the star, and (3) to find the density of the radiating matter.

Not only individual atoms but also molecules, linked groups of atoms, exist in cooler bodies. (Such cooler bodies include the coolest stars, some of the gas between the stars, and the atmospheres of planets.) Molecules also have characteristic sets of absorption lines, and what we say for identifying elements goes for molecules as well.

Where does the radiation that is absorbed go? A very basic physical law called the *law of conservation of energy* says that the radiation cannot simply disappear. The energy of the radiation may be taken up in a collision of the absorbing atom with another atom. Alternatively, the radiation may be emitted again, sometimes at the same wavelength, but even so it could go off in any direction at random and not necessarily toward the observer. Thus, fewer bits of energy proceed straight ahead at that particular wavelength than were originally heading in that direction. This leads to the appearance of an absorption line when we look from that direction.

Moreover, if one element or molecule is present in relatively great abundance, then its characteristic spectral lines will be especially strong. By observing the spectrum of a star or planet one can tell not only which kinds of atoms or molecules are present but also their relative abundances (Appendix 4).

The method of spectral analysis is a powerful tool that can be used to explore the universe from our vantage point on earth. It tells us about the planets, the stars, and the other things in the universe. Further, it has many uses other than those in astronomy. For example, by analyzing the spectrum, one can determine the presence of impurities in an alloy deep inside a blast furnace in a steel plant on earth. One can measure from afar the constituents of lava erupting from a volcano. Sensitive methods developed by astronomers trying to advance our knowledge of the universe often are put to practical uses in fields unrelated to astronomy.

Figure 2–7 We will see in Section 2.3 that telescopes are used primarily to gather light from faint objects. They are also used to increase the resolution, our ability to distinguish details in an image. Photograph *A* shows a pair of point sources (sources so small or far away that they appear as points) that are not quite resolved. In *B*, they have passed Dawes's limit, the condition in which they can barely be resolved. In *C*, they are completely resolved.

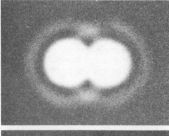

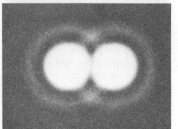

2.3 WHAT A TELESCOPE IS

The question of how to define a telescope has no simple answer, for a "telescope" to observe gamma rays may be a package of electronic sensors launched above the atmosphere, and a radio telescope may be a large number of small aerials strung over acres of landscape. We will begin, nevertheless, by discussing telescopes of the traditional types, which observe the radiation in the visible part of the spectrum. These optical telescopes are important in and of themselves because of the many things we have learned over the years by studying visible radiation. Also, the types of telescopes most used by students and by amateur astronomers (some of whom are quite professional in their approach to the subject) are most likely to be optical. Further, the principles of focusing and detecting electromagnetic radiation that were originally developed through optical observations have widespread use throughout the spectrum.

Contrary to popular belief, (the most important purpose for which most optical telescopes are used is usually to gather light.) (For the next few sections I shall drop the qualifying word "optical.") True, telescopes can be used to magnify as well, but for the most part astronomers are interested in observing fainter and fainter objects and so must collect more light to make these objects detectable. There are certain cases where magnification is important—such as for observations of the sun or for observations of the planets—but stars appear as mere points of light no matter how much magnification is applied.

When we look at the sky with our naked eyes, several limitations come into play in addition to the fact that our eyes see in only one part of the spectrum (the visible). Another limitation is that we can see only the light that passes through an opening of a certain diameter—the pupils of our eyes. In the dark, our pupils dilate so that as much light as possible can enter, but the apertures are still only a few millimeters across. A second additional limitation is one of time. Our brains distinguish a new image about 30 times a second, and so we are unable to store faint images over a long time to accumulate a brighter image. Astronomers overcome both these additional limitations with the combination of a telescope to gather light and a recording device, such as a photographic plate, to store the light. Other equipment may also be used, such as a spectrograph to analyze the spectral content of the light, or simple filters.

(A further advantage of a telescope over the eye, or of a large telescope over a smaller telescope, is that of *resolution*, the ability to distinguish finer details in an image (Fig. 2–7). A telescope is capable of considerably better resolution than the eye, which is limited to a resolution of 1 arc min.) For

For those of you who are photographers and who recognize the use of "f-stops," the eye ranges from f/2.8 to f/22.

Box 2.1 Angular Measure

Since Babylonian times the circle has been divided into parts on a system that uses multiples of 60. If we are at the center of a circle, we consider it to be divided into 360 parts as it extends all around us. Each part is called 1 degree (1°). Each degree, in turn, is divided into 60 minutes of arc (also written 60 arc min), often simply called 60 minutes (60′). And each minute is divided into 60 seconds of arc (also written 60 arc sec), often simply called 60 seconds (60″). Thus in 1°, we have 60 × 60 = 3600 seconds of arc. In this book we usually mention angular measure only to give a sense of appearance. It is useful to realize that your fist at the end of your outstretched arm covers about 5° (we say "subtends 5°") as seen from your eye, and your thumb subtends about ½°. The moon also covers about ½° of the sky; try covering the full moon with your thumb one clear night.

60 is a number that is conveniently divisible by many factors, namely, 2, 3, 4, 5, 6, 10, 12, 15, 20, and 30.

Actually, resolution depends not only on the aperture, but also on the wavelength of radiation being observed; the example we just considered was for figures given in green light, whose wavelength is approximately 5000 Å. As the wavelength of radiation doubles, resolution is halved. For example, a telescope that can resolve two sources emitting 5000 Å (green) light that are 1 arc sec apart could only resolve two 10,000 Å (infrared) sources if they were 2 arc secs apart. The limit of resolution for visible light, called Dawes's *limit, is approximately 2×10^{-3} λ/d arc sec, when the wavelength, λ, is in Å and the diameter of a telescope, d, is in cm.*

example, a telescope can distinguish the two components of a double star from each other in many cases where the unaided eye is unable to do so. If a telescope with a collecting area 10 centimeters across can resolve double stars that are separated by a certain angle, 1 arc sec, a telescope 20 centimeters across, twice the diameter, can resolve stars that are separated by half that angle, ½ arc sec. (In principle, for light of a given wavelength, *resolution is inversely proportional to the diameter of the telescope's primary mirror or lens*.)

We will further consider properties of telescopes in Section 2.9, after we have discussed particular types of telescopes and specific examples.

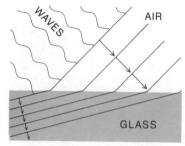

Figure 2–8A Light travels at a velocity in glass that is different from the velocity that it has in air. This leads to refraction.

Figure 2–8B The flower stem appears both bent and displaced at the boundary between the water and the air.

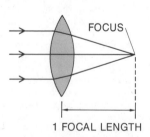

Figure 2–9 The focal length is the distance behind a lens to the point at which objects at infinity are focused. The focal length of the human eye is about 2.5 cm (1 in).

Even for large telescopes, the best resolution that can be achieved on the earth's surface is limited to about 1 arc sec on an average good night by the turbulence in the earth's atmosphere. Thus increasing the diameter of a telescope above a certain size no longer improves the resolution, even on the best of nights, though the advantages in terms of light-gathering power remain.

We will begin by considering the telescopes themselves, and then go on to consider the equally important devices that are used in conjunction with the telescopes.

*2.3a Wave Fronts

Let us consider waves of light hitting a block of glass at an angle, and draw an advancing plane of radiation (for example, the plane representing the first light that would arrive if the source were to suddenly become visible). This advancing plane is called a *wave front* (Fig. 2–8A). (T.V. newscasters, similarly, talk of weather fronts.) The wave front is straight, hence one side meets the glass slightly before the other side. It takes the other side of the wave front a very small additional amount of time to reach the glass, and in that time the first side has traveled a short distance through the glass. The speed of light is lower in the glass than it is in the air, so in that small additional time the part of the wave in the glass travels a shorter distance than the part of the wave still in the air. Because of this, by the time all the wave front has reached the glass, the wave front is moving in a different direction than it was before it started through the glass. This bending of the direction of travel of the light is called *refraction*. Put a straw in a glass of water (Fig. 2–8B), a branch in a lake, or a foot in a bathtub to observe this phenomenon.

2.3b Lenses and Telescopes

A suitably curved piece of glass can be made so that all the light that travels through it is bent, bringing all the rays of each given wavelength to a focus. Such a curved piece of glass is called a *lens.* The lens in your eye does a similar thing to make images of objects of the world on your retina. An eyeglass lens helps the lens in your eye accomplish this task.

The distance of the image from the lens on the side away from the object being observed depends in part on the distance of the object from the front side of the lens. Astronomical objects are all so far away that they are, for the purpose of forming images, as though they were infinitely far away. We say that they are "at infinity." The images of objects at infinity fall at a distance called the *focal length* behind the lens (Fig. 2–9).

The area of which an image is formed is called the *field of view.* A simple lens can be designed to focus light only from directly in front of it or from regions not too far to the side. Objects that are at angles far from the center of the field of view are not focused perfectly.

°Throughout this book, optional sections are starred. They can be skipped without loss of continuity.

The technique of using a lens to focus faraway objects was, we think, developed in Holland in the first decade of the 17th century. It is not clear how Galileo heard of the process, but it is known that Galileo quickly bought or ground a lens and made a simple telescope that he demonstrated in 1610 to the Senate in Venice (Fig. 2–10). He put this small lens and another, smaller lens at opposite ends of a tube. The second lens, called the *eyepiece*, is used to examine and to magnify the image made by the first lens, called the *objective*. Galileo's telescope was very small, but it magnified enough to impress the nobles of Venice, who had assembled to see the new invention.

Galileo turned his simple telescope on the heavens, and what he saw revolutionized not only astronomy but also much of seventeenth century thought. He discovered, for example, that the sun had spots—blemishes— on it, that Jupiter has satellites of its own, and that Venus has phases.

We shall be discussing the fundamental importance of these discoveries later on, in Section 12.5.

2.4 REFRACTING TELESCOPES

Refraction is the bending of light (or other electromagnetic radiation) when the light passes from one medium (e.g., transparent object, air, interstellar space) into another. A lens uses the property of refraction to focus light, and a telescope that has a lens as its major element is called a *refracting telescope*.

The current largest refracting telescope in the world has a lens 1 meter (40 inches) across. It is at the Yerkes Observatory in Williams Bay, Wisconsin (Figs. 2–11 and 2–12), and went into use in 1893.

Refracting telescopes suffer from several problems. For one thing, lenses suffer from *chromatic aberration*, the effect whereby different colors are focused at different points (Fig. 2–13). The speed of light in a sub-

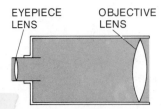

Figure 2–10 A simple refracting telescope consists of an objective lens and an eyepiece. The eyepiece shown here is a different type than the one used by Galileo. (Galileo's was concave—narrower at the center than at the edges— and this one is convex—wider at the center.) Modern eyepieces usually contain several pieces of carefully chosen shapes and materials. The *astronomical refractor* shown here, with a double convex eyepiece lens, was developed later. It gives an inverted image.

Figure 2–11 The Yerkes refractor, still the largest in the world. Note Albert Einstein at right of center in this 1921 picture.

Figure 2–12 The opening of the 1-meter (40-inch) refractor of the Yerkes Observatory was the cause of much notice in the Chicago newspapers in 1893.

Figure 2–11

Figure 2–12

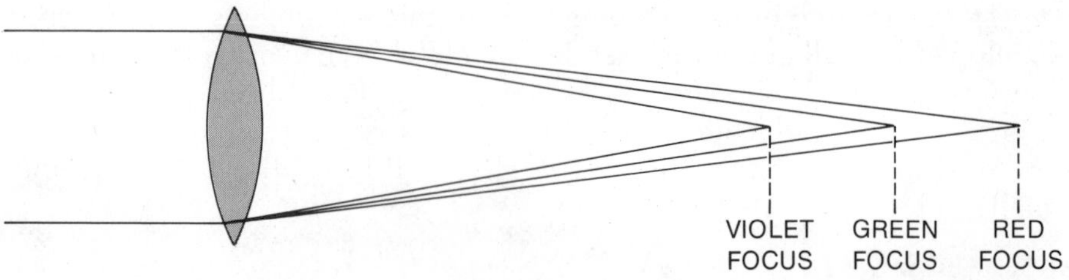

Figure 2–13 The focal length of a lens is different for different wavelengths, which leads to chromatic aberration.

stance—in glass, for example—depends on the wavelength of the light, so light composed of different colors bends different amounts and not all wavelengths can be brought to a focus at the same point. Ingenious methods of making "compound" lenses of different glasses that have slightly differing properties have succeeded in reducing this problem, but the chromatic aberration left over after as much as possible has been eliminated is a fundamental problem for the use of refracting telescopes.

With very large lenses, physical problems arise. It has been difficult to get a pure piece of glass sufficiently free of internal bubbles and sufficiently homogeneous throughout its volume to allow the construction of a lens larger than that at Yerkes. And a lens can be supported only from its rim when it is mounted in a telescope, since nothing may obstruct the aperture. Gravity causes the lens to sag in the middle, and this effect changes as the telescope is pointed in different directions. If the lens is made more rigid by making it thicker, this also makes it heavier and even more difficult to get a sufficiently pure lens blank. The 1-m telescope at Yerkes, for all of these reasons, remains the largest refracting telescope in the world.

2.5 REFLECTING TELESCOPES

Reflecting telescopes are based on the principle of reflection, with which we are so familiar from ordinary household mirrors. A mirror reflects the light that hits it so that the light bounces off at the same angle at which it approached. A flat mirror gives an image the same size as the object being reflected, but funhouse mirrors, which are not flat, cause images to be distorted.

One can construct a mirror in a shape so that it reflects incoming light to a focus. Let us consider a spherical mirror, for example. If you were at the middle of a giant spherical mirror, in whatever direction you looked you

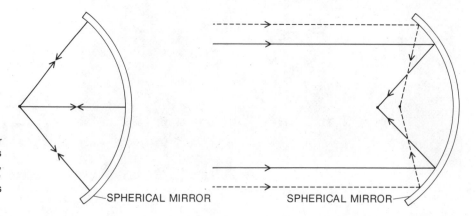

Figure 2–14 A spherical mirror focuses light that originates at its center of curvature back on itself, but suffers *spherical aberration* in that it does not perfectly focus light from infinity.

SPHERICAL MIRROR   SPHERICAL MIRROR

would see your own image. A sphere images whatever is at its center back at the same point, its center (Fig. 2–14). If we were to use just a portion of a sphere, it would still image whatever was at the center of its curvature back at that same point.

But the stars are far away, and so would not be at the center of curvature if we used a spherical mirror. Thus we can use a mirror that makes use of the fact that a parabola focuses *parallel light* to a point (Fig. 2–15). We say that the light from the stars and planets is "parallel light" because the individual light rays are diverging by such an imperceptible amount by the time they reach us on earth that they are practically parallel (Fig. 2–16).

A parabola is a two-dimensional curve that has the property of focusing parallel rays to a point. Telescope mirrors are actually *paraboloids*, which are the three-dimensional curves generated when parabolas are rotated around their axes of symmetry. Over a small area, a paraboloid differs from a sphere only very slightly, and one can ordinarily make a parabolic mirror by first making a spherical mirror and then deepening the center slightly.

The technique of making a small telescope mirror is not excessively arduous. Many amateur astronomers have made mirrors approximately 15 cm (6 inches) across without having had previous experience; it requires perseverance and about 50 hours of time. Making a large telescope mirror is another story, of course, and usually takes years.

Reflecting telescopes have several advantages over refracting telescopes. The angle at which light is reflected does not depend on the color of the light, so chromatic aberration is not a problem for reflectors. One normally deposits a thin coat of a highly reflecting material on the front of a suitably ground and polished surface. Since light does not penetrate the surface, what is inside the telescope mirror does not matter too much. Further, one can have a network of supports all across the back of the telescope mirror, to prevent the mirror from sagging under the force of gravity. All these advantages allow reflecting telescopes to be made much larger than refracting telescopes. (Only reflecting telescopes can be used at wavelengths shorter than those of visible light, because most ultraviolet radiation does not pass through glass.)

A potential problem with small reflecting telescopes could be the fact that one must gather the reflected light without impeding the incoming light. Clearly, if you put your head at the focal point of a small telescope mirror, the back of your head would block the incoming light and you would see nothing at all!

Isaac Newton got around this problem 300 years ago by putting a small diagonal mirror a short distance in front of the focal point to reflect the light so that the focus was outside the tube of the telescope. This type of ap-

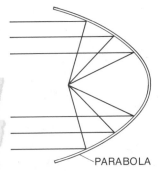

Figure 2–15 A parabola focuses parallel light to a point.

Figure 2–16 Light rays from very distant objects are diverging so slightly by the time they reach us that we speak of *parallel light*.

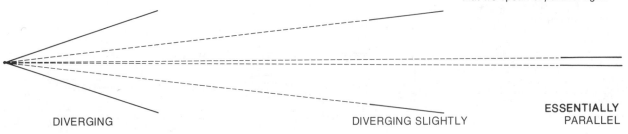

DIVERGING   DIVERGING SLIGHTLY   **ESSENTIALLY** PARALLEL

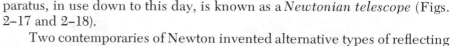

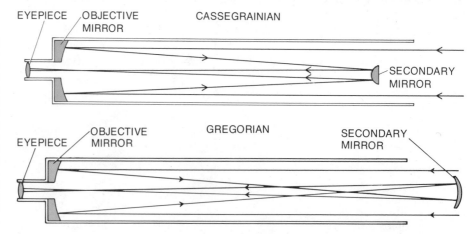

Figure 2-17 A Newtonian reflector.

paratus, in use down to this day, is known as a *Newtonian telescope* (Figs. 2–17 and 2–18).

Two contemporaries of Newton invented alternative types of reflecting telescopes. G. Cassegrain, a French optician, and J. Gregory, a Scottish astronomer, invented types in which the light was reflected by the secondary mirror through a small hole in the center of the primary mirror. These designs are called *Cassegrainian* (or *Cassegrain*) and *Gregorian telescopes* (Fig. 2–19).

The largest telescopes have several interchangeable secondary mirrors built in so that different foci can be used at different times. The very largest telescopes are so huge that the observer can even sit in a "cage" suspended at the end of the telescope at the *prime focus,* the position of the direct reflection of the light from the primary mirror (Fig. 2–20). The largest reflecting telescopes today (See Table 2–2) include the 6-m reflector in the Soviet Union (Color Plate 2), the 5-m reflector on Palomar Mountain in California, and 4-m telescopes at several locations including the Kitt Peak National Observatory in Arizona (Color Plate 1).

The fault of paraboloidal mirrors is that they focus light from only a relatively narrow field of view, limiting astronomers' ability to study extended objects. The narrow field of view also makes it difficult at times for astronomers to find stars bright enough for "guiding," that is, stars to hold steady in view in order to photograph fainter or extended objects near them in the sky. Modern large reflectors use a version of the Cassegrain design that allows a relatively wide field of view compared to that of a paraboloid. The primary mirror has its center deepened from the shape of a paraboloid, and the secondary mirror has a complicated shape to compensate. This

Figure 2-18 The replica of Newton's telescope in the visitor's gallery overlooking the 2.5-meter (98-inch) Isaac Newton telescope when it was at the Royal Greenwich Observatory at Herstmonceux, England, showing the side visible only to the astronomers.

Figure 2-19 A Cassegrainian (often called Cassegrain) telescope has a convex secondary mirror. (The length of the tube is exaggerated in the drawing to show the light paths. Cassegrains normally have short tubes relative to their mirror diameters.) A Gregorian telescope has a concave secondary that would be located to the right of the focus in a diagram oriented like this one. (The Solar Maximum Mission, a NASA spacecraft launched in 1980, has a Gregorian telescope aboard.)

Ritchey-Chrétien design leads to a curved rather than a flat focus, but this problem can be overcome with a small lens near the focus or by curving the film. Almost all the large telescopes constructed in the last 20 years are Ritchey-Chrétiens. So modern reflectors are not usually paraboloids after all!

2.6 SCHMIDT TELESCOPES

As we discussed, a parabolic reflector gives a good focus only for light that strikes the mirror parallel to the axis of the paraboloid. Even Ritchey-Chrétiens are in focus over a field only about 1° across. The fact that such reflectors have only a very small field in good focus limits their usefulness in certain cases.

Bernhard Schmidt, working in Germany in about 1930, invented a type of telescope that combines some of the best features of both refractors and reflectors. In a *Schmidt camera* (Fig. 2–21), the main optical element is a large mirror, but it is spherical instead of being paraboloidal. Before reaching the mirror, the light passes through a thin lens, called a *correcting plate*, that distorts the incoming light in just the way that is necessary to have the spherical mirror focus it over a wide field.

One of the largest Schmidt cameras in the world is on Palomar Mountain, in a dome near that of the 5-m telescope. The Schmidt camera has a correcting plate 1.2 meters (48 inches) across, and a spherical mirror 1.8 meters (72 inches) in diameter. Its field of view is about 7° across, much broader than the field of view, only about 2 minutes of arc across, of the 5-m reflector (Table 2–1).

The field of view of the 5-m is so small that it would be hopeless to try to map the whole sky with it because it would take so long. But the 1.2-meter Schmidt has been used to map the entire sky that is visible from Palomar. The survey was carried out through both red and blue filters; the project took seven years. Sample plates from this National Geographic Society/Palomar Observatory Sky Survey are shown in Figure 2–23. An infrared study is currently under way.

New Schmidt cameras at the European Southern Observatory and at the observatory at Siding Spring, Australia, are now being used in a joint project to extend the survey to the one-quarter of the sky that cannot be seen from Palomar. The 1.2-m Palomar Schmidt is carrying out an infrared survey.

Figure 2–20 The prime-focus cage of the 5-meter Palomar telescope.

The image falls at a location where the eye cannot be put, and thus the image is always recorded on film (Fig. 2–22). Accordingly, this device is often called a Schmidt camera instead of a Schmidt telescope.

Figure 2–21 Edwin P. Hubble, whose observational work is at the basis of modern cosmology, at the guide telescope of the 1.2-meter (48-inch) Schmidt camera on Palomar Mountain.

TABLE 2–1 SCHMIDT TELESCOPES

| Telescope or Institution | Location | Diameter | | Date Completed |
		Corrector (m)/(in)	Mirror (m)/(in)	
Karl Schwarzschild Observatory	Germany	1.3 m/52 in	2.0 m/79 in	1960
Palomar Obs.	California	1.2 m/48 in	1.8 m/72 in	1949
U.K. Schmidt	Australia	1.2 m/48 in	1.8 m/72 in	1973
ESO Schmidt	Chile	1.0 m/40 in	1.6 m/63 in	1973

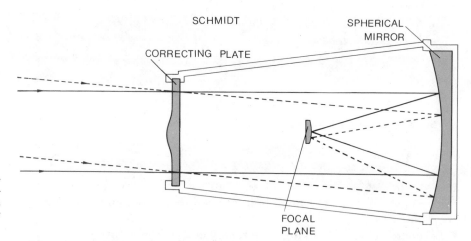

Figure 2–22 By having a non-spherical thin lens called a correcting plate, a Schmidt camera is able to focus a wide angle of sky onto a curved piece of film.

2.7 OPTICAL OBSERVATORIES

Once upon a time, over a hundred years ago, telescopes were put up wherever astronomers happened to be located. Thus all across the country and all around the world, on college campuses and near cities, we find old observatories. Later, sites that take advantage of still, dark mountain skies came into use. Now all major observatories are built at remote sites chosen for the quality of their observing conditions rather than for their proximity to a home campus.

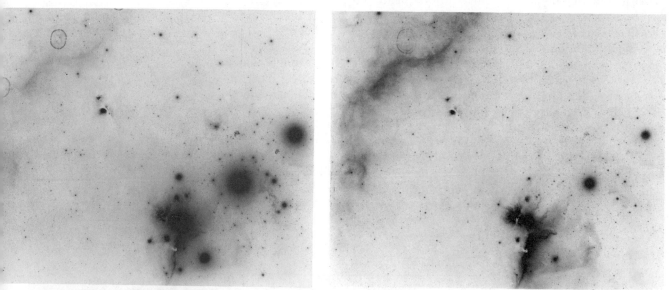

Figure 2–23 A 7° × 7° region of the sky in Orion taken with the 1.2-meter Schmidt camera as part of the National Geographic Society-Palomar Observatory Sky Survey. The left photograph shows the image with a blue-sensitive plate, and the right photograph shows the image with a red filter and a red-sensitive plate. The print is negative, so brighter areas appear blacker. The three bright stars at the lower right are Orion's belt; they are hot stars, which are brighter in the blue than in the red. The bright region to their left radiates mostly the red radiation characteristic of hydrogen, and so is brighter in the red than in the blue. The dust lanes, on the other hand, are more prominent in the blue. The Horsehead Nebula (also shown in Color Plate 59) is a region of dust obscuring part of the hydrogen emission that runs south of this region. Near it is a small nebula that is brighter in the blue than in the red, which leads us to conclude that we are seeing reflected starlight rather than hydrogen emission.

To the upper left is the brightest section of Barnard's ring, a ring of hydrogen-emitting gas that surrounds most of Orion. To the lower left center a large dust cloud causes the number of stars in this region to be less than in the upper right or lower left. This is best seen on the blue print. The three blotches at the upper left are artifacts called "ghost images" caused by internal reflections in the telescope of the three brightest stars.

On Mount Wilson near Los Angeles, George Ellery Hale built a 2.5-m telescope, with a mirror made of plate glass. It began operation in 1917 and was the largest telescope in the world for 30 years.

Box 2.2 Converting Units

In order to know what to multiply and divide when converting from one system of units to another, some people find it convenient to multiply and divide the words representing the units as though they were numbers. They set up a chain of multiplication and division of known conversion factors so that the unit names from which they are trying to convert cancel out. Then they do the same series of multiplications and division with the numbers themselves.

For example, 1 inch = 2.54 cm. If we divide both sides of the equation by 2.54 cm, we have the following equation:

$$\frac{1 \text{ in}}{2.54 \text{ cm}} = 1.$$

Equivalently, we can divide by 1 inch in order to get

$$\frac{2.54 \text{ cm}}{1 \text{ in}} = 1.$$

Now, we can always multiply any number by 1 without changing its value. Thus if we want to put, say, the size of the 200-inch telescope in units of centimeters, we can write:

$$200 \text{ inches} \times \frac{2.54 \text{ cm}}{1 \text{ in}}.$$

The "inches" in the numerator and denominator cancel, and we have 508 cm as the converted figure. Similarly, to convert the size of a 4-meter telescope to inches,

$$4 \text{ m} \times \frac{100 \text{ cm}}{1 \text{ m}} \times \frac{1 \text{ in}}{2.54 \text{ cm}} = \frac{400 \text{ in}}{2.54} = 157.5 \text{ in}.$$

This method works just as well with units of time (months, weeks, days, hours, seconds), units of volume, and so on.

One of the limitations of any telescope, reflecting or refracting, is the length of time that the mirror or lens takes to reach its equilibrium shape

Figure 2–24 Two of Russell W. Porter's drawings of the 5-m (200-inch) telescope on Palomar Mountain.

when exposed to the temperature of the cold night air. In the 1930's, the Corning Glass Works invented Pyrex, a type of glass that is less sensitive to temperature variations than ordinary glass. They cast, with great difficulty, a mirror blank a full 5.08 meters (200 inches) across.

The construction of the telescope was held up by many factors, including the Second World War. Only in 1949 did active observing begin with the telescope (Fig. 2–24). It stands on Palomar Mountain in southern California, and has only just lost its title of largest in the world. It is named the Hale telescope after George Ellery Hale, who was responsible for its construction. Most of the telescopes on Mount Wilson and Palomar Mountain were jointly operated for a time, and the organization was known as the Hale Observatories.

2.7a   New Sites

Other telescopes have been built since, but with a single exception they have all been substantially smaller. The largest telescope in the world is a new 6-m (236-inch) reflector in the Soviet Union, opened in 1976 in the Caucasus (Fig. 2–25).

Newer materials for telescope mirrors have been developed in recent years that are even less sensitive to heat variations than Pyrex. They include fused quartz and ceramic materials. The mirrors of newer telescopes for ground-based observatories as well as for space vehicles are made of these materials.

One of the largest telescopes is the 3-m (120-inch) reflector at the Lick Observatory. The telescope is used by astronomers of several campuses of the University of California. The headquarters of the Lick Observatory are on the main campus at Santa Cruz.

The National Science Foundation has sponsored and built a United

The Hale telescope has been known for years as the "200-inch telescope," but the scientific journals are now converting even this sacred name to metric units. We will round off 5.08 meters to 5 meters in this book. (Note that the mirror is about 17 feet in diameter, the size of an ordinary room.) The hole in the mirror to allow use of the Cassegrain focus is 1 m (40 inches) across; it blocks only $(40/200)^2 = (1/5)^2 = 4$ per cent of the mirror's overall area. Even when the prime focus observing cage, which is 1.8 m (72 inches) across, is installed, only $(1.85)^2 = 13$ per cent of the incoming light is blocked.

Figure 2–25 *(A)* The Soviet 6-meter (236-inch) telescope at the Special Astrophysical Observatory on Mount Pastukhov in the Caucasus in 1979. *(B)* The 6-m mirror during the installation procedure in 1979 of this replacement for the original mirror.

A

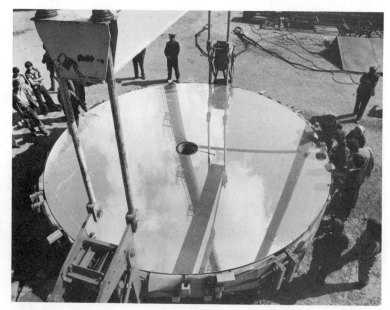

B

Figure 2–26 An aerial view of Kitt Peak. The 4-meter telescope is at the right; the solar telescope is the sloping structure at the lower left. The white dome at extreme left is the 11-meter radio telescope of the National Radio Astronomy Observatory, a separate organization. Sixteen telescopes are now located on Kitt Peak.

States National Observatory. Its telescopes are located on Kitt Peak, a sacred mountain of the Papago Indians, about 80 km (50 miles) southwest of Tucson, Arizona (Fig. 2–26). The headquarters of the Kitt Peak National Observatory are in Tucson, where the astronomers on the staff and the visiting astronomers work except during the specific periods when they are observing with one of the telescopes.

Kitt Peak (as the Kitt Peak National Observatory is usually called) has been set up for use by all the optical astronomers in the United States. Any astronomer can apply for observing time at one of the telescopes at Kitt Peak. A 4-m (158-inch) telescope (shown at the opening of this chapter and in Color Plate 1), opened there in 1973, is the largest of 16 telescopes now on Kitt Peak.

Part of the sky is never visible from sites in the northern hemisphere, and many of the most interesting astronomical objects in the sky are only visible or best visible from the southern hemisphere. Therefore, in the last decade special emphasis has been placed on constructing telescopes at southern hemisphere sites.

Many of the new sites are on coastal mountains west of the Andes range in Chile. The Kitt Peak staff supervised the construction of the Cerro Tololo Inter-American Observatory there (Fig. 2–27), at which a 4-m (158-inch) telescope has been constructed. The staff of the Hale Observatories has supervised, for the Carnegie Institution of Washington, the construction of a 2.5-m (101-inch) telescope on Cerro las Campanas, another Chilean peak. A consortium of European observatories and universities run the European Southern Observatory, where a dozen telescopes including a 3.6-m (142-inch) telescope have been erected on still another Chilean mountain top. It is quite common in the astronomical community to go off to Chile for a while to "observe."

Australia and Great Britain remain at the forefront with an Anglo-Australian Telescope (AAT) of 3.9 meters (153 inches), which was opened at Siding Spring in eastern Australia in 1974. The British 2.5-m (98-inch) Isaac Newton telescope is being moved from the cloudy skies of the south of England to a more favorable site in the Canary Islands. A new mirror, a few

It is not useful to get too exact in specifying the diameters of telescopes. The outer 4 cm (almost 2 inches) of the 4-m mirrors are beveled, thus limiting the usable diameter to 3.96 m (156 inches). Further, the bits of the outermost part of the Kitt Peak mirror do not match the desired shape as accurately as the rest of the mirror; the outermost 19 cm (8 inches) are masked off, leaving a usable diameter of 3.81 m (150 inches). The Anglo-Australian telescope, with 3.89 m (153 inches) of usable aperture, is actually intermediate in size between the Kitt Peak and Cerro Tololo telescopes. We shall refer to them all as "4-meter telescopes."

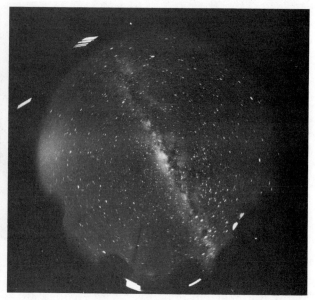

Figure 2–27 Two views of the Cerro Tololo Inter-American Observatory taken from the same location with a fish-eye lens. *(Left)* The camera was held fixed, and star trails and the overexposed image of Venus show above the domes. *(Right)* The camera tracked the stars; the domes are blurred but the Milky Way shows.

centimeters larger, is being installed in it. A 4.2-m (165-inch) telescope is also being constructed alongside.

Another important new site is in Hawaii. A 2.2-m (88-inch) telescope in use on Mauna Kea on the island of Hawaii was joined in 1979 by a 3.8-m (150-inch) United Kingdom Infrared Telescope, a 3.6-m (140-inch) telescope built by a joint Franco-Canadian-Hawaiian group, and by a 3.0-m (120-inch) telescope built by NASA and operated by the University of Hawaii (Fig. 2–28). It is unprecedented for three such major telescopes to be erected at one place at one time. This site is at an especially high altitude,

Figure 2–28 The summit of Mauna Kea on the island of Hawaii now boasts of a variety of large telescopes. The 3.6-m Canada-France-Hawaii Telescope is at far left. Then come the 0.6-m planetary patrol telescope and the 2.2-m telescope of the University of Hawaii, followed by the 3.8-m United Kingdom Infrared Telescope and another 0.6-m telescope. The 3-m NASA/University of Hawaii Infrared Telescope Facility is in the right foreground. The National Science Foundation is planning a 25-m diameter radio telescope for this site to study the millimeter range of the spectrum.

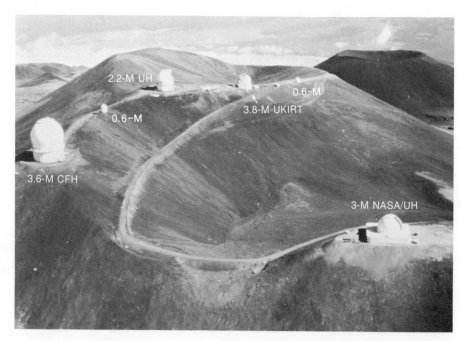

4145 m (13,600 ft), and the air above is especially dry. Since water vapor, which blocks infrared radiation, is minimal, these telescopes are especially suitable for infrared observations.

2.7b New Directions in Telescope Making

Even though the light-gathering power increases rapidly with the size of the telescope, the cost goes up even more rapidly, with the cube of the diameter. Thus the funds have simply not been available in the United States to build a larger telescope than the 5-m (200-inch).

Alternative methods of getting more aperture are now under investigation. The Harvard-Smithsonian Center for Astrophysics and the University of Arizona have jointly built the Multiple Mirror Telescope (MMT) (Fig. 2–29), which has six 1.8-m (72-inch) paraboloids linked together and aligned by lasers, to focus all the light from these several mirrors at the same point. It is much cheaper to build even a half dozen 1.8-m telescopes than it is to build one larger telescope of 4.5 meters (176 inches), the equivalent total aperture. The MMT is on Mount Hopkins, south of Tucson, Arizona. The experiment of trying such an array, which requires constant adjustments of its elements, appears to be a success. Similar techniques may be used in the future for the Next Generation Telescope. Plans for telescopes up to 25 meters across are under consideration (see Section 28.1).

The continued building of optical observatories, with the current unprecedented spate of large telescopes, shows the vitality of ground-based optical astronomical research, even though new techniques involving space telescopes, radio astronomy, and other new techniques are now providing important data at an increasing rate.

Though many of the world's large optical observatories have just been described, the current definition of observatory is no longer the place where the telescopes are located but rather the place where the astronomers are. Most of an astronomer's work is done by studying the data that have been recorded on film, on computer tape or otherwise, during a field trip to an observatory. Other astronomers, theoreticians, carry out their research without ever using a telescope. So it is still possible to have major observatories located in urban areas. The Center for Astrophysics of the Harvard College Observatory and the Smithsonian Astrophysical Observatory (in Cambridge, Massachusetts), and the Observatoire de Paris (in Meudon, France), shown in Figure 2–30, are among many examples. The only thing that you can't do very well at such observatories is observe.

2.8 TELESCOPES IN SPACE

The largest astronomical telescope that has been sent into space has been the 0.9-m (36-inch) telescope aboard the *Copernicus* satellite, the third Orbiting Astronomical Observatory, launched by NASA in 1973. The International Ultraviolet Explorer (IUE), launched by a joint effort of NASA and the British and European space organizations in 1978, carries a 0.45-m (18-inch) telescope with two spectrographs for studies of the ultraviolet. It carries a television-like system that records data much more efficiently than *Copernicus* and hence gives spectra more quickly.

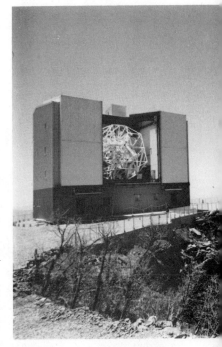

Figure 2–29 The Multiple Mirror Telescope on Mount Hopkins in Arizona became fully operational in 1980.

Figure 2–30 The Paris Observatory has most of its facilities in Meudon, a suburb just to the south of the city. Few stellar observational programs are carried out from this site, though hundreds of astronomers work here.

This brief discussion mentions only optical observatories, and not the many telescopes that are now in existence on the ground and in space to observe other regions of the spectrum, to which we will return in Sections 2.12 and 2.13.

Other laboratories have not been based on observational projects, but nonetheless provide the basic physical and spectroscopic measurements or computer facilities that are important for astrophysical research. They include the Joint Institute for Laboratory Astrophysics in Boulder, Colorado, and the Herzberg Institute in Ottawa.

TABLE 2–2 THE LARGEST OPTICAL TELESCOPES

Telescope or Institution	Location	Diameter (m)	Diameter (in)	Date Completed	Mirror
Soviet Special Astrophysical Observatory	Caucasus	6.0 m	236	1976	Pyrex
Palomar Observatory: Hale Telescope	California	5.0 m	200	1950	Pyrex
Multiple Mirror Telescope (MMT)	Arizona	4.5 m	176	1979	Fused silica
La Palma Observatory: U. K. Telescope	Canary Islands	4.2 m	165	1985	Cer-Vit
Cerro Tololo Inter-American Observatory (CTIO)	Chile	4.0 m	156	1975	Cer-Vit
Anglo-Australian Telescope (AAT)	Australia	3.9 m	153	1975	Cer-Vit
Kitt Peak National Observatory (KNPO): Mayall Telescope	Arizona	3.8 m	150	1974	Quartz
United Kingdom Infrared Telescope (UKIRT)	Hawaii	3.8 m	150	1979	Cer-Vit
European Southern Observatory (ESO)	Chile	3.6 m	142	1976	Fused quartz
Canada-France-Hawaii (CFH)	Hawaii	3.6 m	140	1979	Cer-Vit
Max Planck Institute, Germany	Spain	3.5 m	138	—	
NASA-Hawaii	Hawaii	3.0 m	120	1979	Cer-Vit
Lick Observatory: Shane Telescope	California	3.0 m	120	1959	Pyrex
McDonald Observatory	Texas	2.7 m	107	1968	Fused quartz
Crimean Astrophysical Observatory: Shajn Telescope	Crimea	2.6 m	102	1961	Pyrex
Byurakan Observatory	Armenia	2.6 m	102	1976	Pyrex
Las Campanas Observatory: Irenèe du Pont Telescope	Chile	2.5 m	101	1977	Fused quartz
La Palma Observatory: U. K. Issac Newton Telescope	Canary Islands	2.5 m	101	1982	Zerodur
Mt. Wilson Observatory: Hooker Telescope	California	2.5 m	100	1917	Glass
Wyoming Infrared Observatory (WIO)	Wyoming	2.3 m	92	1977	Cer-Vit
Steward Observatory	Arizona	2.3 m	90	1969	Fused quartz
University of Hawaii	Hawaii	2.2 m	88	1970	Fused silica

Figure 2–31 The Space Telescope is shown in this artist's conception. It will operate from a position above the earth's atmosphere, so it will be able to probe about 7 times deeper into space than ground-based telescopes. The Space Telescope will give images with 0.1 arc second resolution.

The Space Telescope (Fig. 2–31), with a mirror 2.4 meters (95 inches) in diameter, is scheduled for launch in 1983 to concentrate on visible and ultraviolet studies. Free of the twinkling effects caused by turbulence in the earth's atmosphere and located above the atmosphere in perennially black skies, the ST will be able to observe details of objects in space about seven times finer than can be observed with ground-based observatories, partly because the sky is darker and partly because images will be concentrated better on the detectors. This will allow us to see objects about 7 times farther than we can now, which will give us access to a volume of space about 350 times larger than we can now study to the same degree. As a result, our understanding of many astronomical problems will be drastically improved by the capabilities of the Space Telescope. For example, we should be able to improve our knowledge of distances to thousands of galaxies and thus improve our understanding of the overall cosmic scale of distances. We should be able to study quasars so far back in time that conditions in the universe must have been very different from the way they are now.

The Space Telescope's primary mirror (Fig. 2–32), of the Ritchey-Chrétien design, will direct light to one of several analyzing instruments. The major instruments are a Wide Field/Planetary Camera, a Faint Object Spectrograph, a High Resolution Spectrograph, and a High Speed Photometer/Polarimeter.

The Space Telescope, launched from Space Shuttle, will fly freely and will be controlled from earth. Astronauts can visit the Space Telescope with the Space Shuttle to make repairs, carry out maintenance, and even change scientific instruments. Such visits are expected every 2½ years. The ST can even be brought down to earth, refurbished, and launched again. It is designed to work for at least 15 years in orbit.

The Space Shuttle will also be used to launch other telescopes. Infrared telescopes, cooled to 2 K (−271°C), will be among those launched.

Figure 2–32 The 2.4-m (94-inch) mirror blank of the Space Telescope, before it was ground and polished into its final shape.

Work on the faintest objects will be very slow, for at the limit of its observing ability the Space Telescope will receive only 1 photon (i.e., one particle of light energy) per minute. So there will still be plenty for ground-based telescopes to do.

The European Space Agency is cooperating with NASA by providing the solar array, one of the instruments, and some of the staff.

2.9 LIGHT-GATHERING POWER AND SEEING

The principal function of large telescopes is to gather light. The amount of light collected by a telescope mirror is proportional to the area of the mirror, πr^2. (It is often convenient to work with the diameter instead of the

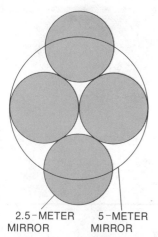

2.5-METER   5-METER
MIRROR   MIRROR

Figure 2–33 A telescope twice the diameter of another has four times the collecting area.

The skies are no longer as dark as they used to be. Much of the problem of light pollution is caused by street lighting. Most street lights, for example, are designed so that some of their light goes up in the air instead of down toward the street. This stray light benefits nobody, and the simple addition of shields on street lights would help the light pollution problem considerably.

radius, since the diameter is the number usually mentioned. The formula for the area of a circle may, of course, also be written as $\pi d^2/4$.)

(The ratio of the areas of two telescopes is thus the ratio of the square of their diameters (d_1^2/d_2^2), which is usually more easily calculated as the square of the ratio of their diameters $(d_1/d_2)^2$. For example, the 5-meter telescope has an area 4 times greater than the area of the 2.5-meter telescope, $(5/2.5)^2$ (Fig. 2–33). Thus during observations of equal duration the 5-m would collect four times as much light as the 2.5-m if all other things were equal, and for the same exposure time would record fainter stars.)

2.9a Seeing and Transparency

(A telescope's light-gathering power (and its ability to concentrate it in as small an area as possible) is really all that is important for the study of individual stars, since the stars are so far away that no details can be discerned no matter how much magnification is used.) The main limitation on the smallest image size that can be detected is caused not in the telescope itself, but rather by turbulence in the earth's atmosphere. The limitations are connected with the twinkling effect of stars. (The steadiness of the earth's atmosphere is called, technically, the *seeing*.) We would say, when the atmosphere is steady, "The seeing is good tonight." "How was the seeing?" is a polite question to ask astronomers about their most recent observing runs. (Another factor of importance is the *transparency*, or how clear the sky is. It is quite possible for the transparency to be good but the seeing to be very bad, or for the transparency to be bad (a hazy sky, for example), but the seeing excellent.)

2.9b Magnification

(For some purposes, including the study of the planets and the resolution of double stars, the magnification provided by the telescope does play a role. Magnification can be important for the study of the sun, as well, and for *extended objects* (as distinguished from *point objects*, objects whose images are points) like nebulae or galaxies.)

(*Magnification* is equal to the focal length of the telescope objective (the primary mirror or lens) divided by the focal length of the eyepiece. By simply substituting eyepieces of different focal lengths one can get different magnifications. However, there comes a point where even though using an eyepiece of short focal length will enlarge the image, it will not give you any more visible detail. When using a telescope one should not exceed the maximum usable magnification, which depends on such factors as the diameter of the objective and the quality of the seeing.)

A telescope of 1 meter focal length will give a magnification of 25 when used with a 40-millimeter eyepiece (that is, 1 m/40 mm = 25), an ordinary combination for amateur observing. If we substitute a 10-millimeter eyepiece, the same telescope has a magnification of 100. We would say that we were observing with "100 power."

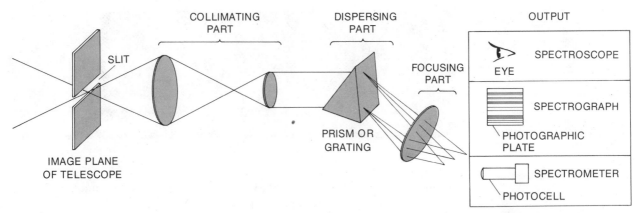

COLLIMATING PART · DISPERSING PART · SLIT · FOCUSING PART · OUTPUT · SPECTROSCOPE · EYE · SPECTROGRAPH · PHOTOGRAPHIC PLATE · PRISM OR GRATING · SPECTROMETER · PHOTOCELL · IMAGE PLANE OF TELESCOPE

Figure 2–34 The light entering the instrument from the left passes through a slit that defines a line of radiation, is made parallel (*collimated;* a double-lens collimator is shown but other alternatives are also used) and directed on the prism (or other dispersing element), and then is focused. The device is called a *spectroscope* if we observe with the eye, a *spectrograph* if we record the data on a photographic plate, and often a *spectrometer* if we record the data electronically.

2.10   SPECTROSCOPY

We have discussed the collection of quantities of light by the use of large mirrors or lenses, and the focusing of this light to a point or suitably small area. Often one wants to break up the light into a spectrum instead of merely photographing the area of sky at which the telescope is pointed.

A prism breaks up light into its spectrum for the same reasons that light focused by a lens has chromatic aberration. A wave front is refracted as it hits the side of the prism, but it is refracted by different amounts at different wavelengths.

When a beam of parallel light falls directly on a prism, the spectral lines that result are often indistinct because the spectrum from one place in the beam overlaps the spectrum from an adjacent spot in the beam. We can limit the blurring of spectral lines that results because of this by allowing only the light that passes through a long, thin opening called a *slit* to fall on the prism. Then the spectral lines are sharp and relatively narrow.

Actually, astronomical spectra are no longer usually made by means of prisms. Light may be broken up into a spectrum without need for a prism by lines that are ruled on a surface very close together. And I do mean close together: one can have 10,000 lines ruled in a given centimeter (25,000 lines in a given inch)! Such a ruled surface is called a *diffraction grating*. Making diffraction gratings is a difficult art.

One can arrange a small telescope so as to examine the spectrum that comes off a prism or grating (Fig. 2–34). A device that makes a spectrum, usually including everything between the slit at one end and the viewing eyepiece at the other, is called a *spectroscope.* When the resultant spectrum is not viewed with the eye but is rather recorded either with a photographic plate or by some other means, the device is called a *spectrograph.* When the resultant spectrum is measured with an electronic device, rather than being recorded on film, the device is called a *spectrometer* or *spectrophotometer.*

2.11   RECORDING THE DATA

When the desired information emerges from a telescope, whether it be an image of a field of stars or a spectrum, astronomers usually want a more

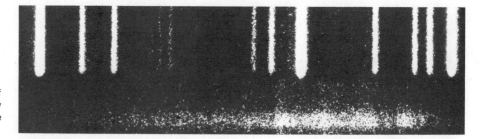

Figure 2–35 This enlargement of the spectrum of a distant galaxy shows clumps of grain on the photographic plate.

Figure 2–36 These three views of Orion (see also Fig. 10–2) show the effect of increased exposure time. The three stars representing the belt, shown also in Fig. 2–23, appear at the center of the photograph. The Orion Nebula (Color Plate 56) is below them. Though the red star Betelgeuse, at the upper left, appears bright to the eye, it does not appear very bright here because these photographs were taken on photographic plates that were sensitive only to the blue part of the spectrum. The longest exposure shows that nebulosity—glowing gas or the reflection of starlight off dust—covers almost the entire constellation.

permanent record than is afforded by merely observing the image and more accuracy than can be guaranteed in a sketch.

Traditionally, one puts a photographic film or "plate" (short for photographic plate, a layer of light-sensitive material on a sheet of glass) instead of an eyepiece at the observer's end of the telescope, so as to get a permanent record of the image or spectrum.

Basically, a photographic plate (or film) consists of a glass or plastic backing covered with an *emulsion*. The emulsion contains grains of a silver compound. When these grains are struck by enough light, they undergo a chemical change. The result, after development, is a "negative" image, with the darkest areas corresponding to the brightest parts of the incident image. Inspection of a photographic plate under high magnification shows the grainy structure of the image (Fig. 2–35).

This use of photographic film also significantly increases our ability to detect signals from faint sources. The eye and brain can process information for only a fraction of a second at a time and do not function cumulatively. A photographic plate, however, can be left exposed to radiation for a long period (Fig. 2–36), sometimes for hours. Just as a long exposure with an ordinary camera can record objects that are only dimly lit, a long exposure with a telescope can record fainter objects than would a shorter exposure.

A

B

C

Other methods of recording data have been developed. Some say that the day of film is past, though new emulsions now available are more sensitive and have finer grain than older emulsions. Still the future seems to lie in electronic devices for many investigations.

The photocell is a basic element device used by astronomers. Certain materials give off an electric current when struck by light. That current can be measured, perhaps by reading a meter or perhaps automatically.

The electrons that become available inside the tube when light hits it can be directed to hit other special elements of the tube. Each of the elements can emit several electrons for each one hitting it. In this way, the original number of electrons can be multiplied, often by a substantial factor (perhaps 10^6). Such a device is called a *photomultiplier.*

Only 15 per cent of the observing time at Palomar is now used for photography.

One advantage of a photocell or photomultiplier is that its response is "linear"; that is, unlike film, it changes by twice as much electric current when twice as much light hits it. This makes it much easier to figure out how bright an object actually is or by what factor of brightness an object or spectrum is changing. One of the disadvantages of photocells and photomultipliers is that they measure only the total intensity of radiation falling on them from whatever direction; they do not form an image in the way that film does. For measuring the intensity of light of a star, photomultipliers are fine. Such measurements of intensity are called *photometry.*

Photomultipliers are much more sensitive to faint signals than is photographic film, and can also be made to be sensitive in a wider region of the spectrum. Photometry has long been carried out with photocells and photomultipliers and can be done very precisely. In some observatories, the procedure is automated so that the strength of the current flowing from the photomultiplier is monitored by a computer. When enough light has been gathered to allow a measurement of a certain accuracy (say, to 1 per cent of the signal level), the computer ends the "exposure" and moves the telescope to the next object.

*2.11a Electronic Imaging Devices

For some years devices called *image tubes* have been used, sometimes to increase the sensitivity of film and sometimes to increase the wavelength coverage. Light from the telescope or spectrograph may fall on the faceplate of a tube and generate internal electrons or light, but ultimately these secondary emissions still fall on a photographic plate or film. Since the light generated need not be at the same wavelength as the incident light, one can, say, have an infrared signal at a wavelength beyond the sensitivity of film fall on the image tube and generate green light to fall on ordinary film or on a television-type camera.

A new series of devices is now able to provide an electronic imaging capability. The technology for making these is related to the new technology that has recently brought electronic calculators to their present versatility and price range. These devices can be used in the place of photographic plates to make either direct images of astronomical objects or of spectra. Rapid advances are being made in this field.

Among several competing imaging devices are "silicon diode vidicons" (Fig. 2–37), for which a series of small crystals, each only a few microns across, can actually be grown on a thin silicon wafer. Each point on the wafer acts as though it were a small photocell.

Figure 2–37 The Cassegrain focus of the 1.5-m telescope at Cerro Tololo, shown here, contains the silicon vidicon spectrometer. The signals are sent to a nearby computer. After astronomer Karen Kwitter finishes her adjustments, she will run the equipment from a computer terminal on the observing floor.

Still more recent electronic detectors are integrated-circuit devices that sense light on a surface and simultaneously scan the image off that surface with internal circuitry. These "self-scanned arrays" come in several types, including Charge-Coupled Devices (CCD's). The name "CCD" refers to the scanning method, which involves shifting the electrical signal from each point to its neighbor one space at a time, analogous to what happens with water in a bucket brigade (for fighting a fire) when buckets of water are passed from one individual to the next. A self-scanned array's picture is created by scanning the picture elements line by line, as occurs in our televisions. CCD's are being used on the Space Telescope.

CCD's are coming into wide-spread use in many fields, such as computer memories.

2.12   OBSERVING AT SHORT WAVELENGTHS

From antiquity until 1930, observations in a tiny fraction of the electro-magnetic spectrum, the visible part, were the only way that observational astronomers could study the universe. Most of the images that we have in our minds of objects in our galaxy (e.g., nebulae) are based on optical studies, since most of us depend on our eyes to discover what is around us.

If they had to make a choice, however, between observing only optical radiation or only everything else, many astronomers might choose "everything else." The rest of the electromagnetic spectrum carries more information in it than do the few thousand angstroms that we call visible light.

It is often useful to think of light as particles of energy called *photons* instead of as waves. Photons that correspond to light of different wavelengths have different energies.

In this section we will first discuss the basic techniques of ultraviolet, x-ray, and gamma-ray astronomy. We will then go on, in the next section, to see how infrared and radio astronomy join with other observing methods to help us investigate space.

2.12a   Ultraviolet and X-Ray Astronomy

Ultraviolet and x-ray photons, which have shorter wavelengths, have greater energies than photons of visible light. Thus they also have enough energy to interact with the film grains. Photographic methods can be used, therefore, throughout the x-ray, ultraviolet, and visible parts of the spectrum. Electronic devices like the ones that work in the visible work in the ultraviolet too.

The International Ultraviolet Explorer (IUE) spacecraft, for example, uses television-type devices (vidicons) to record the spectra that are imaged on them. The data can be read off the vidicons and radioed to earth.

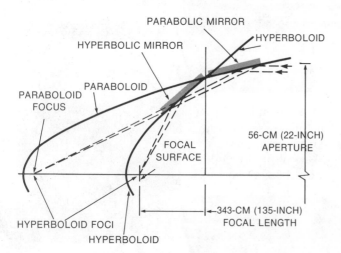

Figure 2–38 X-rays are reflected from mirrors at grazing incidence. The diagram shows the arrangement of reflecting surfaces in HEAO-2, the Einstein Observatory. Two parallel rays are shown entering from the right, and reflecting off a paraboloidal surface. The actual angle of the reflection is less than 1°. The reflected rays are intercepted by a segment of a hyperboloid, which reflects them at grazing incidence to the focus.

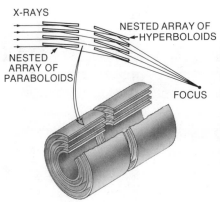

Figure 2–39 Four similar paraboloid/hyperboloid arrangements, all sharing the same focus, are nested within each other to increase the area of telescope surface that intercepts x-rays in NASA's HEAO-2, the Einstein Observatory. Diameters range from 33 to 56 cm (13 to 22 inches). The resolution of the system is 2 arc seconds, within a factor of 4 or 8 of the best ground-based seeing.

At this shorter end of the spectrum, gamma rays, x-rays, and ultraviolet light do not come through the earth's atmosphere. Ozone, a molecule of three atoms of oxygen (O_3), is located in a broad layer between about 20 and 40 kilometers in altitude and prevents all the radiation at wavelengths less than approximately 3000 Å from penetrating. To get above the ozone layer, astronomers have launched telescopes and other detectors in rockets and in orbiting satellites.

Fears that the earth's ozone layer is in danger will be discussed in Section 16.6.

2.12b X-Ray Telescopes

In the shortest wavelength regions, we cannot merely use mirrors to image the incident radiation in ordinary fashion, since the x-rays will pass right through the mirrors! Fortunately, x-rays can still be bounced off a surface if they strike the surface at a very low angle. This is called *grazing incidence*. This principle is similar to that of skipping stones across the water. If you throw a stone straight down at a lake surface, the stone will sink immediately. But if you throw a stone out at the surface some distance in front of you, the stone could bounce up and skip along a few times. By carefully choosing a variety of curved surfaces that suitably allow x-radiation to "skip" along (Fig. 2–38), astronomers can now make telescopes that actually make x-ray images. But the telescopes appear very different from optical telescopes (Figs. 2–39 and 2–40).

Figure 2–40 *(A)* The cylindrical mirror of the Einstein Observatory does not resemble mirrors for optical telescopes because it is used at grazing incidence. This is the outermost of four concentric mirrors. *(B)* Caltech scientist Gordon Garmire looks through a mirror polished for grazing incidence with parabolic and hyperbolic sections.

A

B

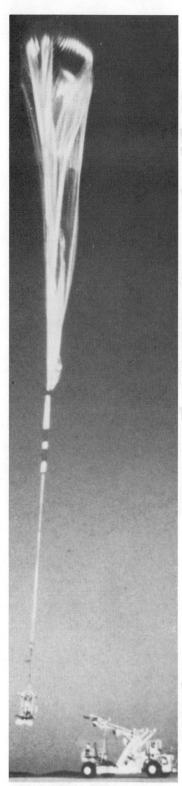

Figure 2-41 A balloon launch carries aloft a 102-cm infrared telescope.

In order to form such high-energy photons, processes must be going on out in space that involve energies very much higher than most ordinary processes that go on at the surface layers of stars. The study of the processes that bring photons or particles of matter to high energies is called *high-energy astrophysics*. A current emphasis in space observations has been in x-ray and gamma-ray spectral regions that give us the most information about these processes. NASA launched a series of three High-Energy Astronomy Observatories (HEAO's) to study high-energy astrophysics. HEAO-1, launched in 1977, surveyed the x-ray sky with high sensitivity and rapid time resolution. HEAO-2, launched in 1978, is able to point at and study in detail the interesting objects mapped by HEAO-1. HEAO-3, launched in 1979, is studying cosmic rays (particles of matter moving with high energies) and gamma rays.

2.13 OBSERVING AT LONG WAVELENGTHS

2.13a Infrared Astronomy

Infrared photons have longer wavelengths and thus lower energies than visible photons. They do not have enough energy to interact with ordinary photographic plates. Some special films can be used at the very shortest infrared wavelengths, but for the most part astronomers have to employ methods of detection involving electronic devices in this region of the spectrum.

Another major limitation in the infrared is that there are very few windows of transparency in the earth's atmosphere. Most of the atmospheric absorption in the infrared is caused by water vapor, which is located at relatively low levels in our atmosphere compared to the ozone that causes the absorption in the ultraviolet. Thus we do not have to go as high to observe infrared as we do to observe ultraviolet. It is sufficient to send up instruments attached to huge balloons (Fig. 2–41).

Many useful observations can be made in the infrared from high mountain tops. This has led to the construction of new infrared telescopes at such sites as Mauna Kea in Hawaii, and to the positioning of the Multiple Mirror Telescope on one of the higher Arizona peaks. The University of Wyoming's new infrared telescope is shown in Figure 2–42.

2.13b Radio Astronomy

At even longer wavelengths, in the radio region of the spectrum, is another window of transparency. The techniques of radio astronomy, now well established as one of the major branches of astronomy, will be discussed briefly here and further in various places in the book, especially in Part V.

One cannot always pinpoint the discovery of a whole field of research as decisively as that of radio astronomy. In 1931, Karl Jansky of the Bell Telephone Laboratories in New Jersey was experimenting with a radio antenna to track down all the sources of noise that might limit the performance of short-wave radiotelephone systems (Figs. 2–43 and 2–44). After a time, he

Figure 2-42 The University of Wyoming's new telescope (*left* and *right*) is optimized for observing in the infrared.

noticed that a certain static appeared at approximately the same time every day. Then he made the key observation: the static was actually appearing four minutes earlier each day. This was the link between the static that he was observing and the rest of the universe. Jansky's static kept sidereal time, just as the stars do (see Section 4.4), and thus was coming from outside our solar system! Jansky was actually receiving radiation from the center of our galaxy.

The principles of radio astronomy are exactly the same as those of optical astronomy: radio waves and light waves are exactly alike—only the wavelengths differ. Both are simply forms of electromagnetic radiation. But different technologies are necessary to detect the signals. Radio waves

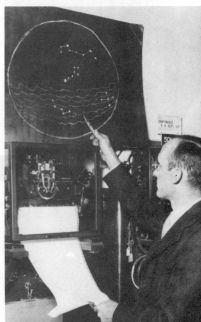

Figure 2-43 Karl Jansky lecturing about his discovery of a source of radio signals moving across the sky according to sidereal time.

Figure 2-44 Karl Jansky with the rotating antenna with which he discovered radio astronomy.

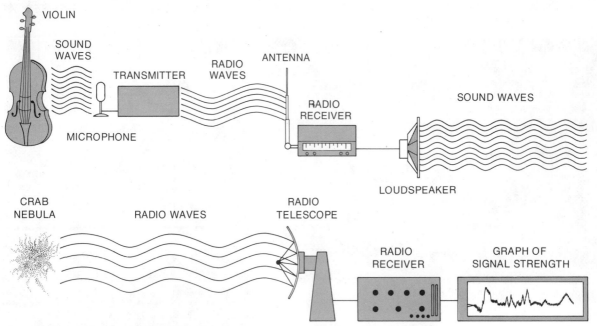

VIOLIN

SOUND WAVES

ANTENNA

TRANSMITTER

RADIO WAVES

MICROPHONE

RADIO RECEIVER

SOUND WAVES

LOUDSPEAKER

CRAB NEBULA

RADIO WAVES

RADIO TELESCOPE

RADIO RECEIVER

GRAPH OF SIGNAL STRENGTH

Figure 2–45 A radio telescope system is similar to a home hi-fi system in that radio waves are electronically amplified. We tend to use the output of a hi-fi, though, to drive a loudspeaker, while astronomers usually record the strength of the signal received by a radio telescope in other ways. In radio astronomy, a reflector is usually used to concentrate the weak radio signals and focus them onto the actual antenna that converts the radio waves into an electric signal.

Figure 2–46 Even though the holes in this radio telescope may look large to the eye, they are much smaller than the wavelength of radio radiation being observed. Thus the telescope appears smooth and shiny to the incoming radio radiation. To appear shiny, the surface of a radio telescope has to be smooth to within a small fraction of the wavelength of radiation, one-twentieth or less.

cause electrical changes in antennas, and these faint electrical signals can be detected with instruments that we call radio receivers (Fig. 2–45).

If we want to collect and focus radio waves just as we collect and focus light waves, we must find a means to concentrate the radio waves at a point at which we can place an aerial. Lenses to focus radio waves are impractically heavy, so refracting radio telescopes are not used. However, radio waves will bounce off metal surfaces. Thus we can make reflecting radio telescopes that work on the same principle as the 5-meter optical telescope on Palomar Mountain. The mirror, called a "dish" in common radio astronomy parlance, is usually made of metal rather than the glass used for optical telescopes. The dish needn't look shiny to our eyes, as long as it looks shiny to incoming radio waves (Fig. 2–46).

We want dishes as big as possible for two reasons: first, a larger surface area means that the telescope will be that much more sensitive. The second reason results from the fact that radio wavelengths are much longer than optical wavelengths. As we saw in Section 2.3, the resolution of any single radio telescope is thus far inferior to that of any single optical telescope. The basic point is that measurements of the size of a telescope are most meaningful when they are in units of the wavelength of the radiation being observed (Fig. 2–47). Thus an optical telescope one meter across used to observe optical light, which is about one two-millionth of a meter in wavelength, is two million wavelengths across. A radio telescope, even one 100 meters across (the size of the currently largest fully steerable dish, at the Max Planck Institute for Radio Astronomy in Bonn, Germany), is only 1000 wavelengths across when used to observe radio waves 10 centimeters in wavelength. The larger the telescope, measured with respect to the wavelength of radiation observed, the narrower is the width of the beam of radiation that the telescope receives. (The reflecting radio telescope dish

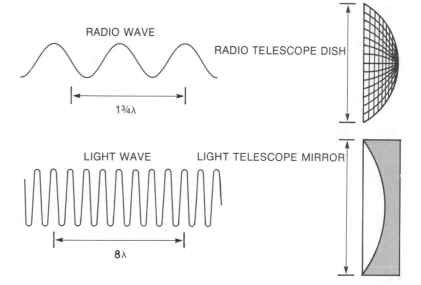

Figure 2–47 It is more meaningful to measure the diameter of telescope mirrors in terms of the wavelength of radiation that is being observed than it is to measure it in terms of units like centimeters that have no particularly relevant significance. The radio dish at top is only 1¾ wavelengths across, while the mirror at bottom is 8 wavelengths across, making it effectively much bigger. The wavelengths are greatly exaggerated in this diagram relative to the size of any actual reflectors. For the Bonn radio telescope, 100 m diameter ÷ 0.1 m per wave = 1000 wavelengths.

Resolution was discussed in Section 2.3, where it was stated that resolution is proportional to wavelength and inversely proportional to telescope diameter.

RADIO WAVE

RADIO TELESCOPE DISH

1¾λ

LIGHT WAVE   LIGHT TELESCOPE MIRROR

8λ

receives radiation only from within a narrow cone. This cone is called the *beam* (Fig. 2–48).) Single-dish radio telescopes have broad beams and cannot resolve any spatial details smaller than the size of their beams. The 100-m telescope at 10 cm has a beam that is four minutes of arc across; at longer wavelengths or with smaller telescopes one cannot even resolve an area as small as the angular size of the moon, 30 minutes of arc. Obviously, this is looking at the sky with less than the optimum resolution, given that large optical telescopes ordinarily resolve 1 arc sec. However, under certain circumstances, one can now use the technique known as interferometry to improve the resolution.) (Interferometry will be discussed in Section 25.6.)

*2.14 A NIGHT AT PALOMAR

In some sense, the opportunity to observe with what has been the largest optical telescope in the world, the 5-m telescope on Palomar Mountain, is the ultimate reward for years of study and research in astronomy. Since Palomar is privately owned, most of the observing time goes to the staff of the Hale Observatories. A dozen individuals on the permanent staff have five nights or as many as 30 nights a year assigned to them. Younger research fellows may have two or three nights each during the years they spend at the Hale Observatories after having received their Ph.D.'s. The remaining one third of the observing time goes to outside applicants from all over the world.

Before using the 5-m Hale telescope, you must have qualified by acquiring experience at some smaller telescope, probably a large telescope in its own right such as the 2.5-m on Mount Wilson. You must then prepare a detailed proposal to a selection committee specifying what it is you want to observe and why you need the 5-m telescope in particular to observe it.

Shortly before your observing time comes, there is a variety of last-minute preparations to make. You might have complicated new equipment

Figure 2–48 The beams of radio telescopes are ordinarily relatively large compared with radio sources. The Crab Nebula looks like a point to the radio telescope shown, even though it looks larger than a point to your eye.

CRAB NEBULA

BEAM WIDTH →

RADIO TELESCOPE

There may be 320 clear nights a year on Palomar Mountain, but many more nights a year would be needed if all requests were to be granted.

to test out, or you might be using one of the standard spectroscopic, photoelectric, or photographic systems permanently installed at the telescope.

Finally, the big day comes and you set off for Palomar. If you are on the staff of the Hale Observatories, your office is in Pasadena, California, and Palomar is about a 2½-hour drive southward and slightly east. If you are at an institution farther away, you would first have to fly to Los Angeles. In any case, you have to have your suitcase packed and be prepared to spend a few days and nights away from home.

As you drive into the mountains from the desert, the foliage changes. By the time you have finished the steep last bit of the climb up Palomar, you find yourself in a pine forest. By noontime of the day you are to start observing, you check into the residence hall, which is called "the Monastery."

Meals are served at the Monastery. The noon meal is for the new arrivals and for the daytime workers; the astronomers who worked the preceding night are usually asleep. After lunch, you set off for the telescope.

The 5-m telescope stands in a huge domed building that is 40 meters (135 ft) high, the equivalent of a 13-story building (Fig. 2–49). If you are doing photographic spectroscopy, the first thing that you would do is to go to the plate vault, a large walk-in refrigerator where special stocks of photographic emulsions are kept. The emulsions are specially chosen for their astronomical characteristics, and are not the ones that you buy in your corner camera store. If you are using electronic devices, you might spend time checking out the computer

You would make certain that the equipment you plan to use is in place. There are so many interchangeable pieces of equipment available for use in

Figure 2–49 A moonlight view of the dome of the 5-m Hale Telescope on Palomar Mountain in California.

Figure 2–50 The control room at the 5-m Hale Telescope with Dave Philip of the Dudley Observatory at the left and the night assistant, Juan Carrasco, at the right. Some of the controls are for pointing the telescope; others are for your specific observations. Note the television screen at right; it is used for guiding.

studying the light reflected by the 5-meter mirror that you usually have to set up your own equipment. The rest of the afternoon might well be spent in carefully focusing the equipment.

It is finally dinner time, just before dark, and you head back to the Monastery. All the daytime personnel have left the mountain, and it is mostly the observers at this meal. There are several other telescopes on the

Figure 2–51 The control room at the 3-m telescope of the Lick Observatory. Astronomers Sandra Faber and Joseph Miller are looking at the video screen, which shows the field of the sky at which the telescope is pointed. Note the computer and other electronic equipment. (Lick Observatory photo)

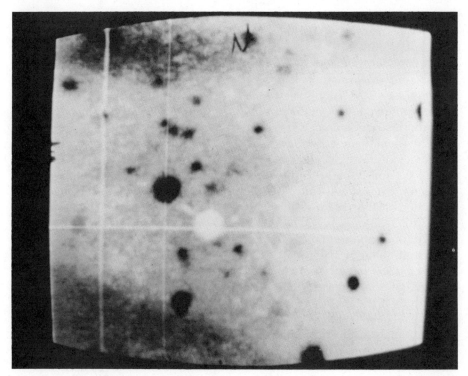

Figure 2–52 The screen of the video guider of the 4-m telescope at Cerro Tololo. The T.V. picture is about 2.5 minutes of arc wide. The white dot represents the hole through which light from the star being studied will fall. The star is the faint object directly above the hole. It is a 19th magnitude star that is being studied for flickering, because it may correspond to an x-ray object. The blackest dot on the screen, due left of the star under study, is a star of magnitude 16.8 that will be used for a guide star. Note that this guide star is 10,000 times fainter than the faintest stars that we can see with the naked eye, yet it can be easily used for guiding. (We discuss the magnitude scale of brightness in Section 4.3.)

Figure 2–53 The Hale telescope.

Figure 2–54 The collecting area of the Hale telescope is large, but the light is focused down to a small area on a spectrographic plate. The spectrum shown is that of the variable star R Coronae Borealis, and was taken by the author. Most of the spectrum visible is that of an iron arc operated in the telescope dome to provide known wavelengths for calibration purposes; the spectrum of the star consists of the few horizontal emission lines between the left and right halves of the calibration spectrum.

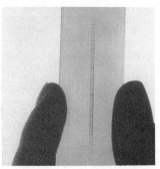

mountain, including a 1.5-m reflecting (60-inch) telescope and the 1.2-m (48-inch) Schmidt telescope, so you have company at dinner.

In the twilight you walk back to the telescope. A night assistant is assigned to the telescope at all times and is responsible for all the mechanical aspects of the telescope. The night assistants know the telescope thoroughly, and also know when a strange noise might mean that the telescope must be shut down, or when high wind or blowing dust means that the telescope dome must be closed for the safety of the mirror.

The telescope is very well lubricated and very well balanced, so even though it weighs over 500 tons it turns very smoothly and needs only a moderate sized motor. But refraction effects in the earth's atmosphere may cause the light from stars to vary in position at a slightly different rate from the normal average, and you must continually check that the telescope is pointing in absolutely the right direction. This is called guiding. Especially in direct photographic work, where you are making an image, observers who are the most careful with their guiding get the best results.

The night assistant points the telescope at the object in the sky that you are ready to observe. If it is a reasonably bright object, then you can check very simply to make sure it is the correct one: you need merely to see if the stars around the objects in the field of view in the telescope have the same position relative to each other as the stars on a chart you have of the area around your object. If you are observing a faint object, on the other hand, you might have to point the telescope a certain fraction of the way between two objects that are bright enough to be visible. You may never see the faint object yourself, though its image will appear on the developed plate. Television-type devices that are sensitive to low levels of light now enable you to see faint objects for alignment or guiding that in years past could not be used (Figs. 2–50, 2–51, and 2–52).

After an hour or so, your exposure is finished. For photographic spectroscopy, you cover the slit and close the film holder. Then you replace the exposed plate with an unexposed plate, and prepare to take another exposure, perhaps of another object. Then you go to the darkroom to develop your plate, gently rocking it back and forth in the developer for several minutes. The night assistant might be able to help you by guiding on the second star while you are developing the plate of the spectrum of the first star. It is important to develop your first plate or to examine your first computer scan right away to make sure that the system is working and that there are no blockages in the light path.

In the course of the night you will probably want to go up to the floor of the dome for a short while just to appreciate the stars and the marvelous monster of an instrument that you are controlling to gather photons from afar (Fig. 2–53).

All too soon the night is over. The dawn is visible in the east, and you must shut down or else the sky light will fog your plate. After developing your last plate (or perhaps leaving it to do at the beginning of the next night), you trudge home to the Monastery to bed.

After a few days on the mountain, you may have a dozen or more plates (Fig. 2–54) to take with you. Your work has just begun. It may take months or years to analyze the data completely, often using measuring devices and computers at your own institution. The thrill of having captured those few photons from a distant star remains long after your night at Palomar.

SUMMARY AND OUTLINE

The spectrum (Section 2.1)
>From short wavelengths to long wavelengths the various types of radiation are known as gamma rays, x-rays, ultraviolet, visible, infrared, radio.

Parts of visible spectrum are ROY G. BIV.

All radiation travels at the "speed of light"; in a vacuum this is 3×10^8 m s^{-1}.

Windows of transparency of the earth's atmosphere exist for visible light and radio waves (plus some parts of the infrared).

Continuous radiation is called the *continuum*.

Spectral lines (Section 2.2)
>Absorption lines are gaps in the continuum shown as narrow wavelength regions where the intensity is diminished from that of the neighboring continuum.

Emission lines are narrow wavelength regions where the intensity is relatively greater than neighboring wavelengths (which are either continuum or zero in intensity).

A given element has a characteristic set of spectral lines under certain conditions of temperature and pressure.

Optical telescopes (Sections 2.3 to 2.9)
>Resolution is inversely proportional to diameter (and proportional to wavelength).

Refractors and reflectors: Ritchey-Chrétiens and Schmidt cameras have relatively wide fields.

Optical observatories and their distribution across the world.

The Space Telescope is to be launched in 1983.

Light-gathering power is proportional to diameter squared.

Spectroscopy and photometry (Sections 2.10 and 2.11)
>Film and electronic devices for gathering photons are described.

Differences in observing in nonvisible parts of the spectrum.

Ultraviolet and x-ray astronomy (Section 2.12)
>Video and grazing incidence techniques used.

Grazing incidence for HEAO-2.

Infrared and Radio Astronomy (Section 2.13)
>Infrared: high altitude sites and electronic devices are necessary.

Radio: single-dish radio telescopes have low resolution.

QUESTIONS

1. What advantages does a reflecting telescope have over a refracting telescope?

2. What limits the resolving power of the 5-meter telescope?

3. Why does it matter whether a telescope is in the northern or southern hemisphere?

4. List the important criteria in choosing a site for an optical observatory meant to study stars and galaxies.

†5. Two reflecting telescopes have primary mirrors 2 m and 4 m in diameter. How many times more light is gathered by the larger telescope in any given interval of time?

6. Why might some stars appear double in blue light though they could not be resolved in red light?

7. For each of the following, identify whether it is a characteristic of a reflecting telescope, a refracting telescope, both, or neither. Give any limitations on the applicability of your answer.
   (a) Free of chromatic aberration.
   (b) Has more severe spherical aberration.
   (c) Requires aluminizing.
   (d) Can be used for photography.
   (e) Has an objective that must be supported only by its rim.
   (f) Can be made in larger sizes.
   (g) Has a prime focus at which a person can work without blocking the incoming light.

___
† This indicates a question requiring a numerical solution.

8. Why can't we use ordinary photographic plates to record infrared images?

9. Why must we observe ultraviolet radiation using rockets or satellites, while balloons are sufficient for infrared observations?

10. Why can radio astronomers observe during the day, while optical astronomers are (for the most part) limited to nighttime observing?

11. What are the advantages of the Multiple Mirror Telescope? How does it compare with the 5-m telescope for studying faint objects?

12. What are the similarities and differences between making radio observations and using a reflector for optical observations? Compare the radiation path, the detection of signals, and limiting factors.

13. Why is it sometimes better to use a small telescope in orbit around the earth than it is to use a large telescope on a mountain top?

14. Why is it better for some purposes to use a medium-size telescope on a mountain instead of a telescope in space?

15. Compare the International Ultraviolet Explorer and the Space Telescope.

16. What are two reasons why the Space Telescope will be able to observe fainter objects than we can now study from the ground?

PART II

When we look up at the sky at night, most of the objects we see are stars. In the daytime, we see the sun, which is itself a star. The moon, the planets, even a comet may give a beautiful show, but, however spectacular, they are only minor actors on the stage of observational astronomy.

From the center of a city, we may not be able to see very many stars, because city light scattered by the earth's atmosphere makes the light level of the sky brighter than most stars. All together, there are about 6000 stars bright enough to be seen with the naked eye under good observing conditions.

Some properties of the stars that can be seen with the naked eye tell us about the natures of the stars themselves. For example, some stars seem blue-white in the sky, while others appear slightly reddish. From information of this nature, astronomers are able to determine the temperatures of the stars. Chapter 3 is devoted to the basic properties of stars and some of the ways we find out such information.

Even a cursory glance at the sky shows that there are patterns in the distribution of stars. The different areas of the sky are called *the constellations*. The names that the ancients gave to these areas, associating them with figures and objects from their myths and religions, are still used down to this day. It takes a great imagination to see Orion (the Hunter) or Draco (the Dragon) in the sky above, but it is nice to be like St. Exupéry's Little Prince (Fig. 3–1), and not lose our sense of wonder at the sky above and the objects in space.

The stars we see defining a constellation may be at very different distances from us. One star may appear bright because it is relatively close to us even though it is intrinsically faint; another star may appear bright even though it is far away because it is intrinsically very luminous. When we look at the constellations, we are observing only the directions of the stars and not whether the stars are physically very close together, as we shall see in Chapter 4. In this chapter we will also discuss the appearance of the sky, and consider time-keeping and calendars.

Aerial view of the Cerro Tololo Inter-American Observatory.

THE STARS

The constellations do not tell us anything about the nature of the stars. Astronomers tend not to be interested in the constellations because the constellations do not give useful information about how the universe works. After all, the constellations merely tell the directions in which stars lie. Most astronomers want to know *why* and *how*. Why is there a star? How does it shine? Such studies of the workings of the universe are called *astrophysics*. Almost all modern day astronomers are astrophysicists as well, for they always not only make observations but also think about their meaning. Chapter 5 describes some of the ingenious ways we use basic information about stars. We can sort stars into categories and tell how far they are away and how they move in space.

Most stars occur in pairs or in larger groups, and we shall discuss types of groupings and what we learn from their study in Chapter 6. Binary stars are important, for example, in finding the masses and sizes of stars. Some stars vary in brightness, a property that we shall use later on to tell us the scale of the universe itself. The study of star clusters leads to our understanding of the ages of stars and how they evolve. Some clusters give off strange bursts of x-rays; such activity probably means that exotic objects are lurking within these clusters.

In Chapter 7 we study an average star in detail. It is the only star we can see close up—the sun. The phenomena we observe on the sun take place on other stars as well, as has recently been verified by telescopes in space.

To do their research, astronomers still often use optical telescopes to gather light from the stars. Following how a stellar biography can be written from the clues that the light carries is like reading a wonderful detective story. But in recent years, the heavens have been studied in other ways besides observing the ordinary light that is given off by many astronomical objects. Radio waves, x-rays, gamma rays, ultraviolet light, infrared light, and cosmic rays are increasingly studied from the earth's surface or from space. Many a contemporary astronomer—even many who consider themselves observers (who mainly carry out observations) rather than theoreticians (who do not make observations, but rather construct theories)—has never looked through an optical telescope. Still, astronomy began with optical studies, and our story begins there.

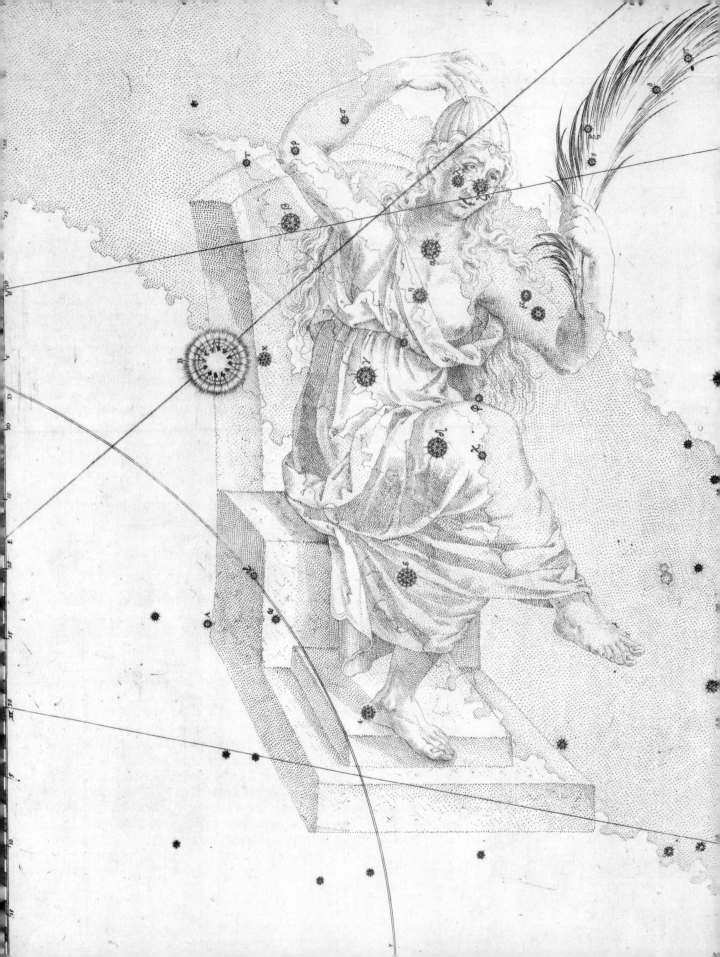

CHAPTER 3

═ORDINARY STARS═

AIMS:

To study the colors and spectra of ordinary stars, and to see how this tells us about the temperatures of the surfaces of the stars

What are these stars that we see as twinkling dots of light in the nighttime sky? The stars are luminous balls of gas scattered throughout space. All the stars we see in the sky are among the 100 billion stellar members of a collection of stars and other matter called the Milky Way Galaxy. The sun, which is the star that controls our planet, is so close to us that we can see detail on its surface, but it is just an ordinary star like the rest.

Stars are balls of gas held together by the force of gravity, the same force that keeps us on the ground no matter where we are on the earth. Long ago, gravity compressed large amounts of gas and dust into dense spheres that became stars. Still, the stars are gaseous through and through; they have no solid surfaces. Energy from the original contraction heated the interiors of the spheres until they were hot enough for nuclear fusion to begin. In this chapter we shall concentrate on how we study the stars.

Stars generate their own energy and light. It has been only 50 years since the realization that the stars shine by nuclear fusion. Now we are looking to the nearest star, our sun, to provide either a direct source of energy in the form of incoming radiation or the answer to the problem of taming the fusion process and recreating it on earth.

We see only the outer layers of the gas of the stars; the interiors are hidden from our view. The outer layers are not generating energy by themselves, but are merely glowing because of the effects of energy transported outward from the stellar interiors. So when we study light from the stars we are not observing the processes of energy generation directly.

We will discuss the Milky Way Galaxy in Part V.

Chapter 7 on the sun can be studied before this chapter if the sections on solar spectroscopy (Sections 7.1a, 7.2a, and 7.3a) are omitted.

In Chapter 8 we will discuss how stars get their energy by this process of nuclear fusion.

One difference between astronomers and scientists of most other fields is that many astronomers are usually observers *rather than* experimentalists *in that they can* **observe** *light from distant objects but cannot ordinarily experiment with the objects directly. Even though we can now land spacecraft on the moon and some of the planets, we cannot seriously hope to even come close to the distant stars.*

Cassiopeia, from Johann Bayer's *Uranometria,* first published in 1603.

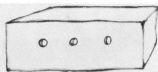

Figure 3–1 *(A)* The Little Prince is shown at top on Asteroid B-612, his home. *(B)* When asked by the Little Prince to draw a sheep, the author drew the above. "This is only his box. The sheep you asked for is inside." The Little Prince replied that the picture was exactly the way he wanted it, and asked for quiet because the sheep had gone to sleep.

From *The Little Prince* by Antoine de Saint Exupéry.

An easy way to convert from degrees Celsius to degrees Fahrenheit is to take the Celsius number, double it, subtract 10 per cent of the result, and add 32. Example: take 30°C, double it to get 60°, subtract 10 per cent of 60° = 6° from 60° to get 54°, and add 32° to get the answer of 86°F.

Box 3.1 Temperature

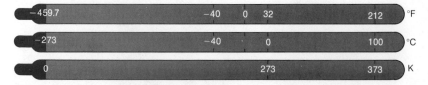

Figure 3–2 Temperature scales.

We must choose the units in which we will discuss temperature for the rest of this book. In America until recently we have used for most everyday purposes the Fahrenheit temperature scale, in which water freezes at 32°F and boils at 212°F, a span of 180°F between freezing and boiling. On this scale, absolute zero (the minimum temperature theoretically possible for any material thing) is −459.7°F (Fig. 3–2). This scale is not used in the physical sciences. Most of the rest of the world uses degrees Centigrade; a modern minor adjustment of the scale is called degrees Celsius. (The Centigrade and Celsius scales differ by less than 0.1° Centigrade or Celsius.) The United States is changing over gradually to this system; most scientists already have for their professional work. On the Centigrade scale, water freezes at 0°C and boils at 100°C, a span of 100°C; a Celsius degree (which is defined in terms of absolute zero) is essentially the same size. Thus 180 Fahrenheit degrees are equivalent to 100 Celsius degrees; simplifying gives 1.8°F per 1°C.

On the Celsius scale, absolute zero is about −273°C. For us on earth the temperatures with which we deal conveniently range around zero or a few tens of degrees. But the freezing and boiling points of water don't have much relevance to the stars. Astronomers choose to use the Kelvin temperature scale, which has its zero point at absolute zero instead of at the freezing point of water. The size of a Kelvin degree (now officially called one kelvin and abbreviated K instead of °K) is the same as the size of a Celsius degree. Thus absolute zero is 0 K, water freezes at +273 K, and water boils at +373 K.

It is simple to change from Celsius degrees to kelvins; simply add 273 to the Celsius temperature. To change from Fahrenheit to Celsius is more complicated. One must first subtract 32°F, to align the freezing points, and then divide by 1.8 to transform the smaller Fahrenheit degrees into the larger Celsius degrees. The formula is °C = (°F − 32) ÷ 1.8.

Stellar temperatures, whether expressed in Fahrenheit, Celsius, or Kelvin degrees, are very high—in the thousands or tens of thousands (Fig. 3–3).

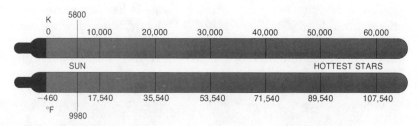

Figure 3–3 Stellar temperature scales.

3.1 THE COLORS OF STARS

In Section 2.1 we saw that the different colors we perceive result from having radiation present at certain ranges of wavelength. For example, if an object appears blue, then most of its radiation (or at least most of its visible radiation) has wavelengths between roughly 4300 and 5000 Å. Shortly, we will see that the distribution of a star's radiation in wavelength depends on the temperature of the surface of that star. Thus, by measuring the color of a star, we can deduce what the star's temperature must be.

When we measure the color of a star, we can graph the intensity of light from the star at a given wavelength on one axis and the wavelength on the other axis. The measurement of intensity can be obtained either by passing the light through a spectrograph and measuring the intensity at each wavelength, or simply by measuring the amount of light that emerges from a set of filters that allows only certain groups of wavelengths to pass and block others.

When we examine a set of these graphs, we can see that, spectral lines aside, the radiation follows a fairly smooth curve (Fig. 3–4). The radiation "peaks" in intensity at a certain wavelength (that is, has a maximum in its intensity at that wavelength), and decreases in amount more slowly on the long wavelength side of the peak than it does on the short wavelength side.

The nature of these graphs can be best understood by first considering the radiation that represents different temperatures. When you first put an iron poker in the fire, it glows faintly red. Then as it gets hotter it becomes redder. If the poker could be even hotter without melting, it would turn yellow, white, and then blue-white. White hot is hotter than red hot (Fig. 3–5).

A set of physical laws governs the sequence of events when material is heated. As the material gets hotter, the peak of radiation shifts toward the blue. This is known as *Wien's displacement law* (Table 3–1).

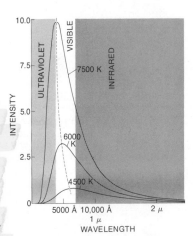

Figure 3–4 The intensity of radiation for different stellar temperatures, according to Planck's law (Section 3.2). The wavelength scale is linear, i.e., equal spaces signify equal wavelength intervals. The dotted line shows how the peak of the curves shifts to shorter wavelengths as temperature increases; this is known as Wien's displacement law.

Figure 3–5 As a rod or poker (perhaps made of iron or carbon) is heated, it first glows red hot and eventually, when it is hotter, appears white hot.

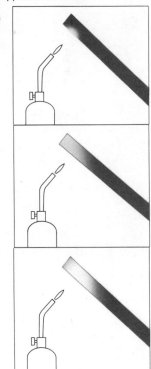

°Box 3.2   Wien's Displacement Law

$$\lambda_{max} \, T = \text{constant},$$

where λ_{max} is the **wavelength** at which the energy given off is at a **maximum,** and T is the temperature. The numerical value of the constant is given in Appendix 2.

TABLE 3–1   WIEN'S DISPLACEMENT LAW

Temperature	Wavelength of Peak	Spectral Region of Wavelength of Peak
3 K	.97 mm	infrared-radio
3000 K	9660 Å	infrared
6000 K	4830 Å	green
12,000 K	2415 Å	ultraviolet
24,000 K	1207 Å	ultraviolet

Note that we are discussing, for the moment, only the continuous part of the spectrum; it is possible to observe both a continuous spectrum and superimposed spectral lines in a single spectrum.

Also, as the material gets hotter, the total energy of the radiation grows quickly. The energy follows the *Stefan-Boltzmann law,* as is shown in Fig. 3–4. This law says that the total energy emitted from each square centimeter of a source in each second grows with the fourth power of the temperature. Thus the ratio of the energies emitted is the fourth power of the ratio of the temperatures. For example, consider the same amount of gas at 10,000 K and at 5000 K. Because 10,000 K is twice 5000 K, the 10,000 K gas gives off $2^4 = 2 \times 2 \times 2 \times 2 = 16$ times more energy than does the same amount of gas at 5000 K. (We must measure our scale of degrees from absolute zero, the minimum temperature possible.)

$E = \sigma T^4$, where E is the flux of energy emitted, the "strength" of the radiation.

---

*Box 3.3 Stefan-Boltzmann Law

$$\text{Energy} = \sigma \, T^4.$$

The numerical value of σ, the Stefan-Boltzmann constant, is given in Appendix 2.

From Wien's displacement law, we can see that the colors of stars in the sky are telling us the stars' temperatures. The reddish star Betelgeuse, thus, is a comparatively cool star, while the blue-white star Sirius is very hot. By simply measuring the intensity of a star with a set of filters at different wavelengths, we can assess the stellar temperature (Fig. 3–6). Astronomers measure colors directly at the telescope with a procedure discussed in Appendix 14.

3.2 PLANCK'S LAW AND BLACK BODIES

The laws of distribution of energy from a heated gas are much more general than their particular application to the stars. In principle, if any gas is heated, its continuous emission follows a certain peaked distribution called *Planck's law,* suggested in 1900 by the German physicist Max Planck (Fig. 3–7). Wien's displacement law and the Stefan-Boltzmann law, which had been discovered earlier from study of experimental data, can be derived from Planck's law. Real gases may not, and usually don't, follow Planck's law exactly. For example, Planck's law governs only the continuous spectrum; any line emission or absorption does not follow Planck's law.

Because of these difficulties, scientists choose to consider an idealization of the actual case. We call this fictional ideal object a *black body,* since we define it as something that absorbs all radiation that falls on it and since black things in our everyday experience do not reflect much light. A black body is an idealized black. Ordinary black paint, on the other hand, does not absorb 100 per cent of the light that hits it. Further, something that looks black in visible light might not seem very black at all in the infrared, and so would not be a real black body.

An important physical law that governs the radiation from a black body says that anything that is a good absorber is a good emitter too. The radiation from a black body follows Planck's law. As a black body is heated, the peak of the radiation shifts toward the short wavelength end of the spectrum (Wien's displacement law) and the total amount of radiation grows rapidly (Stefan-Boltzmann law). The radiation from a black body at a higher

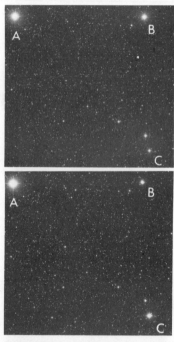

Figure 3–6 The top photograph is a view of several stars in blue light, and the bottom photograph is a view of the same stars in red light. Star A appears about the same brightness in both. Star B, a relatively hot star, appears brighter in the blue photograph. Star C, a relatively cool star, appears brighter in the red photograph.

temperature is greater at every wavelength than the radiation from a cooler black body, but as the temperature increases, the intensity of radiation at shorter wavelengths increases more rapidly than the intensity at longer wavelengths increases. Note that the curves of energies emitted by black bodies at different temperatures in Figures 3–4 and 3–8 do not cross.

To a certain extent, the atmospheres of stars appear as black bodies, in that over a broad region of the spectrum that includes the visible, the continuous radiation from stars follows Planck's law fairly well.

3.3 THE FORMATION OF SPECTRAL LINES

Spectral lines arise when there is a change in the amount of energy that any given atom has. The amount of energy that an atom can have is governed by the laws of *quantum mechanics*, a field of physics that was developed in the 1920's and whose applications to spectra were further worked out in the 1930's. According to the laws of quantum mechanics, light (and other electromagnetic radiation) has the properties of waves in some circumstances and the properties of particles under other circumstances. One cannot understand this dual set of properties of light intuitively and some of the greatest physicists of the time had difficulty adjusting to the idea. Still, quantum mechanics was worked out to explain experimental observations of spectra, and is thoroughly accepted by almost all scientists.

An atom (Fig. 3–9) can be thought of as consisting of particles in its core, which is called the *nucleus*, surrounded by orbiting particles called *elec-*

Figure 3–7 Max Planck *(left)* presented an equation in 1900 that explained the distribution of radiation with wavelength. Five years later, Albert Einstein, in suggesting that light acts as though it is made up of particles, found an important connection between his new theory and the earlier work of Planck. The photograph was taken years later.

Much of what we know about stars comes from studying their spectral lines. First we will discuss spectral lines themselves; we will go on to their appearance in stars in Sections 3.5 and 3.6.

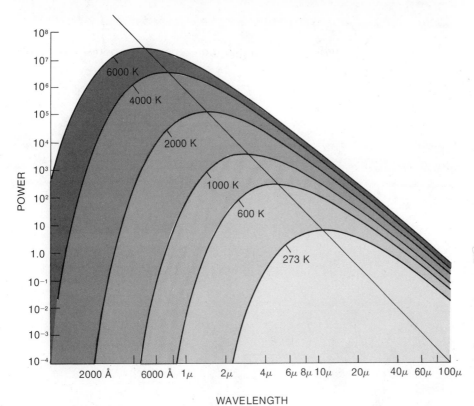

Figure 3–8 Curves for different temperatures representing the intensity of radiation according to Planck's law. The wavelength scale is logarithmic, i.e., equal spaces on the graph signify equal factors multiplying the wavelength or intensity. The peak of the curves shifts to shorter wavelengths, following Wien's displacement law. The total energy being radiated is represented by the area under each curve; it grows rapidly as the temperature increases. The Stefan-Boltzmann law states that the energy grows as the fourth power of the temperature.

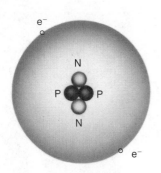

Figure 3–9 A helium atom contains two protons and two neutrons in its nucleus and two orbiting electrons.

An electron volt (eV) is one of the several units in which energy can be expressed. It takes hundreds of billions of eV to equal even a small unit of energy called an erg, and it takes 10^9 ergs per second to light a 100-watt light bulb. Obviously, one eV is pretty small.

Albert Einstein in 1905 first suggested that light can be thought of as quanta of energy and wrote an equation closely related to $E = hc/\lambda$. Planck had previously used the constant h, but in another context.

trons. Examples of nuclear particles are the *proton*, which carries one unit of positive electric charge, and the *neutron*, which has no electric charge. Electrons, each of which has one unit of negative electric charge, are only 1/1800 as massive as protons or neutrons. Normally, an atom has the same number of electrons as it does protons. The negative and positive charges balance, and the atom is electrically neutral.

According to quantum mechanics, an atom can exist only in a specific set of energy states, as opposed to a whole continuum (continuous range) of energy states. An atom can have only discrete values of energy; that is, the energy cannot vary continuously. For example, the energy corresponding to one of the states might be 10.2 electron volts (eV). The next energy state allowable by the quantum mechanical rules might be 12.0 eV. We say that these are *allowed states.* The atom **simply cannot have** an energy intermediate between 10.2 and 12.0 eV. In sum, the energy states are discrete; we say that they are *quantized.* We call them *energy levels.*

When an atom drops from a higher energy state to a lower energy state, without colliding with another atom, the difference in energy is sent off as a bundle of radiation. This bundle of energy, a *quantum*, is also called a *photon*, which may be thought of as a particle of electromagnetic radiation (Fig. 3–10). Photons always travel at the speed of light. Each photon has a specific energy, which does not vary.

A link between the particle version of light and the wave version is seen in the equation that relates the energy of the photon with the wavelength it has: $E = hc/\lambda$. In this equation, E is the energy, h is a constant named after Planck, c is the speed of light (in a vacuum, as always), and λ (lambda) is the wavelength. Since λ is in the denominator on the right hand side of the equation, a small λ corresponds to a photon of great energy. When, on the other hand, λ is large, the photon has less energy. An x-ray photon, for example, of wavelength 1 Å, has much more energy than does a photon of visible light of wavelength 5000 Å.

3.3a Emission and Absorption Lines

When a gas is heated so that many of its atoms are in energy states other than the lowest possible, atoms not only are being raised to higher energy states but also spontaneously drop back to their lowest energy levels, emitting photons as they do so. These emitted photons represent energies at certain wavelengths. As there were not necessarily any photons already

Figure 3–10 Since the hydrogen atom has only a single electron, it is a particularly simple case to study. The lowest possible energy state of an electron in an atom is called its *ground state,* and all energy states are called *excited states.* When an atom in an excited state gives off a photon, it drops back to a lower energy state, perhaps even the ground state. We see the photons as an emission line.

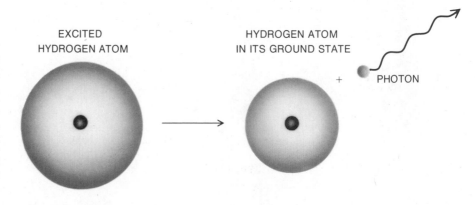

EXCITED
HYDROGEN ATOM

HYDROGEN ATOM
IN ITS GROUND STATE

+ PHOTON

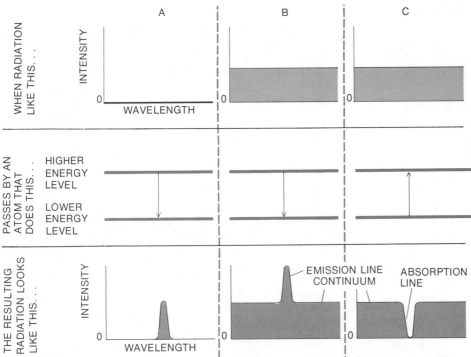

WHEN RADIATION LIKE THIS...

INTENSITY

WAVELENGTH

A   B   C

PASSES BY AN ATOM THAT DOES THIS...

HIGHER ENERGY LEVEL

LOWER ENERGY LEVEL

THE RESULTING RADIATION LOOKS LIKE THIS...

INTENSITY

WAVELENGTH

EMISSION LINE
CONTINUUM

ABSORPTION LINE

Figure 3–11 When photons are emitted, we see an emission line; a continuum may or may not also be present (Cases A and B). An absorption line must absorb radiation from something. Hence an absorption line necessarily appears in a continuum (Case C).

existing at those wavelengths, the new photons appear as bright wavelengths (when they correspond to the visible part of the spectrum). These are the **emission** lines (Fig. 3–11).

(As we saw, when we allow continuous radiation from a body at some relatively high temperature to pass through a cooler gas, the atoms in the gas can take energy out of the continuous radiation at certain discrete wavelengths. We now see that the atoms in the gas are being changed into higher energy states that correspond to these wavelengths. Because energy is being taken up in the gas at these wavelengths, less radiation remains to travel straight ahead at these particular wavelengths. These wavelengths are those of the **absorption** lines (Fig. 3–11). In the visible, these wavelengths appear dark when we look back through the gas. (The energy taken up is soon emitted in other directions, which explains why we can see emission lines when we look from the side, as was shown in Figure 2–5.)

Note that emission lines can appear without a continuum (Case A in Fig. 3–11) or with a continuum (Case B in Fig. 3–11), since they are merely the addition of energy to the radiation field at certain wavelengths. But absorption lines must be absorbed **from** something (Case C in Fig. 3–11), namely the continuum, which is just the name for a part of the spectrum where there is some radiation over a continuous range of wavelengths.

A normal stellar spectrum is a continuum with absorption lines. This simple fact thus leads us to the important conclusion that the temperature in the outer layers of the star is decreasing with height; the mere presence of absorption lines tells us that we have been looking through cooler gas at hotter gas. As a result of this temperature trend with height, emission lines, which result from the presence of gas hotter than any background continuum, are rarely found in stellar spectra. Emission lines can appear, however, in exceptional cases, such as for stars surrounded by hot shells of gas.

We first met emission and absorption spectra in Section 2.2.

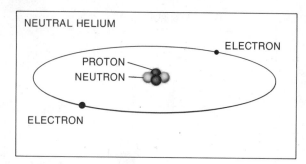
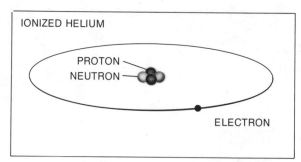

Figure 3–12 An atom missing one or more electrons is *ionized*. Neutral helium is He I and singly ionized helium is He II.

3.3b Excited States and Ions

We can categorize an atom's energy state by an amount of energy and think of a change in energy states as resulting from a change in energy of an atom's electrons. (The electrons are orbiting the nucleus. The production of spectral lines does not involve changes in the nucleus itself.) If a gas were at a temperature of absolute zero, all the electrons in it would be in the lowest possible energy levels. The lowest energy level that an electron can be on is called the *ground state* of that atom.

As we add energy to the atom we can "excite" one or more electrons to higher energy levels; that is, we push the electrons to higher energy levels. If we were to add even more energy, some of the electrons would be given not only sufficient energy to be excited but also enough to escape entirely from the atom. We then say that the atom is *ionized*. The remnant with less than its quota of electrons is called an *ion* (Fig. 3–12). As the remnant has more positive charges in its nucleus than it has negative charges on its electrons, its net charge is positive, and it is sometimes called a *positive ion.*

Before the atom was ionized, it had the same number of protons and electrons and so was electrically neutral. Such an atom is called a neutral atom and is denoted with the Roman numeral I. For example, neutral helium is denoted He I. When the atom has lost one electron (is singly ionized), we use the Roman numeral II. Thus when helium has lost one electron, it is called He II (read "helium two"). If an atom is doubly ionized, that is, has lost two electrons, it is in state III, and so on.

Atoms can be excited or ionized either by collisions or by radiation. In a star the former is the case when the density is sufficiently high that collisions of the atoms with electrons are frequent. When the gas is "heated," this simply means that the constituent atoms are given more energy. The atoms thus collide more frequently and exchange more energy with each other. Thus in a "hotter" gas, more atoms are excited (Fig. 3–13) or ionized.

Figure 3–13 The diagram indicates schematically that as the temperature of a cloud of gas rises, more and more electrons are excited to higher and higher energy levels. Few electrons are excited at low temperatures. The *Boltzmann distribution* tells us how many atoms will be in each of the energy states for a given temperature.

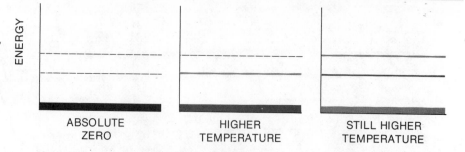

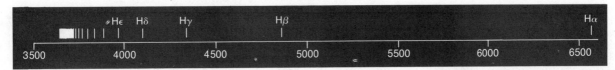

Figure 3–14 The Balmer series, representing transitions down to or up from the second energy state of hydrogen. The strongest line in this series, Hα (H-alpha) is in the red.

3.4 THE HYDROGEN SPECTRUM

Hydrogen emits, and absorbs, a set of spectral lines that fall across the visible spectrum in a distinctive pattern (Fig. 3–14). The strongest line is in the red, the second strongest is in the blue, the third strongest is farther in the blue, and the other lines continue this series, with the spacing between the lines getting smaller and smaller.

It is easy to visualize how the hydrogen spectrum is formed by using the picture (Fig. 3–15) that Niels Bohr (Fig. 3–16), the Danish physicist, laid out in 1913. In the Bohr atom, electrons can have orbits of different sizes that correspond to different energy levels. Only certain orbits are allowable, which is the same as saying that the energy levels are quantized. This simple picture of the atom was superseded by the development of quantum mechanics, but the notion that the energy levels are quantized remains.

The letter n, which labels the energy levels and which appears in the formulas for their energies, is called the principal quantum number. We call the energy level for $n = 1$ the ground level or ground state, as it is the lowest

Figure 3–16 Niels Bohr.

Figure 3–15 The representation of hydrogen energy levels known as the Bohr atom.

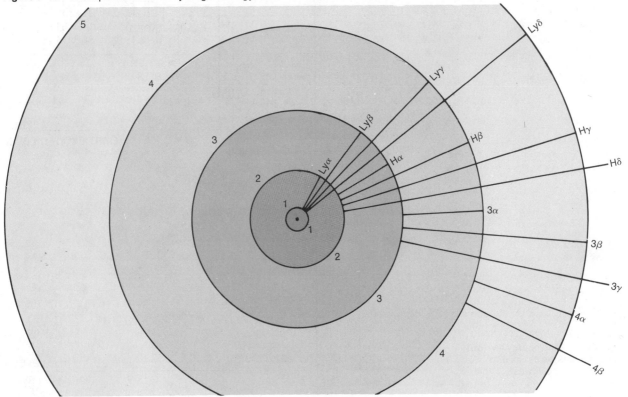

Lyman went so far as to climb Mount McKinley in 1915, hoping that from that great altitude there might happen to be a window of transparency that would pass Lyman alpha. Unfortunately, there is no such window, and his photographs of the solar spectrum were blank in the region where ultraviolet radiation would have appeared.

The hydrogen atom, with its lone electron, is a very simple case. More complicated atoms have more complicated sets of energy levels. Though the principle of determining the wavelengths from the energy differences in transitions remains the same, the sets of spectral lines for atoms and ions with several electrons are much more complicated than the simple series (Lyman, Balmer, etc.) that exist for the hydrogen atom.

possible energy state. The series of transitions from or to the ground level is called the Lyman series, after the American physicist Theodore Lyman. The lines fall in the ultraviolet, at wavelengths far too short to pass through the earth's atmosphere. Lyman's observations of what we now call the Lyman lines were made in a laboratory in a vacuum tank. At present, telescopes aboard earth satellites, such as the International Ultraviolet Explorer (IUE), enable us to observe Lyman lines from the sun and the other stars.

The series with $n = 2$ as the lowest level is in the visible part of the hydrogen spectrum. This series is called the Balmer series. The bright red line in the visible part of the hydrogen spectrum, known as the Hα line, arises from the transition from level 3 to level 2. The transition from the $n = 4$ to the $n = 2$ accounts for the Hβ line, and so on.

There are many other series of lines, each corresponding to transitions to a given lower level (Fig. 3–17).

Since the higher energy levels have greater energy, a spectral line caused by transition from a higher level to a lower level yields an emission line. When, on the other hand, continuous radiation falls on cool hydrogen gas, some of the atoms in the gas can be raised to higher energy levels. Absorption lines result in this case.

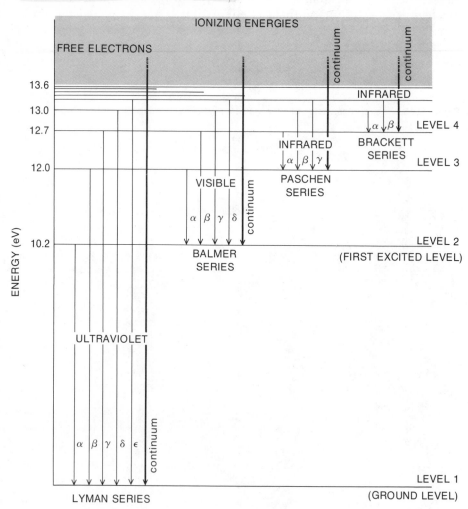

Figure 3–17 The energy levels of hydrogen and the series of transitions among the lowest of these levels.

The concept of the Bohr atom tells us that the energy states of any atom are *discrete* (meaning that they can have only a fixed set of values and not values in between) and that each of the allowable energy states has a fixed energy. Thus a transition from one energy state to another will always result in the same amount of energy being either given off or absorbed. In turn, for a particular element, this transition will always cause the same wavelength to appear no matter whether it is in emission or in absorption.

°Box 3.4   The Balmer Series and the Bohr Atom

The fact that the spectral lines from hydrogen—one in the red, one in the blue, and then others increasing in number farther into the blue and violet—showed a recognizable pattern was long known, but it was not at first obvious how to explain that pattern mathematically. In 1885, a Swiss schoolteacher named Johann Balmer announced that he had found by trial and error a mathematical pattern that fit the Balmer series. If the first line in the series, hydrogen alpha (Hα) in the red, is assigned the number 2; the second line in the series, hydrogen beta (Hβ) in the blue, is assigned the number 3; the third line, hydrogen gamma (Hγ), is assigned 4; etc., then the wavelengths of the lines fit the formulas:

$$1/\text{wavelength (H}\alpha) = \text{constant} \times (1/2^2 - 1/3^2),$$
$$1/\text{wavelength (H}\beta) = \text{constant} \times (1/2^2 - 1/4^2),$$
$$1/\text{wavelength (H}\gamma) = \text{constant} \times (1/2^2 - 1/5^2),$$

and so on for the farther hydrogen spectral lines. These formulas are so simple—they contain only the squares of the smallest integers—that there must be some simple regularity in the hydrogen atom.

Eventually, Theodore Lyman at Harvard, experimenting in the laboratory far in the ultraviolet beyond where the eye can see, discovered another series of hydrogen lines. For this set of lines, now called the *Lyman series*, each line corresponded (with the same constant as for the Balmer series) to

$$1/\text{wavelength (Ly } \alpha) = \text{constant} \times (1 - 1/2^2),$$
$$1/\text{wavelength (Ly } \beta) = \text{constant} \times (1 - 1/3^2).$$

Other scientists soon discovered further series, with $1/3^2$, $1/4^2$, etc., in place of the $1/1^2$ and $1/2^2$ of the Lyman and Balmer series.

Thus a general formula involving small integers explains all the hydrogen lines. If n is a number assigned to each series (1 for the Lyman series, 2 for the Balmer series, etc.) and m signifies all the integers greater than n, the hydrogen lines are all

$$1/\text{wavelength} = \text{constant} \times (1/n^2 - 1/m^2).$$

The fact that each line could be explained by the difference between two such simple numbers indicated that something fundamental was involved. The reason why this simple subtraction explains all the hydrogen lines was found by Niels Bohr in 1913.

In this *Bohr atom,* we can think of each of the allowable energy states as having a fixed energy E_1, E_2, E_3, etc. Each E is equal to a constant times the speed of light *(c)* times 1/(subscript)^2. Thus the Lyman series has 1/wavelength = $E_1 - E_n$, the Balmer series has 1/wavelength = $E_2 - E_n$, etc. (The n represents the integers higher than the subscript in the first term.)

Figure 3–18 The staff of the Harvard College Observatory in about 1917. The most famous worker at the vital task of classifying spectra was Annie Jump Cannon, fifth from the right. In the decades following 1896, she classified over 500,000 spectra. Her catalogue, called the Henry Draper catalogue after the benefactor who made the investigation possible, is still in use today. Many stars are still known by their HD (Henry Draper catalogue) numbers, such as HD 176387. Also in this photograph is Henrietta S. Leavitt, fifth from the left, whose work on variable stars we shall meet in Sections 6.1 and 24.1.

3.5 SPECTRAL TYPES

In the last decades of the 19th century, spectra of thousands of stars were photographed. Differences existed among spectra from different stars, and classifications of the different types of spectra were developed. The most famous worker at the vital task of classifying spectra was Annie Jump Cannon (Fig. 3–18).

One thing that essentially all the stars had in common was the presence of absorption lines in the spectra. (A continuum was present too; of course, one must have a continuum to have something for the spectral lines to absorb from.)

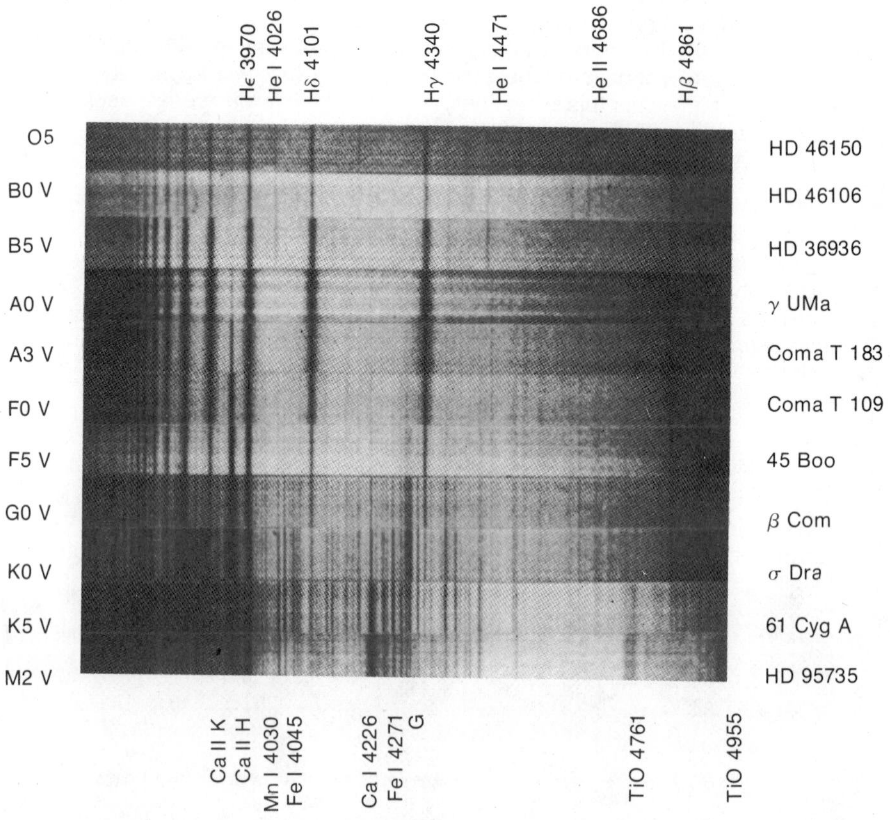

Figure 3–19 Spectral types. Note that all the stellar spectra shown have absorption lines. The Roman numeral "V" indicates that the stars are normal ("dwarf") stars, which will be defined in Section 5.3. The numbers following the letters in the left-hand column represent a subdivision of the spectral types into tenths; the step from B8 to B9 is equal to the adjacent step from B9 to A0. The abbreviations of the elements forming the lines and the individual wavelengths (in angstroms) of the most prominent lines appear at the top and the bottom of the graph.

At first, the stellar spectra were classified only by the strength of certain absorption lines from hydrogen, and were lettered alphabetically: A for stars with the strongest hydrogen lines, B for stars with slightly weaker lines, and so on. These categories are called *spectral types* or *spectral classes* (Fig. 3–19). It was later realized that the types of spectra varied primarily because of differing temperatures of the stellar atmospheres (which, you recall, can independently be measured by observing the color of the star). The hydrogen lines were strongest in stars of a certain temperature and were weaker at both higher and lower temperatures. When we now list the spectral types of stars in order of decreasing temperature, they are no longer in alphabetical order. From hottest to coolest, the spectral types are O B A F G K M.

O stars (that is, stars of type O) are the hottest, and the temperatures of M stars are more than 10 times lower. Generations of American students and teachers have remembered the spectral types by the mnemonic: Oh, Be A Fine Girl, Kiss Me. The spectral types have been subdivided into 10 subcategories each. For example, the hottest B stars are B0, followed by B1, B2, B3, and so on. Spectral type B9 is followed by spectral type A0. It is fairly easy to tell the spectral type of a star by mere inspection of its spectrum.

The actual assignment of spectral types follows the detailed system advanced by W. W. Morgan of the Yerkes Observatory and P. C. Keenan of the Perkins Observatory in 1953 and revised slightly in 1973. They chose a particular set of stellar spectra to serve as standards. All other spectra are compared with these standards to have their spectral types determined.

Historically, the hotter stars are called early types and the cooler stars are called late types. Thus, a B star is "earlier" than an F star, although no evolutionary sequence is now believed to exist, that is, we do not think that stars change from one spectral type to another during the long stable period that takes up most of their lives.

3.5a Stellar Spectra

When we observe the light from a star, we are seeing only the radiation from a thin layer of that star's atmosphere. The interior is hidden within. As we discuss stellar spectra, let us keep this fact in mind.

In parts of Chapters 8 to 11, we will be discussing the processes that go on inside stars.

As we consider stars of different spectral types (Fig. 3–19) ranging from the hottest to the coolest, we can observe the effect of temperature on the nature of the spectra. For example, the hottest stars, those of type O, have temperatures that are sufficiently high so that there is enough energy to remove the outermost electron or electrons from most of the atoms. Since most of the hydrogen is ionized, there are relatively few neutral hydrogen atoms left and the hydrogen spectral lines are weak. Helium is not so easily ionized—it takes a large amount of energy to do so—yet even helium appears not in its neutral state but in its singly-ionized state in an O star. Temperatures range from 30,000 K to 60,000 K in O stars. O stars are relatively rare since they have short lifetimes; none of the nearest stars to us are O stars. The system really begins with spectral types O3 and O4, of which only a handful are known at any distance. No O0, O1, or O2 stars apparently exist. (In the late stages of the evolution of a star, some even hotter objects result, but we are discussing only ordinary stars here.)

Spectral lines arise from transition of electrons between energy states. When hydrogen is ionized, since it has only one electron to begin with, it then has no electrons at all, and thus has no spectral lines.

B stars are somewhat cooler, 10,000 K to 30,000 K. The hydrogen lines are stronger than they are in the O stars, and lines of neutral helium instead of ionized helium are present.

Rigel and Spica are familiar B stars.

A stars are cooler still, 7500 K to 10,000 K. The lines of hydrogen are strongest in this spectral type. Lines of singly ionized elements like magnesium and calcium begin to appear. (All elements other than hydrogen or helium are known as *metals* for this purpose.) O, B, and A stars are all bluish in color.

Sirius, Deneb, and Vega are among the bright A stars in the sky.

K H   D
ULTRAVIOLET   YELLOW

Figure 3–20 The H and K lines of singly ionized calcium are the strongest lines in the visible part of the solar spectrum, and are prominent in spectra of stars of spectral types F, G, K, and M. These classes are known as *late spectral types*, while classes O, A, and B are known as *early spectral types*, although no connotations of age are meant.

Canopus, the second brightest star in the sky (not visible from the latitudes of most of the United States), is a prominent F star, as is Polaris.

F stars have temperatures of 6000 K to 7500 K. Hydrogen lines are weaker in F stars than they are in A stars, but the lines of singly ionized calcium are stronger. Singly ionized calcium has a pair of lines that are particularly conspicuous; they are easy to pick out and recognize in the spectrum. These lines are called H and K (Fig. 3–20).

Besides the sun, Alpha Centauri (actually the brightest of the three stars that together make up the bright point in the sky that we call Alpha Centauri) is also a G star.

G stars, the spectral type of our sun, are 5000 K to 6000 K. They are yellowish in color, as the peak in their spectra falls in the yellow to yellow/green part of the spectrum. The hydrogen lines are visible, but the H and K lines are the strongest in the spectrum. The H and K lines are stronger in G stars than they are in any other spectral type.

Arcturus and Aldebaran, both visible as reddish points in the sky, are K stars.

K stars are relatively cool, only 3500 K to 5000 K. The spectrum is covered with many lines from neutral metals, a strong contrast to the spectra of the hottest stars, which show few spectral lines.

Do not confuse the K in the terminology of the K line of calcium with the K in the name of the spectral type. H was the letter assigned to the strong calcium lines when Fraunhofer made his original list and K was assigned to the shorter wavelength line later on. The H and K lines are clearly shown in Color Plate 5.

Box 3.5

Do not confuse:
    K stars: a spectral class
    K: a kelvin, a unit of temperature
    K: a strong line in many stellar spectra
    K: the element potassium

Betelgeuse is an example of such a reddish star of spectral class M.

M stars are cooler yet, with temperatures less than 3500 K. Their atmospheres are so cool that molecules can exist without being torn apart, and the spectrum shows many molecular lines. (Hotter stars do not usually show molecular lines.) Lines from the molecule titanium oxide are particularly numerous.

For the very coolest stars, there are alternative spectral types that reflect differences in the relative abundances of various elements. For example, stars that are relatively rich in carbon are called *carbon stars* or *C-type*. (These were originally called R and N stars.) *S-type* stars have, among other things, relatively strong zirconium oxide lines, as opposed to titanium oxide lines. Both C- and S-type stars have similar temperatures to M stars.

*The **dispersion** is the measure of how spread out the spectrum is, i.e., how wide the rainbow of color is. Even a low dispersion spectrum, i.e., one that is not spread out very much (we sometimes call this "classification dispersion"), is adequate to tell the spectral type to a fraction of a spectral class (i.e., differentiate F2 from F4).*

We have listed the ordinary spectral types of stars, each of which shows absorption spectrum, that is, a continuum crossed with dark lines. There are many unusual stellar spectra, though. *Wolf-Rayet stars* are O stars that not only show emission lines but also have the strange feature that the emission is particularly broad, that is, a given line may cover several angstroms of spectrum instead of a fraction of an angstrom. The emission in the spectra of Wolf-Rayet stars comes from shells of material that the star has ejected into the space surrounding it.

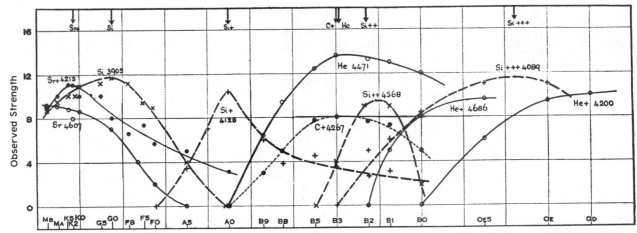

Figure 3-21 In 1925, Cecilia Payne-Gaposchkin (then Cecilia Payne) made this graph of the strength of certain absorption lines versus spectral type. She used it together with theoretical calculations to set the stellar temperature scale for each spectral type.

(Other types of stars, like Oe, Be, Ae, and Me stars (where the "e" stands for emission) also have emission lines, though these are relatively rare.)

The relation of spectral types with the strength of lines in their spectra is shown in Figure 3-21.

*3.5b The Continuous Spectrum

Where does the continuous spectrum come from? In order for there to be an absorption line, there must be some continuous radiation to be absorbed. The simplest picture, one that was believed for many years but was later found to be oversimplified, is that a level just under the surface of a star emits a continuous spectrum, and the light emitted by this layer passes through a layer of gas that absorbs at certain wavelengths. These diminished intensities at certain wavelengths are, of course, the absorption lines. However, we now realize that the processes of continuous emission and spectral-line absorption take place together throughout the outer layers of the stars rather than in separate layers.

The detailed study of the processes of emission and absorption is called the study of *stellar atmospheres*, and nowadays large-scale computers are able to handle complicated sets of equations that describe the transfer of radiation from the outer layers of a star into space. Sometimes the largest computers, such as those at the National Center for Atmospheric Research (NCAR) in Boulder, Colorado, may run for hours to calculate a model of the atmosphere of a single star.)

Figure 3-22 Neutral hydrogen, the hydrogen ion, and the negative hydrogen ion.

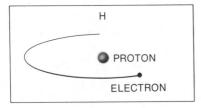

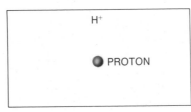

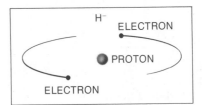

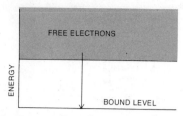

Figure 3–23 A free-bound transition in an atom that has one bound level.

A bound-bound transition would go from one discrete level to another, analogous to jumping on a staircase from one step to another. A free-bound transition would be analogous to jumping from some arbitrary height above the staircase (as from the sky with a parachute) to some particular step.

The particular mechanism that causes the continuous emission is different for stars of different spectral types. Let us consider the situation for the sun, remembering that stars of other spectral types are affected by different processes.

In the sun, the continuous emission is caused by a strange type of ion: the negative hydrogen ion, H⁻ (Fig. 3–22). When second electrons join neutral hydrogen atoms, continuous emission results because the second electrons can have had any amount of energy when they were still "free," that is, outside the atom, as opposed to being restricted to certain energy levels while they are "bound" in the atom. This addition of an electron to an atom is called a "free-bound transition" (Fig. 3–23). Sometimes a free electron is affected by an atom but remains free; such a situation is called a "free-free transition." Since the amount of energy change represented in both these cases is not restricted to a discrete set of values, we get a continuous spectrum in these cases, both of which are important in the sun. In the third situation, a "bound-bound transition," we have a jump between two discrete energy levels, and a spectral line results.

The realization that free-bound and free-free transitions in the negative hydrogen atom cause the solar continuous radiation in the visible and infrared did not come until 1939. Only one atom in 100 million in the sun is in the negative hydrogen ion state, but even this small fraction provides a sufficient number of negative hydrogen atoms to cause the observed effect.

SUMMARY AND OUTLINE

Stars are balls of gas held together by gravitation and generating energy by nuclear fusion.
Colors of stars and of other gas radiating a continuum (Sections 3.1 and 3.2)
As gas gets hotter, it goes from red hot to bluish hot, that is, the peak of the radiation moves to shorter and shorter wavelengths as the temperature increases. This is *Wien's displacement law:*
$\lambda_{max} T$ = constant.
As it gets hotter, the total energy radiated grows rapidly (with the fourth power of the temperature). This is the *Stefan-Boltzmann* law: $E = \sigma T^4$.
Planck's law, discovered after the above laws, gives the amount of radiation at each wavelength, given the temperature. Wien's displacement law and the Stefan-Boltzmann law are really included within Planck's law.
Things whose radiation follows Planck's law precisely are called *black bodies*.

> Wien's displacement law
> $\lambda_{max} T$ = constant
> Stefan-Boltzmann Law
> $E = \sigma T^4$
> Planck's Law relates
> λ, E_λ, and T

The structure of an atom (Section 3.3)
Protons and neutrons in the nucleus with electrons orbiting the nucleus.

Spectral lines (Sections 3.3 and 3.4)
Quantum theory showed that atoms can exist only in discrete states of energy. When an atom drops from a higher energy state to a lower energy state it emits the difference in energy as a photon of a particular wavelength λ and energy hc/λ. Such photons make an emission line. If, on the other hand, an atom takes up enough energy to raise itself from a lower to a higher state, then such atoms lead to the formation of an absorption line in radiation that has been hitting them.
Hydrogen has a series of spectral lines that falls in a distinct pattern across the visible; this is the Balmer series.
The Bohr atom, with each energy level having a different size, is a good way to visualize the hydrogen atom. The Balmer series corresponds to transitions between level 2 and higher levels. The Lyman series, which falls in the ultraviolet (and thus does not pass through the earth's atmosphere), involves transitions to the more basic level, the first level.
Spectral types (Section 3.5)
Hottest to coolest: O B A F G K M
Hydrogen spectra are strongest in type A.
Calcium H and K lines are strongest in type G.
Molecular spectra begin to appear in type M.
All types are absorption spectra. Only a few kinds of stars with emission lines exist.
The sun's continuous spectrum comes from the negative hydrogen ion (H⁻).

QUESTIONS

1. What are two differences between a star and a planet?

†2. Room temperature is now about 65°F. (a) What is this in °C? (b) in kelvins?

3. The sun's surface is about 5800 K. (a) What is this in °C? (b) In °F?

4. The sun's spectrum peaks at 5600 Å. Would the spectrum of a star whose temperature is twice that of the sun peak at a longer or a shorter wavelength? How much more energy than a square centimeter of the sun would a square centimeter of the star give off?

†5. One black body peaks at 2000 Å. Another, of the same size, peaks at 10,000 Å. Which gives out more radiation at 2000 Å? Which gives out more radiation at 10,000 Å? What is the ratio of the total radiation given off by the two bodies?

6. Which contains more information, Wien's displacement law or Planck's law? Explain.

†7. What is the ratio of energy output for a bit of surface of an average O star and a same-sized bit of surface of the sun?

8. Star A appears to have the same brightness through a red and a blue filter. Star B appears brighter in the red than in the blue. Star C appears brighter in the blue than in the red. Rank these stars in order of increasing temperature.

9. (a) From looking at Figure 3–17, draw the Lyman series and the Balmer series on the same wavelength axis. (b) Why is the Balmer series the most observed spectral series of atomic hydrogen?

10. What is the difference between the continuum and an absorption line? The continuum and an emission line? Draw a continuum with absorption lines. Can you draw absorption lines without a continuum? Can you draw emission lines without a continuum? Explain.

†11. Consider a hypothetical atom in which the energy levels are equally spaced from each other, that is, the energy of level n is n. (a) Draw the energy level diagram for the first five levels. (b) Indicate on the diagram the transition from level 4 to level 2 and from level 4 to level 3. (c) What is the ratio of energies of these two transitions? (d) If the $4\rightarrow2$ transition has a wavelength of 4000 Å, what is the wavelength of the $3\rightarrow2$ transition?

12. (a) Why are singly ionized helium lines detectable only in O stars? (b) Why aren't neutral helium lines prominent in the solar spectrum?

13. Compare Planck curves for stars of spectral types O, G, and M.

14. Why don't we see strong molecular lines in the sun?

15. Make up your own mnemonic for the spectral types.

†16. Using Balmer's formula and the fact that the wavelength of Hα is 6563 Å, calculate the wavelength of Hβ. Show your work.

17. What factors determine spectral type? What instrument would you use to determine the spectral type of a star?

18. List spectral types of stars in approximate order of strength, from strongest to weakest, of (a) hydrogen lines, (b) ionized calcium lines, (c) titanium oxide lines.

19. If we take two stones, one twice the diameter of the other, and put them in an oven until they are heated to the same temperature and begin glowing, what will be the relationship between the total energy in the light given off by the stones?

20. Why does the negative hydrogen atom have a continuous rather than a line spectrum?

---

†This indicates a question requiring a numerical solution.

CHAPTER 4

OBSERVING THE SKY

AIMS:
To understand the appearance of
astronomical objects that we see in the
sky, how we describe celestial positions,
and time zones and calendars

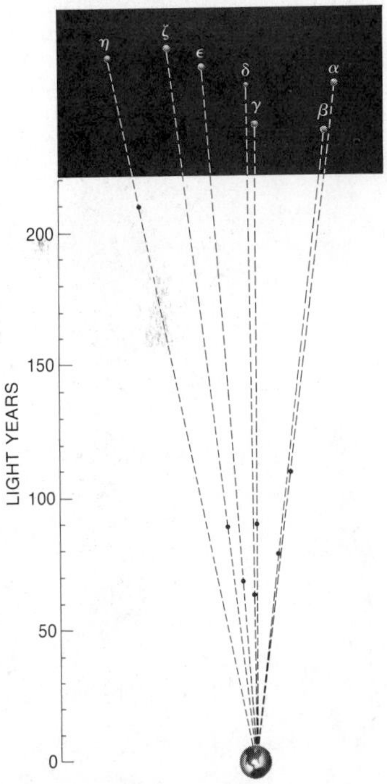

Figure 4-1 The stars we see as a constellation are actually differing distances from us. In this case we see the true distances of the stars in the Big Dipper, part of the constellation Ursa Major, in the lower part of the figure. Their appearance projected on the sky is shown in the upper part.

When we look up at the sky on a dark night, from a location outside a city, we see a fantastic sight. If the moon is up, its splendor can steal the show; not only does its pearly appearance draw our attention but also the light it gives off makes the sky so bright that we cannot see the other objects well. But when the moon is down, we see bright jewels in the inky sky. Generally, the brightest few shine steadily, which is how we can tell that they are planets. The rest twinkle—sometimes gently, sometimes fiercely—which reveals them to be stars.

The Milky Way arches across the sky, and if we are in a good location on a dark night, it is quite obvious to the naked eye. (City inhabitants may never see the Milky Way at all.) If you know where to look, from the northern hemisphere you can see a hazy spot that is actually a galaxy, rather than individual stars. It appears in the constellation Andromeda. From the southern hemisphere, two other galaxies, the Magellanic Clouds, can be easily seen.

Other things are sometimes seen in the sky. If we are lucky, a bright comet may be there, but this happens rarely. More often, meteors—shooting stars—will dart across the sky above. And, these days, we may see the steady light of a spacecraft move in a stately fashion across the sky. The lights of airplanes can be seen quite regularly.

This 10½-hour exposure shows southern star trails over the Anglo-Australian telescope in Siding Spring, Australia.

4.1 THE CONSTELLATIONS

The International Astronomical Union put the scheme of constellations on a definite system. The sky was officially divided into 88 constellations (see Appendix 10) with definite boundaries, and every star is now associated with one and only one constellation.

Long, long ago, when Egyptian and other ancient astronomers were beginning to study and understand the sky, they divided the sky into regions containing fairly distinct groups of stars. The groups, called *constellations*, were given names and stories were attached, perhaps to aid in remembering.

Actually, the constellations are merely areas in the sky that happen to have stars in particular directions as seen from our vantage point on the earth. There is no significance to the apparent groupings, nor are the stars in a given constellation necessarily associated with each other in any direct manner (Fig. 4–1).

The stories we now associate with the constellations come from Greek mythology (Fig. 4–2), though the names may have been associated with particular constellations more to honor Greek heroes than because the constellations actually looked like these people. Other civilizations (American Indians, for example) attached their own names, pictures, and stories to the stars.

Ursa Major was originally the princess Callisto, an attendant of the goddess Juno, who became jealous of her. To protect Callisto, Jupiter turned her into a bear. However, when Callisto's son was about to kill the bear one day, Jupiter turned him into another bear and placed both of them in the sky.

Sometimes familiar groupings in the sky do not make up a complete constellation. Such groupings are called *asterisms;* the Big and Little Dippers are examples, because they are really parts of the constellation Ursa Major (the Big Bear) and Ursa Minor (the Little Bear), respectively.

Box 4.1   The Sky Maps

A set of four sky maps are bound into this book, one for each of the seasons. Choose the map that corresponds to your season of the year; the evening hours to which it corresponds are given at the top of the page. Hold the map over your head so that you are looking up at it, with the top of the page to the north. This will put the other directions marked on the map—south, east, and west—in the positions corresponding with those directions on earth.

The maps are drawn for observers at a latitude of 40 degrees, which corresponds to a line through Philadelphia, Indianapolis, Denver, and halfway between San Francisco and the California-Oregon border. If you are farther north than that, Polaris will appear higher in your sky. If you are farther south, on the other hand, Polaris will appear lower in the sky.

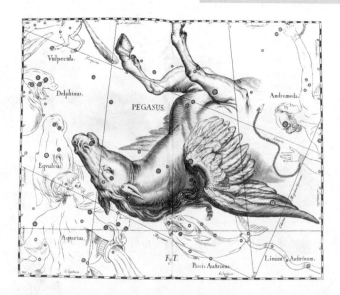

Figure 4–2 Pegasus, from the atlas of Hevelius. The horse appears reversed from what we see in the sky (in addition to being upside down), because the atlas is drawn as though we are looking at a globe of the sky from the outside instead of from the inside.

4.2   TWINKLING

What about the twinkling? It is not a property of the stars themselves, but merely an effect of our earth's atmosphere. The starlight is always being bent by moving volumes of air in our atmosphere, making the images of the stars appear to be larger than points, to dance around slightly (a property called *seeing*) and to change rapidly in intensity when observed with the naked eye (a property called *scintillation*). The change in intensity, scintillation, is what we non-technically call "twinkling." The planets, on the other hand, do not usually seem to twinkle. This is because they are close enough to earth that they appear as tiny disks large enough to be seen through telescopes. Though each point of the disk of light representing a planet may change slightly in intensity, the disk is made of so many points of light that the total intensity doesn't change. To the naked eye the planets thus appear to shine more steadily than the stars.

Even the stars don't twinkle when seen through a sufficiently large telescope, because the changes in intensity average out over the surface of the telescope.

4.3   LIGHT FROM THE STARS

The most obvious thing we notice about the stars is that they have different brightnesses. Over 2000 years ago, the Greek astronomer Hipparchus divided the stars that he could see into classes of brightness. His work was extended by Ptolemy in Alexandria in about A.D. 140. In their classification, the brightest stars were said to be of the first magnitude, somewhat fainter stars were of the second magnitude, and so on down to the sixth magnitude, which represented the faintest stars that could be seen with the naked eye.

4.3a   Apparent Magnitude

In the nineteenth century, when astronomers became able to make quantitative measurements of the brightnesses of stars, the magnitude scale was placed on an accurate basis. It was discovered that the brightest stars that could be seen with the naked eye were about 100 times brighter than the faintest stars. This corresponded to a difference of 5 magnitudes on Hipparchus' scale, a number that was used to set up the present magnitude system. Since it tells how bright a star **appears**, it is called *apparent magnitude.*

The new system is based on the definition that a difference of five magnitudes means a ratio of intensity of exactly 100 times. Thus a star of first magnitude is exactly 100 times brighter than a star of the sixth magnitude. The brightnesses of the stars that had been classified by the Greeks were placed on this new magnitude scale (Fig. 4–3). Many of the stars that had been of the first magnitude in the old system were indeed "first magnitude stars" in the new system. But a few stars were much brighter, and thus corresponded to "zeroth" magnitude, that is, a number less than one. A couple, like Sirius, the brightest star in the sky, were brighter still. The numerical magnitude scale was easily extended to negative numbers. On this scale, Sirius is brighter than magnitude −1. The new scale, being numerical, admits fractional magnitudes. Sirius is actually magnitude −1.4, for example. Physiologically, this type of measurement makes sense because the human eye happens to perceive equal **ratios** of luminosity (such a ratio is

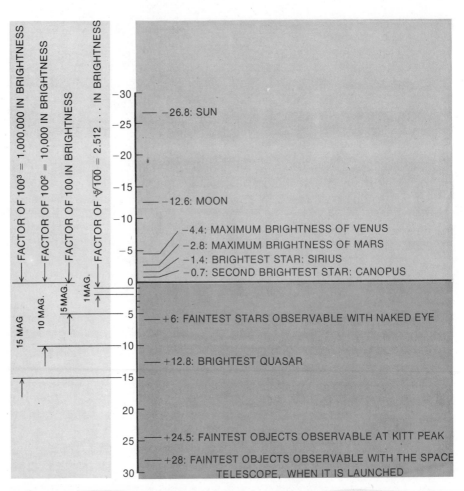

Figure 4–3 The apparent magnitude scale is shown on the vertical axis. At the left, sample intervals of 1, 5, 10, and 15 magnitudes are marked and translated into multiplicative factors.

The interval from magnitude −1.4 to magnitude 23.5 is nearly 25 magnitudes. Since each 5 magnitudes represents a multiplication of 100 times in brightness, 25 magnitudes is a factor of 100 × 100 × 100 × 100 × 100 (since it is 100 multiplied by itself 5 times, we abbreviate this by writing $100^5 = (10^2)^5 = 10^{10}$), or 10,000,000,000.

the energy coming from one object divided by the energy coming from another) as roughly equal **intervals** (additive steps) of brightness, which is just how the magnitude scale is set up.)

The scale was extended not only to stars brighter than first magnitude, but also to stars fainter than 6th magnitude. Since 5 magnitudes corresponds to a hundred times in brightness, 11th magnitude is 100 times fainter than 6th magnitude (that is, 1/100 times as bright); 16th magnitude, in turn, is 100 times fainter than 11th magnitude. The faintest stars that can be photographed with the 5-meter (200-inch) telescope on Palomar Mountain in California, the largest optical telescope in the United States and until recently the largest optical telescope in the world, are fainter than 23rd magnitude.

*4.3b Working with the Magnitude Scale

We have seen how a difference of 5 magnitudes corresponds to a factor of 100 in brightness. What does that indicate about a difference of 1 magnitude? Since an increase of 1 in the magnitude scale corresponds to a decrease in brightness by a certain factor, we need a number that, when multiplied by itself 5 times, will equal 100. This number is just $\sqrt[5]{100}$. Its value is 2.512 . . . , with the dots representing an infinite string of other digits. For many practical purposes, it is sufficient to know that it is nearly 2.5.)

Thus a second magnitude star is about 2.5 times fainter than a first magnitude star. A third magnitude star is about 2.5 times fainter than a second magnitude star. Thus, a third magnitude star is approximately $(2.5)^2$ (which is about 6) times fainter than a first magnitude star. (We could easily give more decimal places, writing 6.3 or 6.31, but the additional figures would not be meaningful. We started with only one digit being significant when we wrote down "2nd magnitude," so our data were given to only 1 *significant figure.* Our result is thus only accurate to 1 significant figure.)

In similar fashion, using a factor of 2.512 for the single magnitudes and a factor of 100 for each group of 5 magnitudes, we can very simply find the ratio of intensities of stars of any brightness.

Example: What is the difference between magnitude -1 and magnitude 6, the ratio between the brightest and faintest stars that are seen with the naked eye?

Answer: The difference is 7 magnitudes, and the factor is thus $(100)(2.5)(2.5)$, which is about 600. The star of -1st magnitude is 600 times brighter than the star of 6th magnitude.

One unfortunate thing about the magnitude scale is that it operates in the opposite sense from a direct measure of brightness. Thus the brighter the star, the lower the magnitude (a second magnitude star is brighter than a third magnitude star), which is sometimes a confusing convention. This kind of problem comes up occasionally in an old science like astronomy, for at each stage astronomers have made their new definitions so as to have continuity with the past.

*4.4 COORDINATE SYSTEMS

The stars and other astronomical objects that we study are at a wide range of distances from the earth, and it is much harder to determine these distances than it is to determine the direction to these objects. Though we

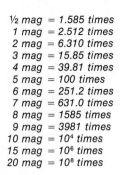

½ mag	= 1.585 times
1 mag	= 2.512 times
2 mag	= 6.310 times
3 mag	= 15.85 times
4 mag	= 39.81 times
5 mag	= 100 times
6 mag	= 251.2 times
7 mag	= 631.0 times
8 mag	= 1585 times
9 mag	= 3981 times
10 mag	= 10^4 times
15 mag	= 10^6 times
20 mag	= 10^8 times

We must be careful not to fool ourselves that we can gain in accuracy in an arithmetical process; the number of significant figures we put in is the number of significant figures in the result.

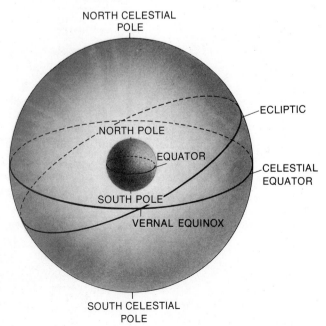

Figure 4-4 The celestial equator is the projection of the earth's equator onto the sky, and the ecliptic is the sun's yearly path through the stars. The vernal equinox is one of the intersections of the ecliptic and the celestial equator. From a given location at the latitude of the United States, the stars nearest the north celestial pole never set and the stars nearest the south celestial pole never rise above the horizon.

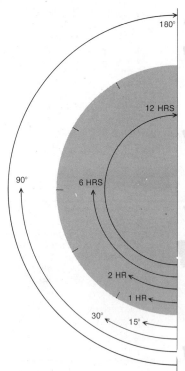

Figure 4-5 Right ascension is measured in units of time. 24 hours of right ascension make up a full 360° circle, so each hour of right ascension corresponds to 15° of arc. Your sidereal time is the right ascension of the stars on your meridian at that instant.

The zero-point of right ascension is the vernal equinox, *one of the two locations where the ecliptic– the sun's path in the sky–and the celestial equator cross.*

may have to know the distance to an object to understand how it works, to make observations of an object, we need know only its direction. For this reason, for observational purposes only, it is useful to think of the astronomical objects as being at a common distance, all hung on the inside of a large (imaginary) sphere, which we call the *celestial sphere.* In this section, we will see how we describe the positions of objects on the celestial sphere.

Astronomers and geographers share the need to set up systems of coordinates—*coordinate systems*—to designate the positions of places in the sky or on the earth.

The geographers' system is familiar to most of us: longitude and latitude. Lines (actually half-circles) of longitude called *meridians* run from the north pole to the south pole. The zero circle of longitude has been adopted, by international convention, to run through the former site of the Royal Greenwich Observatory in England. We measure longitude by the number of degrees east or west an object is from the meridian that passes through Greenwich.

Latitudes are defined by parallel circles that run around the earth, all parallel to the equator. Zero degrees of latitude corresponds to the equator; ninety degrees of latitude corresponds to the poles (90° of north latitude for the north pole and 90° of south latitude for the south pole).

The astronomers' system for the sky corresponds exactly to the geographers' system for the earth, except that the astronomers use the names *right ascension* for the equivalent of celestial longitude and *declination* for celestial latitude. They are measured with respect to a *celestial equator,* which is the extension of the earth's equator into space, and *celestial poles,* which are on the extensions of the earth's axis of spin into space and thus lie above the earth's poles (Fig. 4–4).

Right ascension and declination form a coordinate system fixed to the stars. To observers on earth, the stars appear to revolve every 23 hours 56 minutes. The coordinate system thus appears to revolve at the same rate. Actually, of course, the earth is rotating and the stars and celestial coordinate system remain fixed.

Astronomers set up a timekeeping system called *sidereal time* (sidereal means "by the stars"). The sidereal time is the right ascension of any star on that place's meridian, the great circle linking the north and south poles and passing through the *zenith,* the point directly overhead (Fig. 4–5). Thus each location on the earth has a different sidereal time at each instant. Astronomers find this system convenient because they can consult clocks that are set to run on sidereal time. Thus by merely knowing the sidereal time, they can tell whether a star is favorably placed for observing.

Sidereal time and solar time are not exactly the same. They are both caused by the rotation of the earth on its axis. A *sidereal day* is the length of time that a given right ascension takes to return to the same position in the sky. A *solar day* is the length of time that the sun takes to return to the same position in the sky. Since the earth is revolving around the sun every year, by the time a day has gone by, the earth has moved 1/365 of the way around the sun. Thus after the stars have returned to their same positions in the sky, it takes 1/365 of approximately 24 hours=4 minutes for the sun to return to its position. A solar day is thus approximately 4 minutes (actually 3 minutes 56 seconds of time) longer than a sidereal day (Fig. 4–6).

Every observatory has both sidereal clocks, for the astronomers to tell when to observe their stars, and solar clocks, for the astronomers to gauge when sunrise will come and to know when to go to dinner. A solar clock and a sidereal clock show the same time (on a 24-hour system) on only one day each year. The next day the sidereal clock is 4 minutes ahead, the second day afterward it is 8 minutes ahead, and so on. Six months later the two clocks differ by 12 hours and the stars that were formerly at their highest at midnight are then at their highest at noon, when the sun is out. As a result, they may not be visible at all at that season.

The coordinate in the sky that is perpendicular to the celestial equator marks the declination. Declination, latitude in the sky, is measured in degrees north or south of the celestial equator in exactly the same manner that latitude on the earth is measured with respect to the terrestrial equator.

Actually, there are minor effects that cause the celestial coordinates of the stars to change slightly. One such effect is *precession*, which causes a slow drift in the coordinate system with a 26,000-year period (Fig. 4–7). Precession takes place because the earth's axis doesn't always point exactly at the same spot in the sky; it rather traces out a small circle and takes approximately 26,000 years to return to the same orientation. As a result of precession, one has to make small corrections in any catalogue of celestial positions to update them to the present time. Precession is a small effect—the change in celestial coordinates is less than one minute of arc (one sixtieth of a degree) per year. Don't worry, Polaris will be the North Star again in about 26,000 years, but in between—in, say, A.D. 14,980—the North Star will be Vega.

Another minor effect that causes changes in the right ascensions and declinations of celestial objects is "proper motion," an actual motion of individual objects across the sky (see Section 5.6), but this effect is so minuscule that even when astronomers work at it, the proper motion is difficult to measure.

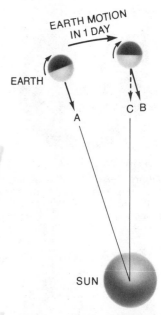

Figure 4–6 While the earth rotates once on its own axis with respect to the stars (one sidereal day), it also moves slightly in its orbit around the sun. Thus after one sidereal day, arrow A becomes arrow B. But one solar day has passed only when arrow B rotates a little farther and becomes arrow C. This takes an additional 4 minutes, making a solar day 4 minutes longer than a sidereal day.

4.5 MOTIONS IN THE SKY

At the latitudes of the United States, which range from +25 degrees for the tip of the Florida Keys up to +49 degrees for the Canadian border, and

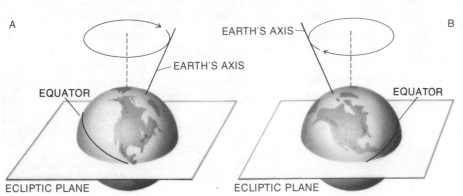

Figure 4–7 The earth's axis precesses with a period of 26,000 years. The two positions shown are separated by 13,000 years.

As the earth's pole precesses, the equator moves with it (since the earth is a rigid body). The celestial equator and the ecliptic will always maintain the 23½° angle between them, but the points of intersection, the equinoxes, will change. Thus, over the 26,000-year precession cycle, the vernal equinox will move through all the signs of the zodiac. It is now in the constellation Pisces and approaching Aquarius (and thus the celebration in the musical "Hair" of the "Age of Aquarius").

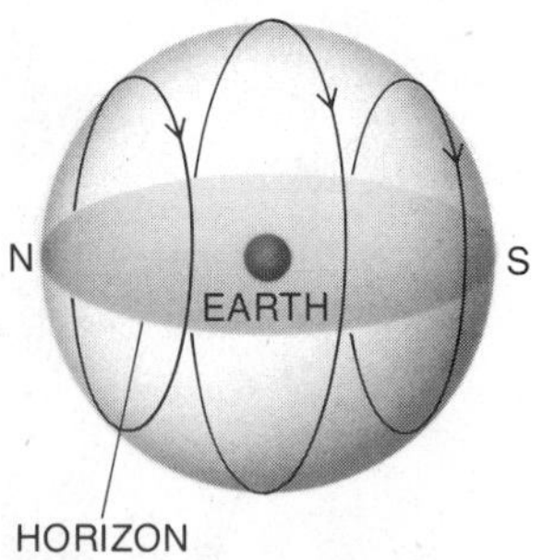

Figure 4–8 From the equator, the stars rise straight up, pass right across the sky, and set straight down.

ZENITH

HORIZON

Figure 4–9 From the pole, the stars move around the sky in circles parallel to the horizon, never rising or setting.

These concepts involve spherical geometry, which many individuals find difficult to visualize without practical experience. But a little experience with a telescope or in a planetarium can make right ascension and declination seem very easy.

A large alt-azimuth mount is often less expensive to construct than an equatorial mount of the same size, and can be housed in a smaller and thus less expensive dome. The 6-m Soviet telescope and the Multiple Mirror Telescope in Arizona have alt-azimuth mounts for these reasons.

down to +19 degrees in Hawaii and up to +67 degrees in Alaska, the stars rise and set at angles to the horizon. In order to understand the situation, it is best first to visualize simpler cases.

If we were standing on the equator, the stars would rise perpendicularly to the horizon (Fig. 4–8). The north celestial pole would lie exactly on the horizon in the north, and the south celestial pole would lie exactly on the horizon in the south. Each star would rise somewhere on the eastern half of the horizon; each would remain "up" for twelve hours, and then would set. We would be able to see all stars, no matter what their declination. The sun, no matter what its declination, would also rise and set twelve hours apart, so the day and night would each last twelve hours.

If, on the other hand, we were standing on the north pole, the north celestial pole would be directly overhead, and the celestial equator would be on the horizon (Fig. 4–9). All the stars would move around the sky in circles parallel to the horizon. Since the celestial equator is on the horizon, we could see only the stars with northern declinations. The stars with southern declinations would never be visible.

Let us consider in this paragraph a latitude between the equator and the north pole, say 40 degrees north latitude. Then the stars seem to rise out of the horizon at oblique angles. The north celestial pole (with the north star nearby it) is always visible in the northern sky, and is at an *altitude* of 40 degrees above the horizon. The star Polaris, a 2nd magnitude star, happens to be located within one degree of the north celestial pole, and so is called the *pole star.*

If you point a camera at the sky and leave its lens open for a long time, many minutes or hours, the stars appear as trails. Those near the south celestial pole (see the figure opening this chapter) or the north pole (Fig. 4–10) move in relatively small and obvious circles around the poles. The circles followed by stars farther from the celestial poles are so large that sections of them seem straight (Fig. 4–11).

When astronomers want to know if a star is favorably placed for observing, they must know both its right ascension and declination. By seeing if the right ascension is reasonably close to the sidereal time, they can tell if it is the best time of year at which to observe (Fig. 4–12). But they must also know the star's declination to know how long it will be above the horizon each day.

Telescopes are often mounted at an angle such that one axis points directly at the north celestial pole, i.e., at a position in the sky very close to that of Polaris. (In the southern hemisphere, the axis would point at the south celestial pole.) This axis is called the *polar axis.* Thus since all stars move across the sky in circles centered nearly at Polaris, the telescope must merely turn about that axis to keep up with the stellar motions. The other axis of the telescope is used to point the telescope in declination.

Since motion around only one axis is necessary to track the stars, one need have only a single small motor set to rotate once every 24 sidereal hours. This motor turns the polar axis in the direction opposite to the rotation of the earth. The principle is the same for a small telescope in your backyard as for the 5-meter telescope at Palomar. Arrangements of this type are called *equatorial mounts,* since one axis rotates perpendicularly to the celestial equator (Fig. 4–13).

The alternative to this system is to mount a telescope such that one axis

Figure 4–10

Figure 4–11

Figure 4–10 Star trails over the 4-m telescope of the Kitt Peak National Observatory. (Richard and Dolores Hill)

Figure 4–11 Near the celestial equator, the star circles are so large that they appear almost straight in this view past the Yerkes Observatory in Wisconsin.

goes up-down (altitude) and the other goes around (azimuth). The azimuth motion is parallel to the horizon and the altitude motion is perpendicular. Binoculars on stands at scenic overlooks are mounted this way. To track a star, continual adjustments have to be made in both axes, which was formerly very inconvenient. However, the availability of computers to make the necessary calculations allows the mounting of large new telescopes in this system, which is called *alt-azimuth*.

Figure 4–12 For each time of the year, we can see the constellations that are in the direction away from the sun.

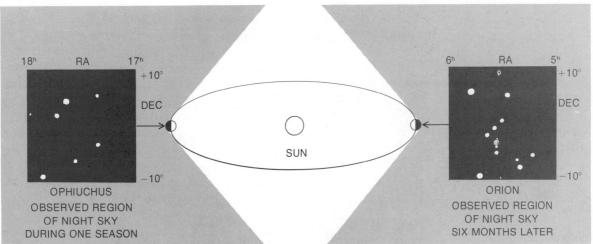

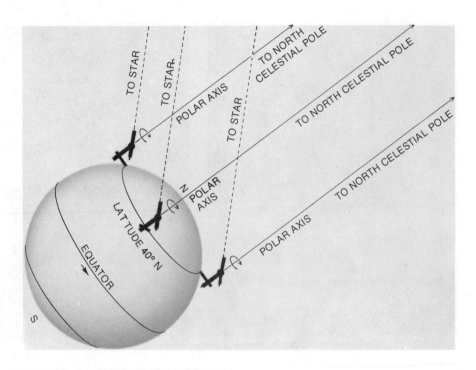

Figure 4–13 An equatorially mounted telescope need rotate on only one axis to keep pointing at a star. This axis is called the *polar axis;* it is fixed for telescopes in the northern hemisphere so that it points at the north celestial pole. Rotation of the telescope around this axis keeps up with the rotation of the earth on its axis. The axis perpendicular to the polar axis sweeps out a circle of declination in the sky and is called the *declination axis.*

*4.5a Positions of the Sun, Moon, and Planets

Although the stars are fixed in their positions in the sky, the sun's position varies through the whole range of right ascension each year. The path of the sun in the sky with respect to the stars is called the *ecliptic.* The earth's axis is inclined by 23½ degrees with respect to the celestial equator. The ecliptic and the celestial equator cross at two points. The sun crosses one of those points, called the *vernal equinox,* on the first day of spring. The sun crosses the other intersection, the *autumnal equinox,* on the first day of autumn. In the northern hemisphere, these occur on approximately March 21st and September 23rd, respectively. On these days, the sun's declination is 0°; the sun's declination varies over the year from +23½° to −23½° (Fig. 4–14).

These points are called *equinoxes* ("equal nights"; *nox* is Latin for night) because the daytime and the nighttime are supposedly equal on these days.

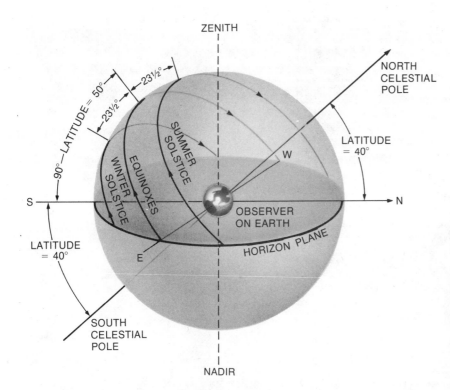

ZENITH

NORTH CELESTIAL POLE

LATITUDE = 50°

23½°

23½°

90° — LATITUDE = 40°

SUMMER SOLSTICE

WINTER SOLSTICE

EQUINOXES

LATITUDE = 40°

W

S

OBSERVER ON EARTH

N

LATITUDE = 40°

E

HORIZON PLANE

SOUTH CELESTIAL POLE

NADIR

Figure 4–14 The path of the sun at different times of the year. Around the summer solstice, June 21, the sun is at its highest declination, rises highest in the sky, stays up longer (because, as shown, more of its path is above the horizon), and rises and sets farthest to the north. The opposite is true near the winter solstice, December 21. The diagram is drawn for a latitude of 40 degrees.

Actually, because refraction (bending) of light by the earth's atmosphere makes the sun appear to rise a little early and set a little late, and the fact that the top of the sun rises ahead of the middle of the sun, the daytime exceeds the nighttime at U.S. latitudes by about 10 minutes on the days of the equinoxes. The days of equal daytime and nighttime precede the vernal equinox and follow the autumnal equinox by a few days.

If we were at the north pole, whenever the sun had a northern declination, it would be above our horizon and we would have daytime. The sun would move in a circle all around us, moving essentially parallel to the horizon. From day to day it would appear slightly higher in the sky for 3 months, and then move lower. The date when it is highest in the sky is called the *summer solstice*. It occurs approximately on June 21st each year. On that

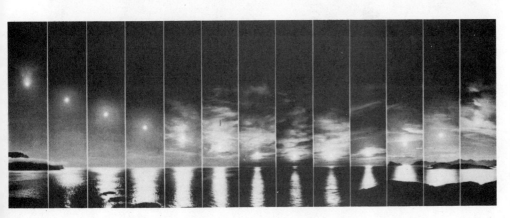

Figure 4–15 In this series taken in June from northern Norway, above the Arctic Circle, one photograph was taken each hour for an entire day. The sun never set, a phenomenon known as the *midnight sun*. Since the site was not at the north pole, the sun and stars move somewhat higher and lower in the sky in the course of a day.

date the sun is 23½° above the horizon because its declination is +23½°. The time of the year when the sun never sets is known as the time of the *midnight sun* (Fig. 4–15), which lasts six months at the poles and shorter times at locations other than the poles. The sun crosses the celestial equator and begins to have northern declination at the vernal equinox. The sun crosses the celestial equator in the other direction six months later at the autumnal equinox. Since the sun goes 23½° above the celestial equator, the midnight sun is visible anywhere within 23½° of the north pole, a boundary at 66½° latitude known as the Arctic Circle. The midnight sun within 23½° of the south pole, within the Antarctic Circle, is six months out of phase with that near the north pole.

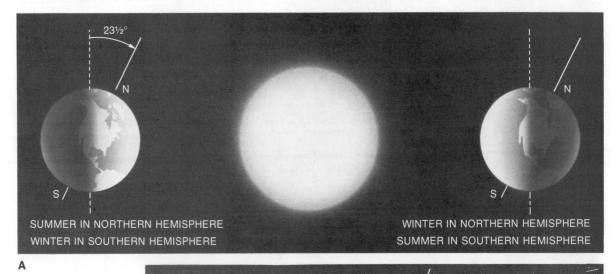

A

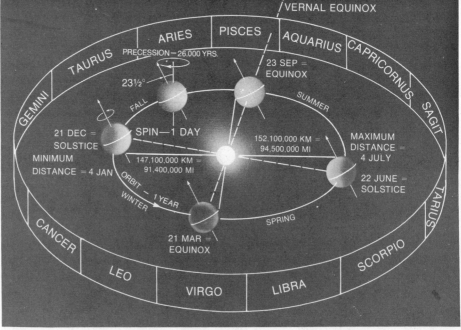

Figure 4–16 *(A)* The seasons occur because the earth's axis is tipped with respect to the plane of its revolution around the sun. When the northern hemisphere is tilted toward the sun, it has its summertime; at the same time, the southern hemisphere is having its winter. The diagram is not to scale. *(B)* The motions of the sun in the sky in the course of a year.

B

© 1970 United Feature Syndicate, Inc.

© 1970 United Feature Syndicate, Inc.

Figure 4–17 Cartoons by Charles Schulz.

When the sun is at the summer solstice, it is at its greatest northern declination and is above the horizon of all northern hemisphere observers for the longest time each day. Thus daytimes in the summer are longer than daytimes in the winter, when the sun is at its lowest declinations. In the winter, the sun not only is above the horizon for a shorter period each day but also never rises very high in the sky. The time of its lowest declination is called the *winter solstice*. Winter in the northern hemisphere corresponds to summer in the southern hemisphere.

The seasons (Fig. 4–16A), thus, are caused by the variation of declination of the sun, which, in turn, is caused by the fact that the earth's axis of spin is tipped by 23½° with respect to the perpendicular to the plane of the earth's orbit around the sun. This cycle, of course, repeats once each year. A summary of the motions of the sun in the sky appears as Figure 4–16B.

The moon goes around the earth once each month, and thus the moon's right ascension changes through the entire range of ascension once each month. Since the moon's orbit is inclined to the celestial equator, the moon's declination also varies.

The planets' motions in the sky are less easy to categorize, but they also change their right ascension and declination from day to day.

Tables of the daily positions of the sun, moon, and planets are published each year by the U.S. Naval Observatory in Washington, D.C., in a book entitled The American Ephemeris and Nautical Almanac.

*4.6 TIME AND THE INTERNATIONAL DATE LINE

Every city and town on earth used to have its own time system, based on the sun, until widespread railroad travel made this inconvenient. In 1884, an international conference agreed on a series of longitudinal time zones. Now all localities in the same zone have a standard time (Fig. 4–18). Since there are twenty-four hours in a day, the 360 degrees of longitude around the earth are divided into 24 standard time zones, each 15 degrees wide. Each time zone is centered on a meridian of longitude exactly divisible by 15. Because the time is the same throughout each zone, the sun is not directly overhead at noon at each point in a given time zone, but in principle is less than about half an hour off. Standard time is based on a mean solar day, the average length of a solar day.

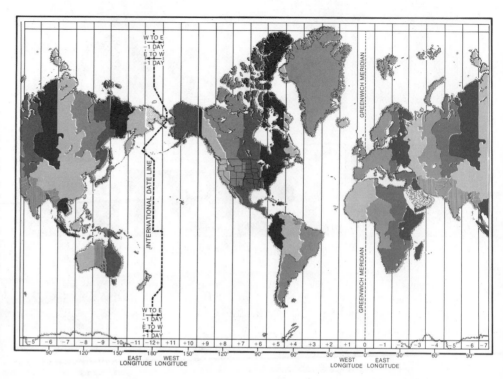

Figure 4–18 Although in principle the earth is neatly divided into 24 time zones, in practice, political and geographic boundaries have made the system much less regular. The existence of daylight-saving time in only some places confuses the time zone system. Some countries even have double-daylight saving time. Other countries, shown with cross hatching, have time zones that differ by one-half hour and even 40 minutes. Nepal even differs by 10 minutes from neighboring India. In the U.S., most states have daylight-saving time for six months a year—from the last Sunday in April until the last Sunday in October. But Arizona, Hawaii, and parts of Indiana have standard time year round.

While the sun seems to move in the sky from east to west, the time in any one place gets later. We can visualize noon, and each hour, moving around the world from east to west, minute by minute. We would get a particular time back 24 hours later, but if the hours circled the world continuously the date would not have changed. So we specify a north-south line and have the date change there. We call it the *international date line.* England won for Greenwich, then the site of the Royal Greenwich Observatory, the distinction of having the basic line of longitude, zero degrees. Realizing that the international date line would disrupt the calendars of those who crossed it, that line was thus put as far away from the populated areas of Europe as possible: near or along the 180 degree longitude line. The international date line passes from north to south through the Pacific Ocean and actually bends to avoid cutting through continents or groups of islands, thus providing them with the same date as their nearest neighbor, as shown in Figure 4–19.

In the summer, in order to make the daylight last into later hours, many countries have adopted daylight-saving time. Clocks are set ahead 1 hour on a certain date in the spring. Thus if darkness falls at 6 P.M. E.S.T., that time is called 7 P.M. E.D.T., and most people have an extra hour of daylight after work. In most places, that hour is taken away in the fall, though some places have adopted daylight-saving time all year. The phrase to remember to help you set your clocks is "fall back, spring ahead." Of course, daylight-saving time is just a bookkeeping change in how we name the hours, and doesn't result from any astronomical changes.

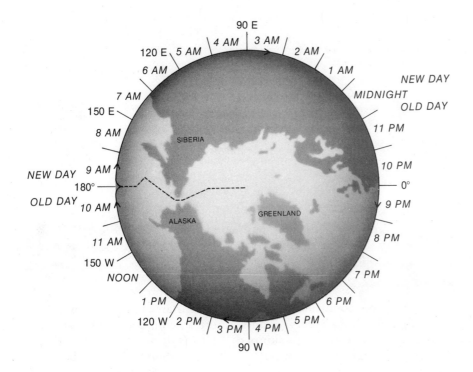

*4.7 CALENDARS

The period of time that the earth takes to revolve once around the sun is called, of course, a *year*. This period is about 365¼ mean solar days. A *sidereal year* is the interval of time that it takes the sun to return to a given position with respect to the stars. A *solar year* (in particular, a *tropical year*) is the interval between passages of the sun through the vernal equinox, the point where the ecliptic crosses the celestial equator. The standard tropical year was 1900. Tropical years are growing shorter by about half a second per century.

Roman calendars had, at different times, different numbers of days in a year, so the dates rapidly drifted out of synchronization with the seasons (which follow solar years). Julius Caesar decreed that 46 B.C. would be a 445-day year in order to catch up, and defined a calendar, the *Julian calendar*, that would be more accurate. This calendar had years that were normally 365 days in length, with an extra day inserted every fourth year in order to bring the average year to 365¼ days in length. The fourth years were, and are, called *leap years*.

The Julian calendar was much more accurate than its predecessors, but the actual solar year 1985 will be 365 days 5 hours 45.6 seconds long, some 11 minutes 14 seconds shorter than 365¼ days (365 days 6 hours). By 1582, the calendar was about 10 days out of phase with the date at which Easter had occurred at the time of a religious council 1250 years earlier, and Pope Gregory XIII issued a bull—a proclamation—to correct the situation. He dropped 10 days from 1582.

A sidereal year differs from a solar year by 20 minutes 24 seconds because of the earth's precession, discussed in Section 4.4.

The name of the fifth month, formerly Quintilis, was changed to honor Julius Caesar; in English we call it July. The year then began in March; the last four months of our year still bear names from this system of numbering. Augustus Caesar, who carried out subsequent calendar reforms, renamed August after himself. He also transferred a day from February in order to make August last as long as July.

Many citizens of that time objected to the supposed loss of the time from their lives and to the commercial complications. Does one pay a full month's rent for the month from which the days were omitted, for example? "Give us back our fortnight," they cried.

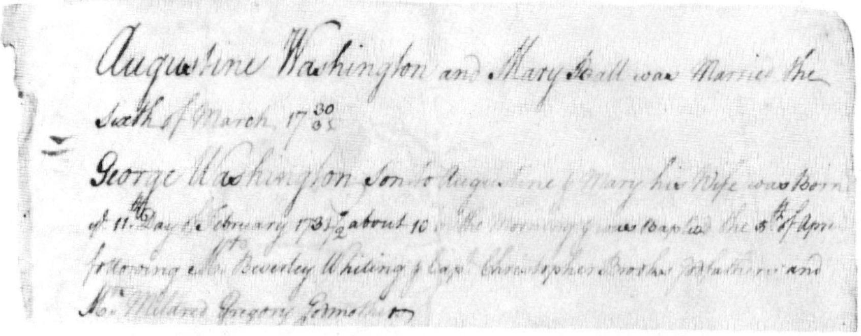

Augustine Washington and Mary Ball was Married the
Sixth of March 1730/31
George Washington Son to Augustine & Mary his Wife was Born
ye 11th Day of February 1731/2 about 10 in the Morning & was Baptiz'd the 5th of April
following Mr. Beverley Whiting & Capt. Christopher Brooks Godfathers and
Mrs. Mildred Gregory Godmother

Figure 4–20 In George Washington's family bible his date of birth is given as 1731/2. Some contemporaries would have said 1731; we now say 1732.

In the Gregorian calendar, years that are evenly divisible by four are leap years, except that three out of every four century years, the ones not divisible evenly by 400, have only 365 days along with all the other years. Thus 1600 was a leap year, 1700, 1800, and 1900 were not, and 2000 will again be a leap year.

Although many countries adopted the Gregorian calendar as soon as it was promulgated, Great Britain (and its American colonies) did not adopt it until 1752, when 11 days were skipped. As a result, we celebrate George Washington's birthday on February 22nd, even though he was born on February 11. Actually, since the year had begun in March instead of January, 1752 was cut short. Washington was born in February 1731, often then written February 1731/32, but we now refer to his date of birth as February 22nd, 1732 (Fig. 4–20). The Gregorian calendar is the one in current use. It will be over 3000 years before this calendar is as much as one day out of step with the seasons.

*4.8 ASTRONOMY AND ASTROLOGY

Astrology is not at all connected with astronomy, except in a historical context, so does not really deserve a place in a text on contemporary astronomy. But since so many people associate astrology with astronomy, and since astrologers claim to be using astronomical objects to make their predictions, let us use our astronomical knowledge to assess astrology's validity.

Most professional astronomers would privately agree to sentiments something like those expressed in this section, or at least reach the same conclusion. Many astronomers don't even think that astrology is worth discussing, or at best, that discussions have no effect on those who "believe" and so are a waste of time. We will not take this latter position here.

Astrology is an attempt to predict or explain our actions and personalities on the basis of the positions of the stars and planets now and at the instants of our births. Astrology has been around for a long time, but it has never been shown to work. Believers may cite incidents that reinforce their faith in astrology, but no successful scientific tests have ever been carried out. If something happens to you that you had expected because of an astrological prediction, you would more certainly notice that this event occurred than you would notice the thousands of other unpredicted things that happened to you that day. And likewise with hindsight. We do enough things, have sufficiently varied thoughts, and interact with enough people that if we make many predictions in the morning, some of them are likely to be at least partially fulfilled during the day. We simply forget that the rest ever existed. Besides, we always have the easy out of being able to say, "well, we just didn't have a good astrologer."

The constellations or the signs of the zodiac (Fig. 4–21) don't actually exist as physical objects. They are merely projections of the positions of stars that may be located at very different distances from us.

Figure 4–21 Twelve constellations through which the sun, moon, and planets pass make up the zodiac. Drawing by Handelsman; © 1978 The New Yorker Magazine, Inc.

Studies have shown that superstition actively constricts the progress of science and technology in various countries around the world and is therefore not merely an innocent force. It is not just that some people harmlessly believe in astrology. Their lack of understanding of scientific structure may actually impede the proper scientific training of people needed to work on the problems of our age, including the problems of pollution and the energy crisis. A recent paper even shows how widespread superstitious beliefs can impede smallpox-prevention programs. Thus many scientists are not content to ignore astrology, but actively oppose its dissemination. Further, if large numbers of citizens do not understand the scientific method and the difference between science and pseudoscience, how can they intelligently vote on or respond to scientific questions that have societal implications?

A major reason why scientists in general and astronomers in particular don't believe in astrology is that they cannot conceive of a way in which it would work. The human brain is so complex that it seems most improbable that any celestial alignment can affect people, including newborns, in an overall way. The celestial forces that are known cannot be sufficient to set personalities nor influence day to day events. And even if people reply that they do not think astrology is true but merely find it interesting, many scientists feel that so many strange and exciting explainable things are going on in the universe that we wonder why anybody should waste time with far-fetched astrological concerns.

After all, we will be discussing such fascinating things as neutrinos from the sun, pulsars, and black holes. Later in this book we will consider complex molecules that have spontaneously formed in interstellar

Gloria: Oh Ma, you don't believe in astrology, do you?
Edith: No, but it's fun to know what's going to happen to you, even if it don't.

from All in the Family

Figure 4–22 The ecliptic is divided into the signs of the zodiac in this illustration from the *Epitoma In Almagestum Ptolemaei (Venice, 1496)* by Regiomontanus. This was the first translation into Latin of Ptolemy's *Almagest,* which contained the standard description of Ptolemy's earth-centered universe. The *Almagest* had been formerly available only in Arabic. "Almagest" means "the greatest."

space, and try to decide whether the universe will expand forever. We have sent a rocket into interstellar space bearing a portrait of humans, and have beamed a radio message toward a group of stars 24,000 light years away. These topics and actions are part of modern astronomy, what contemporary, often conservative, scientists are doing and thinking about the universe. How prosaic and fruitless it thus seems to spend time pondering celestial alignments and wondering whether they can affect individuals. In fact, even the alignments that most astrologers use are not accurately calculated, for the precession of the earth's pole has changed the stars that are overhead at a given time of year from what they were millennia ago when astrological tables that are often still in current use were computed, as we described in Section 4.6a.

Moreover, astrology doesn't work. Bernie I. Silverman, a psychologist then at Michigan State University, tested specific values: Do Libras and Aquarians rank "Equality" highly? Do Sagittarians especially value "Honesty"? Do Virgos, Geminis, and Capricorns (Fig. 4–22) treasure the value "Intellectual"? Several astrology books agreed that these and other similar examples are values typical of those signs. Although believers often criticize objections on the ground that these group horoscopes are not as valuable or accurate as individualized charts, surely some general assumptions and rules hold in common.

The subjects, 1600 psychology graduate students, did not know in advance just what was being tested. They gave their birthdates, and the questioners determined their astrological signs. The results: no special correlation was apparent for any of the signs with the values they were supposed to hold. This was confirmed by statistical tests. When asked to what extent they shared the qualities of each given sign, as many subjects ranked themselves above average as below, regardless of their astrological signs.

Furthermore, Silverman tested whether individuals from particular pairs of signs were especially compatible or incompatible by sampling marriage and divorce rates. He compared these rates with the predictions of two astrologers as to compatible or incompatible combinations. No correlation existed. Those born under "compatible" signs neither married more frequently nor divorced less frequently than would be expected by chance, nor did those born under "incompatible" signs marry less frequently or divorce more frequently.

Several such investigations have been carried out. Astronomers George Abell and colleagues have concluded, for example, that there is no correlation between the birth dates of sports champions and suitable astrological signs, a discussion that has led to extended statistical wrangling.

But if astrology is so meaningless, why does it still have so many adherents? Well, it could be the bandwagon effect, and Silverman had an ingenious test that endorsed this idea. He took twelve personality descriptions from astrology books, one for each astrological sign, and displayed them to two groups of individuals. The first group, composed of 51 subjects, was told to which astrological signs the descriptions pertained, and was asked to write their own signs on the covers of the questionnaires. More than half the members of this group thought that the descriptions listed under their own signs were, for each individual, among the four best descriptions of themselves out of the twelve choices. It would seem, if one considered only this phase of the experiment, that astrology was working.

Yet when the second group was given the twelve descriptions without mention of astrology, being told that they came from a book entitled "Twelve Ways of Life," their choices were random. Only 30 per cent chose their own sign's description as being in the group most closely describing them. So the idea that astrology can predict personality types seems to be the result of self-delusion. When people know what they are expected to be like, they tend to identify themselves with the description. But that doesn't mean that they actually satisfy the description that astrology predicts for them better than any other description.

From an astronomer's view, astrology is meaningless, unnecessary, and impossible to explain if we accept the broad set of physical laws we have conceived over the years to explain what happens on the earth and in the sky. Astrology snipes at the roots of all pure science. Moreover, astrology patently doesn't work. If people want to believe in it on an *a priori* basis, as a religion, or have a personal astrologer act as a psychologist, let them not try to cloak their beliefs in scientific astronomical gloss. The only reason people may believe that they have seen astrology work is that it is a self-fulfillling means of prophecy, conceived of long ago in times when we knew less about the exciting things that are going on in the universe. Let's all learn from the stars, but let's learn the truth.

A historian of science has suggested that the bandwagon effect by itself is insufficient to explain belief in astrology, because astrological beliefs have persisted for so long. Astrology, following this line of reasoning, joins other pseudosciences in satisfying intuitional needs of individuals who do not clearly understand what science is, on what it is based, and what it has to offer us.

Figure 4–23 The signs of the zodiac are shown on this map dating from 1700, when it was drawn by Frederik de Wit. (Courtesy of the Royal Library of Copenhagen)

SUMMARY AND OUTLINE

The constellations (Section 4.1)
  Stars in them are not necessarily physically grouped.
Twinkling (Section 4.2)
  Stars appear to twinkle; planets usually do not.
  Twinkling is caused in the earth's atmosphere.
Magnitudes (Section 4.3)
  Lower (or negative) numbers are the brightest objects.
  5 magnitudes difference = 100 times in brightness.
  Brightest star is magnitude −1.4; faintest stars visible to the naked eye are about sixth magnitude.
Coordinate systems (Section 4.4)
  Celestial longitude and latitude are right ascension and declination.
  The celestial equator and celestial poles are the points in the sky that are on the extensions of the earth's equator and poles into space.
  Sidereal time is time by the stars.
  Sidereal day: a given right ascension returns to the same position.
  Solar day: the sun returns to the same position in the sky.
  Precession is a slow drift in the coordinate system; 26,000-year period.
  The ecliptic and the celestial equator cross at the equinoxes.
Motions in the sky (Section 4.5)
  Stars rise and set; at the earth's equator we would see them do so perpendicularly to the horizon.
  At the earth's poles, we would see the stars move around parallel to the horizon.

At or near the poles, the sun and moon rise and set only when they change sufficiently in declination; the sun goes through this sequence once a year—thus "the midnight sun."
The seasons are caused by the sun's variations in declination.
Telescopes can be mounted so that motion on one axis allows the earth's rotation to be counteracted; such mounts are called "equatorial."
Alt-azimuth mounts have separate motions in altitude and azimuth; it takes both motions to follow the stars.
Time (Section 4.6)
  Standard time is based on a mean solar day.
  The date changes at the international date line.
  Daylight-saving time: "fall back, spring ahead."
Calendars (Section 4.7)
  Leap years are needed to keep up with the 365¼ day year.
  The Julian calendar, introduced by Julius Caesar, was a reasonably good calendar, but, over the centuries, the days drifted.
  We now use the Gregorian calendar, a corrected calendar set up in 1582.
Astronomy and astrology (Section 4.8)
  No scientific basis known for astrology.
  Belief in such pseudoscience impedes the advance of science and technology.
  Statistical arguments show that astrology doesn't work.

QUESTIONS

1. Explain why we cannot tell by merely looking in the sky that stars in a given constellation are at different distances, while in a room we can easily tell that objects are at different distances from us. What is the difference between the two situations?

2. If you look toward the horizon, are the stars you see likely to be twinkling more or less than the stars overhead? Explain.

3. Is the planet Uranus, which is in the outskirts of the solar system, likely to twinkle more or less than the planet Venus?

4. Explain how it is that some stars never rise in our sky, while others never set.

†5. Venus can reach magnitude −4.4, while the brightest star, Sirius, is magnitude −1.4. How many times brighter is Venus at its maximum than is Sirius?

†6. Pluto is about 14th magnitude at most. How many times fainter is it than Venus, which can surpass magnitude −4?

7. Why is it that planets change in apparent brightness?

†8. If a variable star brightens by a factor of 15, by how many magnitudes does it change?

†9. If a variable star starts at 5th magnitude and brightens by a factor of 60, at what magnitude does it appear?

†10. Venus can be brighter than magnitude −4. Betelgeuse is a first magnitude star (m = +1). How many times brighter is Venus at magnitude −4 than Betelgeuse?

†11. Star A has magnitude +11. Star B appears 10,000 times brighter. What is the magnitude of star B? Star C appears 10,000 times fainter than star A. What is its magnitude?

†12. Star A has magnitude +10. The magnitude of star B is +5 and of star C is +3. How much brighter does star B appear than star A? How much brighter does C appear than A? How much brighter does C appear than B?

13. Why do the stars twinkle?

†14. How much brighter is a 0th magnitude star than a +3rd magnitude star?

15. Between the vernal equinox, March 21st, and the autumnal equinox, 6 months later,

   (a) by how much does the right ascension of the sun change;

   (b) by how much does the declination of the sun change;

   (c) by how much does the right ascension of Sirius change;

   (d) by how much does the declination of Sirius change?

16. (a) How does the declination of the sun vary over the year?

   (b) Does its right ascension increase or decrease from day to day? Justify your answer.

17. We normally express longitude on the earth in degrees. Why would it make sense to express longitude in units of time?

18. If a planet always keeps the same side toward the sun, how many sidereal days are there in a year on that planet?

†19. When it is 6 P.M. on October 1st in New York City, what time of day and date is it in Tokyo?

20. What is the advantage of an equatorial mount? Why did large telescopes used to be made with equatorial mounts while they are now sometimes made with alt-azimuth mounts?

---

†This indicates a question requiring a numerical solution.

CHAPTER 5

STELLAR DISTANCES AND MOTIONS

AIMS:
To describe the method of trigonometric parallax and how it tells us the distances to the stars.
To describe the Hertzsprung-Russell diagram, and what it tells us about types of stars and their distances.
To describe the Doppler effect, and how it tells us about stellar motions.

We have seen in Chapter 3 how we can analyze the spectrum of an individual star to tell us the state of the outer layer of that star; for example, we can tell how hot it is. In this chapter, we see how information about the spectra of many stars can be put together to tell us about the overall properties of stars.

We also see how we can fairly directly measure the distances to the nearest stars, and that these distances, together with classification by spectral type, can give us distances to farther stars. We see how graphing the temperatures and brightnesses of stars gives us an important tool: the Hertzsprung-Russell diagram. Then we study the Doppler effect, which enables us to tell how stars are moving.

Later in this book we shall use this same method to discover that indeed the whole universe is expanding, with galaxies moving away from us in all directions.

5.1 STELLAR DISTANCES

We can tell a lot by looking at a star or by examining its radiation through a spectrograph. But such observations do not tell us directly how far away the

---

Stars in the Milky Way in the constellation Sagittarius, looking toward the center of our galaxy. Two globular clusters are visible.

Figure 5–2 The nearer stars seem to be slightly displaced with respect to the farther stars when viewed from different locations in the earth's orbit.

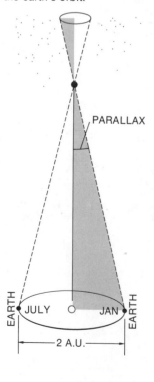

star is. Since all the stars are but points of light in the sky even when observed through the biggest telescopes, we have no reference scale to give us their distances (Fig. 5–1).

The best way to find the distance to a star is to use the principle that is also used in rangefinders in cameras. Sight toward the star from different locations, and see how the direction toward the star changes. You can see this effect by holding out your thumb at arm's length. Examine it first with one eye closed and then with the other eye closed. Your thumb seems to change in position as projected against a distant background; it appears to move across the background by a certain angle, about 2°. This is because your eyes are a few centimeters apart from each other so each eye has a different point of view.

Now hold your thumb up closer to your face, just a few centimeters away. Note that the angle your thumb seems to jump across the background as you look through first one eye and then through the other is greater than it was before. Exactly the same effect can be used for finding the distances to nearby stars. The nearer the star is to us, the farther it will appear to move across the background of distant stars, which are so far away that they do not appear to move with respect to each other.

To maximize this effect, we want to observe from two places that are separated from each other as much as possible. For us on earth, that turns out to be the position of the earth at intervals of six months. In that six-month period, the earth moves to the opposite side of the sun. In principle, we observe the position of a star in the sky (photographically), with respect to the background stars, and several months later, when the earth has moved part way around the sun, we repeat the observation. (Six months later, the star would be up in the daytime and could not be seen.) The straight line joining the points in space from which we observe is called the *baseline*. The distance from the earth to the sun is called an *Astronomical Unit* (A.U.). In the case above, we have observed with a baseline 2 A.U. in length. The angle across the sky that a star seems to move (with respect to the background of other stars) between two observations made from the ends of a baseline of 1 A.U. (half the maximum possible baseline) is called the *parallax* of the star (Fig. 5–2). From the parallax, we can use simple trigonometry to calculate the distance to the star. This is called the method of *trigonometric parallax.*

The basic limitation to the method of trigonometric parallax is that the farther away the star is, the smaller is its parallax. It turns out that this method can only be used for about ten thousand of the stars closest to us; most other stars have parallaxes too small for us to measure. (These other stars, with negligible parallaxes, are the distant stars that provide the unmoving background against which we compare.) Even Proxima Centauri, the star nearest to us, has a parallax of only ¾ arc sec. This is the angle subtended (Fig. 5–3) by a dime at a distance of 2 km. Clearly we must find other methods to measure the distances to most of the stars.

Parallaxes are measured by comparing pairs of photographs taken several months apart. Measurements are repeated over several years to determine the effect of actual motions of stars and to reduce observational errors. Trigonometric parallaxes can now be measured for stars as far as 300 light years from the sun, which is less than 1 per cent of the diameter of our galaxy, and as faint as 20th magnitude.

The Space Telescope will be able to use its very high resolution to

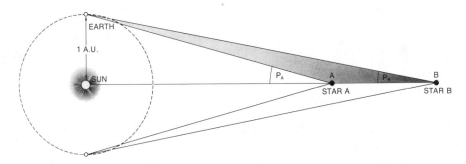

Figure 5–3 Star B is farther from the sun than star A and thus has a smaller parallax. The parallax angles are marked P_A and P_B. To *subtend* is to take up an angle; as seen from point A, 1 A.U. subtends an angle P_A. As seen from point B, the same 1 A.U. subtends an angle P_B.

pinpoint the positions of stars extremely accurately when it is launched in 1983. It will thus lead to great improvement in *astrometry*, the study of the positions of stars and their apparent motion across the sky. Since the resolution of the Space Telescope will be improved by a factor of 7 over current capabilities, we will be able to measure parallaxes about 7 times farther out into space than we can now. Also, the parallaxes remeasured for closer objects will be much more accurate. Among the many projects that will become possible is a thorough search for dark companions (that is, companions that are not shining or are shining very faintly)—perhaps even planets—of nearby stars. New astrometric measurements are also necessary for providing accurate positions for faint optical objects so they can be identified with emitters in other parts of the spectrum, such as those identified in radio astronomy or with x-ray satellites, especially the Einstein Observatory.

5.1a   Parsecs

Since parallax measures have such an important place in the history of astronomy, a unit of distance was defined in terms that relate to these measurements. If we were outside the solar system, and looked back at the earth and the sun, they would appear to us to be separated in the sky, and we could measure the amount by which they are separated as an angle. From twice as far away as Pluto, for example, when the earth and the sun were separated by the maximum amount, they would be approximately 1° apart. As we go farther and farther away from the solar system, 1 A.U. subtends a smaller and smaller angle. When we go about 60 times farther away, 1 A.U. subtends only 1 arc min (60 arc min = 1°). From 60 times still farther, 1 A.U. subtends 1 arc sec (60 arc sec = 1 arc min). Note that the **angle** subtended by the astronomical unit is a measure of the **distance** we have gone (Fig. 5–3).

When we are at the distance at which 1 A.U. subtends only 1 arc sec, we call that distance *1 parsec* (Fig. 5–4). A star that is 1 parsec from the sun has a parallax of one arc **sec.** One parsec is a long distance. Most astronomers tend to use parsecs instead of light years, the distance that light travels in a year (= 9.5×10^{12} km), when talking about stellar distances. It takes light 3.26 years to travel 1 parsec; thus 1 parsec = 3.26 light years. Note that parsecs and light years are **distances,** just like kilometers or miles, even though their names contain references to their definitions in terms of angles or time.

Devices to measure the positions of stars on photographic plates are called measuring engines; *the recent development of automatic measuring engines has enabled trigonometric parallaxes to be measured for stars five times more distant than could previously be determined. New ground-based techniques using electronic detectors will soon improve the situation further.*

A European space satellite that would be devoted entirely to astrometry has been proposed for launch in the mid 1980's. It would be named Hipparchos after the Greek astronomer (whose name we have already used in this book in the Americanized form Hipparchus) who in the second century B.C. *first divided the stars into classes of brightness and who made a catalogue of star positions.*

*Hipparchos stands for **Hi**gh* ***Pr**ecision **Par**allax **Co**llecting **S**atellite.*

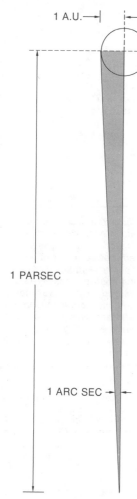

1 A.U.→|

1 PARSEC

1 ARC SEC →|←

Figure 5–4 A *parsec* is the distance at which 1 A.U. subtends an angle of 1 arc sec. The drawing is not to scale. 1 arc sec is actually a tiny angle.

*5.1b   Computing with Parsecs

The advantage of using "parsecs" instead of "light years," "kilometers," or any other distance unit is that the distance in parsecs is equal to the inverse of the parallax angle in seconds of arc:

$$d(parsecs) = 1/p \text{ (seconds of arc).}$$

For example, a star with a parallax angle of 0.5 arc sec is 2 parsecs away, and would thus be one of the nearest stars to the sun. Since the parallax angle is the quantity that is measured directly at the telescope, it was convenient to choose a distance unit very closely related to the parallax angle.

5.2   ABSOLUTE MAGNITUDES

We have thus far defined only the *apparent magnitudes* of stars, how bright the stars **appear** to us. However, stars can appear relatively bright or faint for either of two reasons: they could be intrinsically bright or faint, or they could be relatively close or far away. We can remove the distance effect, giving us a measure of how actually bright a star is, by choosing a standard distance and considering how bright all stars would appear if they were at that standard distance. The standard distance that we choose is 10 parsecs. We define the *absolute magnitude* of a star to be the magnitude that the star would appear to have if it were at a distance of 10 parsecs. We normally write absolute magnitude with a capital M, and apparent magnitude with a small m.

Thus if a star happens to be exactly 10 parsecs away from us, its absolute magnitude is exactly the same as its apparent magnitude. If we were to take a star that is farther than 10 parsecs away from us and somehow move it to be 10 parsecs away, then it would appear brighter to us than it does at its actual position. Since the star would be brighter, its absolute magnitude would be a lower number (for example, 2 rather than 6) than its apparent magnitude.

On the other hand, if we were to take a star that is closer to us than 10 parsecs, and move it to 10 parsecs away, it would then be farther away and therefore fainter. Its absolute magnitude would be higher (more positive, for example, 10 instead of 6) than its apparent magnitude.

To assess just how much brighter or fainter that star would appear, we must realize that the intensity of light from a star follows the *inverse square law* (Fig. 5–5). That is, the intensity of a star varies inversely with the square of the distance of the star from us. (This law holds for all point sources of radiation, that is, sources that appear as points without length or breadth.) If we could move a star twice as far away from us, it would appear four times fainter. If we could move a star 9 times as far away, it would grow 81 times fainter. Of course, we are not physically moving stars (we would burn our shoulders while pushing), but merely considering how they would appear at different distances. In all this, we are assuming that there is no matter in space between the stars to absorb light. Unfortunately, this is not always a good assumption (as we shall see in Chapter 23).

For many purposes, it is sufficient just to understand how the inverse square law affects brightness. Astronomers, however, normally speak in terms of magnitudes. Remember that a difference of one magnitude is a factor of approximately 2.5 in brightness. Thus, if a star is 2.5 times fainter

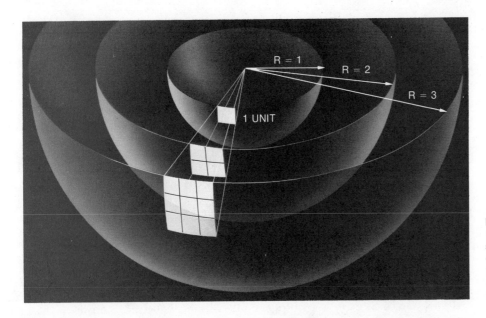

Figure 5-5 The inverse square law. Radiation passing through a sphere twice as far away as another sphere has spread out so that it covers $2^2 = 4$ times the area; n times farther away it covers n^2 times the area.

(that is, gives off 2.5 times less light) than another star, it is 1 magnitude fainter. If it is 3 times fainter, it is slightly more than 1 magnitude fainter. If it is 6 times fainter, it is slightly less than 2 magnitudes fainter.

Example: A star is 20 parsecs away from us, and its apparent magnitude is +4. What is its absolute magnitude?

Answer: If the star were moved to the standard distance of 10 parsecs away, it would be twice as close as it was at 20 parsecs and, therefore, by the inverse-square law, would appear four times brighter. Since 2.5 times is one magnitude, and $(2.5)^2 = 6.25$ is two magnitudes, it would be approximately 1½ magnitudes brighter. Since its apparent magnitude is 4, its absolute magnitude would be $4 - 1½ = 2½$.

Astronomers have a formula to carry out this calculation, but the formula merely does numerically what we have just carried out logically. The formula is $m - M = 5 \log_{10} r/10$, where m is the apparent magnitude, M is the absolute magnitude, and r is the distance in parsecs. Note that if $r = 10$, then $r/10 = 1$, $\log 1 = 0$, and $m = M$. If $r = 20$, as in our example above, $m - M = 5 \log 2 = 5 \times 0.3 = 1.5$ magnitudes. Be careful not to get carried away using the formula without understanding the point of the manipulations.

*5.2a Photometry

Note that the actual value of the magnitude, either absolute or apparent, can depend on the wavelength region in which we are observing. Let us consider a blue star and a red star, each of which gives off the same amount of energy. The blue star can seem much brighter than the red star if we observe them through a blue filter, and the red star can seem much brighter than the blue if we observe them through a red filter. Thus we must specify the wavelength range in which we are observing. For example, one often sees magnitudes listed as M_{pg} (for photographic) and M_v (for visual). The sen-

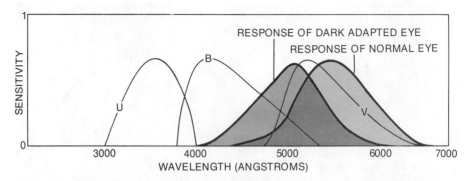

Figure 5–6 The U, B, and V curves represent the standard set of filters used by many astronomers. The response of the eye under normal conditions and the response of the dark adapted eye are also shown.

sitivity of photographic plates is greatest farther toward the violet than is the sensitivity of the human eye, which has its peak sensitivity in the green, so use of m_{pg} is equivalent to observing through a filter passing shorter wavelengths than the visual range.

A standard set of filter wavelengths has been defined with one filter in the ultraviolet passing a broad band of wavelengths centered at 3700 Å (called U), one filter in the blue passing a broad band centered at 4400 Å (called B), and one filter in the yellow passing a broad band centered at 5500 Å (called V for visual). Sets of equivalent filters exist at observatories all over the world. Hundreds of thousands of stars have had their colors measured with this *UBV set of filters;* we call the process *three-color photometry.* Graphs of the sensitivities of the UBV filters appear in Figure 5–6. U, B, and V are the **apparent** ultraviolet, blue, and visual magnitudes, respectively (and are unfortunately and inconsistently written with capital letters); the corresponding **absolute** magnitudes are written $M_U, M_B,$ and M_V. A four-color system, uvby, has ultraviolet, violet, blue, and yellow filters. This is called *four-color photometry.*

Other filters exist for other spectral ranges; a star's magnitude measured far in the infrared, for example, may be very different from magnitudes measured in the blue. For example, R (red) and I (infrared) filters are now sometimes used together with UBV filters.

*5.2b   Bolometric Magnitude

"Bolometer," the instrument used to measure the total amount of energy arriving in all spectral regions, derives its name from "boli," Greek for "beam of light."

The physical quantity of fundamental importance is not the magnitude as observed through any given filter but rather the magnitude corresponding to the total amount of energy given off by the star over all spectral ranges. This is called the **bolometric magnitude,** M_{bol}. We cannot, however, measure bolometric magnitudes directly; we would have to measure the energy given off by the star over the entire spectrum and add these contributions. A major uncertainty in the past has been how much energy was contributed by the ultraviolet part of the spectrum that was blocked out by the earth's atmosphere. Only recently, with spacecraft (especially the International Ultraviolet Explorer, IUE), can we measure the ultraviolet contribution directly. (Note that this is radiation farther in the ultraviolet than that measured by the U magnitude.)

5.3   THE HERTZSPRUNG-RUSSELL DIAGRAM

In about 1910, Ejnar Hertzsprung (Fig. 5–7) in Denmark and Henry Norris Russell (Fig. 5–8) at Princeton University in the United States independently plotted a new kind of graph (Fig. 5–9). On one axis, the abscissa (x-axis), each graphed a quantity that measured the temperature of stars. On the other axis, the ordinate (y-axis), each graphed a quantity that measured the brightness of stars. They found that all the points that they plotted fell in limited regions of the graph rather than being widely distributed over the graph.

There are several possible ways to plot a measure of the temperature without plotting the temperature itself. One can plot the spectral type, for example, or even the color of a star (usually in the form of color index, as described in Appendix 13).

There are also several possible ways to plot the brightness. The simplest is merely to plot the apparent magnitudes of stars, with the brightest stars (i.e., those with the most negative magnitudes) at the top. But remember that a star can appear bright either by really being intrinsically bright or alternatively by being very close to us. One way to get around this problem is to plot only stars that are at the same distance away from us.

How do we find a group of stars at the same distance? Luckily, there are clusters of stars in the sky (described in more detail in Chapter 6) that are

Figure 5–7 Ejnar Hertzsprung in the 1930's.

Figure 5–8 Henry Norris Russell and his family.

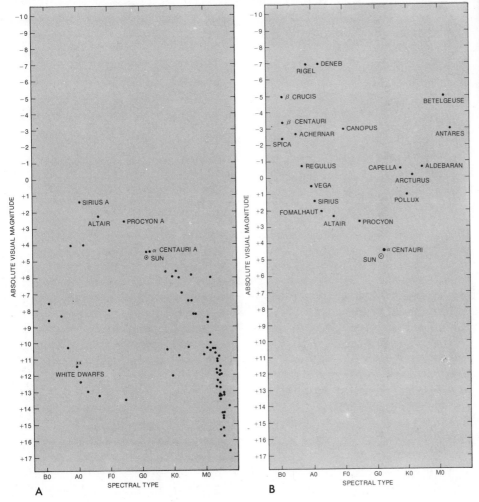

A

B

Figure 5–9 Hertzsprung-Russell diagrams *(A)* for the nearest stars in the sky and *(B)* for the brightest stars in the sky. The brightness scale is given in *absolute magnitude*. Because the effect of distance has been removed, the intrinsic properties of the stars can be compared directly on such a diagram.

Note that none of the nearest stars is intrinsically very bright. Also, the brightest stars in the sky are, for the most part, intrinsically very luminous, even though they are not usually the very closest to us.

Figure 5-10 A physical grouping of stars in space. The globular cluster ω (Omega) Centauri, photographed with the Anglo-Australian telescope.

really clusters of stars in space at approximately the same distance away from us. We do not have to know what the distance is to plot a diagram of the type plotted by Hertzsprung; all we have to know is that the stars are really clustering in space (Fig. 5–10).

If we somehow knew the absolute magnitudes of the stars, that would also be a good thing to plot. When we know the distances (for relatively close stars, we can get this from trigonometric parallax measurements, which is what Russell originally did), we can calculate the absolute magnitudes, which are actually what is plotted in Figure 5–9.

A plot of temperature versus brightness is known as a *Hertzsprung-Russell diagram*, or simply as an *H-R diagram*. Note that since H-R diagrams were sometimes originally plotted by spectral type, from O to M, the hottest stars are on the left side of the graph. Thus temperature increases from right to left. Also, since the brightest stars are on the top, magnitude decreases toward the top. In some sense, thus, both axes are plotted backwards from the way a reasonable person might choose to do it if there were no historical reasons for doing it otherwise.

When plotted on a Hertzsprung-Russell diagram, the stars lie mainly on a diagonal band from upper left to lower right (Fig. 5–11). Thus the hottest stars are normally brighter than the cooler stars. Most stars fall very close to this band, which is called the *main sequence*. Stars on the main sequence are called *dwarf stars*, or *dwarfs*. There is nothing strange about dwarfs; they

Figure 5-11 The Hertzsprung-Russell diagram, with the stars of Figure 5–9 included. The spectral type axis *(bottom)* is equivalent to the temperature axis *(top)*. The absolute magnitude axis *(left)* is equivalent to the luminosity axis *(right)*.

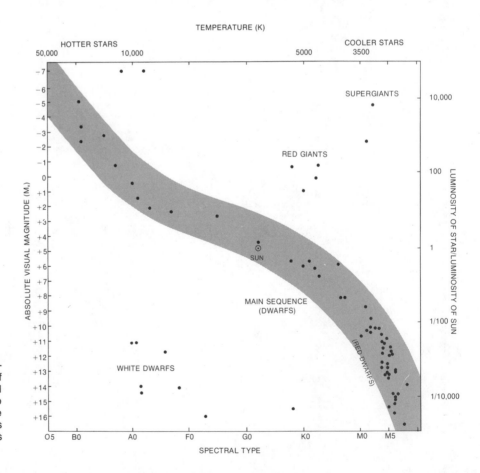

are the normal kind of stars. The sun is a type G dwarf. Some dwarfs are quite large and bright; the word "dwarf" is used only in the sense that these stars are not a larger, brighter kind of star that we will define below as giant.

Some stars lie above and to the right of the main sequence. That is, for a given spectral type the star is intrinsically brighter than a main-sequence star. These stars are called *giants*, because their luminosities are large compared to dwarfs of their spectral type. Some stars, like Betelgeuse, are even brighter than normal giants, and are called *supergiants*. Since two stars of the same spectral type have the same temperature and, according to the Stefan-Boltzmann law, the same amount of emission from each area of their surfaces, the brighter star must be bigger than the fainter star of the same spectral type.

The Stefan-Boltzmann law, Section 3.1, linked the energy of a volume of gas to the fourth power of its temperature.

Stars in a class of faint hot objects, called *white dwarfs*, are located below and to the left of the main sequence. They are smaller and fainter than main-sequence stars of the same spectral type.

The use of the Hertzsprung-Russell diagram to link the spectrum of a star with its brightness is a very important tool for stellar astronomers. The H-R diagram provides, for example, another way of measuring the distances to stars, as we shall see below.

5.4 SPECTROSCOPIC PARALLAX

When we take a group of stars whose distances we can measure directly, by some method like that of trigonometric parallax, we can plot a Hertzsprung-Russell diagram for all these stars. We have thus established a "standard" Hertzsprung-Russell diagram, and from this standard diagram can read off the absolute magnitude that corresponds to a given spectral type for any star on the main sequence.

We can apply the standard Hertzsprung-Russell diagram to find the distance to any star whose spectrum we can observe, no matter how faint, if we know that the star is on the main sequence. We first find the absolute magnitude that corresponds to its spectral type. We must also know the apparent magnitude of the star, but we can get that easily and directly by simply observing it. Once we know both the apparent magnitude and the absolute magnitude, we have merely to figure out how far the star has to be from the standard distance of 10 parsecs to account for the difference $m - M$. (We get this from the inverse square law.)

One can tell whether a star is a dwarf, giant, or supergiant by looking closely at its spectrum, because slight differences exist for a given spectral type. This procedure is called luminosity classification. *Once we know a star's* luminosity class, *we can place the star on the H-R diagram and find its spectroscopic parallax.*

Example: We see a G2 star on the main sequence. Its apparent magnitude is +8. How far away is it?

Answer: From the H-R diagram, we see that $M = +5$. The star appears fainter than it would be if it were at 10 parsecs. It is approximately $(2.5)^3 = 6 \times 2.5 = 15$ times fainter. (More accuracy in carrying out this calculation would not be helpful because the original data were not more accurate.) By the inverse square law, this means that it is approximately 4 times farther away (exactly 4 times would be a factor of $4^2 = 16$ in brightness). Thus the star is 4 times 10 parsecs, or 40 parsecs, away.

We are measuring a distance, and not actually a parallax, but by analogy with the method of trigonometric parallax for finding distance, the method using the H-R diagram is called finding the *spectroscopic parallax.*

5.5 THE DOPPLER EFFECT

The Doppler effect is one of the most important tools that astronomers can use to understand the universe. This is because without having to measure the distance to an object they can use the Doppler effect to determine its *radial velocity*, its speed measured with respect to us in the direction toward or away from us (the radius). The Doppler effect in sound is familiar to most of us, and its analogue in electromagnetic radiation, including light, is very similar.

The Doppler shift allows us to measure only velocities toward or away from us and not side to side. Since these velocities are along a radius extending outward from earth, we call them radial veloci- *ties. If the object is moving at some angle to the radius, then the Doppler shift measures only the part of the velocity (technically, the* component *of the velocity) in the radial direction.*

5.5a Explanation of the Doppler Effect

You may be familiar with how the sound of a train whistle, a jet engine, or a motorcycle motor changes in pitch as the train, plane, or motorcycle first approaches you and then passes you and begins to recede. The object that is emitting the sound waves is approaching you at first. By the time it emits a second wave, it has moved closer to you than it had been when it emitted the first wave. Thus the waves pass more frequently than they would if the source were not moving. The wavelengths seem compressed, and the pitch is higher. When the emitting source passes you, the wavelengths become stretched, and the pitch of the sound is lower (Fig. 5–12).

With light waves, the physical mechanism is not as straightforward as is the stretching or compressing of the wavelengths of sound, but the effect is similar. As a body emitting light, or other electromagnetic radiation, approaches you, the wavelengths become slightly shorter than they would be if the body were at rest. Visible radiation is thus shifted slightly in the direction of the blue (it doesn't actually have to become blue, only be shifted in that direction). We say that the radiation is *blueshifted*. Conversely, when the emitting object is receding, the radiation is said to be *redshifted* (Fig. 5–13). We generalize these terms to types of radiation other than light, and say that radiation is blueshifted whenever it changes to shorter wavelengths, and redshifted whenever it changes to longer wavelengths.

The point of all this is that we can measure a Doppler shift for any object we can see that has something in its spectrum for which we can measure an accurate wavelength, such as a spectral line. We can then tell how fast the object is moving along a radius toward or away from us. If we were moving at the same speed and in the same direction as a source of radiation or sound, it would have no net velocity with respect to us. Thus no Doppler shift would be observed.

*5.5b Working with Doppler Shifts

The wavelength when the emitter is at rest is called the *rest wavelength*. Let us consider a moving emitter. The fraction of the rest wavelength that the wavelength of light is shifted is the same as the fraction of the speed of light at which the body is traveling (or, for a sound wave, the fraction of the speed of sound).

We can write

$$\frac{\text{change in wavelength}}{\text{original wavelength}} = \frac{\text{velocity of emitter}}{\text{speed of light}}$$

or,

$$\frac{\Delta\lambda}{\lambda_0} = \frac{v}{c},$$

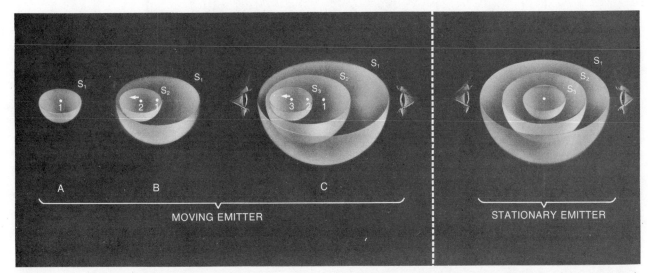

Figure 5–12 An object emits waves of radiation that can be represented by spheres representing the peaks of the wave, each centered on the object and expanding. In the left side of the drawing, the emitter is moving in the direction of the arrow. In part *A*, we see that the peak emitted when the emitter was at point 1 becomes a sphere (labeled S_1) around point 1, though the emitter has since moved toward the left. In part *B*, some time later, sphere S_1 has continued to grow around point 1 (even though the emitter is no longer there), and we also see sphere S_2, which shows the position of the peaks emitted when the emitter had moved to point 2. Sphere S_2 is thus centered on point 2, even though the emitter has continued to move on. In *C*, still later, yet a third peak of the wave has been emitted, S_3, this time centered on point 3, while spheres S_1 and S_2 have continued to expand.

Contrast the case at the extreme right, in which the emitter does not move and so all the peaks are centered around the same point. No redshifts or blueshifts arise.

For the case of the moving emitter, observers who are being approached by the emitting source (those on the left side of the emitter, as shown on the left side of Case *C*) see the three peaks drawn coming past them relatively bunched together, (that is, at shorter intervals of time, which is the same as saying at a higher frequency) as though the wavelength were shorter, since we measure the wavelength from one peak to the next. This corresponds to a wavelength farther to the blue than the original, a *blueshift*. Observers from whom the emitter is receding (those on the right side of the emitter, as shown on the right side of Case *C*), see the three peaks coming past them with decreased frequency (at increased intervals of time), as though the wavelength were longer. This corresponds to a color farther to the red, a *redshift*.

Note that once the light wave is emitted, it travels at a constant speed ("the speed of light," 3×10^{10} cm s^{-1}, equivalent to seven times around the earth in a second), so that a shorter (lower) wavelength corresponds to a higher frequency, and a longer (higher) wavelength corresponds to a lower frequency.

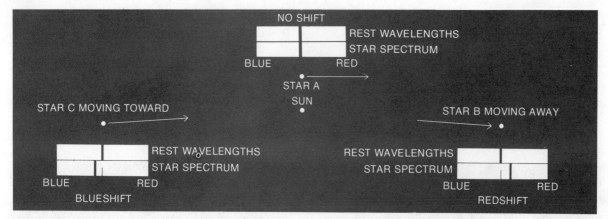

NO SHIFT
REST WAVELENGTHS
STAR SPECTRUM
BLUE   RED
STAR A
SUN
STAR C MOVING TOWARD   STAR B MOVING AWAY
REST WAVELENGTHS   REST WAVELENGTHS
STAR SPECTRUM   STAR SPECTRUM
BLUE   RED   BLUE   RED
BLUESHIFT   REDSHIFT

Figure 5–13 The Doppler effect. In each pair of spectra, the position of the spectral line in the laboratory is shown on top and the position observed in the spectrum of the star is shown below it. Lines from approaching stars appear blueshifted, lines from receding stars appear redshifted, and lines from stars that are moving transverse to us are not shifted because the star has no velocity toward or away from us. A short vertical line marks the unshifted position on the spectra showing shifts.

where $\Delta\lambda$ is the change in wavelength (the Greek delta, Δ, usually stands for the change), λ_0 is the original (rest) wavelength, v is the velocity of the emitting body, and c is the velocity of light ($= 3 \times 10^5$ km s^{-1}). We define positive velocities as velocities of recession and negative velocities as velocities of approach. The new wavelength, λ, is equal to the old wavelength plus the change in wavelength, $\lambda_0 + \Delta\lambda$.

Example: A star is approaching at 30 km s^{-1} (i.e., $v = -30$ km s^{-1}). At what wavelength do we see a spectral line that was at 6000 Å (which is in the red part of the spectrum) when the radiation left the star?

Answer:

$$\frac{\Delta\lambda}{\lambda_0} = \frac{v}{c} = \frac{-30 \text{ km s}^{-1} \times 10^5 \text{ cm km}^{-1}}{3 \times 10^{10} \text{ cm s}^{-1}} = \frac{-3 \times 10^6 \text{ cm s}^{-1}}{3 \times 10^{10} \text{ cm s}^{-1}} = -10^{-4}.$$

$\Delta\lambda = -10^{-4} \lambda_0 = -10^{-4} \times 6000$ Å $= -0.6$ Å. The change in wavelength, $\Delta\lambda$, is thus -0.6 Å. (Since the star is approaching, this is a blueshift, and the new wavelength is slightly shorter than the original wavelength.) The new wavelength, λ, is thus $\lambda_0 + \Delta\lambda = 6000$ Å $- 0.6$ Å $= 5999.4$ Å. It is still in the red part of the spectrum. The change would be easily measurable, but would not be apparent to the eye.

In this example, you were given the rest wavelength. In a real situation, astronomers identify the shifted line and the element it is from by the line's position in the overall pattern of lines in the spectrum. They can then easily measure the position of the line in a spectrum of the relevant element taken in a laboratory on earth (which, of course, has no Doppler shift).

Note that since there are proportions on both sides of the equation, if we take care to use v and c in the same units (e.g., km s^{-1}, cm s^{-1}, or whatever), the $\Delta\lambda$ will be in the same units that λ is in, no matter whether that is angstroms, centimeters, or whatever.

The 30 km s^{-1} given in the example is typical of the random velocities that stars have with respect to each other. These velocities are small on a universal scale, much too small to change the overall color that the eye perceives, but large compared to terrestrial velocities (30 km s^{-1} = 30 km s^{-1} × 3600 s hr^{-1} = 108,000 km hr^{-1}).

Even the Doppler shift of a car's 30 km hr^{-1} velocity, 1/3600 times less than the speed of the star in the example, is easily measurable by police radar, as many have unfortunately discovered.

Doppler shifts resulting from these random velocities are relatively small. We had to go outside our own galaxy to find larger Doppler shifts, until a weird object in our galaxy called SS433 (Section 10.14) was discovered in 1979 to have large Doppler shifts corresponding to velocities one-quarter the speed of light. More generally, distant galaxies and quasars have large Doppler shifts.

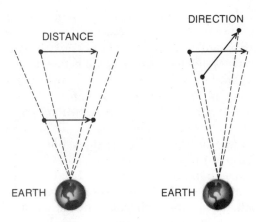

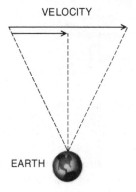

Figure 5–14 Stars can have different proper motions because *(left)* they are at differing distances from us even though they have the same speed through space oriented in the same direction *(middle)*, their directions of motion are oriented differently even though they have the same speed through space and are at the same distance from us, or *(right)* they are actually traveling at different speeds through space.

5.6   STELLAR MOTIONS

On the whole, the network of stars in the sky is fixed. But radial velocities can be measured for all stars, and some of the stars are seen to move slightly across the sky with respect to the more distant stars. The actual velocity of a star in 3-dimensional space, with respect to the sun, is called its *space velocity*, but when we detect a star's change in position in the sky, we know only through what angle it moved. We cannot tell how far it moved in linear units (like km or light years) unless we also happen to know the distance to the star.

We usually deal separately with the part of the velocity of a star that is toward or away from us (the radial velocity), and the angular velocity of the star across the sky (how fast the object is moving across the sky in units of

The stars that show proper motions are generally the closest stars to us. If more distant stars were moving at the same velocities as measured in km s⁻¹, their angular movement would be imperceptible (Fig. 5–14).

Figure 5–15 Two photographs taken 11 months apart. In making this combined print, one of the negatives was shifted slightly. The proper motion of Barnard's star is clearly visible.

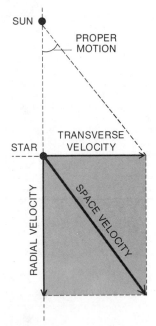

Figure 5–16 If we know a star's proper motion and its distance, we can compute its linear velocity through space in the direction across our field of view. The Doppler effect gives us its linear velocity toward or away from us. These two velocities can be combined to tell us the star's actual velocity through space, its *space velocity*.

angle). The angular velocity is called the *proper motion*. These motions are perpendicular to each other.

Proper motion is ordinarily very small, and is accordingly very difficult to measure. Usually we must compare the positions of stars taken at intervals of decades to detect a sufficient amount of proper motion to allow a measurement of any accuracy to be made. Only a few hundred stars have proper motions as large as 1 arc sec per year; proper motions of other stars are sometimes measured in seconds of arc per century! The Space Telescope will soon improve the quality of our measurement by a considerable factor.

The star in the sky with the largest proper motion was discovered in 1916 by E. E. Barnard, and is known as Barnard's star (Fig. 5–15). It moves across the sky by 10¼ arc sec per year. Some call it "Barnard's runaway star" because of this proper motion, which, though the highest for any known star, is still minuscule. Barnard's star is seen to move so rapidly across the sky because it is relatively close to us. Only 1.8 parsecs away, it is the nearest star to us after the sun, Proxima Centauri and the rest of the Alpha Centauri system (of which Proxima is a part). Since the moon appears half a degree across, Barnard's star moves across the sky by the equivalent of the diameter of the moon in only 180 years.

If we know the distance to a star, the proper motion and the distance together give us the star's actual velocity across space from side to side, its *transverse velocity*. From both the transverse velocity and the radial velocity (obtained from the Doppler shift), we can calculate the star's *space velocity* (Fig. 5–16), which is its real velocity in linear units.

SUMMARY AND OUTLINE

Trigonometric parallax (Section 5.1)
We triangulate, using the earth's orbit as a baseline.
We find the angle through which a nearby star appears to shift; the more the shift, the nearer the star.
This method works only for the nearest stars; the Space Telescope will improve our capabilities.
A *parsec* is the distance of a star from the earth when the radius of the earth's orbit subtends 1 arc sec when viewed from the star.
Absolute magnitude (Section 5.2)
If we know how bright a star *appears*, and how far away it is, we can calculate how intrinsically bright it is. This can be placed on the same scale as that of apparent magnitudes, and is known as the *absolute magnitude*. The absolute magnitude is defined as the magnitude a star would appear to have if it were at a distance of 10 parsecs.

The apparent brightness of a point object follows the inverse square law: brightness decreases with the square of the distance.
We can use the inverse square law to relate apparent magnitude, absolute magnitude, and distance.
The Hertzsprung-Russell diagram (Section 5.3)
A plot of temperature versus brightness.
Stars fall in only limited regions of the graph.
Main sequence contains most of the stars; these stars are called *dwarfs*.
Giants are brighter (and bigger) than dwarfs; *supergiants* are brighter and bigger still.
White dwarfs are fainter than dwarfs, and so fall below the main sequence.
Method of *spectroscopic parallax* (Section 5.4)
Observing the spectrum of a star tells us where it falls on an H-R diagram; this tells us its absolute magnitude. Since we can easily observe its apparent magnitude, we can derive its distance.

The Doppler effect (Section 5.5)
Radiation from objects that are receding is shifted to longer wavelengths: redshifted. Radiation from objects that are approaching is shifted to shorter wavelengths: blueshifted.
We can measure only the *radial velocity,* not the velocity from side to side, in this way.
The angular velocity from side to side is measured by observing the *proper motion*—the motion of the star across the sky. This can be observed only for the nearest stars. If we know the distance to the star, from both distance and proper motion we can derive the star's *transverse velocity.*
The radial velocity and the transverse velocity together give us a star's *space velocity,* its actual motion through space with respect to the sun.
Angular velocity is usually measured in arc sec year^{-1} and radial velocity in km s^{-1}.

QUESTIONS

1. You are driving a car, and the speedometer shows that you are going 80 km hr^{-1}. The person next to you asks why you are going only 75 km hr^{-1}. Explain why you each saw different values on the speedometer.

2. Would the parallax of a nearby star be larger or smaller than the parallax of a more distant star? Explain.

†3. A star has an observed parallax of 0.2 arc second. Another star has a parallax of 0.02 arc sec. (a) Which star is farther away? (b) How much farther away is it?

†4. (a) What is the distance in parsecs to a star whose parallax is 0.05 arc sec? (b) What is the distance in light years?

†5. Estimate the farthest distance for which you can detect parallax by alternately blinking your eyes. To what angle does this correspond?

†6. Vega is about 8 parsecs away from us. What is its parallax?

7. What is the significance of the existence of the main sequence?

8. What is the observational difference between a dwarf and a white dwarf?

†9. Two stars have the same apparent magnitude and are the same spectral type. One is twice as far away as the other. What is the relative size of the two stars?

†10. Two stars have the same absolute magnitude. One is ten times farther away than the other. What is the difference in apparent magnitudes?

†11. A star has apparent magnitude of +5, and is 100 parsecs away from the sun. If it is a main sequence star, what is its spectral type? (Hint: refer to Fig. 5–11.)

†12. A star is 30 parsecs from the sun and has apparent magnitude +2. What is its absolute magnitude?

†13. A star has apparent magnitude +9 and absolute magnitude +4. How far away is it?

14. If a star is moving away from the earth at very high speed, will the star have a continuous spectrum that appears hotter or cooler than it would if the star were at rest? Explain.

15. (a) What does the proper motion of Barnard's star indicate about its distance from us? (b) What would Barnard's star's proper motion be if it were twice as far away from us as it is?

†16. A star has a proper motion of 10 arc sec per century. Can you tell how fast it is moving in space in km s^{-1}?

†17. Two stars have the same space velocity in the same direction, but star A is 10 parsecs from the earth and star B is 30 parsecs from the earth. (a) Which has the larger radial velocity? (b) Which has the larger proper motion?

18. How might the Space Telescope help in determining proper motions?

---

†This indicates a question requiring a numerical solution.

CHAPTER 6

VARIABLE STARS AND STELLAR GROUPINGS

AIMS:
To discuss multiple stars, variable stars,
and stellar clusters, and to draw
conclusions about the distances to stars
and about the evolution of stars

We often think of stars as individual objects that shine steadily, but many stars vary in brightness and most stars actually have companions close by. Also, many stars appear as members of groupings called associations or clusters. By studying the effects that the stars have on each other, or by studying the nature of all the stars in a stellar cluster, we can derive much information that we could not discover by studying the stars one at a time. This information even leads us to a general understanding of the life history of the stars.

6.1 BINARY STARS

Most of the objects in the sky that we see as single "stars" really contain two or more component stars. Sometimes a star appears double merely because two stars that are located at different distances from the sun appear in the same line of sight. Such systems are called *optical doubles,* and will not concern us here. We are more interested in stars that are physically associated with each other. We will use the terms *double star* or *binary star* interchangeably to mean two or more stars held together by the gravity they have for each other. (Even systems with three or more stars are usually called "double stars.")

The easiest way to tell that more than one star is present is by looking through a telescope of sufficiently large aperture. Stars that appear double when observed directly are called *visual binaries.* The resolution of small

Albireo (β Cygni) contains a B star and a K star, which make a particularly beautiful pair because of their different colors.

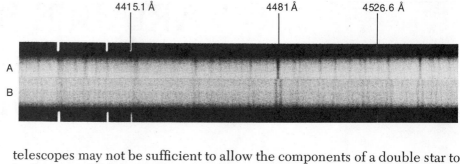

A

B

Figure 6–1 Two spectra of Mizar (ζ Ursae Majoris) taken 2 days apart show that it is a spectroscopic binary. The lines of both stars are superimposed in the upper stellar absorption spectrum *(A)*, but are separated in the lower spectrum *(B)* by 2 Å, which corresponds to a relative velocity of 140 kms⁻¹. Emission lines from a laboratory source are shown at the extreme top and extreme bottom to provide a comparison with a source at rest that has lines at known wavelengths.

telescopes may not be sufficient to allow the components of a double star to be "separated" from each other, and larger telescopes can thus distinguish more double stars. When the stars have different colors, they form particularly beautiful objects to observe in even small telescopes. Five to 10 per cent of the stars in the sky are visual binaries.

We are not always fortunate enough, however, to have the two components of a double star appear far enough apart from each other to be observed visually. Sometimes a star appears as a single object through a telescope, but one can see that the spectrum of that "object" actually consists of the overlapping spectra of at least two objects. If we can detect the presence of two spectra of different types—of, say, one hot star and one cool star—then we say that the object has a *composite spectrum.*

We can tell that a second star is present even when an image appears single if the spectral lines we observe change in wavelength with time. We know of no other way such wavelength changes can occur except for Doppler shifts, which indicate that the object we see is moving. We deduce that two or more stars are present, and are revolving around each other. Such an object is called a *spectroscopic binary* (Fig. 6–1). As the stars in the system orbit each other, unless we are looking straight down on the orbit from above, each spends half of its orbit approaching us and the other half receding from us, relative to its average space motion. The velocity variations in spectroscopic binaries are periodic and, of course, the spectrum of each component varies separately in wavelength. Note that even if the spectrum of one of the stars is too faint to be seen, we can still tell if the star is a spectroscopic binary from the variations in radial velocity of the visible component (Fig. 6–2).

A double star may fall in more than one category of double stars. For example, two stars that revolve around each other in an orbit that makes them eclipse would compose a spectroscopic binary too. Only nearby binary stars are close enough to be seen as visual binaries, but we can detect binaries with composite spectra at much greater distances, even in other galaxies.

Careful spectroscopic studies have shown that two thirds of all solar-type stars have stellar companions. Though the presence of companions of stars of other spectral types has not been studied in such detail, it seems that about 85 per cent of the stars are members of double star systems. Few stars are single, an idea that is verified by noticing that many of the nearest stars to our sun (Appendix 8), stars we can study in detail, are double.

We can detect visual binaries most easily when the stars are relatively far apart. The period of the orbit is relatively long in such cases, over 100,000 years in most cases. So we know little about the motions in such systems because of our short human lifetimes, even though most double

Figure 6–2 Two spectra of α Geminorum B taken at different times show a Doppler shift. Thus the star is a spectroscopic binary, even though lines from only one of the components can be seen. The comparison spectrum of a laboratory source appears at the top and bottom. Note that emission lines from this laboratory source are in the same horizontal positions at extreme top and bottom, while the absorption lines of the stellar spectra are shifted laterally (that is, in wavelength) with respect to each other.

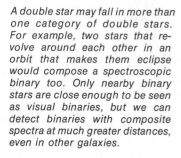

stars discovered are visual binaries. It is harder to detect spectroscopic binaries so we know fewer of them. But for this group, most of whose periods are between one day and one year, we are able to determine such important details of the orbit as size and period (Fig. 6–3).

Sometimes the components of a double star pass in front of each other, as seen from our viewpoint on the earth. The "double star" then changes in brightness periodically, as one star cuts off the light from the other. Such a pair of stars is called an *eclipsing binary* (Fig. 6–4). The easiest to observe is Algol, β Persei (beta of Perseus), in which the eclipses take place every 69 hours, dropping the total brightness of the system from magnitude 2.3 to 3.5. (A third star is present in the Algol system. It orbits the other two every 1.86 years, and does not participate in the eclipse.)

Note that the way that a binary star appears to us depends on the orientation of the two stars not only with respect to each other, but also with respect to the earth. If we are looking down at the plane of their mutual orbit (Fig. 6–5), then we might see a composite spectrum; under the most favorable conditions the star might be seen as a visual binary. But in this orientation we would never be able to see the stars eclipse. We would also not be able to see the Doppler shifts typical of spectroscopic binaries, since only radial velocities contribute to the Doppler shift.

It is possible that a star could be a "double," but still not be detectable by any of the above methods. There are cases where the existence of a double star shows up only as a deviation from a straight line in the proper motion of the "star" across the sky. Such stars are called *astrometric binaries* (Fig. 6–6).

Many of the celestial sources of x-rays that have been observed in recent years from orbiting telescopes turn out to be binary systems. Matter from one member of the pair falls upon the other member, heats up, and radiates x-rays.

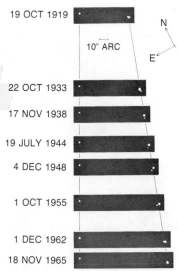

Figure 6–3 The two components of the visual binary Kruger 60 are seen to orbit each other with a period of about 44 years in this series of photographs taken at the Leander McCormick Observatory in Virginia and at the Sproul Observatory in Pennsylvania. We also see the proper motion of the Kruger 60 system, as it moves farther away from the single star at the left over the years.

Astrometric techniques are important in the discovery of white dwarfs (see Section 9.4) and of very cool ordinary dwarfs, and for the discussion of the chances of finding extraterrestrial life (see Section 21.2).

Binary x-ray sources will be discussed in Sections 6.6 and 10.13. In Section 9.8 we will discuss more generally how the transfer of some of the mass from one member of a binary system to the other affects the evolution of each of the stars, possibly in very significant ways.

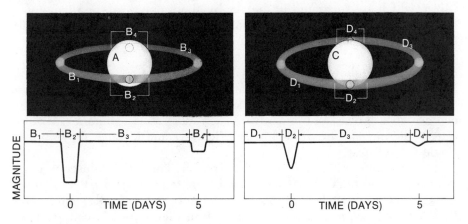

Figure 6–4 The shape of the light curve of an eclipsing binary depends on the sizes of the components and the angle from which we view them. At lower left, we see the light curve that would result for star B orbiting star A, as pictured at upper left. When star B is in the positions shown with subscripts at top, the regions of the light curve marked with the same subscripts result. The eclipse at B_4 is total. At right, we see the appearance of the orbit and the light curve for star D orbiting star C, with the orbit inclined at a greater angle than at left. The eclipse at D_4 is partial. From earth, we observe only the light curves, and use them to determine what the binary system is really like, including the inclination of the orbit and the sizes of the objects.

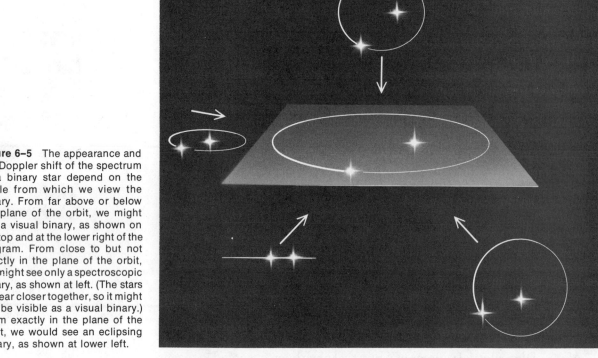

Figure 6-5 The appearance and the Doppler shift of the spectrum of a binary star depend on the angle from which we view the binary. From far above or below the plane of the orbit, we might see a visual binary, as shown on the top and at the lower right of the diagram. From close to but not exactly in the plane of the orbit, we might see only a spectroscopic binary, as shown at left. (The stars appear closer together, so it might not be visible as a visual binary.) From exactly in the plane of the orbit, we would see an eclipsing binary, as shown at lower left.

Figure 6-6 Sirius A and B, an astrometric binary. From studying the motion of Sirius A (often called, simply, Sirius) astronomers deduced the presence of Sirius B before it was seen directly. Sirius B's orbit is larger because Sirius A is a more massive star. They contain 2.14 and 1.05 times as much mass as the sun, respectively.

Though one usually speaks of "double stars," there is actually nothing to prevent stars from existing in threes, or fours, or with even more components. For example, the second star from the end of the handle of the Big Dipper, Mizar, has an apparent fainter companion called Alcor. Alcor and Mizar, which are separated in the sky by about one-third the diameter of the moon, are an optical double. Many people can see them as double with the naked eye. The American Indians knew these two stars as a horse and rider.

A telescope shows that two stars are present in Mizar, that is, that Mizar is a binary, with the components separated by 1/50 the angular distance from Alcor to Mizar. The two components of Mizar A (often known simply and confusingly as Mizar) and Mizar B are each spectroscopic binaries. Mizar A is a double-line spectroscopic binary (Fig. 6-1), and Mizar B shows only a single set of lines.

Another interesting star is Castor, α Gem. Through a telescope it appears as a triple star: Castor A and Castor B are relatively close to each other and Castor C is somewhat farther off. Castor A, B, and C are all spectroscopic binaries; we saw the spectrum of Castor B in Figure 6-2. Thus Castor is a sextuple star.

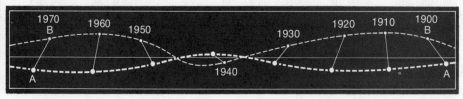

6.2 STELLAR MASSES

The study of binary stars is of fundamental importance in astronomy because it allows us to determine stellar masses. If we can determine the orbits of the stars around each other, we can calculate theoretically the masses of the stars necessary to produce the gravitational effects that lead to those orbits. For a star that is a visual binary with a sufficiently short period (only 20 or even 100 years, for example), we are able to determine the masses of both of the components. If a star is a spectroscopic binary, even one with spectral lines of both stars present, but is not a visual binary, we can find only the lower limits for the masses—that is, we can say that the masses must be larger than certain values. This limitation occurs because we cannot usually tell the inclination of the plane of the orbit of the stars around each other. Thus in this case we know only the radial velocities and not the actual velocities of the stars in their orbits. If the spectroscopic binary is also an eclipsing binary then we know the angle of inclination and can find the individual masses.

Our intuitive notion of gravity tending to pull two objects together, with the strength of gravity greater when the masses of the stars are greater, is sufficient to illustrate the concept that studying the orbits of binary stars can tell us their masses.

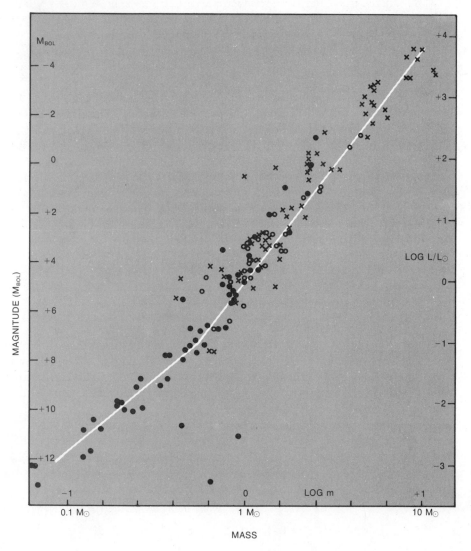

Figure 6–7 The mass-luminosity relation, measured from binary stars. Most of the stars fit within a narrow band. Data from visual components are shown as circles (the filled circles are higher quality data than the open circles) and data from eclipsing components are shown as crosses.

The relation is defined by straight lines of different slopes for stars brighter than and fainter than magnitude 7.5. We do not know if there is a real difference between the brighter and fainter stars that causes this or whether it is an effect introduced when the data are reduced, at which stage we try to take account of the energy that the stars radiate outside of the visible spectrum.

Note that white dwarfs, which are relatively faint for their masses, lie below the mass-luminosity relation. Red giants lie above it. Thus we must realize that the mass-luminosity relation holds only for main-sequence stars. (Courtesy of W. D. Heintz, updated to 1978)

So we do not always know as much as we would like about the masses of the stars in binary systems. Of course, we are better off than we are for stars outside binary systems, for we cannot directly measure their masses at all! From studies of the several dozen binaries for which we can accurately tell the masses, astronomers have graphed the luminosities (intrinsic brightnesses of the stars against the masses. ("Against" in this context means that the luminosities are plotted on one axis and the masses on the other.) Stars lie on a narrow band, the *mass-luminosity relation* (Fig. 6–7), which shows that the masses and luminosities (measured from the apparent brightnesses and distances) of stars are closely related. The mass-luminosity relation in this form is valid only for stars on the main sequence, that is, for normal dwarfs. The more massive a star is, the brighter it is. The mass, in fact, is the prime characteristic that determines where on the main sequence a star will settle down to live its lifetime.

About 90 per cent of stars, those on the main sequence, follow the mass-luminosity relation. The other stars, such as giants or white dwarfs, are special cases that we will discuss later on.

The most massive stars we know are about 50 times more massive than the sun, and the least massive stars are about 25 times less massive than the sun. A dwarf star (i.e., main-sequence star) of 10 solar masses is a B star. A dwarf star of 2 solar masses is an A star. The sun, which of course has 1 solar mass, is a G star. A dwarf of half a solar mass is a K star, and a dwarf of four tenths of a solar mass or less is an M star.

There are many more stars of low mass than there are of high mass. In fact, the number of stars of a given mass in any sufficiently large volume of space increases as we consider stars of lower and lower mass. Thus most of the mass in a cluster of stars, or in a galaxy, comes from stars at the lower end of the H-R diagram. Even though the many faint stars contain most of the total mass in a group of stars, most of the light that we receive from a group of stars, on the contrary, comes from the few very brightest stars. A single O star can outshine a thousand K main-sequence stars. Thus the majority of the stars in a star system together give off a minor fraction of the light; the few brightest stars in that system give off most of the radiation.

Example: *Castor C is not only a spectroscopic binary but also an eclipsing binary, and is sometimes known as YY Gem. From the fact that eclipses occur, we know that the orbits of the stars lie in a plane parallel to our line of sight, as in the bottom left case of Figure 6–5. Thus we can use measurement of the Doppler shift to derive the speed in its orbit of each component, and from this and its measured period can find the size of its orbit. From this we can derive (we use a form of Kepler's third law, which we discuss in Sections 12.7 and 12.8) the sum of the masses of the components. Since we can tell from the spectra that the masses of each are the same, we simply divide this sum by two to get the mass of each star by itself. Each turns out to contain 0.6 times as much mass as the sun.*

6.3 STELLAR SIZES

Stars appear as points to the naked eye and as small, fuzzy disks through large telescopes. The blurring of the stellar images into disks results from distortion by the earth's atmosphere, and hides the actual size and structure of the surface of the stars. The sun is the only star for which we can measure its angular diameter easily and directly.

Astronomers use an indirect method to find the sizes of most stars. If we know the absolute magnitude of a star (from some type of parallax measurement) and the temperature of the surface of the star (from measuring its spectrum), then we can tell the amount of surface area the star must have. The extent of surface area, of course, depends on the radius.

One type of direct measurement works only for eclipsing binary stars.

As the more distant star is hidden behind the nearer star, one can follow the rate at which the intensity of radiation from the further star declines. From this information together with Doppler measurements of the velocities of the components, one can calculate the sizes of the stars. One can sometimes even tell how the brightness varies across a star's disk.

It is much more difficult to measure the size of a single star directly. One way of measuring the diameter is by *lunar occultation*. As the moon moves with respect to the star background, it occults (hides) stars. By studying the light from a star in the fraction of a second it takes for the moon to completely block it, we can deduce the size of the stellar disk. Unfortunately, the moon only passes over 10 per cent of the sky in the course of a year.

Other methods for measuring the diameters of single stars use a principle called *interferometry*, a technique that measures incoming radiation at two different locations and then combines the two signals. This gives the effect, for the purpose of determining the resolution, of a single very large telescope (Figure 6–8). The method overcomes many of the problems of blurring by the earth's atmosphere.

The first such measurements were carried out at Mt. Wilson 50 years ago; 7 of the stars that subtended the largest angles in the sky had their diameters measured. These were the largest of the closest stars, that is, nearby red giants and supergiants. A modern variation of this technique has been worked out at the University of Maryland.

The above interferometric methods work best for large, cool stars. For the last 20 years, R. Hanbury Brown and his associates in Australia have used another type of stellar interferometer (Fig. 6–9) that works best for hot, bright stars. They have determined the diameters of three dozen stars whose diameters cannot be measured with other techniques.

Only a few dozen stellar diameters have been measured directly, and they confirm the indirect measurements (Fig. 6–10). Direct and indirect measurements show that the diameters of main-sequence stars decrease as we go from hotter to cooler; that is, O dwarfs are relatively large and F dwarfs are relatively small. Indirect measurements alone show that this trend continues for K and M stars, which are smaller still. As for stars that are not on the main sequence, red giants are indeed giant in size as well as in brightness.

Details will be discussed in the context of radio observations in Section 25.6.

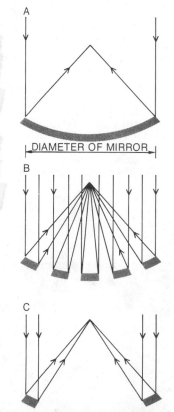

Figure 6–8 A large single mirror *(A)* can be thought of as a set of smaller mirrors *(B)*. Since the resolution of a telescope for light of a certain wavelength depends only on the telescope's aperture, as we saw in Section 2.3, retaining only the outermost segments *(C)* matches the resolution of a full-aperture mirror. We use a property of light called *interference* to analyze the incoming radiation so the device is called an *interferometer*. The parts need not be physically connected to each other, allowing them to be separated by a great distance.

We shall discuss interference in Section 27.4 in the context of radio observations.

Figure 6–9 Hanbury-Brown's intensity interferometer in Australia, used to measure the diameters of hot stars. The two sets of mirrors can be placed up to 188 meters apart from each other.

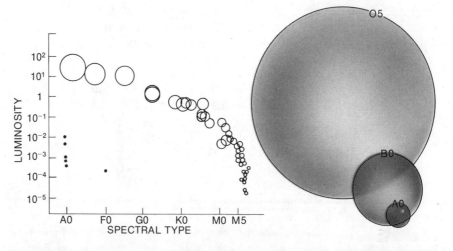

Figure 6–10 The H-R diagram for stars within 17 light years of the sun. The open circles indicate approximately the relative diameters of the main-sequence stars; the white dwarfs are shown as dots. In addition, relative sizes of O, B, and A stars are shown at right.

---

*Box 6.1   Speckle Interferometry

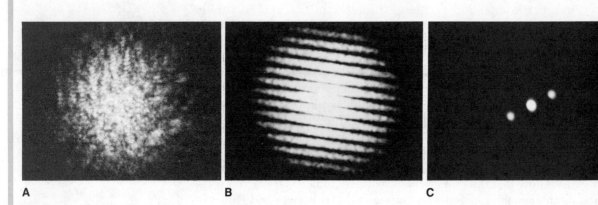

A   **B**   **C**

Figure 6–11 *See legend on opposite page.*

The surface of a star is darker at its limb (edge) than at the center of the disk, a phenomenon known as limb darkening. *You can see this phenomenon for the sun on the photograph opening Chapter 7.*

Over the last decade, a new procedure called *speckle interferometry* has come into use. At any one instant, the image of a star through a large telescope looks speckled (as shown in Fig. 6–11) because different parts of the image are affected by different small turbulent areas in the earth's atmosphere. A long exposure blurs these speckles, and gives us the fuzzy disk that normally results when we photograph a star. The speckle interferometry technique involves taking photographs or electronic images of the speckle pattern with very short exposures—on the order of 1/50 second—and then using mathematical techniques and computer assistance to deduce the properties of the starlight that entered the telescope. Thus, in this manner, from many short exposures of a stellar image, scientists can measure the diameter and the limb darkening.

Interferometric techniques have been used not only to measure the diameters of individual stars but also to resolve the components of close binary stars. Essentially, the speckle pattern of a double star consists of two overlapping speckle patterns, one from each component. Each individual speckle is paired with one from the other star. We can measure the separation and orientation of each pair of speckles. Optical or computer methods can then be used to produce a single photograph on which measurements that give information about the separation and orientation of the binary star system can be made. Speckle interferometry has already resolved the components of dozens of close binary stars (Fig. 6–11C). Several of the systems had not been previously known to be binary. The determination of the separation of double-line spectroscopic binaries allows the masses of the components to be determined.

6.4 VARIABLE STARS

Some stars can be seen to vary in brightness with respect to time. One basic parameter that characterizes the variation of a variable star is the *period*. The period is the time it takes for a star to go through its entire cycle of variation and to return to the original degree of brightness.

One way for a star to vary in brightness, as we have seen, is for the star to be an eclipsing binary. Sometimes we get the light simultaneously from both members of the binary, and sometimes the light from one of the members is at least partially blocked by the other member. But it is possible for individual stars to vary in brightness all by themselves. Thousands of such stars are known in the sky. The periods of the variations can range from seconds for some types of stars to years for others. Plots of the brightness of a star (usually in terms of its magnitude) versus time are called *light curves*. The light curves of some types of variable stars, as we shall see next, can be used to provide vital information about the distance scale.

---

*Box 6.2   Naming Variable Stars

A naming system has been adopted for variable stars (often called simply "variables") that helps us recognize them on any list of stars. The first variable to be discovered in a constellation is named R, followed by the genitival form ("of the . . .") of the Latin name of the constellation (see Appendix 10), e.g., R Coronae Borealis (R of Corona Borealis). One continues with S, T, U, V, W, X, Y, Z, then RR, RS, etc., up to RZ, then SS (not SR) up to SZ, and so on up to ZZ. Then the system starts over with AA up to AZ, BB (not BA) and so on up to QZ. The letter J is omitted to avoid confusion with I. This system covers the first 334 variables; after that one numbers the stars beginning with V for variable (V335 . . .). It should be noted that variable stars with more commonly known names, such as Polaris and δ Cephei, retain their common names instead of being included in the lettering system.

We will limit our discussion to three types of variables, one of which is especially numerous, and the other two of which have provided important information about the scale of distance in the universe.

6.4a Mira Variables

A type of variable star is often named after its best-known or brightest example. A star named Mira in the constellation Cetus (*o* Ceti, omicron of the Whale) fluctuates in brightness with a long period (Fig. 6–12). Red stars that share this characteristic are called *Mira variables*. The period of a given Mira-type star can be from three months up to about two years. The period of an individual star is not strictly regular; it can vary from time to time from the average period.

Mira itself is sometimes of apparent magnitude 9, and is thus invisible to the naked eye. However, it brightens fairly regularly by about six magnitudes, a factor of 250, with a period of about 11 months. At maximum brightness it is quite noticeable in the sky. As it brightens, its spectral type changes from M5 to M9. Thus real changes are taking place at the surface of the star that result in a change of temperature.

Figure 6–11 A single short (1/50 sec) exposure of a speckle pattern of the double star κ (Kappa) Ursae Majoris is shown in *A*. Adding 50 of them together and using a special photographic technique (the equivalent of computer processing) gives a composite photograph *(B)* on which bright fringes separated by dark bands appear. The spacing of the bright fringes from each other and their orientation can be measured and interpreted in terms of the separation of the components of the binary and the angle at which the stars are oriented in the sky as seen from earth. (We are not going into the mathematical method of transformation here; the photographs are merely to illustrate how the technique is carried out.) Another optical (or computer) process of transformation gives the form of part *C*, in which the fainter of the stars appears twice, symmetrically to upper right and lower left of the image of the brighter of the stars. Even coarse measurements on *C* made with a ruler would tell you information that can be used to find the separation and orientation of the stars. A major advantage of this method is that it works in many cases that cannot be measured in other ways. Even without understanding the technique in general, the ease of measuring *C* clearly shows the power of the method. (Photographs by Harold A. McAlister, taken at the Kitt Peak National Observatory)

Finding the distance scale in the universe, the distances to astronomical objects in linear units like km or light years, is one of the basic problems that astronomers attack.

Figure 6–12 The light curve for Mira, the prototype of the class of long-period variables.

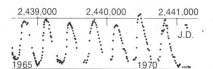

Mira stars are giants of spectral type M, and are about 700 times the diameter of the sun. This would extend beyond Mars if one were in our solar system. Mira stars emit most of their radiation in the infrared. They are also the source of strong radio spectral lines from water vapor.

These stars, which are the most numerous type of variable star in the sky, are also known as *long-period variables.*

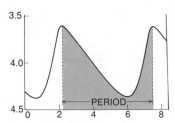

Figure 6–13 The light curve for δ Cephei, the prototype of the class of Cepheid variables.

In Greek mythology, Cepheus was a king of Ethiopia, Cassiopeia's husband, and Andromeda's father.

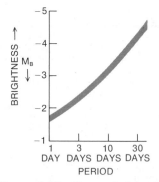

Figure 6–14 The period luminosity relation for Cepheid variables.

Figure 6–15 Henrietta S. Leavitt, who worked out the period-luminosity relation, at her desk at the Harvard College Observatory in 1916.

6.4b Cepheid Variables

The most important variable stars in astronomy are the *Cepheid variables* (cef'e-id). The prototype is δ Cephei (delta of Cepheus) (Fig. 6–13). Cepheid variables have very regular periods that, for individual Cepheids (as they are called), can be from 1 to 100 days.

Cepheids are relatively rare stars in our galaxy; only about 700 are known. Some have been detected in other galaxies. δ Cephei itself, as an example from our own galaxy, varies between apparent magnitudes 3.6 and 4.3 with a period of 5.4 days (Fig. 6–13).

Cepheids are important because a relation has been found that links the periods of the light changes, which are simple to measure, with their absolute magnitudes. For example, if we measure that the period of a Cepheid is 10 days, we need only look at Figure 6–14 to see that the star is of absolute magnitude −3. We can then compare its absolute magnitude to its apparent magnitude, which gives us its distance from us.

The study of Cepheids is the key to our current understanding of the distance scale of the universe and has allowed us to determine that the objects in the sky that we now call galaxies (discussed in Chapter 25) are giant systems comparable to that of our own galaxy (discussed in Chapter 22).

The story started about 75 years ago with Henrietta Leavitt (Fig. 6–15) who, at the Harvard College Observatory, was studying the light curves of variable stars in the southern sky. In particular, she was studying the variables in the Large and Small Magellanic Clouds (Fig. 6–16), two hazy areas in the sky that were discovered by the crew of Magellan's expedition around the world when they sailed far south. Whatever the Magellanic Clouds were—we now know them to be galaxies, but Leavitt did not know this—it looked like a concentrated cloud of material. It thus seemed clear that for each cloud all its stars were at approximately the same distance away from the earth. Thus even though she could only plot the apparent magnitudes, the relation of the absolute magnitudes to each other was exactly the same as the relation of the apparent magnitudes.

By 1912, Leavitt had established the light curves and determined the periods for two dozen stars in the small Magellanic Cloud. She plotted the magnitude of the stars (actually a median brightness, a value between the maximum and minimum brightness) against the period. She realized that there was a fairly strict relation between the two quantities, and that the Cepheids with longer periods were brighter than the Cepheids with shorter periods. By simply measuring the periods, she could determine the magnitude of one star relative to another; each period uniquely corresponds to a magnitude.

Henrietta Leavitt could measure **apparent** magnitude but she did not know the distance to the Magellanic Clouds so could not determine the **absolute** magnitude (intrinsic brightness) of the Cepheids. To find the distance to the Magellanic Clouds, we first had to be able to find the distance to *any* Cepheid—even one not in the Magellanic Clouds—to tell its absolute magnitude. A Cepheid of the same period but located in the Magellanic Clouds would presumably have the same absolute magnitude.

To find the absolute magnitude of a Cepheid it would seem easiest to start with the one nearest to us. Unfortunately, not a single Cepheid is close enough to the sun to allow its distance to be determined by the method of trigonometric parallax. More complex, statistical methods had to be used to study the relationships between stellar motions and distances. This gave the distance to a nearby Cepheid. Once we have the distance to and thus the absolute magnitude of a nearby Cepheid, we know the absolute magnitude of all stars of that same period in the Magellanic Clouds, since all stars of the same period have the same absolute magnitude.

From that point on it is easy to tell the absolute magnitude of Cepheids of **any** period in the Magellanic Clouds. After all, if a Cepheid has **apparent** magnitude that is, say, 2 magnitudes brighter than the apparent magnitude of our Cepheid of known intrinsic brightness, then its **absolute** magnitude is also 2 magnitudes brighter. After this process, we have the period-luminosity relation in a more useful form—period vs. absolute magnitude. We call this process "the calibration" of the period-luminosity relation.

Thus Cepheids can be employed as indicators of distance: First we identify the star as a Cepheid (by studying its spectrum and the shape of its light curve). Then we measure its period. Third, the period-luminosity relation gives us the absolute magnitude of the Cepheid. And last, we calculate how far a star of that absolute magnitude would have to be moved from the standard distance of 10 parsecs to appear as a star of the apparent magnitude that we observe.

When the calibration of the period-luminosity relation was worked out quantitatively by the American astronomer Harlow Shapley in 1917, the distance to the Magellanic Clouds could be calculated. They were very far away, a distance that we now know means that they are not even in our galaxy! Instead, they are galaxies by themselves, two small irregular galaxies that are companions of our own, larger, galaxy. (Shapley's values have been superseded by later measurements, but the conclusion remains valid.)

Cepheids can also be seen not only in our own galaxy and in the Magellanic clouds but also in more distant galaxies. They are bright stars (giants and supergiants, in fact) and so can be seen at quite a distance. The use of Cepheids is the prime method of establishing the distance to all the nearer galaxies.

As the Cepheids were further studied, astronomers could make a model for their variations on the basis of their light curves and of their spectra. The absorption lines in the spectra of a Cepheid show Doppler shifts that prove that the star is actually pulsating—expanding and contracting—as it changes in brightness. The size of a Cepheid variable may change by 5 to 10 per cent as it goes through its pulsations. The surface temperature also varies. The star is brightest roughly at its hottest phase. A detailed theory of stellar pulsations that explains these effects has been worked out.

Figure 6–16 From the southern hemisphere, the Magellanic Clouds are high in the sky. In this view from Perth, Australia, the Large Magellanic Cloud is about 20° above the horizon and the Small Magellanic Cloud is twice as high. They are not quite this obvious to the naked eye. Canopus, the second brightest star in the sky, appears at the lower left.

At present, the Cepheid period-luminosity relation is calibrated for the open cluster the Hyades, which is now agreed to be 44 pc distant.

In the 1950's, when overlapping methods of finding distances were applied to some relatively nearby stars and clusters of stars, it was realized that there was a second type of Cepheid whose light curves looked the same but which were about 1.5 magnitudes fainter at each given period than the original type. The distances to many galaxies had to be recalculated because this distinction between ordinary Cepheids and Type II Cepheids had previously not been made. Our estimates of the distances to many distant galaxies doubled.

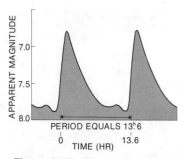

Figure 6–17 The light curve of RR Lyrae, the prototype of the class known as RR Lyrae variables or cluster variables.

6.4c RR Lyrae Variables

Many stars are known to have short regular periods, less than one day in duration. Certain of these stars, no matter what their period, have light curves of a certain distinctive shape (Fig. 6–17). All these stars have the same average absolute magnitude.

Such stars are called *RR Lyrae stars* after the prototype of the class. Since many of these stars appear in globular clusters (which will be described in Section 6.5), RR Lyrae stars are also called *cluster variables*.

Once we detect an RR Lyrae star by the shape of its light curve, we immediately know its absolute magnitude, since all the absolute magnitudes are the same (about 0.6). Just as before, we measure the star's apparent magnitude and can thus easily calculate its distance, which is also the distance to the cluster.

6.5 CLUSTERS AND STELLAR POPULATIONS

Even aside from the hazy band of the Milky Way, the distribution of stars in our sky is not uniform. There are certain areas where the number of stars is very much higher than the number in adjacent areas. Such sections of the sky are called *star clusters* (Color Plates 47 and 48).

One type of star cluster appears as just an increase in the number of stars in that limited area of sky. Such clusters are called *open clusters*, or *galactic clusters* (Fig. 6–18). The most familiar example of a galactic cluster is the Pleiades, a group of stars visible in the winter sky. The eye sees at least six stars very close together. With binoculars or the smallest telescopes, dozens more can be seen. A larger telescope reveals hundreds of stars. Another galactic cluster is called the Hyades, which forms the "v" that outlines the

The Pleiades are often known as the Seven Sisters, *after the seven daughters of Atlas who were pursued by Orion and who were given refuge in the sky. That one is missing—the Lost Pleiad—has long been noticed. Of course, a seventh star is present (and hundreds of others as well), although too faint to be plainly seen with the naked eye. The Pleiades seem to be riding on the back of Taurus.*

Figure 6–18 The Pleiades, a galactic cluster. The six brightest stars are visible to the naked eye.

Figure 6–19 The Hyades are a galactic cluster that outlines the face of Taurus, the bull, pictured here in a plate from the atlas of Hevelius published in 1690.

face of Taurus, the bull, a constellation best visible in the winter sky (Fig. 6–19). More than a thousand such clusters are known, most of them too faint to be seen except with telescopes.

All the stars in a galactic cluster are packed into a volume not more than 10 parsecs across.

Stars in galactic clusters seem to be representative of stars in the spiral arms of our galaxy and of other galaxies. When the spectra of stars in galactic clusters are analyzed to find the relative abundances of the chemical elements in their atmospheres, we find that over 90 per cent of the atoms are hydrogen, most of the rest are helium, and less than 1 per cent are elements heavier than helium.* This is similar to the composition of the sun. Such stars are said to belong to stellar *Population I*.

The structure of our galaxy is described in Chapter 22.

The second major type of star cluster appears in a small telescope as a small, hazy area in the sky. Observation with larger telescopes can distinguish individual stars, and reveals that these clusters are really composed of many thousands of stars packed together in a very limited space. The clusters take spherical forms, and are known as *globular clusters* (Fig. 6–20).

Globular clusters can contain 10,000 to one million stars, in contrast to the 20 to several hundred stars in a galactic cluster. A globular cluster can fill

---

*The structure of the nuclei of the elements is discussed in Section 8.3. For the present, we need only know that hydrogen, whose most abundant form contains only one nuclear particle, is the least massive element, and helium, whose most abundant form contains 4 nuclear particles, is the second least massive element.

Figure 6–20 The globular cluster M3 in the constellation Canes Venatici, the hunting dogs.

a volume up to about 30 parsecs across. It has fewer stars at its periphery and more closely packed stars toward the center.

Globular clusters are typically found only in regions above and below the galactic plane. We say that they are in the galactic *halo*. (We do not see many in the plane of the galaxy because they are hidden by interstellar dust there.) The abundances of the elements heavier than helium in globular clusters are much lower, by a factor of 10 or more, than their abundances in the sun. Such stars are said to belong to *Population II*.

The fact that galactic clusters and globular clusters are distributed throughout our galaxy in different fashions has been used to determine the makeup and structure of our galaxy. We shall return to this point in more detail in Chapter 22.

6.5a H-R Diagrams and the Ages of Galactic Clusters

The Hertzsprung-Russell diagram for several galactic clusters is shown in Figure 6–21. For the purposes of this discussion it is most important to note that the horizontal axis is a measure of temperature and the vertical axis is a measure of brightness.

The H-R diagrams for different galactic clusters appear to be similar over the lower part of the main sequence but diverge at the upper part. The stars at the upper left are the more massive ones. They are more luminous and use up their nuclear fuel at a faster rate than the cooler, more numerous, ordinary stars like the sun. Stars of all masses are on the main sequence in the stable, middle-age part of their lifetimes. Later in their lives they become larger and thus brighter, since they then have more surface area to radiate light. Since the y-axis of the H-R diagram shows brightness, the stars thus move upward from the main sequence. Their outer layers may also, after a time, become cooler, and the stars move toward the right on the H-R dia-

At the telescope, we measure temperatures as a "color index" (see Appendix 13).

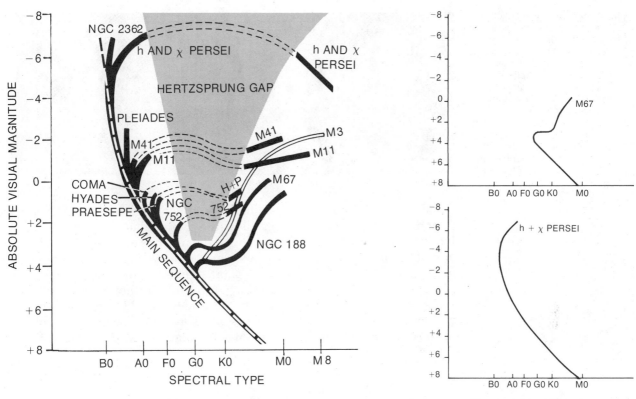

gram. Since the more massive stars finish their main-sequence lifetimes relatively quickly, they "move off" the main sequence before the less massive stars.

When we examine the H-R diagram of a cluster, we are observing a group of stars all of which were presumably formed at the same time in the same region of space and so have similar chemical compositions. We presume that all the stars "fell" on the main sequence during most of their lives (that is, the points representing their magnitudes and temperatures were on the main sequence). Points representing the fainter stars usually fall on the main sequence of the H-R diagram. But as we see in Figure 6–21, for some galactic clusters the points representing the brighter, more massive stars are to the right side of the main sequence. We interpret this to mean that these stars have passed through the phase of their lives in which they fell on the main sequence and have "moved off" the main sequence.

Remember, the more massive the star, the faster it uses up its fuel and so the shorter its main-sequence lifetime. From seeing how high on the main sequence the H-R diagram of a galactic cluster turns off, we can deduce the age of a cluster. After all, the longer we wait, the farther down the turnoff moves. We can read the turnoff point as we read a clock, based on theoretical calculations of how long the main-sequence phase lasts for stars of given masses. Calculations indicate that the cluster is older than the main-sequence lifetimes of those stars in that cluster that have moved off the main sequence. On the other hand, the fact that stars of a certain temperature and absolute magnitude, and thus a certain mass, are still on the main sequence indicates that the cluster is not yet old enough for stars of this mass to have finished their main-sequence lifetimes.

Figure 6–21 *(A)* Hertzsprung-Russell diagrams of several galactic clusters, showing the overlay of several individual diagrams. We see that the fainter stars of galactic clusters are on the main sequence, while the brighter stars are above and to the right of the main sequence. Almost all stars in the younger clusters, like h and χ Persei, follow the main sequence, while the hotter and more massive members of older clusters, like M67, have had time to evolve toward the red giant region. By observing the point where the cluster turns off the main sequence, we deduce the length of time since its stars were formed, i.e., the age of the cluster. The presence of the *Hertzsprung gap* (shaded), a region in which few stars are found, indicates that stars evolve rapidly through this part of the diagram.

Part of the H-R diagram for the globular cluster M3 (hollow bar) is graphed for comparison. *(B)* Two of the H-R diagrams of individual galactic clusters are shown separately here to illustrate how several of these are put together to make the composite diagram shown in *A*.

Figure 6–22 The double cluster in Perseus, h and χ Persei, a pair of galactic clusters that are readily visible in a small telescope and close enough together that they appear in the same field of view. Study of the H-R diagram for stars in this cluster reveals that they are relatively young. Perseus is a northern constellation that is most prominent in the winter sky. In Greek mythology, Perseus slew the Gorgon Medusa and saved Andromeda from a sea monster.

|◄─DIAMETER OF MOON─►|

Figure 6–21 shows that stars in the double cluster in Perseus, h and χ (h and chi) Persei (Fig. 6–22), follow the main sequence for most of its length. The line of points representing the stars turns off the main sequence only at its upper end. Thus h and χ Persei are very young clusters. They may have formed only 10^7 years ago, a short time scale from an astronomical point of view. The Pleiades, which turns off further down the main sequence, is still a young cluster but is somewhat older than the double cluster in Perseus. It may be 10^8 years old. The Hyades, which turns off the main sequence still further down, may be much older, possibly 10^9 years.

°**Box 6.3   Main-Sequence Fitting**

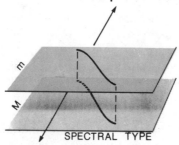

Figure 6–23 The main sequence measured in terms of apparent magnitude, drawn on the top sheet, can be slid up and down in the direction of the arrows, varying the magnitude but not the spectral type. When the measured main-sequence matches a standard main sequence drawn in terms of absolute magnitude, bottom sheet, the difference in magnitudes gives the distance to the cluster.

If we measure apparent magnitudes for a large number of stars in a cluster, we can deduce the absolute magnitude by the following method. We already know that all the stars in the cluster are at the same distance. Since we know that the lower part of the main sequence will agree for all galactic clusters, we can take an H-R diagram measured in apparent magnitudes and merely slide it up and down along lines of constant spectral type (Fig. 6–23) until we see the best agreement with respect to a standard H-R diagram drawn in terms of absolute magnitudes. Then we have found the number of magnitudes by which the absolute and apparent magnitude scales differ, and thus *from the H-R diagram we can easily calculate the distance to the cluster.*

For example, let us consider just one point on the main sequence. A-type stars are known to have absolute magnitude of about +2. If we see an A star of apparent magnitude +7, we know that the cluster is farther enough away than 10 parsecs that its stars appear 5 magnitudes = 100 times fainter. Because of the inverse square law of brightness, we know that the star and its cluster are 10 times further away than 10 parsecs, which is 100 parsecs (see Section 5.2). Actually, this procedure is more precise when we *fit* (superimpose) a whole segment of a main sequence, or some other part of an H-R diagram. The procedure is thus called *fitting,* as in "fitting the main sequence."

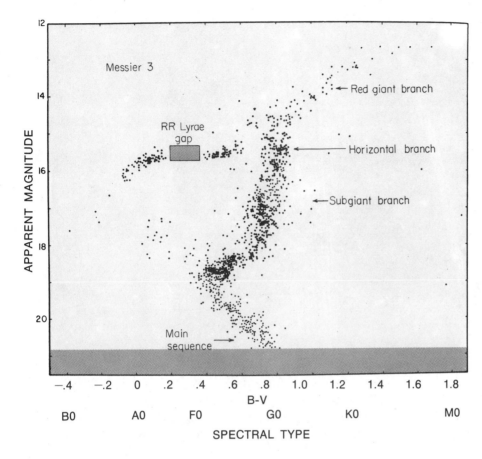

Figure 6–24 The Hertzsprung-Russell diagram for the globular cluster M3, part of which was included in Figure 6–21A. There is a gap in the horizontal branch where no stars of constant brightness are found. The RR Lyrae variables fall here, some 200 for this cluster (though most clusters contain few of them). We can observe only the members of the cluster that are brighter than a certain limit. Thus no stars appear in the shaded region at the bottom.

The positions of stars on the horizontal axis were actually determined from measurements of the colors of the stars made at the telescope by comparing their brightnesses in the blue and in the yellow (which gives a *color index*). The transformation to spectral type is approximate. Color index is described in Appendix 13.

6.5b   H-R Diagrams for Globular Clusters

The H-R diagram for a globular cluster is shown in Figure 6–24. H-R diagrams for all globular clusters look very similar in form; all have stubby main sequences like this one.

From the fact that all globular clusters have H-R diagrams with short main sequences, we can conclude that the globular clusters must be very old. Detailed studies assign ages from about 8 to 12.5 billion years. If we assume that the globular clusters formed when or soon after our galaxy formed, then our galaxy must be about 12.5 billion years old.

The H-R diagrams for globular clusters have prominent *horizontal branches*. The horizontal branch goes leftward from the stars on the right side of the diagram that have long since turned off the main sequence. It represents very old stars that have evolved past their giant or supergiant phases and are returning leftward. No galactic cluster has stars that old. In the horizontal branch there is a gap where stars of constant brightness are not found. Only the RR Lyrae stars fall in this gap.

*6.6   X-RAY BURSTERS AND GLOBULAR CLUSTERS

For the last few years, since scientists have been able to place telescopes in orbit, they have been able to observe x-rays from celestial sources. The

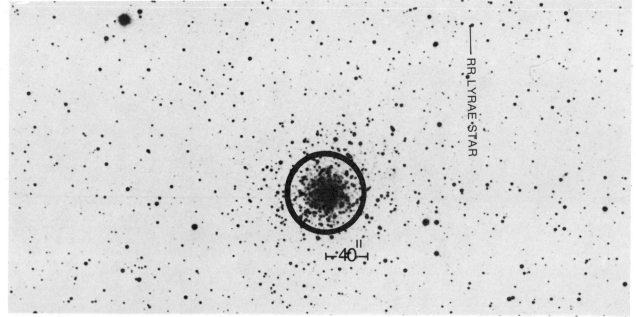

RR LYRAE STAR

⊢40″⊣

Figure 6-25 The circle shows an early measurement of the position of the source of x-ray pulses discussed at right. The size of the circle reflects our uncertainty of the position as measured from the SAS-3 satellite. When drawn on a photograph of the sky, globular cluster NGC 6624 was included in it. More recent measurements from the Einstein Observatory, with an accuracy of only 2 arc seconds, has localized the source still further—to the very center of the cluster. It is difficult to tell from inspection of the photograph whether the lone RR Lyrae star that has been detected in this field of view is indeed part of the cluster (and thus the cluster is much bigger than is apparent on the photograph); the star's distance is about 2 kiloparsecs smaller than the 8 kiloparsecs distance to the cluster that has been derived from fitting its horizontal branch to a standard H-R diagram.

Bursts in the infrared have been discovered from the cluster—Liller 1—in which the rapid burster is located.

strength of the x-radiation varies strongly. In one globular cluster, the x-radiation sometimes changes in intensity by a factor of 5 from one minute to the next. In 1975, intense bursts of x-rays from this globular cluster were discovered, each presenting an increase in intensity of about 25 times in half a second, followed by a slower decay of the intensity (Fig. 6–26). The bursts occur every few hours in this cluster, though the interval between bursts varies somewhat. In each burst, the cluster emits about a million times the energy radiated by the sun in the same amount of time! Of the eight globular clusters now known to be x-ray sources, six—a very high percentage—are known to give off such bursts.

Recent studies, including observations with the Einstein Observatory (Fig. 6–27), have shown that an additional two dozen of these sources of x-ray bursts exist scattered across the sky, in addition to the eight in globular clusters. All these sources are called *bursters*. The bursts recur every few hours in most sources. One such source gives off bursts very frequently—once every 10 to 100 seconds. For obvious reasons, this unique source is called the "rapid burster." Most bursters seem to be "on" for a few weeks and then "off," a time when no bursts are detectable.

The leading explanations of the bursters involve the most unusual and exciting objects in astronomy. The most likely interpretation at present is that the bursters are examples of binary x-ray sources, that is, a binary star system that gives off x-rays. In this case, one of the two objects is a neutron star, a highly compact star that is the result of the collapse of a star after it has exhausted its nuclear fuel, left the main sequence, and become a supergiant. We shall be studying neutron stars further in Chapter 10, where we will see that they also show up as pulsars.

The explanation of bursters by this model is that the neutron star's gravity pulls some of its companion's matter away from the companion (which is presumably a normal main-sequence star) and onto itself. When the gas hits the surface of the neutron star it detonates, quite like a hydrogen bomb going off. This causes the x-rays that we detect.

(An alternative way that a neutron star can cause bursts of x-rays is if it gathers up—*accretes*—from its companion a disk of matter spinning around with it as it rotates. When the matter from this accretion disk drops onto the neutron star, the detonation and resulting flash of x-rays occur. The off/on nature of the bursts would be caused, in this model, by something interrupting the flow of matter in the accretion disk. This version, when detailed theoretical analysis is performed, allows the bursts to take place more frequently than the other version, and is probably the only way that the rapid burster can be explained.)

In the neutron star model, we can understand why so many bursters are in globular clusters. After all, the high density of stars in a globular cluster makes it relatively likely that one star will capture another to form a binary system. But the neutron star model does not explain where bursters' binary companions are hidden; no trace of periodic motion of bursters has been detected.

The other main line of explanation for bursters involves a very strange notion that had been advanced on theoretical grounds even prior to the discovery of bursters: a very massive black hole may exist at the center of many globular clusters. We discuss at length the properties of black holes in Chapter 11, but suffice it to say here that if a supergiant star dies and has a relatively large mass left as it collapses, then the collapse will not stop at a neutron star stage. The collapse will continue forever, and the mass of the star would become so concentrated that everything—even light and all other radiation—would be pulled in. The effect would be as though light were affected by gravity, and gravity were too strong to allow the light to escape. Matter that is falling into a black hole can become very hot—after it enters the black hole itself the matter is heard from no more—and x-rays (which are photons of very high energy) result.

It has been predicted on theoretical grounds that the observed kind of x-rays—with rapid variations plus sharp bursts—would result if a black hole containing between 10 and 100 times the mass of the sun existed in the center of a globular cluster. And since the gravitational pull is stronger in the middle of a globular cluster than elsewhere in the cluster, that would be a reasonable place for the hulks of dead stars to accumulate and combine into such a giant black hole.

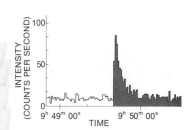

Figure 6–26 An x-ray burst (shaded) from the globular cluster NGC 6624 observed on September 28, 1975. The burst lasted about 10 seconds.

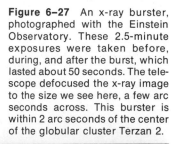

Figure 6–27 An x-ray burster, photographed with the Einstein Observatory. These 2.5-minute exposures were taken before, during, and after the burst, which lasted about 50 seconds. The telescope defocused the x-ray image to the size we see here, a few arc seconds across. This burster is within 2 arc seconds of the center of the globular cluster Terzan 2.

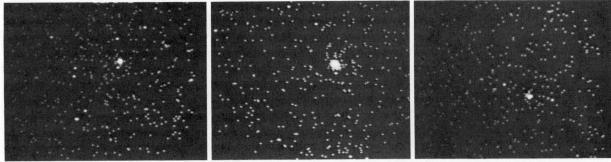

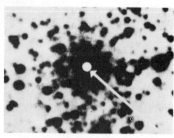

Figure 6–28 The position of the x-ray burster in the globular cluster Terzan 2 has been narrowed by the Einstein Observatory to 2 arc sec, and is right in the middle of the cluster. Its position is shown with a dot. The x-ray data were obtained by Jonathan Grindlay of the Harvard-Smithsonian Center for Astrophysics, and the optical picture was obtained by Grindlay at the Cerro Tololo Inter-American Observatory.

One difficulty with the black hole theory is that it does not explain how bursters can exist outside of globular clusters. Dozens of such bursters are known. Also, though it seemed at first that the x-ray burst sources were exactly at the center of the globular clusters, which is more reasonable for a black hole than for a binary source, Einstein Observatory observations are of higher precision (Fig. 6–28) and indicate that some of the x-ray sources are slightly displaced from the cluster centers.

In any case, the deduction that the members of a binary system may well interact so dramatically or that giant black holes may exist is an interesting merger of modern and traditional astronomy. For example, many of the basic equations we need to interpret the mass and other properties of the binary system and so understand the neutron star have been developed in the course of classical studies over many decades. Also, to know how much energy the cluster is emitting in the form of x-rays requires us to know the distance to the cluster. We find this distance by fitting the horizontal branch of its Hertzsprung-Russell diagram to a standard H-R diagram.

Globular clusters, far from being uninteresting relics of old stars, turn out to be fascinating places that are earning increased attention. And many of the tried and true methods of classical astronomy are proving useful on the forefront of contemporary research.

SUMMARY AND OUTLINE

Binary stars (Section 6.1)
  Optical doubles
  Visual binaries
  Binaries with composite spectra
  Spectroscopic binaries
  Eclipsing binaries
  Astrometric binaries
Determination of stellar masses (Section 6.2)
  Mass-luminosity relation
Determination of stellar sizes (Section 6.3)
  Indirect: calculate from absolute magnitude or measure in eclipsing binary system
  Direct: interferometry
Variable stars (Section 6.4)
  Mira variables (Section 6.4a)
  Cepheid variables (Section 6.4b)
    Period-luminosity relation
    Uses for determining distances
  RR Lyrae variables (Section 6.4c)
    All of approximately the same absolute magnitude
    Uses for determining distances

Clusters and stellar populations (Section 6.5)
  Galactic (open) clusters (Section 6.5a)
    Population I: relatively high abundance of elements heavier than helium
    Representative of spiral arms
    20 to several hundred members
    Turn-off on H-R diagram gives age
  Globular clusters (Section 6.5b)
    Population II: relatively low abundance of metals
    Representative of galactic halo
    10^4 to 10^6 members
    Old enough to have H-R diagrams with horizontal branches
Main-sequence fitting (Box 6.3)
  Compare measured H-R diagram with one calibrated in terms of absolute magnitude
X-ray bursts (Section 6.6)
  A higher percentage than expected are in globular clusters
  Neutron stars or massive black holes may be the cause of the bursts

QUESTIONS

1. Sketch the orbit of a double star that is simultaneously a visual, an eclipsing, and a spectroscopic binary.

†2. How much brighter than the sun is a main-sequence star whose mass is 10 times that of sun?

3. (a) Assume that an eclipsing binary contains two identical stars. Sketch the intensity of light received as a function of time.
   (b) Sketch to the same scale another curve to show the result if both stars were much larger while the orbit stayed the same.

4. Describe two methods of determining the sizes of stars.

†5. A Cepheid variable has a period of 30 days. What is its absolute magnitude?

†6. An RR Lyrae star has an apparent magnitude of 6. How far is it from the sun?

7. Explain briefly how observations of a Cepheid variable in a distant galaxy can be used to find the distance to the galaxy.

8. Briefly distinguish Population I from Population II stars.

†9. Star A has color index 0.0. Star B has color index 1.0. Which has a higher surface temperature? (See Appendix 13.)

†10. A main-sequence star is 3 times the mass of the sun. What is its luminosity relative to that of the sun?

†11. (a) Use the mass-luminosity relation to determine about how many times brighter than the sun are the most massive main-sequence stars of which we know.
   (b) How many times fainter are the least massive main-sequence stars?

12. When we see the light from a distant galaxy, are we seeing mostly low or high mass stars?

13. When we consider the gravitational effects of a distant galaxy on its neighbors, are we measuring the effects of mostly low or high mass stars?

†14. What is the absolute magnitude of an RR Lyrae star with an 18-hour period?

15. Cluster X has a higher fraction of main-sequence stars than cluster Y. Which cluster is probably older?

16. What is the advantage of studying the H-R diagram of a cluster, compared to that of the stars in the general field?

17. How do bursters get their name?

18. What is the difference between the variations of a burster and a Cepheid?

---

†This indicates a question requiring a numerical answer.

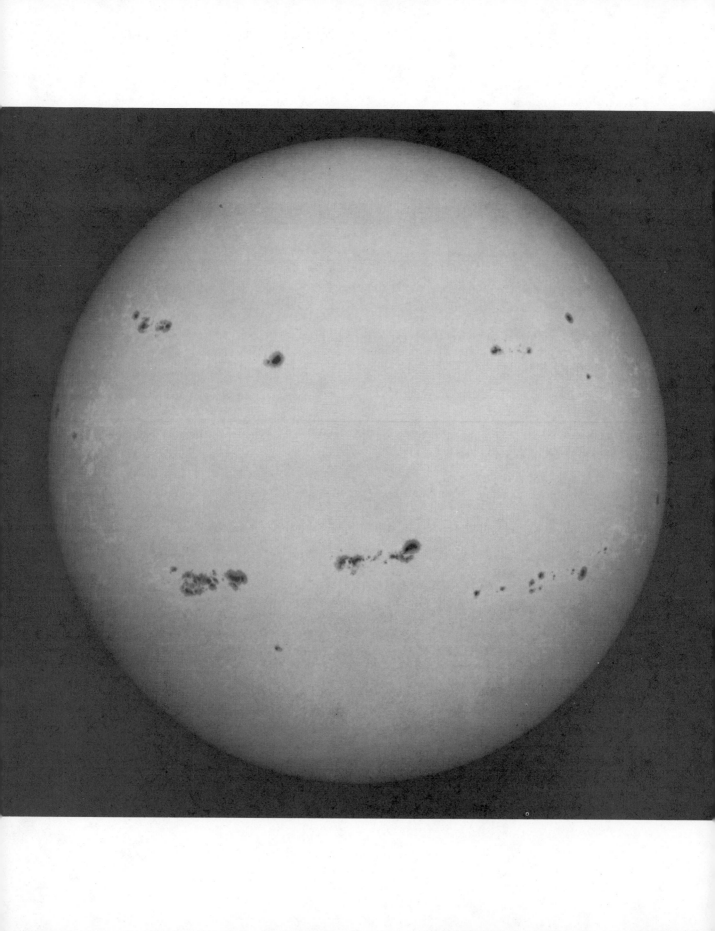

CHAPTER 7

THE SUN

AIMS:
To study the sun, which is the nearest star
and an example of all the other stars
whose surfaces we cannot observe in such
detail

We have thus far discussed a range of individual stars of different spectral classes and have discussed groupings of stars in close physical proximity to each other. Studying these distant stars has allowed us to learn a lot about the properties of stars and how they evolve. But not all stars are far away; one is close at hand. By studying the sun, we not only learn about the properties of a particular star but also can study the details of processes that undoubtedly take place in more distant stars as well. We will first discuss the *quiet sun*, the solar phenomena that appear every day. Afterwards, we will discuss the *active sun*, solar phenomena that appear non-uniformly on the sun and vary over time.

This chapter can be treated at any point in the course after Section 3.2.

7.1 BASIC STRUCTURE OF THE SUN

We think of the sun as the bright ball of gas that appears to travel across our sky every day. We are seeing only one layer of the sun; the properties of the solar interior below that layer and of the solar atmosphere above that layer are very different. The outermost parts of the solar atmosphere even extend through interplanetary space beyond the orbit of the earth.

The edge of the sun or any star is known as the limb.

The layer that we see is called the *photosphere*, which simply means the sphere from which the light comes (from the Greek *photos*, light). As is typical of many stars, about 94 per cent of the atoms and nuclei in the outer parts are hydrogen, about 5.9 per cent are helium, and a mixture of all the other elements make up the remaining one tenth of one per cent. The overall composition of the interior is not very different.

The sun is an average star, since stars much hotter and much cooler, and intrinsically much brighter and much fainter, exist. Radiation from the pho-

The sun, observed in white light on November 9, 1979, near solar maximum.

tosphere peaks in the middle of the visible spectrum; after all, our eyes evolved over time to be sensitive to that region of the spectrum because the greatest amount of the solar radiation occurred there. If we lived on a planet orbiting an object that emitted mostly x-rays, we, like Superman, might have x-ray vision.

The disk of the sun takes up about one-half a degree across the sky; we say that it *subtends* one-half degree. This is large enough for us to see detail on the solar surface, and we shall describe that detail in subsequent sections.

Beneath the photosphere is the solar *interior*. All the solar energy is generated there at the solar *core*, which is about 10 per cent of the solar diameter at this stage of the sun's life. The temperature there is about 15,000,000 K. In Chapter 8 we will discuss how energy is generated in the sun and other stars.

The photosphere is the lowest level of the *solar atmosphere* (Fig. 7–1). The parts of the atmosphere above the photosphere are very tenuous, and contribute only a small fraction to the total mass of the sun. These upper layers are very much fainter than the photosphere, and cannot be seen with the naked eye except during a solar eclipse, when the moon blocks the photospheric radiation from reaching our eyes directly. Now we can also study these upper layers with special instruments on the ground and in orbit around the earth.

Just above the photosphere is a jagged, spiky layer about 10,000 km thick, only about 1.5 per cent of the solar radius. This layer glows colorfully pinkish when seen at an eclipse, and is thus called the *chromosphere* (from the Greek *chromos*, color). Above the chromosphere, a pearly-white halo called the *corona* (from the Latin, crown) extends tens of millions of kilometers into space. The corona is continually expanding into interplanetary space and in this form is called the *solar wind*. We shall discuss all these phenomena in succeeding sections.

Figure 7–1 The parts of the solar atmosphere and interior. The solar surface is depicted as it appears through unfiltered light (called *white light*), and through filters that pass only light of certain elements in certain temperature stages. These specific wavelengths, counter-clockwise from the top, are the Hα line of hydrogen, which appears in the red, the K line of ionized calcium, which appears in the part of the ultraviolet that passes through the earth's atmosphere and can be seen with the naked eye, and the line of ionized helium at 304 Å in the extreme ultraviolet, which can be observed only from rockets and satellites.

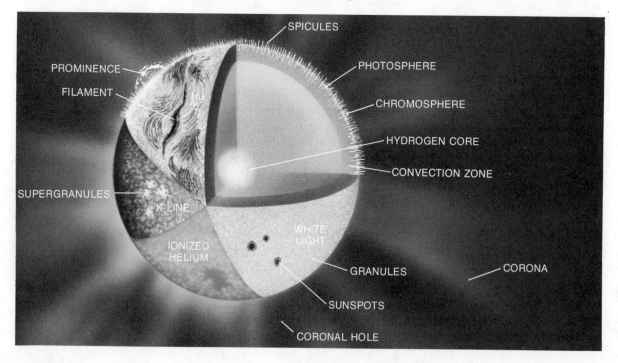

7.2 THE PHOTOSPHERE

The sun is the only star close enough to allow us to study its surface in detail. The phenomena we discover on the sun presumably also exist on other stars.

7.2a High-Resolution Observations of the Photosphere

One major limitation in observing the sun is the turbulence in the earth's atmosphere. We have seen how turbulence is a limiting factor in stellar observing, as it causes the twinkling of the stars. The problem is even more serious for studies of the sun, since the sun is up in the daytime when the atmosphere is heated by the solar radiation, and so is more turbulent than it is at night.

Only at the very best observing sites, specially chosen for their steady solar observing characteristics, can one see detail on the sun subtending an angle as small as 1 second of arc. This corresponds to about 700 km on the solar surface, the distance from Boston to Washington, D.C. Occasionally, objects ½ arc second across can be seen. Since atmospheric turbulence causes bad "seeing" that limits our ability to observe small-scale detail (see Section 2.9), it does little good to build solar telescopes larger than about 50 cm (20 inches) in order to increase resolution, even though larger telescopes are inherently capable of resolving finer detail.

When one studies the solar surface with 1 arc second resolution, one sees a salt-and-pepper texture called *granulation* (Fig. 7–2). The effect is similar to that shown by boiling liquids on earth, which are undergoing what is called *convection*. Convection, the transport of energy by moving masses of matter, is one of the ways in which energy can be transported. Conduction and radiation are the other major methods.

Granulation on the sun seems to be an effect of convection. Each granule is only about 1000 km across, and represents a volume of gas that is rising from and falling to a shell of convection, called the *convection zone*, located below the photosphere. The granules are convectively carrying energy from the hot solar interior to the base of the photosphere.

It was discovered in the early 1960's that areas of the upper photosphere are oscillating up and down with a 5-minute period. This result, which was totally unexpected, is now thought to be caused by waves of energy coming from the convection zone.

7.2b The Photospheric Spectrum

The spectrum of the solar photosphere, as are spectra of all similar stars, is a continuous spectrum with absorption lines (Fig. 7–3). Tens of thousands of these absorption lines, which are also called Fraunhofer lines, have been photographed and catalogued. They come from most of the chemical elements, although some of the elements have many lines in their spectra and some have very few. The majority of the lines in the solar spectrum come from iron and such other elements as magnesium, aluminum, calcium, titanium, chromium, nickel, and sodium. The hydrogen Balmer lines are strong but few in number. Helium is not excited at the low temperatures of the photosphere, and so its spectral lines do not appear.

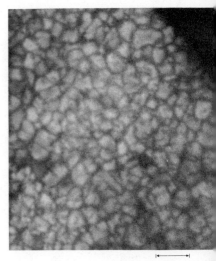

10 ARC SEC

Figure 7–2 The picture was taken during the partial phase of the 1973 solar eclipse so that the sharp edge of the moon, visible at upper right, shows the blurring caused by the earth's atmosphere. The smallest features visible are about 1 arc sec across. The picture was taken with the 40-cm vacuum telescope located in the Canary Islands of the West German Kiepenheuer Institute.

The temperature of the gas in the photosphere is about 5800 K. The sun has spectral type G2.

Granulation can be seen in white light—all the radiation from the visible part of the spectrum taken together without being filtered— as distinguished from, say, red light or light of the Hα line of hydrogen.

The sun also seems to vibrate— ring like a gong—with a 2 hr 40 m period. This has been confirmed by observations made at the earth's South Pole when the sun was up around the clock. The vibrations may teach us about the solar interior, just as seismic waves on earth tell us about earth's interior.

The set of hydrogen's spectral lines in the visible are the Balmer series.

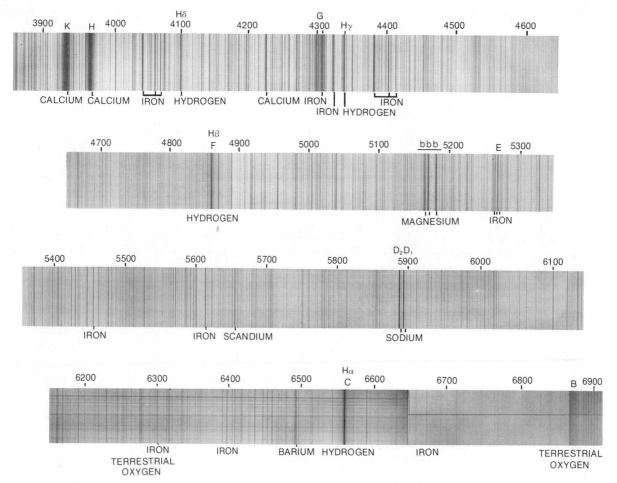

Figure 7-3 The visible part of the solar spectrum.

Fraunhofer's original spectrum is shown in Color Plate 5.

*We are discussing here only the part of the solar spectrum in the visible. The formation of the Fraunhofer lines was discussed in Section 3.3. Note that to absor**b** is spelled with a "b," but ab-sor**p**tion is spelled with a "p."*

Fraunhofer, in 1814, labeled the strongest of the absorption lines in the solar spectrum with letters from A through H. His C line, in the red, is now known to be the first line in the Balmer series of hydrogen, and is called Hα (H alpha). We still use Fraunhofer's notation for some of the strong lines. The D lines, a pair of lines close together in the yellow part of the spectrum, are caused by neutral sodium (Na I). The H line and the K line, both in the part of the violet spectrum that is barely visible to the eye, are caused by singly ionized calcium (Ca II). None of the elements other than helium makes up as much as one tenth of one per cent of the number of hydrogen atoms (Table 7–1). Nevertheless, the D, H, and K lines are among the strongest in the visible spectrum. Of the elements other than hydrogen and helium, sodium and calcium are among the more abundant. Also, most of the absorbing power of these elements at the temperatures of the photosphere and chromosphere is concentrated in just a few lines. Furthermore, these lines occur in the visible part of the spectrum.

The places where the continuous spectrum (called the *continuum*) and most of the absorption lines are formed are mixed together throughout the photospheric layers. The energy emitted in them is not formed in these layers, but has rather been transported upward from the solar interior.

Box 7.1   Solar Observatories

Figure 7–4

Figure 7–5

Figure 7–6

Figure 7–7

Figure 7–8

Figure 7–4   The Sacramento Peak Observatory in Sunspot, New Mexico, has a solar tower. Light passes into a window at the top, and the entire light path inside is a vacuum in order to eliminate the effects of air currents.

Figure 7–5   A large solar telescope is located at the Kitt Peak National Observatory, alongside the nighttime telescopes. A mirror at the top the size of a person (note the figure on top) reflects light down a long sloping tube.

Figure 7–6   The Big Bear Solar Observatory, part of the Hale Observatories, is located on an artificial island in the middle of a lake in southern California. Because air flows smoothly over bodies of water, this leads to exceptionally good "seeing."

Figure 7–7   The Mount Wilson Observatory has a solar tower 50 meters tall, which is often used to make magnetic maps of the sun.

Figure 7–8   The Haleakala Observatory, on Maui, Hawaii, is at the top of a volcano, high enough to be above the atmospheric inversion layer that keeps turbulence and weather below. This is their coronagraph, a telescope that allows the corona to be studied without an eclipse.

TABLE 7–1 SOLAR ABUNDANCES OF THE MOST COMMON ELEMENTS

	Symbol	Atomic Number
For each 1,000,000 atoms of hydrogen, there are	H	1
63,000 atoms of helium	He	2
690 atoms of oxygen	O	8
420 atoms of carbon	C	6
87 atoms of nitrogen	N	7
45 atoms of silicon	Si	14
40 atoms of magnesium	Mg	12
37 atoms of neon	Ne	10
32 atoms of iron	Fe	26
16 atoms of sulfur	S	16
3 atoms of aluminum	Al	13
2 atoms of calcium	Ca	20
2 atoms of sodium	Na	11
2 atoms of nickel	Ni	28
1 atom of argon	Ar	18

7.3 THE CHROMOSPHERE

7.3a The Appearance of the Chromosphere

Under high resolution, we see that the chromosphere is not a spherical shell around the sun but rather is composed of small spikes that rise and fall. They have been compared in appearance to blades of grass or burning prairies. These small spikes are called *spicules* (Fig. 7–9).

Spicules are more-or-less cylinders of about 1 arc second in diameter and perhaps ten times that in height, which corresponds to about 700 km across and 7000 km tall. They seem to have lifetimes of about 5 to 15 minutes, and there may be approximately half a million of them on the surface of the sun at any given moment.

Studies of velocities on the sun showed the existence of large organized convection cells of matter on the surface of the sun called *supergranulation*. Supergranulation cells look somewhat like polygons of approximately 30,000 km diameter. Supergranulation is an entirely different phenomenon from granulation. Each supergranulation cell may contain hundreds of individual granules. Supergranules can be seen in Figure 7–10.

Matter seems to well up in the middle of a supergranule and then slowly move horizontally across the solar surface to the supergranule boundaries. The matter then sinks back down at the boundaries. This slow circulation of matter seems to be a basic process of the lower part of the solar atmosphere. The network of supergranulation boundaries, called the *chromospheric network*, is visible in the radiation of hydrogen alpha or the H and K lines of calcium.

Figure 7–9 Spicules at the solar limb, observed through a filter passing a narrow band of wavelengths centered 1 Å redward of Hα.

(Chromospheric matter appears to be at a temperature of approximately 15,000 K, somewhat higher than the temperature of the photosphere.) New ultraviolet spectra of distant stars recorded by the International Ultraviolet Explorer have shown unmistakable signs of chromospheres in stars of spectral types like the sun. Thus by studying the solar chromosphere we are also learning what the chromospheres of other stars are like.

Figure 7–10 Supergranulation is best visible in an image like this one, in which the velocity field of the sun is shown. Dark areas are receding and bright areas are approaching us. The super-granules are each less than 1 mm across in this reproduction. Because of the flow of matter from the center to the edge of each supergranule, the Doppler shift makes one side appear dark and the other bright on this velocity image.

Box 7.2 The Heating of the Solar Chromosphere and Corona

The temperature rises in the region above the photosphere. This must happen because additional energy is being deposited there; after all, the temperature does not rise without reason. Waves originate in the convection zone and turn into *shock waves*—waves that compress matter in the same way that waves in a sonic boom on earth compress the air. The shock waves pass through the photosphere and dissipate their energy in the chromosphere and, even higher, in the corona. The granules and spicules, and the 5-minute oscillation, may be manifestations of these shock waves. But nobody knows exactly how the waves transfer their energy to the chromospheric and coronal gas. Several models have been proposed, but there is no agreement. This is an important field of current research.

The answers have important implications not only for astronomy but also for energy research. The gas in the sun is a mixture of ions and electrons, called a *plasma,* in a magnetic field. In developing controlled nuclear fusion on earth as a source of energy, we must learn how to contain plasmas in a magnetic field in the laboratory. For the moment, we can best study plasmas and their properties in the sun and the stars.

7.3b *The Chromospheric Spectrum at the Limb*

During the few seconds at an eclipse of the sun that the chromosphere is visible, its spectrum can be taken. This type of observation has been performed ever since the first spectroscopes were taken to eclipses in 1868.

Figure 7–11 The flash spectrum 0.6 sec after totality at the African solar eclipse of 1973. We see the chromospheric spectrum.

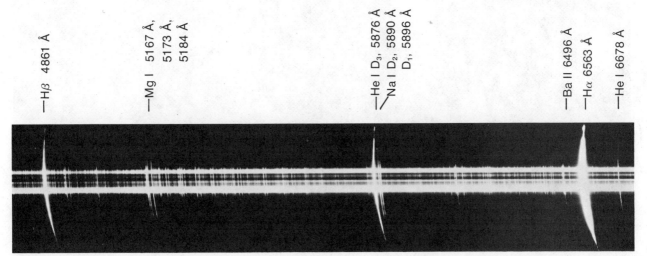

We have been discussing the visible chromospheric spectrum. In the far ultraviolet, as studied from satellites, the chromospheric and coronal lines are emission lines even when viewed at the center of the disk (Color Plates 14 and 15).

Since the chromosphere then appears as hot gas silhouetted against dark sky, the chromospheric spectrum consists primarily of emission lines. The chromospheric emission lines appear to flash into view at the beginning and at the end of totality, so the spectrum of the chromosphere is known as the *flash spectrum* (Fig. 7–11 and Color Plate 13).

7.4 THE CORONA

7.4a The Structure of the Corona

During total solar eclipses, when first the photosphere and then the chromosphere are completely hidden from view, a faint white halo around the sun becomes visible. This *corona* (Fig. 7–12) is the outermost part of the solar atmosphere, and is the link between the sun and interplanetary and interstellar space. The corona is at a temperature of about 2,000,000 K.

Even though the temperature of the corona is so high, the actual amount of energy in the solar corona is not large. The temperature quoted is actually a measure of how fast individual particles (electrons, in particular) are moving. There aren't very many coronal particles, even though each particle has a high velocity. The corona has less than one billionth the density of the earth's atmosphere, and would be considered to be a very good vacuum in

Figure 7–12 The total solar eclipse of February 16, 1980, photographed from Palem, India. The photograph was taken in red light through a filter that is denser at its center than at its edges so as to reduce the bright inner corona. This allows us to see the much fainter streamers of the outer corona in the same photograph, a factor of 10,000 in brightness. Bright prominences are visible on the limb.

a laboratory on earth. For this reason, the corona serves as a unique and valuable celestial laboratory in which we may study gaseous plasmas in a near vacuum.

Photographs of the corona show that it is very irregular in form. Beautiful long *streamers* extend away from the sun in the equatorial regions. At the poles, delicate thin *plumes* are suspended above the surface. The shape of the corona varies continuously and is thus different at each successive eclipse. The structure of the corona is maintained by the magnetic field of the sun.

*7.4b   The Coronal Spectrum

The visible region of the coronal spectrum, when observed at eclipses, shows a continuum, and also both absorption lines and emission lines. The emission lines do not correspond to any normal spectral lines known in laboratories on earth or on other stars, and for many years their identification was one of the major problems in solar astronomy. In the late 1930's, it was discovered that they arose in atoms that were multiply ionized. This was the major indication that the corona was very hot. In the photosphere we find atoms that are neutral, singly ionized, or doubly ionized (Ca I, Ca II, and Ca III, for example). In the corona, on the other hand, we find ions that are ionized approximately a dozen times (Fe XIV, for example, iron that has lost 13 of its normal quota of 26 electrons). The corona must be very hot indeed, millions of degrees, to have enough energy to strip that many electrons off atoms.

By multiply (mult-i-plē) we mean more than once, that is, twice, three times, and so on.

Figure 7–13 The graph illustrates why we need an eclipse to see the corona. The brightness of the sun's surface is shown at far left, with the edge (limb) of the sun indicated. Outside the limb, you can readily see that the sky is much brighter than the corona. Even the purest blue sky (from certain high mountains) allows viewing of only the innnermost corona.

The sky brightness during an eclipse drops drastically, as shown by the screened horizontal band at the bottom, and allows the corona to be seen.

The continuous spectrum of the corona has two parts. The K-corona (from the scattering of photospheric light by electrons in the corona) falls off in intensity more rapidly than does the F-corona (from the scattering of photospheric light by interplanetary dust). Thus the inner part of the corona we see is primarily light bounced off electrons, and the outer part of the corona we see is primarily light bounced off the dust.

The E-corona is the set of emission lines. Although the total energy of all the emission lines is less than the energy received from the continuous spectrum of the corona, a given emission line is stronger than the K- and F-coronas at its particular wavelength.

The dark side of the moon is illuminated only by earthshine, sunlight reflected off the earth to the moon.

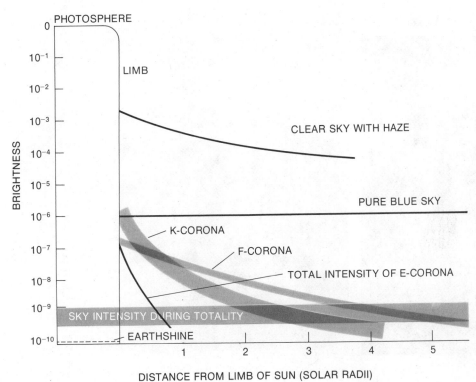

DISTANCE FROM LIMB OF SUN (SOLAR RADII)

The continuum is caused partly by electrons located near the sun, as they scatter (bounce) photospheric light to us, and partly by a similar scattering by dust particles near the orbit of the planet Mercury. The absorption lines disappear from the part that is scattered by the electrons because Doppler shifts to and fro blur the lines so much. The absorption lines are retained in the part of the photospheric radiation that is scattered by the dust (Fig. 7–13).

*7.4c Other Ground-Based Observations of the Corona

(The corona is normally too faint to be seen except at an eclipse of the sun because it is fainter than the everyday blue sky. But at certain locations on mountain peaks on the surface of the earth, the sky is especially clear and dust-free, and the innermost part of the corona can be seen (Fig. 7–14). Special telescopes called *coronagraphs* have been built to study the corona from such sites. Coronagraphs are built with special attention to detail, since their object is not to gather a lot of light but to prevent the strong solar radiation from being scattered about within the telescope.)

Since so many of the problems in observing the faint solar corona are caused by the earth's atmosphere, it is obvious that we want to observe the corona from above the atmosphere. If we stood on the moon, which has no air, we would see the corona rising each lunar morning a bit ahead of the sun.

But it is much less expensive to observe the corona from a satellite in orbit around the earth than it is to do so from the moon. The unmanned seventh Orbiting Solar Observatory, OSO-7 (1971–1974), the manned Skylab missions (1973–74), and the unmanned Solar Maximum Mission, launched in 1980, used coronagraphs to photograph the corona hour by hour

Figure 7–14 From a few mountain sites, the innermost corona can be photographed without need for an eclipse. The coronagraph at the Haleakala Observatory of the University of Hawaii on Maui took these observations on March 8, 1970, the day after a total eclipse, with this coronagraph. Five exposures were superimposed to improve the image.

Figure 7–15 This example of solar ejection was photographed from Skylab. Here we see two pictures superimposed. The eruption at the extreme right, photographed with the coronagraph, is the response of the corona to a lower-level eruption, photographed in the ultraviolet and superimposed on the central dark region occulted by the coronagraph.

Figure 7–14

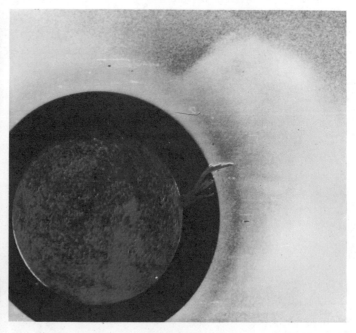

Figure 7–15

in visible light (Color Plates 16 to 19). These satellites could study the corona to much greater distances from the solar surface than can be studied with coronagraphs on earth. Among the major conclusions of the research is that the corona is much more dynamic than we had thought. For example, many blobs of matter were seen to be ejected from the corona into interplanetary space (Fig. 7–15), often in connection with solar flare activity (which will be discussed in Section 7.7).

7.5 SPACE OBSERVATIONS OF THE CHROMOSPHERE AND CORONA

The gas in the corona is so hot that it emits mainly x-rays, photons of high energy. The photosphere, on the other hand, is too cool to emit x-rays. As a result, when photographs of the sun are taken in the x-ray region of the spectrum, they show the corona and its structure (Fig. 7–16).

X-ray astronomy, which is now able to study distant stars, galaxies, and quasars, began with studies of the sun. Herbert Friedman and Richard Tousey of the Naval Research Laboratories were pioneers who, starting in the 1940's, launched rockets with x-ray devices for observing the sun. The techniques of making high-quality x-ray images were developed in the 1960's and 1970's by Riccardo Giacconi, Giuseppe Vaiana, and others at the American Science & Engineering Company. The technique reached fruition with the x-ray imaging telescopes sent aloft in rockets in the early 1970's and shortly thereafter with an orbiting x-ray telescope in Skylab in 1973 and 1974. Skylab was a manned mission that NASA sent aloft to make use of leftover equipment from the Apollo moon-landing program. A battery of solar telescopes was included as part of Skylab. Three successive crews of astronauts visited Skylab, for periods of 1 to 3 months.

Figure 7–16 shows, most obviously, that the corona is very structured. This structure varies with the 11-year cycle of solar activity, which we shall discuss in Section 7.7. The brightest regions visible in x-rays are part of *active regions* that can also be seen in white light. (In white light, these active regions include sunspots.) The x-ray images show a higher layer of the solar atmosphere than the white-light images show.

Detailed examination of the x-ray images shows that most, if not all, the radiation appears in the form of loops of gas joining points separated from each other on the solar surface. The latest thinking is that the corona is composed entirely of these loops. One important line of thought that follows from this new idea is that we must understand the physics of coronal loops in order to understand how the corona is heated. It is not sufficient to think in terms of a uniform corona, since the corona is obviously so non-uniform.

The x-ray image also shows a very dark area at the sun's north pole and extending downward across the center of the solar disk. (As the sun rotates from side to side, obviously, the part extending across the center of the disk will not usually be facing us.) These dark locations are *coronal holes* (Fig. 7–17), regions of the corona that are particularly cool and quiet. The density of gas in those areas is lower than the density in adjacent areas.

There is usually a coronal hole at one or both of the solar poles. Less often, we find additional coronal holes at lower solar latitudes. The regions of the coronal holes seem very different from other parts of the sun. Some

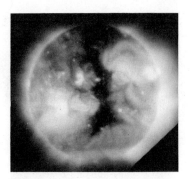

Figure 7–16 An x-ray photograph of the sun taken from Skylab on June 1, 1973. The filter passed wavelengths 2–32 Å and 44–54 Å. The dark region across the center is a coronal hole.

The High Energy Astronomy Observatories detect x-rays from the distant stars. These x-rays no doubt originate in the stars' coronas. The International Ultraviolet Explorer is studying spectral lines in the chromospheres of the stars. Thus within the last few years, we have become able to study the outer atmospheres of many more stars than the sun, and the sun has gained new importance as a primary example that we can study in detail.

A   He II 304 Å

B   Ne VII 465 Å

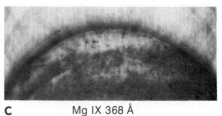

C   Mg IX 368 Å

Figure 7–17 This sequence, taken on August 14, 1973, shows the coronal hole, a phenomenon that usually appears over one or both poles of the sun and often at other latitudes as well. Ionized helium *(A)* shows chromospheric temperatures, Ne VII *(B)* shows the transition zone, and Mg IX *(C)* shows the corona. A macrospicule is visible as the jet on top in the He II photograph.

features are found only there, such as jets of gas resembling chromospheric spicules but on a larger scale, called macrospicules (Color Plate 14 and Fig. 7–17).

Skylab also carried a coronagraph to allow observation of the corona in visible light. Because of limitations of space technology, not only the photosphere but also the inner part of the corona had to be blocked in order to limit the amount of light scattering around in the telescope, but the series of regular hour by hour observations of the outer corona are unsurpassed. The Skylab observations revealed that the corona is much more active than had been thought, with large eruptions occurring much more often than had been expected. The Solar Maximum Mission is now continuing this work.

Solar physicists used to think that the corona was heated to millions of degrees by shock waves that carried energy upward from underneath the photosphere. Though this theory had seemed very well established and had been worked on continuously since 1946, it was discarded in 1980 on the basis of new evidence. Some of the important new evidence was a set of observations from the OSO-8 spacecraft. The data showed directly that the shock waves carried too little energy to heat the corona by a factor of 1000.

It was also discovered from the Einstein Observatory x-ray spacecraft that even type O and B stars give off enough x-rays that they must have hot coronas, even though their internal structure is such that they cannot generate shock waves. The x-ray observations show that different mechanisms are necessary to heat the coronas of other stars. Some of these different mechanisms probably heat the solar corona, too. Though the old ideas about coronal heating have been discarded, there is no agreement about what the real mechanism is for heating the corona. Possibly energy is generated by a twisting and untwisting of the magnetic field in active regions. This would also explain the Skylab x-ray pictures, which show that the corona is hotter above active regions.

*7.6 SOLAR ECLIPSES

Since the solar chromosphere and corona are visible to the eye only at the time of a total solar eclipse, eclipses have played a major role in solar physics. Let us discuss them in detail.

The sun (more precisely, the photosphere) is very large and very far from us. The moon is 400 times smaller in diameter than the sun, but it is also 400 times closer to the earth. Because of this, the sun and the moon subtend almost exactly the same angle in the sky—about ½°—which is a happy coincidence (Fig. 7–18).

The moon's position in the sky, at certain points in its orbit around the

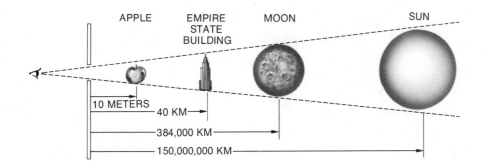

APPLE EMPIRE STATE BUILDING MOON SUN

10 METERS

40 KM

384,000 KM

150,000,000 KM

Figure 7–18 An apple, the Empire State Building, the moon and the sun are very different from each other in size, but here they subtend the same angle because they are different distances from us.

earth, comes close to the position of the sun. This happens approximately once a month at the time of the new moon. Since the lunar orbit is inclined with respect to the earth's orbit, the moon usually passes above or below the line joining the earth and the sun. But occasionally the moon passes close enough to the earth-sun line that the moon's shadow falls upon the surface of the earth.

At a total solar eclipse, the lunar shadow barely reaches the earth's surface (Fig. 7–19). As the moon moves through space on its orbit, and as the earth rotates, this lunar shadow sweeps across the earth's surface in a band up to 300 km wide. Only observers stationed within this narrow band can see the total eclipse.

From anywhere outside this band, one sees only a partial eclipse. Sometimes the moon, sun, and earth are not precisely aligned and the darkest part of the shadow—called the *umbra*—never hits the earth. We are in the intermediate part of the shadow, which is called the *penumbra*. Only a partial eclipse is visible on earth under these circumstances. As long as the slightest bit of photosphere is visible, even as little as 1 per cent, one cannot see the important eclipse phenomena—the chromosphere and corona. Thus partial eclipses are of little value for most serious scientific purposes. After all, the photosphere is 1,000,000 times brighter than the corona; if one per cent is showing then we still have 10,000 times more light from the photosphere than from the corona, which is enough to ruin our opportunity to see the corona.

If you are fortunate enough to be standing in the zone of a total eclipse, you will find excitement all around you as the approaching eclipse is anticipated. Even uneducated people have formed an impression of the cause of an eclipse: it immediately follows the arrival of a horde of astronomers. If the astronomers come, can the eclipse be far behind?

About an hour or an hour and a half before totality begins, the partial phase of the eclipse starts. Nothing is visible to the naked eye immediately, but if you were to look at the sun through a special filter you would see that the moon was encroaching on the sun (Fig. 7–20). At this stage of the eclipse, it is necessary to look through a special filter, as in Figure 7–21 (or to project the image of the sun with a telescope or a pinhole camera onto a surface such as a piece of cardboard), in order to protect your eyes, for the photosphere is visible. Its direct image on your retina for an extended time could cause burning and blindness, though even in the few cases where eye damage is done, the damage is usually minor.

The eclipse progresses gradually, and by 15 minutes before totality the sky grows strangely dark, as though a storm were gathering. During the

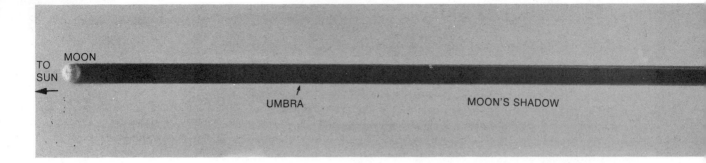

MOON

UMBRA

MOON'S SHADOW

Figure 7–19 A solar eclipse with the earth, moon, and the distance between them shown to actual scale.

Figure 7–20 The partially eclipsed sun observed from Africa in 1973.

minute or two before totality begins, bands of shadow race across the ground. These *shadow bands* are caused in the earth's atmosphere, and thus tell us about the terrestrial rather than about the solar atmosphere.

You still need the special filter to watch the final seconds of the partial phase. Only a thin sliver of photosphere is visible. Then the sliver breaks up into a chain of beads along the rim of the moon. These *Baily's beads* are the photosphere shining through valleys that happen to be located at the edge of the moon and oriented so as to make the lunar rim irregular.

The last Baily's bead seems especially bright to the eye. It glistens and dazzles for a few seconds, sparkling like a diamond. This is called the *diamond ring effect* (Color Plate 11).

With the passage of the diamond ring effect, the total phase of the eclipse has begun. For a few seconds, a reddish glow is visible in a narrow band around the leading edge of the moon. This is the chromosphere and perhaps one or more prominences. Scientists studying the chromosphere have to work fast and have their photographic devices operating at a rapid pace in order to make their observations in these few seconds.

Then the corona is visible in all its glory (Figs. 7–21B and 7–25 and Color Plate 12). One can see the equatorial streamers and the polar plumes. At this stage, the photosphere is totally hidden so it is perfectly safe to stare at the corona with the naked eye and without filters. It has approximately the same brightness as the full moon, and is equally safe to look at. Unfortunately, many people are not adequately informed about this most important phase of the total eclipse, and miss the spectacle.

The total phase may last a few seconds, or it may last as long as a little over 7 minutes. Spectrographs are operated, photographs are taken through special filters, rockets photograph the spectrum from above the earth's atmosphere, and tons of equipment are brought into operation to study the corona during this brief time of totality.

It is as though chemists were told that they could study a certain chemical reaction for only 5 minutes a year. They might have to go to Africa to study the reaction; they could bring all their equipment and even bring colleagues with their own equipment, but like it or not the reaction would take place for only five minutes. Then they would have to wait a year or more for another chance.

At the end of the eclipse, the diamond ring appears on the other side of the sun, then Baily's beads, and then the final partial phases. All too soon, the eclipse is over.

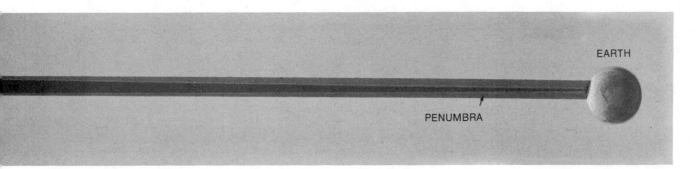

EARTH

PENUMBRA

7.6a Observing Eclipses

On the average, a solar eclipse occurs somewhere in the world every year and a half. The band of totality usually does not cross populated areas of the earth, and astronomers often have to travel great distances to carry out their observations.

The last total eclipse to cross the continental U.S. and Canada occurred in 1979. The band of totality crossed the northwestern and north central United States and central Canada. The rest of North America saw only a partial eclipse, and so missed all the most interesting eclipse phenomena

Figure 7–21A These children, who visited our eclipse site in Loiengalani, Kenya, just prior to the 1973 total solar eclipse, were practicing the use of special filters to protect their eyes when observing the partial phases.

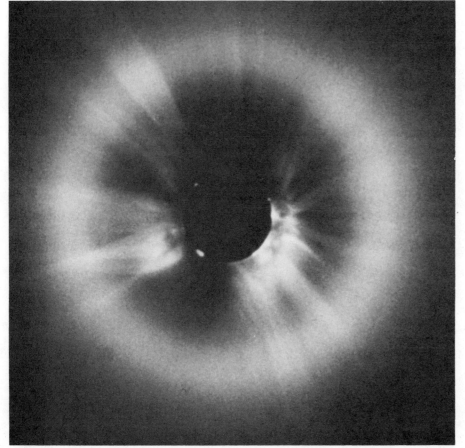

Figure 7–21B A coronal photograph from the February 26, 1979 eclipse. The sun was photographed from an airplane over North Dakota. A radially-graded filter was used. Much streamer detail can be seen, and the south polar region seems to be a coronal hole. (Courtesy of Dr. Charles Keller, William H. Regan, and Maxwell T. Sandford of the University of California Los Alamos Scientific Laboratory)

TABLE 7–2   TOTAL SOLAR ECLIPSES

Date	Maximum Duration	Location
1981 July 31	2:03	U.S.S.R.
1983 June 11	5:11	Indian Ocean, Indonesia, New Guinea
1984 November 22	1:59	New Guinea, South Pacific Ocean
1985 November 12	1:59	South Pacific Ocean
1988 March 18	3:46	Indonesia, Philippines
1990 July 22	2:33	Finland, U.S.S.R., U.S.A. (Alaska)
1991 July 11	6:54	U.S.A. (Hawaii), Central and South America

Figure 7–22   This map shows the paths that the moon's shadow will take during those total solar eclipses occurring between 1979 and 2017. Anyone standing along such a path on the dates indicated will see a total solar eclipse. Outside of the path a partial eclipse will be observed. Note that an annular eclipse of the sun will be visible in the southeastern United States in 1984.

(diamond ring, corona, etc.). The next total eclipse in the U.S. will be visible in Alaska in 1990, and another total eclipse will be visible from Hawaii in 1991. The next total eclipse in the continental United States won't be until 2017. Canada won't see another total eclipse until 2024. Table 7–2 and Figure 7–22 show future eclipses.

Sometimes the moon subtends a slightly smaller angle in the sky than the sun, because the moon is on the part of its orbit that is relatively far from the earth. When a well-aligned eclipse occurs in such a circumstance, the moon doesn't quite cover the sun. An annulus—a ring—of photospheric light remains visible, so we call this an *annular eclipse* (Fig. 7–23). The

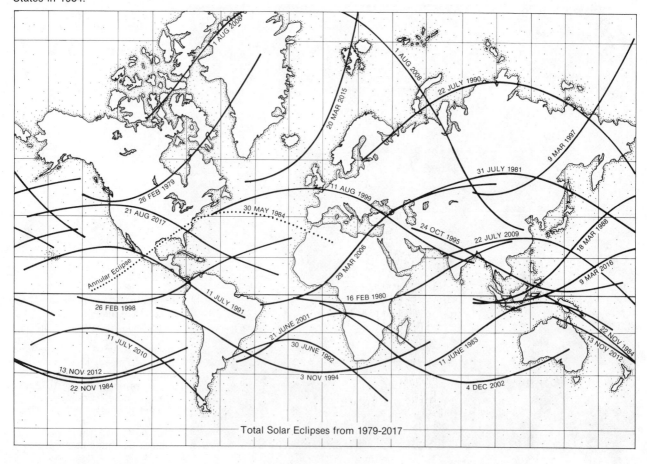

Total Solar Eclipses from 1979-2017

remaining light is so bright that we cannot see the corona. Such an annular eclipse will cross the southeastern United States on May 30, 1984. Most of the rest of the United States will see a partial eclipse.

In these days of orbiting satellites that can study ultraviolet spectral lines from the corona or carry coronagraphs above the earth's atmosphere, is it worth travelling to observe eclipses? There is much to be said for the benefits of eclipse observing. Eclipse observations are a relatively inexpensive way, compared to space research, of observing the chromosphere and corona to find out such quantities as the temperature, density, and magnetic field structure. Even realizing that some eclipse experiments may not work out because of the pressure of time or because of bad weather, eclipse observations are still very cost effective. And for some kinds of observations, those at the highest resolutions, space techniques have not yet matched ground-based eclipse capabilities. Space coronagraphs, even though they can study the outer part of the corona day to day, are not able to observe the inner and middle corona that we observe at eclipses.

Figure 7–23 A member of the author's expedition using a dense filter to observe the annular eclipse of December 24, 1974 in the Andes Mountains near Bogotá, Colombia. In an annular eclipse, the moon appears smaller than the sun and a ring (annulus) of photosphere remains visible.

7.6b An Eclipse Expedition

An eclipse expedition is unlike most other kinds of observing experiences that astronomers have. A lengthy period of preparation is followed by a tense and pressure-packed but fascinating few weeks at the expedition site, all leading up to the inexorable deadline of the eclipse itself. The few minutes of totality are glorious, and pass all too quickly. Then, if the experiment is successful, it may take many months or years to study the data and draw conclusions.

The most recent major expedition took place to observe the total solar eclipse of February 16, 1980. The band of totality crossed Africa and India. We began planning a year before, when we sat down to think about the existing state of solar astronomy and considered which experiment would be most worthwhile. Recent theoretical studies of the sun and new technology led us to choose our experiment. Other groups of scientists went through similar processes. There is still so much to learn about the sun that everyone quite independently chooses different topics for investigation.

When the time for the eclipse came, a joint expedition of American scientists, sponsored by the National Science Foundation, went to a site in India, where the weather forecasts were especially favorable. The Sacramento Peak Observatory made images of part of the corona in one of its emission lines in order to study motions of gas in the corona. The Kitt Peak National Observatory took coronal spectra, also to study motions in the corona. The University of Hawaii studied the spectra of the corona and of prominences. The High Altitude Observatory and Southwestern College studied the intensity and polarization of the corona. The University of Minnesota observed the corona in the infrared to study interplanetary dust, an investigation that included a search for a ring around the sun analogous to the rings around Jupiter, Saturn, and Uranus. The Johns Hopkins University and Gettysburg College studied shadow bands.

My own group from Williams observed the coronal spectrum in the ultraviolet and in the infrared to study the density and temperature of the corona. A second Williams experiment, with additional support from the National Geographic Society, studied the possibility that parts of the corona

A

3

Figure 7–24 *(A)* One of the Indian scientists, Dr. Arvind Bhatnagar of the Udaipur Solar Observatory, setting up his equipment. The dome of the 1.2-m telescope is visible in the background. The American teams were set up on a roof next to the dome. *(B)* The Williams group setting up part of its apparatus, with visiting school children in the background. The 30-cm flat mirror being rotated is used to reflect the sun into the apparatus; it tracks the sun in the sky so that the rest of the apparatus doesn't have to move.

are oscillating with a very short—1 second—period, which is a test of one group of theories of how the corona is heated to millions of degrees. Scientists from India had experiments at nearby sites on the observatory grounds.

The main site was about 60 kilometers south of the city of Hyderabad in southern India, a drive of an hour and three quarters because of the mixture of bicycle rickshaws, three-wheeler taxis, cars, buses, and trucks sharing the road with pedestrians and ox-carts, a common mode of transportation in that part of the world (Fig. 7–24). The University in Hyderabad had an observatory at this site, though their 1.2-m telescope, a regular nighttime telescope (and thus not suitable for looking at the sun), was not used during the eclipse. We did, however, take advantage of the sturdy buildings there both for setting up equipment and for lodging. Two of the groups had to be on the center-line of totality, and they set up in another small town somewhat farther from Hyderabad. The rest of us sacrificed a few seconds of totality for the better facilities of the observatory buildings.

Fourteen tons of equipment, including 1100 kg (2500 pounds) from my own group, were shipped ahead to India, where they cleared customs and were transported to the site by the time we got there three weeks in advance of the day of totality. Local workers constructed sturdy piers of brick and cement to our specifications. Pouring concrete is always an important part of eclipse expeditions, and it helps on an eclipse expedition to be not only an astronomer but also a combination carpenter, mason, machinist, electronics engineer, optical physicist, and more.

Working under the 32°C (90°F) sun was physically draining, and the heat was not good for our electronic measuring devices or computers (Fig. 7–25). We arranged (we thought) for a 3 foot by 4 foot by 5 foot canvas covering to shield the electronics, but wound up with a 3 meter by 4 meter by 5 meter tent that was so large that it prevented us from seeing the sun, an experience that points to the advantage of standardized systems of measurement. We had to take it down.

Most of the scientists and students stayed in the rooms at the Observatory, each furnished with cots, mosquito netting, and a table. We were able to travel to a hotel in Hyderabad, which also catered the food under a tent down at the observing site.

One by one, the experiments were made to work. As is usual, we needed last-minute on-the-spot drilling of holes and other construction. A generator provided American-style power (110 volts 60 cycles) for everybody. Tracking motors on telescopes, motor-driven camera shutters and film advances, electronic instruments, computers, and other equipment all depended on the power. The local power was not only of the wrong type but also very unreliable, and we couldn't take a chance on a power failure during the eclipse.

The weather was, of course, a major topic of discussion because everything depended on having clear skies for the eclipse. It was clear every morning, though clouds tended to form in the afternoon, and the eclipse was to take place in the late afternoon. On the days previous to the eclipse, sometimes the sun was covered with clouds at eclipse time and sometimes not. We hoped that even if clouds started to form on eclipse day, the cooling of the atmosphere caused by the partial phases of the eclipse itself would impede further cloud formation or even break up clouds that had already formed. Two days before the eclipse, the day was completely cloudy for the first time, and nervousness grew. The day before the eclipse was pretty clear, though there were some small clouds in the sky in the afternoon.

Eclipse morning was completely clear, and we all made our final preparations. A sudden power failure made us abruptly anxious, but it turned out that somebody on another experiment had plugged in a new machine to erase computer tapes (never, never plug in anything on eclipse day that you haven't plugged in on the same electric line before), and after we found the proper circuit breaker, all was ok again. In the early afternoon, as we went through final rehearsals of what we would do in the 2 minutes and 10 seconds of totality, a few small clouds appeared in the sky.

The computer programs and data used to predict the times and locations of eclipses are now so accurate that the eclipse can be predicted to better than a second. It is always impressive to me that just at the proper time for the first contact of the moon and the sun, one can really see (using special filters to reduce the solar intensity by a factor of 100,000) the dark disk of the moon blocking part of the surface.

A few minutes later, about an hour before totality, we weren't quite so pleased, as a big dark cloud covered the partially eclipsed sun and started growing. It was larger across than the angle my hand appeared to cover at the end of my outstretched arm. Fortunately, after fifteen agonizing minutes, the cloud moved away and then dispersed, perhaps because of the eclipse's cooling of the atmosphere. A clear period gave us hope, but then, starting a half hour before totality, it was cloudy about half the time. Only at 8 minutes before totality did the last cloud pass from in front of the sun. By this time we had even received by radio a report on how the corona had appeared during the eclipse observations made from an airplane off the coast of Kenya, where totality was already over.

As the sliver of sun diminished rapidly, shadow bands flickered over the landscape very prominently. Finally the countdown came over a loudspeaker: 10, 9, 8 . . . . The last crescent of sun, viewed through our filters and imaged through filters on our equipment onto a television screen, disappeared. (Figure 7–26 shows the progress of the eclipse phases.) At the

Figure 7–25 The Williams College expedition in mid-eclipse. The corona is visible at upper left. The sun was in the clear, although there were clouds around it in the sky. The sky at the horizon is bright because there we are seeing light from outside the cone of the moon's shadow. The corona is also seen on a television screen showing an image used for maintaining the accurate tracking of the telescopes. Other scientific equipment is barely visible in silhouette.

A

B

Figure 7–26 *(A)* The diamond ring effect, photographed by the Williams College—Hopkins Observatory expedition to the Japal-Rangapur Observatory in India. *(B)* Totality observed from India.

count of 1 the diamond ring dazzled us as we pulled all the filters away. Our computer was taking data at 30 times per second, and its tape drive could be heard faintly clicking away. We moved around the image of the sun on our instruments as we had planned in order to study different regions of the corona. Suddenly 70 seconds had passed and exactly one minute was left, though it seemed that the eclipse had just begun. We each stole glances at the corona, which was hovering beautifully in the sky with the extremely spiky structure typical of this maximum phase of the sunspot cycle. Only a faint prominence was visible, much less obvious an object than we had hoped for at solar maximum.

Then a countdown resumed—15, 14, 13 . . . . "What is this?" shouted one of my team, hoping that it was only mid-eclipse. But no, the eclipse was over already, and "lens caps on" was the next command.

We still had many minutes of additional data to take—we recorded special lamps of known spectra we had brought with us in order to provide a wavelength scale and observed the solar photosphere to compare with our coronal observations. But eventually we could pause and reflect on the glory of the eclipse we had just seen.

We took away with us from India computer and video tapes of data and of the eclipse itself. It will take quite some time to put the data into our computers at home and to study them. Those who observed on film took home the film to develop under carefully controlled conditions. Eventually, we will give papers at meetings of the American Astronomical Society and publish results in scientific journals such as the *Astrophysical Journal* and *Solar Physics*, so that our results from the eclipse of 1980 can become part of astronomers' understanding of the sun in particular and the stars in general.

7.7 SUNSPOTS AND OTHER SOLAR ACTIVITY

We have discussed the components of the sun as though it were a static object: we described the photosphere, the chromosphere, and the corona, and will describe the interiors of the sun and stars of similar mass in Chapter 8. But a host of other time-varying phenomena are superimposed on the basic structure of the sun. Many of them, notably the sunspots, vary with an 11-year cycle, which is called the *solar activity cycle.*

7.7a Sunspots

Sunspots (Fig. 7–27) are the most obvious manifestation of solar activity. They are areas of the sun that appear relatively dark when seen in white light. Sunspots appear dark because they are giving off less radiation than the photosphere that surrounds them. This implies that they are cooler areas of the solar surface, since cooler gas radiates less than hotter gas. Actually, if we could somehow remove a sunspot from the solar surface and put it off in space, it would appear bright against the dark sky; a large one would give off as much light as the full moon.

A sunspot includes a very dark central region, called the *umbra* from the Latin for "shadow" (pl: *umbrae*). The umbra is surrounded by a *penumbra* (pl: *penumbrae*), which is not as dark (just as during an eclipse the umbra of the shadow is the darkest part and the penumbra is less dark).

To understand sunspots we must understand magnetic fields. When iron filings are put near a simple bar magnet on earth, the filings show a pattern that is illustrated in Figure 7–28. The magnet is said to have a north pole and a south pole, and the magnetic field linking them is characterized by what we call *magnetic lines of force,* or *magnetic field lines* (after all, the iron filings are spread out in what look like lines). The earth (as well as some other planets) has a magnetic field that has many characteristics in common with that of a bar magnet. The structure seen in the solar corona, including polar plumes and equatorial streamers, results from matter being constrained by the solar magnetic field.

In this section we discuss the active sun.

Figure 7–27 *(A)* A sunspot, showing the dark *umbra* surrounded by the lighter *penumbra.* Granulation is visible in the surrounding photosphere. *(B)* A pair of sunspots within a common penumbra.

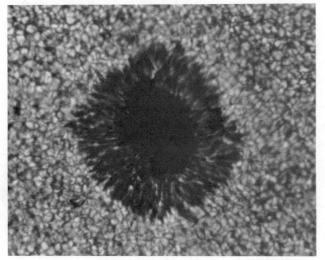

A

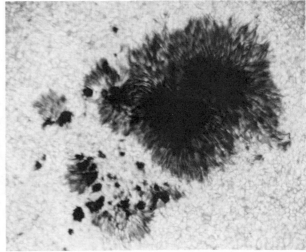

B

We can measure magnetic fields on the sun by using a spectroscopic method. In the presence of a magnetic field, certain spectral lines are split into a number of components and the amount of the splitting depends on the strength of the magnetic field. This is called the *Zeeman effect.*

Measurements of the solar magnetic field were first made by George Ellery Hale. He showed, in 1908, that the sunspots are regions of very high magnetic field strength on the sun, thousands of times more powerful than the earth's magnetic field or than the average solar magnetic field. Sunspots usually occur in pairs, and often these pairs are part of larger groups. In each pair, one sunspot will have a polarity typical of a north magnetic pole and the other will have a polarity typical of a south magnetic pole.

Magnetic fields are able to restrain matter—this is the property we are trying to exploit on earth to contain superheated matter sufficiently long to allow nuclear fusion for energy production to take place. The strongest magnetic fields in the sun occur in sunspots. The magnetic fields in sunspots restrain the motions of the matter there, and in particular keep convection from carrying energy to photospheric heights from lower, hotter levels. This results in sunspots being cooler and darker, though exactly why they remain so for weeks is not known. The parts of the corona above active regions are hotter and denser than the normal corona. Presumably the energy is guided upward by magnetic fields. These locations are prominent in radio or x-ray maps of the sun.

Sunspots were discovered in 1610, independently by Galileo in Italy, Fabricius and Christopher Scheiner in Germany, and Thomas Harriot in England. In about 1850, it was realized that the number of sunspots varies with an 11-year-cycle, as is shown in Figure 7–29. This is called the *sunspot cycle,* although we now realize that many related signs of solar activity vary with the same period.

Besides the specific magnetic fields in sunspots, the sun seems to have a weak overall magnetic field with a north magnetic pole and a south magnetic pole, which may entirely result from the sum of the weak magnetic fields from vanished sunspots. Every 11-year cycle, the north magnetic pole and south magnetic pole on the sun reverse polarity; what had been a north

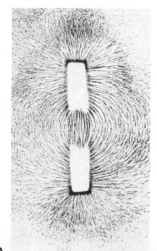

A

B

Figure 7–28 *(A)* Lines of force from a bar magnet are outlined by iron filings. One end of the magnet is called a north pole and the other is called a south pole. Similar poles ("like poles")—2 norths or 2 souths—repel each other and unlike poles (1 north and 1 south) attract each other. Lines of force go between opposite poles. *(B)* The magnetic field on November 9, 1980, the time that the white light photograph that opens this chapter was taken. One polarity appears as relatively dark and the opposite polarity as relatively light. Note how the regions of strong fields correspond to the positions of sunspots. Notice also how the two members of a pair of sunspots have polarity opposite to each other, and how which polarity comes first as the sun rotates is different above and below the equator.

Figure 7–29 The 11-year sunspot cycle is but one manifestation of the solar activity cycle.

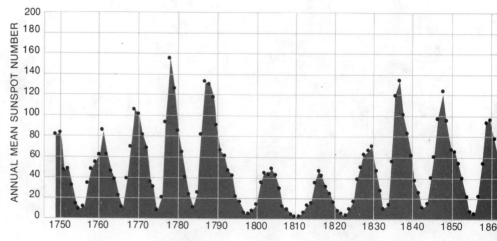

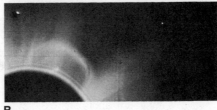

A **B**

Figure 7–30 *(A)* NASA's Solar Maximum Mission (SMM) is shown mounted on top of the Multi-Mission Spacecraft (MMS) prior to launching. The Solar Maximum Mission has already provided astronomers with new data on flares and other solar activity. Launched to study the sun during this current period of maximum activity, the satellite is scheduled to be retrieved from space with the Space Shuttle sometime in the future. *(B)* The eruption of a large prominence led to this transient event in the corona, observed from the coronagraph aboard Solar Maximum Mission. 56 minutes elapsed between the two frames. (Courtesy of Lewis House, Ernest Hilder, William Wagner, and Constance Sawyer—HAO/NCAR/ NSF and NASA)

magnetic pole is then a south magnetic pole and vice versa. This occurs a year or two after the number of sunspots has reached its maximum. For a time during the changeover, the sun may even have two north magnetic poles or two south magnetic poles! But the sun is not a simple bar magnet, so this strange-sounding occurrence is not prohibited. Because of the changeover, it is 22 years before the sun returns to its original configuration, so the real period of the solar activity cycle is 22 years.

A dozen years of daily observations of the solar surface from the Mount Wilson Observatory have revealed that the sun's surface consists of wide bands whose rotation periods vary. Studies by Robert F. Howard and Barry LaBonte showed in 1979 that the faster bands rotate abut 20 kms^{-1} more rapidly than the slower bands. The bands drift toward the equator over a 22-year period. High magnetic fields appear at the boundaries between some of the fast and slow bands, and spots began to appear when these boundaries reached a region about halfway between the pole and the equator. Perhaps the magnetic fields become twisted and deformed at the boundaries.

The last maximum of the sunspot cycle—the time when there are the greatest number of sunspots—took place in late 1979. Shortly thereafter, NASA launched a satellite called the Solar Maximum Mission to study solar activity at that time (Fig. 7–30).

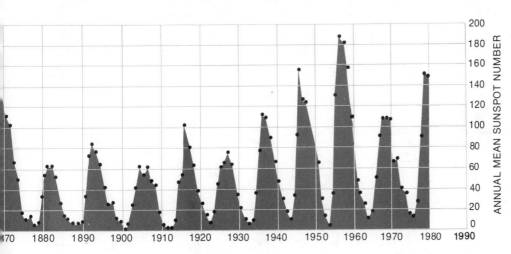

Figure 7–31 Maunder's butterfly diagram. When the latitude of each spot on the visible surface of the sun is graphed each month, the graph eventually resembles a butterfly, showing that the latitude of sunspots grows smaller as the sunspot cycle advances. One can also see that a new cycle begins at higher latitudes even before the old cycle has ended at lower latitudes.

°Box 7.3 The Butterfly Diagram

In 1904, E. Walter Maunder plotted, month by month, the latitude of each sunspot on the sun. He found that his diagram looked like that in Figure 7–31. Since the pattern resembles a butterfly, this is called *Maunder's butterfly diagram.* It shows that early in a sunspot cycle, new spots form close to latitude 30°. As the cycle progresses, new spots are formed closer and closer to the equator. At the end of a cycle, just before the time when the number of sunspots is at a minimum, spots of the old cycle may be appearing at the equator while the first spots of the new cycle may already be appearing at higher latitudes.

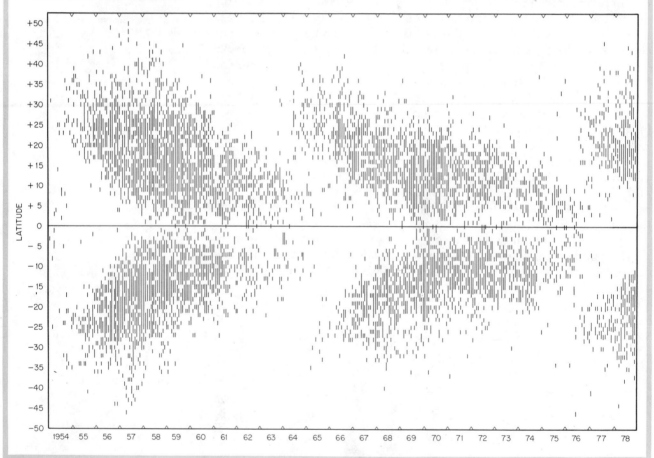

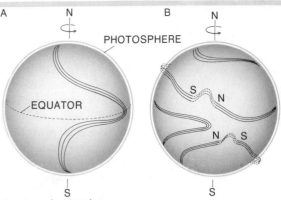

Figure 7–32 A leading model to explain sunspots suggests that the solar differential rotation winds tubes of magnetic flux around and around the sun. When the tubes kink and penetrate the solar surface, we see the sunspots that occur in the areas of strong magnetic field.

Box 7.4 The Origin of Sunspots

Although the details of the formation of sunspots are not yet understood, a general picture was suggested in 1961 by Horace W. Babcock of the Hale Observatories, elaborating on ideas that had been advanced by E. N. Parker of the University of Chicago. In the Babcock model, just under the solar photosphere the magnetic field lines are bunched in tubes of magnetic field that wind around the sun. These tubes are formed by a mechanism (Fig. 7–32) that will be discussed next.

The sun rotates approximately once each earth month. Different latitudes on the sun rotate at different speeds. Though a solid ball like the earth rotates at a constant rate at all latitudes (both Tampa and New York rotate in 24 hours), a gaseous ball like the sun can rotate differentially (Fig. 7–33). Because of this *differential rotation,* if a line of sunspots started at the same longitude on the sun at a given time, the ones closest to the equator would make a full revolution faster than the ones farthest from the equator. Gas at the equator rotates in 25 days, gas at 40 degrees latitude rotates in about 28 days, and gas nearer the poles rotates even more slowly. The differential rotation can also be measured spectrographically by studying Doppler shifts. (Not only does the surface of the sun rotate differentially, but also the sun probably rotates at different speeds at different distances from its core, though we cannot directly observe the interior. Different atmospheric heights appear to rotate at slightly different speeds.)

A line of force that may have started out in the direction from north to south on the solar surface is wrapped around the sun by the action of the differential rotation. These lines collect in the equivalent of tubes located not far beneath the solar photosphere. Under some circumstances, buoyant forces carry part of a tube upward until the tube sticks up through the solar photosphere. Where the tube emerges we see a sunspot of one magnetic polarity, and where the tube returns through the surface we see a sunspot of the other polarity. The north and south polarities are connected by magnetic lines of force that extend above the sun.

Because of the differential rotation, the spiral winding of the magnetic lines of force is tighter at higher latitudes on the sun than at lower latitudes. Thus the instability that allows part of a tube to be carried to the surface arises first at higher solar latitudes. As the solar cycle wears on, the differential rotation continues and the tubes rise to the surface at lower and lower latitudes. This explains the latitude effect that we have seen illustrated in the butterfly diagram. The picture must be modified by the effect of the alternating bands of high and low velocity discovered recently. This discovery shows that the basic structure that causes sunspots is located deep in the solar interior, because only through such a deep connection could bands in the northern and southern hemispheres be linked with each other, as the observations show they must be.

The effects that lead to the formation of sunspots arise from the interaction of the solar differential rotation with turbulence and motions in the convective zone. Dynamos in factories on earth also depend on the interaction of rotation and magnetic fields; thus these theories for the solar activity are called *dynamo theories,* and one says that the sunspots are generated by the *solar dynamo.* The dynamo theories can explain such important observations as the existence of the sunspot cycle and its 22-year period, the connection of sunspots to regions of high magnetic field, and Maunder's butterfly diagram.

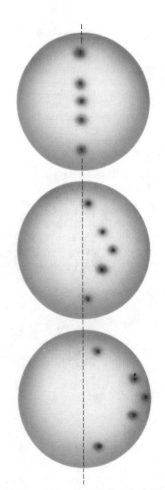

Figure 7–33 The notion that the sun rotates differentially is illustrated by this series, which shows the progress of a schematic line of sunspots month by month. The equator rotates faster than the poles by about 4 days per month.

7.7b Flares

Violent activity sometimes occurs in the regions around sunspots. Tremendous eruptions called *solar flares* (Figs. 7–34 and 7–35) can eject particles and emit radiation from all parts of the spectrum into space. These solar storms begin in a few seconds and can last up to four hours. A typical flare lifetime is 20 minutes. Temperatures in the flare can reach 5 million kelvins, even hotter than the quiet corona. Flare particles that are ejected reach the earth in a few hours or days and can cause disruptions in radio transmission, cause the aurorae (Fig. 7–36 and Color Plate 36)—the *aurora borealis* is the "northern lights" and the *aurora australis* is the "southern lights"—and even cause surges on power lines. Because of these solar-terrestrial relationships, high priority is placed on understanding solar activity and being able to predict it. The U.S. government even has a solar weather bureau to forecast solar storms, just as it has a terrestrial weather bureau.

Solar flares also emit x-rays, which have been studied from Skylab, from OSO's, and from meteorological satellites. Observing flares in the ultraviolet and x-ray region was one of the Solar Maximum Mission's major goals. The radio emission of the sun also increases at the time of a solar flare.

No specific model is accepted as explaining the eruption of solar flares. But it seems that a tremendous amount of energy is stored in the solar magnetic fields in sunspot regions. Something unknown triggers the release of the energy.

7.7c Plages, Filaments, and Prominences

Studies of the solar atmosphere in Hα radiation also reveal other types of solar activity. Bright areas called *plages* (from the French word for beach,

Figure 7–34 One of the largest solar flares in decades occurred on August 7, 1972, and led to power blackouts, short wave radio blackouts, and aurorae. *(A)* The whole sun is shown at the peak of the flare, 15:30 U.T. (Universal Time, approximately corresponding to Greenwich Time). *(B)* The development of the flare with time and its subsidence is shown in this sequence from the Big Bear Solar Observatory. It was hours before the flare completely died down. Other strong flares also occurred in this active region during that week. Labels show Universal Time in hours and minutes. All photographs are through a filter passing only Hα.

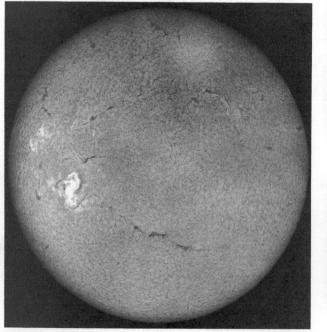

A

15:16

15:31

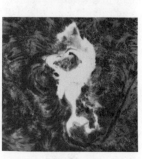
15:23

15:55

B

Figure 7–35 A flare near the solar limb, photographed at Big Bear on August 20, 1971.

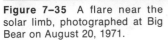
Figure 7–35   Figure 7–36

Figure 7–36 An aurora borealis, beautiful patterns of color in the sky, photographed in Alaska. The aurorae are phenomena of the earth's atmosphere, and are caused by incoming particles from the sun.

Effects in the upper atmosphere of the earth during solar maximum may cause so much radio "skip"—which makes even very distant stations audible, as the signal skips or bounces off the earth's ionosphere (an upper atmospheric level)—that CB radios become harder to use.

Figure 7–37 The sun, photographed in hydrogen radiation at the last solar maximum, shows bright areas called plages and many dark filaments around the many sunspot regions.

pronounced "plahges") surround the entire sunspot region (Fig. 7–37). Dark *filaments* are seen threading their way across the sun in the vicinity of sunspots. The longest filaments can extend for 100,000 km. When these filaments happen to be on the limb of the sun, they can be seen to project into space, often in beautiful shapes. Then they are called *prominences* (Fig. 7–38). Prominences can be seen with the eye at solar eclipses, and glow pinkish at that time because of their emission in Hα and a few other spectral lines. They can be observed from the ground even without an eclipse, if an Hα filter is used. Prominences appear to be composed of matter in a similar condition of temperature and density to matter in the quiet chromosphere. Space observations can lead to maps of the temperature structure inside a prominence (Fig. 7–39).

Sometimes prominences can hover above the sun, supported by magnetic fields, for weeks or months. They are then called *quiescent prominences*, and can extend tens of thousands of kilometers above the limb. Other prominences can seem to undergo rapid changes (Color Plates 14 and 15 and Fig. 7–30).

7.8 SOLAR-TERRESTRIAL RELATIONS

Careful studies of the solar activity cycle are now increasing our understanding of how the sun affects the earth. Although for many years scientists were skeptical of the idea that solar activity could have a direct effect on the earth's weather, scientists presently seem to be accepting more and more the possibility of such a relationship.

Figure 7-38 Many prominences appear around the sun, and a giant eruptive prominence ejects matter in this hydrogen-light photograph from the Haleakala Observatory in Hawaii.

Figure 7-39 Observations in the ultraviolet from the Harvard College Observatory experiment aboard Skylab reveal the temperature structure in and around the prominences, which are themselves at about 15,000 K. (A) Triply ionized oxygen gas (O IV) is at a temperature of about 130,000 K and represents the transition zone from the prominence to the corona. (B) Five-times ionized oxygen gas (O VI) is somewhat hotter (300,000 K) and represents both gas in the transition zone and gas in the corona. (C) Nine-times ionized magnesium gas (Mg X) represents gas in the corona at a temperature of 1,500,000 K.

An extreme test of the interaction may be provided by the interesting probability that there most likely were no sunspots at all on the sun from 1645 to 1715! The sunspot cycle may not have been operating for that 70-year period (Fig. 7-40). This was known to Maunder and others in the early years of this century but was largely forgotten until its importance was recently noted and stressed by John A. Eddy of the High Altitude Observatory. Although no counts of sunspots exist for most of that period, there is evidence that people were looking for sunspots; it seems reasonable that there were no counts of sunspots because they were not there and not just because nobody was observing. A variety of indirect evidence has also been brought to bear on the question. For example, the solar corona appeared very weak when observed at eclipses. It may be significant that the anomalous sunspotless period coincided with a "Little Ice Age" in Europe and with a drought in the southwestern United States. Another important conclusion from the existence of this sunspotless period is that the solar activity cycle may be much less regular than we had thought.

The evidence for the Maunder minimum is indirect, and has been challenged. For example, the Little Ice Age could have as much or more to do with volcanic dust in the air or with changing patterns of land use than with sunspots. Old auroral records may show activity, though this is controversial. It should come as no particular surprise that several mechanisms affect the earth's climate on this time scale, rather than only one.

*7.9 THE SOLAR WIND

At about the time of the launch of the first earth satellites, in 1957, it was realized that the corona must be expanding into space. This phenomenon is called the *solar wind*. The expansion causes comet tails (Fig. 7-41) always to point away from the sun, but the definite existence of a solar wind had not been deduced from this fact.

The solar wind extends into space far beyond the orbit of the earth, possibly even beyond the orbit of Pluto. The density of particles, always low, decreases with distance from the sun. There are only about 5 particles

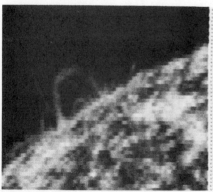

A

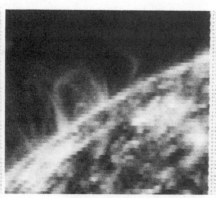

B

C

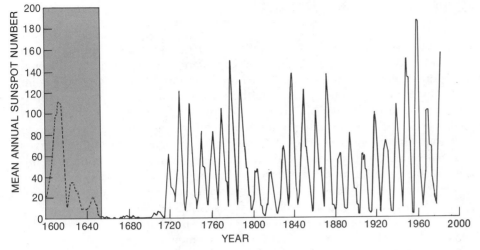

Figure 7–40 The Maunder minimum (1645–1715), when sunspot activity was negligible for decades, may indicate that the sun does not have as regular a cycle of activity as we had thought. Activity is extrapolated into the shaded regions with measurements made by indirect means (such as the frequency of auroras).

in each cubic centimeter at the earth's orbit. The particles are a mixture of ions and electrons.

It is now thought by many astronomers that the solar wind emanates from coronal holes (Section 7.5). The earth's outer atmosphere is bathed in the solar wind. Thus research on the nature of the solar wind and on the structure of the corona is necessary to understand our environment in the solar system.

In 1975 and 1976, two joint German-American satellites named Helios were sent into the solar wind to go as close to the sun as possible, within about 44,000,000 km from the solar center. Other satellites are in orbit around the earth, and even in orbit around the sun at points quite distant from the earth, in order to give as complete a picture of the solar wind from as widely spaced points of view as possible. No observations, however, have ever been made outside the plane in which all the planetary orbits lie. The Solar Polar Mission, a pair of spacecraft scheduled to be launched in 1985, should provide such a view. The spacecraft will be sent out to Jupiter, and take advantage of that planet's gravitational pull to send them out of the earth's orbital plane and back over the poles of the sun.

Gas in the solar wind takes about 10 days to reach us from the bottom of the corona, by which time it has accelerated to 400 km s⁻¹. It takes only two more weeks to reach Jupiter at its ever increasing speed.

Three NASA satellites, International Sun-Earth Explorers, were launched in 1977 and 1978 to study the solar wind. Two are in earth orbit and the third is hovering at a libration point, a place in space where gravity from the sun just balances gravity from the earth-moon system.

Figure 7–41 The solar wind causes the wavy streaming of the tails of comets, as in this view of Comet Mrkos in 1957.

*7.10   THE SOLAR CONSTANT

The solar constant is the amount of energy per second that would hit each square centimeter of the earth at its average distance from the sun if the earth had no atmosphere. 99% of solar energy is in the range from 2760 Å to 49,000 Å (nearly 5 microns), and 99.9% is between 217 Å and 10.94 microns.

Calories are a unit of energy, and watts are a unit of power in the metric system.

Solar Maximum Mission has detected short-term variations of almost 0.2 per cent. A long-term accuracy of 0.1 per cent is expected.

Every second a certain amount of solar energy passes through each square centimeter of space at the average distance of the earth from the sun. This quantity is called the *solar constant*. Accurate knowledge of the solar constant is necessary to understand the terrestrial atmosphere, for to interpret our atmosphere completely we must know all the ways in which it can gain and lose energy. Further, knowledge of the solar constant enables us to calculate the amount of energy that the sun itself is giving off, and thus gives us an accurate measurement on which to base our quantitative understanding of the radiation of all the stars.

Ground-based measurements give a value of about 2 calories/cm²/min, which is equivalent to about 135 milliwatts/cm². The value is determined to an accuracy of 1.5 per cent; we have not yet been able to determine the solar constant to higher accuracy. Higher accuracy measurements are necessary, and an experiment for this purpose flew on the Solar Maximum Mission. One interesting aim is to determine if there are long-term changes in the solar constant, perhaps arising from changes in the luminosity of the sun. Certainly our climate would be profoundly affected by small long-term changes. Of course, if the solar constant changed with time, it wouldn't really be a constant after all.

7.11   THE SUN AND THE THEORY OF RELATIVITY

The intuitive notion we have of gravity corresponds to the theory of gravity advanced by Isaac Newton in 1687. We now know, however, that Newton's theory and our intuitive ideas are not sufficient to explain the universe in detail. Theories advanced by Albert Einstein in the first decades of this century now provide us with a more accurate understanding.

The sun, as the nearest star to the earth, has been very important for testing some of the predictions of Albert Einstein's theory of gravitation, which is known as the general theory of relativity. The theory, which Einstein advanced in final form in 1916, made three predictions that depended on the presence of a large mass like the sun for experimental verification. These predictions were (1) the gravitational deflection of light, (2) the advance of the perihelion of Mercury, and (3) the gravitational redshift. We shall discuss the gravitational redshift in Section 9.5 and discuss the other two tests here.

Figure 7–42A The prediction in Einstein's own handwriting of the deflection of starlight by the sun, taken from a letter from Einstein to Hale at Mt. Wilson. "Lichtstrahl" is "light ray." Einstein asked if the effect could be measured without an eclipse, to which Hale replied negatively. The drawing came from a date when Einstein had developed only an early version of the theory, which gave a predicted value half of his later predicted value.

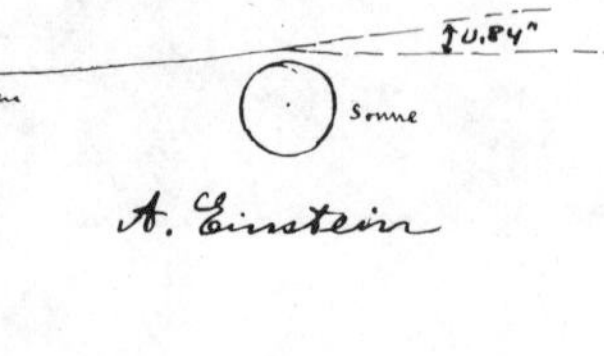

Einstein's theory predicts that the light from a star would act as though it were bent toward the sun by a very small amount (Fig. 7–42). We on earth, looking back, would see the star from which the light was emitted as though it were shifted slightly away from the sun. Only a star whose radiation grazed the edge of the sun would seem to undergo the full deflection; the effect diminishes as one considers stars farther away from the solar limb. To see the effect, one has to look near the sun at a time when the stars are visible, and this could be done only at a total solar eclipse.

The British astronomer Arthur Eddington and other scientists observed the total solar eclipse of 1919 from sites in Africa and South America. The

Box 7.5 The Special and General Theories of Relativity

In 1905, Einstein advanced a theory of relative motion that is called the *special theory of relativity*. A basic postulate is that nothing can go faster than the speed of light and, strangely, that the speed of light is the same even if we are moving toward or away from the object that is emitting the light that we are observing. Einstein's theory shows that the values that we measure for length, mass, and the rate at which time advances depend on how fast we are moving relative to the object we are observing. One consequence of the special theory, published by Einstein a year later, is that mass and energy are equivalent and that they can be transformed into each other, following a relation $E = mc^2$. The consequences of the special theory of relativity have been tested experimentally in many ways, and the theory has long been well established.

"Special relativity" was limited in that it did not take the effect of gravity into account. Einstein proceeded to work on a more general theory that would explain gravity. Isaac Newton in 1667 had shown the equations that explain motion caused by the force of gravity, but nobody had ever said how gravity actually worked. Einstein's *general theory of relativity* links three dimensions of space and one dimension of time to describe a four-dimensional space-time. It explains gravity as the effect we detect when we observe objects moving in curved space. Imagine the two-dimensional analogy to curved space of a billiard ball rolling on a warped table. The ball would tend to curve one way or the other. The effect would be the same if the table were flat but there were objects with the equivalent of gravity spaced around the table. In four-dimensional space-time, masses warp the space near them and objects or light change their path as a result, an effect that we call gravity. Thus Einstein's theory of general relativity explains gravity as an observational artifact of the curvature of space.

In Einstein's own words, "If you will not take the answer too seriously, and consider it only as a kind of joke, then I can explain it as follows. It was formerly believed that if all material things disappeared out of the universe, time and space would be left. According to the relativity theory, however, time and space disappear together with the things."

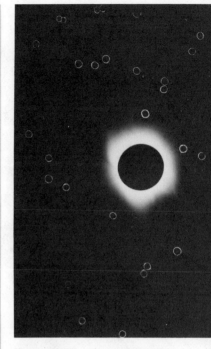

Figure 7–42B The photographic plate from the 1922 eclipse expedition to test Einstein's theory. The circles mark the positions of the stars used in the data reduction; the stars themselves are too faint to see in the reproduction.

effect for which they were looking was a very delicate one, and it was not enough merely to observe the stars at the moment of eclipse. One had to know what their positions were when the sun was not present in their midst, so the astronomers had already made photographs of the same field of stars six months earlier when the same stars were in the nighttime sky. The duration of totality was especially long, and the sun was in a rich field of stars, the Hyades, making it a particularly desirable eclipse for this experiment. Even though his observations had been limited by clouds, Eddington detected that light was deflected by an amount that agreed with Einstein's revised predictions. The scientists hailed this confirmation of Einstein's theory, and from the moment of its official announcement, Einstein was recognized by scientists and the general public alike as the world's greatest scientist.

Although the experiment has been repeated, most recently in Mexico in 1970, Africa in 1973, and India in 1980, the precision of even the best results is only about 5 per cent (that is, we cannot tell from the data whether the actual value is as much as 5 per cent more or less than the value we specify as the "measured" value). The data agree with Einstein's prediction,

Actually, according to the general theory of relativity, the presence of the mass warps the space. Thus the light travels in a straight line but space itself is curved. The effect is the same as it would be if space were flat and the light were bent.

When Einstein eventually heard of the result, he was so pleased that he wrote his mother a postcard to tell her of his success. "Dear Mother: Joyful news today," Einstein wrote. "H. A. Lorentz has telegraphed me that the English expeditions have actually proved the deflection of light near the sun."

Figure 7–43   Albert Einstein visiting California in the 1930's.

Figure 7–44   The advance of the perihelion of Mercury, with the amount of the advance considerably exaggerated.

but are not accurate enough to distinguish between Einstein's theory and newer, more complicated, rival theories of gravitation. Fortunately, the effect of gravitational deflection is constant through the electromagnetic spectrum and the test can now be performed more accurately by observing how the sun bends radiation from radio sources, especially quasars. The results agree with Einstein's theory to within 1 per cent, enough to make the competing theories very unlikely.

A related test involves not deflection but a delay in time of signals passing near the sun. This can now be performed with signals from interplanetary spacecraft, most recently with a Viking orbiter near Mars, and these data also agree with Einstein's theory.

Another of the triumphs of Einstein's theory, even in its earliest versions, was that it explained the "advance of the perihelion of Mercury." The orbit of Mercury, as are the orbits of all the planets, is elliptical; the point at which the orbit comes closest to the sun is called the *perihelion* (Fig. 7–44). The elliptical orbit is pulled around the sun over the years, mostly by the gravitational attraction on Mercury by the other planets, so that the perihelion point is at a different orientation in space. Each century (!) the perihelion point appears to move around the sun by approximately 5600 seconds of arc (which is less than 2°). Subtracting the effects of precession (the changing orientation of the earth's axis in space, described in Section 4.5) and the gravitational effects of the other planets leaves 43 seconds of arc per century whose origin had not been understood on the basis of Newton's theory of gravitation.

Einstein's general theory predicts that the presence of the sun's mass warps the space near the sun (Fig. 7–45). Since Mercury is sometimes closer

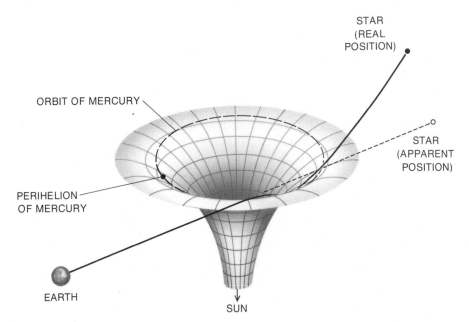

STAR
(REAL
POSITION)

ORBIT OF MERCURY

STAR
(APPARENT
POSITION)

PERIHELION
OF MERCURY

EARTH

SUN

Figure 7–45 Under the general theory of relativity, the presence of a massive body essentially warps the space nearby. This can account for both the bending of light near the sun and the advance of the perihelion point of Mercury by 43 arc sec per century more than would otherwise be expected. The diagram shows how a two-dimensional surface warped into three dimensions can change the direction of a "straight" line that is constrained to its surface; the warping of space is analogous, although with a greater number of dimensions to consider.

to and sometimes farther away from the sun, it is sometimes traveling in space that is warped more than the space it is in at other times. This should have the effect of changing the point of perihelion by 43 seconds of arc per century. The agreement of this prediction with the measured value was an important observational confirmation of Einstein's theory. More recent refinements of solar system observations have shown that the perihelions of Venus and of the earth also advance by the even smaller amounts predicted by the theory of relativity.

Historically, in explaining the perihelion advance, general relativity was explaining an observation that had been known. On the other hand, in the bending of light it actually predicted a previously unobserved phenomenon.

As a general rule, scientists try to find theories that not only explain the data that are at hand, but also make predictions that can be tested. This is an important part of the *scientific method*. Because the bending of electromagnetic radiation by a certain amount was a prediction of the general theory of relativity that had not been anticipated, the verification of the prediction was a more convincing proof of the theory's validity than the theory's ability to explain the perihelion advance.

We shall see later on how observations of a pulsar in a binary system (Section 10.11), a system in which the perihelion advance should be and is 4° per year, has provided even more decisive proof.

Figure 7–46 Einstein lecturing at the College de France in Paris in 1922.

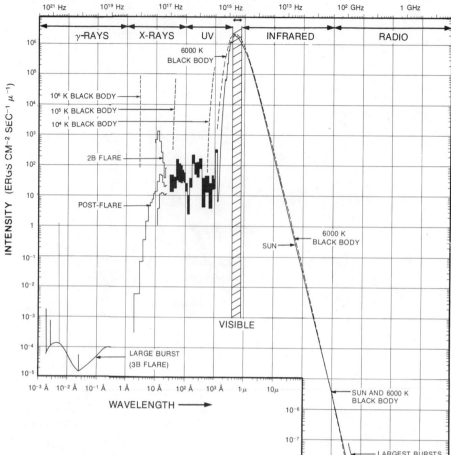

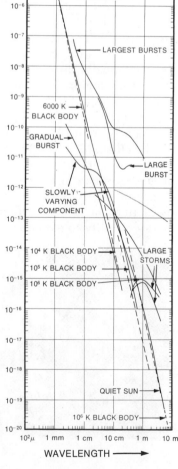

Figure 7–47 The solar spectrum has been studied over its entire range, from gamma rays on the short wavelength end to radio waves on the long wavelength end. The visible part of the spectrum, shown with cross-hatching, makes up only a small part of the total. The solar intensity is in near agreement through the visible and the infrared with the black body curve for a temperature of 6000 K.

In the part of the ultraviolet above about 1000 Å, the spectrum is slightly depressed from the 6000 K curve because of the presence of many absorption lines. At wavelengths shorter than 1000 Å or longer than 1 cm, the sun does not agree at all with the 6000 K black body curve. At the longest and shortest wavelengths, the solar spectrum agrees with the 1,000,000 K black body curves representative of the temperatures of the corona and of solar flares. At those very short or very long wavelengths, we see hotter radiation from higher in the solar atmosphere than the photosphere, and do not see through to the photosphere.

While the central part of the solar spectrum does not vary significantly with solar activity, the short wavelength and long wavelength ends vary greatly. The radio spectrum, for example, has both a component that varies slowly (over the solar activity cycle) and a rapidly varying flare component. Ranges of intensities, which depend on the amount of activity and the presence of flares, are shown. The area and brightness class of flares is shown; a 3B flare is the largest and brightest. In the γ-ray region, the strong emission lines that appear in flares in addition to continuous radiation are shown. (Individual emission or absorption lines are not shown in other parts of the spectrum.)

Note the wide range of intensity that the observed solar spectrum has: over 25 orders of magnitude.

The tick marks on the horizontal axes represent wavelength. The conversion to frequency units is shown at top; for example, 3 mm corresponds to 10^2 GHz (the wavelength tick marks that correspond to these powers of ten in frequency units are elongated).

The data graphed here were compiled by Harriet H. Malitson of the Goddard Space Flight Center, mostly from data included in *The Physical Output of the Sun,* edited by O. R. White (Boulder: University of Colorado Press, 1976).

SUMMARY AND OUTLINE

Definitions: parts of the sun (interior, photosphere, chromosphere, corona, spicules, granulation, supergranulation, coronal holes)

Spectra:
Photospheric spectrum: continuum with Fraunhofer lines; 5800 K; abundances of the elements (Section 7.2b)

Chromospheric spectrum: emission at the limb during eclipses in the visible part of the spectrum (flash spectrum); typical temperature is 15,000 K (Section 7.3b)

Coronal spectrum: high temperature of 2,000,000 K (Section 7.4b)

Space observations: Orbiting Solar Observatories, Skylab, Solar Maximum Mission (Section 7.5)

Eclipse phenomena: band of totality, Baily's beads, diamond ring (Section 7.6)

Solar activity cycle: sunspots, magnetic fields, flares, plages, filaments, prominences (Section 7.7)

Solar-terrestrial relations (Section 7.8)
Possible effect of solar activity on terrestrial weather
No solar activity during the Maunder minimum

Solar wind flows from coronal holes (Section 7.9)

Solar constant and its possible variability (Section 7.10)

Using the sun to test Einstein's general theory of relativity (Section 7.11)
Gravitational deflection
Predicted in advance of detection
First detected at eclipses
Now also tested in radio region of spectrum
Advance of the perihelion of Mercury
Phenomenon was known in advance of Einstein's theory
Now also verified for the binary pulsar

QUESTIONS

1. Sketch the sun, labeling the interior, the photosphere, the chromosphere, the corona, sunspots, and prominences. Give the approximate temperature of each.

2. What elements make most of the lines in the Fraunhofer spectrum? What elements make the strongest lines? Why?

3. Explain why the photospheric spectrum is an absorption spectrum and the chromospheric spectrum seen at an eclipse is an emission spectrum.

4. Graph the temperature of the interior and atmosphere of the sun as a function of distance from the center.

5. (a) Why isn't there a solar eclipse once a month?
(b) Whenever there is a total solar eclipse, a lunar eclipse occurs either two weeks before or two weeks after. Explain.

6. Why can't we observe the corona every day from any location on earth?

7. Describe the series of phenomena one observes at a total solar eclipse.

8. Why does the chromosphere appear pinkish at an eclipse?

9. How do we know that the corona is hot?

10. Describe the sunspot cycle and a mechanism that explains it.

11. What are the differences between the solar wind and the normal "wind" on earth?

12. (a) What is the solar constant?
(b) Why is it difficult to measure precisely?

13. (a) Of what part of the spectrum is it most important to make accurate measurements in order to determine the solar constant?
(b) What would the solar constant be if we lived on Mars?

14. In what tests of general relativity does the sun play an important role? Describe the current status of these investigations.

TOPICS FOR DISCUSSION

1. Discuss the relative importance of solar observations (a) from the ground, (b) at eclipses, (c) from unmanned satellites, and (d) from manned satellites.

We have seen how the Hertzsprung-Russell diagram for a cluster of stars allows us to deduce the age of the cluster and the ages of the stars themselves (see Section 6.5). Our human lifetimes are very short compared to the billions of years that a typical star takes to form, live its life, and die. Thus our hope of understanding the life history of an individual star depends on studying large numbers of stars, for presumably we will see them at different stages of their lives.

Though we can't follow an individual star from cradle to grave, by studying many different stars we can study enough stages of development to write out a stellar biography. Chapters 8 through 11 are devoted to such life stories. Similarly, we could study the stages of human life not by watching someone's aging, but rather by studying people of all ages who are present in, say, a city on a given day.

In Chapter 8, we shall study the birth of stars, and observe some places in the sky where we expect stars to be born very soon, possibly even in our own lifetimes. We shall then consider the properties of stars during the long, stable phase in which they spend most of their histories. Then we go on to consider the ways in which stars end their lives. Stars like the sun sometimes eject shells of gas that glow beautifully; the remainders then contract until they are as small as the earth (see Chapter 9). Sometimes when a star is newly visible in a location where no star was previously known to exist, we may be seeing the spectacular death throe of a star that had earlier been too faint to observe (see Chapter 10). The study of what happens to stars after that explosive event is currently a topic of tremendous interest to many astronomers. Some of the stars become pulsars (also discussed in Chapter 10). Other stars, we think, may even wind up as black holes, objects that are invisible and therefore difficult—though not impossible—to observe (see Chapter 11).

Following how a stellar biography can be written from the clues that we get from observation is like reading a wonderful detective story. As new methods of observation become available, we are able to make better

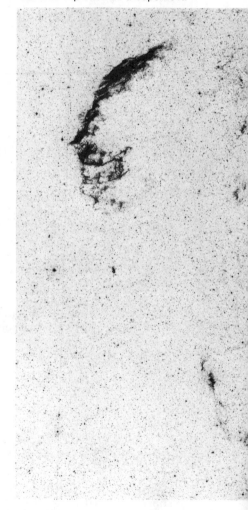

The Cygnus loop, the remnant of a supernova. The remainder of the star that explodes as a supernova

STELLAR EVOLUTION

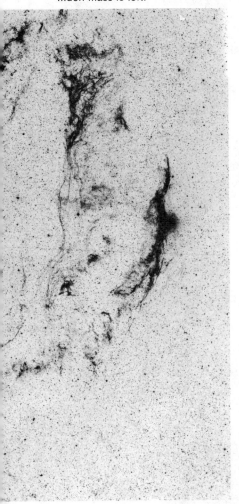

contracts to become either a pulsar or a black hole, depending on how much mass is left.

deductions, so the extension of our senses throughout the electromagnetic spectrum has led directly to a better understanding of stellar evolution. Our work in the x-ray part of the spectrum, for example, is intimately connected with our current exciting search for a black hole. The High-Energy Astronomy Observatories, one of which is shown on this page, represented a major advance.

In these chapters, we shall see that an important new tool has also been added: the computer. Calculations that would have taken years or centuries to carry out—and indeed which never would have been carried out because of the time involved or the probability of error—can now be routinely made in minutes. As computers grow faster and cheaper, as they do every year, our capabilities grow.

The importance of this Part of the book is based on a theorem that states that all stars of the same mass and of the same chemical composition evolve in the same way, if we neglect the effects of rotation and magnetic field. Since the chemical differences among stars are usually not overwhelming, it is largely the mass that is the chief determinant of stellar evolution. Because of this fact, we need only study a few groups of stars to gain pictures of the evolution of most stars.

We can set up divisions, using terminology from the sport of boxing, depending on the mass of stars. We use the name *featherweight star* for collapsing gas that is less than about 7 per cent as massive as the sun; it never reaches stardom, and becomes a black dwarf. Stars more massive than that but containing less than about four solar masses eventually become white dwarfs; we call these *lightweight stars*. *Middleweight stars*, which contain between four and eight solar masses when they are on the main sequence, explode as supernovae and then wind up as neutron stars. We now think that these are the pulsars. The *heavyweight stars*, containing greater than about 8 solar masses, may also explode as supernovae, but wind up withdrawing from the universe in the form of black holes. The next four chapters deal with these possibilities.

CHAPTER 8

YOUNG AND MIDDLE-AGED STARS

AIMS:
To study the formation of stars, their
main-sequence lifetimes, and the
mechanisms of stellar energy generation

Even though individual stars shine for a relatively long time, they are not eternal. Stars are born out of gas and dust that may exist within a galaxy; they then begin to shine brightly on their own. Though we can observe only the outer layers of stars, we can deduce that the temperatures at their centers must be millions of kelvins. We can even deduce what it is deep down inside that makes the stars shine.

In this chapter we will discuss the birth of stars, then the processes that go on in a stellar interior during a star's life on the main sequence, and then begin the story of the final stages of their lives. The following three chapters will continue the story.

8.1 STARS IN FORMATION

The process of star formation starts with a region of gas and dust of slightly higher density than its surroundings. Perhaps this results only from a random fluctuation in density. If the density is high enough, the gas and dust begin to contract under the force of gravity. As they contract, energy is released, and it turns out (as a consequence of the more general physical law known as the *virial theorem*) that half of that energy heats the matter. As the gas and dust are heated, they begin to give off an appreciable amount of radiation.

The "track" of this protostar, the path it takes on a Hertzsprung-Russell

The gas and dust from which stars are forming is best observed in the infrared and radio regions of the spectrum. We shall have more to say about these regions when we discuss star formation in the Milky Way Galaxy in Section 24.5.

A photograph in red light of the region near Herbig-Haro object number 100 in the dark cloud in the constellation Corona Australis. *Herbig-Haro objects* are bright nebulous regions of gas and dust that may be stars in formation. The Herbig-Haro object is just to the lower left of the center of the photograph, near the center of the black patch that is the high-density cloud from which the H–H object was formed.

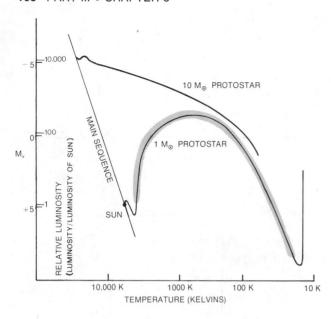

Figure 8–1 Evolutionary tracks for two protostars, one of 1 solar mass and the other of 10 solar masses. ($M_\odot$ is the symbol for the mass of the sun.) The evolution of the shaded portion of the track for the 1 solar mass star is described in the text. These stages of its life last for about 50 million years. More massive stars whip through their protostar stages more rapidly; the corresponding stages for a 10 solar mass star may last only 200,000 years. These very massive stars wind up being very luminous.

"Proto-" is a prefix of Greek origin meaning "primitive."

diagram, is shown in Figure 8–1. At first, the protostar brightens, and the track moves upward and toward the left on the diagram. While this is occurring, the central part of the protostar continues to contract and the temperature rises. The higher temperature results in a higher pressure, which pushes outward more and more strongly. Eventually, a stage is reached where this outward force balances the inward force of gravity of the central region.

Box 8.1   Evolutionary Track

We can plot the position of a star on the Hertzsprung-Russell diagram for any particular instant of its life. As a protostar, for example, it would have a specific luminosity and temperature a million years before it began nuclear fusion, and another specific luminosity and temperature half a million years later. Each of the pairs of luminosity and temperature would correspond to a point on the H-R diagram. When the hydrogen burning began, the star would also correspond to a specific point on the H-R diagram. If we connect the points representing the entire lifetime of the star, we have an *evolutionary track*. A star, of course, is only at one point of its track at any given time.

By this time, the dust has vaporized and the gas has become opaque, so that energy emitted from the central region does not escape directly. The outer layers continue to contract. Since the surface area is decreasing, the luminosity decreases and the track of the protostar begins to move downward on the H-R diagram. As the protostar continues to heat up, it also continues to move toward the left on the H-R diagram.

The time it takes for this gravitational contraction depends on the mass of gas in the protostar. Very massive stars contract to approximately the size of our solar system in only 10 thousand years or so. These massive objects become O and B stars, and are located at the top of the main sequence.

They are sometimes found in groups, which are called *O and B associations*.

Less massive stars contract much more leisurely. A star of the same mass as the sun may take tens of millions of years to contract, and a less massive star may take hundreds of millions of years to pass through this stage.

Theoretical analysis shows that the dust surrounding the stellar embryo we call a protostar should absorb much of the radiation that the protostar emits. Do we detect any objects in the sky that meet the characteristics we expect from this scenario?

Theoretically, the radiation from the protostars should heat the dust to temperatures that produce primarily infrared radiation. Indeed, infrared astronomers have found several objects that are especially bright in the infrared but that have no known optical counterparts. These objects seem to be located in regions where the presence of a lot of dust and gas and other young stars indicates that star formation might be going on. Some of these infrared objects may contain stars in formation.

In the visible part of the spectrum, several classes of stars that vary erratically are found. One of these classes, called *T Tauri stars*, includes stars that have a wide range of spectral types. Their visible radiation can vary by as much as several magnitudes, and they are known to be strong infrared emitters. Presumably these are stars that have not quite settled down to a steady and reliable existence.

T Tauri stars are found in close proximity to each other; these groupings are known as *T associations*. Besides the stars in the T associations, bright nebulous regions of gas and dust called *Herbig-Haro objects* are also present (Fig. 8–2). Herbig-Haro objects may be the sites of star formation.

We discussed how the wavelength at which gas gives off the most radiation depends on the temperature of the gas in Section 3.1.

Figure 8–2 T Tauri itself is embedded in a Herbig-Haro object. It has changed in brightness considerably in the last century, first fading from view and then brightening considerably. This probably results from changing illumination from T Tauri on a fixed dust cloud. Herbig-Haro objects are named after George Herbig of the Lick Observatory and Guillermo Haro of the Mexican National Observatory.

8.2 STELLAR ENERGY GENERATION

All the heat energy in stars that are still contracting toward the main sequence results from the gravitational contraction itself. If this were the only source of energy, though, stars would not shine for very long on an astronomical time—only about 30 million years. Yet we know that even rocks on earth are older than that, since rocks billions of years old have been found. We must find some other source of energy to hold the stars up against their own gravitational pull.

The gas in the protostar will continue to heat up until the central portions become hot enough for *nuclear fusion* to take place. Using this process, which we will soon discuss in detail, the star can generate enough energy inside itself to support it during its entire lifetime on the main sequence.

The basic fusion processes in most stars fuse four hydrogen nuclei into one helium nucleus, just as hydrogen atoms are combined into helium in a hydrogen bomb here on earth. In the process, tremendous amounts of energy are released.

A hydrogen nucleus is but a single proton. A helium nucleus is more complex. It consists of two protons and two neutrons (Fig. 8–3). The mass of the helium that is the final product of the fusion process is slightly less than the sum of the masses of the four hydrogen nuclei that went into it. A small amount of the mass "disappears" in the process: 0.007 (a number that is easy to remember for James Bond aficionados) of the mass of the four hydrogen nuclei.

In the interiors of stars, we are dealing with nuclei instead of the atoms we have discussed in stellar atmospheres, because the high temperature strips the electrons off the nuclei. The electrons are mixed in with the nuclei through the center of the star; there can be no large imbalance of positive and negative charges, or strong repulsive forces would arise inside the star.

Featherweight stars may never become hot enough for fusion processes to take place.

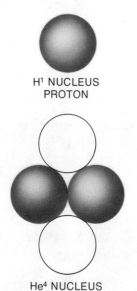

H¹ NUCLEUS
PROTON

He⁴ NUCLEUS

Figure 8-3 The nucleus of hydrogen's most common form is a single proton, while the nucleus of helium's most common form consists of two protons and two neutrons.

Many other nuclear particles have been detected–some only show up for tiny fractions of a second under extreme conditions in giant particle accelerators ("atom smashers"). We need not discuss the other particles in detail for most astronomical contexts. The theory that the nuclear particles are all made up of even more fundamental particles called quarks is being more and more accepted as experimental evidence in its favor mounts.

Figure 8-4 Hydrogen and helium ions. The sizes of the nuclei are greatly exaggerated with respect to the sizes of the orbits of the electrons.

The mass does not really simply disappear, but is rather converted into energy according to Albert Einstein's famous formula $E = mc^2$. Now c, the speed of light, is a large number, and c^2 is even larger. Thus even though m is only a small fraction of the original mass, the amount of energy released is prodigious. The loss of only 0.007 of the mass of the sun, for example, is enough to allow the sun to radiate as much as it does at its present rate for a period of at least 10 billion (10^{10}) years. This fact, not realized until 1920, solved the long-standing problem of where the sun and the other stars got their energy.

All the main-sequence stars are approximately 90 per cent hydrogen (that is, 90 per cent of the atoms are hydrogen), so there is lots of raw material to stoke the nuclear "fires." We speak colloquially of *nuclear burning*, although, of course, the processes are quite different from the chemical processes that are involved in the "burning" of logs or of autumn leaves. In order to be able to discuss these processes, we must first discuss the general structure of nuclei and atoms.

8.3 ATOMS

An atom consists of a small *nucleus* surrounded by *electrons*. Most of the mass of the atom is in the nucleus, which takes up a very small volume in the center of the atom. The effective size of the atom, the chemical interactions of atoms to form molecules, and the spectra are determined by the electrons.

The nuclear particles with which we need be most familiar are the *proton* and *neutron*. The neutron has no electric charge and the proton has one unit of positive electric charge. The electrons, which surround the nucleus, have one unit each of negative electric charge. When an atom loses an electron, it has a net positive charge of 1 unit for each electron lost. The atom is now a type of ion (Fig. 8–4).

The definition of ions does not concern the nucleus; for the nucleus, we need keep track only of the numbers of protons and neutrons and not of orbiting electrons. The number of protons in the nucleus determines the quota of electrons that the neutral state of the atom must have, since it determines the charge of the nucleus. Each element is defined by the specific number of protons in its nucleus. The element with one proton is hydrogen, that with two protons is helium, that with three protons is lithium, and so on. (A table of the elements appears in Appendix 4.)

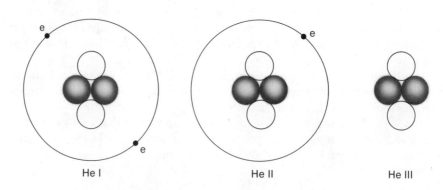

H I   H II

He I   He II   He III

$_1H^1$ $_1H^2 = D$ = DEUTERIUM $_1H^3 = T$ = TRITIUM $_2He^3$ $_2He^4$

Figure 8–5 Isotopes of hydrogen and helium. $_1H^2$ (deuterium) and $_1H^3$ (tritium) are much rarer than the normal isotope, $_1H^1$. $_2He^3$ is much rarer than $_2He^4$.

The number of neutrons in a nucleus is not fixed for a given element, although the number is always somewhere between 1 and 2 times the number of protons. (Hydrogen, which need have no neutrons, and helium are the only exceptions.) The possible different forms of the same element are called *isotopes*. Isotopes of an element have only slightly different numbers of neutrons.

For example, the nucleus of ordinary hydrogen contains one proton and no neutrons. An isotope of hydrogen (Fig. 8–5) called deuterium (and sometimes "heavy hydrogen") has one proton and one neutron. Another isotope of hydrogen called tritium has one proton and two neutrons.

Most isotopes do not have specific names, and we keep track of the numbers of protons and neutrons with a system of superscripts and subscripts. The subscript **before** the symbol denoting the element is the number of protons, and a superscript **following** the symbol is the total number of protons and neutrons together. The number of protons is called the *atomic number*, and the number of protons and neutrons together is called the *mass number*. For example, $_1H^2$ is deuterium, since deuterium has one proton, which gives the subscript, and an atomic mass of 2 which gives the superscript. This means that the atomic number of deuterium equals 1 and its mass number is 2. Similarly, $_{92}U^{238}$ is an isotope of uranium with 92 protons (atomic number = 92) and a mass of 238, which is divided into 92 protons and $238 - 92 = 146$ neutrons.

Each element exists only in certain isotopic forms. For example, most naturally occurring helium is in the form $_2H^4$, with a lesser amount as $_2H^3$. Sometimes an isotope is not stable, in that after a time it will spontaneously change into another isotope or element; we say that such an isotope is *radioactive*.

During certain types of radioactive decay, a particle called a *neutrino* is given off (Fig. 8–6). A neutrino is a neutral particle (its name comes from the Italian for "little neutral one").

We have generally thought that neutrinos, interestingly, travel at the speed of light. Now, most matter cannot travel at the speed of light, according to Einstein's special theory of relativity. (Its mass gets larger and larger and approaches infinity as its speed approaches the speed of light). Relativity theory shows that the mass of an object is equal to a constant quantity called the *rest mass* divided by a quantity that approaches zero as the object goes faster and faster in approaching the speed of light. So the only way that an object can travel at the speed of light is if its rest mass is zero, because then the quotient is always zero, which is the only case in which it does not become infinite. Thus if the neutrinos travel at the speed of light, they have no rest mass. That is, the neutrino, like the photon, would have no mass if it were at rest (which it never is), so there would be no rest mass to grow larger.

New results from atomic accelerators—atom smashers—were reported in 1980 that indicated that the neutrino may have some small rest mass after all. The mass measured in these experiments is tiny, less than .0001 of the mass of an electron, but if this result is confirmed then it can have profound consequences for physics and

In a nuclear reaction one starts out with certain nuclear particles (protons and neutrons, for example) and possibly radiation and winds up with other particles and whatever radiation may be emitted. The particles that come out of a reaction are different from the particles that go into a reaction, just as though a Datsun and a Cadillac collided and after an x-ray zapped a passer-by, a Pinto and a Mercedes were found when the smoke cleared.

Figure 8–6 Enrico Fermi, Werner Heisenberg, and Wolfgang Pauli enjoy an outing on Lake Como in 1927. Pauli predicted on theoretical grounds that the neutrino would exist. Fermi further developed the theory and gave the particle its name. Heisenberg, one of the originators of quantum mechanics, is most widely known for his *uncertainty principle*. The principle shows that there are fundamental limits in how accurately we can pinpoint the position or momentum of a nuclear particle (or other object) because the very act of observing it disturbs it.

astrophysics. Since each cubic centimeter of the universe apparently contains about 100 neutrinos, even a small rest mass for each one would mean that most of the matter in the universe was in the form of neutrinos.

Further, we already know that there are three different forms of neutrinos. The new experiments indicate that neutrinos may change back and forth from one form to another, changing slightly in mass as they do so. If this is established, then neutrinos are even harder to detect than we had thought, because they might have changed into a form other than the one our equipment is searching for.

The experimental results are still preliminary, and are not accepted by all scientists. We must await further research.

Neutrinos have a very useful property for the purpose of astronomy: they do not interact very much with matter. Thus when a neutrino is formed deep inside a star, it can usually escape to the outside without interacting with any of the matter in the star. This means that the neutrino heads right out of the star, without bumping into any of the matter in the star along the way. Electromagnetic radiation, on the other hand, does not escape from inside a star so easily. A photon of radiation can travel only about 1 cm in a stellar interior before it is absorbed, and it is millions of years before a photon zigs and zags its way to the surface.

The elusiveness of neutrinos makes them valuable probes—indeed, the only possible direct probes—of the conditions inside the sun at the present time. This same elusiveness also makes them difficult to detect. Later in this chapter we shall discuss the major experiment that is being performed to study solar neutrinos.

8.4 STELLAR ENERGY CYCLES

Several chains of reactions have been proposed to account for the fusion of four hydrogen atoms into a single helium atom. Hans Bethe, now at Cornell University, suggested some of these chains during the 1930's. The different possible chains that have been proposed are important at different temperatures, so chains that are dominant in the centers of very hot stars may be different from the ones that are dominant in the centers of cooler stars.

When the center of a star is at a temperature less than 15×10^6 K, the

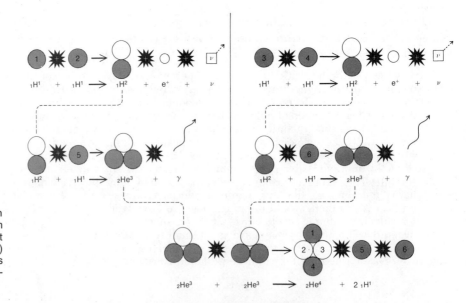

Figure 8–7 The proton-proton chain; e^+ stands for a positron (a particle like an electron except for having positive charge), ν (nu) is a neutrino, and γ (gamma) is radiation at a very short wavelength.

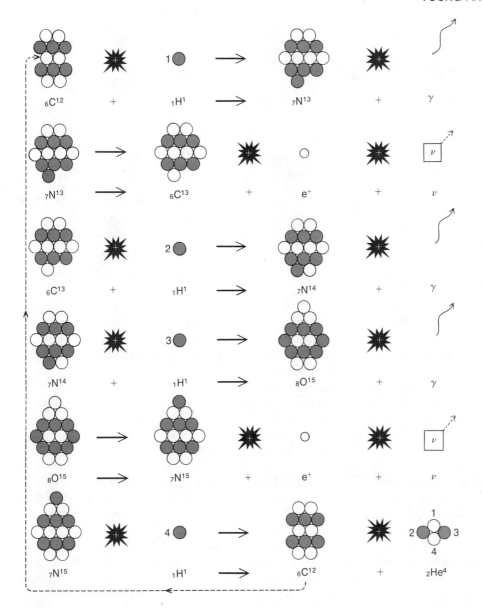

Figure 8-8 The carbon cycle. The 4 hydrogen atoms are numbered. Note that the carbon is left over at the end, ready to enter into another cycle.

proton-proton chain is dominant (Fig. 8–7). In the first stage, two ordinary nuclei of hydrogen fuse to become a deuterium nucleus, a positron, and a neutrino. The neutrino immediately escapes from the star, but the positron soon collides with an electron. They annihilate each other, forming gamma rays.

Next, the deuterium nucleus fuses with yet another nucleus of ordinary hydrogen to become an isotope of helium with two protons and one neutron. High-energy radiation is released at this stage in the form of more gamma rays.

Finally, two of these helium isotopes fuse to make one nucleus of ordinary helium plus two nuclei of ordinary hydrogen. Six hydrogens have been put in and the result is one helium plus two hydrogens, a net transformation of four hydrogens into one helium. The small fraction of mass that disappears in the process is converted into energy according to the formula $E = mc^2$.

For stellar interiors hotter than that of the sun, the *carbon cycle* would

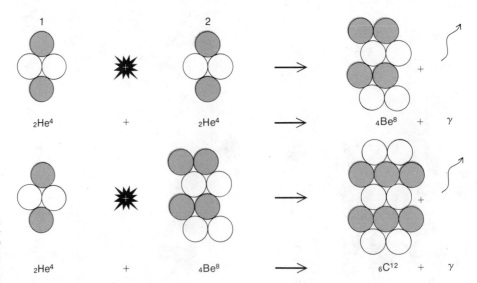

Figure 8–9 The triple-alpha process, which takes place only at temperatures above about 10^8 K. Beryllium, $_4Be^8$, is but an intermediate step.

The Greek characterizations derive from a former confusion of radiation and particles. We know now that α particles, formerly called α rays, are helium nuclei; β particles, formerly called β rays, are electrons; and γ rays, as we have seen, are radiation of short wavelength.

dominate (Fig. 8–8). The carbon cycle begins with the fusion of a hydrogen nucleus with a carbon nucleus. After many steps, and the insertion of four hydrogen nuclei, we are left with one helium nucleus plus a carbon nucleus. Thus as much carbon remains at the end as there was at the beginning, and the carbon can start the cycle again. Again, four hydrogens have been converted into one helium, 0.007 of the mass has been transformed, and an equivalent amount of energy has been released according to $E = mc^2$. Since the cycle can also start with nitrogen and oxygen, the process is sometimes called the *carbon-nitrogen cycle* or the *CNO cycle.*

Stars with even higher interior temperatures, above 10^8 K, can fuse helium nuclei to make carbon nuclei. The nucleus of a helium atom is called an "alpha particle" for historical reasons. Since three helium nuclei ($_2He^4$) go into making a single carbon nucleus ($_6C^{12}$), the procedure is known as the *triple-alpha process* (Fig. 8–9). A series of other processes can build still heavier elements inside stars. This is the study of *nucleosynthesis.*

The theory of nucleosynthesis can account for the abundances we observe of the elements heavier than helium. Currently, we think that the synthesis of isotopes of hydrogen and helium took place in the first few minutes after the origin of the universe (see Section 27.6), and that the heavier elements were formed, along with additional helium, in stellar processes.

8.5 THE STELLAR PRIME OF LIFE

Now that we have discussed basic nuclear processes, let us return to an astronomical situation. We last discussed a protostar in a collapsing phase, with its internal temperature rapidly rising.

One of the most common definitions of temperature describes the temperature in terms of the velocities of individual atoms or other particles. Since this type of temperature depends on kinetics (motions) of particles, it is called the *kinetic temperature*. A higher kinetic temperature (we shall call it simply "temperature" from now on) corresponds to higher particle velocities.

For a collapsing protostar, the energy from the gravitational collapse goes into giving the individual particles greater velocities; that is, the temperature rises. For nuclear fusion to begin, atomic nuclei must get close enough to each other so that the force that holds nuclei together, the *nuclear force* (technically called the *strong force*) can play its part. But all nuclei have positive charges, because they are composed of protons and neutrons. The positive charges on any two nuclei cause an electrical repulsion between them, and this force tends to prevent fusion from taking place.

However, at the high temperatures typical of a stellar interior, some nuclei have enough energy to overcome this electrical repulsion and to come sufficiently close to each other for the strong force to take over. The electrical repulsion that must be overcome is the reason why hydrogen nuclei, which have net positive charges of 1, will fuse at lower temperatures than will helium nuclei, which have net positive charges of 2.

Box 8.2 Forces in the Universe

There are four known types of forces in the universe:

(1) The *strong force,* also known as the *nuclear force,* is the strongest. It is the binding force of nuclei. Although it is very strong close up, it grows weaker rapidly with distance.

Actually, particles that are subject to the strong force, including protons and neutrons, are composed of particles called *quarks* (from a line of James Joyce's *Finnegan's Wake*). There are six kinds ("flavors") of quarks, called "up," "down," "strange," "charmed," "truth," and "beauty." (These names are whimsical and do not have the same meaning that the words have in general speech.) Each kind of quark can have one of three properties called *colors.* Each of the six flavors in three colors makes 18 quarks in all; since each quark has an antiquark, there are really 36. This number is so large that even quarks may not be truly basic.

The strong force is carried between quarks by a particle; since this particle provides the "glue" that holds nuclear particles together, it is known as a "gluon." Recent work with atomic accelerators has reported the discovery of evidence for the existence of the first five of the quarks and of the gluon.

Murray Gell-Mann and George Zweig independently "invented" quarks.

(2) The *electromagnetic force,* 1/137 the strength of the strong force, leads to electromagnetic radiation in the form of photons. It is the force involved in chemical reactions.

(3) The *weak force,* important only in the decay of certain elementary particles, is currently under careful study by particle physicists. It is very weak, only 10^{-13} the strength of the strong force, and it also has a very short range.

(4) The *gravitational force* is the weakest of all over short distances, only 10^{-39} the strength of the strong force. But the effect from the masses of all the particles is cumulative—they add up—so that on the scale of the universe gravity dominates the other forces.

A theory worked out in the 1960's simultaneously explained the electromagnetic force and the weak force. These two forces are thus different aspects of a single force in this "unified" theory. An analogous situation was the discovery in 1877 by James Clerk Maxwell that two forces then thought of as separate—electricity and magnetism—were actually unified as "electromagnetism."

Sheldon Glashow and Steven Weinberg of Harvard and Abdus Salam of Imperial College, London, and Trieste received the 1979 Nobel Prize in Physics for this work.

Theoretical physicists are making substantial progress in finding a "grand unification" theory that explains the strong force as well as the electromagnetic and weak forces in a unified way. Theories that incorporate gravity as well are also under study, but progress is less far along.

Box 8.3 Stellar Interiors

The study of stellar interiors was for many years purely the domain of the theoretician working with equations. Sets of equations were found that accounted for the interior structure of stars, and sets of theoretical solutions were worked out. With the advent of high-speed computers that can perform in minutes calculations that would have taken years to do by hand, the study of stellar interiors is now carried much further. Fewer simplifying assumptions need to be made. Theoreticians can now make a "model" of a stellar interior, that is, a set of values of temperature, density, and pressure for each position in the star, which they consider as a series of concentric shells. Then they calculate shell by shell what must be going on in the interior, making certain that the values of temperature, density and pressure that are derived for the bottom of one layer match the values that are derived for the top of the layer immediately below (Fig. 8–10). They thus make certain that these quantities change continuously from each height to the next of the stellar model in the computer, as they do in real life.

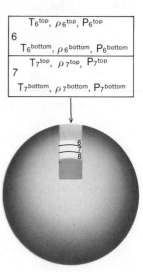

Figure 8–10 For theoretical calculations of stellar interiors, the star is often divided up into thin shells. Physical parameters such as temperature, density, and pressure must be the same at the top of one shell as they are at the bottom of the next higher shell. Two of the layers shown in white in the cutaway are enlarged at the top. T stands for temperature, ρ (the Greek letter rho) stands for density, and P stands for pressure.

*Note that a star does not move **along** the main sequence if its mass doesn't change; it stays for a long time at essentially the same place on the H-R diagram.*

In the center of a star, the fusion process is a self-regulating one. If the nuclear energy production rate increases, then an excess pressure is generated that would tend to make the star grow larger if it were not for the fact that this expansion in turn would cool down the gas. Thus the star finds a temperature and size at which it can remain roughly stable for a very long time. This is the main-sequence phase of a stellar lifetime. Fusion processes are well regulated in stars. When we learn how to control fusion in power-generating stations on earth, which currently seems a long way off, our energy crisis will be over.

The more mass a star has, the hotter its core becomes before it generates enough pressure to counteract gravity. The hotter core leads to a higher surface luminosity, so this explains the mass-luminosity relation (Section

6.2). Thus more massive stars use their nuclear fuel at a much higher rate than less massive stars, and even though the more massive stars have more fuel to burn, they go through it relatively quickly. The next three chapters discuss the fate of stars when they have used up their hydrogen.

Depending on the temperature and density inside the star, and thus on how massive the star is, the helium that is formed inside the star may itself fuse into carbon via the triple-alpha process. In the interiors of the most massive stars, the carbon itself burns, and then the products of the carbon combustion burn, and so on, until eventually an iron core is built up.

From theoretical analysis of basic properties of nuclei we have discovered that iron is the heaviest element that can form in a stellar interior. All the fusion processes that form nuclei lighter than and including iron are processes that release energy. On the other hand, one must add energy to the system in order to form nuclei heavier than iron. Since there is no source for such extra energy, iron actually accumulates at the center of the star.

8.6 THE NEUTRINO EXPERIMENT

One can make a perfectly good model for what goes on inside a star, and it can look quite satisfactory when it comes from the computer, but nonetheless one would like to find some experimental confirmation of the model. Happily, the models are consistent with the conclusions on stellar evolution that one can make from studying different types of Hertzsprung-Russell diagrams. Still, it would be nice to observe a stellar interior directly.

Since the stellar interior lies under opaque layers of gases, we cannot observe directly any electromagnetic radiation it might emit. All the gamma rays, x-rays, ultraviolet, visible, infrared, and radio radiation that we receive come from the surface layers or atmospheres of stars.

Only the neutrino, a nuclear particle mentioned in Section 8.3, can escape directly from a stellar interior. It interacts so weakly with matter that it is hardly affected by the presence of the rest of the solar mass, and zips right out into space at the speed of light.

Raymond Davis, Jr., a chemist at the Brookhaven National Laboratory, has spent the last 12 years trying to detect the neutrinos from the solar interior that on theoretical grounds should be created by nuclear reactions there.

How do we detect the neutrinos? Neutrinos, after all, pass through the earth and sun, barely affected by their mass. At this instant, neutrinos are passing through your body. Davis makes use of the fact that very occasionally a neutrino will interact with the nucleus of an atom of chlorine and transform it into an isotope of argon.

This transformation takes place very rarely, so Davis needs a large number of chlorine atoms. He found it best to do this by filling a large tank with liquid cleaning fluid, C_2Cl_4, where the subscripts represent the number of atoms of carbon and chlorine in the molecule, which is called perchloroethylene. One fourth of the chlorine is the Cl^{37} isotope, which is able to interact with a neutrino. Davis now has a large tank containing 400,000 liters (100,000 gallons) of this cleaning fluid.

Figure 8-11 The neutrino telescope, deep underground in the Homestake Gold Mine in Lead, South Dakota, consists mainly of a tank containing 400,000 liters of perchloroethylene. The tank is now surrounded by water.

Box 8.4 Fusion Chains for the Neutrino Experiment

Neutrinos formed in the sun's normal proton-proton (p-p) chain do not have enough energy for Davis' apparatus to detect them. The normal p-p chain begins with two hydrogen nuclei (protons), and results in one deuteron, one positron, and a weak neutrino. One quarter of one per cent of the time, one time in 400, an alternative reaction occurs in which two hydrogens plus an electron (p + e + p) result in a deuteron and a neutrino. This *pep* neutrino is somewhat more energetic (has more energy) than the p-p neutrino, and can barely be detected by Davis' apparatus (Fig. 8–11).

Neutrinos that are even more detectable are formed further along in the p-p and pep chains. Again, these higher energy neutrinos are not formed in the most usual branch (in which deuterium and hydrogen become He^3 and two He^3's make He^4 plus 2 H^2's), but in a minor branch involving boron and beryllium.

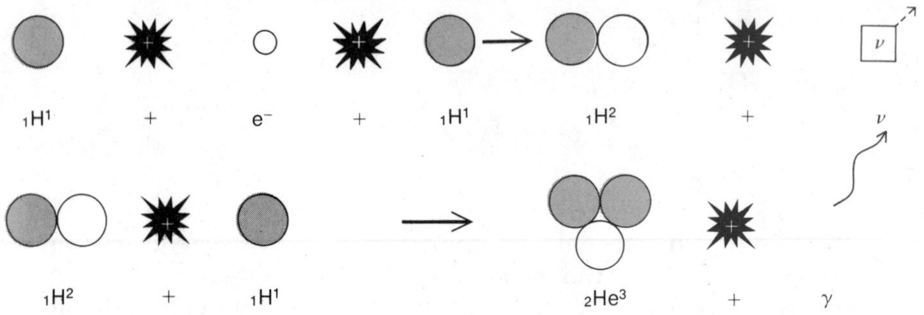

$$_1H^1 \qquad + \qquad e^- \qquad + \qquad _1H^1 \qquad \rightarrow \qquad _1H^2 \qquad + \qquad \nu$$

$$_1H^2 \qquad + \qquad _1H^1 \qquad \longrightarrow \qquad _2He^3 \qquad + \qquad \gamma$$

Figure 8–12 The proton-electron-proton (pep) chain.

Since his experiment is our only direct opportunity to test the theoretical work on stellar interiors, Davis' results are exceedingly important.

At first Davis reported: No neutrinos. Theoreticians went back to work and tried to carry out complicated calculations including more realistic assumptions than they had previously used. They included, for example, the fact that the sun is not just a stationary ball of gas but that it is rotating. Still the new calculations did not give predictions of appreciably lower value. As time went on, Davis refined his experiment and increased the total amount of observing time. His latest value is not zero, but is less than one third of the best theoretical prediction.

Even with this huge tankful of chlorine atoms, calculations show that Davis should expect only about one neutrino-chlorine interaction every day. This experiment is surely one of the most difficult ever attempted.

When an interaction occurs, a radioactive argon atom is formed. Any such argon atoms can be removed from the tank by standard chemical techniques. Then the radioactivity of the few resulting argon atoms can be measured. Davis runs the experiment for 1- to 3-month intervals by leaving the tank undisturbed for that length of time.

Davis set up his apparatus far underground, in order to shield it from other particles from space that could also interact with the chlorine. Thus 1.5 km underground, deep in the Homestake Gold Mine in Lead, South Dakota, you will find the most incredible of the world's "telescopes"—this 400,000-liter tank of cleaning fluid (Figs. 8–11 and 8–13). (Davis denies the story that after he bought all his cleaning fluid he was besieged by wire-coat-hanger salesmen.) The tank is even submerged in water, to further prevent contamination by outside particles (Fig. 8–14).

It is fair to say that Davis' results have astounded the scientific community, which has confidently expected to hear reports of a small but measurable number of neutrino interactions. But Davis is finding fewer neutrinos than expected. His measurement is at the minimum level that standard theories are able to predict.

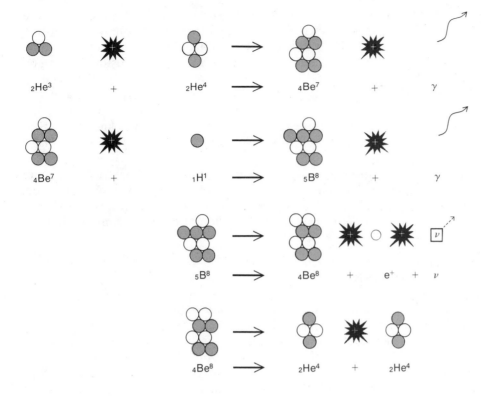

$$_2He^3 \quad + \quad _2He^4 \quad \longrightarrow \quad _4Be^7 \quad + \quad \gamma$$

$$_4Be^7 \quad + \quad _1H^1 \quad \longrightarrow \quad _5B^8 \quad + \quad \gamma$$

$$_5B^8 \quad \longrightarrow \quad _4Be^8 \quad + \quad e^+ \quad + \quad \nu$$

$$_4Be^8 \quad \longrightarrow \quad _2He^4 \quad + \quad _2He^4$$

Figure 8–13 The chain of fusion via the boron-beryllium route, through a version of the proton-proton chain, leads to the ejection of neutrinos of an energy that should be detectable by Davis' apparatus.

Several questions immediately come to mind. The first deals with Davis' apparatus, and whether there are some experimental effects that could explain the results in some normal manner. Davis has carried out a series of careful checks of his apparatus, and most astronomers and chemists are convinced that the experimental setup is not the cause of the problem.

Next, perhaps we do not understand the neutrino as well as we thought. Perhaps it disintegrates during the 8 minutes it takes to travel from the sun to the earth. That would indeed account for the failure to detect neutrinos, but would itself be an important result for physics. There is no experimental evidence for this situation, and not many scientists are still seriously considering it.

Another possibility is that we really don't understand the basic physics of nuclear reactions as well as we had thought. For example, the new experiments mentioned in Section 8.3 that the neutrino may have a small mass and may change from one state to another would drastically change our predictions for the number of neutrinos we would expect. Davis' chlorine would detect only the type of neutrinos we think the sun emits. If these neutrinos changed to another type by the time they reached earth, then he wouldn't detect them.

Another possibility now being considered with increasing respect is the idea that maybe we are doing everything right but that the sun is simply not generating neutrinos at the moment! The neutrino experiment tells us what the sun is doing now, or really what it did eight minutes earlier. The other way we can detect the effect of the nuclear furnace in the solar interior is to observe the solar surface, but it takes millions of years for energy

Perhaps ice ages in the past resulted from other periods long ago when the sun was turned "off"; perhaps the current episode will lead to ice ages hundreds of thousands of years from now.

Figure 8–14 Davis tried out the water that helps the kilometer of earth overhead shield his neutrino telescope from particles other than neutrinos.

Gallium is sensitive to neutrinos of lower energy than those with which chlorine interacts. Davis would need 50 tons of gallium, roughly equivalent to a year's production. Nonetheless, gallium production is increasing since this element is used in many solid state devices including the light-emitting diodes (LED's) that form the digits on many pocket calculators and digital watches. Gallium turns into germanium from an interaction with a neutrino, and Davis is now working on chemical means to separate the few germanium atoms from the gallium, which is liquid at room temperature.

generated inside to work its way up to the surface. We have thought of main-sequence stars as sedentary objects, regularly transforming hydrogen into helium at a steady rate. But they may be actually cyclic, and we would be merely observing the sun at a period when little nuclear fusion is going on in its interior.

Research is continuing to try to produce theoretical explanations of Davis' neutrino results. In the meantime, Davis continues to run his apparatus. The uncertainties involved in both the experimental and theoretical values must be reduced before we can definitely say that there is a major problem.

Davis plans to extend his search by using gallium, a material even more sensitive to neutrinos than is chlorine. Another experiment, using lithium, is also being developed. Parallel searches are starting in the Soviet Union. Observational results in this field take a long time to accumulate but are worth waiting for.

8.7 DYING STARS

We have been considering what we know and how much we may not really know about stars as they spend their middle age on the main sequence as dwarfs. Their main-sequence lifetimes, temperatures, and brightnesses depend on how much mass they contain. After this phase of life, we have seen that the stars evolve upward on the H-R diagram. Eventually they become brighter. Also, their surfaces become cooler and they move to the right on the H-R diagram. Once the stars evolve off the main sequence, mass differences play an even more important role.

We shall devote the next three chapters to the various end stages of stellar evolution. First we shall discuss the less massive stars, such as the sun. In Chapter 9, we shall see how such stars swell in size to become giants, possibly become planetary nebulae, and then end their lives as white dwarfs.

More massive stars come to more explosive ends. In Chapter 10, we shall see how some stars are blown to smithereens, and how strange objects called "pulsars" may be remnants. Other remnants appear to be part of x-ray–emitting binary systems.

In Chapter 11, we shall discuss the strangest kind of stellar death of all. The most massive stars may become "black holes," and effectively disappear from view.

SUMMARY AND OUTLINE

Protostars (Section 8.1)
   Their evolutionary tracks are on the H-R diagram
   They may form in dark clouds
   Infrared observations indicate we may be observing
      dust shells around protostars

T Tauri stars (Section 8.1)
   Irregular variations in magnitude exist
   They may not have reached the main sequence
Stellar energy generation (Sections 8.2, 8.3, and 8.4)
   Nuclear fusion

Definition of nuclear particles, isotopes
The proton-proton chain dominates in cooler stars, including the sun
The carbon cycle is for hotter stars
The triple-alpha process takes place after helium has been formed and at high temperatures
Stellar nucleosynthesis
Stars on the main sequence maintain a balance between pressure and gravity (Section 8.5)

A test of the theory: the neutrino experiment (Section 8.6)
The mechanism described
The result has been too few neutrinos
The possible explanations
The consequences for our faith in the theory
Mass as the determining factor in evolution (Section 8.7)

QUESTIONS

1. Since individual stars can live for billions of years, how can observations taken at the current time tell us about stellar evolution?

2. What is the source of energy in a protostar? At what point does a protostar become a star?

3. What is the *evolutionary track* of a star?

4. Arrange the following in order of development: OB associations; T Tauri stars; interstellar gas and dust; pulsars.

†5. What is the net charge (in terms of proton charges) of Ca I, Mg II, Fe II, Fe XII? (Consult Appendix 4 for a list of the elements. The number of protons an element has is called its atomic number, and is tabulated there.)

6. (a) If you remove one neutron from helium, the remainder is what element?
   (b) Now remove one proton. What is left?
   (c) Why is He IV not observed?

7. (a) Explain why nuclear fusion takes place only in the centers of stars rather than on their surfaces as well.
   (b) What is the major fusion process that takes place in the sun?

8. If you didn't know about nuclear energy, what possible energy source would you suggest for stars? What is wrong with these alternative explanations?

†9. (a) If all the hydrogen in the sun were converted to helium, what fraction of the solar mass would be lost?
   (b) How many times the mass of the earth would that be?

10. (a) How does the temperature in a stellar core determine which nuclear reactions will take place?
    (b) Why do more massive stars have shorter main-sequence lifetimes?

11. Why is the gravitational attraction between you and the earth stronger than the electrical attraction?

12. In what form is energy carried away in the proton-proton chain?

13. In the proton-proton chain, the products of some reactions serve as input for the next and therefore don't show up in the final result. Identify these intermediate products. What are the **net** input and output of the proton-proton chain?

14. What do you think would happen if nuclear reactions in the sun stopped? How long would it be before we noticed?

15. Why do neutrinos give us different information about the sun than does light?

16. Why are the results of the solar neutrino experiment so important?

———————————

†This indicates a question requiring a numerical solution.

CHAPTER 9

THE DEATH OF STARS LIKE THE SUN

AIMS:
To understand what happens to stars of up
to 4 solar masses when they have finished
their time on the main sequence of the
H-R diagram

The sun is just an average star, in that it falls in the middle of the main sequence. But the majority of stars have less mass than the sun. So when we discuss the end of the main-sequence lifetimes of low-mass dwarfs, including not only the sun but also all stars containing up to a few times its mass, we are discussing the future of most of the stars in the universe. In this chapter we will discuss the late—(post–main-sequence—stages of evolution of all stars that, when they are on the main sequence, contain up to about 4 solar masses. We shall call these *lightweight stars.*)Let us consider a star like the sun, in particular, remembering that all stars up to 4 solar masses go through similar stages but at different rates.

9.1 RED GIANTS

During the main-sequence phase of lightweight stars, as with all stars, hydrogen in the core gradually fuses into helium. By about 10^{10} years after a one-solar-mass star first reaches the main sequence, no hydrogen is left in its core, which is thus composed almost entirely of helium. Though there is no longer any hydrogen in the core, hydrogen is still undergoing fusion in a shell around the core.

Since no fusion is taking place in the core, there is no ongoing nuclear process to replace the heat that flows out of this hot central region. The core

Figure 9–1 An H-R diagram, showing the evolutionary tracks of 0.8 and 1.5 solar mass stars, as the stars evolve from the main sequence to become red giants. For each star, the first dot represents the point where hydrogen burning starts and the second dot the point where helium burning starts. The vertical axis is the luminosity of the star, divided by that of the sun; the horizontal axis is the effective temperature at which the star radiates.

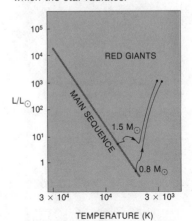

The Bubble Nebula, NGC 7635, often described as a planetary nebula (although some recent work indicates that it might receive its energy from outside rather than from inside, unlike ordinary planetary nebulae).

Half the energy from gravitational contraction always goes into kinetic motion in the interior. An example of such increased kinetic motion is the rise in temperature that we call heat. This is an important theorem, the virial theorem, which we met before in Section 8.1 in discussing the contraction of protostars.

Figure 9–2 A red giant swells so much it can be the diameter of the earth's orbit.

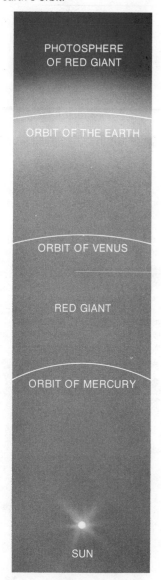

PHOTOSPHERE
OF RED GIANT

ORBIT OF THE EARTH

ORBIT OF VENUS

RED GIANT

ORBIT OF MERCURY

SUN

no longer has enough pressure to hold up both itself and the overlying layers against the force of gravity. As a result, the core then begins to contract under the force of gravity (we say it contracts *gravitationally*). This gravitational contraction not only replaces the heat lost by the core, but also in fact heats the core up further. Thus, paradoxically, soon after the hydrogen burning in the core stops, the core becomes hotter than it was before because of the gravitational contraction.

As the core becomes hotter, the hydrogen-burning shell around the core becomes hotter too, and the reactions proceed at a higher rate. Part of the increasing energy production goes into expanding the outer parts of the star. In fact, the surface temperature of the star decreases. Thus, as soon as a main-sequence star forms a substantial core of helium, the outer layers grow slowly larger and redder. The star winds up giving off more energy (because the nuclear reaction rates are increasing). The total luminosity, which is the emission per unit area times the total area, is increased. The luminosity increases even though each bit of the surface is less luminous than before (as a result of its decreasing temperature) because the total surface area that is emitting increases rapidly.

The process proceeds at an ever accelerating pace, simply because the hotter the core gets, the faster heat flows from it and the more rapidly it contracts and heats up further. The hydrogen shell gives off more and more energy. The layers outside this hydrogen-burning shell continue to expand.

The sun is presently still at the earlier stage where changes are stately and slow. But in a few billion years, when the hydrogen in its core is exhausted, the time will come for the sun to brighten and redden faster and faster until it eventually swells and engulfs Mercury and Venus. At this point, the sun will no longer be a dwarf but will rather be what is called a *red giant* (Fig. 9–1). Its surface will be so close that it will sear and char everything on earth. When the sun is sufficiently bloated to be nearly the size of the earth's orbit (Fig. 9–2), we won't be around to admire the show.

While the outer layers are expanding, the inner layers continue to heat up, and eventually reach 10^8 K. At this point, the triple-alpha process begins for all but the least massive lightweight stars, as groups of three helium nuclei fuse into single carbon nuclei. (This triple-alpha process was described in Section 8.4.) For stars about the mass of the sun, the onset of the triple-alpha process happens rapidly. The development of this *helium flash* produces a very large amount of energy in the core, but only for a few years. The consequences of this input of energy at the center are to reverse the evolution that took place prior to the helium flash: now the core expands and the outer portion of the star contracts. As the star adjusts to this situation it becomes smaller and less luminous, moving back down and to the left on the H-R diagram. When the effect of the helium flash is stabilized, the star is burning hydrogen and helium in separate shells; its core is made of carbon.

9.2 PLANETARY NEBULAE

When the carbon core forms in a red giant, it contracts and heats up as did the helium core before it. This drives up the hydrogen-to-helium reaction rates in the shell surrounding the core. The star then moves up and to the right on the H-R diagram, becoming a luminous red giant once again. This time, however, the swelling and cooling continue on and on until, after a time, the outer layers grow sufficiently cool that the nuclei and electrons combine and form neutral atoms. As electrons that are free have more energy than electrons that are bound up as parts of atoms, the system as a whole experiences an increase in energy when the previously free electrons recombine with the nuclei.

The excess energy is given off as photons, which are absorbed by the gas and blow the outer layers outward. But as the outer layers expand, they in turn become cooler, so that still more nuclei and electrons recombine and still more energy is released. In this way a vicious circle quickly sets in, which is broken in many cases when the outer layers of the star are blasted off. By the time the blown-off layers have moved a few astronomical units away from the star, they have spread into a shell thin and cool enough to be transparent.

We know of a thousand such objects in our galaxy that can be explained by the above theoretical picture, and there may be 10,000 more. They were named *planetary nebulae* because hundreds of years ago, when they were discovered, they appeared similar to the planets Uranus and Neptune when viewed in a small telescope. Both planets and "planetary nebulae" appeared as small, greenish disks. We have since discovered what planetary nebulae really are; they have nothing at all to do with planets, but they retain their old name for historical reasons. Their greenish color is caused by the presence of certain strong emission lines of multiply ionized oxygen (that is, oxygen that has lost more than one electron) and other elements. These lines happen to fall in the green region of the visible spectrum.

Planetary nebulae are exceedingly beautiful objects (Figs. 9–3, 9–4, and 9–5, and Color Plates 49, 50, and 51). These nebulae are actually semi-transparent spheres or cylinders of gas. When we look at their edges, we are looking obliquely through the shells of gas (Fig. 9–6), and there is enough gas along our line of sight to be visible. But when we look through the center of the spherical shell, there is less gas along our line of sight, and it appears transparent.

Figure 9–3 NGC 2392, a planetary nebula in the constellation Gemini, the Twins. This object is sometimes known as the Eskimo Nebula.

Figure 9–4 The Dumbbell Nebula, M27, a planetary nebula in the constellation Vulpecula, the Fox.
Figure 9–5 NGC 7293, a planetary nebula in the constellation Aquarius, the Water Bearer.

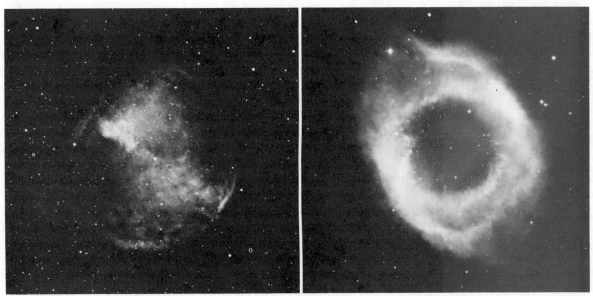

Figure 9–4

Figure 9–5

Figure 9–6 A hollow shell, when viewed from outside, can give the appearance of a ring, since when looking at the center we see perpendicularly through nearly-transparent gas. Only near the edges of the shell, where we are looking through the gas at a very oblique angle, does the gas become readily visible.

1.4 solar masses is a theoretically derived limit.

Do not confuse the term "white dwarf" with the term "dwarf." The former refers to the dead hulks of stars in the lower left of the H-R diagram, while the latter refers to normal stars on the main sequence.

Astronomers are particularly interested in planetary nebulae because they want to study the various means by which stars can eject mass into interstellar space. This tells them both how a star can reduce its own mass and also about the origin of some of the interstellar matter. Each planetary nebula represents the ejection of 10 or 20 per cent of a solar mass, which is only a small fraction of the mass of the star.

We can measure the ages of planetary nebulae by tracing back the shells at their current rate of expansion and calculating when they would have been ejected from the star. Most of the planetary nebulae we can observe appear to be less than 50,000 years old. This is consistent with our picture of planetary nebulae as a transient phase in the lifetimes of stars of about one solar mass; after about 100,000 years the nebulae will have expanded so much that the gas will be invisible.

9.3 WHITE DWARFS

Consider what happens later on for a star that on the main sequence has up to approximately four solar masses. We have seen it transform all the hydrogen in its core into helium and then all the helium in its core into carbon. The star does not then heat up sufficiently to allow the carbon to fuse into still heavier elements. Accordingly, there comes a time when nuclear reactions are no longer generating energy to maintain or increase the internal pressure that balances the force of gravity.

All stars that at this stage in their lives contain less than 1.4 solar masses have the same fate. Many or all stars in this low-mass group come from the red giant phase, and may pass through the phase of being central stars of planetary nebulae. Those that do not go through a planetary nebula stage lose parts of their mass in some other way, perhaps during the giant stage. In any case, they lose large fractions of their original masses. Stars that contained one solar mass when they were on the main sequence now contain 0.5 solar mass. Stars that contained 4 solar masses when they were on the main sequence probably now contain only 1.4 solar masses.

When their nuclear fires die out for good, the stars with less than 1.4 solar masses remaining shrink in size, and reach a stable condition that we shall describe below. As they shrink they grow very faint (in the opposite manner to that in which red giants grew brighter as they grew bigger). Because the Planck curve of radiation from many of the stars peaks to the left of the center of the visible part of the spectrum, most appear whitish in color. (Since the intensity falls off slowly on the long wavelength side of the Planck curve, even those stars whose radiation peaks in the uv give off amounts of radiation that vary only slightly across the visible, and so also appear whitish.) Some stars of this type are cooler, though, and so appear yellow or even red. Whatever their actual color, all these stars are called *white dwarfs*. The white dwarfs occupy a region of the Hertzsprung-Russell diagram that is below and to the left of the main sequence (Fig. 9–7).

White dwarfs represent a stable phase in which stars of less than 1.4 solar masses live out their old age. Obviously, something must be holding up the material in the white dwarfs against the force of gravity; nuclear reactions generating thermal pressure no longer take place in their interiors. The property that holds up the white dwarfs is a condition called *electron degeneracy;* we thus speak of "degenerate white dwarfs."

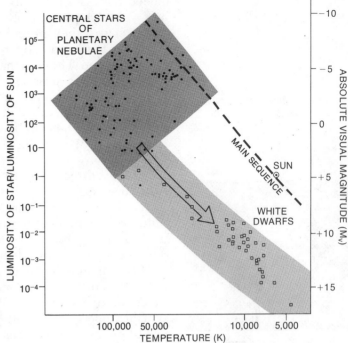

Figure 9–7 When the planetary nebula blows off a red giant, the core of the star remains visible and can be detected at the center of the nebula. Since this *central star* has a high temperature, it appears to the left of the main sequence. These stars eventually shrink and cool to become white dwarfs, as shown on this H-R diagram.

Electron degeneracy is a condition that arises in accordance with certain laws of quantum mechanics (and is not something that is intuitively obvious). As the star contracts and the electrons get closer together, there is a continued increase in their resistance to being pushed even closer. This manifests itself as a pressure. There comes a time when the pressure generated in this way exceeds the normal thermal pressure. This only happens at very great densities. When this pressure from the degenerate electrons is sufficiently great, it balances the force of gravity and the star stops contracting.

Thus, the effect of the degenerate electron pressure is to stop the white dwarf from contracting; the gas is then in a very compressed state. In a white dwarf, a mass approximately that of the sun is compressed into a volume only the size of the earth (Fig. 9–8). A single teaspoonful of a white dwarf weighs 5 tons; it would collapse a table if you somehow tried to put some there. A white dwarf contains matter so dense that it is in a truly incredible state.

What will happen to the white dwarfs with the passage of time? Because of their electron degeneracy, they can never contract further. Still, they have some heat stored in the nuclear particles present, and that heat will be radiated away over the next billions of years. Then the star will be a burned-out hulk called a *black dwarf*, though it is probable that no white dwarfs have yet lived long enough to reach that final stage. It will be billions of years before the sun becomes a white dwarf and then many billions more before it reaches the black dwarf stage.

Some black dwarfs come from stars that were not massive enough to begin hydrogen burning; others are cooled white dwarfs.

Figure 9–8 The sizes of the white dwarfs are not very different from that of the earth. A white dwarf contains about 300,000 times more mass than does the earth, however.

9.4 OBSERVING WHITE DWARFS

White dwarfs, as we can see from their positions on the H-R diagram, are very faint and thus should be correspondingly difficult to detect. This is

The center of mass *of a system is the point around which each individual mass appears to orbit.*

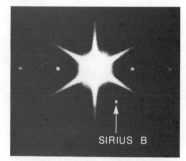

SIRIUS B

Figure 9–9 Sirius A appears as a bright overexposed image with six spikes caused in the telescope system; fainter additional images of Sirius A appear at regular spacing to both sides. Sirius B appears to the lower right of the brightest image of Sirius A; its additional images are too faint to see. Sirius A and B differ by 10 magnitudes, a factor of 10,000, and are only a few arc seconds apart, making them very difficult to photograph on the same picture. This led to the use of the special technique in which much of the overexposed light from Sirius A is caused to fall in the six spikes so that Sirius B can be seen between the spikes. (Photographed with the 66-cm U.S. Naval Observatory refractor by Irving Lindenblad)

A gravitational redshift *is a redshift caused by gravity, as predicted by the general theory of relativity. Einstein's theory predicts that light leaving a mass would be slightly redshifted. The greater the amount of gravity at the place where the light is emitted, the greater the redshift.*

indeed the case. White dwarfs are discovered by looking for bluish (hot) stars with high proper motions or by their gravitational effect on their companion stars if they are in binary systems.

In the former case, by studying proper motions we find the stars that are close to the sun. The ones that are fainter than main-sequence stars would be at their distances must be white dwarfs. To understand the latter case, we must realize that for any system of masses orbiting each other, we can define an imaginary point, called the *center of mass*, which moves in a straight line across the sky. The path in the sky of any of the individual bodies, though, appears wavy. We have already discussed such astrometric binaries in Chapter 6.1.

At least three of the 40 stars within 5 parsecs of the sun—Sirius (Fig. 9–9), 40 Eridani, and Procyon—have white dwarf companions. Another nearby object, known as van Maanen's star, is a white dwarf, although not in a multiple system. So even though we are not able to detect white dwarfs at great distances from the sun, there seems to be a great number of them. Hundreds of white dwarfs have been discovered.

Even though the theory that explains white dwarfs has seemed consistent with actual observations, their size had never before been measured directly until 1975 when a set of observations was made in the ultraviolet from the Copernicus satellite. The ultraviolet part of the spectrum of Sirius B could be distinguished from that of its much brighter neighbor, and was found to peak at 1100 Å. This corresponds to a temperature of under 30,000 K. Since the absolute magnitude of Sirius B is known, its surface area and thus its radius can be found. The radius of Sirius B is indeed only 4200 km, smaller than the earth's. Many white dwarfs are now being observed with the International Ultraviolet Explorer spacecraft. IUE observations confirm that the temperature of Sirius B is about 26,000 K. The observations were taken in the 1150 to 2000 Å region by Erika Bohm-Vitense, Terry Dettmann, and Stelios Kapranidis of the University of Washington in Seattle. This temperature is too low to explain why we detect x-rays from the Sirius system. We shall discuss observing with the IUE in Section 9.7.

*9.5 WHITE DWARFS AND THE THEORY OF RELATIVITY

One special reason to study white dwarfs is to make use of them as a laboratory to test extreme physical conditions. It is impossible to create such strong gravity in a laboratory on earth. We have already discussed two tests of general relativity in Section 7.11: the deflection of starlight by the sun, and the advance of the perihelion of the planet Mercury. A third test is the *gravitational redshift* of light.

The sun is bright enough that we can spread out its spectrum and measure the wavelengths of spectral lines from the photosphere with tremendous accuracy. The redshift that Einstein predicted has been successfully detected, though the effect is minuscule. Furthermore, turbulence in the solar photosphere, such as the rise and fall of granules, distorts and confuses the results. One would like to find a body with stronger gravity than the sun, and white dwarfs are just such objects.

Unfortunately, one must know the mass and size of the white dwarf to perform the test accurately, and the masses and sizes of white dwarfs are not known very well. The observational part of the test is not difficult, because strong redshifts are found. The results agree with the predictions of the theory of relativity to within the possible error that results from our uncertain knowledge of the masses and sizes of white dwarfs. The test has best been carried out with 40 Eridani B.

9.6 NOVAE

Although the stars are generally thought to be unchanging on a human time scale, occasionally a "new star," a *nova* (pl: *novae*), becomes visible. Such occurrences have been noted for thousands of years; ancient Oriental chronicles report many such events.

A nova is a newly visible star rather than actually a new star. A nova, rather, represents a brightening of a star by 5 to 15 magnitudes or more, which is equivalent to a brightening by hundreds or millions of times. Often, while brightening by this large factor, a nova passes from the realm of objects that cannot be seen with the naked eye to that of objects that can be seen with the naked eye, or from objects that are too faint to be seen even with the largest telescopes to objects that can be observed.

A nova (Fig. 9–10) may brighten within a few days or weeks. It ordinarily fades drastically within months, and then continues to fade gradually over the years. Several "recurrent" novae that appear to brighten at intervals of years or decades are known.

The spectra of novae show absorption lines that are Doppler shifted to the blue. This implies that gas between the star and earth is moving toward the earth. From this we deduce that gas is being thrown off by the star. Soon the photospheric material that has been ejected thins out enough to become transparent, and we begin to see emission lines from the gas that has expanded in all directions from the star. Months or years after the outburst of light, the shell of gas sometimes becomes detectable through optical telescopes (Fig. 9–12).

Only 10^{-4} or so of a solar mass is thrown off in a nova outburst, so there is no reason that recurrent novae can't repeat their outbursts many times. Statistics from another galaxy show that all novae probably recur eventually.

Many astronomers believe that most if not all novae occur when a binary system has one member that has evolved into a white dwarf and another

In this century, exceptionally bright novae appeared in 1901, 1918, 1925, 1934, 1942, and 1975 (Fig. 9–11).

Nova shells expand at 100 times the rate of planetary nebulae.

Planetary nebulae, in contrast, give off perhaps 20% of the star's mass. They happen only once per star.

Figure 9–10 Nova Herculis 1934, showing its rapid fading from 3rd magnitude on March 10, 1935 to below 12th magnitude on May 6, 1935.

August 28, 1975   11:30 U.T.   August 30, 1975   6:45 U.T.

Figure 9–11 These photographs are part of a unique series of observations covering the eruption of Nova Cygni 1975 during the period of its brightening. A Los Angeles amateur astronomer, Ben Mayer, was repeatedly photographing this area of the sky at the crucial times to search for meteors. Never before had a nova's brightening been so well observed. On August 28 *(left)*, no star was visible at the arrow. By August 30 *(right)*, the nova had reached 2nd magnitude, the brightness of Deneb, which is seen at the right.

member that is en route to becoming a red giant. Though the details are not certain, the nova might occur through the following processes. We have mentioned that the outer layers of a red giant are not held very strongly by the star's gravity. If a white dwarf is nearby, some of the matter originally from the red giant can surround one or both of the stars. Some may fall on the surface of the white dwarf and perhaps trigger the nuclear reactions on its surface for a brief time. The energy thus produced blows off the material. Obviously, this mechanism could recur every few years as more new matter falls on the white dwarf. /

*9.7 OBSERVING WITH THE IUE SPACECRAFT

The most widely used telescope now available to astronomers is not on a mountain top, but is rather hovering over the earth. Known as IUE, the International Ultraviolet Explorer, it has supplied data to hundreds of astronomers in the United States and around the world since its launch in 1978.

IUE (Fig. 9–13) carries a 45-cm (18-inch) telescope (Fig. 9–14) of the Ritchey-Chrétien design. It uses a vidicon (that is, a television-type device) for a detector, and records the data in a particularly efficient way so that it can survey the spectrum of a star through the entire ultraviolet in an hour or so, depending on the brightness of the star. IUE is so sensitive that it can also be used to observe much fainter objects, including galaxies and even quasars.

IUE is in an orbit that carries it around the earth once every 24 hours. Since the earth rotates underneath at the same rate, the spacecraft hovers within view of NASA's Goddard Space Flight Center in Greenbelt, Maryland, all the time. It actually traces out an ellipse on the ground that brings it within view of a European data center in Spain for 8 hours a day.

Astronomers at telescopes on earth have to be careful not to point their telescopes too close to either the sun or the moon. Astronomers (and Telescope Operators) working with IUE have to be careful as well not to point the telescope too close to the earth, which is at different directions as seen from the spacecraft at different times.

The spacecraft, which was built by NASA, is controlled from Goddard for two thirds of every day. In acknowledgment of the European contribution, the spacecraft is controlled from Spain by European astronomers for the other third of each day.

American astronomers prepare proposals that describe what they want to observe with IUE and why, and send them periodically to NASA. For the successful proposers, the observing time follows some months later. Three days or a week of 16-hour shifts, for example, might be available for a given project. Let us say that it is our proposal to observe the outer atmospheres of stars. We take our leave from home for a week or so and set out for the Goddard Space Flight Center, which is in a suburb of Washington, D.C.

Figure 9–12 In 1951, a shell of gas could be seen to surround Nova Herculis 1934.

We arrive at an office building, not at all like a telescope dome. In a small room on an ordinary office corridor, we find the American control room for this fabulous telescope in space.

Before we arrive, we have carefully figured out exactly what objects we want to observe, and indeed have a list of specific objects approved by the IUE team. We meet the Telescope Operator, the person who actually sends commands from the control room. We are really there in an advisory capacity.

Acquiring the star in the field of view might take fifteen minutes. Then an image of a field of view 16 minutes of arc across—about half the diameter of the moon—appears on a television screen before us (Fig. 9–15). Our star is probably within the field of view, because the telescope points pretty accurately where we tell it to look.

If there are several stars close together in the field, however, we can have the computer automatically display on the same screen a star map in which each star is represented by a box, the size of the box representing the brightness of a star. We then simply have the telescope slide around until the stars in the field of view fit within the boxes. This is certainly a far cry from having to locate stars from their appearance in the sky!

Now we tell the Telescope Operator to take the first exposure. We have to choose which of two possible wavelength bands to use: a relatively short one from 1150 Å, below the wavelength of hydrogen's Lyman α line, up to 2000 Å, or else a relatively long ultraviolet range from 1900 Å up to 3200 Å. Only the upper tip of the upper range comes through the earth's atmosphere, so just what is to be found in these spectra is often a tremendous surprise.

The spectrum is displayed both on a television screen on the console and also on a large screen overhead (Fig. 9–16), for the convenience of both observers and colleagues who may watch through a glass wall.

After our hour-long exposure, the data are radioed back from the spacecraft to an antenna at Goddard, and soon appear on our video screen. We have the Telescope Operator begin the next exposure, perhaps the other wavelength band for the same star.

As soon as these instructions are radioed to the spacecraft, we can make use of one of the nicest features of IUE—an interactive computer. We can tell the computer to display an enlarged region of the spectrum on the screen.

Since white dwarfs have the peak of their radiation in the ultraviolet part of the spectrum, they are especially appropriate for observation with IUE. Indeed, observations with IUE, by determining the location of the peak more precisely, help improve our understanding of white dwarf temperatures.

Figure 9–13 The International Ultraviolet Explorer spacecraft, IUE, before its launch. It is 4.3 meters high and weighs 671 kg. The telescope extends out the top, and solar cells to provide power are visible at both sides. The spectrographs are in the middle section.

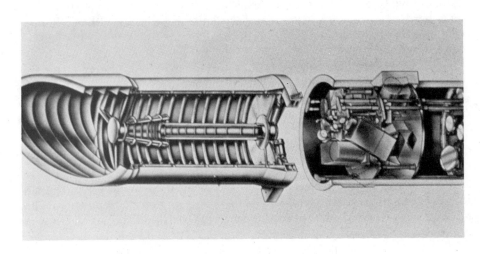

Figure 9–14 IUE's telescope, a 45-cm Ritchey-Chrétien, has a mirror made of beryllium. Baffles, devices that keep light from bouncing around inside the telescope, are used.

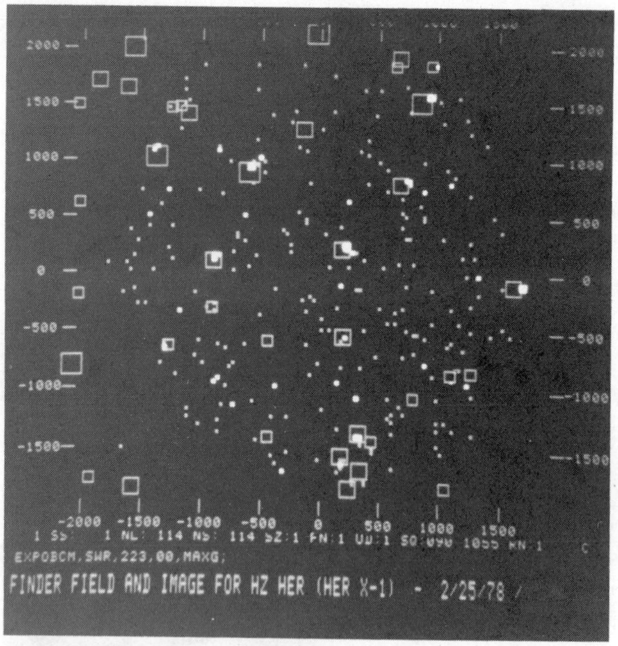

Figure 9-15

Figure 19-16

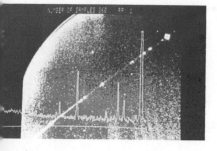

Figure 9-15 Within a couple of minutes, IUE displays its field of view so that the astronomers can specify which object to observe and to use as a guide star.

Figure 9-16 The control room contains video screens to display the data *(left)* and spacecraft housekeeping equipment *(right)*, plus a large projection tv screen *(top)* for a second view of the data. The high dispersion spectrum is dispersed by a kind of spectrograph called an *echelle,* in which many short strips of spectrum, each only 20 Å wide, are displayed one above the other. This is particularly convenient for a vidicon detector, which has a round face, because it fills up much of the viewing surface, making observing pretty efficient.

Figure 9-17 The computer can plot on its screen within seconds a graph of any part of the spectrum that the astronomer designates.

We can even have the computer display, in a few seconds, a graph of a part of the spectrum (Fig. 9–17). No more do we have to develop photographic plates and then have the densities on the emulsion traced laboriously. Within seconds, we have a spectrum in view, and can even lay a plastic overlay on it to make a few measurements right away. So much of the spectrum of objects is completely unknown in the ultraviolet that important results have regularly been found in this way.

Another advantage of the interactive computer system is that by the time our next exposure is complete, we have already gained a good idea about the kind of results we might hope for from the first spectrum. We would also know whether it was well exposed or has to be retaken.

A rest at our motel—our shifts may well run through the night—brings us up to the time to observe again. We see where the telescope is pointing, figure out what to observe first, and start again. A week of this is exhausting but rewarding. We even have preliminary data to take home with us, though computer tapes with final treatments of the data will follow us by mail a few weeks later.

IUE—the first international observatory in space—has been a tremendous success. And it should last aloft for decades. It has already sent back data on stars, planets, nebulae, galaxies, and quasars, and we look for continued productivity in the years to come.

9.8 THE EVOLUTION OF BINARY STARS

A star evolves in an orderly fashion and lives on the main sequence for a length of time that depends only on its mass and chemical composition. The effect of mass showed particularly clearly in our earlier discussion of how to determine the ages of star clusters (Section 6.5). A G2 dwarf like the sun, for example, lives on the main sequence for about 10 billion years, while a B star, which contains about 10 times as much mass as the sun, has a main-sequence lifetime of only a few million years.

But steady evolution takes place only for stars whose mass does not change substantially during their main-sequence lifetimes. The realization that many of the x-ray sources observed with new space satellites are binary systems has spurred new interest in studying the evolution of binaries. The importance of the investigation is underscored by the long-known fact that most stars are members of binary systems.

In a binary system, the mutual gravitational attraction the two stars have for each other can sometimes pull off mass, which then flows from one star to the other. This, in turn, can have a drastic effect on the evolution of both stars.

In order for one star to pull mass off another star, the force of gravity from the first star must be attracting some part of the surface of the second star more strongly than that star's own gravity is attracting it. This happens most readily when the second star is very large (for example, a red giant, which has swelled considerably over its main-sequence size). Then its outer layers are much further from its center. Since a star's gravity acts as though all the star's mass were concentrated at the center of the star, gravity does not bind the outer layers very strongly when the star has grown so large.

The effect of the first star's gravity on the second star varies from point to point because the different points are at different distances from the center of the first star. The second star may be distorted (Fig. 9–18) because of this *differential effect*, which is called a *tidal force* because the tides of the earth's oceans also arise from such a differential effect, in that case chiefly from the moon's gravity.

Let us consider a binary system made up of a main-sequence star (a dwarf) and a red giant. The parts of the red giant that are closer to the dwarf feel a stronger pull of gravity from the dwarf than the parts of the red giant that are farther from the dwarf. The tidal force pulls the outer layers of the giant component until it is more egg-shaped than round. In fact, we can draw a figure-8 shaped curve (Fig. 9–19) that marks the edge of the volumes in which each star's own gravity dominates, which we call the *Roche lobes* of the two stars. We categorize double stars according to the relation of each of

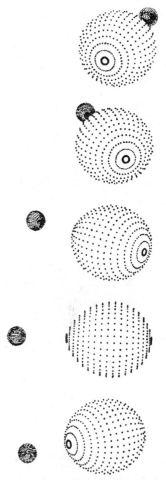

Figure 9–18 A computer simulation of an eclipse in a short period binary. The five phases shown here are separated by 1/10 of a period. The light we would measure from such a system would appear to vary in part because of the eclipses and in part because of the distortion of the components from being round.

Figure 9–19 The Roche lobes for a binary system in which the primary starts with 80 per cent of the mass and the secondary starts with 20 per cent of the mass. As the mass is transferred, the separation between the two stars diminishes until the two are equal in mass and then the separation increases. By the end of the transfer, the relation of the masses is reversed. Numbers below each frame of the sequence show the fraction of the total mass in the left component.

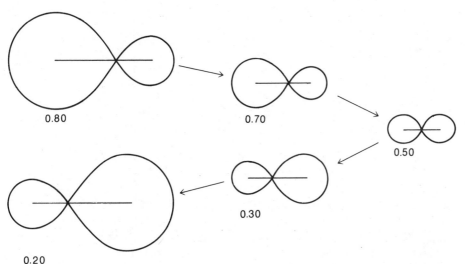

0.80

0.70

0.50

0.30

0.20

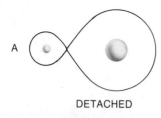

A

DETACHED

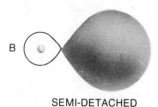

B

SEMI-DETACHED

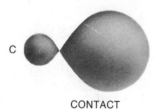

C

CONTACT

D

COMMON ENVELOPE
(OVERCONTACT)

Figure 9–20 Depending on whether one or both stars fills its Roche lobe, we can categorize double stars as *(A)* detached, *(B)* semi-detached, and *(C)* contact binaries. When both stars overfill their Roche lobes, we have an overcontact binary *(D)*.

the components to its Roche lobe (Fig. 9–20). The size of the Roche lobes is calculated theoretically from knowledge of the stars' masses. In *contact binaries*, the stars fill their Roche lobes; in *detached binaries*, they do not.

A dwarf is much smaller than its Roche lobe. But a giant may be so large that it fills its Roche lobe. When this happens, material can flow freely through the "neck" of the figure-8 and fall onto the other star. In this way, the dwarf gains mass while the giant star loses mass. This takes place rapidly at first. Sometimes the fact that the mass is newly transferred shows up as unusual abundances of the chemical elements. Angular momentum, the quantity that describes the tendency to keep spinning, is transferred too. The angular momentum causes the transferred mass to spiral down onto the dwarf. The distribution of the stars' masses thus becomes asymmetrical, leading to an asymmetrical light curve as one star occults the other (Fig. 9–21).

We have already seen how the phenomenon of mass transfer leads to novae when the smaller star is a white dwarf instead of the ordinary dwarf we have been discussing. In recent years, various types of binary stars have been studied in newly accessible parts of the spectrum, such as the ultraviolet with IUE and the x-ray with the HEAO's.

If the recipient star gains a lot of mass, its evolution can speed up very considerably. It can carry on nuclear fusion in its interior at a greatly increased rate, and will change spectral type. The relative brightness of the two stars in the binary system can change completely, or even invert.

Since it is very difficult or even impossible to gain an accurate idea of the amount of mass flowing between stars in a binary system, or the amount which has already been transferred in the previous million or so years, stellar evolution is considerably more complicated to understand for a star in a binary system than it is for an isolated star.

Figure 9–21 The bright eclipsing binary star β Lyrae may actually be in the stage of exchanging mass at a rapid rate. Here we see that its light curve is asymmetric with respect to the time of its minimum light, undoubtedly because of the mass flowing into the system. Its spectral lines are asymmetric as well.

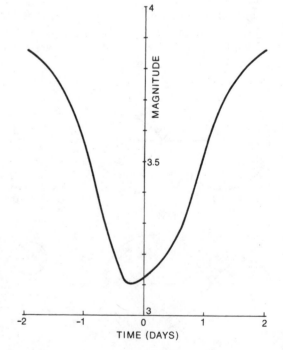

SUMMARY AND OUTLINE

Red giants (Section 9.1)
  They follow the end of hydrogen burning in the core
    Core contracts gravitationally
    Core and hydrogen-burning shell become hotter
    The increased energy expands the outer layers
    The total luminosity is the luminosity of each bit of
      surface times the total surface area
    Star is red, so each bit of surface has relatively low
      luminosity
    Surface area is very greatly increased, so total
      luminosity is greatly increased
  Helium flash
    Rapid onset of triple-alpha process
Planetary nebulae (Section 9.2)
  Their appearance
  Ages: less than 50,000 years old
  The gas is blown off when it absorbs photons
  Mass loss: 0.1 or 0.2 solar mass
White dwarfs (Sections 9.3, 9.4, and 9.5)
  End result of lightweight stars

Less than 1.4 solar masses remaining
Electron degeneracy
Detecting white dwarfs via proper motion or by their
  presence in a binary system
Use in testing the gravitational redshift
Einstein's general theory of relativity endorsed
Novae (Section 9.6)
  Mass loss: only 10^{-4} solar mass
  Binary model: interaction of a red giant and a white
    dwarf
  X-ray novae
International Ultraviolet Explorer (IUE) (Section 9.7)
  A NASA/European spacecraft with an international
    set of observers can study faint objects in the
    ultraviolet
Evolution of binary stars (Section 9.8)
  Exchange of mass between components of binary
    stars can change their evolution drastically
  A star gives off mass to its companion when it fills its
    Roche lobe

QUESTIONS

1. What event signals the end of the main-sequence life of a star?

2. When hydrogen burning in the core stops, the core contracts and heats up again. Why doesn't hydrogen burning start again?

3. When the core starts contracting, what eventually halts the collapse?

4. If you are outside a spherical mass, the force of gravity varies inversely as the square of the distance from the center. What is the ratio of the force of gravity at the surface of the sun to what it will be when the sun has a radius of one astronomical unit?

5. Why is helium "flash" an appropriate name?

6. If you compare a photograph of a nearby planetary nebula taken 80 years ago with one taken now, how would you expect them to differ?

7. Why is the surface of a star hotter after the star sheds a planetary nebula?

8. What keeps a white dwarf from collapsing further?

9. What are the differences between the sun and a one-solar-mass white dwarf?

10. When the sun becomes a white dwarf, approximately how much mass will it have? Where will the rest of the mass have gone?

11. Which has a higher surface temperature, the sun or a white dwarf?

12. Sketch an H-R diagram indicating the main sequence, and the region of white dwarfs. Now indicate the position of an ordinary dwarf that appears yellowish.

13. Why does the study of the gravitational effects of a system often give a better estimate of the mass of the system than the direct observation of all the components?

14. What evidence do we have that material is being ejected from novae?

15. When the proton-proton chain starts at the center of a star, it continues for billions of years. When it starts at the surface (as in a nova) it only lasts a few weeks. How can you explain the difference?

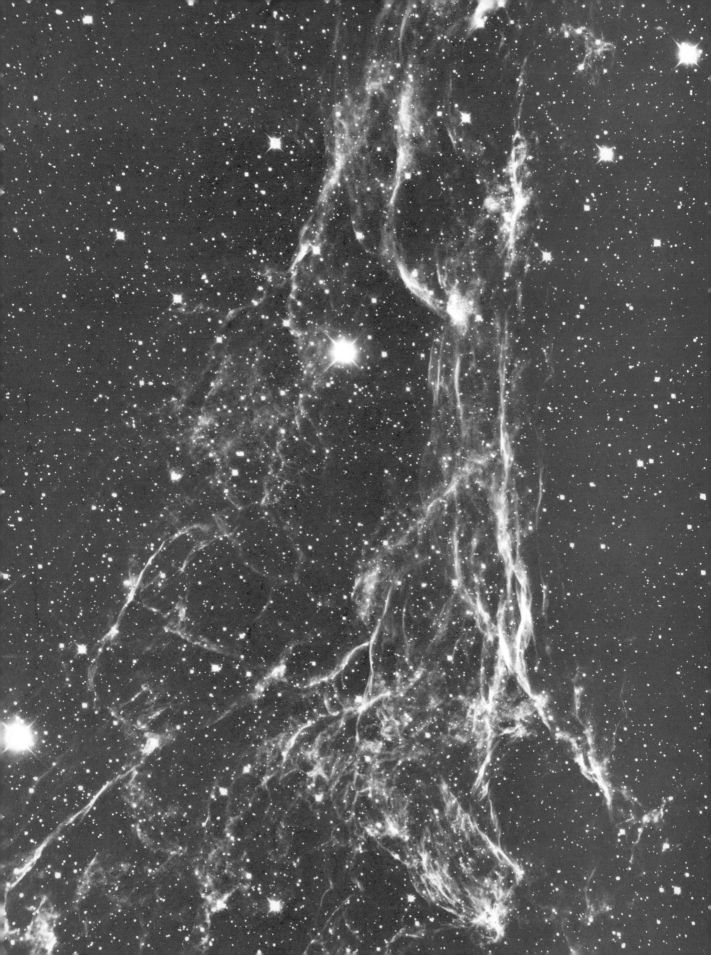

CHAPTER 10

SUPERNOVAE, NEUTRON STARS, AND PULSARS

AIMS:
To see how stars of medium and heavy mass become supernovae, with the cores of the stars of medium mass condensing to become neutron stars. We observe the neutron stars as pulsars

We have seen how the run-of-the-mill lightweight stars end, not with a bang but a whimper. The most massive stars, stars that contain more than 8 solar masses when they are on the main sequence, put on a more dazzling display. These objects, which we can call *heavyweight stars,* cook the heavy elements deep inside themselves. They then blow themselves almost to bits, forming still more heavy elements in the process. The matter that is left behind settles down to even stranger states of existence than that of a white dwarf.

In this chapter and the next one we shall discuss the fate of stars more than 8 times as massive as the sun, many of which explode as *supernovae.* We shall see that some of the massive stars, after the supernova stage, become neutron stars, which we detect as pulsars and x-ray binaries. In Chapter 11, we shall discuss the death of other heavyweight stars, which become black holes.

10.1 RED SUPERGIANTS

Stars that are much more massive than the sun whip through their main-sequence lifetimes at a rapid pace. These prodigal stars use up their store of hydrogen very quickly. A star of 15 solar masses may take only 10 million

A division using the graphic terms of lightweight (Chapter 9), middleweight, and heavyweight stars was suggested by Martin Schwarzschild of Princeton. At present, how middleweight stars evolve is uncertain; they probably just lose mass and become white dwarfs like the lightweight stars.

Note that we are really discussing mass rather than weight. Mass is an intrinsic property of matter; weight, on the other hand, is a less fundamental quantity. The weight of any mass is the force of gravity on it.

Filaments at the north end of the supernova remnant known as the Cygnus Loop.

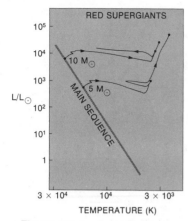

Figure 10–1 An H-R diagram (as in Fig. 9–1) showing the evolutionary tracks of 5 and 10 solar mass stars as they evolve from the main sequence to become red supergiants. For each star, the first dot represents the point where hydrogen burning starts, the second dot represents the point where helium burning starts, and the third dot the point where carbon burning starts.

years from the time it first reaches the main sequence until the time when it has exhausted all the hydrogen in its core. This is a lifetime a thousand times shorter than that of the sun. When the star exhausts the hydrogen in its core, the outer layers expand and the star becomes a red giant.

For these massive stars, the core can then gradually heat up to 100 million degrees, and the triple-alpha process begins to transform helium into carbon. However, for more massive stars the helium burns steadily after the helium ignition, unlike the helium flash of less massive stars.

By the time helium burning is concluded, the outer layers have expanded even further, and the star has become much brighter than even a red giant. We call it a *red supergiant* (Fig. 10–1); Betelgeuse, the star that marks the shoulder of Orion, is the best-known example (Fig. 10–2). Supergiants are inherently very luminous stars, with absolute magnitudes of up to -10. The sun, with an absolute magnitude of $+5$, is only one-millionth as luminous as the most brilliant of the red supergiants. A supergiant's mass is spread out over such a tremendous volume that its average density is less than one millionth that of the sun.

Some scientists think that the carbon core can contract, heat up, and begin fusing into still heavier elements. Two C^{12} nuclei can fuse into magnesium, Mg^{24}, for example. Eventually, in some cases, even elements in the iron group (iron, cobalt, and nickel) build up. The core, containing the elements with the highest mass numbers, is surrounded by layers of elements of different mass, with the lightest toward the periphery and the heaviest toward the center.

10.2 SUPERNOVAE

After reaching supergiant status, a very massive star will eventually

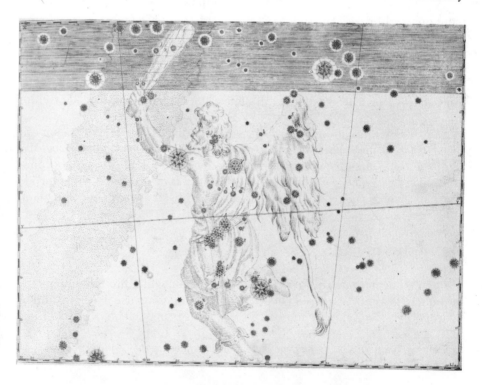

Figure 10–2 Betelgeuse is the star labeled α in the shoulder of Orion, the Hunter, shown here in Johann Bayer's *Uranometria*, first published in 1603.

explode in a glorious burst called a *supernova* (pl: *supernovae*). The conditions in the center of the star change so quickly that it becomes very difficult to model them satisfactorily in sets of equations or on computers. Many different models have been advanced to describe what happens in a supernova; none is universally accepted.

Let us describe an example of a specific set of events that could lead to a supernova explosion in order to illustrate how the advance of stellar evolution can run away with itself and go out of control.

In the case we are considering, a substantial iron core has been formed and begins to shrink and heat up. The iron represents the ashes of the previous stage of nuclear burning. Eventually the temperature becomes high enough for the iron to undergo nuclear reactions. The stage is now set for disaster because iron nuclei have a fundamentally different property from other nuclei when undergoing nuclear reactions. Unlike other nuclei, iron absorbs rather than produces energy in order to undergo either fusion (the merging of nuclei) or fission (the splitting of nuclei). Thus iron takes up energy when it is being transformed to other elements, either heavier or lighter. This energy is no longer available to heat the core.

The core responds to this loss of energy by shrinking and heating up still more. The iron can be broken up into lighter nuclei by the high-energy photons (gamma rays) of radiation that are generated in the fusion process, and in doing so absorbs still more energy. The process goes out of control. Within seconds—a fantastically short time for a star that has lived for millions and millions of years—the core collapses and heats up catastrophically. It was formerly thought that matter falling in upon the collapsing core from the layers above would undergo violent nuclear reactions in the intense heat of the core, but this process is now thought to be less likely and less important than the following process:

As matter falls in upon the collapsing core from the layers above, electrons and protons combine to make neutrons. In the process, huge numbers of neutrinos, the elusive particles we discussed in Sections 8.3 and 8.6, are given off. The core collapses; theoretical calculations show that it then bounces outward.

The rebounding core meets the star's outermost material, which is rapidly falling inward. The collision causes heavy elements to form and throws off the outer layers. With this tremendous explosion, the star is destroyed. Only the core is left behind. This explosion of the star is called its supernova phase. To summarize, the essential part of the story is that once the core is made up of iron, the star is likely to explode.

The heavy elements formed either near the center of the star or in the supernova explosion are spread out into space, where they enrich the interstellar gas. When a star forms out of such enriched gas, these heavy elements are present, so the abundances of elements heavier than hydrogen are higher than they are in older stars. Also, this is the source of heavy elements in planets. Thus supernovae provide the heavy elements that are necessary for life to arise. Most of the atoms in each of us have been through such supernova explosions. The third High-Energy Astronomy Observatory (HEAO-3) carries a gamma-ray spectrometer that can detect gamma-ray spectral lines from the formation of heavy elements in supernovae.

In the days or weeks following the explosion, the amount of radiation emitted by the supernova can equal that emitted by the rest of its entire

This model explains Type II supernovae, which are the explosion of massive stars. Type I supernovae, on the other hand, may take place in less massive stars, perhaps from the addition of mass to a white dwarf, which leads to the star's collapse. Type I and II supernovae can be distinguished from each other by their spectra and the rate at which their brightness changes.

It was, until recently, believed that the outer layers of the star are so dense that they stop the neutrinos. In so doing they would be pushed away from the core with great force. But recent work in nuclear physics has indicated that the neutrinos are delayed by 1 second or so in escaping from the core, so by the time they escape, it may be too late for them to blow off the outer layers.

Figure 10–3 Views of the central part of the galaxy NGC 5252, taken in 1959 and in 1972. The supernova that appeared in 1972 was nearly as bright as the rest of the galaxy.

Figure 10-4 Views of the supernova that erupted in 1979 in the galaxy M100. Less spiral structure appears in the later view, chiefly because of a different exposure time, yet the supernova appears prominently at lower left. (Photo by Paul Griboval with an electronic camera and the 0.76-m telescope of the McDonald Observatory, University of Texas)

galaxy. Most of the supernovae that we observe (Figs. 10–3 and 10–4) are in distant galaxies, and we can see a single star outshine its galaxy for a period of weeks. It may brighten by over 20 magnitudes, a factor of 10^8 in luminosity.

10.2a Supernova Remnants

The gas we see as supernova remnants includes not only the gas from the exploded star but also interstellar gas swept up and compressed by shock waves from the supernova explosion. The Cygnus loop, whose picture opened this Part, is mostly matter that was compressed by such shock waves.

Tycho's supernova, observed by the great astronomer in 1572, is the bright object at the left edge of Cassiopeia's chair as shown in the drawing opening Chapter 3. Kepler's supernova, observed by that great astronomer in 1604, was the last to be seen from earth.

Optical astronomers have photographed two dozen of the stellar shreds that are left behind, which are known as *supernova remnants* (Fig. 10–5, the photograph that opens this chapter, and Color Plates 54 and 55), in our galaxy alone. The gas gives off not only optical radiation but also strong radio radiation, and the supernova remnants are best studied by means of radio astronomy (which was described in Section 2.13). About 100 supernova remnants are known from observations made in the radio part of the spectrum. Some of these have now also been observed in the x-ray part of the spectrum (Fig. 10–6 and Color Plates 52 and 53).

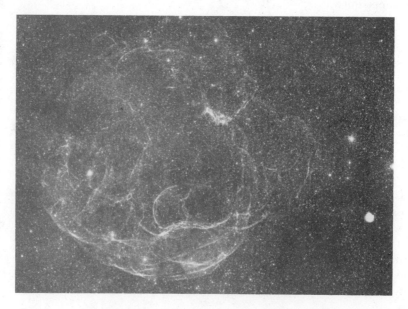

Figure 10-5 S147, the remnant of a supernova explosion in our galaxy. The long delicate filaments shown cover an area over $3° \times 3°$, about 40 times the area of the moon.

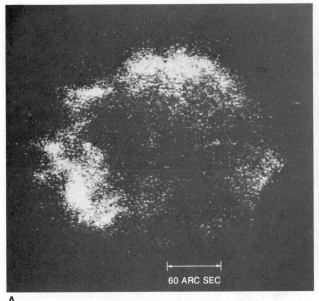

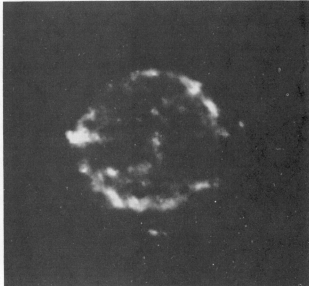

60 ARC SEC

A   B

Figure 10–6 The Cassiopeia A supernova remnant. *(A)* An x-ray image made with the Einstein Observatory. *(B)* A radio image of Cas A made with the Very Large Array. The source is a relatively recent supernova remnant, but the supernova itself was not observed. We can date it to 1667 by studying its expansion. Both images show a bright shell of emission, presumably the supernova matter encountering the interstellar medium. Analysis of the x-ray spectrum shows that the abundances of the elements silicon, sulfur, and argon are enhanced over the normal solar abundance, demonstrating that heavy elements are indeed formed in supernovae. Other supernova remnants have even greater enhancements of these elements. (The x-ray image is courtesy of Steven Murray et al. of the Harvard-Smithsonian Center for Astrophysics. The radio image was made by A. R. Thompson et al., and is courtesy of R. Hjellming.)

Supernovae were named at a time when it was thought that they were merely unusually bright novae. But now we know that novae and supernovae are very different phenomena. (A supernova explosion represents the death of a star and the scattering of most of its material, while a nova uses up only a small fraction of a stellar mass and can recur.)

In our own galaxy, we have historical records of only five supernovae in the last thousand years, so we can conclude that many of the 24 optical remnants and over 100 radio remnants known must have come from stars that exploded before such events were recorded.

In 1054 A.D., Chinese chronicles recorded the appearance of a "guest star" in the sky that was sufficiently bright that it could be seen in the daytime. No one is certain why no Western European records of the supernova were made, for the Bayeux tapestry illustrates a comet and thus shows that even in the Middle Ages people were aware of celestial events. A reference to a sighting of the supernova in Constantinople has recently been discovered. Certain cave and rock paintings made by Indians in the American southwest may show the supernova (Fig. 10–7), though this interpretation is controversial.

When we look at the reported position in the sky, in the constellation Taurus, we see an object that clearly looks as though it is a star torn to shreds (see Fig. 10–17 and Color Plate 54). It is called the Crab Nebula. The Crab appears to be about 1/5 of the distance from the sun to the center of our galaxy. The age, the location, and the current appearance of the object leave us no doubt that the Crab is the remnant of the supernova whose radiation reached earth in 1054 A.D., twelve years before William the Conqueror invaded England. We find its age by noting the rate at which the filaments in the Crab Nebula are expanding. Tracing the filaments back in time shows that the filaments were at a single point in approximately 1140 A.D. The small discrepancy between this and 1054 A.D. can be accounted for by the reasonable assumption that the expansion has been slowing down.

Figure 10–7 An American Indian cave painting discovered in northern Arizona that may depict the supernova explosion 900 years ago that led to the Crab Nebula. It had been wondered why only Oriental astronomers reported the supernova, so searches have been made in other parts of the world for additional observations.

Figure 10–8 HEAO-3, launched in 1979 to study cosmic rays and gamma rays.

From a study of the rate at which supernovae seem to appear in distant galaxies, it is estimated that supernovae should appear in our galaxy about once every 15 to 50 years. There have been only a few known supernovae in our galaxy in the past 900 years, though any that occurred on the far side of our galaxy may have been hidden from our view by all the interstellar matter. There could be a supernova in our part of the galaxy any day. We've already been waiting for hundreds of years, since the last one was observed in 1604.

10.3 COSMIC RAYS

Most of the information we have discussed thus far in this book has been gleaned from the study of the electromagnetic radiation, which can be thought of as either waves or as particles called photons. But certain high-energy particles of matter have been discovered to be traveling through space in addition to the photons. These particles are nuclei of atoms moving at tremendous velocities. They are called *cosmic rays*.

The cosmic rays formed deep in space, called *primary cosmic rays*, often interact with atoms in the earth's atmosphere, at which time *secondary cosmic rays* are given off. It is possible to detect many such secondary rays without leaving the earth's surface.

To best capture cosmic rays from outer space, one must travel above most of the earth's atmosphere in a rocket, satellite, or balloon. (Only a few of the most energetic cosmic rays reach the ground.) Balloons bear stacks of suitable plastics to altitudes of 50 kilometers; the cosmic rays leave marks as they travel through the plastic, and a three-dimensional picture of the track can be built up because of the three-dimensional nature of the stack. Similar experiments will soon be carried out aboard Space Shuttle. The third High-Energy Astronomy Observatory (HEAO-3) carried two cosmic ray experiments into space in 1979 (Fig. 10–8). These experiments are measuring the abundances of the different isotopes of elements from beryllium up to uranium, the heaviest element that occurs naturally on earth. One of the experiments can even detect nuclei heavier than uranium, if any such exist.

The origin of the primary cosmic rays has long been a matter of controversy. It seems likely that these cosmic rays are particles accelerated to tremendous velocities in supernova explosions, and that they have been traveling through space since they were ejected.

10.4 NEUTRON STARS

We have discussed the fate of the outer layers of a star that explodes as a supernova. Now let us discuss the fate of the core.

As iron fills the core of a massive star, the temperatures are so high that the iron nuclei begin to break apart into smaller units like alpha particles (helium nuclei). The pressure is no longer high enough to counteract gravity, and the core collapses.

As the density increases, the electrons are squeezed into the nuclei and react with the protons there to produce neutrons and neutrinos. The neutrinos escape, thus stealing still more energy from the core, and may also help eject the outer layers if enough neutrinos interact as they pass through. In any case, a gas composed mainly of neutrons is left behind in the dense core as the outer layers explode as a supernova.

Following the explosion, the core may contain as little as a few tenths of a solar mass or possibly as much as two or three solar masses. This remainder is at an even higher density than that at which electron degeneracy holds up a white dwarf. At this density a condition called *neutron degeneracy*, in which the neutrons cannot be packed any more tightly, appears. This condition is completely analogous to electron degeneracy. The pressure caused by neutron degeneracy balances the gravitational force that tends to collapse the core, and as a result the core reaches equilibrium as a *neutron star.*

Whereas a white dwarf packs the mass of the sun into a volume the size of the earth, the density of a neutron star is even more extreme. A neutron star may be only 20 kilometers or so across (Fig. 10–9), in which space it may contain the mass of about two suns. In its high density, it is like a single, giant nucleus. A teaspoonful of a neutron star could weigh a billion tons. This density of matter may seem simply inconceivable, but on theoretical grounds it is possible for it to exist even on such a large scale.

Before it collapses, the core (like the sun) has only a weak magnetic field. But as the core collapses, the magnetic field is concentrated, and grows stronger as a result. By the time the core shrinks to neutron star size, it has an extremely powerful magnetic field, much stronger than any we can produce on earth.

A neutron star may be the strangest type of star of which we can conceive. When neutron stars were discussed in theoretical analyses in the 1930's, there seemed to be no hope of actually observing one. Nobody had a good idea of how to look for a neutron star. Let us leave this story here for a moment, and jump to consider events of 1967, which will later prove to be related to the search for neutron stars.

If more than two or three solar masses remain, then the force of gravity will overwhelm even neutron degeneracy. We shall discuss that case in Chapter 11.

Figure 10–9 A neutron star may be only 20 kilometers in diameter (10 km in radius), the size of a city, even though it may contain a solar mass or more. A neutron star might have a solid, crystalline crust about a hundred meters thick. Above these outer layers, its atmosphere probably takes up only another few **centimeters.** Since the crust is crystalline, there may be irregular structures like mountains, which would only poke up a few centimeters through the atmosphere.

10.5   THE DISCOVERY OF PULSARS

By 1967, radio astronomy had become a flourishing science. Dozens of large radio telescopes were in existence all around the world, and were being used to observe radio emission from objects in space. Radio telescopes, just like radio receivers in our living rooms, are subject to static— rapid variations in the strength of the signal. It is difficult to measure the average intensity of a signal if the signal strength is jumping up and down many times a second, so radio astronomers usually adjust their instruments so that they do not record any variation in signal shorter than a second or so.

This adjustment should have had the effect of merely "smoothing" the incoming signal, while not distorting it. Of course it would mask any rapid variations in signal, but none were expected to arise in the source itself, as most astronomical sources were thought to be relatively steady in their emission of electromagnetic radiation. The only rapid variations expected besides terrestrial static were those that correspond to the twinkling of stars.

Electrons in the earth's ionosphere (Section 16.5) and in interstellar space contribute lesser amounts to the observed scintillation of radio sources.

Figure 10–10 The radio telescope—actually a field of aerials—with which pulsars were discovered at Cambridge, England. The total collecting area was large, so that it could detect faint sources, and the electronics was set so that rapid variations could be observed.

Figure 10–11 This pulsar, PSR 0329 + 54, has a period of 0.7145 second. (Observations made in 1979 at the 91-m (300-ft) radio telescope of the National Radio Astronomy Observatory by Joseph H. Taylor, Marc Damashek and Peter Backus of the University of Massachusetts at Amherst.)

An astronomer at Cambridge University in England, Antony Hewish, wanted to study this "twinkling" of radio sources, which is called *scintillation*. The light from stars twinkles because of effects of our earth's atmosphere. The radio waves from radio sources scintillate not because of any terrestrial effects but rather because radio signals are affected by clouds of electrons in the solar wind. To study scintillations, Hewish built a radio telescope (Fig. 10–10) that was uniquely able to study rapid variations of faint signals.

One day in 1967, Jocelyn Bell (now Jocelyn Burnell), then a graduate student working on the project, noticed that a set of especially strong variations in the signal appeared in the middle of the night, when scintillations caused by the solar wind are usually weak. After a month of observation, it became clear to her that the position of the source of the signals remained fixed with respect to the stars (that is, according to sidereal time, which was discussed in Section 4.5), a sure sign that the object was not terrestrial or solar.

Detailed examination of the signal showed that, surprisingly, it was a rapid set of pulses, with one pulse every 1.3373011 seconds. The pulses were very regularly spaced (Fig. 10–11).

One immediate thought was that the signal represented an interstellar beacon sent out by extraterrestrial life on another star. For a time the source was called an LGM, for Little Green Men. Soon, Burnell located three other sources, pulsing with regular periods of 0.253065, 1.187911, and 1.2737635 seconds, respectively.

They were briefly called LGM 1, LGM 2, LGM 3, and LGM 4, but by this time it was obvious that the signals had not been sent out by extraterrestrial life. For one thing, it seemed unlikely that there would be four such beacons at widely spaced locations in our galaxy. Besides, any beings would probably be on a planet orbiting a star, and no effect of a Doppler shift from any orbital motion was detected.

When the discovery was announced to an astonished astronomical community, it was immediately apparent that the discovery of these pulsating radio sources, called *pulsars*, was one of the most important astronomical discoveries of the decade. But what were they?

Other observatories turned their radio telescopes to search the heavens for new pulsars. This often required the purchase of new equipment to be able to follow signals that varied in time so quickly. Dozens of new pulsars were found. The periods range from hundredths of a second to four seconds.

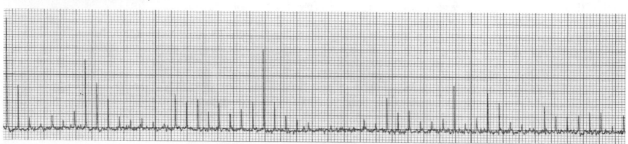

10.6 PROPERTIES OF PULSAR SIGNALS

A typical pulse shape is shown in Figure 10–12. The time from one pulse to the next pulse is called the *period*. The pulse itself lasts only a small fraction of the period.

When the positions of all the known pulsars are plotted on a chart of the heavens, it can easily be seen that they are concentrated along the plane of our galaxy (Fig. 10–13) rather than being uniformly distributed over the sky. Thus they are clearly objects in our galaxy, for if they were extragalactic objects (objects located outside our galaxy), we would expect them to be distributed uniformly.

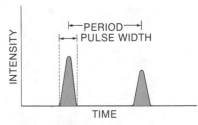

Figure 10–12 A schematic series of pulses.

10.7 WHAT ARE PULSARS?

Theoreticians went to work to try to explain the source of the pulsars' signal. The problem had essentially two parts. The first part was to explain what supplies the energy of the signal. The second part was to explain what causes the signal to be regularly timed, that is, what the clock mechanism is.

From studies of the second part of the problem, the choices were quickly narrowed down. The signals could not be coming from a pulsation of a whole normal main-sequence star because such a star would be too big. If the sun, for example, were to turn off all at one instant, we on earth would not see it go dark all at once. The point nearest to us would disappear first, and in the next seconds the darkening would spread farther back around the side of the sun (Fig. 10–14). Pulsars pulsed too rapidly to be the size of a normal dwarf star.

That left white dwarfs and neutron stars as the likely candidates. Remember, at that time, neutron stars were merely objects that had been

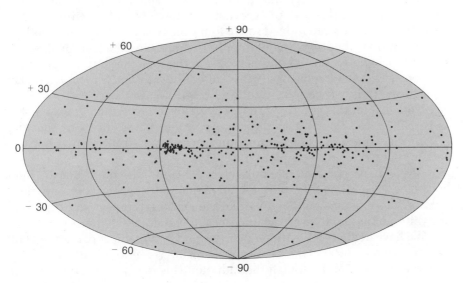

Figure 10–13 The distribution of the 328 known pulsars on a projection that maps the entire sky, with the plane of the Milky Way along the zero degree line on the map. The concentration of pulsars near 60° galactic longitude on this map merely represents the fact that this section of the sky has been especially carefully searched for pulsars because it is the area of the Milky Way best visible from the Arecibo Observatory. From the concentration of pulsars along the plane of our galaxy, we can conclude that pulsars are members of our galaxy; had they been extragalactic, we would have expected to see as many near the poles of this map. We can extrapolate that there are at least 100,000 pulsars in our galaxy.

Figure 10–14 Even if a large star were to turn off all at one time, the size of the star is such that it would appear to us to darken over a measurable period of time because radiation travels at the finite speed of light. Pulsars pulse so rapidly that we know that they cannot be large objects flashing.

The first pulsars discovered all had periods longer than ¼ second. Later, a shorter period pulsar was discovered (as will be described in Section 10.8); others with periods shorter than ¼ second are now also known.

predicted theoretically but had never been detected observationally. Could the pulsars be the more ordinary objects, namely, special kinds of white dwarfs?

Astronomers could conceive of two basic mechanisms as possibilities for the clock. The pulses might be coming from a star that was actually oscillating in size and brightness, or they might be from a star that was rotating. A third alternative was that the emitting mechanism involved two stars orbiting around each other in a binary system. However, it was soon shown that systems of orbiting stars would not have the characteristics observed for pulsars. This left the rotation or oscillation of white dwarfs or neutron stars.

On theoretical grounds, the denser a star the more rapidly it oscillates. A white dwarf oscillates in less than a minute. However, it could not oscillate as rapidly as once every second, as would be required if pulsars were oscillating white dwarfs. Neutron stars, however, would indeed oscillate more rapidly, but they would oscillate once every 1/1000 second or so, too rapidly to account for the pulsars. Thus oscillating neutron stars are ruled out as well.

What about rotation? Consider a lighthouse whose beacon casts its powerful beam many miles out to sea. As the beacon goes around, the beam sweeps past any ship very quickly, and returns again to illuminate that ship after it has made a complete rotation. Perhaps pulsars do the same thing: they emit a beam of radio waves that sweeps out a path in space. We see a pulse each time the beam passes the earth (Fig. 10–15).

But what type of star could rotate at the speed required to account for the pulsations? Could a white dwarf be rotating fast enough? Recall that a white dwarf is approximately the size of the earth. If an object that size were to rotate once every 1/4 second, the centrifugal force, the outward force resulting from the rotation, would overcome even the immense inward gravitational force of a white dwarf. The outer layers would begin to be torn off. Thus pulsars were probably not white dwarfs; if a pulsar with a period shorter than 1/4 second were to be found, then the white dwarf model would be completely ruled out.

Neutron stars, on the other hand, are much smaller, so the centrifugal force would be weaker and the gravitational force would be stronger. As a result, neutron stars can indeed rotate four times a second. There is nothing to rule out their identification with pulsars. Since no other reasonable possibility has been found, astronomers accept the idea that pulsars are in fact neutron stars that are rotating. This is called the *lighthouse model*. Thus by discovering pulsars, we have also discovered neutron stars.

Note that for this argument to be complete and convincing, we must be as certain as we can be that there are no other possibilities to consider. Whenever we argue from elimination of other possibilities, we must be certain that we have a complete list (Table 10–1).

At the moment, about 300 pulsars have been discovered. But the lighthouse model implies that there are many more, for we can see only those pulsars whose beams happen to strike the earth. If a pulsar were spinning at a different angle, then we would not know that it was there.

We have dealt above with only the second part of the explanation of the emission from pulsars: the clock mechanism. We understand much less well the mechanism by which the radiation is actually emitted in a beam.

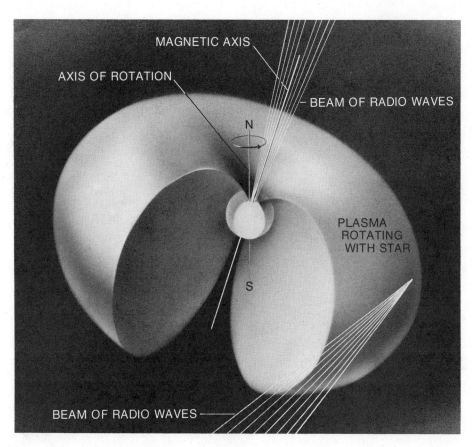

MAGNETIC AXIS

AXIS OF ROTATION

BEAM OF RADIO WAVES

N

S

PLASMA ROTATING WITH STAR

BEAM OF RADIO WAVES

Figure 10–15 In the light-house model for pulsars, which is now commonly accepted, a beam of radiation flashes by us once each pulsar period, just as a light-house beam appears to flash by a ship at sea. It is believed that the generation of a pulsar beam is related to the neutron star's magnetic axis being aligned in a different direction than the neutron star's axis of rotation. The mechanism by which the beam is generated is not currently understood. Two possible variations of the lighthouse theory are shown here. The beam may be emitted along the magnetic axis, as shown in the top half of the figure. Alternatively, the beam could be generated on the surface of a doughnut of magnetic field (shown cut away) that surrounds and rotates with the neutron star, as shown in the bottom half of the figure.

Presumably it has something to do with the extremely powerful magnetic field of the neutron star. Energy may be generated far above the surface of the neutron star in the magnetic field, perhaps near the location where the magnetic field lines, which are swept along with the star's rotation, attain velocities that approach that of light. The radio waves may be generated by charged particles that have escaped from the neutron star near its magnetic poles. These magnetic poles may not coincide with the neutron star's poles of rotation. After all, the earth's north magnetic pole is not at the north pole but near Hudson's Bay, Canada (Fig. 10–16).

Figure 10–16 The magnetic north pole is the point on the earth's surface, about 1400 km from the north geographic pole, to which a magnetic compass would lead you starting from any other point. Interestingly, the magnetic pole is in continuous motion, and its mean position moves about 0.1° per year. It is believed that the same fluid motions of the earth's core that create the geomagnetic field are responsible for the movement of the magnetic pole.

TABLE 10–1   POSSIBLE EXPLANATIONS OF PULSARS

Hypothesis	Probability
Regular dwarf or giant	Ruled out
System of orbiting white dwarfs	Ruled out
System of orbiting neutron stars	Ruled out
Oscillating white dwarf	Ruled out
Oscillating neutron star	Ruled out
Rotating white dwarf	Ruled out
Rotating neutron star	Most likely

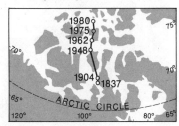

10.8 THE PULSAR IN THE CRAB NEBULA

Several months after the first pulsars had been discovered, strong bursts of radio energy were discovered coming from the direction of the Crab Nebula (Fig. 10–17). The source was named NP 0532 (for **N**ational Radio Astronomy Observatory **P**ulsar at 5 hours and 32 minutes of right ascension). The bursts were sporadic rather than periodic, but there was hope that the astronomers were seeing only the strongest pulses and that a periodicity could be found.

Within two weeks of feverish activity at several radio observatories, it was discovered that the pulsar in the Crab Nebula had a period of 0.033 second, the shortest period by far of all the known pulsars. The fastest pulsar previously known pulsed four times a second, while NP 0532 pulsed at the very rapid rate of thirty times a second. No white dwarf could possibly rotate that fast.

Furthermore, the Crab Nebula is a supernova remnant, and theory predicts that neutron stars should exist at the centers of supernova remnants. Thus the discovery of a pulsar there, exactly where a neutron star would be expected, was the discovery that clinched the identification of pulsars with neutron stars. Presumably, we do not detect pulsars in the positions of other supernovae because their spin axes are oriented such that their beams of radiation do not strike the earth.

Optical astronomers, of course, had searched very carefully at the reported positions of pulsars in order to try to discover optical objects there. But no optical objects had been found, and at this time the position of the Crab Nebula had not been examined since it was not yet known to contain a pulsar. The search had been abandoned.

Soon after NP 0532 was discovered, three astronomers at the University of Arizona decided to examine its position to look for an optical pulsar. They turned their telescope towards a faint star in the midst of the Crab Nebula, and it very soon became apparent (Fig. 10–18) that the star was pulsing! Essentially they did something that many astronomers had come to think impossible—they found an optical pulsar.

Figure 10–17 *(A)* The Crab Nebula, the remnant of a supernova explosion that became visible on earth in 1054 A.D. The pulsar, the first pulsar detected to be blinking on and off in the visible part of the spectrum, is marked with an arrow. In the long exposure necessary to take this photograph, the star turns on and off so many times that it appears to be a normal star. Only after its radio pulsation had been detected was it observed to be blinking in optical light. It had long been suspected, however, of being the leftover core of the supernova because of its unusual spectrum, which has only a continuum and no spectral lines. *(B)* An x-ray image of the Crab Nebula, made with the Einstein Observatory, shows the pulsar as the bright object at its center. (Image by Harvey Tananbaum et al., Harvard–Smithsonian Center for Astrophysics)

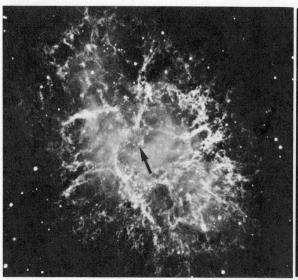

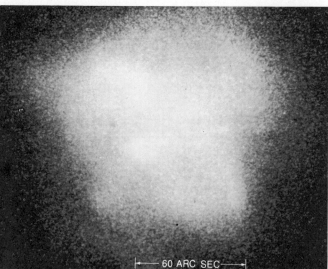

60 ARC SEC

A

B

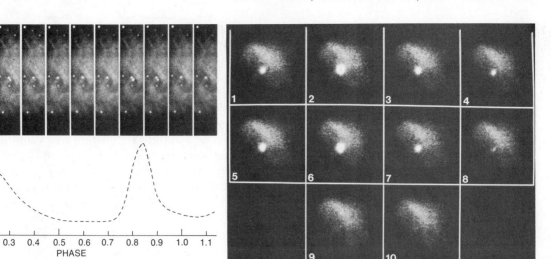

A

B

Figure 10–18 *(A)* A sequence of pictures taken by adding light from different phases of the pulsar's period shows the pulsar apparently blink on and off. Other stars in the field remain constant in intensity. Observations like this one clinched the identification of this star with the pulsar. (Observation by S. P. Maran and J. C. Brandt at the Kitt Peak National Observatory) *(B)* High-resolution x-ray images of the Crab Nebula and its pulsar, displayed in 10 time intervals covering the 33 millisecond pulse period. A total of 10 hours of observation is included. The main pulse occurs in the second interval, and an in-between pulse called the "interpulse" falls in the sixth. (These Einstein Observatory data have been provided by F. R. Harnden, Jr., High Energy Astrophysics Division of the Harvard-Smithsonian Center for Astrophysics)

For about ten years after that discovery, no other optical pulsar was found. During that time there were 150 objects known to be pulsing in the radio spectrum, and only one—which has the shortest period yet discovered and which is thus presumably the youngest known pulsar—in the optical spectrum. In 1977, flashes from the Vela pulsar, the third fastest, were discovered. Several objects are known that pulse x-rays or gamma rays, but the mechanism by which these pulses are generated is different from the run-of-the-mill radio-pulsing pulsar.

10.9 THE CLOCK SLOWS DOWN

When pulsars were first discovered, their most prominent feature was the extreme regularity of the pulses. It was hoped that they could be used as precise timekeepers, perhaps even more exact than atomic clocks on earth.

After they had been observed for some time, however, it was noticed that the pulsars were slowing down very slightly. Their periods were all gradually increasing.

The filaments in the Crab Nebula still glow brightly, even though over 900 years have passed since the explosion. Furthermore, the Crab gives off tremendous amounts of energy across the spectrum from x-rays to radio waves. It had long been wondered where the Crab got this energy. Theoretical calculations show that the amount of rotational energy lost by the Crab's neutron star as its rotation slows down is just the right amount to provide the energy radiated by the entire nebula. Thus the discovery of the Crab pulsar solved a long-standing problem in astrophysics: where the energy originates that keeps the Crab Nebula shining.

*10.10 DISTANCES AND DISPERSION

It is easy to tell the direction to a pulsar, but it is difficult to tell exactly how far away it is. We must know the distance in order to tell what the intrinsic luminosity of the pulsar is—that is, exactly how much energy it is giving off—just as we must know the distance to an ordinary star in order to determine its absolute magnitude.

The radio signals from a pulsar have a property that allows us to tell how far away the pulsar is. When a pulsar emits a pulse of radiation, signals of all frequencies are emitted at the same instant. But signals of different frequencies travel at different velocities through space.

One may ask, doesn't all electromagnetic radiation travel at the speed of light? Actually, the speed of light is constant in a vacuum, and signals of all frequencies travel at the same speed in a vacuum. But whenever ionized matter is present, then radiation travels somewhat more slowly than "the speed of light" and different frequencies travel at slightly different velocities. In interstellar space, it is principally the free electrons there that affect the speed of radiation.

The time delay of a pulse from one frequency to the next is known as dispersion. The measurement of the dispersion of the arrival times of pulses is the major method astronomers have of estimating the distance to pulsars.

Electrons are scattered throughout interstellar space, about 1 electron for every 30 cubic centimeters. This is an exceedingly low density of matter, for there are over 10^{19} atoms in one cubic centimeter of air in the room in which you are sitting. Yet before reaching us the radiation from pulsars has to travel large distances—thousands of parsecs in some cases—so the total amount of material it traverses adds up.

Thus by the time the signal from a pulsar has passed through interstellar space and reaches us, pulses at different frequencies arrive at different times (Fig. 10–19). By merely measuring the delay in arrival time from frequency to frequency we can determine the total number of electrons that the signal has traversed. Then, if we know from other studies the average density of electrons in interstellar space, we can calculate the approximate distance to the pulsar.

Consider the analogous situation of the dispersion of hurdlers arriving at the finish line of a race. They all set off together. Let us say that we know that Leslie jumps slightly faster than Hilary by a constant amount (and so that without hurdles their speeds would be equal). If we observe the hurdlers after 50 meters of the race, Leslie will be a certain distance ahead. If we observe the hurdlers after 100 meters, Leslie will be twice as far ahead. At any later time we can tell how far the hurdlers have gone by how far Leslie has pulled ahead. (We must know if Leslie has lapped Hilary, that is, made an extra revolution of the track.) The arrival time of the hurdlers has dispersed.

If the distance to a pulsar happens to be known by some other method —for example, the distance to the Crab Nebula is known to be 2000 parsecs from optical studies—then one can use the observed dispersion to find the average electron density in space. This value can then be used to find the distances to other pulsars. The currently accepted value for the electron density of 1 electron per 30 cubic centimeters is somewhat lower than had been previously thought.

This method of determining distances is not foolproof, for it depends on the assumption that the electron density of space is the same everywhere. We know that there are clouds of high-density gas scattered throughout the plane of our galaxy, and the electron density is apt to be slightly higher in the plane of our galaxy than when we look out of the plane. We can take these effects into account in part. We must simply be aware of the uncertainties in our estimates of distance based on the dispersion method.

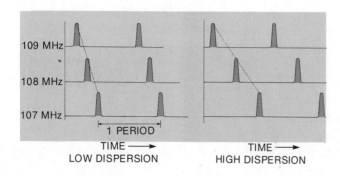

Figure 10–19 Because of the presence of electrons in the interstellar medium, the arrival time of pulses at lower frequencies are slightly delayed over the arrival time of pulses at slightly higher frequencies. The slope of the line showing the time delay as a function of frequency can be directly interpreted to give the number of electrons between the pulsar and us. Thus the pulsar whose pulses are shown on the right is farther away from us than the pulsar whose pulses are shown on the left.

*10.11 THE BINARY PULSARS

We recall that astronomers can only determine masses for stars that are in binary systems. All the pulsars that had been found prior to 1974 were lone objects in space. Thus although it seemed likely that pulsars would have the masses that were predicted theoretically for neutron stars, we could not confirm this directly.

Over the past years, Joseph H. Taylor has searched for new pulsars, with the detection of unusual specimens as one of his goals. In 1974, Taylor and Russell Hulse, both of the University of Massachusetts at Amherst, found a pulsar whose period of pulsation (approximately 0.059 sec) did not seem very regular. They finally discovered that its variation in pulse times could be explained if the pulsar were orbiting another star with a period of 8 hours. If so, when the pulsar was approaching us the pulses would be jammed together a bit, and when it was receding the pulses would be spread apart in time in a type of Doppler effect.

No optical object has been found at the location of this *binary pulsar*, so it can be concluded that both objects are too faint to be seen from the earth. The diameter of the orbit of the pulsing component is approximately the diameter of the sun. Thus the accompanying object cannot be a star of normal size; it may be a neutron star too. No pulses have been detected from it, but it could simply be oriented at a different angle.

Unfortunately, when only one star of a binary pair is detectable, we cannot uniquely determine the mass of the stars, that is, determine one and only one value that is consistent with the observations. We can only calculate limits on the masses, and a quantity, also mentioned in Section 6.2, that is a mixture of sums and products of the two masses (called the *mass function*). But at least it can be concluded that the results are consistent with the interpretation of pulsars as neutron stars.

Further, the main interest in binary pulsars has come about because of their ability to provide powerful tests of the general theory of relativity. In Section 7.11, we discussed the excess advance of the perihelion of Mercury, which is only 43 seconds of arc per **century**. It can be calculated that the Hulse-Taylor binary pulsar, which is in a much stronger gravitational field than is Mercury, should have its periastron, the closest point of its orbit to its companion (corresponding to the "perihelion" point, the closest approach of a planet's orbit to the sun), advance by 4° per **year** (Fig. 10–20). (Converting 4° to arc sec shows that the advance is by 14,400 arc sec year^{-1} or 1,440,000 arc sec century^{-1}. We can see that 4° per year is about 35,000 times the size of the effect for Mercury.) A change of 4° is large enough to measure easily. Observations of the binary pulsar clearly show the advance and so provide a strong confirmation of Einstein's general theory of relativity.

Binary pulsars are discovered when astronomers notice that a pulsar's period does not seem to remain constant over a lengthy period of time. In 1979 and 1980, the binary nature of two more pulsars was realized. One had been slowing down for some time, but then began to speed up. Its orbital period is 4 years. Unfortunately, relativistic effects with such a long period are much less than those for the original binary pulsar, with its 8-hour orbital period. The third binary pulsar's period of just under 25 hours is not so much greater than the original's. However, its orbit is almost circular, and with such a small deviation from an ellipse the interesting relativistic effects will also be small.

All Taylor's conclusions are based on careful analysis of the variations in the time of arrival of the pulses.

Figure 10–20 The periastron of the binary pulsar PSR 1913 + 16 advances by 4° per year, which can be interpreted as a strong endorsement of Einstein's general theory of relativity. The angle is shown for the apastron, the farthest point of the pulsar from the other object.

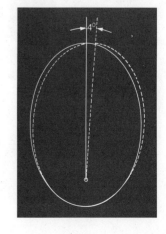

*10.12 GRAVITATIONAL WAVES

Most of the data we have thus far discussed in this book have come from studies of the electromagnetic spectrum. But Einstein's general theory of relativity predicts the existence of still another type of signal that we might be able to detect: gravitational waves.

Whereas electromagnetic waves affect only charged particles, we would expect gravitational waves to affect all matter. But gravitational waves, if they exist, would probably be very weak. They would be emitted only in situations where large masses were being accelerated rapidly. In the previous section, we mentioned one possible example: two neutron stars revolving in close orbits around each other. Another possible example would be a supernova or the collapse of a massive star to become a black hole. Even a simple close binary system containing a white dwarf or a neutron star will give out a reasonable amount of gravitational radiation.

Joseph Weber of the University of Maryland was the first to build sensitive apparatus to try to search for gravitational waves. His detector is a large metal cylinder, delicately suspended from wires so as to insulate it from restraint and from local motion. When a gravitational wave hits the cylinder, it should begin vibrating. Weber has sensitive apparatus to measure the vibrations. Starting in 1969, Weber reported success in detecting gravitational waves, but the frequency of events was much higher than we would have expected.

Eventually other scientists built other gravitational wave apparatus, some in configurations that they had calculated would be more sensitive than Weber's. None of the other scientists detected any gravitational waves, and most astronomers feel that Weber's detectors must have been affected by other kinds of events, perhaps of terrestrial origin. Nonetheless, Weber gets credit for having stimulated this exciting field. New and still more sensitive apparatus is being built.

In the meantime the Hulse-Taylor binary pulsar turned out to be a quicker and a better way to detect gravity waves. If gravity waves exist, this binary system should emit them strongly enough that there should be detectable consequences. In particular, the energy carried away would make the

Figure 10–21 *(A)* Joseph Weber and his gravitational wave detector, a large aluminum cylinder, weighing 4 tons, delicately suspended so that gravitational waves would set it vibrating 1660 times per second. Weber's results have not been verified, and his experiment has drawn criticism from other scientists. However, Weber's efforts have stimulated interest in detecting gravity waves, and newer, more sensitive detectors are currently under construction. *(B)* MIT's laser interferometer to measure gravity waves. A laser beam enters the vertical tube A. It is then split into 2 beams that bounce back and forth between A and B and between A and C. The beams are then combined, which allows us to determine if the 2 different paths have changed in length. A passing gravity wave would change the effective length of one of the paths. The technique, which is similar to that used in the 1880's by A. A. Michelson to study the speed of light in various directions, can measure changes of a fraction of the wavelength of the laser light. It is believed that future larger versions would be sensitive enough to detect gravity waves from supernovae in nearby galaxies.

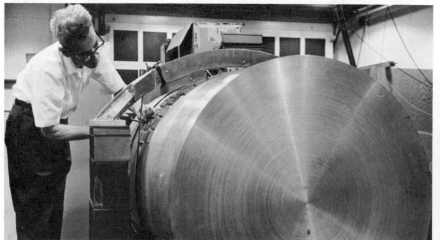

A

B

orbits of the stars around each other shrink, which would result in a slight speedup of the period of revolution. In late 1978, it was reported that such a slight speedup—1/10,000 second per year—has been found. This is strong evidence for the existence of gravity waves. Gravity waves, in turn, are confirmation of a prediction of the general theory of relativity.

*10.13 X-RAY BINARIES

Telescopes in orbit that are sensitive to x-rays have detected a number of strong x-ray sources, some of which are pulsating. Several of these objects pulse only in the x-ray region of the spectrum. One of the most interesting is Hercules X-1 (the first x-ray source to be discovered in the constellation Hercules). It is pulsating with a 1.24-second period.

Hercules X-1 and the other pulsating x-ray sources are apparently examples of the type of evolution of binary systems that we discussed in Section 9.8. Theoreticians think that the x-ray sources are radiating because mass from the companion is being funneled toward the poles of the neutron star by the neutron star's strong magnetic field. Unlike the slowing down of the pulse rate of pulsars (which give off pulses in the radio region of the spectrum), the pulse rate of the binary x-ray sources usually speeds up. The period of Hercules X-1, for example, is growing shorter.

Hercules X-1 is the companion to a long-observed variable star (a single-line spectroscopic binary, a type of double star described in Section 6.1). The visible star is named HZ Herculis (variable star HZ in the constellation Hercules). The two components orbit each other every 1.7 days. Observations indicate that the companion has 1.3 solar masses, within the limit that neutron degeneracy can support. With the aid of the latest electronic techniques, we have even been able to detect individual photons emitted from the optically visible star when the beam of x-rays from the dark companion, according to the lighthouse theory, strikes it every 1.24 seconds. These observations are consistent with the theory that Hercules X-1 contains a neutron star.

Thus we have in recent years discovered two types of systems in which neutron stars are located. First, isolated neutron stars may give off pulses of radio waves, and we detect these stars as pulsars. Second, neutron stars in binary systems tend to give off pulses of x-rays.

*10.14 SS433: THE STAR THAT IS COMING AND GOING

Probably the most exciting single object now known in astronomy is SS433 (Fig. 10–22), an object in our galaxy brought to the astronomical community's attention in 1979. SS433 contains spectral lines that change in wavelength by a tremendous amount, hundreds of angstroms. This corresponds to the Doppler shift from a velocity of over 40,000 kms^{-1}, about 15 per cent of the speed of light, and is entirely unprecedented for an object in our galaxy. Furthermore, spectral lines appear simultaneously that are shifted to the red and to the blue by that same tremendous amount. Something seems to be coming and going at the same time.

The present phase of the story began when, following radio and x-ray studies of the object, Bruce Margon of UCLA observed that a set of SS433's

"SS" indicates that it is from a catalogue of stars with hydrogen emission lines compiled in the 1960's by C. Bruce Stephenson and Nicholas Sanduleak of Case Western Reserve University.

Figure 10–22 SS433 is shown with an arrow on this negative print of a photograph taken with the 1.2-m Palomar Schmidt telescope. A vertical strip across the center of the photograph shows relatively few stars, indicating the presence of dust lanes there. SS433, which is 14th magnitude, is certainly not very prominent in this photograph. Bright filaments show at the left in this Hα picture and a radio source, known as W50, that is presumed to be a supernova remnant also appears. The picture shows a field about 1° wide in the constellation Aquila.

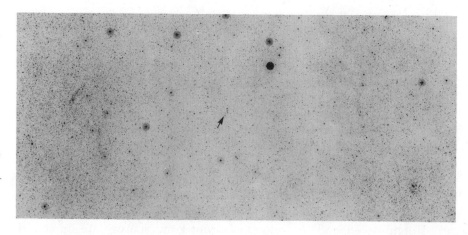

spectral lines seemed to change wildly in wavelength from night to night (Fig. 10–23). A change of wavelength of 150 Å occurred over a 4-day interval, for example! It was difficult at first to see what was changing, and it took some time and a lengthy series of observations before it could be first realized and then proved that the wavelengths varied in a regular fashion with a period of 164 days (Fig. 10–24).

It was later discovered at the Dominion Astrophysical Observatory in Victoria, Canada, that SS433 is in orbit with a 13-day period around another object, making it a binary system. So it can be considered a type of x-ray binary, though it is only a weak x-ray source (Fig. 10–25). It has been detected in the radio spectrum as well (Fig. 10–26).

Figure 10–23 Three spectra of SS433 taken over a four-day period, showing variation of wavelength of spectral features even over that short time. The Hα line at center remains stationary, while images of the Hα line are offset to both left and right.

Figure 10–24 The redshifts and blueshifts of spectral lines in SS433 vary in a regular fashion with a 164-day period.

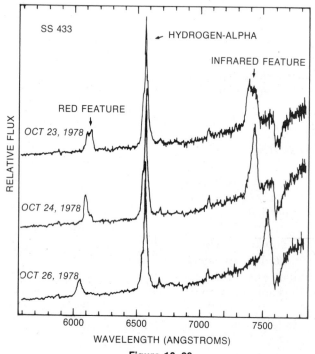

Figure 10–23

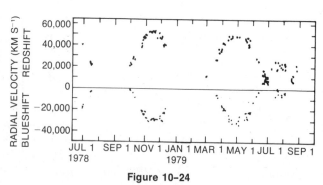

Figure 10–24

Figure 10–25 An x-ray image of SS433 from the Einstein Observatory. Two jets (or, at least, features that resemble two jets) can be seen extending to the sides of the bright central image. This central peak is actually a point source, though it is spread on this image because of the exposure necessary to emphasize the jets. SS433 is not an especially strong source; some of the other binary x-ray sources are 100 times stronger. (Einstein image courtesy of Jonathan E. Grindlay and Fred Seward of the Center for Astrophysics, and Ernest Seaquist and William Gilmore of the University of Toronto)

15 ARC–MIN

SS433 is probably a collapsed object such as a neutron star (or possibly a black hole), surrounded by an accretion disk of material gathered from the companion. In a leading model (Fig. 10–27), material is being ejected from the accretion disk along a line which, because of the precession of the accretion disk, traces out a cone. (Precession is the wobbling motion, as of a child's top, that we have already met for the earth's rotation in Section 4.5.) If we assume that SS433 is oriented with its spin axis more-or-less but not completely perpendicular to the direction in which we are looking, we see radiation given off by the material being ejected upward as redshifted and radiation given off by the material being ejected downward as blueshifted. The redshift and blueshift vary as the line traces out the cone. Other models have also been suggested.

From the interstellar absorption lines in SS433's spectrum, Margon and his colleagues estimated that the object is at least 3500 parsecs from us, about a third of the sun's distance from the center of our galaxy.

The original group included UCLA scientists Margon, Holland Ford, Jonathan Katz, Karen Kwitter (now at Williams College), and Roger Ulrich, and Lick scientists Remington Stone and Arnold Klemola.

FEBRUARY 1980

DECLINATION

1 ARC SEC

RIGHT ASCENSION
0.1 SEC

A

MAY 1980

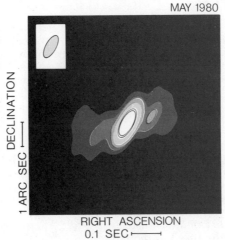

DECLINATION

1 ARC SEC

RIGHT ASCENSION
0.1 SEC

B

Figure 10–26 SS433 observed at a radio wavelength of 6 cm with the Very Large Array (VLA). SS433 is in the center of a supernova remnant known as W50. We see two images, one from February 1980 and the other from May 1980. Slight changes are visible in the structure and orientation of what appear to be the jets, in accordance with the model we show below. The ellipse in the upper left shows the size and shape of the VLA's resolution. (Courtesy of Ernest Seaquist and William Gilmore of the University of Toronto and John Stocke of the University of Arizona)

SS433 is a weird and still unique object in astronomy. Since both components seem to be about 1.5 solar masses, it may well contain one neutron star and one ordinary star, and be an extreme case of an x-ray binary system. Whatever it is, SS433 is undergoing extensive study in all regions of the spectrum from observatories all over the world and in space.

Figure 10–27 A model of SS433 in which the radiation emanates from two narrow beams of matter that are given off by the accretion disk. The model follows ideas of Mordechai Milgrom of the Weizmann Institute in Israel, Bruce Margon, now at the University of Washington, and Jonathan Katz and George Abell of UCLA. The velocity of gas emitted is 80,000 km s⁻¹ in each direction, though we see a somewhat lower velocity because we do not see the jets head on. In some models, variations in pressure and density of the disk can cause the emission.

SUMMARY AND OUTLINE

This entire chapter concerns the evolution of stars of more than 8 solar masses.

When they exhaust their hydrogen in the core, they become red giants.

Later, heavier elements are built up in the core, and the stars become supergiants (Section 10.1).

After iron cores form, the stars explode as supernovae. Only a few optical supernovae have been seen in our galaxy in the last 900 years. Supernova remnants can be studied with techniques of radio astronomy and x-ray astronomy (Section 10.2).

Cosmic rays, high-energy particles in space, probably come from supernovae (Section 10.3).

After a star's fusion stops, neutron degeneracy can support a remaining mass of up to 2 or 3 solar masses; more massive stars will be discussed in the next chapter.

A neutron star may be only 10 km in radius, and so is fantastically dense (Section 10.4).

Pulsars were discovered when a radio telescope was built that did not mask rapid time variations in the signal (Section 10.5).

Pulsars have very regular signals, although they have been discovered to be slowing down very slightly. Pulse periods range from 0.033 to 4 sec (Section 10.6).

Scientists have identified pulsars with rotating neutron stars by a process of elimination (Section 10.7):

They can't be full-size stars, because the pulse width is too small.

They can't be a binary pair of white dwarfs or neutron stars.

They can't be oscillating white dwarfs or neutron stars, because the periods of oscillation would be too long or too short, respectively.

They can't be rotating white dwarfs, because white dwarfs can barely rotate fast enough to account for the first known pulsars and surely can't rotate fast enough to account for the pulsar in the Crab Nebula.

Rotating neutron stars are all that are left, and could satisfy the observations. Thus we accept the lighthouse model.

A pulsar is observed in the center of the Crab Nebula. a supernova remnant, where we would expect to observe a neutron star (Section 10.8).

Pulsars are considered to be objects that emit in the radio spectrum, although the Crab and the Vela pulsars also pulse light, x-rays, and γ-rays. Other sources pulse only x-rays (Section 10.8).

Pulsars are gradually slowing down as they grow older (Section 10.9).

The distance to a pulsar can be determined by observing the delays in arrival time from frequency to frequency if you know the electron density in space (Section 10.10).

One pulsar has been discovered to be in a binary system. Its periastron advances 4°/year, confirming general relativity (Section 10.11).

General relativity predicts the existence of gravitational waves. Attempts to detect them directly by their effect on cylinders of material on earth have probably been unsuccessful. The orbital period of the binary pulsar is slowing down by just the amount that is expected to correspond to the emission of gravitational waves from that system, confirming Einstein's theory (Section 10.12).

Many x-ray sources are caused by the infall of material on a neutron-star member of a binary system from its companion (Section 10.13).

An object in our galaxy, SS433, has spectral lines that shift periodically to the red and to the blue by about 15 per cent of the speed of light, an unprecedented amount. The Doppler-shifted radiation may come from beams ejected from an accretion disk about the neutron star in an x-ray binary and changing their orientation to us because the disk is precessing (Section 10.14).

QUESTIONS

1. Why are red supergiants so bright?

2. What are the basic differences between a nova and a supernova?

3. If we see a massive main-sequence star (a heavyweight star), what can we assume about its age, relative to most stars? Why?

4. What do we know about the core of a star when it leaves the main sequence?

5. In a supernova explosion of a 20 solar mass star, about how much material is blown away?

†6. A supernova can brighten by 20 magnitudes. By what factor of brightness is this? Show your calculation.

†7. A typical galaxy has a luminosity 10^{11} times that of the sun. If a supernova equals this luminosity, and if it began as a B star, by what factor did it brighten? By how many magnitudes?

8. Would you expect the appearance of the Crab Nebula to change in the next 500 years? How?

†9. If the light from a supernova 2000 parsecs from us was received in 1054, when did the star actually explode?

10. What are cosmic rays?

11. In your own words, fill in Table 10–1 with the reasons why each of the explanations for pulsars involving single stars was ruled out except for the case of rotating neutron stars.

12. How do we find the distances to pulsars?

†13. If there is one electron every 30 cm³ in interstellar space, how many electrons are in a "tube" 1 cm in diameter and 2000 parsecs long?

14. In view of current theories about supernovae and pulsars, list the pieces of evidence that indicate that the Crab pulsar is a young one.

15. Explain two important consequences of the discovery of the binary pulsar.

16. Why is the discovery of gravity waves a test of the general theory of relativity?

.17. Sketch a model for an x-ray binary and explain how it leads to x-ray emission.

†18. A shift of the Hα line in SS433 of 40,000 km s⁻¹ corresponds to a shift of how many Å in wavelength? (Ignore any effects of relativity.)

19. Describe why the precession of an accretion disk in SS433 can cause varying Doppler shifts.

20. Explain why the true velocities of the jets in SS433 may be even greater than the velocities we measure directly from Doppler shifts.

---

†This indicates a question requiring a numerical solution.

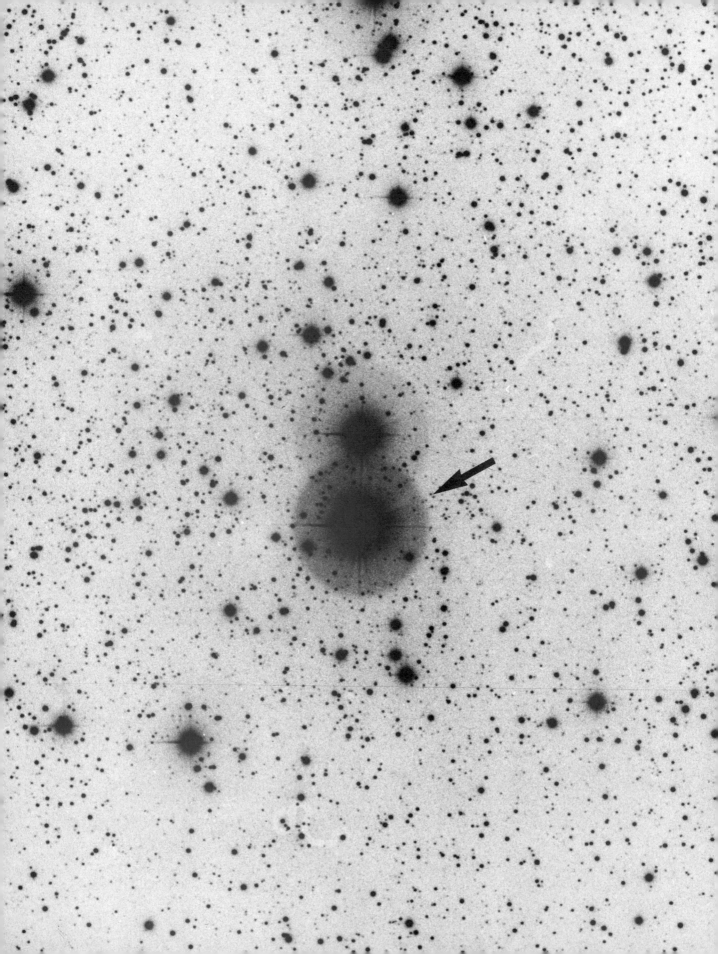

CHAPTER 11

BLACK HOLES

AIMS:
To understand what black holes are, how
they form, and how they might be
detected

The strange forces of electron and neutron degeneracy support dying lightweight, middleweight, and some heavyweight stars against gravity. The strangest case of all occurs at the death of most massive stars, which contained up to about 60 or more solar masses when they were on the main sequence. They generally undergo supernova explosions like less massive heavyweight stars, but some of these most massive stars may retain cores of over 2 or 3 solar masses. Nothing in the universe is strong enough to hold up the remaining masses against the force of gravity. The masses collapse, and continue to collapse forever. We call the result of such a collapse a *black hole* Later, we shall discuss the formation of black holes in processes other than those that result in the collapse of a star.

11.1 THE FORMATION OF A STELLAR BLACK HOLE

When nuclear fusion ends in the core of a star, gravity causes the star to contract. We have seen in Chapter 9 that a star that has a final mass of less than 1.4 solar masses will end its life as a white dwarf. In Chapter 10 we saw that a heavier star will become a supernova, and that if the remaining mass of the core is less than 2 or 3 solar masses it will wind up as a neutron star. We are now able to observe both white dwarfs and neutron stars and so can study their properties directly.

It seems reasonable that in some cases more than 2 or 3 solar masses remain after the supernova explosion. The star collapses through the neutron star stage, and we know of no force—not even neutron degeneracy—that can stop the collapse. In some cases, the matter may have become so dense as the

Do not confuse a black hole with a black body. A black body is merely a radiating body whose radiation follows Planck's law. A black hole, essentially, does not radiate (though an exception to this rule—really an alternate way of looking at the whole picture—will be discussed in Section 11.6).

The overexposed dark object in the center of this negative print is a blue supergiant star HDE 226868, which is thought to be the companion of the first black hole to be discovered, Cygnus X-1.

star collapsed that it passed through the neutron star stage before any explosion took place; in these cases, no supernova resulted.)

For a long time, astronomers assumed that the most massive stars would somehow lose most of their mass before or in the process of collapsing, and end up as white dwarfs. That is one reason why astronomers have long been interested in studying the various mechanisms of *mass loss*. Several such are known: mass is known to flow gradually into space from the outer layers of giants and supergiants, and mass is ejected into space as the shell of a planetary nebula or in a supernova explosion. The solar wind represents such a net loss of mass from our sun. Even though stars with masses as large as 60 solar masses are known, it was long assumed that somehow almost all of the mass would be lost as the star evolved so that all stars could wind up as white dwarfs.

But astronomers no longer make that assumption. Pulsars have been identified with neutron stars, and some seem to have masses greater than 1.4 times that of the sun. Further, the image that most astronomers had decades ago of a placid universe was radically altered by the discovery of violent events that give rise to bursts of radiation in the x-ray or radio regions of the spectrum. No longer was there a preconception against the idea that very massive stars can collapse without losing most of their mass, winding up in exotic states.

The value for the maximum mass that a neutron star can have is a result of theoretical calculation. We must always be aware of the limits of accuracy of any calculation, and particularly a calculation that deals with matter in a state that is very different from states that we have been able to study experimentally. (Still, the best modern calculations show that the limit of a neutron star mass is about 2 solar masses. Even allowing a margin for error, the limit is probably less than 3 solar masses.)

(We may then ask what happens to a 5 or 10 or 50 solar mass star as it collapses, if it retains more than 3 solar masses. It must keep collapsing, getting denser and denser. We have seen that Einstein's general theory of relativity predicts that a strong gravitational field will redshift radiation, and that this prediction has been verified both for the sun and for white dwarfs (as discussed in Section 9.5). Also, radiation will be bent by a gravitational field, or at least appear to us on earth as though it were bent. This prediction has been verified for the sun (see Section 7.11).)

(As the mass contracts, radiation is continuously redshifted more and more, and radiation leaving the star other than perpendicularly to the surface is bent more and more. Eventually, when the mass has been compressed to a certain size, radiation from the star can no longer escape into space. The star has withdrawn from our observable universe, in that we can no longer receive radiation from it. We say that the star has become a *black hole*.)

Why do we call it a black hole? We think of a black surface as a surface that reflects none of the light that hits it. Similarly, any radiation that hits the surface of a black hole continues into the black hole and is not reflected. In this sense, the object is perfectly black.

11.2   THE PHOTON SPHERE

Let us consider what happens to radiation emitted by the surface of a star as it contracts. Although what we will discuss affects radiation of all

Any ordinary surface is not truly "black." A black piece of paper, for example, may not reflect light and certainly doesn't emit any light of its own, but it does radiate infrared radiation corresponding to its temperature. A black hole, on the other hand, devours radiation at all wavelengths and emits nothing; no radiation would be seen coming from the black hole itself.

Figure 11–1 As the star contracts, a light beam emitted other than radially outward will be bent.

wavelengths, let us simply visualize standing on the surface of the collapsing star with a flashlight.

If we stand on the surface of a supergiant star, we note only very small effects of gravity on the light from our flashlight. We can shine the beam straight up, or at any angle, and it seems to go straight out into space.

As the star collapses, two effects begin to occur. Although we on the surface of the star cannot notice them ourselves, a friend of ours on a planet revolving around the star can detect the effects and radio back information to us about them. For one thing, our friend will see that our flashlight beam is redshifted; this could be told either by observing a spectral line, if any, or by noting the shift of the peak of the Planck curve of the radiation toward the red. Second, our flashlight beam would be bent by the gravitational field of the star (Fig. 11–1). If we shined the beam straight up, it would continue to go straight up. But if we shined it away from the vertical, the beam would be bent even farther away from the vertical. If bent enough, it will not be able to escape from the star (Fig. 11–2).

Only if the flashlight is pointed within a certain angle of the vertical does the light continue outward. This angle forms a cone, with its apex at the flashlight, and is called the *exit cone* (Fig. 11–3). As the star grows smaller yet, we find that the flashlight has to be pointed more directly upward in order for its light to escape. The exit cone grows smaller as the star continues to shrink.

When we shine our flashlight upward in the exit cone, the light escapes. When we shine our flashlight in a direction outside the exit cone, the light is bent sufficiently that it falls back to the surface of the star. When we shine our flashlight exactly along the side of the exit cone, the light goes into orbit around the star, neither escaping nor returning to the surface (Fig. 11–3).

The sphere around the star in which the light can orbit is called the *photon sphere*. Its size can be calculated theoretically. It is 13.5 km in radius for a star of 3 solar masses, and its size varies linearly with the mass; that is, it is 27 km in radius for a star of 6 solar masses, and so on.

As the star continues to contract, the theory shows that the exit cone gets narrower and narrower. Light emitted within the exit cone edge still escapes. The photon sphere remains at the same height even though the matter inside it has contracted further, since the total amount of matter within has not changed.

11.3 THE EVENT HORIZON

If we were to depend on our intuition, which is based on classical physics as advanced by Newton, we might think that the exit cone would simply continue to get narrower. But when we apply the general theory of relativity, we find that at a certain radius the cone vanishes. Light no longer can escape into space, even when it is traveling straight up, away from the center of the gravitational mass.

The solution to Einstein's equations that predicts this was worked out by Karl Schwarzschild in 1916, shortly after Einstein advanced his general theory. The radius of the star at the time at which light can no longer escape is called the *Schwarzschild radius* or the *gravitational radius,* and the spherical surface at that radius is called the *event horizon* (Fig. 11–4). The radius of the photon sphere is exactly $3/2$ times this Schwarzschild radius.

Figure 11–2 Light can be bent so that it falls back onto the star.

"Now, here, you see, it takes all the running you can do, to keep in the same place. If you want to get somewhere else, you must run twice as fast as that."
The Red Queen,
Through the Looking Glass
by Lewis Carroll, 1871

Figure 11–3 When the star has contracted enough (the inner sphere), only light emitted within the exit cone escapes. Light emitted on the exit cone goes into the photon sphere. The further the star contracts within the photon sphere, the narrower the exit cone becomes.

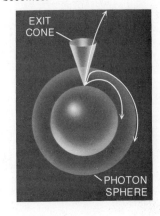

EXIT CONE

PHOTON SPHERE

Figure 11-4 When the star becomes smaller than its Schwarzschild radius, we can no longer observe it. We say that it has passed its *event horizon*, by analogy to the statement that we cannot see a thing on earth once it has passed our horizon.

The sun's Schwarzschild radius is only 3 km. The Schwarzschild radius for the earth is only 9 mm; that is, the earth would have to be compressed to a sphere only 9 mm in radius in order to form an event horizon and be a black hole.

The Schwarzschild solutions *of equations from Einstein's general theory of relativity are the theoretical basis for the existence of black holes.*

We can visualize the event horizon in another way, by considering a classical picture (that is, one based on the Newtonian theory of gravitation). The picture is essentially that conceived in 1796 by Laplace (who was also responsible for the nebular hypothesis of planetary formation as described in Section 12.11). (A body must have a certain velocity, called the *escape velocity*, to escape from the gravitational pull of another body.) For example, we have to launch rockets at 11 km s⁻¹ (40,000 km hr⁻¹) in order for them to escape from the earth's gravity. (For a more massive body of the same size, the escape velocity would be higher. Now imagine that this body contracts, and we are drawn closer to the center of the mass with all the mass concentrated in a sphere below us. As this happens, the escape velocity would rise. As the body crosses its Schwarzschild radius, the escape velocity becomes equal to the speed of light. Thus even light cannot escape. If we then begin to apply the special theory of relativity, we might then reason that since nothing can go faster than the speed of light, nothing can escape. Now let us return to the picture according to the general theory of relativity.)

(The size of the Schwarzschild radius depends linearly on the amount of mass that is collapsing. A star of three solar masses, for example, would have a Schwarzschild radius of 9 km. A star of six solar masses would have a Schwarzschild radius of twice 9 km, or 18 km. One can calculate the Schwarzschild radii for less massive stars as well, although the less massive stars would be held up in the white dwarf or neutron star stages and not collapse to their Schwarzschild radii (although conceivably some lesser mass object could be sufficiently compressed in a supernova).)

(Note that anyone or anything on the surface of a star as it passed its event horizon would not be able to survive. An observer would be torn apart by the tremendous difference in gravity between head and foot. (This is called a *tidal force*, since this kind of difference in gravity also causes the tides on earth.) If the tidal force could be ignored, though, the observer on the surface of the star would not notice anything particularly wrong as the star passed its event horizon. If we chose to stay with the observer at that time, we could still be pointing our flashlights up into space trying to signal our friends on earth. But once we passed the event horizon, no answer would ever come because our signal would never get out.)

(Once the star passes inside its event horizon, it continues to contract. Nothing can ever stop its contraction. In fact, the mathematical theory predicts that it will contract to zero radius, a situation that seems physically

Figure 11-5 This drawing by Charles Addams is reprinted with permission of *The New Yorker Magazine, Inc.,* © 1974.

impossible to conceive of. The point at which it will have zero radius (it has infinite density there) is called a *singularity*. Strange as it seems, theory predicts that a black hole contains a singularity.

Even though the mass that causes the black hole has contracted further, the event horizon doesn't change. It remains at the same radius forever, as long as the amount of mass inside doesn't change. Any matter that ventures too close to the black hole will be pulled in by the gravity, and the mass inside the event horizon will become larger. Thus, since the event horizon depends on the amount of mass, the event horizon will become slightly larger. As the black hole sits in space, or even moves through space as it continues the proper motion across space that the original star had, it may encounter material. Its gravity would pull this material in and the black hole would grow larger. It will consume whatever it encounters, and nothing can stop it.

*11.4 ROTATING BLACK HOLES

Once matter is inside a black hole, it loses its identity in the sense that from outside a black hole, all we can tell is the mass of the black hole, the rate at which it is spinning, and what electric charge it has. These three quantities are sufficient to completely describe the black hole. Thus, in a sense, black holes are simple objects to describe physically, because we only have to know three numbers to characterize each one. By contrast, for the earth we have to know shapes, sizes, densities, motions, and other parameters for the interior, the surface, and the atmosphere. The theorem that describes the simplicity of black holes is often colloquially stated by astronomers active in the field as "a black hole has no hair"; that is, a black hole has no basic properties at all that can be described aside from mass, spin, and charge. All other properties, such as size, can be derived from these three basic properties.

Most of the theoretical calculations about black holes, and the Schwarzschild solutions in particular, are based on the assumption that black holes do not rotate. But this assumption is only a convenience; we think, in fact, that the rotation of a black hole is one of its important properties. It was not until 1963 that Roy P. Kerr solved Einstein's equations for a situation that was later interpreted in terms of the notion that a black hole is rotating. In this more general case, a second event horizon, with somewhat different properties, appears. It is called the *stationary limit*. Unlike the situation for a non-rotating black hole, in a rotating black hole (Fig. 11–6) it is not necessary that matter inside this second event horizon reach the singularity. The matter might be able to reappear at some other point in our universe, or even reappear in our universe at some other time in the past or future!

Also unlike the case of a non-rotating black hole, for which the singularity is always unreachably hidden within an event horizon, when the rotation is very fast Kerr's solutions allow the existence of a singularity out in the open, not surrounded by an event horizon. Such a point is called a *naked singularity* and, if one exists, we would have no warning before we ran into it. Most theoreticians assume the existence of a law of "cosmic censorship," which requires all singularities to be "clothed" in event horizons, that is, not naked.

Figure 11–6 The description of a black hole so far in this chapter has assumed that the black hole is not rotating. But it seems reasonable, or even likely, that a black hole would rotate (just as a neutron star is rotating rapidly); a set of solutions to Einstein's equations discovered by Roy P. Kerr can be interpreted to describe a rotating black hole. From the top, the object looks like an ordinary non-rotating black hole. But in a non-rotating black hole we can never have any contact with matter inside the event horizon. In the rotating black hole, there is a region inside one of the event horizons (the *stationary limit*) with which we can have some contact. This region is called the *ergosphere* because, in principle, work (which in Greek is *ergon*) can be extracted from it. ("Work" is a technical term in physics closely related to the ability to supply energy.) The stationary limit is so named because within it, nothing can be at rest. Light directed in the sense (that is, around with it rather than against it) of the black hole's rotation will escape from the ergosphere, and an astronaut could too.

The drawing shows a cross-section of a rotating black hole; actually, the ergosphere wraps completely around the equator and diminishes to zero thickness only at the poles.

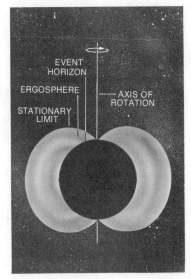

Rotating black holes have two types of event horizons with a region called the *ergosphere* between them (Fig. 11–6). An object can be shot into the ergosphere of a rotating black hole at such an angle that, after it is made to split into two, one part continues into the event horizon while the other part is ejected from the ergosphere with more energy than the incoming particle had. This energy must have come from somewhere; it could only have come from the rotational energy of the black hole because there are no other possibilities. Thus by judicious use of the ergosphere, we can tap the energy of the black hole. We are always looking for new energy sources and this is the most efficient known, although it is obviously very far from being practical.

11.5 DETECTING A BLACK HOLE

What if we were watching the black hole from the outside? Photons that were just barely inside the exit cone when emitted would leak very slowly away from the black hole. These few photons would be spread out in time, and the image would be very faint. As the star went through its event horizon, the last photons to reach us would be redshifted, so the star would get very red and then blink out. It would do this in a fraction of a second, so the odds are unfavorable that we would actually see a collapsing star as it went through the event horizon.

But all hope is not lost for detecting a black hole, even though we can't hope to see the photons from the moments of collapse. The black hole disappears, but it leaves its gravity behind. It is a bit like the Cheshire Cat from *Alice's Adventures in Wonderland*, which fades away, leaving only its grin behind (Fig. 11–7).

The black hole attracts matter, and the matter accelerates toward it. Some of the matter will be pulled directly into the black hole, never to be seen again. But other matter will go into orbit around the black hole, and will orbit at a high velocity.

It seems likely that the gas in orbit will be heated. Theoretical calculations of the friction that will take place between adjacent filaments of gas in orbit around the black hole show that the heating will be so great that the gas will radiate strongly in the x-ray region of the spectrum. Thus, though we cannot observe the black hole itself, we can hope to observe x-rays from the gas surrounding it.

In fact, a large number of x-ray sources are known in the sky. They have been surveyed from NASA satellites called Uhuru and, more recently, HEAO's (High-Energy Astronomy Observatories). Some of these sources have been identified with galaxies or quasars. Others pulse regularly, like Hercules X-1, and are undoubtedly neutron stars (see Section 10.13). Some of the rest, which pulse sporadically, may be related to black holes.

It is not enough to find an x-ray source that gives off sporadic pulses, for one can think of other mechanisms besides revolution of matter around a black hole that can lead to such pulses. One would like to show that a collapsed star of greater than 3 solar masses is present.

In Section 6.2, we discussed how masses are determined from studies of binary stars. Our hope of pinning down the mass of a collapsed object, and thus identifying it with a black hole, rests on such a study.

Figure 11–7 Lewis Carroll's Cheshire Cat, from *Alice's Adventures in Wonderland*, shown here in John Tenniel's drawing, is analogous to a black hole in that it left its grin behind when it disappeared while a black hole leaves its gravity behind when its mass disappears. Alice thought that the Cheshire Cat's persisting grin was "the most curious thing I ever saw in all my life!" We might say the same about the black hole and its persisting gravity.

looked goodnatured, she thought: still it had *very* long claws and a great many teeth, so she felt it ought to be treated with respect.

"Cheshire Puss," she began, rather timidly, as she did not at all know whether it would like the

When we search the position of the x-ray sources, we are interested in finding a spectroscopic binary (that is, a star whose spectrum shows a Doppler shift that indicates the presence of an invisible companion). Then, if we can show that the companion is too faint to be a normal, main-sequence star, it must be a collapsed star. If, further, the mass of the unobservable companion is greater than 3 solar masses, it must be a black hole.

The most persuasive case is named Cygnus X-1, and was the first x-ray source that was found to vary in intensity on a time scale of several milliseconds. In 1971, radio radiation was found to come from the same direction. A 9th magnitude star called HDE 226868 was found at its location (Fig. 11–8).

HDE 226868 has the spectrum of a blue supergiant, and thus has a mass of about 15 times that of the sun. Its spectrum is observed to vary in radial velocity with a period of 5.6 days, indicating that the supergiant and the invisible companion are orbiting each other with that period. From the orbit, it is deduced that the invisible companion must certainly have a mass greater than 4 solar masses; the best estimate is 8 solar masses. Because this is so much greater than the limit of 2 or 3 solar masses above which neutron stars cannot exist, it seems that even allowing for possible errors in measurement or in the theoretical calculations, too much mass is present to allow the matter to settle down as a neutron star. Thus many, and probably most, astronomers believe that a black hole has been found in Cygnus X-1.

Calculations have been carried out, especially by Kip Thorne of Caltech, to predict how the supergiant and the black hole would interact if we accept the binary theory. Thorne finds that the binary theory predicts that the gravity of the black hole would attract matter from the outer layers of the supergiant, and that this matter would go into orbit around the black hole. Because of the orbital motion of the matter, it would gather up—accrete—a disk around the black hole, not unlike a large-scale version of Saturn's rings (Fig. 11–9). As the matter is heated, it gives off x-rays.

Matter in the accretion disk around the black hole would orbit very quickly. If a hot spot developed somewhere on the disk, it might beam a cone of radiation into space, similar to the lighthouse model of pulsars. If the hot spot lasted for several rotations, we could detect a pulse of x-rays every time the cone swept past the earth. Thorne predicts that the period of the x-ray pulses would be extremely short, only a few milliseconds.

The Uhuru satellite was launched in 1970 in an international effort by an Italian team from a site in Kenya; Uhuru means "freedom" in Swahili, and is the Kenyan national motto.

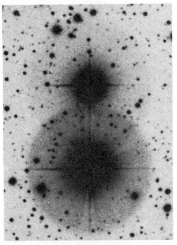

Figure 11–8 A section of the image reproduced in the photograph opening this chapter of the blue supergiant star HDE 226868. The black hole Cygnus X-1 that is thought to be orbiting the supergiant star is not visible. Note that the fact that the image of the supergiant appears so large is entirely an overexposure effect in the film and does not represent the actual angular diameter subtended by the star, which is too small to resolve from the earth.

Circinus X-1 and V861 Scorpii are other prime candidates.

Figure 11–9 An artist's conception of the disk of swirling gas that would develop around a black hole like Cygnus X-1 *(right)* as its gravity pulled matter off the companion supergiant *(left)*. The x-radiation would arise in the disk. (Painting by Lois Cohen, used courtesy of the Griffith Observatory)

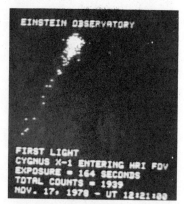

Figure 11-10 An x-ray view of Cygnus X-1, the first picture taken with HEAO-2, which is called the Einstein Observatory. The resolution of the telescope system aboard is about 4 arc sec, whereas the orbiting gas around the black hole—the "accretion disk"—is 100,000 times smaller, according to calculations, so that a single photograph like this one mainly shows that the telescope optics are focusing properly. The object was allowed to drift into the field of view for this first image, hence the appearance of a "tail." (Courtesy of Riccardo Giacconi, Harvard-Smithsonian Center for Astrophysics.)

The Schwarzschild radius of a black hole varies linearly with the mass and so the volume contained inside the event horizon varies with the cube of the mass. The density, which is the mass divided by the volume, is thus proportional to one over the square of the mass. Density = mass ÷ volume $\propto m \div d^3 = m \div m^3 = 1/m^2$, where m is mass and d is diameter. For very large masses, therefore, the density can be very low. Example: If we consider a black hole of (3×10^8) $M_\odot$, then the density is reduced by 10^{16} over the density of a 3 solar mass $(3\ M_\odot)$ black hole, where $M_\odot$ stands for the mass of the sun.

Searches for the short period pulses have been made with several x-ray satellites, most recently the Einstein Observatory (HEAO-2). Cygnus X-1 (Fig. 11-10) and a couple of other sources show pulses on a time scale of 10 milliseconds but not on a scale of 1 millisecond. Detailed observations and analysis continue.

It still looks as though Cygnus X-1 represents the first observational detection of a black hole.

*11.6 NON-STELLAR BLACK HOLES

We have discussed how black holes can form by the collapse of a star of greater than 3 solar masses. But theoretically a black hole will result if a mass of any amount is sufficiently compressed. No mass less than 3 solar masses will contract sufficiently under the force of its own gravity in the course of stellar evolution. But the density of matter was so high at the time of the origin of the universe (see Chapter 27) that smaller masses may have been sufficiently compressed to form what are called mini black holes.

There is no good evidence for a mini hole, but some scientists predict that there could be some the size of pinheads (and thus masses equivalent to those of asteroids) floating around in space.

Stephen Hawking, the English astrophysicist who suggested the existence of mini black holes, has further deduced that small black holes can seem to emit energy in the form of elementary particles (neutrinos and so forth). The mini holes would thus evaporate and disappear. This may seem to be a contradiction to the concept that mass can't escape from a black hole. And indeed, when we consider effects of quantum mechanics, the simple picture of a black hole that we have discussed up to this point is not sufficient. Hawking suggests that a black hole so affects space near it that a pair of particles—a nuclear particle and its antiparticle—can form simultaneously. The antiparticle disappears into the black hole, and the remaining particle reaches us. This situation is important only for the smallest mini black holes, for the amount of radiation decreases sharply as we consider black holes of greater and greater masses. Only mini black holes up to the mass of an asteroid—far short of stellar masses—would have had time to disappear since the origin of the universe. Hawking's ideas set a lower limit on the size of black holes now in existence, since we think the mini black holes were formed only at the time of the big bang (which is discussed in Chapter 27).

On the other extreme of mass, we can consider what a black hole would be like if it contained a very large number, i.e., thousands or millions, of solar masses. Thus far, we have considered only black holes the mass of a star or smaller. Such black holes form after a stage of high density. But the more mass involved, the lower the density needed for a black hole to form. For a very massive black hole, one containing hundreds of millions or billions of solar masses, the density would be fairly low when the event horizon formed, approaching the density of water. For even higher masses, the density would be lower yet.

Thus if we were traveling through the universe in a spaceship, we couldn't count on detecting a black hole by noticing a volume of high density. We could pass through the event horizon of a high-mass black hole without even noticing. We would never be able to get out, but it would be hours before we would notice that we were being drawn into the center at an accelerating rate.

Where could such a supermassive black hole be located? The existence of bursters (Section 6.6) may be evidence that black holes of about 100 solar masses are at the center of globular clusters. The center of our galaxy is one possibility for an even more massive black hole. Strange bursts of x-ray emission have been detected, and it is reasonable that there could be a concentration of mass there (see Section 22.2). Though we do not expect to observe radiation from the black hole itself, we hope to observe radiation from the gas surrounding it. The core of the galaxy M87 is another prime suspect (see Box 25.1). Quasars (Section 26.4) are other possible locations for massive black holes.

SUMMARY AND OUTLINE

Gravitational collapse occurs if more than 2 or 3 solar masses remain (Section 11.1)
Critical radii of a black hole
   Photon sphere: exit cones form (Section 11.2)
   Event horizon: exit cones close (Section 11.3)
   Schwarzschild radius (gravitational radius) defines the limit of the black hole, the event horizon
   Singularity
Rotating black holes (Section 11.4)
   Can get energy out of the ergosphere

Detecting a black hole (Section 11.5)
   X-radiation expected
   Detection in a spectroscopic binary, invisible high-mass companion
   Cygnus X-1: a black hole?
   Observations with HEAO's
Non-stellar black holes (Section 11.6)
   Mini black holes could have formed in the big bang
   Very massive black holes do not have high densities

QUESTIONS

1. Why doesn't electron or neutron degeneracy prevent a star from becoming a black hole?

2. Why is a black hole blacker than a black piece of paper?

3. (a) Is light acting more like a particle or more like a wave when it is bent by gravity?
   (b) Can you explain the bending of light as a property of a warping of space, as discussed in Section 7.11?

4. (a) How does the escape velocity of the moon compare with the escape velocity of the earth? Would a larger rocket engine be necessary to escape from the gravity of the earth or from the gravity of the moon?
   (b) How will the velocity of escape from the surface of the sun change when the sun becomes a red giant? A white dwarf?

†5. (a) What is the Schwarzschild radius for a 10 solar mass star?
   (b) What is your Schwarzschild radius?

6. What is the relation in size of the photon sphere and the event horizon? If you were an astronaut in space, could you escape from within the photon sphere of a rotating black hole? From within its ergosphere? From within its inner event horizon?

7. (a) Why does the mass of a black hole that results from a collapsed star tend to increase rather than decrease with time?
   (b) What property would show up in mini black holes, if they exist, that would allow them to eventually lose mass?

8. Would we always notice when we reached a black hole by its high density? Explain.

9. Could we detect a black hole that was not part of a binary system?

10. (a) Under what circumstances does the presence of an x-ray source associated with a spectroscopic binary suggest to many astronomers the possible presence of a black hole? Why?
    (b) For what additional properties of the objects in the binary system do we search?

---

†This indicates a question requiring a numerical solution.

PART IV

The Earth and the rest of the solar system may be important to us, but they are only minor companions to the stars. In *Captain Stormfield's Visit to Heaven,* by Mark Twain, the Captain races with a comet and gets off course. He comes into heaven by a wrong gate, and finds that nobody there has heard of "the world" ("**the** world, there's billions of them!" says a gatekeeper). Finally, the gatekeepers send a man up in a balloon to try to detect "the world" on a huge map. The balloonist has to travel so far that he rises into clouds and after a day or two of searching he comes back to report that he has found it: an unimportant planet, more properly named "the Wart."

We too must learn humility as we ponder the other objects in space. And while it is no doubt the case that the heavens are filled with a vast assortment of suns and planets more spectacular than our own, still, the solar system is our own local environment. We would like to understand it and come to terms with it as best we can. Besides, in understanding our own solar system, we may even find some keys to understanding the rest of the Universe.

To get an idea of its scale, imagine that the solar system is scaled down and placed on a map of the United States. Let us say that the sun is a hot ball of gas taking up all of Rockefeller Center, more than a kilometer across, in the center of New York City.

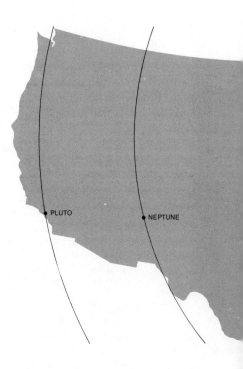

We would then find Mercury to be a ball 4 meters across at the distance of mid–Long Island, and Venus to be a 10-meter ball one and a half times farther away. The Earth is only slightly bigger and is located at the distance of Trenton, New Jersey. Mars is half that size, 5 meters across, located past Philadelphia.

Only for the planets beyond Mars would the planets be much different in size from the Earth, and the separations become much greater. Jupiter is 100 meters across, the size of a baseball stadium, past Pittsburgh at the Ohio line. Saturn without its rings is a little smaller than Jupiter (including the rings it is a little larger), and is past Cincinnati toward the Indiana line. Uranus and Neptune are each about 30 meters across, about the size of a baseball infield, and are at the distance of Topeka and Santa Fe, respectively. And Pluto, as small as the inner planets, is as far away as Los Angeles, 40 times farther away from the sun than the Earth is. Occasionally a comet sweeps in from Alaska, or some other random direc-

THE SOLAR SYSTEM

tion, passes around the Sun, and returns in the general direction from which it came.

The planets fall naturally into two groups. The first group, the *terrestrial planets,* consists of Mercury, Venus, Earth, and Mars. All are rocky in nature. The terrestrial planets are not very large, and have densities about five times that of water. (When the metric system was set up, the gram was defined so that water would have a convenient density of exactly 1 gram/cubic centimeter. Thus the terrestrial planets have densities of about 5 g cm^{-3}.)

The second group, the *giant planets,* consists of Jupiter, Saturn, Uranus, and Neptune. All these planets are much larger than the terrestrial planets, and also much less dense, ranging down to slightly below the density of water. Jupiter and Saturn are largely gaseous in nature, similar to the Sun in composition. Uranus and Neptune have large gaseous atmospheres but have rocky or icy surfaces and interiors. Pluto, planet number nine, is anomalous in several of its properties, and so may have had a very different history from the other planets.

Between the orbits of the terrestrial planets and the orbits of the giant planets are the orbits of thousands of chunks of small "minor planets." These *asteroids* range up to 1000 kilometers across. Sometimes much smaller chunks of interplanetary rock penetrate the Earth's atmosphere and hit the ground. We shall discuss these *meteorites* and where they came from together with asteroids and comets.

Many people are interested in the planets in order to study their history—how they formed, how they have evolved since, and how they will change in the future. Others are more interested in what the planets are like today. Still others are interested in the planets mainly to consider whether they are harboring intelligent life.

Because the Moon and some other objects in the solar system have undergone less erosion on their surfaces than has the Earth, in observing these celestial bodies we actually see them as they appeared eons ago. In this way the study of the planets and other objects in the solar system as they are now provides information about the solar system's early stages and its origin. Other objects, like Jupiter's moon Io, change even more rapidly in their surface appearance than does the Earth. We shall see in the following chapters how the study of the solar system is varied in nature, and how much these studies tell us about our own planet.

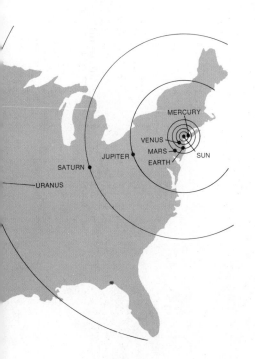

CHAPTER 12

THE EARLY HISTORY OF ASTRONOMY

AIMS:
To discuss the scale and structure of the
solar system, the history of our
understanding of the geocentric and
the heliocentric theories, Kepler's laws
about the motions of the planets,
Newton's law of gravity, and the origin of
the solar system

In this chapter we describe some general ideas of the solar system and its structure. First we describe such commonly observed phenomena as phases of the moon and eclipses. Then we discuss the origins of astronomy, which began with such practical things as keeping time and marking the arrival of the seasons. We trace the development of early astronomy through the work of some of the principal scientists who contributed to the field.

In this chapter we consider the period up to 1687, when Isaac Newton published his major work. This date marks the origin of what we might call the modern era of astronomy. We also consider the historical development of our ideas of the structure and origin of the solar system.

Other historical developments since the time of Newton are treated throughout this book in the context of the various separate topics.

12.1 THE PHASES OF THE MOON AND PLANETS

From the simple observation that the apparent shapes of the Moon and planets change, we can draw conclusions that are important for our understanding of the mechanics of the solar system. The fact that the Moon goes through a set of *phases* approximately once every month is perhaps the most familiar everyday astronomical observation (Fig. 12–1). In fact, the

"The Moon" is often capitalized to distinguish it from moons of other planets; in general writing, it is usually written with a small "m." In Part IV, we shall capitalize "Earth" to put it on a par with the other planets and the Moon. For consistency, we shall also capitalize "Sun" and "Universe."

The relative sizes of the planets and the Sun. The relative sizes of the planets are exaggerated with respect to the Sun by a factor of 2 in this drawing.

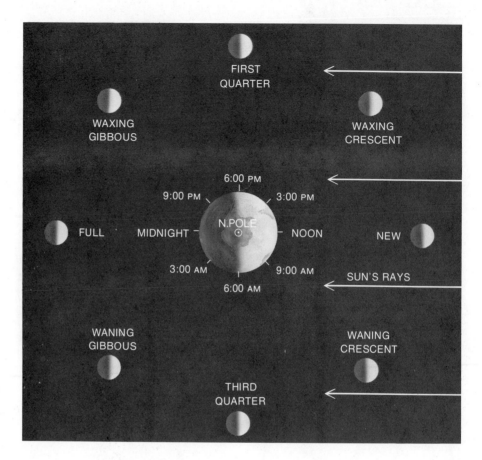

Figure 12–1 The phases of the Moon depend on the Moon's position in its orbit around the Earth. Here we visualize the situation as if we could be high above our north pole, looking down.

The phases of moons or planets are the shapes of the sunlighted areas as seen from our vantage point.

Figure 12–2 The phases of the Moon.

name "month" comes from the word "moon." The actual period of the phases, the interval between a particular phase of the Moon and its next repetition, is approximately 29½ Earth days (Fig. 12–2). This period can vary by as much as 13 hours. The explanation of the phases is quite simple.

The Moon is a sphere, and at all times the side that faces the Sun is lighted and the side that faces away from the Sun is dark. The phase of the Moon that we see from the Earth, as the Moon revolves around us, depends on the relative orientation of the three bodies: Sun, Moon, and Earth. The situation is simplified by the fact that the plane of the Moon's revolution

4 days

7 days
1st quarter

10 days

14 days
full

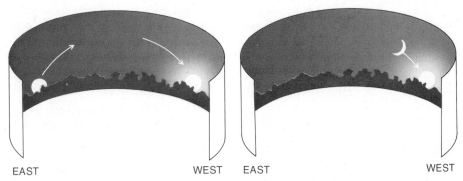

EAST   WEST   EAST   WEST

Figure 12–3 Because the phase of the Moon depends on its position in the sky with respect to the Sun, we can see why a full moon is always rising at sunset while a crescent moon is either setting at sunset, as shown here, or rising at sunrise.

around the Earth is nearly, although not quite, the plane of the Earth's revolution around the Sun.

Basically, when the Moon is almost exactly between the Earth and the Sun, the dark side of the Moon faces us. We call this a "new moon." A few days earlier or later we see a sliver of the lighted side of the Moon, and call this a "crescent." As the month wears on, the crescent gets bigger, and about 7 days after new moon, half the face of the Moon that is visible to us is lighted. We sometimes call this a "half moon." Since this occurs one fourth of the way through the phases, the situation is also called a "first-quarter moon." (Instead of apologizing for the fact that astronomers call the same phase both "quarter" and "half," I'll just continue with a straight face and try to pretend that there is nothing strange about it.)

When over half the Moon's disk is visible, we have a "gibbous" moon. One week after the first-quarter moon, the Moon is on the opposite side of the Earth from the Sun, and the entire face visible to us is lighted. This is called a "full moon." One week later, we have a "third-quarter moon." Then we go back to "new moon" again and repeat the cycle of phases.

Note that since the phase of the Moon is related to the position of the Moon with respect to the Sun, one can tell when the Moon will rise at a certain phase. For example, since the Moon is 180° across the sky from the Sun when it is full, a full moon is always rising just as the Sun sets (Fig. 12–3). Each day thereafter, the moon rises approximately one hour later (24 hours divided by the period of revolution of the Moon around the Earth). The third-quarter moon, then, rises near midnight, and is high in the sky at sunrise. The new moon rises with the Sun in the east at dawn. The first-quarter moon rises near noon and is high in the sky at sunset. It is often visible in the late afternoon.

From To Jane (The Keen Stars Were Twinkling) *by Percy Bysshe Shelley (written in 1822)*

The stars will awaken,
Though the moon sleep a full hour later,
Tonight. . . .

20 days

22 days
third quarter

24 days

26 days

Figure 12–4 The crescent of Earth seen from a distance of 450,000 km by the Pioneer Venus spacecraft, whose trajectory took it briefly outside the Earth's orbit.

Of the seven eclipses of the Sun and the Moon that can occur in a single year, most are partial eclipses, where the Sun or the Moon is only partially covered. The total eclipses, which are rarer, are much more spectacular.

Figure 12–5 The plane of the Moon's orbit is tipped with respect to the plane of the Earth's orbit, so the Moon usually passes above or below the Earth's shadow.

The Moon is not the only object in the solar system that is seen to go through phases. Mercury and Venus both orbit inside the Earth's orbit, and so sometimes we see the side that faces away from the Sun and sometimes we see the side that faces toward the Sun. Thus at times Mercury and Venus are seen as crescents, though it takes a telescope to observe their shapes. Spacecraft to the outer planets have looked back and seen the Earth as a crescent (Fig. 12–4).

The outer planets, though, from Mars on outward, are never between the Sun and the Earth. Thus they never appear as crescents, although sometimes they can appear gibbous.

12.2 ECLIPSES

Because the Moon's orbit around the Earth, and the Earth's orbit around the Sun are not precisely in the same plane (Fig. 12–5), the Moon usually passes slightly above or below the Earth's shadow at full moon, and the Earth usually passes slightly above or below the Moon's shadow at new moon. But every once in a while, up to seven times a year, the Moon is at the part of its orbit that crosses the Earth's orbital plane at full moon or new moon. When that happens, we have a lunar or a solar eclipse.

We have already discussed eclipses of the Sun, when the Moon comes directly between the Earth and the Sun. Many more people see a lunar eclipse than a solar eclipse when one occurs. At a lunar eclipse, the Moon lies entirely in the Earth's shadow and sunlight is entirely cut off from it (Fig. 12–6). So anywhere on the Earth that the Moon has risen, the eclipse is visible. In a solar eclipse, on the other hand, the alignment of the Moon between the Sun and the Earth must be precise, and only those people in a narrow band on the surface of the Earth see the eclipse.

A lunar eclipse is a much more leisurely event to watch than a solar eclipse. The partial phase, when the Earth's shadow gradually covers the Moon, lasts for hours. And then the total phase, when the Moon is entirely within the Earth's shadow, can itself last for over an hour. During this time, the sunlight is not entirely shut off from the Moon (Fig. 12–7). A small

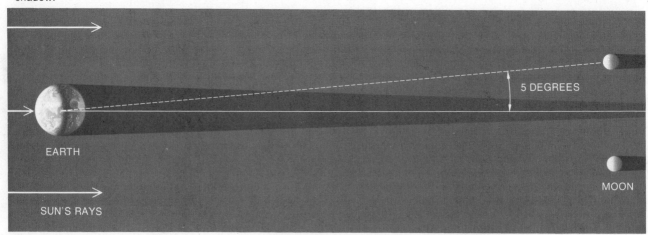

EARTH

5 DEGREES

MOON

SUN'S RAYS

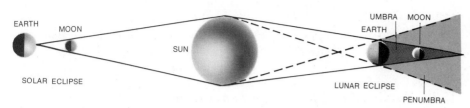

Figure 12–6 When the Moon is between the Earth and Sun, we observe an eclipse of the Sun. When the Moon is on the far side of the Earth from the Sun, we see a lunar eclipse. The part of the Earth's shadow from which the Sun is only partially shielded from the Moon's view is called the *penumbra;* the part of the Earth's shadow from which the Sun is entirely shielded from the Moon's view is called the *umbra.*

amount is refracted around the edge of the Earth by our atmosphere. Most of the blue light is taken out during the sunlight's passage through our atmosphere; this is how blue skies are made for the people part way around the globe from the point at which the Sun is overhead. The remaining light is reddish, and this is the light that falls on the Moon. Thus, the eclipsed Moon appears reddish.

12.3 ANCIENT ASTRONOMY—THE MAIN LINE

*12.3a Egyptian Astronomy

In these days of digital watches, electronic calculators, and accurate time on the radio and television, it is hard to imagine 1000 B.C. or earlier, when there were no mechanical clocks and even the idea of a calendar was vague. Yet each spring, the Nile River flooded the neighboring farmlands, and the Egyptian farmers wanted to know how to predict when this would happen. Farmers everywhere, even though not dependent on annual flooding, wanted to know when to expect the spring and the spring rains so that they could plant their crops in ample time.

Eventually, some people noticed that, rising in the eastern sky, Sirius, the brightest star in the sky, became visible before sunrise just at the time of year when the crops should be planted and when the Nile was about to flood. This first visibility of Sirius in the morning sky—the *heliacal rising* of Sirius—served as a marker for Egyptian astronomers and priests and enabled them to let everybody know that the springtime had come. After all, the Egyptian climate does not show the changing of the seasons as obviously as does New England's.

Figures in the sky—constellations—had been recorded by the Sumerians in at least 2000 B.C., and maybe earlier. The bull, the lion, and the scorpion are constellations that date from this time. Other constellations that we have in the present time were recorded by a Greek, Thales of

It has long been known that Egyptian pyramids and temples erected since 1500 B.C. have astronomical alignments. And even the Great Pyramid of Giza from 2600 B.C. has its four faces accurately aligned north, east, south, and west. The main axis of the temple of Karnak is aligned with mid-winter sunrise. And at the equinoxes, the Sun shone directly through the temple at Abu Simbel so as to light up the sanctuary inside. Also, the Sun god, Ra, was pictured widely. Astronomy was used for religious as well as practical purposes.

Figure 12–7 The Earth's atmosphere scatters shorter wavelengths much more than it scatters longer wavelengths, so red light is scattered much less than blue light. Some of the red light survives its passage through the Earth's atmosphere and is bent, or refracted, toward the Moon. Thus the Moon often appears reddish during the total phase of a lunar eclipse.

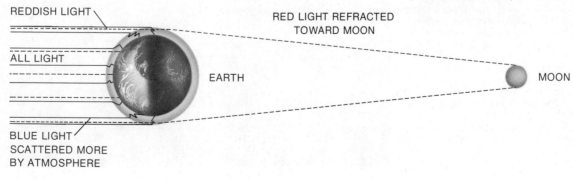

REDDISH LIGHT

RED LIGHT REFRACTED TOWARD MOON

ALL LIGHT

EARTH

MOON

BLUE LIGHT SCATTERED MORE BY ATMOSPHERE

Miletus, in about 600 B.C., at the dawn of Greek astronomy. Also, certain constellations were common to several civilizations. The Chinese also had a lion, a scorpion, a hunter, and a dipper, for example.

12.3b   Greek Astronomy: The Earth at the Center

When we observe the planets in the sky, we notice that their positions vary from night to night with respect to each other and with respect to the stars. The stars, on the other hand, are so far away that their positions are relatively fixed with respect to each other. The fact that the planets appear as "wandering stars" was known to the ancients; our word "planet" comes from the Greek word for "wanderer." Of course, both stars and planets revolve together across our sky essentially once every 24 hours; by the wandering of the planets we mean that the planets appear to move at a slightly different rate so that over a period of weeks or months they change position with respect to the fixed stars.

The motion of the planets in the sky is not continuously in the same direction with respect to the stars. Most of the time they appear to drift eastward, with respect to the background stars, but sometimes they drift backwards, that is, westward. We call the backward motion *retrograde motion* (Fig. 12–8).

The ancient Greeks began to explain the motions of the planets through making theoretical models of the geometry of the solar system. For example, by studying which of the known planets had the longest periods of retrograde motion, they were able to discover the order of distance of the planets.

One of the earliest and greatest philosophers, Aristotle, who lived in Greece about 350 B.C., summarized the astronomical knowledge of his day into a qualitative cosmology that remained dominant for 1800 years. On the basis of what seemed to be very good evidence—what he saw—Aristotle thought, and actually believed that he knew, that the Earth was at the center of the universe and that the planets, the Sun, and the stars revolved around it (Fig. 12–9). The Universe was made up of a set of 55 celestial spheres that fit around each other and that had rotation as their natural motion. Each of the heavenly bodies was carried around the heavens by a sphere, whose motion was compounded by the motion of other spheres. These motions combined to account for the various observed motions, including revolution around the Earth, retrograde motion, and motion above and below the ecliptic. Each planet had several spheres to account for its various motions.

In ancient times, five planets were known—Mercury, Venus, Mars, Jupiter, and Saturn, plus the Sun and the Moon.

The planets and Moon appear to move on or near the ecliptic, *the path of the Sun in the sky.*

We give dates in our conventional B.C. *and* A.D. *notation, but obviously the ancient Greeks couldn't do that. They actually dated events from Olympiads, just as the Romans mostly dated events from the accession of consuls, their rulers. By the late Roman period, a traditional epoch of the foundation of Rome was used, just as we use 1* A.D. *as the origin of our system of dates. (Don't buy any coins dated 43* B.C.*)*

Figure 12–8 The retrograde loop of Mars in 1977 and 1978. Opposition, when Mars was on the side of the Earth directly opposite the sun, occurred in the constellation Cancer in January 1978; Gemini and Canis Minor are among the other constellations visible in this photograph of a planetarium simulation. Pollux is the bright star just above and to the right side of the retrograde loop. Regulus is the bright star to the left that Mars passed closely. The next opposition of Mars will be in March 1982. (Photograph by George Lovi, Vanderbilt Planetarium)

The outermost sphere was that of the fixed stars, beyond which lay the prime mover, *primum mobile,* that caused the general rotation of the stars overhead.

Aristotle's theories ranged through much of science. He held that all bodies below the sphere of the Moon were made of four basic "elements": earth, air, fire, and water. The fifth "essence"—the quintessence—was a perfect, unchanging transparent element of which the celestial spheres around that of the Moon are formed.

Aristotle's theories dominated scientific thinking for almost two millennia, until the Renaissance. Unfortunately, most of his theories were far from what we now consider to be correct, so we tend to think that the widespread acceptance of Aristotelian physics impeded the development of science.

In about 140 A.D., almost 500 years after Aristotle, the Greek astronomer Claudius Ptolemy (Fig. 12–10), in Alexandria, presented a detailed theory of the Universe that explained the retrograde motion. Ptolemy's model was, as was Aristotle's, Earth-centered. To account for the retrograde motion of the planet's, the planets had to be moving not simply on large circles around the Earth but rather on smaller circles, called *epicycles,* whose centers moved around the Earth on larger circles, called *deferents* (Fig. 12–11). (The notion of epicycles and deferents had been advanced earlier by such astronomers as Hipparchus, whose work on star catalogues we mentioned in Section 4.3.) It seemed natural that the planets should follow circles in their motion, since circles were thought to be "perfect" figures. Sometimes the center of the deferent was not centered at the Earth (and thus the circle was an *eccentric*). The planets moved at a constant rate of angular motion (that is, the angle through which they moved was the same for each identical period of time), but another complication was that the point around which the planet's angular motion moved uniformly was neither at the center of the Earth nor at the center of the deferent. The epicycle moved at a uniform angular rate about still another point, the *equant.* The equant was on the far side of the center of the deferent with respect to the Earth. Ptolemy's views were very influential in the study of astronomy, because versions of his ideas and of the tables of planetary motions that he computed were accepted for nearly 15 centuries. His major work, the *Almagest,* contained both his ideas and a summary of the ideas of his predecessors (especially those of Hipparchus), and is the major source of our knowledge of Greek astronomy.

Figure 12–9 Aristotle's Earth-centered theory. The Earth is at the center, orbited in larger and larger circles by the Moon, Mercury, Venus, the Sun, Mars, Jupiter, Saturn, and the stars.

Figure 12–10 Ptolemy. (Burndy Library, photograph by Owen Gingerich)

Figure 12–11 *(A)* In the Ptolemaic system, a planet would move around on an epicycle which, in turn, moved on a deferent. Variations in the apparent speed of the planet's motion in the sky could be accounted for by having the epicycle move uniformly around a point called the equant, instead of moving uniformly around the Earth or the center of the deferent. The Earth and the equant were on opposite sides of the center of the deferent and equally spaced from it. When the planet was in the position shown, it would be in retrograde motion. *(B)* In the Ptolemaic system, the projected path in the sky of a planet in retrograde motion is shown.

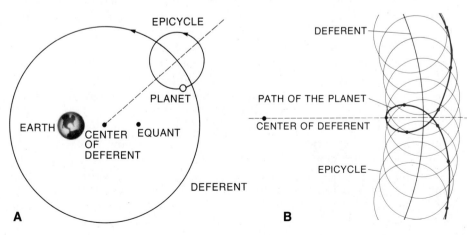

EPICYCLE

PLANET

EARTH   CENTER   EQUANT
OF
DEFERENT

DEFERENT

A

DEFERENT

PATH OF THE PLANET

CENTER OF DEFERENT

EPICYCLE

B

It was known that the lengths of the seasons are not equal. The present-day values for the northern hemisphere are

spring	92^{d}19^h
summer	93^{d}15^h
autumn	89^{d}20^h
winter	89^{d}0^h.

If the Sun goes in a circular orbit at uniform speed, then the orbit cannot be centered at the Earth.

Parallaxes are not visible to the naked eye nor could they be observed with the instruments available to the ancient Greeks.

°Box 12.1   Stellar Parallaxes and the Motion of the Earth

In addition to objecting to the notion that the Earth moved, Aristotle, Aristarchus, and Ptolemy had also reasoned (correctly) that if the Earth revolved around the Sun, then during the course of a year the stars would show slight displacements in the sky because of parallax, the effect of our changing vantage point. The parallaxes of even the nearby stars are too small for these parallaxes to have been observed before the invention of the telescope (and were not discovered until 1838). The Greek astronomers deserve credit, at least, for the fact that their theories agreed in this sense better with observations than did the heliocentric theory. It was only much later that accurate observations changed the weight of the evidence on this point.

---

*12.3c   Other Ancient Astronomy

At the same time that Greek astronomers were developing their ideas, Babylonian priests were computing positions of the Moon and planets. The major work was carried out during the period from about 700 B.C. up until about 50 A.D.

This work is sometimes referred to as Chaldean—pronounced something like kal-dē-an in English. Chaldea was the southern part of Babylonia.

The Babylonian tablets that have survived show not only lists and tables of planetary positions and eclipses but also predictions of such quantities as the times when planets would be closest to (in *conjunction* with) and farthest from (in *opposition* to) the Sun in the sky and when objects would be visible for the first or the last time in a year.

Babylonian methods were communicated to the Greeks, and it is through the Greeks that Babylonian astronomy had its influence on western thought.

Elsewhere in the world, other civilizations were also developing astronomical ideas. China, for example, had astronomers who recorded the occurrences not only of eclipses but also of novae, and who discussed the possible motion of the Earth. And they had known basic ideas, like that of the calendar, for centuries. India had observatories thousands of years ago. In Polynesia, the stars were long used for accurate navigation at sea. Stonehenge and other sites are evidence of prehistoric observations in England.

Novae to the ancients were newly visible stars.

12.4   NICOLAUS COPERNICUS: THE SUN AT THE CENTER

The credit for the breakthrough in our understanding of the solar system belongs to Nicolaus Copernicus (Fig. 12–12), a Polish astronomer whose 500th birthday was celebrated by the astronomical community in 1973. Copernicus advanced a *heliocentric*—sun-centered—theory (Fig. 12–13). He suggested that the retrograde motion of the planets could be readily explained if the Sun rather than the Earth was at the center of the Universe, that the Earth is a planet, and that the planets move around the Sun in circles.

Helios was the sun god in Greek mythology.

Aristarchus of Samos, a Greek scientist, had also suggested a heliocentric theory 18 centuries earlier, though we do not know how detailed a picture of planetary motions he presented. His heliocentric suggestion required the apparently ridiculous notion that the Earth itself moved, in contradiction to our senses and to the theories of Aristotle. If the Earth is rotating, for example, why aren't birds and clouds left behind the moving Earth? Only the 17th century discovery by Isaac Newton of laws of motion

Figure 12–12

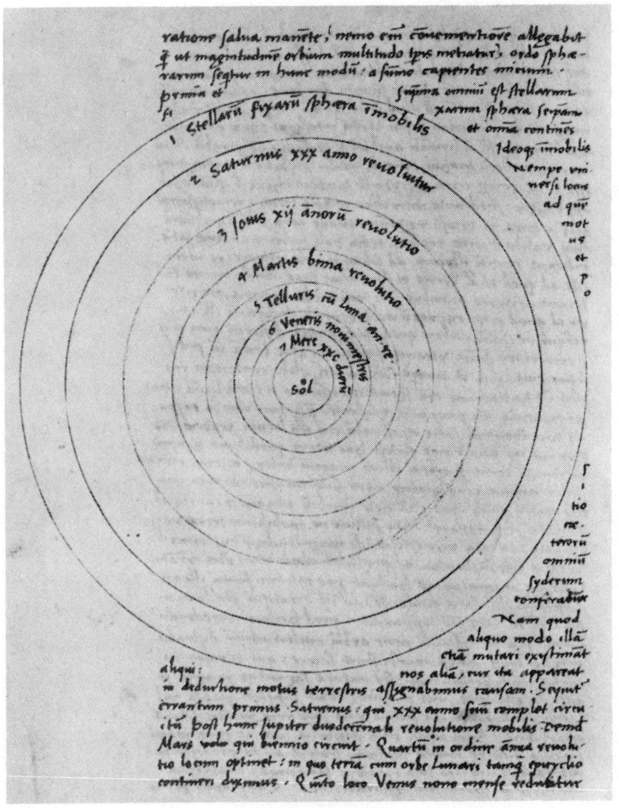

Figure 12–13

substantially different from Aristotle's solved this dilemma. Aristarchus' heliocentric idea had long been overwhelmed by the *geocentric*—Earth-centered—theories of Aristotle and Ptolemy.

Copernicus' theory, although it put the Sun instead of the Earth at the center of the solar system (and, for then, the Universe), still assumed that the orbits of celestial objects were circles. As a result, in order to improve agreement between theory and observation, Copernicus still invoked the presence of some epicycles, though he eliminated the equant. (Since they were not as important as the epicycles that had to be used in the Ptolemaic theory to provide retrograde motion, we may call Copernicus' smaller circles "epicyclets.") Further, the detailed predictions that Copernicus himself computed on the basis of his theory were not in much better agreement with the existing observations than tables based on Ptolemy's model, because in many cases Copernicus still used Ptolemy's observations. These observations were deficient in that the uncertainty with which a result was known was relatively large and also in that the values themselves were sometimes wrong. The heliocentric theory appealed to Copernicus and to many of his contemporaries on philosophical grounds rather than because direct comparison of observations with theory showed the new theory to be better.

Copernicus' heliocentric theory, published in 1543 in the book he called *De Revolutionibus, On the Revolutions*, (Fig. 12–15), explained the retrograde motion of the planets as follows (Fig. 12–16):

Figure 12–12 Copernicus, in a 16th century woodcut.

Figure 12–13 The page of Copernicus' original manuscript in in which he drew his heliocentric system. The sun (sol) is at the center surrounded by Mercury (Merc), Venus (Veneris), Earth (Telluris), Mars (Martis), Jupiter (Jovis), Saturn (Saturnus), and the fixed stars. The manuscript is in the University library in Cracow.

Figure 12–14 Copernicus' signature from the Uppsala University Library in Sweden, photographed by Charles Eames and reproduced courtesy of Owen Gingerich.

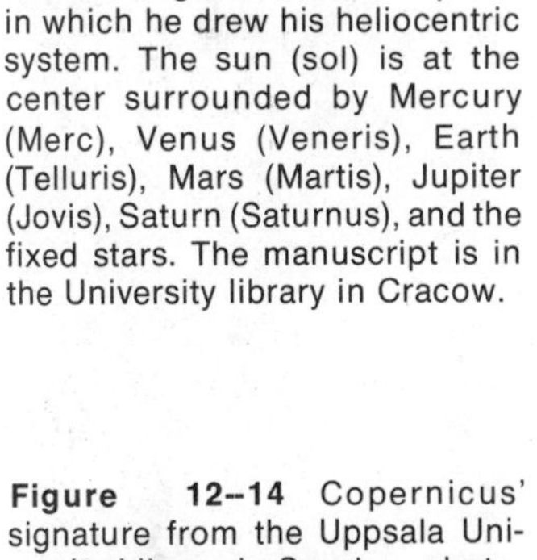

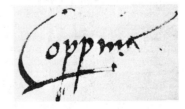

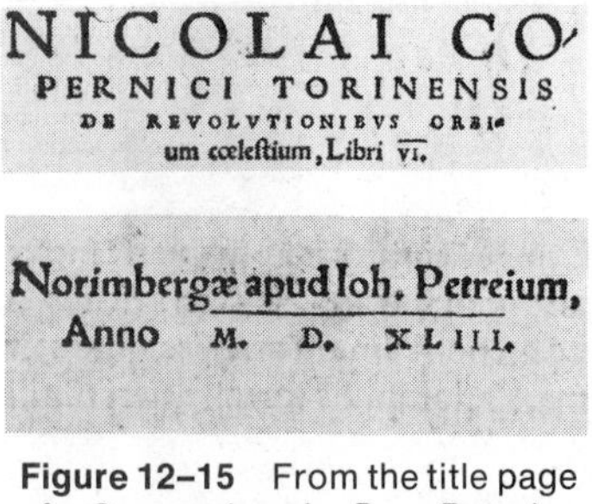

NICOLAI CO
PERNICI TORINENSIS
DE REVOLVTIONIBVS ORBI-
um cœlestium, Libri VI.

Norimbergæ apud Ioh. Petreium,
Anno M. D. XLIII.

Figure 12–15 From the title page of Copernicus' *De Revolutionibus*. About 200 copies of this work are currently known to be extant.

Leonardo da Vinci, for example, deduced that when we are able to see the dark part of the Moon faintly lighted, it is because of earthshine, the reflection of sunlight off the Earth and over to the Moon. His work, astronomical and otherwise, both artistic and scientific, took place about 1500.

Let us consider, first, an outer planet like Mars as seen from the Earth. As the Earth approaches the part of its orbit that is closest to Mars (which is orbiting the Sun much more slowly than does the Earth), the projection of the Earth–Mars line to the stars (which are essentially infinitely far away compared to the planets) moves slightly against the stellar background. As the Earth in its orbit comes closest to Mars, and then passes it, the projection of the line joining the two planets can actually seem to go backward, since the Earth is going at a greater speed than Mars. Then, as the Earth continues around its orbit, Mars appears to go forward again. A similar explanation can be demonstrated for the retrograde loops of the inner planets.

12.5 GALILEO GALILEI

As art, music, and architecture began to flourish after the Middle Ages, astronomy also developed. Copernicus' work can be considered to be the beginning of the astronomical renaissance, and from that time on there was continual development.

The Italian scientist Galileo Galilei began to believe in the Copernican system in the 1590's, and later provided important observational confirmation of the theory. In 1610, simultaneously with the first settlements in the American colonies, Galileo was the first to use a telescope for astronomical observation.

In his book *Sidereus Nuncius, The Starry Messenger*, published in 1610, he reported that with his telescope he could see many more stars than he could with the naked eye, and could see that the Milky Way and certain other hazy-appearing regions of the sky actually contained individual stars. He described views of the Moon, including the discovery of mountains, craters, and the relatively dark regions he called (and that we still call) *maria* (pronounced mar' ē-a), seas. And he discovered that small bodies revolved around Jupiter (Fig. 12–17). This discovery proved that all bodies did not

Figure 12–16 The Copernican theory explains retrograde motion as an effect of projection. The drawing shows the explanation for Mars or for the other *superior planets,* that is, planets whose orbits lie outside that of the Earth. Similar drawings can explain retrograde motion for *inferior planets,* that is, planets whose orbits lie inside that of the Earth (namely, Mercury and Venus).

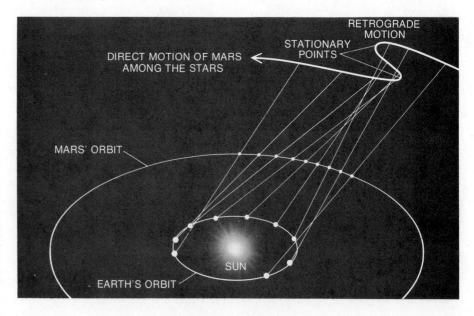
DIRECT MOTION OF MARS AMONG THE STARS — STATIONARY POINTS — RETROGRADE MOTION — MARS' ORBIT — SUN — EARTH'S ORBIT

Color Plate 22 (top, left): Copernicus' original manuscript of *De Revolutionibus*, the sixteenth-century work in which he set out his heliocentric theory. (Photo by Charles Eames)

Color Plate 23 (top right): Comet West, a bright comet visible in 1976. Broad bands are visible in the 50 million-kilometer-long dust trail. The faint blue gas tail extends straight up past the bright star Epsilon Pegasi. (Photo by Martin Grossmann)

Color Plate 24 (center): Apollo 17 view of the Taurus-Littrow Valley on the Moon. This huge, fragmented boulder had rolled a kilometer down the side of the North Massif to here. Scientist-astronaut Harrison Schmitt is at the left. The Lunar Rover is on the right. (NASA photo)

Color Plate 25 (bottom, left): The blast-off of Apollo 17, a night launch. (Photo by the author)

Color Plate 26 (bottom, center): An Apollo 17 astronaut with an instrument package. (NASA photo)

Color Plate 27 (bottom, right): The crescent Earth seen from Apollo 11 in orbit around the Moon. (NASA photo)

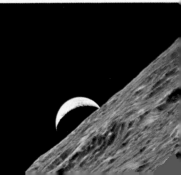

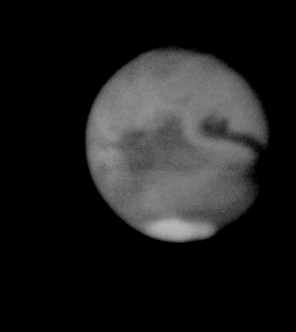

Color Plate 28 (top, left): Earth photographed from Apollo 11. Africa and the Middle East are clearly visible. (NASA photo)

Color Plate 29 (top, right): Mars, photographed from Earth as part of the International Planetary Patrol.

Color Plate 30 (bottom, left): Mars, photographed from the Viking 1 spacecraft as it approached in June 1976. Olympus Mons, the large volcano, is toward the top right of the picture. The Tharsis Mountains, a row of three other volcanoes, are also visible. To the left of these volcanoes, the irregular white area may be surface frost or ground fog. The large impact basin, Argyre, is the circular feature at the bottom of the disk. (NASA photo)

Color Plate 31 (bottom, right): The other side of Mars, photographed from Viking 1. The giant canyon in the upper hemisphere has a diameter larger than the coast-to-coast diameter of the United States.

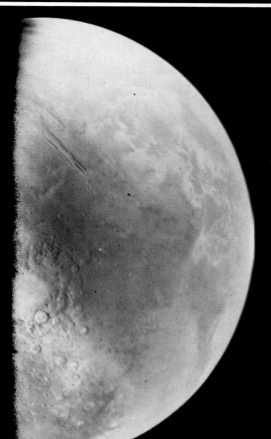

Color Plate 32 (top): Viking 1's sampler scoop is in the foreground of this view from the Martian surface. Angular rocks of various types can be seen. Large blocks one to two meters across can be seen on the horizon, which is about 100 meters from the spacecraft. The horizon may be the rim of a crater. (NASA photo)

Color Plate 33 (bottom, left): Mars' Utopia Planitia as seen from Viking 2, with the Lander's boom in the foreground. Many of the rocks are porous and sponge-like, similar to some of the Earth's volcanic rocks. (NASA photo; color-corrected version courtesy of Friedrich O. Huck)

Color Plate 34 (bottom, right): Mars' Utopia Planitia as seen from Viking 2. This winter scene reveals frost under some of the rocks (NASA photo; color-corrected version courtesy of Friedrich O. Huck)

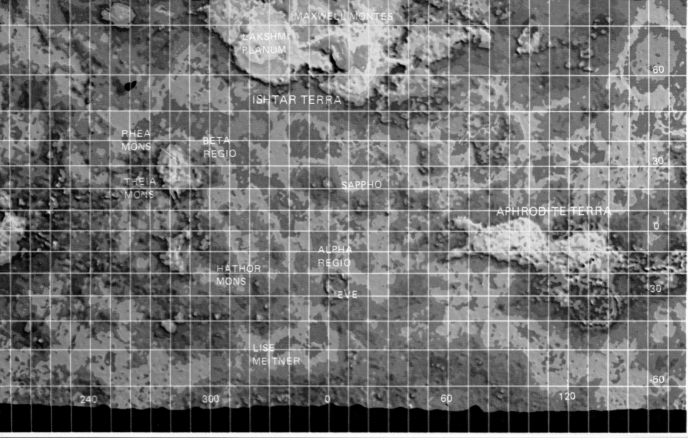

Color Plate 35 (top): A radar map of Venus from Pioneer Venus Orbiter. Most of Venus is covered by a rolling plain, shown in green and blue. The highlands (yellow and brown contours) sit atop the plain, like continents. Aphrodite Terra is half as big as Africa. Ishtar Terra is the size of Australia, but looks relatively large in this Mercator projection. The lowlands, though resembling Earth's ocean basin, cover only 16 percent of the planet. (Experiment and data—MIT; maps—U.S. Geological Survey; NASA/Ames spacecraft)

Color Plate 36 (bottom): An aurora borealis photographed in the Goldstream Valley, Alaska. (Gustav Lamprecht photo)

Color Plate 37 (top): Jupiter, its Great Red Spot, and 2 of its moons, in this Voyager 1 photograph. The satellite Io, can be seen against Jupiter's disk. The satellite Europa is visible off the limb at the right.

Color Plate 38 (bottom): Jupiter's Great Red Spot, photographed from Voyager 2. Gas in both the Spot and the adjacent white oval is circulating in the same direction.

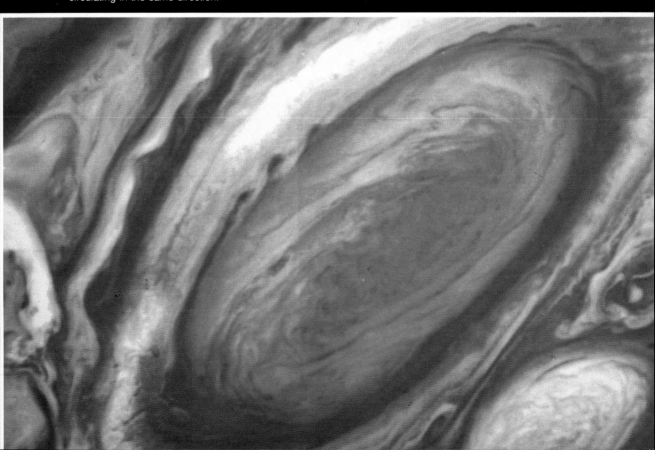

Color Plate 39 (top): Jupiter, photographed from Voyager 1. Io is visible above the Great Red Spot. Europa is also visible.

Color Plate 40 (bottom): (A) Io, photographed at a range of 826,000 km from Voyager 1. (B) Volcanoes erupting on Io, photographed from Voyager 2. Two volcanic eruption plumes that are about 100 km high and strongly scatter blue light appear on the limb.

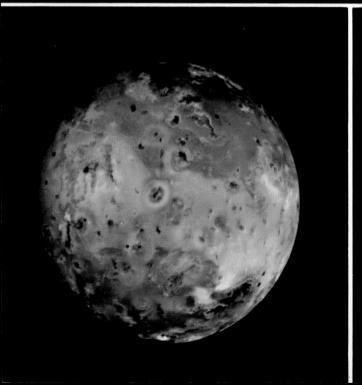

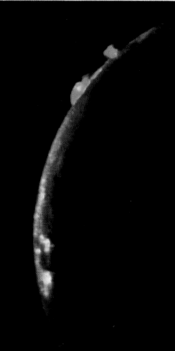

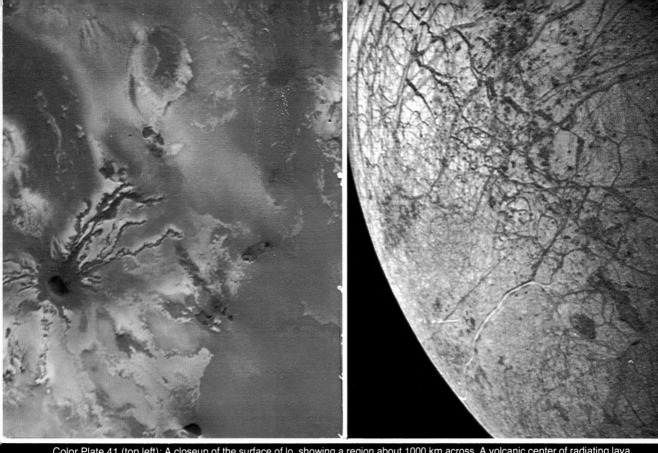

Color Plate 41 (top left): A closeup of the surface of Io, showing a region about 1000 km across. A volcanic center of radiating lava flows is visible at left center.

Color Plate 42 (top right): Europa, photographed from Voyager 2, showing its very smooth and fractured crust.

Color Plate 43 (bottom left): Ganymede, photographed from Voyager 2. The dark, cratered, circular feature is about 3200 km in diameter.

Color Plate 44 (bottom right): Callisto, photographed from Voyager 1. The bull's-eye, named Valhalla, is a large impact basin. The outer ring is about 2600 km across.

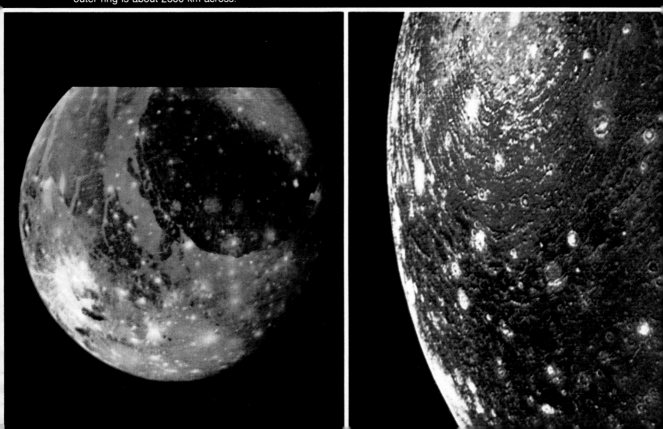

Color Plate 45 (top): Saturn, on March 11, 1974, showing bands on its disk and Cassini's division in its rings. (Lunar and Planetary Laboratory, University of Arizona photo)

Color Plate 46: Saturn and two of its moons, Tethys and Dione, were photographed by Voyager November 3, 1980, from 13 million km (8 million (JPL/NASA photo)

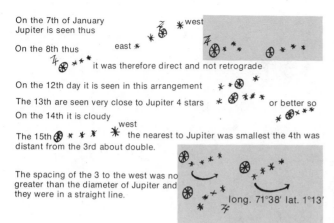

On the 7th of January Jupiter is seen thus

On the 8th thus east ✳

it was therefore direct and not retrograde

On the 12th day it is seen in this arrangement

The 13th are seen very close to Jupiter 4 stars — or better so

On the 14th it is cloudy

The 15th ✳ ✳ ✳ ✳ the nearest to Jupiter was smallest the 4th was distant from the 3rd about double.

The spacing of the 3 to the west was no greater than the diameter of Jupiter and they were in a straight line.

long. 71°38′ lat. 1°13′

Figure 12–17 A translation *(left)* of Galileo's original notes *(right)* summarizing his first observations of Jupiter's moons in January 1610. The shaded areas were probably added later. It had not yet occurred to Galileo that the objects were moons in revolution around Jupiter.

revolve around the Earth, and also, by displaying something that Aristotle and Ptolemy obviously had not known about, showed that the science of Aristotle and Ptolemy was not all there was to know (Fig. 12–18).

Subsequently, Galileo found that Saturn had a more complex shape than that of a sphere (though it took better telescopes to actually show the rings). He was one of several people to almost simultaneously discover sunspots, and he studied their motions on the surface of the Sun. Galileo also discovered that Venus went through an entire series of phases (Fig. 12–19). This could not be explained on the basis of Ptolemaic theory, because if Venus traveled in an epicycle located between the Earth and the Sun, Venus should always appear as a crescent (Fig. 12–20). Also, more generally, Galileo's observations showed that Venus was a body similar to the Earth and the Moon in that it received light from the Sun rather than generating its own.

But while the Roman Catholic Church was not concerned with anyone's making models to explain observations, it was concerned with assertions that those models represented physical truth. It is difficult to say how much they feared that Galileo would generalize his ideas to a philosophical and religious level. In his old age, Galileo was forced by the Inquisition to recant his belief in the Copernican theory.

Figure 12–18 The frontispiece to Galileo's *Dialogue Concerning the Two World Systems,* 1632, which led to his being taken before the Inquisition. Copernicus is at the right. Aristotle and Ptolemy are at the left.

Figure 12–19 The phases of Venus. Note that Venus is a crescent only when it is in a part of its orbit that is relatively close to the Earth, and so it looks larger at those times.

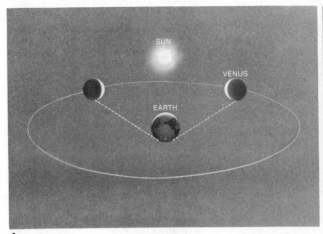

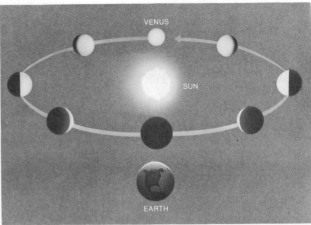

A   **B**

Figure 12–20 In the Ptolemaic theory *(A)*, Venus and the Sun both orbited the Earth, but because it is known that Venus never gets far from the Sun in the sky, the center of Venus' epicycle was restricted to always fall on the line joining the center of the deferent and the Sun. In this diagram Venus could never get farther from the Sun than the region restricted by the dotted lines. Thus, Venus would always appear as a crescent, though before the telescope was invented, this could not be verified. In the heliocentric theory *(B)* Venus is sometimes on the near side of the Sun, where it appears as a crescent, and it is sometimes on the far side, where we can see half or more of Venus illuminated. This agrees with Galileo's observations, a modern version of which appears as Figure 12–19.

12.6 TYCHO BRAHE

In the last part of the 16th century, not long after Copernicus' death and when Galileo was a child, Tycho Brahe, a Danish nobleman, began a series of observations of Mars and other planets. He set up an observatory on an island off the mainland of Denmark (Fig. 12–21). The building was called Uraniborg (after Urania, the muse of astronomy). Tycho's positional observations were considerably more precise than any observations that had been made up to that time, even though the telescope had not yet been invented. In 1597, Tycho lost his financial support in Denmark and moved to Prague, arriving two years later. A young assistant, Johannes Kepler, came there to work with him. At Tycho's death, in 1601, Kepler—who had worked with Tycho for only 10 months—was left to analyze all the observations that Tycho and his assistants had made (though Kepler first had to get access to the data from Tycho's family, which proved troublesome).

Figure 12–21 *(A)* Tycho's observatory at Uraniborg, Denmark. Here Tycho is seen showing the mural quandrant that he used to measure the altitudes at which stars and planets crossed the meridian. *(B)* Tycho, known throughout Europe, advanced his own, Earth-centered cosmology in which the Sun and Moon revolved around the Earth while the other planets revolved around the Sun. Though his cosmology was shortly forgotten, Tycho's current reputation results from his having provided the observational base for Kepler's research. The Comet of 1577 also appears on this plate reproduced from Tycho's book about that comet, published in 1588.

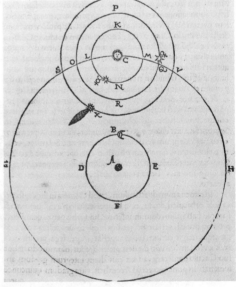

A   **B**

12.7 JOHANNES KEPLER

Tycho's observational data showed that the tables then in use did not adequately predict planetary positions. Kepler started to study them in 1600, and carried out detailed numerical calculations. (Nowadays we could use a computer to calculate in minutes results that took Kepler weeks to work out.) Kepler was eventually able to make sense out of the observations of Mars that Tycho had made, and thus clear up the discrepancies between the predictions and the observations of Mars' position in the sky. In 1609 and 1618, Kepler published three laws based on his empirical analyses. To this day, we consider these laws to be the basis of our understanding of the motions of the solar system.

Box 12.2 Kepler's Laws of Planetary Motion

(1) The planets orbit the sun in ellipses, with the Sun at one focus.
(2) The line joining the Sun and a planet sweeps through equal areas in equal times.
(3) The squares of the periods are proportional to the cubes of the planets' distances from the Sun. (More accurately: the square of the period of a planet is proportional to the cube of the semimajor axis of its orbit—half the longest dimension of the ellipse.)

12.7a Kepler's First Law (1609)

Until Kepler worked out these laws, even the heliocentric calculations assumed that the planets followed "perfect" orbits, namely, circles. The discovery by Kepler that the orbits were in fact ellipses greatly improved the accuracy of the calculations.

An ellipse is a curve that is defined in the following way. First choose any two points on a plane; these points are called the *foci* (each is a *focus*). From any point on the ellipse, we can draw two lines, one to each focus. The sum of the lengths of these two lines is the same for each point on the ellipse.

It is easy to draw an ellipse (Fig. 12–22). Put two nails or thumbtacks in a piece of paper, and link them with a piece of string that has some slack in it. If you pull a pen around while the pen keeps the string taut, the pen will necessarily trace out an ellipse; the string doesn't change in length, so the sum of the lengths of the lines from the pen to the foci equals the length of the string, which is constant. The shape of the ellipse will change if you

Figure 12–22 How to draw an ellipse.

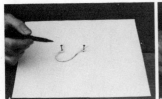   

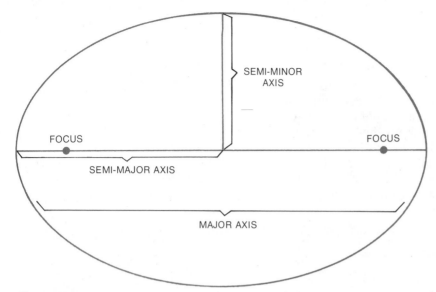

Figure 12–23 The parts of an ellipse.

Figure 12–24 The ellipse shown has the same *perihelion* distance (closest approach to the Sun) as does the circle. Its eccentricity, the distance between its foci divided by its major axis, is 0.5. If the perihelion distance is kept constant but the eccentricity allowed to reach 1, then we have a parabola. For eccentricities greater than 1 we have hyperbolas.

Figure 12–25 An ellipse is a conic section, in that the intersection of a cone and a plane that passes through the sides of a cone (and not the bottom) is an ellipse. If the plane is parallel to an edge of the cone, a parabola results. If the plane is tipped further over so that it is neither parallel to the edge nor intersects the side, then hyperbolas result. So parabolas and hyperbolas are conic sections as well.

change the length of the string, or if you change the distance between the foci.

The *major axis* of the ellipse is the line within the ellipse that passes through the two foci, or the length of that line (Fig. 12–23). We often speak of the *semimajor axis*, which is just half the length of the major axis (semi- is a prefix from the Greek, and means "half"). The *minor axis* is the part of the line lying within the ellipse that is drawn perpendicular to the major axis and bisects it, or the length of that line (Fig. 12–24). When one of the foci lies on top of the other, the major and minor axes are the same length, and the ellipse is the special case that we call a *circle*.

Circles, ellipses, parabolas, and hyperbolas are *conic sections*, that is, sections of a cone (Fig. 12–25). If you slice off the top of a cone, the shape of the slice is an ellipse, as long as your slice goes through the sides of the cone (Fig. 12–25). If the slice is parallel to the side of the cone, you get a parabola. If the slice is tipped still further over, a hyperbola results.

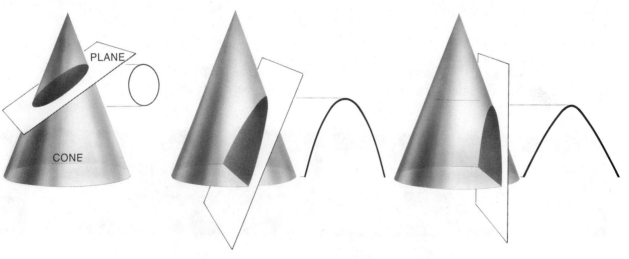

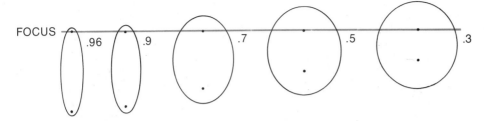

Figure 12–26 A series of ellipses of the same major axis but different eccentricities. The *foci* are marked; these are the two points inside with the property that the sum of the distances from any point on the circumference to the foci is constant. As the eccentricity—distance between the foci divided by the major (longer) axis—approaches 1, the ellipse approaches a straight line. As the eccentricity approaches zero, the foci come closer and closer together. A circle is an ellipse of zero eccentricity.

We often describe an ellipse by giving its *eccentricity* (Fig. 12–26) which is just the distance between the foci divided by the length of the major axis. For a circle, the distance between the foci is zero, so the eccentricity is zero.

If one focus is allowed to go off infinitely far, we have parabolas and hyperbolas.

For a planet orbiting the Sun, the Sun is at one focus of the elliptical orbit; nothing special marks the other focus (we say that it is "empty").

12.7b Kepler's Second Law (1609)

The second law, also known as the *law of equal areas,* governs the speeds with which the planets travel in their orbits. When a planet is at its greatest distance from the Sun in its elliptical orbit, the line joining it with the Sun sweeps out a long, skinny sector. (A *sector* is the area bounded by two straight lines from a focus of an ellipse [or, for a circle, two radii] and the part of the ellipse joining their outer ends.) This sector has the same area as the short, fat sector formed in the same period of time when the planet is closer to the sun, when the planet travels faster in its orbit (Fig. 12–27).

12.7c Kepler's Third Law (1618)

To understand the third law (it is ordinarily referred to as *Kepler's third law* and is demonstrated in Figure 12–28), let us compare the orbit of, say, Jupiter with that of the Earth. We can choose to work in units that are convenient for us on the Earth. We call the average distance from the sun to the Earth *1 Astronomical Unit* (1 A.U.). Similarly, the unit of time that the Earth takes to revolve around the Sun is defined as *one year*. Using these values, if we know from observation that Jupiter's period of revolution around the Sun is 11.86 years, we can use Kepler's third law to find the semimajor axis of Jupiter's orbit. If P is the period of revolution of a planet and R is the radius of its orbit (we often speak loosely in terms of R, the radius, though we really mean *a*, the semi-major axis),

$$\frac{P\ (\text{Jupiter})^2}{P\ (\text{Earth})^2} = \frac{R\ (\text{Jupiter})^3}{R\ (\text{Earth})^3}.$$

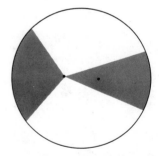

Figure 12–27 Kepler's second law states that the two shaded sectors, which represent the areas covered by a line drawn from a focus of the ellipse to an orbiting planet in a given length of time, are equal in area. The Sun is at this focus; nothing is at the other focus.

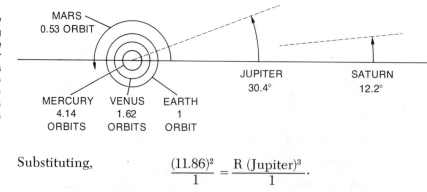

Figure 12–28 Kepler's third law relates the period of an orbiting body to the size of its orbit. The outer planets orbit at much slower velocities than the inner planets and also have a longer path to follow in order to complete one orbit. The distances the planets travel in their orbits in 1 year is shown here.

Substituting,

$$\frac{(11.86)^2}{1} = \frac{R\ (\text{Jupiter})^3}{1}.$$

Now $(11.86)^2$ can be rounded off to 12^2, which is approximately 150. So the radius of Jupiter's orbit around the Sun is approximately the cube root of 150, or approximately 5 A.U. The acutal measured value is 5.2 A.U.—the small difference was almost entirely introduced in the rough estimation process. So it checks.

We have used Kepler's third law to determine how the period of a planet revolving around the Sun is related to the size of its orbit. The constant of proportionality between P^2 and R^3 (that is, the number by which R^3 must be multiplied to get P^2 if we don't do the simplification of comparing with values for the Earth) depends on the mass of the central body, which is the Sun in this case. We shall see in the next section that Isaac Newton derived this mathematically.

The period of revolution of satellites around other bodies follows Kepler's laws as well (Fig. 12–29).

12.8 ISAAC NEWTON

Isaac Newton was born in England in 1642, the year of Galileo's death. He was to become the greatest scientist of his time and perhaps of all time. At other places in this book we describe his discovery that visible light can

Figure 12–29 Most orbiting satellites are only 200 km or so above the Earth's surface, and orbit the Earth in about 90 minutes. From Kepler's third law we can see that the velocity in orbit of a satellite decreases as the satellite gets higher and higher. At about 6½ Earth radii, the satellite orbits at the same velocity as that at which the Earth's surface rotates underneath. Thus the satellite is in *synchronous rotation,* and always remains over the same location on Earth. This property makes such synchronous satellites useful for relaying communications. The International Ultraviolet Explorer is in synchronous orbit, allowing constant command of its telescope.

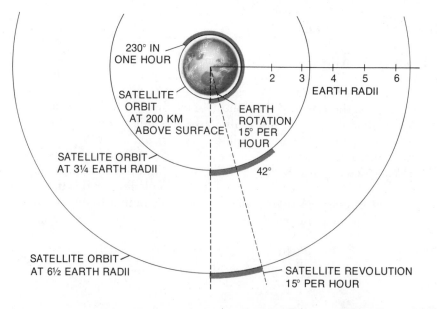

URANUS	NEPTUNE	PLUTO
4.29°	2.18°	1.5°

be broken down into a spectrum and his invention of the reflecting telescope. Here we shall speak only of his ideas about motion and gravitation.

For many years he developed his ideas about the nature of motion and about gravitation. In order to derive them mathematically, he invented calculus. Newton long withheld publishing his results, possibly out of shyness. Finally, his friend Edmond Halley, whose name we associate with the famous comet, persuaded him to publish his work. The *Philosophiae Naturalis Principia Mathematica (Mathematical Principles of Natural Philosophy)*, known as *The Principia*, appeared in 1687 (Fig. 12–30). In it, Newton showed that the motions of the planets and comets could all be explained by the same law of gravitation that governed bodies on Earth. In fact, he derived Kepler's laws on theoretical grounds.

In the derivations, Newton used the law of gravitation that he had discovered; Newton—whether or not you believe that an apple fell on his head—was the first to realize the universality of gravity. He formulated the law that the force of gravity (F) between two bodies varies directly with the product of the masses (m_1 and m_2) of the two bodies and inversely with the square of the distance between them:

$$\text{force of gravity} \propto m_1 m_2 / d^2.$$

$\propto$ means "is proportional to"; it means that the left hand side is equal to a constant times the right hand side. The constant, whose value is given in Appendix 2, is called G or the universal gravitational constant *because to the best of our knowledge it is constant throughout the Universe.*

We can write $F = G m_1 m_2 / d^2$. This is known as the law of universal gravitation; it is a universal law in that it works all over the universe rather than being limited to local applicability (on Earth or even in the solar system).

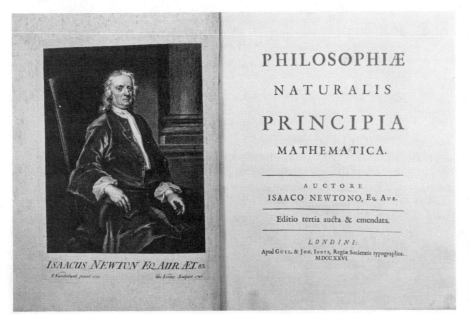

Figure 12–30 The title page and frontispiece of the third edition of Isaac Newton's *Principia Mathematica,* published in 1726. The first edition had appeared in 1687.

*Box 12.3 Kepler's Third Law, Newton, and Planetary Masses

Astronomers nowadays often make rough calculations to test whether physical processes under consideration could conceivably be valid. Astronomy has also had a long tradition of exceedingly accurate calculations. Pushing accuracy to yet one more decimal place sometimes leads to important results.

For example, Kepler's third law, in its original form—the period of a planet squared is proportional to its distance from the Sun cubed (P^2 = constant × R^3)—holds to a reasonably high degree of accuracy and seemed completely accurate when Kepler did his work. But now we have more accurate observations. If we consider each of the planets in turn, and if a term involving the sum of the masses of the Sun and the planet under consideration is included in the equation

$$P^2 = \frac{\text{constant}}{m_{\text{sun}} + m_{\text{planet}}} \times R^3,$$

the agreement with observation is improved in the fourth decimal place for the planet Jupiter and to a lesser extent for Saturn. The masses of the other planets are too small to have a detectable effect; even the effects caused by Jupiter and Saturn are tiny.

Isaac Newton derived the equation in its general form:

$$P^2 = \frac{4\pi^2}{G\,(m_1 + m_2)}\,a^3$$

for a body with mass m_1 revolving in an elliptical orbit with semimajor axis a around a body with mass m_2. The constant G is the *universal gravitational constant*.

Newton's formula shows that the planet's mass contributes to the value of the proportionality constant, but its effect is very small. (The effect can be detected even today only for the most massive planets—Jupiter and Saturn.)

Kepler's third law, and its subsequent generalization by Newton, applies not only to planets orbiting the Sun but also to any bodies orbiting other bodies under the control of gravity. Thus it also applies to satellites orbiting planets. We determine the mass of the Earth by studying the orbit of our Moon, and determine the mass of Jupiter by studying the orbits of its moons. Until recently, we were unable to reliably determine the mass of Pluto because we could not observe a moon in orbit around it. The discovery of a moon of Pluto in 1978 has finally allowed us to determine Pluto's mass, and we discovered that our previously best estimates (based on Pluto's gravitational effects on Uranus) were way off. The same formula can be applied to binary stars to find their masses.

12.9 THE REVOLUTION AND ROTATION OF THE PLANETS

The motion of the planets around the Sun in their orbits is called *revolution*. The spinning of a planet is called *rotation*. The Earth, for

Figure 12–31 It is all too easy to "misspeak" when discussing rotation, which refers to a body turning on its axis, and revolution, which refers to a body orbiting around another body.

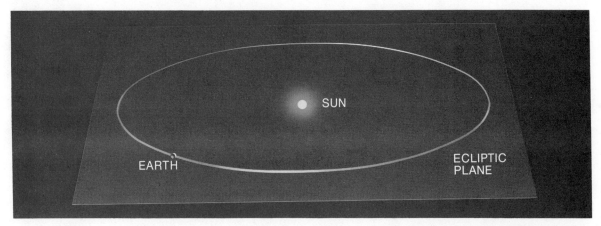

Figure 12–32 The ecliptic plane is the plane of the Earth's orbit around the Sun.

example, revolves around the Sun in 365 days and rotates on its axis in 1 day (Fig. 12–31).

The orbits of all the planets lie in approximately the same plane. They thus take up only a disk whose center is at the Sun, rather than a full sphere. Little is known of the parts of the solar system away from this disk, although the comets may originate in a spherical cloud around the Sun.

The plane of the Earth's orbit around the Sun is called the *ecliptic plane* (Fig. 12–32). Of course, since we are on the Earth rather than outside the solar system looking in, the position of the Sun appears to move across the sky with respect to the stars. The path that the Sun takes among the stars, as seen from the Earth, is called *the ecliptic* (see also Section 4.5a).

The *inclinations* of the orbits of the other planets with respect to the ecliptic are small, with the exception of Pluto (Figs. 12–33 and 12–34). Of the other planets, Mercury has an inclination of 7°; the remainder, except for Pluto, have inclinations of less than 4°. Pluto's much larger inclination of 17° is discrepant and is just one of the pieces of evidence suggesting that Pluto may not have formed under the same circumstances as the other planets in the solar system.

The fact that the planets all orbit the Sun in essentially the same plane is one of the most important facts that we know about the solar system. Its explanation is at the base of most cosmogonical models, and central to that explanation is a property that astronomers and physicists use in analyzing spinning or revolving objects: *angular momentum.* The amount of angular

The inclination *of the orbit of a planet is the angle that the plane in which its orbit lies makes with the plane in which the Earth's orbit lies.*

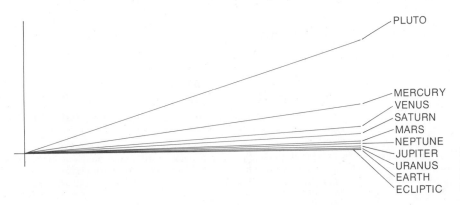

Figure 12–33 The orbits of the planets, with the exception of Pluto, have only small inclinations to the ecliptic plane.

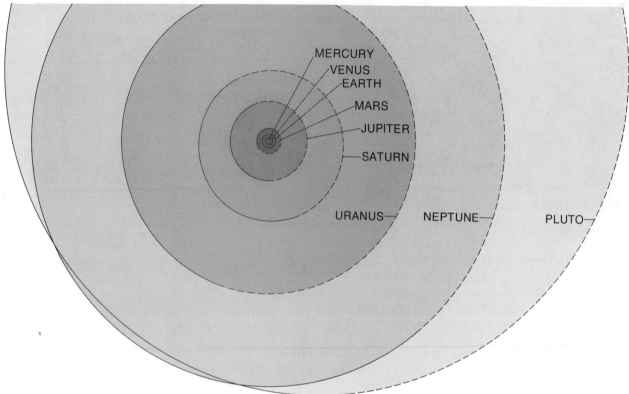

MERCURY
VENUS
EARTH
MARS
JUPITER
SATURN
URANUS⎯ NEPTUNE⎯ PLUTO⎯

Figure 12–34 The planetary orbits, with the exception of Pluto's, are of relatively low eccentricity and appear at regular spacings. The portion of the orbits that are below the ecliptic plane are shown with dotted lines.

momentum of a small body revolving around a large central body is (distance from the center) × (velocity) × (mass). The importance of angular momentum lies in the fact that it is conserved; that is, the total angular momentum of the system (the sum of the angular momenta of the different parts of the system) doesn't change even though the distribution of angular momentum among the parts may change. The total angular momentum of the solar system should thus be the same as it was in the past, unless some mechanism is carrying angular momentum away.

The most familiar example of angular momentum is an ice skater. She may start herself spinning by exerting force on the ice with her skates, thus giving herself a certain angular momentum. When she wants to spin faster, she draws her arms in closer to her (Fig. 12–35). This changes the distribution of her mass so that it is effectively closer to her center (the axis around which she is spinning), and to compensate for this she starts to rotate more quickly so that her angular momentum remains the same.

Thus from the fact that every planet is revolving in the same plane and in the same direction, we deduce that the solar system was formed of primordial material that was rotating in that direction. We would be very surprised to find a planet revolving around the Sun in a different direction— and we do not—or, to a lesser extent, to find a planet rotating in a direction

Figure 12–35 An ice skater draws her arms in, in order to redistribute her mass. Because her angular momentum is conserved, she begins to spin faster in order to compensate for the new mass distribution. Dorothy Hamill is shown here during her gold medal performance in the free style event at the Innsbruck Olympics in 1976. You can tell she is spinning rapidly because her skirt and hair are thrown outward.

opposite to that of its fellows. There are, however, two examples of such backwards rotation, which must each be carefully considered.

Such rotation in the opposite sense is called retrograde rotation. *Do not confuse it with* retrograde motion, *which has to do with the apparent motion of the position of a planet in the sky as seen from the Earth.*

12.10 BODE'S "LAW"

For over two hundred years, a numerical relation called "Bode's law" has been known to give the approximate distances of the planets from the Sun, though it has never been theoretically understood.

Bode's law says to write down first 0, then 3, and then keep doubling the previous number: 6, 12, 24, etc. Now add 4 to each number to get the series 4, 7, 10, 16, 28, and so on. Then divide each number by 10. The results: .4, .7, 1.0, etc., give the approximate radii of the planetary orbits in astronomical units (Table 12–1).

1 A.U., the average distance from the Earth to the Sun, equals 150,000,000 km.

The observational tests of Bode's law are connected with the outer planets and asteroids. When Uranus was discovered by William Herschel in England in 1781, its orbital distance fit Bode's law and the discovery spurred theoretical work to derive the "law." Bode's law also predicted a planet between Mars and Jupiter. The asteroids, some of which are at this distance, as predicted by Bode's law, began to be discovered twenty years after Uranus. The fact that there were many asteroids (minor planets), instead of one major planet, was disturbing to many. To have a perspective in time, let us realize that this active discussion of Bode's law was going on in Europe at the time of the American Revolution.

The relation was first published anonymously by Titius of Wittenberg in 1766 in a translation of a book by a Swiss scientist, but Bode's name became attached to the relation a half-dozen years later when Bode, director of the Berlin Observatory, popularized it. Sometimes we now call it the Titius-Bode law.

But Neptune and Pluto, which were discovered later, do not fit Bode's law, and no mathematical derivation of Bode's law has ever been found. (Some numerical calculations with large computers do show planetary spacings that follow different rules.)

At present, most astronomers think of Bode's law as merely a numerical coincidence and a historical curiosity. It is, at best, a way of memorizing planetary distances (should one want to do so).

It is no surprise that the planetary orbits have spaces between them such that we don't find two planets in orbits very close to each other. Most matter that would exist between the planets would be swept up by the gravitational forces of the planets as they go around the Sun.

TABLE 12–1 BODE'S LAW

Planet	Distance by Bode's Law (A.U.)	Actual Mean Distance (A.U.)
Mercury	0.4	0.39
Venus	0.7	0.72
Earth	1.0	1.00
Mars	1.6	1.52
Asteroids	2.8	—
Jupiter	5.2	5.20
Saturn	10.0	9.54
Uranus	19.6	19.2
Neptune	38.8	30.6
Pluto	77.2	39.4

12.11 THEORIES OF COSMOGONY

The solar system exhibits many regularities, and theories of its formation must account for them. The orbits of the planets are almost, but not quite, circular, and all lie in essentially the same plane. All the planets revolve around the Sun in the same direction, which is the same direction in which the Sun rotates. Moreover, almost all the planets and planetary satellites rotate in that same direction. Some planets have families of satellites that revolve around them in a manner similar to the way that the planets revolve around the Sun. And cosmogonical theories must explain the spacing of the planetary orbits and the distribution of planetary sizes and compositions.

René Descartes, the French philosopher, was one of the first to consider the origin of the solar system in what we would call a scientific manner. In his theory, proposed in 1644, circular eddies of all sizes called vortices were formed in a primordial gas at the beginning of the solar system and eventually settled down to become the various celestial bodies.

After Newton proved that Descartes' vortex theory was invalid, it was 60 years before the next major developments. The Comte de Buffon suggested in France in 1745 that the planets were formed by material ejected from the Sun when what he called a "comet" hit it. (At that time, the composition of comets was unknown, and it was thought that comets were objects as massive as the Sun itself.) Later versions of Buffon's theory, called *catastrophe theories*, followed a similar line of reasoning although they spoke explicitly of a collision with another star. The requirement of a collision could be loosened by postulating that the material for the planets was drawn out of the Sun by the gravitational attraction of a passing star. This latter possibility is called a *tidal theory*.

But catastrophe and tidal theories are currently out of fashion, for they predict that only very few planetary systems would exist, since calculations have been made that show that only very few stellar collisions or near-collisions would have taken place in the lifetime of the galaxy. There is some observational evidence that many stars have planets, which would require a more common method of formation. (There is new evidence contradictory to some of these observational results, though, as we shall see in Section 21.2.) Also, theoretical calculations show that gas drawn out of a star in a collision or by a tidal force would not condense into planets, but would rather disperse.

The study of the origin of the solar system is called cosmogony. *Note that "cosmogony" is pronounced with a hard "g."*

The theories of cosmogony that astronomers now tend to accept stem from another 18th century idea. Immanuel Kant, the noted German philosopher, suggested in 1755 that the Sun and the planets were formed by the same type of process. In 1796, the Marquis Pierre Simon de Laplace, the French mathematician, independently advanced a similar kind of theory to Kant's when he postulated that the Sun and the planets all formed from a spinning cloud of gas called a *nebula.* Laplace called this the *nebular hypothesis,* using the word "hypothesis" because he had no proof that it was correct. The spinning gas supposedly threw off rings that eventually condensed to become the planets. But though these beginnings of modern cosmogony were laid down in the time of Benjamin Franklin and George Washington, the theory still has not been completely understood or quantified, for not all the stages that the primordial gas would have had to follow are understood.

Further, some of the details of Laplace's theory were later thought for a time to be impossible. For example, it was calculated in the last century, by the great British physicist James Clerk Maxwell, that the planets would not

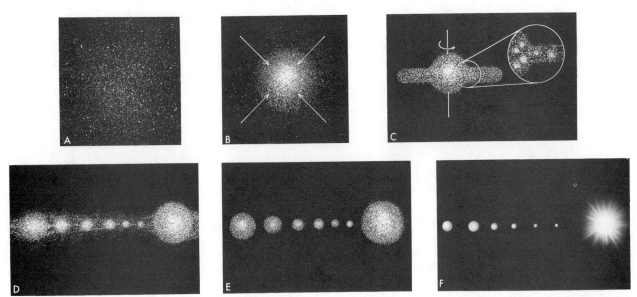

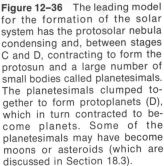

Figure 12–36 The leading model for the formation of the solar system has the protosolar nebula condensing and, between stages C and D, contracting to form the protosun and a large number of small bodies called planetesimals. The planetesimals clumped together to form protoplanets (D), which in turn contracted to become planets. Some of the planetesimals may have become moons or asteroids (which are discussed in Section 18.3).

have been able to condense out of the rings of gas, and also that they could not have been set to rotate as fast as they do.

So for a time, the nebular hypothesis was not accepted. But the current theories of cosmogony again follow Kant and Laplace (Fig. 12–36). In these *nebular theories* the Sun and the planets condensed out of what is called a *primeval solar nebula*. Some five billion years ago, billions of years after the galaxies began to form, smaller clouds of gas and dust began to contract out of interstellar space. Similar interstellar clouds of gas and dust can now be detected at many locations in our galaxy.

There is increasing evidence that the collapse of gas to form the solar system was set off by shock waves from a nearby supernova (Fig. 12–37). The

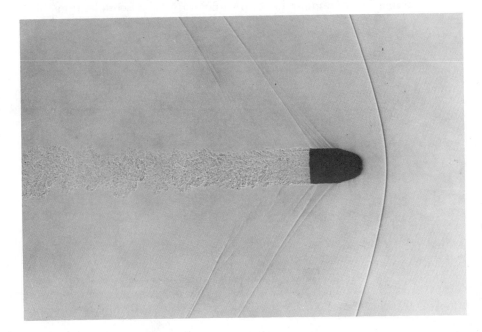

Figure 12–37 This photograph shows a shock wave in air made by a bullet travelling faster than the speed of sound. The sharp curved line at the right is the shock wave; it represents a sharp change in pressure. (Courtesy of Harold E. Edgerton, MIT)

evidence concerns the existence of unusual abundances of certain isotopes that we expect to be formed in such supernova explosions.

Because of random fluctuations in the gas and dust from which it formed, the primeval solar nebula probably would have had a small net spin from the beginning. As it contracted, it would have begun to spin faster because of the conservation of angular momentum (the same reason that the ice skater spins faster). Gravity would have contracted the spinning nebula into a disk, for in the directions perpendicular to the plane in which the nebula is rotating there was no force to oppose gravity's pull.

Angular momentum is a measure of the amount of spin.

Perhaps rings of material were left behind by the solar nebula as it contracted towards its center. Perhaps there were additional agglomerations beside the *protosun*, the part of the solar nebula that collapsed to become the Sun itself. These additional agglomerations, which may have been formed from interstellar dust, would have grown larger and larger. Millimeter-size particles would have clumped into larger bodies and then larger still; the size of the clumps would have increased until the bodies were kilometers and then thousands of kilometers across. The intermediate bodies, hundreds of kilometers across, are called *planetesimals*. (The idea that planetesimals exist and are an intermediate stage in the formation of the planets was borrowed from collision theories.) Laplace's old idea that the protosun had thrown off rings that condensed to become planets, an idea that had been rejected by Maxwell, was not necessary to provide planets after all.

The planetesimals combined under the force of gravity to form *protoplanets*. These protoplanets may have been larger than the planets that resulted from them because they had not yet contracted, though gravity would ultimately cause them to do so. Some of the larger planetesimals may themselves have become moons.

As the Sun condensed, some of the energy it gained from its contraction went to heat it until its center reached the temperature at which nuclear fusion reactions began taking place. The planets, on the other hand, were simply not massive enough to heat up sufficiently to have nuclear reactions start.

Several modifications of this basic theory have been worked out to accommodate particular observational facts that are known about the solar system. For example, one has to explain the fact that the inner planets are small, rocky, and dense, while the next group of planets out are large and are made of light elements. Since the protosun would have been made primarily out of hydrogen and helium, with just traces of the heavier elements formed in earlier cyclings of the material through stars and supernovae, one simply has to provide a mechanism for ridding the inner regions of the solar system of this lighter material. One possible way is to say that the Sun flared fiercely and/or often in its younger days (similarly to a T Tauri star, as described in Section 8.1), and that the lighter elements were blown out of the inner part of the solar system. But proto-Jupiter and the other outer protoplanets, which were much farther away from the Sun, would have retained thick atmospheres of hydrogen and helium because of their high gravity.

A major modern method of calculation considers how the temperature decreases with distance from the center of the solar nebula, ranging from almost 2000 K close to the protosun, down to only about 20 K at the distance of Pluto. Various elements are able to condense—solidify—at different locations because of these differing temperatures. For example, in the positions occupied by the terrestrial planets, it was too hot for icy substances to form, though such ices are present in the giant planets. Such calculations are perhaps the dominant consideration in much current cosmogonic research.

The cosmogonical ideas we have discussed are all speculative, for the nature of the formation of the solar system is not definitively understood, nor are the details known. Professional astronomers join students of astronomy in wishing that the theory could be laid out neatly and conclusively for all to see, but we are not yet able to do that.

There are still more possibilities that bedevil cosmogonical research. Some of the planets and moons may not have been formed in their present configurations. Perhaps the Earth captured another protoplanet, which became the Moon. Perhaps Pluto was ejected from an orbit around Neptune into an orbit as an independent planet. Perhaps gravitational encounters put some of the planets and moons into retrograde rotation.

SUMMARY AND OUTLINE

The scale of the solar system and the planets themselves
The phases of the Moon and planets (Section 12.1)
Lunar and solar eclipses (Section 12.2)
Ancient astronomy (Section 12.3)
  Egyptian astronomy (Section 12.3a)
    Astronomy served the practical purposes of the farmers
  Greek astronomy (Section 12.3b)
    The apparent motion of the planets
      Retrograde motion
    Geocentric theory of Aristotle and Ptolemy
    Epicycles, deferents, equants needed to explain the planetary motions
  Babylonian, Chinese, and early British astronomy (Section 12.3c)
Heliocentric theory (Section 12.4)
  Copernicus, *De Revolutionibus* published in 1543
  Earlier heliocentric idea of Aristarchus of Samos was not widely accepted
  Retrograde motion explained as a projection effect
  Copernican system still used circular orbits and epicycles
  Copernicus preferred his system on philosophical grounds rather than superiority of its predictions, because ancient observations were still used and Copernican system used circular orbits which did not always fit the data
Galileo (Section 12.5)
  First to use a telescope for astronomical observations
  His observations supported the heliocentric theory and further showed that Aristotle and Ptolemy had not known everything

Tycho Brahe (Section 12.6)
  Amassed the best set of observations that had ever been obtained
Kepler (Section 12.7)
  Studied Tycho's observations and discovered three laws of planetary motion
  Kepler's laws: (1) orbits are ellipses; (2) equal areas are swept out in equal times; (3) period squared is proportional to distance cubed
Newton (Section 12.8)
  Newton worked out laws of gravity and of motion
  He was able to derive Kepler's laws and find the masses of the planets
The ecliptic; inclinations of the orbits of the planets (Section 12.9)
  Pluto has largest inclination by far
  Other planets have only slight inclinations
Conservation of angular momentum (Section 12.9)
  All planets revolve in the same direction
  Most planets rotate in the same sense
Bode's "law" (Section 12.10)
  Write down 0, 3, 6, 12, and so on; add 4 and divide by 10
  The "law" probably does not have significance
Theories of cosmogony (Section 12.11)
  Catastrophe theories
  Nebular theories
    Kant and Laplace
    Newer versions: protosun, planetesimals, and protoplanets
    A nearby supernova may have caused our solar system to come into being

QUESTIONS

1. What are the features that distinguish the terrestrial from the giant planets? What features does Pluto have in common with either group?

2. Suppose that you live on the Moon. Sketch the phases of the Earth that you would observe for various times during the Earth's month.

3. If you lived on the Moon, would the motion of the planets appear any different than from Earth?

4. If you lived on the Moon, how would the position of the Earth change in your sky over time?

5. If you lived on the Moon, what would you observe during an eclipse of the Moon? How would an eclipse of the Sun by the Earth differ from an eclipse of the Sun by the Moon that we observe from Earth?

6. Sketch what you would see if you were on Mars and the Earth passed between you and the Sun. Would you see an eclipse? Why?

7. (a) If you lived on Saturn, describe the phases that the Earth would seem to go through on the Ptolemaic system.
   (b) Now describe the phases that the Earth would seem to go through on the Copernican system.

8. (a) If you lived on Saturn, describe the phases that Uranus would seem to go through on the Ptolemaic system.
   (b) Now describe the phases that Uranus would seem to go through on the Copernican system.

9. Discuss the velocity that a planet must have around its epicycle in the Ptolemaic theory with respect to the velocity that the epicycle has around the deferent if we are to observe retrograde motion.

10. Discuss the following statement: "With the addition of epicycles, the geocentric theory of the solar system could be made to agree with observations. Since it was around first, and therefore better known, it should have been kept."

11. How do the predictions of the Ptolemaic and Copernican systems differ for the variation in size of Venus?

12. If we double the height of the orbit of a satellite that has been in a synchronous orbit around the Earth, what will its new period be?

13. Discuss four new observations made by Galileo, and show how they tended to support, tended to oppose, or were irrelevant to the Copernican theory.

14. Do Kepler's laws permit circular orbits?

15. At what point in its orbit is the Earth moving fastest?

†16. Use Kepler's third law and the fact that Mercury's orbit has a semimajor axis of 0.4 A.U. to deduce the period of Mercury. Show your work.

†17. What is the period of an object orbiting 1 A.U. from a 9 solar mass star?

18. What was the difference in the approach of Kepler and Newton to the discovery of laws that controlled planetary orbits?

†19. (a) Use the data in Appendix 5 to show for which planets the square of the period does not equal the cube of the semimajor axis to the accuracy given.
   (b) Use the formula in Box 12.3 to show that including the effects of planetary masses removes the discrepancy.

†20. The occupants of Planet X note that they are 1 greel from their sun (a greel is the Planet X unit of length), and they orbit with a period of 1 fleel. They observe Planet Y which orbits in 8 fleels. How far is Planet Y from their sun (in greels)?

21. How can Pluto have a longer period than Neptune, even though Pluto is and will be closer to the Sun until the year 2000?

22. Explain how conservation of angular momentum applies to a diver doing somersaults or twists. What can divers do to make sure they are vertical when they hit the water?

23. If two planets are of the same mass but different distances from the Sun, which will have the higher angular momentum around the Sun? (Hint: Use Kepler's third law.)

24. If the planets condensed out of the same primeval nebula as the Sun, why didn't they become stars?

25. Which planets are likely to have their original atmospheres? Explain.

---

†This indicates a question requiring a numerical answer.

PREVIEW OF THE PLANETS AND THEIR MOONS

When you have completed Chapters 13 through 19, you should be able to identify the following pictures, and understand the features to look for in order to make the identifications:

A.   Which of the following has (a) a surface covered with craters (b) a giant canyon; (c) erupting volcanoes? Choose from the Moon, Mars, and Jupiter's moon Io.

B.   Which of the following has (a) an atmosphere; (b) lines of cliffs called scarps; (c) rayed craters? Choose from the Moon, Mercury, and Mars.

C.   Which of the following has (a) one ring; (b) several prominent rings; (c) several narrow rings very difficult to observe from Earth? Choose from Saturn, Jupiter, and Uranus.

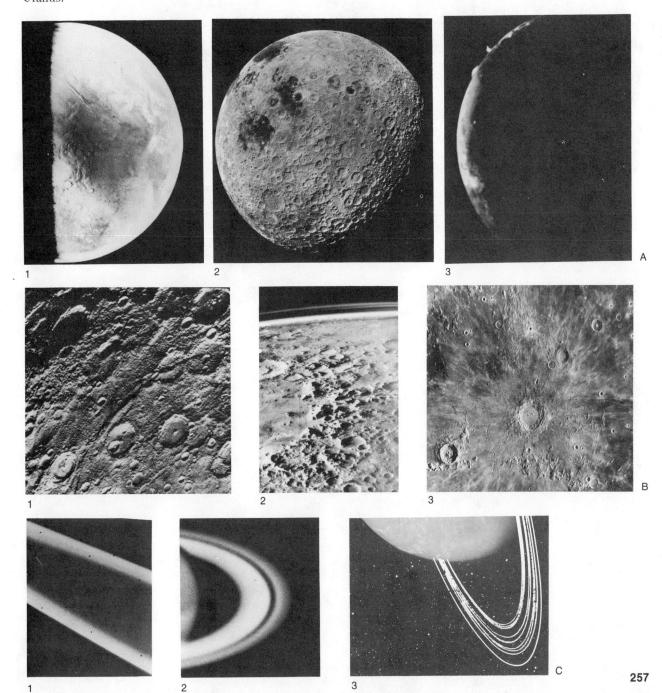

CHAPTER 13

THE MOON

AIMS:
To see how direct exploration of the Moon has increased our knowledge manyfold, although such fundamental questions as how the Moon was formed remain unanswered

The Earth's nearest celestial neighbor—the Moon—is only 380,000 km (238,000 miles) away from us on the average, close enough that it appears sufficiently large and bright to dominate our nighttime sky. The Moon's stark beauty has called attention to it since the beginning of history, and studies of the Moon's position and motion led to the earliest consideration of the solar system, to the prediction of tides, and to the establishment of the calendar.

13.1 THE APPEARANCE OF THE MOON

The fact that the Moon's surface has different kinds of areas on it is obvious to the naked eye. Even a small telescope—Galileo's, for example—reveals a surface pockmarked with craters. The *highlands* are heavily cratered; other areas, called *mare* (pronounced mar′-ā; plural, *maria*, pronounced mar′ē-a), are relatively smooth, and indeed the name comes from the Latin word for sea (Fig. 13–1). But there are no ships sailing on the lunar seas and no water in them; the Moon is a dry, airless, barren place. The Moon's mass is only 1/81 that of the Earth, and the gravity at its surface is only 1/6 that of the Earth. Any atmosphere and any water that may once have been present would long since have escaped into space. The Moon is 3476 km (2160 miles) in diameter, about one-fourth the diameter of the Earth, a very large fraction for a moon.

Many types of structures are visible on the Moon besides the smooth maria and the cratered highlands. In the highlands, there are *mountain ranges* and *valleys*. Lunar *rilles* are clefts that can extend for hundreds of kilometers along the surface (Fig. 13–2). Some are relatively straight while others are sinuous. On the other hand, raised *ridges* (Fig. 13–2) also occur. The craters themselves come in all sizes, ranging from as much as 295 km

Figure 13–1 This Earth-based photograph shows Mare Imbrium at the upper left, the Apennine Mountains at the lower right, and many craters, some of which have central peaks.

Scientist-astronaut Harrison Schmitt collecting small rocks and rock chips with a lunar rake during the Apollo 17 mission to the Taurus-Littrow region of the Moon.

Figure 13–2 Ridges with the Aridaeus Rille.

Figure 13–3 The full Moon. Note the dark maria and the lighter, heavily cratered highlands. The positions of the 6 American Apollo (A) and Soviet Luna (L) missions from which material was returned to Earth for analysis are marked (Table 13–1).

across for one called Bailly down to tiny fractions of a millimeter. Crater *rims* can be as much as several kilometers high, much higher than the rim of the Grand Canyon stands above the Colorado River. Examples of all these features will be seen in the photographs that illustrate this chapter.

When the Moon is full, it is especially bright in the nighttime sky—bright enough, in fact, to cast shadows or even to read by. But full moon is a bad time to try to observe surface structure, for we are looking along the same line that is being traveled by the sunlight, and any shadows we see are short. When the Moon is in a crescent phase or even a half-moon, however, the part of the lunar surface where the shadows are long faces us. At those times, the lunar features stand out in bold relief.

Shadows are longest near the *terminator*, the line separating day from night. Since the Moon goes around the Earth and returns to the same position with respect to the Earth and Sun once every 29½ days (this is the *synodic revolution period* of the Moon), the terminator moves around the Moon in that time. The phases repeat with this period, which is called the *synodic month*. Most locations on the Moon are thus in about 15 days of sunlight, during which time they become very hot, and then 15 days of darkness, during which time their temperature drops to as low as 160 K.

Box 13.1   Sidereal and Synodic Periods

Sidereal: with respect to the stars. For example, the Moon revolves around the Earth in about 27⅓ days with respect to the stars.

Synodic: with respect to some other body, usually the Earth. For example, by the time that the Moon has made one sidereal revolution around the Earth, the Earth has moved about ¹/₁₃ of the way around the Sun (about 27°). Thus the Moon must travel a little farther to catch up (Fig. 13–4). The synodic period of the Moon, the synodic month, is about 29½ days.

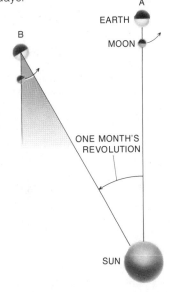

Figure 13–4   After the Moon has completed one revolution around the Earth with respect to the stars, which it does in 27⅓ days, it has moved from A to B. The Moon has still not swung far enough around to again be in the same position with respect to the Sun because the Earth has revolved one month's worth around the Sun. It takes about an extra two days for the Moon to complete its revolution with respect to the Sun, which gives a synodic rotation period of about 29½ days. The extra angle it must cover is shaded. In the extra two days, both Earth and Moon have continued to move around the Sun (toward the left in the diagram).

The Moon rotates on its axis at the same rate as it revolves around the Earth, always keeping the same face in our direction. The Earth's gravity has locked the Moon in this pattern, interacting with a bulge in the distribution of the lunar mass to prevent the Moon from rotating freely. As a result of this interlock, we always see essentially the same side of the Moon from our vantage point on Earth.

In some sense, before the period of exploration by the Apollo program we knew more about almost any star than we did about the Moon. As a solid body, the Moon reflects the solar spectrum rather than emitting one of its own, so we were hard pressed to determine even the composition or the physical properties of its surface.

Another debate that raged was the origin of the craters. One side thought that they were formed by the impact of *meteorites* (Section 20.2). Others thought they were volcanic in nature and were the result of internal lunar activity. This has been finally settled (see Section 13.3c) in favor of impacts, but we also see numerous signs of volcanic activity.

13.2   LUNAR EXPLORATION

The space age began on October 4, 1957, when the U.S.S.R. launched its first Sputnik (the Russian word for *travelling companion*) into orbit. The shock galvanized the American space program and within months American spacecraft were also in Earth orbit. The ability to observe from space has benefited all phases of astronomy.

Figure 13–5 Neil Armstrong, the first person to set foot on the Moon, took this photograph of his fellow astronaut Buzz Aldrin climbing down from the Lunar Module of Apollo 11 on July 20, 1969. The site is called Tranquility Base, as it is in the Sea of Tranquility (Mare Tranquillitatis).

In the case of the Moon and planets, the ability to travel into space did not merely free us from the obscuration of the Earth's atmosphere but also allowed us to explore the planets directly. The Moon, as the closest celestial body, was obviously the place to begin.

In 1959, the Soviet Union sent its Luna 3 spacecraft around the Moon; it radioed back the first murky photographs of the Moon's far side. Now that we have high-resolution maps of most of the lunar surface, it is easy to forget how big an advance that was.

In 1961, President John F. Kennedy announced that it would be a national goal of the United States to put a man on the Moon, and bring him safely back to Earth, by 1970. This grandiose goal led to the largest coordinated program in the history of the world, and had an outstandingly successful conclusion.

The American lunar program, under the direction of the National Aeronautics and Space Administration (NASA), proceeded in gentle stages. On one hand, the ability to carry out manned space flight was developed with single-astronaut sub-orbital and orbital capsules, called Project Mercury, and then with two-astronaut orbital spacecraft, called Project Gemini. Simultaneously, on the other hand, a series of unmanned spacecraft were sent to the Moon.

The manned and unmanned trains of development came together with Apollo 8, which circled the Moon on Christmas Eve, 1968, and returned to Earth. The next year, Apollo 11 brought humans to land on the Moon for the first time. It went into orbit around the Moon after a three-day journey from Earth, and a small spacecraft called the Lunar Module (LM) separated from the larger Command Module. On July 20, 1969—a date that from the long-range standard of history may be the most significant of the last millennium—Neil Armstrong and Buzz Aldrin left Michael Collins orbiting in the Command Module and landed on the Moon (Fig. 13–5). In the preceding days there had been much discussion of what Armstrong's historic first words should be, and millions listened as he said "One small step for man, one giant leap for mankind." (He meant to say "for a man.")

The Lunar Module carried a variety of experiments, including devices to test the soil, a camera to secure close-up stereo photos of the lunar soil, a sheet of aluminum with which to capture particles from the solar wind, and a seismometer. Later Lunar Modules carried additional experiments, some even including a vehicle (Fig. 13–6). In all, six Apollo missions carried people to the Moon. Unfortunately, the missions seemed to be confused in the popular mind with just one of the experiments—the collection of rocks.

TABLE 13–1 MISSIONS TO THE LUNAR SURFACE AND BACK TO EARTH

Apollo 11	U.S.A.	1969	manned
Apollo 12	U.S.A.	1969	manned
Luna 16	U.S.S.R.	1970	unmanned
Apollo 14	U.S.A.	1971	manned
Apollo 15	U.S.A.	1971	manned
Luna 20	U.S.S.R.	1972	unmanned
Apollo 16	U.S.A.	1972	manned
Apollo 17	U.S.A.	1972	manned
Luna 24	U.S.S.R.	1976	unmanned

A

B

These rocks and dust, returned to Earth for detailed analysis, were indeed important, but represented only a fraction of the purpose of each mission.

The Soviet Union, which has sent three unmanned spacecraft to land on the lunar surface, collect lunar soil, and return it to Earth, continued lunar work longer. Its first two round-trip spacecraft each collected a few grams of lunar soil. Luna 24, which went to the Moon in August 1976, drilled to a depth of ½ meters (2 feet) below the lunar surface and brought a long, thin cylinder of material back to Earth. In 1970 and 1973, two other Soviet spacecraft had carried remote-controlled rovers, Lunokhods 1 and 2, that traveled over 10 or so kilometers of the lunar surface over a period of many months each.

13.3 THE RESULTS FROM APOLLO

The kilometers of film exposed by the astronauts, the 382 kilograms (843 pounds) of rock brought back to Earth (Fig. 13–7), the rolls and rolls of magnetic tape recording the results from lunar seismographs, and other data, all studied by hundreds and hundreds of scientists from countries all over the Earth, have led to new views of several basic questions. They have raised many new questions about the Moon and the solar system as a whole. Let us consider (1) the composition of the lunar surface, (2) the chronology of the Moon, (3) the origin of the lunar craters, (4) the structure of the lunar interior and whether it is hot or cold, and (5) how and where the Moon was formed.

13.3a The Composition of the Lunar Surface

The types of rocks that were encountered on the Moon are types that are familiar to terrestrial geologists. All the rocks are *igneous,* which means that they were formed by the cooling of lava. The moon has no *sedimentary* rocks, which are formed by deposit in water. (Earth's sedimentary rocks include limestone and shale.)

In the maria, the rocks are mainly *basalts* (a type of rock that results from the cooling of molten material like lava) (Fig. 13–8A). The highland rocks are *anorthosites.* (Anorthosites—defined by their particular combination of minerals—are rare on Earth, though the Adirondack Mountains are made of

Figure 13–6 *(A)* Eugene Cernan riding on the Lunar Rover during the Apollo 17 mission. The mountain in the background is the east end of the South Massif. *(B)* Drawing by Alan Dunn; © 1971 *The New Yorker Magazine, Inc.*

Figure 13–7 A moon rock from Apollo 14 being handled in the Lunar Receiving Laboratory at the Johnson Space Center in Houston.

A   B

Figure 13–8 *(A)* A basalt returned to Earth by Apollo 15. It
weighs 1.5 kg. Note the many
vesicles, spherical cavities that
arise in volcanic rock because of
gas trapped at the time of the
rock's formation. *(B)* A breccia,
dark gray and white in color, returned to Earth by Apollo 15.

them.) Anorthosites, though they have also cooled from molten material,
have done so under different conditions than basalts and have taken longer
to cool.

Some of the rocks in each type of location are *breccias* (Fig. 13–8B),
mixtures of fragments of several different types of rock that have been
compacted and welded together. The ratio of breccias to crystalline rocks is
much higher in the highlands than in the maria because there have been
many more impacts in the visible highland surface, as shown by the greater
number of craters.

Under a microscope, one can easily see the contrast between lunar and
terrestrial rocks. The lunar rocks contain no water, and so never underwent
the reactions that terrestrial rocks undergo. The lunar rocks also show that no
oxygen was present when they were formed.

The astronauts also collected some *lunar soils,* bits of dust plus larger
fragments from the Moon's surface. This *regolith* was built up by bombardment of the lunar surface by meteorites of all sizes over a period of billions of
years. Some of these proved, on examination through the microscope, to
contain a type of structure that is not common on Earth. Small glassy globules
(Fig. 13–9) are mixed in. In some of the soils brought back glass coatings
and glass globules were present; the bubbly globules in these cases undoubtedly resulted from the melting of rock during its ejection from the site
of a meteorite impact and the subsequent cooling of the molten material. In
some cases, where most of the soil consisted of glass beads, a volcanic origin
is possible.

Almost all of these rocks and soils have lower proportions of elements
with low melting points (*volatile* elements) than does the Earth. On the other
hand, there are relatively high proportions of elements with high melting
points (*refractory* elements) like calcium, aluminum, and titanium compared to terrestrial abundances. Titanium, which is only a minor constituent
of Earth rocks, amounted to 10 per cent of some lunar samples. Of elements
that are even rarer on Earth, the abundances of uranium, thorium, and the
rare-earth elements are also increased.

In particular, mare basalts, which were probably formed in the lunar
interior instead of on the surface, are enriched in refractory elements and
depleted in volatile ones compared to similar basalts on the Earth. Similarly,
abundances in anorthosites, which probably cooled from molten material

Figure 13–9 An enlargement of
a glassy spherule from the lunar
dust collected by the Apollo 11
mission. The shape and composition of the spherule indicate
that it was created when a
meteorite crashed into the Moon,
melting lunar material and splash-
ing it long distances. The glassy
bead is enlarged 3200 times in this
photograph taken with a scanning
electron microscope.

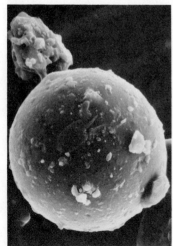

after it floated to the surface, differ from abundances in terrestrial anorthosites. In sum, the Moon and the Earth seem to be similar chemically, though real differences in overall composition exist. For example, the Moon as a whole is probably enriched in calcium and aluminum relative to the Earth.

Something that seems fairly definite is that none of the lunar rocks contain any trace of water bound inside their minerals. This ends all hope that water existed on the Moon at any time in the past, and so seems to eliminate the possibility that life evolved there.

13.3b Lunar Chronology

One way of dating the structures on the surface of a moon or planet is to observe the number of craters on them, a method that was used before Apollo. If we assume that the events that cause craters—whether they are impacts of meteors or the eruptions of volcanoes—continue over a long period of time and that there is no strong erosion, surely those locations with the greatest number of craters must be the oldest. Relatively smooth areas—like the maria—must have been covered over with volcanic material at some relatively recent time (which is still billions of years ago). Even from the Earth we can count the larger craters. And when one crater is superimposed on another (Fig. 13–10), we can be certain that the superimposed crater is the younger one.

A few craters on the Moon, notably Copernicus (Figs. 13–3 and 13–11), have thrown out obvious rays of lighter-colored matter. Since these rays extend over other craters, they are younger than these other craters. The youngest rayed craters may be very young indeed—perhaps only a few

Figure 13–10 This view from the orbiting Apollo 15 Command Module shows a smaller crater, Krieger B, superimposed on a larger crater, Krieger. Obviously, the smaller crater is younger than the larger one. Several *rilles* (clefts along the lunar surface that can be hundreds of kilometers in length) and *ridges* are also visible. Sometimes the areas in the centers of craters, the *floors,* are smooth but sometimes craters have *central peaks.*

Figure 13–11 The crater Copernicus, seen in this ground-based photograph, has rays of light material emanating from it. This light material was thrown out radially when the meteorite that formed the crater impacted.

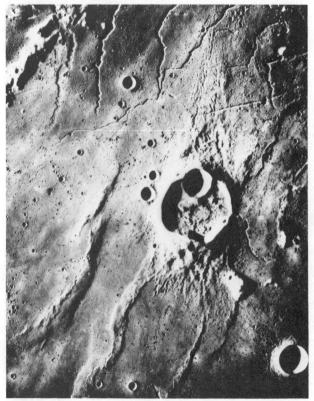

Figure 13–10

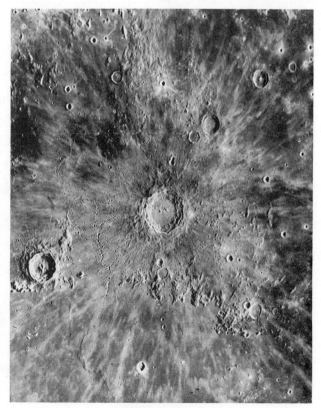

Figure 13–11

hundred million years. The rays darken with time, so rays that may have once existed near other craters are now indistinguishable from the rest of the surface. Rays are visible for about 3 billion years.

Crater counts, or the superposition of one crater on another, give only relative ages. We could find the absolute ages only when rocks were physically returned to Earth, which allowed us to study them in laboratories. Dates are found from study of the fraction of radioactive isotopes (particular forms of a chemical element, as described in Section 8.3) to nonradioactive isotopes. The oldest rocks that were found at the locations sampled on the Moon were formed 4.42 billion years ago. The youngest rocks were formed 3.1 billion years ago. (Ages are given to the accuracy to which they can be measured; the 4.42-billion year age was especially precise and represented a catastrophic event on the Moon.)

There are major overall differences between the ages of highland and maria rocks. The highland rocks were formed between about 3.9 and 4.4 billion years ago, and the maria between 3.1 and 3.8 billion years ago. One highland rock even showed signs of having been melted and re-formed twice—mainly at 3.93 billion years but with some crystals that were 4.4 billion years old included. Several highland rocks are exactly the same age, all 4.42 billion years old. So 4.42 billion years ago may therefore have been the origin of the lunar surface material, that is, the time when it last cooled.

All the observations can be explained on the basis of the following general picture. The Moon formed 4.6 billion years ago. We know that the top 100 km or so of the surface was molten after about 200 million years. This could have been caused by the original heat or by an intense bombardment of meteorites (or debris from the period of formation of the Moon and the Earth) that melted the surface entirely. The decay of radioactive elements also provided heat. Then the surface cooled definitively. From 4.2 to 3.9 billion years ago, bombardment (perhaps by planetesimals) caused most of the craters we see today. About 3.8 billion years ago, the interior of the Moon heated up sufficiently (from radioactive elements inside) that vulcanism began; lava flowed on the lunar surface and filled the largest basins that resulted from the earlier bombardment, thus forming the maria (Fig. 13–12). By 3.1 billion years ago, the era of vulcanism was over, and the Moon has been geologically pretty quiet since then.

Up to this time, the Earth and the Moon shared similar histories. But

Radioactive isotopes are those that decay spontaneously; that is, they change into other isotopes even when left alone. Stable isotopes remain unchanged. For certain pairs of isotopes–one radioactive and one stable–we know the proportion of the two when the rock was formed. Since we know the rate at which the radioactive one is decaying, we can calculate how long it has been decaying from a measurement of what fraction is left.

Figure 13–12 A series of artist's views of the formation of the lunar surface compared with a recent photographic map. *(A)* The Moon before the formation of the present mare surface material about 4 billion years ago. The concentric rings of the Imbrium basin, now Mare Imbrium, show prominently at upper left. The last mare basin, Mare Oriental (partly in view on the left—west—limb) has just been formed. The artist has removed most mare material and late craters, and freshened certain early craters. The geological dominance of the structures and blankets of the multiringed basins is clear. *(B)* The Moon soon after the formation of most of the mare material approximately 3.3 billion years ago. Young craters such as Tycho and Copernicus are absent. The Moon looks much like it does now, though surface details differ. *(C)* A modern lunar map from the U.S. Geological Survey. (Drawings by Donald E. Davis under the guidance of Don E. Wilhelms of the U.S. Geological Survey)

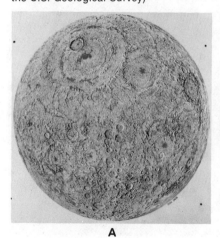

A   B   C

Figure 13–13 This view from Apollo 15 shows the Oceanus Procellarum region looking north of the crater Aristarchus. The lunar mountains and ridges, including those shown here, are formed of debris. This makes them fundamentally different from mountains on the Earth, which are formed as a result of internal processes relating to vulcanism and the butting together of continental plates according to the theory of continental drift. (Motions of the crust of a planet are called *tectonic*.)

Figure 13–14A Mare Orientale, shown here, is at the edge of the surface of the Moon we see from Earth so was little known before space exploration of the Moon. One of the unmanned pre-Apollo spacecraft radioed back this picture, which shows concentric rings extending outward over 450 km in radius. They were probably caused by a meteor impact. The stripes are artifacts in the process of transmitting data. The origin of most other craters not surrounded by such rings is more ambiguous.

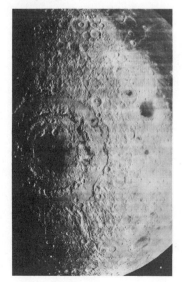

active lunar history stops about 3 billion years ago, while the Earth continued to be geologically active (Fig. 13–13). Further, because the Earth's interior continued to send gas into the atmosphere and because the Earth's higher gravity retained that atmosphere, the Earth developed conditions in which life evolved. The Moon, because it is smaller than the Earth, presumably lost its heat more quickly and also generated a thicker crust.

Almost all the rocks on the Earth are younger than 3 billion years of age; erosion and the remolding of the continents as they move slowly over the Earth's surface, according to the theory of plate tectonics (Section 16.3), have taken their toll. The oldest single rock ever discovered on Earth has an age of 3.7 billion years. So we must look to extraterrestrial bodies—the Moon or meteorites—that have not suffered the effects of plate tectonics or erosion (which occurs in the presence of water or an atmosphere) to study the first billion years of the solar system.

13.3c   The Origin of the Craters

The debate over whether the craters were formed by meteoritic impact or by volcanic action began in pre-Apollo times (Fig. 13–14). The results from Apollo indicate that most craters resulted from meteoritic impact, even though some very small fraction may represent the effects of vulcanism. It is almost as though the wrong question was being asked all along. The question

Figure 13–14B This series, made with a very short-exposure "strobe" light, shows the result of a falling milk drop. The formation of a lunar crater by a meteorite is similar, because the energy of the impact makes the surface material flow like a liquid. (Courtesy of Harold E. Edgerton, MIT)

Meteorites hit the Moon with such high velocities that huge amounts of energy are released at the impact. The effect is that of an explosion, as though it had been TNT or an H-bomb exploding.

should not have been, "Are the craters volcanic?" but rather, "Is there evidence of vulcanism?" The answer to the latter question is "yes." Only a very few craters result from vulcanism, but there are many other signs of volcanic activity, including the lava flows that filled the maria and modifications of many impact craters by subsequent vulcanism. In any case, over 99 per cent of the lunar craters that we see with our telescopes were caused by the impacts of meteorites.

The photographs of the far side of the Moon have shown us that the near and far hemispheres are quite different in overall appearance. The maria that are so conspicuous on the near side are almost absent from the far side, which is cratered all over (Figs. 13–15 and 13–16). The asymmetry in the distribution of maria may arise inside the Moon itself by an uneven distribution of mass. Once any asymmetry is set up, then the Earth's gravity would lock one side toward us. It is interesting to note that not only the Moon but also the

Figure 13–15 The far side of the Moon looks very different from the near side in that there are few maria (compare with Fig. 13–3). This photograph was taken from Apollo 16, and shows some of the near-side maria at the upper left.

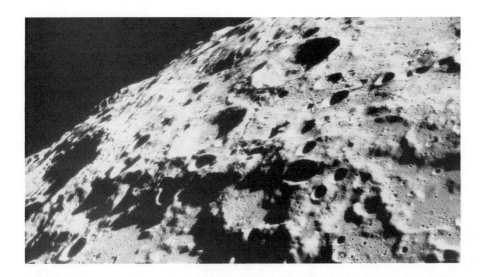

Figure 13-16 An oblique view of the lunar far side, showing how rough it looks.

Earth, Mars, and probably Mercury have asymmetric hemispheres. The cause of this is unknown.

13.3d The Lunar Interior

Before the Moon landings, it was widely thought that the Moon was a simple body, with the same composition throughout. But we now know it to be a differentiated body (Fig. 13–17) like the planets. It has a *crust* of relatively light material at its surface and a silica-rich *mantle* making up most or all of the interior. It may also have a metallic (iron-rich) *core* at its center, though any such core would take up a much smaller fraction of the lunar interior than the Earth's core does of the Earth's interior.

Differentiated bodies have layers of materials in different states and of different compositions, as opposed to bodies that are homogeneous throughout.

The lunar crust is perhaps 65 kilometers thick on the near side and twice as thick on the far side. This asymmetry may explain the different appearances of the near and far sides, because lava would be less likely to flow through the thicker part of the crust.

The interior of a moon or planet cannot be observed directly, though there are many indirect methods of deduction that can be brought to bear. One of the best methods has scientists using the seismograph, the device employed on Earth to detect earthquakes. The speeds at which different types of waves are transmitted through the lunar interior tell us of the condition of the interior.

Working seismometers were left by Apollo astronauts at four widely spaced locations on the Moon. The seismometers enabled us to locate the origins of the thousands of weak moonquakes that occur each year—almost all of which are magnitudes 1 to 2 on the Richter scale used on Earth. Perhaps three moonquakes per year reach Richter 4, a strength which on Earth could be felt but which would not cause damage. On one occasion (July 21, 1972) a meteorite hit the far side of the Moon and generated seismic waves strong enough that we would have expected them to travel through to the near side, where the seismometers are all located. From the fact that one type of seismic wave—so-called *shear waves*—did not travel through the core while the other types did, most researchers deduce that the core is molten or at least plastic in consistency.

The instruments that were left on the Moon were shut off by NASA as an economy measure in 1977.

We think that magnetic fields are formed in molten planetary interiors, for then the material inside the crust can circulate. If the interior had always been solid, no magnetic field would have resulted.

On the other hand, the Moon has no general magnetic field. If the

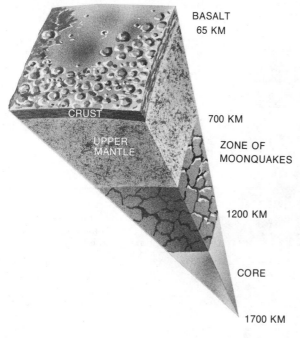

BASALT
65 KM

CRUST

UPPER MANTLE

700 KM

ZONE OF
MOONQUAKES

1200 KM

CORE

1700 KM

Figure 13–17 The Moon's interior. The depth of mare basalts is greater under maria, which are largely on the side of the Moon nearest the Earth. Almost all the 10,000 moonquakes observed originated in a zone halfway down toward the center of the Moon, a distance ten times deeper than most terrestrial earthquakes. This fact can be used to interpret conditions in the lunar interior. If too much of the interior of the Moon were molten, the source of moonquakes probably could not have remained suspended there. The deep moonquakes came from about 80 locations, and were triggered at each location twice a month by tidal forces resulting from the variation in the Earth-Moon distance.

interior of the Moon were molten, as is the Earth's interior, it might be expected that a more intense magnetic field than is detected would be set up. The weak magnetic field we do detect frozen into lunar rocks could be left over from a core that could have been molten long ago but has since cooled. We shall see in subsequent chapters that the unexpected detection of a magnetic field on Mercury has shown us that we do not understand as well as we had thought how planetary magnetic fields are formed.

Apollo astronauts made direct measurements of the rate at which heat flows upward through the top of the lunar crust. The rate is one-third that of the heat flow on Earth. The value is important for checking theories of the lunar interior.

Tracking the orbits of the Apollo Command Modules and other satellites that orbited the Moon also told us about the lunar interior. If the Moon were a perfect, uniform sphere of mass, the spacecraft orbits would have been perfect ellipses. Any deviation of the orbit from an ellipse can be interpreted as an effect of an asymmetric distribution of lunar mass. (The effects of the Earth, Sun, or other planets are relatively small.)

It might be noted that although the finding of mascons on the Moon was unexpected, there are two dozen locations on Earth where terrestrial "mascons" have even larger gravitational effects than does the largest lunar mascon.

One of the major surprises of the lunar missions was the discovery of *mascons*, regions of **mass con**centrations near and under most maria that lead to anomalies in the gravitational field, that is, deviations of the gravitational field from being spherical. The mascons may be lava that is denser than the surrounding matter. The existence of the mascons is evidence that the whole lunar interior is not molten, for if it were, then these mascons could not remain near the surface. However, the mascons could be supported by a crust of sufficient thickness.

In sum, we have contradictory evidence as to whether the lunar interior is hot or cold, molten or not. Still, as a result of the meteorite impact detected with the seismic experiment, most scientists believe that the Moon's core is molten. If such a small body as the Moon formed with a hot interior, then we know that the planetesimals (Section 12.11) would have had to coalesce very rapidly into planets, since the heat did not have time to be radiated away.

The study of the Moon gives us insight into the formation of the whole solar system.

13.3e The Origin of the Moon

Among the models that have been considered in recent years for the origin of the Moon are the following:

(1) *Fission:* the Moon was separated from the material that formed the Earth;

(2) *Capture:* the Moon was formed far from the Earth in another part of the solar system, and was later captured by the Earth's gravity; and

(3) *Condensation:* the Moon was formed near to and simultaneously with the Earth in the solar system.

Comparison of the chemical composition of the lunar surface with the composition of the terrestrial surface has been important in narrowing down the possibilities. The mean lunar density of 3.3 grams/cm³ is close to the average density of the Earth's major upper region (the mantle), which had led to some belief in the fission hypothesis. However, detailed examination of the lunar rocks and soils indicates that the abundances of elements on Moon and Earth are sufficiently different to indicate that the Moon did not form directly from the Earth. (Though some minerals (Fig. 13–18) that do not exist on Earth have been discovered on the Moon, this results from different conditions of formation rather than from abundance differences.) Also, the acceptance of the theory of continental drift (Section 16.3) for the formation of the Pacific Ocean basin makes it seem less likely that the Pacific Ocean was the hole left behind when the Moon was ripped from the Earth.

Still, the fission hypothesis, at least in the case that the Moon separated from the Earth very early on, cannot be excluded. The length of the interval in which fission could have happened, however—given the extreme ages of lunar soil and some of the rocks—was only a short one. New evidence that the ratios of three oxygen isotopes are the same on the Moon as they are on the Earth but different from the ratio for meteorites indicates that the fission hypothesis is still to be reckoned with.

The capture model also has evidence against it. The conditions necessary for the Earth to capture such a massive body by gravity seem too unlikely for this to have occurred. However, the evidence against the model is not conclusive—one can't apply statistical methods to one example—and several possible ways have been suggested in which the Moon could have been captured by the Earth. Perhaps a larger body was broken up by tidal forces and only part was captured.

Although there are chemical differences between the Moon and the Earth, they are not so overwhelming as to exclude the condensation possibility, and most astronomers currently believe in some version of this model. The idea that the Moon has an iron core adds backing to this theory. In sum, it seems most probable that the Earth and the Moon formed near each other as a double planet, probably by accretion of planetesimals. The condensation model thus closely connects the question of the origin of the Moon to the larger question of the origin of the solar system, which we also think involved the accretion of planetesimals.

It had been hoped that landing on the Moon would enable us to clear up the problem of the lunar origin. But though the lunar programs have led to modifications and updating of the models described above, none of these models has been entirely ruled out. Nobody can say definitely which, if any, of the three is correct.

Figure 13–18 A crystal of armalcolite (**Arm**strong-**Ald**rin-**Col**lins, the crew of Apollo 11), examined under a polarization microscope. This mineral has been found only on the Moon.

13.4 THE VALUE OF MANNED LUNAR RESEARCH

Color Plates 24 to 27 are devoted to the Moon and lunar exploration.

The issue of manned versus unmanned space research has long been controversial among astronomers, who tend to favor unmanned research because its substantially lower cost would allow more work to be carried out. The success of the Apollo missions plus the ability of the Skylab astronauts to rescue a damaged spacecraft and operate complex equipment requiring on-the-spot judgment has tempered this opinion for many. The cost overruns on the Space Shuttle program, however, have drained funds from NASA's research programs, again leaving many scientists wishing that they had unmanned spacecraft.

We must recall that the decision to start the program of manned lunar exploration was largely the result of political factors. Sending a person to the Moon was a necessary part of the project. While it is true that much of the cost of the project was related to the cost of keeping humans alive and safe, it is also true that this part of the cost was justified on this ground and the equivalent amount of money would not have been put into straight astronomical research.

We have also succeeded through the Apollo program (and in the succeeding Skylab program, which used rockets and spacecraft developed and built for Apollo) in showing that humans can function in space and carry out useful work over long periods of time. Weightlessness and other effects of spaceflight (such as calcium depletion in bones) do not seem to be major barriers to space exploration. Perhaps this knowledge of our human capabilities will be the longest-lasting benefit of the Apollo and Skylab programs.

The Moon Treaty, "The Agreement Governing Activities of States on the Moon and Other Celestial Bodies," was opened for signing at the United Nations in 1980. It is an attempt to provide international law in advance of when it is needed to control competition for the Moon's wealth. The treaty is controversial because its phrase that the moon and its resources are "the common heritage of mankind" leaves ambiguous whether the Moon (and other celestial bodies such as asteroids) would be common property or open to private enterprise (mining, etc.).

For those of us who remember the shock of Sputnik, and followed each small step farther and farther off the Earth's surface and into space, it is hard to believe that the era of manned lunar exploration not only has begun but also has already ended. At present we have no plans to send more people to the Moon. The United States does not even have any unmanned missions planned, while the Soviet Union has indicated they may send another unmanned spacecraft to the Moon within the next few years. Perhaps by the turn of the 21st century, manned lunar exploration will resume. Twenty or thirty years from now, we may each be able to visit the Moon as researchers or even as tourists.

SUMMARY AND OUTLINE

Lunar features (Section 13.1)
  Fixed, maria, highlands, craters, mountains, valleys, rilles, ridges
  Dependent on sun angle: terminator
Revolution and rotation (Box 13.1)
  Sidereal: with respect to the stars
  Synodic: with respect to another body
Lunar exploration (Section 13.2)
  Unmanned series
  Manned series in Earth orbit: Mercury, Gemini
  Manned and unmanned lines of development merge in Apollo
  Six Apollo landings: 1969 to 1972

Three Soviet unmanned spacecraft brought samples back to Earth
Two Soviet Lunokhods roved many kilometers over the lunar surface
Composition of the lunar surface (Section 13.3a)
  Mare basalts: from cooling of lava
  Highland anorthosites
    More breccias—broken up and re-formed rock—in highlands
  Soils
Chronology (Section 13.3b)
  Relative dating by crater counting
  Radioactive dating gives absolute ages (Fig. 13–19)
  Highland rocks (3.9–4.4 billion years) older than maria (3.1–3.8 billion years)

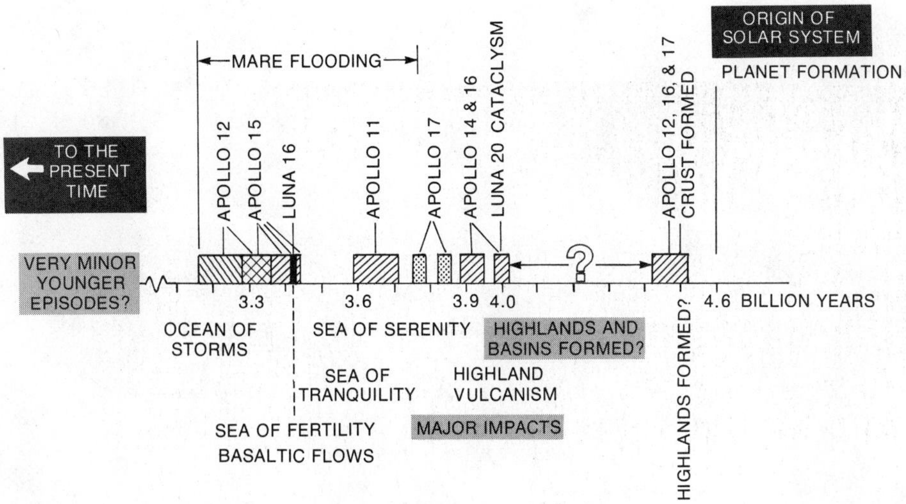

Figure 13–19 The chronology of the lunar surface, based on work carried out at the Lunatic Asylum, as the Caltech laboratory of Gerald Wasserburg is called. The ages of rocks found in seven missions are shown in boxes. The names and descriptions below the line took place at the times indicated by the positions of the words.

Probable model: Moon formed 4.6 b.y. At 4.4 b.y., the surface, which had been molten, cooled. Meteorite bombardment from 4.2 to 3.9 b.y. made most of the craters. Interior heated up from radioactive elements; the resulting vulcanism from 3.8 to 3.1 b.y. caused lava flows that formed the maria. Then lunar activity stopped.

Craters (Section 13.3c)
  Almost all formed in meteoritic impact
  Signs of vulcanism also present on surface, though only very few craters are volcanic
  Far side has no maria, quite different in appearance from near side
Interior (Section 13.3d)

Moon is differentiated into a core, a mantle, and a crust
Seismographs left by Apollo missions reveal that interior is molten
Lack of lunar magnetic field, on the other hand, is an indication that interior is not molten, though seismic evidence seems definitive
Weak moonquakes occur regularly
Mascons discovered from their gravitational effects
Origin (Section 13.3e)
  Condensation, fission, and capture theories
  All still viable, with most astronomers tending to believe in a condensation theory

QUESTIONS

1. Compare the lengths of sidereal and synodic months. Which is longer? Why?

2. If the mass of the Moon is $^1/_{81}$ that of the Earth, why is the gravity at the Moon's surface as great as $^1/_6$ that at the Earth's surface?

3. To what location on Earth does the terminator on the Moon correspond?

4. Why is the heat flow rate related to the radioactive material content in the lunar surface?

5. What does cratering tell you about the age of the surface of the Moon, compared to that of the Earth's surface?

6. Why is it not surprising that the rocks in the lunar highlands are older than those in the maria?

7. Why are we more likely to learn about the early history of the Earth by studying the rocks from the Moon than those on the Earth?

8. What do the mascons tell us about the interior of the Moon?

9. Choose one of the proposed theories to describe the origin of the Moon and discuss the evidence pro and con.

10. (a) Describe the American and Soviet lunar explorations.
    (b) What can you say about future plans?

TOPIC FOR DISCUSSION

Discuss the scientific, political, and financial arguments for resuming manned exploration of the Moon.

CHAPTER 14

MERCURY

AIMS:
To discuss the difficulties in studying
Mercury from the Earth, and what
space observations have told us

Mercury is the innermost planet, and until very recently has been one of the least understood. Except for distant Pluto, its orbit around the Sun is the most elliptical (the difference between the maximum and minimum distances of Mercury from the Sun is as much as 40 per cent of the average distance, compared with less than 4 per cent for the Earth). Its average distance from the Sun is 58 million kilometers (36 million miles), which is 4/10 of the Earth's average distance. Thus Mercury is 0.4 A.U. from the Sun.

Since we on the Earth are outside Mercury's orbit looking in at it, Mercury always appears close to the Sun in the sky (Fig. 14–1). At times it

In order to know when Mercury will be favorably located in the sky for you to observe it, you should read the sky descriptions published in monthly magazines like Sky and Telescope or Astronomy and in some local newspapers.

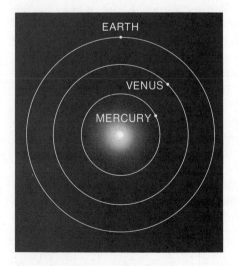

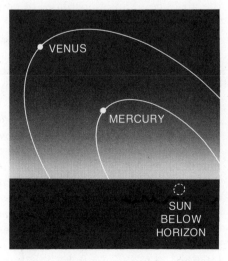

Figure 14–1 Since Mercury's orbit is inside that of the Earth, Mercury is never seen against a really dark sky. A view from the Earth appears at right, showing Mercury and Venus at their greatest respective distances from the Sun. The open view of their orbits is an exaggeration; we actually see the orbits nearly edge on.

Mercury, photographed at a distance of 200,000 km from Mariner 10, revealed a cratered surface. The symbol for Mercury appears at the top of this page.

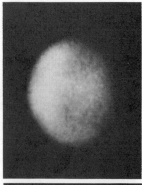

Figure 14–2 From the Earth, we cannot see much surface detail on Mercury. These views, among the best ever taken on Earth, were made by the New Mexico State University Observatory. Mercury was only 7.1 *(top)* and 5.1 *(bottom)* seconds of arc across.

It seemed reasonable that there could be a bulge in the distribution of mass of Mercury. The side that was bulging would be attracted to the Sun by gravity, locking the rotation to the revolution, just as the Moon is locked to the Earth. This is called synchronous rotation. *Synchronous rotation implies that the periods of rotation and revolution are equal, and so the less massive body would always keep the same face toward the more massive body.*

If Mercury were in a circular orbit, it would be in synchronous rotation. But because its orbit is elliptical, its bulge is locked toward the Sun mainly when it is at perihelion.

rises just before sunrise, and at times it sets just after sunset, but it is never up when the sky is really dark. The maximum angle from the Sun at which we can see it is 28°, which means that the Sun always rises or sets within about two hours of Mercury's rising or setting. Of course, the difference in time is usually even less than this maximum. Because of Mercury's closeness to the Sun in the sky, astronomers have never gotten a really good view of Mercury from the Earth, even with the largest telescopes. Many people have never seen it at all. (Copernicus' deathbed regret was that he had never seen Mercury.) The best photographs taken from the Earth show Mercury as only a fuzzy ball with faint, indistinct markings (Fig. 14–2). These photographs could be taken only when Mercury was low in the sky near sunrise and sunset. Consequently the light from Mercury had to pass obliquely through the Earth's atmosphere, making a long path through turbulent air. Hence the photographs are blurred.

14.1 THE ROTATION OF MERCURY

Astronomers, from studies of drawings and photographs, did as well as they could to describe Mercury's surface. A few features could barely be distinguished, and the astronomers watched to see how long those features took to rotate around the planet. From these observations they decided that Mercury rotated in the same amount of time that it took to revolve around the Sun. An 88-day period of rotation for Mercury, matching its known 88-day period of revolution around the Sun, thus appeared in all the reference works and textbooks. This led to the fascinating conclusion that Mercury could be both the hottest planet and the coldest planet in the solar system. The subsolar point, the surface that is closest to the Sun, would always be the same region of Mercury's surface since it would never rotate around. The point on the opposite side of Mercury would never receive sunlight and, in the absence of an atmosphere to conduct and convect heat, would be the coldest point.

Radio astronomy studies of Mercury indicated that the dark side of Mercury was too hot for a surface that was always in the shade. Simply, the signal strength depends on the temperature.

Later, we became able not only to receive radio signals emitted by Mercury, as above, but also to transmit radio signals from Earth and detect the echo. This technique is called *radar*. Since Mercury is rotating, one side of the planet is always receding relative to the other. Studies of how the radio signals are changed as they bounce off the planet tell us the speed of the planet's rotation.

The results were a surprise: scientists had been wrong about the period of Mercury's rotation. It actually rotates in 59 days.

Mercury's 59-day period of rotation is exactly ⅔ of the period of its revolution, so the planet rotates three times for each two times it revolves around the Sun. Although the inertia of its rotation—its tendency to continue rotating—is too strong to be overcome by the gravitational grip of the Sun, the Sun's steadying pull is strongest every 1½ rotations. At those times, Mercury is in its perihelion position, so the gravitational bulge on Mercury is as near to the Sun as it can be. Mercury's spin was probably once much faster, and was slowed down by the fact that the Sun's gravity attracted the bulge

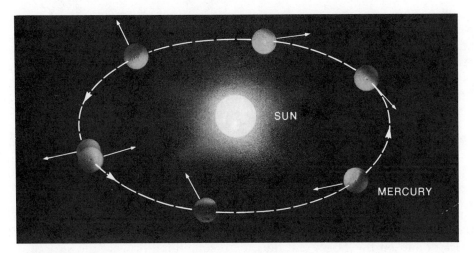

Figure 14–3 Follow the arrow that starts facing rightward toward Sun in the image of Mercury at the left of the figure, as Mercury revolves along the dotted line. Mercury, and thus the arrow, rotate once with respect to the stars in 59 days, when Mercury has moved only 2/3 of the way around the Sun. (Our view is as though we were watching from a distant star.) Note that after one full revolution of Mercury around the Sun, the arrow is facing away from the Sun. It takes another full revolution, a second 88 days, for the arrow to be again facing the Sun. Thus the rotation period with respect to the Sun is twice 88, or 176, days.

more than it attracted the rest of the planet. A *gravitational interlock* is acting.

This rotation period is measured with respect to the stars; that is, the period is one sidereal day, the interval between successive returns of the stars to the same position in the sky. Mercury's rotation and revolution combine to give a value for the rotation of Mercury relative to the Sun (that is, a Mercurian solar day) that is neither the 59-day *sidereal rotation period* nor the 88-day period of revolution. As can be seen from careful analysis of Figure 14–3, if we lived on Mercury we would measure each day and each night to be 88 Earth days long. We would alternately be fried and frozen for 88 Earth days at a time. Mercury's *solar rotation period* is thus 176 days long, twice the period of Mercury's revolution.

Though the subsolar point is not always at the same place on the surface and so is not eternally heated, we still measure a temperature of about 700 K there.

We know from Kepler's second law that Mercury travels around the Sun at different speeds at different times in its eccentric orbit. This effect, coupled with its slow rotation on its axis, would lead to an interesting effect if we could stand on Mercury's surface. From some locations we would see the Sun rise for an Earth day or two, and then retreat below the horizon from which it had just come, when the speed of Mercury's revolution around the Sun dropped below the speed of Mercury's rotation on its own axis. Later, the Sun would rise again, and then continue across the sky.

No harm was done by the scientists' misconception of Mercury's rotational period for all those years, but the story teaches all of us a lesson: we should not be too sure of so-called facts, even when they are stated in all the textbooks. Don't you believe everything you read here, either.

14.2 OTHER KNOWLEDGE FROM GROUND-BASED OBSERVATIONS

Even though the details of the surface of Mercury can't be studied very well from the Earth, there are other properties of the planet that can be better studied. For example, we can measure Mercury's *albedo*, the fraction of sunlight hitting Mercury that is reflected from it (Fig. 14–4). We can measure

On rare occasions, Mercury goes into transit across the Sun; that is, we see it as a black dot crossing the Sun. The next transit is on December 11, 1986. The following transits will be in 1993, 2003, and 2006.

The albedo (from the Latin for whiteness) is the ratio of light reflected from a body to light received from it.

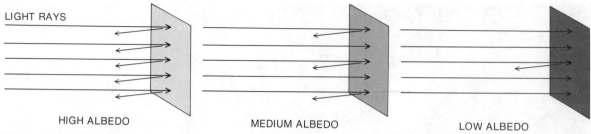

LIGHT RAYS

HIGH ALBEDO

MEDIUM ALBEDO

LOW ALBEDO

Figure 14–4 *Albedo* is the fraction of radiation reflected. A surface of low albedo looks dark.

this because we know how much sunlight hits Mercury (we know how bright the Sun is and how far away Mercury is from it). Then we can easily calculate at any given time how much light Mercury reflects, from both (1) how bright Mercury looks to us and (2) its distance from the Earth. Once we have a measure of the albedo, we can compare it with the albedo of materials on the Earth and on the Moon and thus learn something of what the surface of Mercury is like.

Let us consider some examples of albedo. An ideal mirror reflects all the light that hits it; its albedo is thus 100 per cent (the very best real mirrors have albedos of as much as 96 per cent). A black cloth reflects essentially none of the light; its albedo is almost 0 per cent. Mercury's overall albedo is only about 6 per cent. Its surface, therefore, must be made of a dark—that is, poorly reflecting—material. The albedo of the Moon is similar. In fact, Mercury (or the Moon) appears bright to us only because it is contrasted against a relatively dark sky; if it were silhouetted against a bedsheet, it would look relatively dark, as if it had been washed in Brand X instead of Tide.

Mercury's density turns out to be 5.5 $g\,cm^{-3}$, about the same density as that of Venus and the Earth. Thus Mercury's core, like the Earth's, must be heavy; it too is made of iron. Since Mercury has less mass than the Earth and is therefore less compressed, Mercury's core must have even more iron than the Earth to give the planets the same density.

From Mercury's apparent angular size and its distance from the Earth—which can be determined from knowledge of its orbit—we have determined that Mercury is less than half the diameter of the Earth. Since Mercury has no moon, we can determine its mass only from its gravitational effects on bodies that pass near it, such as occasional asteroids and comets. The most accurate mass now comes from tracking the Mariner 10 spacecraft flyby. The mass is five times greater than that of our Moon and 5½ per cent that of Earth. Its density (its mass divided by its volume) can thus be calculated, and is roughly the same as the Earth's. (An accurate value is given in Appendix 5.)

14.3 MARINER 10

So astronomers had, typically, deduced a lot from limited data. In 1974, we learned much more about Mercury in a brief time. We flew right by. The tenth in the series of Mariner spacecraft launched by the United States went to Mercury. First it passed by Venus and then had its orbit changed by Venus' gravity to direct it to Mercury. Tracking the orbits improved our measurements of the gravity of these planets and thus of their masses. Further, the 475-kg (1042-pound) spacecraft had a variety of instruments on board. One was a device to measure the magnetic fields in space and near the two planets. Another measured the infrared emission of the planets and thus their temperatures. Two others of these instruments—a pair of television cameras—provided not only the greatest popular interest but also many important data.

Mariner 10's flyby of Venus will be described in the next chapter.

14.3a Photographic Results

When Mariner 10 flew by Mercury the first time (yes, it went back again),

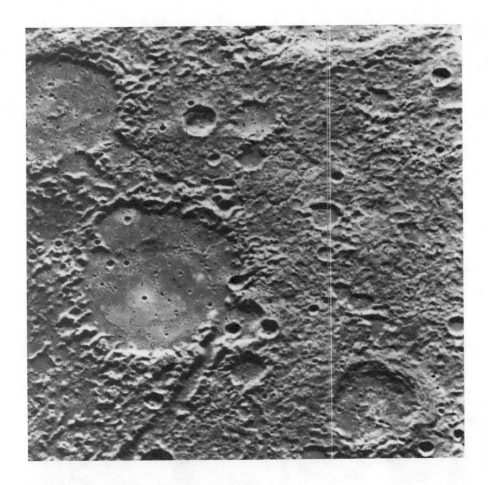

Figure 14–5 This 150 × 300 km area of Mercury, photographed from a distance of 35,000 km as Mariner 10 approached the planet for the first time, shows a heavily cratered surface with many low hills. The valley at bottom is 7 km wide and over 100 km long. The large flat-floored crater is about 80 km in diameter.

it took 1800 photographs that were transmitted to Earth. It came as close as 750 km to Mercury's surface.

The most striking overall impression is that Mercury is heavily cratered (see the photograph that opened this chapter). At first glance, it looks like the Moon! But there are several basic differences between the features on the surface of Mercury and those on the lunar surface. We can compare how the mass and location in the solar system of these two bodies affected the evolution of their surfaces.

Mercury's craters seem flatter than those on the Moon, and have thinner rims (Fig. 14–5). This is largely an effect of Mercury's higher gravity. The craters may have been eroded by any of a number of methods, such as the impacts of meteorites or micrometeorites (large or small bits of interplanetary rock). Alternatively, erosion may have occurred during a much earlier period when Mercury may have had an atmosphere, or internal activity, or been flooded by lava.

Most of the craters themselves seem to have been formed by impacts of meteorites. In many areas they appear superimposed on relatively smooth plains. A class of smaller, brighter craters are sometimes, in turn, superimposed on the larger craters and thus must have been made afterwards (Figs. 14–6 and 14–7). The rate at which objects hit and formed craters is probably about the same for all the terrestrial planets.

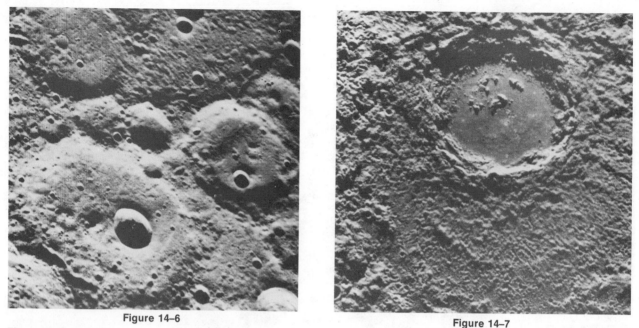

Figure 14–6

Figure 14–7

Figure 14–6 A fresh new crater, about 12 km across, in the center of an older crater basin.
Figure 14–7 Another fresh crater on Mercury, about 120 km across. Because Mercury's gravitational field is higher than that of the Moon, material ejected by an impact on Mercury does not travel as far across the surface.

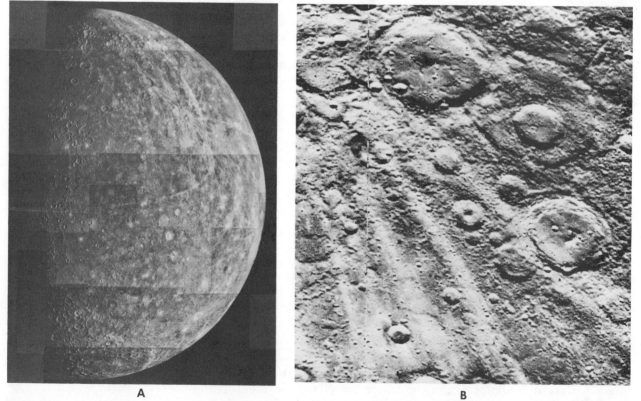

A

B

Figure 14–8 (A) Many rayed craters can be seen on this photograph of Mercury, taken six hours after Mariner 10's closest approach on its first pass. The north pole is at the top and the equator extends from left to right about 2/3 of the way down from the top. (B) A field of rays radiating from a crater off to the top left, photographed on the second pass of Mariner 10. The crater at top is 100 km in diameter.

Some craters have rays of higher albedo emanating from them (Fig. 14–8), just as some lunar craters do. The ray material represents relatively recent crater formation (that is, within the last hundred million years). The ray material must have been tossed out in the impact that formed the crater.

One interesting kind of feature that is visible on Mercury is a line of cliffs hundreds of miles long; on Mercury as on Earth such lines of cliffs are called *scarps*. The scarps are particularly apparent in the region of Mercury's south pole (Figs. 14–9 and 14–10). Unlike fault lines we know on the Earth, such as the San Andreas fault in California, there are no signs of geologic tensions like rifts or fissures nearby. The more that Mercury has been photographed and the photographs analyzed, the more we realize that these cliffs or scarps are global in scale, and not just isolated occurrences.

It seems that these scarps may actually be wrinkles in the crust of the planet. Mercury's core, judging by the fact that its density is about the same as the Earth's, is probably iron and takes up much of the central volume, perhaps 50 per cent of the volume or 70 per cent of the mass. At one time, perhaps the core was molten, and shrank by 1 or 2 km as it cooled. The crust would have settled down with it, making the scarps in the quantity that is in fact observed. Alternatively, it seems possible that Mercury formerly rotated more rapidly. As the rotation slowed down, the shape of the planet would have become less oblate (that is, closer to round instead of bulging at the equator). The crust might have cracked and wrinkled as it tried to match the new shape of the interior.

One part of the Mercurian landscape seems particularly different from the rest (Fig. 14–11). It seems to be grooved, with relatively smooth areas

Figure 14–9 A scarp *(arrow)* more than 300 km long extends from top to bottom in this second-pass picture.

The ground level on one side of a scarp is different from that on the other side; a scarp is not merely a line of mountains in a flat plain.

Figure 14–10 This first-pass view of Mercury's northern limb shows a prominent scarp *(arrow)* extending from the limb near the middle of the photograph. The photograph shows an area 580 km from side to side.

Figure 14–11 This mosaic of pictures from the first pass of Mariner 10 shows a terrain unique to Mercury—hills and ridges cut across many of the craters and the inter-crater areas.

By "high resolution," we mean that smaller details are visible than would be visible under "low resolution."

Figure 14–12 A high resolution view of the fractured and ridged plains of the Caloris Basin, photographed from a distance of 19,000 km on Mariner 10's third pass, 34 minutes after the spacecraft's closest approach.

between the grooves. It is called the "weird terrain." Just a couple of areas of this type have been found on the Moon, and no others are known on Mercury. The weird terrain is 180° around Mercury from the Caloris Basin, the site of a major meteorite impact. Shock waves from that impact may have been focused halfway around the planet.

Box 14.1   Naming the Features of Mercury

The mapping of the surface of Mercury leads to a need for names. The scarps are being named for historical ships of discovery and exploration, such as Endeavour (Captain Cook's ship), Santa Maria (Columbus' ship), and Victoria (the first ship to sail around the world, which it did in 1519–22 under Magellan and his successors). Some plains are being given the name of Mercury in different languages, such as Tir (in ancient Persian), Odin (an ancient Norse god), and Suisei (Japanese). Craters are being named for non-scientific authors, composers, and artists, in order to complement the lunar naming system, which honors scientists.

The Mariner 10 mission was a navigational coup not only because it used the gravity of Venus to get the spacecraft to Mercury, but also because scientists and engineers were able to find an orbit around the Sun that brought the spacecraft back to Mercury several times over. Every six months Mariner 10 and Mercury returned to the same place at the same time. As long as the gas jets for adjusting and positioning Mariner functioned, it was able to make additional measurements and to send back additional pictures in order to increase the photographic coverage. On its second visit, for example, in September 1974, Mariner 10 was able to study the south pole and the region around it for the first time. This pass was devoted to photographic studies. The spacecraft came within 48,000 kilometers of Mercury, farther away than the 750 kilometer minimum of the first pass, but the data were still very valuable. On its third visit, in March 1975, it had the closest encounter

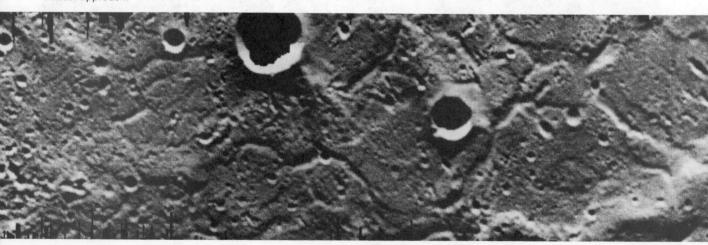

ever—only 300 km above the surface (Fig. 14–12). Thus it was able to photograph part of the surface with a high resolution of only 50 meters. Then the spacecraft ran out of gas for the small jets that control its pointing, so even though it still passes close to Mercury every few months, it can no longer take clear photographs or send them back to Earth.

14.3b Infrared Results

The cameras were not the only instruments on board Mariner. The infrared radiometer, for example, gave data that indicate that the surface of Mercury is covered with fine dust, as is the surface of the Moon, to a depth of at least several centimeters. Astronauts sent to Mercury, whenever they go, will leave footprints behind them.

The Mariner 10 mission gave more accurate measurements of the temperature changes across Mercury than had been determined from the Earth. In a few hundred kilometers at the terminator, the line between Mercury's day and night, the temperature falls about 775 K (500°C) to about 425 K (150°C), and then drops even lower farther across into the dark side of the planet.

14.3c Results from Other Types of Observations

The ability to study Mercury from so close up has led to the discovery that Mercury even has an atmosphere, although an all-but-negligible one. It is only a few billionths as dense as the Earth's, so slight that someone standing on Mercury would need special instruments to detect it. Traces of helium, oxygen, carbon, argon, nitrogen, and xenon have been detected with a spectrometer that operated in the ultraviolet. The presence of helium is a particular surprise, because helium is a light element and would be expected to escape from Mercury's weak gravity within a few hours. So there must be a constant source to replace it. Either it comes from the Sun in the solar wind, or, more likely, it comes from radioactive decay of uranium and thorium on Mercury. Hydrogen from the solar wind is also present.

One more surprise—perhaps the biggest of the mission—was that a magnetic field was detected in space near Mercury. It was discovered on Mariner 10's first pass, and then confirmed on the third pass. The field is weak; extrapolated down to the surface it is about 1 per cent of the Earth's. It had been thought that magnetic fields were generated by the rapid rotation of molten iron cores in planets, but Mercury is so small that the core would have quickly solidified. Perhaps the magnetic field generated has been frozen into Mercury since the core was molten.

There are reasonable arguments against both the leading explanations for Mercury's magnetic field. It would seem that it has to be either generated at present or else a relic of a past field. If it were now being generated, you might expect that we would see some evidence on Mercury's surface of internal heat or vulcanism younger than 3 billion years. But the alternative is that the magnetic field has been generated in the past, and we think that high temperatures are required to generate a field. These high temperatures would have destroyed the field once the generation stopped. Scientists do not agree on the reason for Mercury's magnetic field.

Mariner 10 detected lots of electrons near Mercury. Perhaps they are trapped in some sort of belt by the magnetic field, similar to the Van Allen belts around the Earth. But perhaps they are bound to Mercury for shorter times than electrons trapped by the Earth's magnetic field.

No moons of Mercury have ever been detected, and Mariner 10 did an especially careful job of searching. The amount of light that any moon reflects depends on both its size and its albedo. After all, a moon is a certain brightness because it reflects a certain amount of sunlight. A relatively small size could be compensated for with a relatively high albedo. We know now that there is no moon any bigger than 5 kilometers across with the same low albedo as Mercury itself. A higher albedo would correspond to a still smaller upper limit on size.

Spacecraft encounters like that of Mariner 10 have revolutionized our knowledge of Mercury. The kinds of information that we can bring to bear on the basic questions of the formation of the solar system and of the evolution of the planets are much more varied now than they were just a short time ago.

The Van Allen belts *are belts of charged particles surrounding the Earth. They were perhaps the first major discovery of orbiting Earth satellites, the earliest of which was Sputnik, launched on October 4, 1957. American satellites soon followed; James Van Allen of the University of Iowa analyzed data from them and discovered that the satellites regularly passed through belts of particles around the Earth. We will discuss the Earth's Van Allen belts in Section 16.7.*

SUMMARY AND OUTLINE

Difficult to observe from the Earth; never far from the
    Sun in the sky
Radio astronomy (Section 14.1)
    Surface temperatures
    Rotation period linked to orbit: 1 day is 2 years long
Albedo (Section 14.2)
    Low albedo means a poor reflector
    Mercury has a low albedo, only 6 per cent
Mass and density (Section 14.2)
Mariner 10 observations (Section 14.3)
    Photographic results

Types of objects
    Craters, maria, scarps, "weird terrain"
Mechanisms
    Impacts, vulcanism, shrinkage of the crust
Results from infrared observations
    Dust on the surface
    Temperature measurements
Results from other types of observations
    A small atmosphere
        Where does the helium come from?
    A big surprise: a magnetic field, which tells us
        about the histories of Mercury and of the Earth

QUESTIONS

1. Assume that on a given day, Mercury sets after the Sun. Draw a diagram, or a few diagrams, to show that the height of Mercury above the horizon depends on the angle that the Sun's path in the sky makes with the horizon as the Sun sets. Discuss how this depends on the latitude or longitude of the observer.

2. If Mercury did always keep the same side toward the Sun, does that mean that the night side would always face the same stars? Draw a diagram to illustrate your answer.

3. Explain why a day on Mercury is 176 Earth days long.

†4. Using information from Appendix 5, calculate the maximum Doppler shift that results from Mercury's rotation.

5. If you increased the albedo of Mercury, would its temperature increase or decrease? Explain.

6. List those properties of Mercury that could best be measured by spacecraft observations.

7. Think of how you would design a system that transmits still photographs taken by a spacecraft back to Earth and then produces the picture. The system can be simpler than your TV since still pictures are involved.

8. How would you distinguish an old crater from a new one?

9. What evidence is there for erosion on Mercury? Does this mean there must have been water on the surface?

10. List three major findings of Mariner 10.

†This indicates a question requiring a numerical solution.

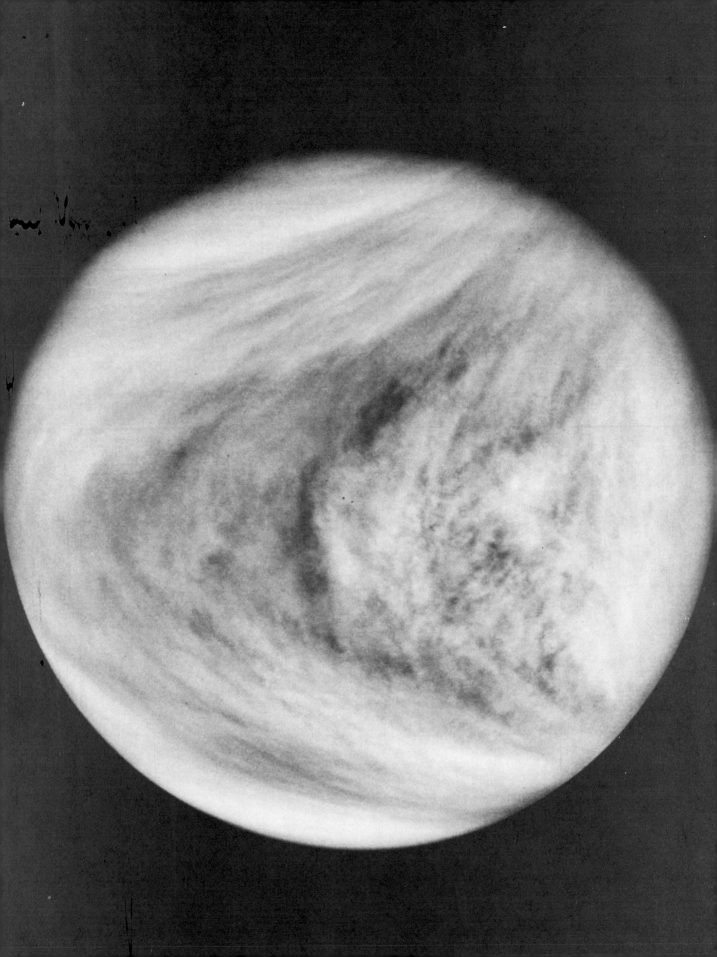

CHAPTER 15

VENUS

AIMS:
To discuss Venus' clouds, the composition of its atmosphere, and the structure of its surface, and to use this knowledge to improve our understanding of the Earth's structure and atmosphere

Venus and the Earth are sister planets; their sizes, masses, and densities are about the same. But they are as different from each other as the wicked sisters were from Cinderella. The Earth is lush, has oceans and rainstorms of water, an atmosphere containing oxygen, and creatures swimming in the sea, flying in the air, and walking on the ground. On the other hand, Venus is a hot, foreboding planet with temperatures constantly about 750 K (900°F), a planet on which life seems unlikely to develop. Why is Venus like that? How did these harsh conditions come about? Can it happen to us here on Earth?

Venus orbits the Sun at a distance of 0.7 A.U. Although it comes closer to us than any other planet—it can approach as close as 45 million kilometers (30 million miles)—we still do not know much about it because it is always shrouded in heavy clouds (Fig. 15–1). Observers in the past saw faint hints of structure in the clouds, which seemed to indicate that these clouds might circle the planet in about 4 days, rotating in the opposite sense from Venus' orbital revolution, but the clouds never parted sufficiently to allow us to see the surface.

15.1 THE ATMOSPHERE OF VENUS

The tops of Venus' clouds have been studied from the Earth. The clouds turn out to be primarily composed of droplets of sulfuric acid, H_2SO_4, with water droplets mixed in. Sulfuric acid may sound strange as a cloud constituent, but it is interesting to note that the Earth too has a significant

Figure 15–1 A crescent Venus, observed with the 5-m Hale telescope in blue light. We see only a layer of clouds.

Venus, photographed from the Pioneer-Venus Orbiter spacecraft in 1979. The picture shows the cloud tops; we see a turbulent atmosphere. The small mottled features near the center appear to be cells of convection caused by the heating of the atmosphere by solar radiation.

Figure 15–2 An infrared spectrum of Venus, showing the region from 8600 Å to 8820 Å, which includes spectral bands from carbon dioxide.

Astronomers often use the adjective Cytherean *to describe Venus. Cythera was the island home of Aphrodite, the Greek goddess who corresponded to the Roman goddess Venus.*

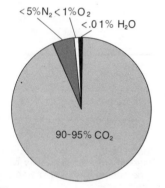

Figure 15–3 The composition of Venus' atmosphere.

layer of sulfuric acid droplets in its stratosphere, a higher layer of the atmosphere that we shall discuss in the next chapter. However, the water in the lower layers of the Earth's atmosphere, circulating because of weather, washes out the sulfur compounds in them, whereas Venus has sulfur compounds in the lower layers of its atmosphere in addition to those in its clouds.

Sulfuric acid takes up water very efficiently, so there is little water vapor above Venus' clouds. It took especially careful work from high-altitude sites on the Earth to detect the presence of the small amount of Cytherean water vapor there. This observation was difficult, because the spectral lines of water vapor from Venus were masked by the spectral water vapor lines that arose from the Earth's own atmosphere.

Spectra taken on Earth (Fig. 15–2) show that carbon dioxide is present in the atmosphere of Venus with a high concentration. In fact, carbon dioxide makes up over 90 per cent of Venus' atmosphere (Fig. 15–3). The Earth's atmosphere, for comparison, is mainly made of nitrogen, and has a fair amount of oxygen as well. Carbon dioxide makes up less than 1 per cent of the terrestrial atmosphere.

The surface pressure of Venus' atmosphere is 90 times higher than the pressure of Earth's atmosphere, and the large amount of carbon dioxide is the reason why. Carbon dioxide on Earth dissolved in sea water and eventually formed our terrestrial rocks. If this carbon dioxide were to be released from the Earth's rocks, along with other carbon dioxide trapped in sea water, our atmosphere would become as dense and have as high a pressure as that of Venus. Venus, slightly closer to the Sun than Earth and thus hotter, was too hot for the reactions that capture carbon dioxide to take place. And Venus had no liquid water to help put the carbon into rocks.

15.2 THE ROTATION OF VENUS

In 1961, radar astronomy penetrated the clouds and provided an accurate rotation period for the surface of Venus. Venus, because of its relative proximity to Earth, is an easier target for radar than Mercury. Venus rotates in 243 days with respect to the stars in the direction opposite from the other planets; this backward motion is called retrograde rotation, to distinguish it from forward (direct) rotation. Venus revolves around the Sun in 225 Earth days. These periods combine, in a way similar to that in which Mercury's sidereal day and year combine (see Section 14.1), so that a solar day on Venus would correspond to 127 Earth days; that is, the planet's rotation would bring the Sun back to the same position in the sky every 127 days.

The notion that Venus is in retrograde rotation seems very strange to astronomers, since the other known planets revolve around the Sun in the same direction, and almost all the planets and satellites also rotate in that same direction. Because of the conservation of angular momentum (which was described in Section 12.9), and since the original material from which

A sidereal day on Venus is 243 Earth days long.

the planets coalesced was undoubtedly rotating, we expect that its original angular momentum should now be divided up among the Sun and the planets. We thus expect all the planets to revolve and rotate in the same sense.

It is thus a problem to explain why Venus rotates "the wrong way." Nobody knows the answer. One possibility depends on the idea that when Venus was in the process of forming, the planetesimals formed clumps of different sizes. Perhaps the second largest clump struck the largest clump at such an angle as to cause the result to rotate backwards. Scientists do not like *ad hoc* ("for this special purpose") explanations like this one, because they have been constructed to explain specific situations and do not permit generalization. Nevertheless, that's all we can do in this case.

The slow rotation of the solid surface contrasts with the rapid rotation of the clouds. The clouds rotate in the same sense as the surface of Venus rotates but much more rapidly, once every 4 days.

The rotation of Uranus, as we shall see in Section 19.2, is also a problem.

15.3 THE TEMPERATURE OF VENUS

The surface of Venus can be detected from Earth with radio telescopes. Radio waves emitted by the surface penetrate the clouds. Just as we can determine the surface temperature of a star from studying the distribution of its radiation over optical wavelengths and by the strength of the emission, so can we also determine the temperature of the Cytherean surface by studying the radio emission. (The optical radiation doesn't pass through the clouds.) The surface is very hot about 750 K (475°C).

15.3a *The Greenhouse Effect*

In addition to directly measuring the temperature on Venus, we can

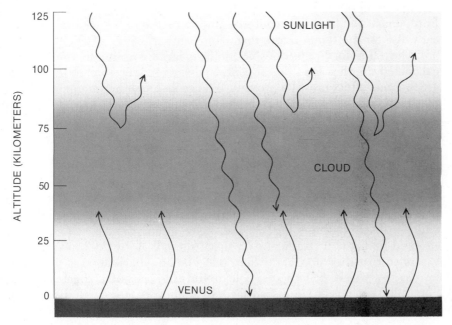

Figure 15–4 Sunlight can penetrate Venus' clouds, so the surface is illuminated with radiation in the visible part of the spectrum. Venus' own radiation is mostly in the infrared, and the presence of carbon dioxide in the atmosphere blocks the transmission of most of this radiation. This is called the greenhouse effect. Water vapor and sulfuric acid particles block the rest of the infrared. The Earth's atmosphere is not as opaque as is Venus' in the infrared, so the Earth's greenhouse effect is much smaller, about 35 K of heating.

calculate theoretically approximately what it should be if the Cytherean atmosphere did not impede the radiation. This value—less than 375 K (100°C)—is much lower than the measured values. The high temperatures derived from radio measurements indicate that energy does not leave Venus as readily as it leaves the Earth.

What accounts for this difference? What is happening on Venus is similar to the process that is generally—though incorrectly—thought to occur in greenhouses here on the Earth. The process is thus called the *greenhouse effect* (Fig. 15–4).

Sunlight passes through the Cytherean atmosphere in the form of radiation in the visible part of the spectrum. The sunlight is absorbed by, and so heats up, the surface of Venus. At the temperatures that result, the radiation given off is mostly in the infrared. But the carbon dioxide in Venus' atmosphere is opaque to infrared radiation, so the energy is trapped. Thus Venus heats up above the temperature it would have if the atmosphere were transparent; the surface radiates more and more energy. Understanding such processes involving the transfer of energy is but one of the practical results of the study of astronomy.

Greenhouses on Earth don't work quite this way. The closed glass of greenhouses on Earth prevents convection and the mixing in of cold outside air. The trapping of energy by the "greenhouse effect" is a less important process in an actual greenhouse. Try not to be bothered by the fact that the greenhouse in your backyard is not heated by the "greenhouse effect."

15.3b *Calculating the Temperatures of the Planets*

We know that the amount of radiation that Venus receives from the Sun must be in balance with the amount of radiation that Venus gives off; we say that the situation is in equilibrium (Fig. 15–5). Since the amount of radiation that an object gives off depends on the temperature of that object, we can calculate the amount of radiation that Venus would give off if it were at various temperatures and if it reacted to radiation in the same ways that the Earth does. Matching these output values with the input, we find a theoretical value for the temperature that Venus should have.

The value for Venus does not, in fact, agree with the theoretical value calculated in this method. It is this disagreement that tells us that a greenhouse effect must be present. We use a similar argument to calculate that a greenhouse effect may be present, for example, on Saturn's moon Titan, though the effect is not as dramatically large as it is for Venus.

15.4 RADAR MAPPING OF VENUS

We can study the surface of Venus by using radar to penetrate Venus' clouds. "Radar" is an acronym for **radio detection and ranging.** In this section, we discuss ground-based radar observations. Beams of radio signals we send from a radio telescope on Earth spread out by the time they reach Venus so that they are larger than that planet. The following section describes how we are nonetheless able to study the parts of the beams reflected from small regions on Venus' surface. In Section 15.7, we will discuss radar observations recently made by a satellite in orbit around Venus.

The carbon dioxide, plus water vapor and sulfuric acid droplets, makes the atmosphere of Venus opaque to infrared radiation and leads to a large greenhouse effect. The Earth's atmosphere is not so opaque in the infrared, and its greenhouse effect is much smaller, about 35 K.

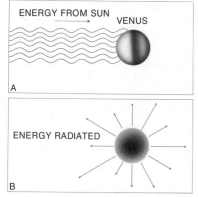

Figure 15–5 Venus receives energy from only one direction, and heats up and radiates energy (mostly in the infrared) in all directions. From balancing the energy input and output, astronomers can calculate what Venus' temperature would be if Venus had no atmosphere. This type of calculation is typical of those often made by astronomers.

Since light travels 300,000 km s⁻¹, and Venus' radius is only 6000 km, it takes only about 1/50 s between the time the radiation strikes the point on Venus nearest us and the time it reaches the edge we can see, which is halfway around. Thus the echo we receive back on Earth from a brief pulse we send out is spread out over a very brief time interval. It must be recorded very accurately on tape or in a computer to allow careful analysis of the variations of the signal.

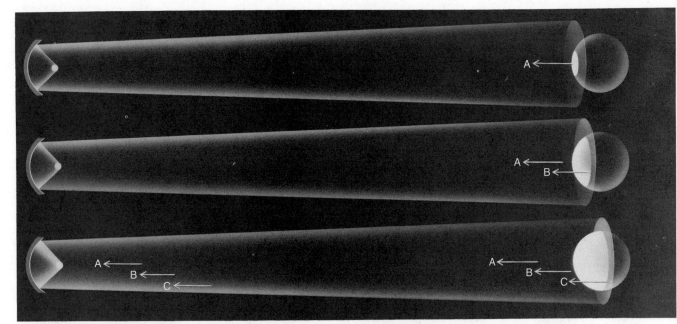

Figure 15–6 A pulse of radio waves sent from the telescope on Earth, shown at left, is reflected from Venus, shown on the right. The radar pulse first hits the point on Venus nearest to the Earth (A), and then spreads as time goes on in concentric circles (B and C). The reflected pulse is thus spread out over a longer time interval than was the transmitted pulse. By observing the time of arrival of each part of the reflected pulse back at the Earth, we can tell from which concentric circle that part of the returned pulse was reflected.

*15.4a   How Radar Mapping Works

Radar involves sending radio waves and studying the echoes that return. The point of Venus nearest Earth reflects the signal back a fraction of a second before points farther away (Fig. 15–6). Thus by following how the signal that returns differs from the signal that was sent out, aside from the obvious fact that the echo will be weaker than the original signal, we can tell something about the surface of Venus.

The radio waves we send out first hit Venus at the point of Venus that is nearest to the Earth. In the instants following, the incoming waves spread out in concentric circles around that point (Fig. 15–6). Thus the signals that return from Venus in following instants represent areas on different individual circles of the concentric circles we have drawn on the surface of Venus.

But this technique has narrowed down the region of Venus from which we receive a signal at a given time to a circle on the planet, and not to a specific point. We can make use of the planet's rotation to give us further information. Since Venus is rotating, one side of the planet is receding relative to the other. (We can ignore its orbital motion, which contributes equally to the Doppler shift of each side of the planet.) As a result of the rotation, the frequency of the signal received from one side of Venus has a Doppler shift that is different from the Doppler shift of the frequency of the signal from the other side (Fig. 15–7).

Thus the time analysis of the signal gives a concentric circle on Venus,

Figure 15–7 Everything the same distance from the axis of rotation of a planet, i.e., the surface of the cylinder, rotates at the same velocity. Our radar signal from Earth travels parallel to the plane shown at right in the direction shown by the arrow. Any signal reflected on the part of Venus that intersects the plane has the same Doppler shift. We see only the side of the circle shown that is facing us; from our vantage point it looks like a straight line. Thus by observing the signal at a given frequency, which has undergone a certain Doppler shift, we define a line from top to bottom across the surface of the planet. This line intersects the circle shown here and in Fig. 15–6 at the two marked points.

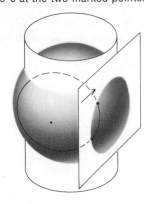

The 305-m dish at Arecibo in Puerto Rico (shown in Fig. 21–6), the 36-m "Haystack" dish in Massachusetts, and the 64-m dish at Goldstone, California, are equipped for radar. Using one of these large dishes together with smaller radio telescopes a few kilometers away as "interferometers" (which are discussed in Section 25.6) allows us to have the high resolution that a single telescope that same few kilometers across would have. With such high resolution, the ambiguity of two areas on the surface can be resolved.

and the Doppler analysis of the signal gives a straight line across the surface of Venus. The two intersect at two points (Fig. 15–7). Using two different radio telescopes together allows us to have a field of view small enough to distinguish between the two points, though still not as small as the regions distinguished by the radar technique. Once the two-point ambiguity is resolved, we know the position on the planet to which a given signal we receive corresponds. From the strength of the echo and a comparison of its variation with time to that of a sphere, we can tell the roughness of the region of Venus' surface reflecting the signal and the altitude of that region.

15.4b The Surface of Venus

The ground-based radar maps of Venus (Fig. 15–8) show large-scale surface features, some very rough and others relatively smooth, similar to the

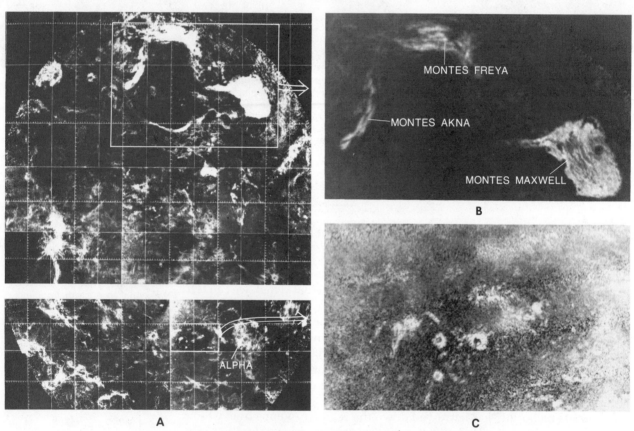

Figure 15–8 *(A)* A montage of radar maps showing one fourth of the surface of Venus. The observations were made from the Arecibo Observatory. Increasing brightness indicates increasing levels of reflected power. In general, contrast changes are determined by differences in the degree of roughness on a scale of a few centimeters. High inclinations of surface features can also give a bright return. The resolution varies over the image between 10 and 20 km. *(B)* The low reflectivity region Planum Lakshmi (a plain) and the high reflectivity region Montes Maxwell (mountains) make up most of a continental sized region, Terra Ishtar. This image has been enhanced to show details of the structure of the three mountains and has a resolution of 10 km. *(C)* The craters are 30 to 65 km in diameter. They are relatively smooth and are surrounded by rough areas of ejected material. The craters are probably the result of impacts on the surface, but a volcanic origin cannot be ruled out as yet. (Courtesy of D. B. Campbell, Arecibo Observatory)

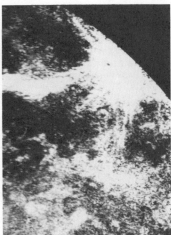

A

B

Figure 15–9 *(A)* An area 2400 km × 4600 km at an average resolution of 17 km, shows the Alpha region, the large bright region at the lower right. Whiter areas correspond to rougher terrain than the darker areas. Alpha is 1100 km across, but it is not clear from the ground-based radar to what geological type of structure it corresponds. The region at lower left shows many craters. *(B)* The large bright feature on the top of this ground-based radar image is the region Beta, which is about 10 km high. This image covers an area 2400 × 4600 km at an average resolution of 10 km. In Section 15.7 we will see how spacecraft radar observations have given us a more complete view. (Courtesy of D. B. Campbell, Arecibo Observatory)

variation we find of surfaces on the Moon. Many craters have been found, all relatively shallow, but ranging up to 1000 km across. Most have probably been formed by meteoritic impact.

From Venus' size and from the fact that its mean density is similar to that of the Earth, we conclude that its interior is also probably similar to that of the Earth. This means that we might expect to find volcanoes and mountains on Venus, and that venusquakes probably occur too. The ground-based radar results endorse this point of view. In Section 15.7 we will discuss some current radar observations from a space satellite.

The region Alpha, long known for its very high radar reflectivity, is circular and 1100 km across. It appears to have no counterpart on Earth. A central dark region may be a volcano. The region contains a very large number of roughly parallel ridges about 19 km apart. Some of the ridges can be traced for distances of hundreds of kilometers.

A huge peak, Beta, whose base is over 750 km in radius and which has a depression 40 km in radius at its summit has been detected by radar, and is probably a giant volcano (Fig. 15–9). It has long tongues of rough material extending as far as 480 km from it in an irregular fashion. A cluster of some 20 smaller peaks resembles the configuration of clusters of volcanoes on Earth. At present, we cannot tell whether any of the volcanoes are currently active. A long trough at Venus' equator, 1500 km in length, seems to resemble the Rift Valley in East Africa, the Earth's largest canyon. It may similarly result from movements of Venus' crust, just as the Rift Valley resulted from movements of the Earth's crust. Recent improvements have enabled radars to reach resolutions better than 20 km on Venus. This has led, for example, to the discovery of a large area that is probably a lava flow, which tells us that volcanic conditions have existed in the interior.

The quality of ground-based radar maps improves continually as the sensitivity and power of radar facilities improve.

*15.5 TRANSITS OF VENUS

Every so often, as seen from Earth, Venus passes in front of the surface of

Of course, only Mercury and Venus can have transits, as seen from the Earth. The other planets never pass between the Earth and the Sun.

The atmosphere of Venus at the transit was visible farther around the edge of the planet than the crescent extended.

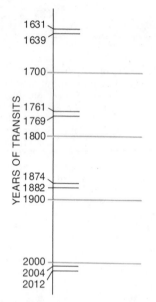

Figure 15–10 Transits of Venus.

the Sun. These *transits* of Venus are rare events, much rarer than transits of Mercury. Transits of Venus occur in pairs, the second following the first by 8 years. The pairs are spaced at very long intervals. Transits occurred in 1631 and 1639, in 1761 and 1769, and then in 1874 and 1882 (Fig. 15–10). Since Venus appears just as a small black dot crossing the face of the sun, as does Mercury when it transits across the Sun every few years, transits of Venus are not spectacular events to watch in the same way that eclipses are.

In times past, astronomers used transits of Mercury and Venus to get information about the size of the solar system (by triangulation). Venus' atmosphere, in addition, was discovered at the transit of 1761 when it appeared silhouetted against the Sun. International cooperation in scientific pursuits was fostered by the many expeditions to observe that transit. In hope of further success, Captain Cook sailed to Tahiti from England on one of many expeditions that observed the transit of 1769. Captain Cook mapped Australia and a few other places as bonuses along the way; it is not often that the side benefits of astronomical research are so apparent. Travelling off in a jet plane to see a solar eclipse anywhere in the world is certainly less arduous than was this lengthy expedition by ship.

Because of the results from space exploration, the transits of the future will be of little scientific importance. But nonetheless, when the next pair of transits comes in 2004 and 2012, their rarity will make them worth seeing.

15.6 SPACE OBSERVATIONS

Venus was an early target of both American and Soviet space missions (Table 15–1). Let us first discuss the basic results that have been found over the years. Then we shall go on to discuss our current picture of Venus more completely in the context of the set of Soviet and American spacecraft that reached Venus in 1978.

15.6a Early Spacecraft Observations

American spacecraft flew by Venus in 1962 and 1967. The Soviet spacecraft concentrated more on landings. The Soviet *Venera* are double and separate into two when they reach the vicinity of Venus. One part attempts a landing, and the other makes observations as it flies by. Each Venera lander radios information concerning temperatures and pressures back to Earth as it descends through the atmosphere of Venus. The spacecraft are designed to withstand the tremendous pressures expected at the base of Venus' atmosphere. Unfortunately, even so, the earlier Soviet spacecraft didn't transmit from the surface. Later, Venera 7 to 10 were designed to withstand the higher pressures that we now know are typical of Venus: 90 times that of Earth.

In 1970, the Venera 7 spacecraft radioed 23 minutes of data back from the surface of Venus. Two years later, the lander from Venera 8 survived on the surface of Venus for 50 minutes. They confirmed the ground-based results of high temperatures and found high pressures. They measured that the Cytherean atmosphere contains over 90 per cent carbon dioxide (CO_2). The fact that the concentration of carbon dioxide was high had been known from Earth-based observations.

The United States spacecraft Mariner 10 took thousands of photographs

TABLE 15–1   UNMANNED PROBES LAUNCHED TO VENUS

Spacecraft	Launched by	Arrival Date	Comments
Venera 1	USSR	1961	Failed en route
Mariner 2	USA	1962	Flyby
Venera 2	USSR	1966	Flyby
Venera 3	USSR	1966	Crash-landed
Venera 4	USSR	1967	Probed atmosphere
Mariner 5	USA	1967	Flyby
Venera 5 and 6	USSR	1969	Probed atmosphere
Venera 7	USSR	1970	23 minutes of operation on surface
Venera 8	USSR	1972	50 minutes of operation on surface
Mariner 10	USA	1974	Flyby
Venera 9	USSR	1975	53 minutes of operation on surface; photograph
Venera 10	USSR	1975	65 minutes of operation on surface; photograph
Pioneer Venus 1	USA	1978	Orbiter
Pioneer Venus 2	USA	1978	Probes dropped through atmosphere
Venera 11	USSR	1978	Lander (95 min of operation on surface) and flyby
Venera 12	USSR	1978	Lander (110 min of operation on surface) and flyby

of Venus in 1974 as it passed by en route to Mercury. It went sufficiently close and its imaging optics were good enough that it was able to study the structure in the clouds with resolution of up to 200 meters, about an eighth of a mile (Fig. 15–11). The structure shows only when viewed in ultraviolet light. We could observe much finer details from the spacecraft than we can see from the Earth.

The clouds appear as long, delicate streaks, as do cirrus clouds viewed in visible or ultraviolet light on Earth. In the Cytherean tropics the clouds also show a mottling, which suggests that convection, the boiling phenomenon, is going on. A big "eye" of convection is visible just downwind from the point at which the Sun is overhead and, therefore, at which its heating

No magnetic field has been detected. This indicates that either Venus does not have a liquid core or that the core does not have a high conductivity.

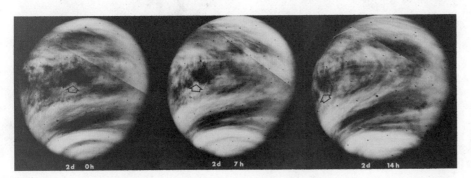

Figure 15–11   A series of photographs of the circulation of the clouds of Venus, photographed at 7-hour intervals from Mariner 10 in the ultraviolet near 3550 Å. The contrast has been electronically enhanced so that small actual differences in contrast are made apparent. The size of the feature indicated by the arrows is about 1000 km.

actions are at the maximum. Having an "eye" downwind from the subsolar point is a perfect example of what atmospheric scientists would expect, since hot air rises and starts the process of convection. Strong winds blow the clouds at these upper levels around the planet at 300 km hr^{-1}, as rapidly as the jet stream blows on Earth; the surface is rotating only very slowly. The raging winds and clouds observed on Venus stay at high levels above the surface, without much interaction with atmospheric regions below.

Studies of Venus like these have great practical value. The better we understand the interaction of solar heating, planetary rotation, and chemical composition in setting up an atmospheric circulation, the better we will understand our Earth's atmosphere. We then may be better able to predict the weather and discover jet routes that would aid air travel, for example. The potential savings of large amounts of money from this knowledge is enormous: it would be many times the investment we have made in planetary exploration.

15.6b Pictures of Venus' Surface

In 1975 Soviet scientists succeeded in landing a spacecraft on Venus and having it survive long enough to send back photographs. Venera 9 and Venera 10 each sent down a lander (Fig. 15–12). The single photograph that the Venera 9 lander took in the 53 minutes before it succumbed to the tremendous temperature and pressure showed a clear image of sharp-edged, angular rocks (Fig. 15–13). This came as a surprise to some scientists, who had thought erosion would be rapid in the dense Cytherean atmosphere, and that the rocks should therefore have become smooth or disintegrated into sand.

Three days later, the Venera 10 lander reached the surface and transmitted data for 65 minutes, including a photograph (Fig. 15–14). The landing site was 2000 kilometers distant from the Venera 9 site. The rocks in the Venera 10 site were not sharp; they resembled huge pancakes, and between them were sections of cooled lava or debris of weathered rock.

Thus we have two photographs of the surface of Venus, the first photographs ever taken on the surface of another planet. The one from Venera 9 looked typical of a "young mountainscape," in the words of Soviet scientists, with sharp rocks possibly ejected from volcanoes. On the other hand, the one from Venera 10 "showed as a landscape typical of old mountain formation."

Figure 15–12 A model of the Soviet Venera 9 lander.

Figure 15–13 The surface of Venus photographed from Venera 9. The photograph is reproduced in a flat strip, but was actually a scan from upper left, down toward the center, and up toward the right. Think of looking to your left at the horizon, then down at your feet, and then up to the horizon at the right. That is why the horizon is visible as tilted lines at upper left and at upper right.

Figure 15–14 The surface of Venus photographed from Venera 10.

The Veneras measured the surface wind velocity to be only 1 to 4 km hr⁻¹, although, since the atmosphere is much denser than that of the Earth, more material would hit a rock in a given time on Venus than on Earth at the same wind velocity. Nevertheless, the low wind speed we see makes it seem likely that erosion is caused not by sand blasting but by melting, temperature changes, chemical changes (which can be very efficient at the high temperature of Venus), and other mechanisms.

The Venera landers also made measurements of the soil, determining that its chemical composition and density correspond to that of a differentiated body, like the Earth, the Moon, and Mars. They measured temperatures of 760 K and 740 K at the two sites, and a pressure over 90 times that of the Earth's atmosphere, confirming the earlier measurements.

15.7 CURRENT VIEWS OF VENUS

1978 marked a new round of space exploration of Venus, with both the United States and the Soviet Union sending elaborate missions. Each launched a pair of spacecraft.

NASA's Pioneer Venus 1 (Fig. 15–15) was the first spacecraft to go into orbit around Venus, allowing it to make observations over a lengthy time period. Its elliptical orbit ranges between 400 km and 65,000 km in altitude.

Figure 15–15 The launch of NASA's Pioneer Venus 1.

Pioneer Venus 1's dozen experiments include cameras to study Venus' weather by photographing the planet regularly in ultraviolet light (Fig. 15–16 and the figure opening this chapter). Following its orbit over a period of months is allowing detailed determination of Venus' gravitational field.

Pioneer Venus 1 is carrying a small radar to study the topography of Venus' surface. The resolution is comparable to that of Earth-based radars, but the orbiting radar has the advantage of being able to map a wider area. The observations show that Venus has a wide variety of terrains, some similar to those on Earth and some resembling those on the Moon (Fig. 15–17 and Color Plate 35). Examples are both large volcanic areas and large highland regions. One highland region is equivalent in size and elevation to Africa and another mountain range to Australia. The giant canyon, 1500 km long, is 5 km deep and 400 km wide. It is the largest in the solar system. Venus, however, has much less continental area, respectively, than the Earth.

Aphrodite, the Greek goddess of love, was the equivalent of the Roman goddess Venus. Ishtar was the Babylonian goddess of love and war. Major features on Venus are being named after mythical goddesses, minor features after other mythical female figures, and still smaller circular features after famous women.

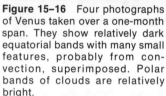

Figure 15–16 Four photographs of Venus taken over a one-month span. They show relatively dark equatorial bands with many small features, probably from convection, superimposed. Polar bands of clouds are relatively bright.

Many large craters, some 320 km in diameter, have been found (Fig. 15–18). Most have prominent central peaks and, similar to lunar craters, seem to be the result of meteoritic impacts.

The prominent radar feature known as Beta (Fig. 15–19) has a pair of gently-sloping mountains shaped like volcanoes. A giant mountain massif known as Maxwell (Fig. 15–20) is 2 km taller than Mt. Everest. The sum of the evidence seems to be that volcanic and mountain-building processes

Figure 15–17 An artist's conception of Venus' surface based on the Pioneer Venus Orbiter's radar map (Color Plate 35). Two continents exist: Aphrodite Terra *(foreground),* which is comparable in scale with Africa, and Ishtar Terra *(top),* which is comparable in scale with the continental United States or Australia. The volcano Maxwell, 10.8 km above Venusian "sea level" (the average radius of Venus), is on the right of Ishtar Terra and is the highest point on Venus. Beta Regio appears to be a huge double shield volcano, and is on the left edge of the disk we see in the picture. Sixty per cent of Venus' surface is covered with a huge rolling plain. Only about 16 per cent of Venus' surface is covered with lowlands, compared with over two thirds of Earth's surface covered with oceans. (NASA/Ames)

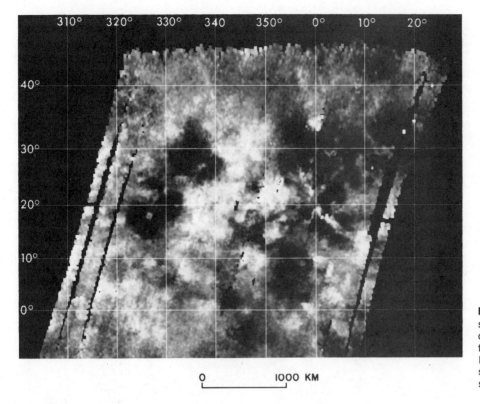

0   1000 KM

Figure 15–18 A radar image showing two impact craters with central peaks. It is unusual that these craters have been found in a lowland region because all other such cratered terrain in the solar system is in highlands.

similar to the processes on Earth, plus meteoritic impacts, which play a very minor role on Earth, have been important in shaping Venus' surface. But Venus' crust is apparently thicker than the Earth's, and has choked off further continent building.

Pioneer Venus 2 arrived at Venus at about the same time as Pioneer Venus 1. Pioneer Venus 2 was a "multiprobe," several spacecraft (Fig. 15–21) that travelled to the vicinity of Venus together and then separated. A basic cylinder was called the "bus"; a large probe and three smaller probes were mounted on it. They entered Venus' atmosphere and radioed data back as they plummeted to the surface at widely separate locations. The bus studied the composition of the upper atmosphere, and the probes were designed to survive the entire trip through the atmosphere in order to radio back data throughout the trip. They were not, however, designed to survive impact on the surface, though one did radio back data for an additional 67 minutes. Unfortunately, there were no special instruments on board to take proper advantage of this unexpected opportunity. One of the probes also measured that about 2 per cent of the sunlight reaching Venus filters down to the surface.

The probes confirmed that carbon dioxide makes up 96 per cent of the atmosphere. The probes found less water vapor but more sulfur dioxide than they had expected in Venus' lower atmosphere. This is particularly important since one of the possible flaws in the theory that Venus' atmosphere is heated so substantially by the greenhouse effect is that the presence of carbon dioxide alone is not enough to cause the greenhouse effect. The spectrum of carbon dioxide has gaps that would let out so much infrared

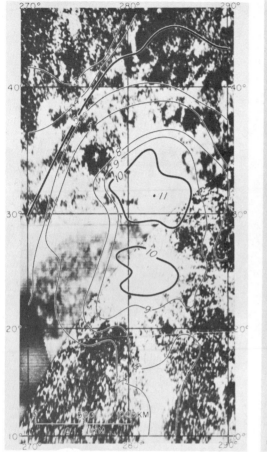

Figure 15–19

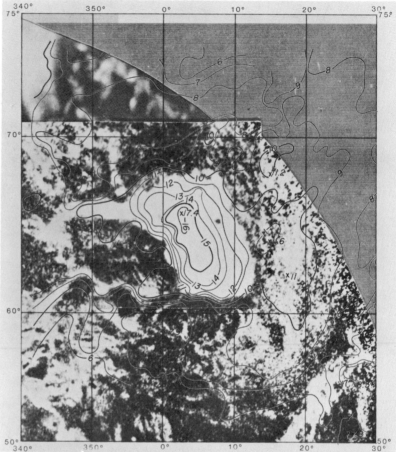

Figure 15–20

Figure 15–19 A radar image showing Beta, a region uplifted by volcanic action. It is larger than the whole Hawaiian Island chain. Contours of equal elevation are shown. The numbers are elevations in km above a base radius of 6045 km.

Figure 15–20 Maxwell Montes, observed by the Pioneer Venus radar, is the highest region of the northern continent, Ishtar Terra. Maxwell Montes rises high above the plain, Planum Lakshmi, that takes up the other end of the continent. The highest places on Maxwell Montes are taller than Earth's Mount Everest. Maxwell's slopes are steep and are covered with rocks.

radiation that the greenhouse effect would not take place. The spectra of water vapor or sulfur dioxide in sufficient amounts, however, can plug those gaps. Enough sulfur dioxide is now thought to be present for this purpose, which endorses the greenhouse effect theory.

Venus' clouds start 48 km above its surface, and extend upward about 30 km. This is much higher than terrestrial clouds, which rarely go above 10 km. The Pioneer probes detected three distinct layers of Cytherean clouds, separated from each other by regions of relatively low density. The lowest layer, relatively dense, is only about 2 km thick. Over it is the second layer, about 7 km thick. The uppermost layer does not have a sharp top.

Reactions among sulfur, hydrogen, and oxygen are important all through Venus' atmosphere. Sulfur dioxide, SO_2, is an important constituent of the remainder of the atmosphere below the clouds that is not carbon dioxide. The sulfuric acid, H_2SO_4, particles of which make up the clouds, comes from interactions of the same three elements.

The infrared sensing equipment aboard the Orbiter has revealed that the cloud tops are over 20°C warmer at the poles than at the equator (Fig. 15–22). Cloud-top temperatures are slightly higher on the night side than on the day side, which could mean that the cloud tops drop lower at night. There is a depression in the clouds, 3000 km wide, at the north pole and a 1100 km

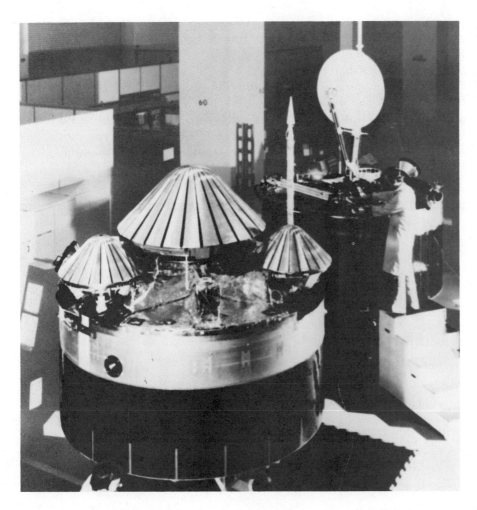

Figure 15–21 Pioneer Venus 1, the orbiter, is in the background and Pioneer Venus 2, which contains the basic bus, a large probe, and three smaller probes (of which one is hidden) is in the foreground.

Figure 15–22 An infrared polar view from the Pioneer Venus Orbiter. The north pole is at the center, and the outer circle represents 45° latitude. The subsolar point is at the bottom. Brighter regions correspond to warmer cloud-top temperatures in these measurements at a wavelength of 11 microns. Two bright "eyes" straddle the pole. A dark cold "collar" surrounds the pole. (NASA photo, courtesy of David J. Diner)

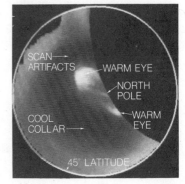

hole where there were few or no clouds. One of the probes penetrated near the north pole and found that below the clouds the temperature at the pole is cooler than the temperature at the equator measured by another probe. These phenomena are different from what we have on Earth. They seem to support the theory that on Venus the warm atmosphere at the equator rises, cools as it expands, moves to the poles and sinks. The atmosphere heats up as it compresses while sinking. Thus the lower atmosphere circulates from equator to pole. Above it, the upper atmosphere seems to circulate from pole to equator. On Earth, our planet's faster rotation stretches out the weather patterns.

The two Soviet missions each consisted of a flyby, which acted as a communications relay, and a lander. They reached Venus shortly after the American missions, and the landers parachuted softly to the surface. On the way, they measured the chemical composition of the atmosphere. One lander sent back data from the surface for almost two hours and the other for an hour and a half. The data indicated a temperature of 735 K. No pictures were sent.

Both American and Soviet missions reported lightning at a frequent rate. The Soviet landers detected up to 25 pulses of energy per second between 5 and 11 km above the surface, which explains the constant glow that was recorded by the two Pioneer probes that descended into Venus' night side. Some scientists, though, think that the glow is caused by the temperature of the surface rather than by lightning.

NASA is planning a Venus Orbiting Imaging Radar (VOIR) spacecraft to be launched by the Space Shuttle in 1986. It will circle in a near polar orbit at an altitude of 300 km and will map nearly the whole surface of Venus with 1 km resolution. It will also map a few selected areas at high resolution, 1/10 km. A joint Soviet-French probe carried around in Venus' atmosphere by a balloon is planned as well.

SUMMARY AND OUTLINE

Venus' atmosphere (Section 15.1)
  Mainly composed of carbon dioxide
  Clouds primarily composed of sulfuric acid droplets
A slow rotation period, in retrograde (Section 15.2)
The temperature of Venus (Section 15.3)
  A high temperature, 750 K, caused by the greenhouse effect
  Greenhouse effect: visible light in, turns to infrared, infrared can't escape
Radar mapping (Section 15.4)
  Radar maps show volcanoes and mountains
  Time study of radar echo gives concentric circles; Doppler study gives a vertical line; the two intersect in two points
Transits of Venus (Section 15.5)
Early space observations (Section 15.6)
  U.S. Mariner 10 observations of clouds and atmosphere
  Soviet Venera 9 and 10 landers: temperature measurements and photographs of the surface

Recent space exploration (Section 15.7)
  U.S. Pioneer Venus 1 and Pioneer Venus 2
    Orbiter studied Venus for many months
      Followed evolution of clouds
      Mapped surface with radar; further evidence of volcanoes
      Temperature maps show circulation of atmosphere
    Multiprobe contained bus and 3 probes
      Measured composition of atmosphere and temperatures
      Observed lightning
      Measured enough sulfur dioxide to confirm greenhouse theory
  U.S.S.R. Venera 11 and 12
    Radioed data from surface
    Observed temperature, lightning, composition of atmosphere

QUESTIONS

1. Make a table displaying the major similarities and differences between the Earth and Venus.

2. Why does Venus have more carbon dioxide in its atmosphere than does the Earth?

3. Why do we think that there have been significant external effects on the rotation of Venus?

4. If observers from another planet tried to gauge the rotation of the Earth by watching the clouds, what would they find?

5. Suppose a planet had an atmosphere that was opaque in the visible but transparent in the infrared. Describe how the effect of this type of atmosphere on the planet's temperature differs from the greenhouse effect.

6. Why do radar observations of Venus provide more data about the surface structure than a Mariner flyby?

7. If one removed all the CO_2 from the atmosphere of Venus, the pressure of the remaining constituents would be how many times the pressure of the Earth's atmosphere?

8. The Earth's magnetic field protects us from the solar wind. What does this tell you about whether or not the solar wind particles are charged?

9. Some scientists have argued that if the Earth had been slightly closer to the Sun, it would have turned out like Venus; outline the logic behind this conclusion.

10. Outline what you think Venus would be like if we sent an expedition to Venus that resulted in the loss of almost all of the CO_2 from its atmosphere.

†11. Estimate the angular size of Venus as viewed from Earth (at its closest approach). In a radar experiment, the radar pulse is spread out over a cone whose point (technically, apex) has an angle of 1 arc min. What fraction of this radiation strikes Venus?

†12. How big would the beam of a radio telescope or array of telescopes have to be to distinguish be-

tween the radar return from a point at 45° north latitude from a point at 45° south latitude on Venus?

13. Compare the ground-based and Orbiter-based radar observations of Venus.

14. Do radar observations of Venus study the surface or the clouds? Explain.

15. Besides the radar observations, briefly discuss three other types of observations of Venus.

---

†This indicates a question requiring a numerical solution.

CHAPTER 16

OUR EARTH

AIMS:
To describe and understand the Earth
on a planetary scale, and to realize
the limitations of our studies of other
planets from seeing how someone on
another planet would interpret
observations of the Earth

We have learned quite a bit about the solar system; some of our knowledge has come from comparing the properties of various planets with known properties of the Earth. On the other hand, the study of other planets and the Moon has taught us a lot about the history and evolution of the Earth. Elsewhere in this book, we discuss various features of the Earth as a basis for comparison. In this chapter, we will summarize some of our knowledge of the Earth's structure and history. Some of the concepts are also mentioned elsewhere.

16.1 THE VIEW FROM SPACE

Since it doesn't seem fair to know so much more about the Earth than the other planets simply because we live on it, let us consider our own planet as though it were being described by an inhabitant of Mars. Other material about the Earth—on continental drift and the composition of the atmosphere, for example—will be covered in later sections of this chapter.

REPORT OF THE MARTIAN ACADEMY OF SCIENCE

For centuries we have known Earth as an interesting object in our sky; sometimes it is the morning star and sometimes it is the evening star. Since it

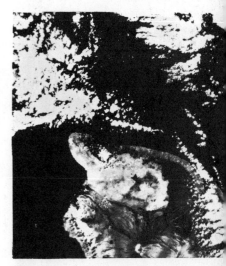

Figure 16–1 A negative print of Hawaii, showing the shield volcano Mauna Kea. Measured from the ocean floor, it is the highest mountain on Earth. (Everest rises from a high plain). The largest canyon-like feature on Earth is the Great Rift Valley in Africa.

The Earth, photographed from Apollo 17. The symbol for the Earth appears at the top of the page.

Figure 16-2 The Martian Terra 5 presumably landed in the ocean, shown here off the coast of Massachusetts and New Hampshire.

Figure 16-3 At first, the Martians sighted the Great Wall of China. This view over Chicago and Lake Michigan shows more distinct signs of civilization.

Figure 16-4 A drive-in movie must look strange to a visiting alien.

is an inner planet for us, Earth is always within a few hours of rising or setting, so we can never observe it high overhead in a dark sky and through the minimum amount of our atmosphere.

The major problem we have observing Earth is not our atmosphere but its own. Often major portions of Earth are covered with white clouds that prevent us from seeing the surface.

When we can see through the clouds, we see that Earth is mostly covered with a blue-greenish dark substance of much lower albedo than the clouds (as shown on the photograph on the facing page and on Color Plate 28). A smaller fraction of the surface is covered with lighter-colored material, and this lighter material changes in color somewhat with the seasons. As springtime comes to each hemisphere of Earth, the lighter material becomes greenish.

There is much less of the life-giving carbon dioxide on Earth than there is in our own atmosphere. But oxygen, a gas that is deadly to us, does exist by itself in Earth's atmosphere, both as O_2 and O_3. Earth has polar caps, one in the north and another in the south, and their sizes change with the season. Perhaps they are tremendous reservoirs of carbon dioxide in the form of ice (frozen CO_2), as are our polar caps. Some Martian scientists, but a minority, think that the polar caps on Earth might partially be made of frozen dihydrogen oxide (H_2O), which we call "wet ice," and are planning to take better spectra to study this possibility. They claim that their infrared measurements of Earth's temperature indicate that the planet is too warm for ordinary carbon dioxide ice.

Earth is accompanied by a remarkable moon, called Selene. Selene is not too much smaller than the planet Mercury, and ranks in size with the giant moons of Jupiter and Saturn. This is unusual because Earth is a martian planet rather than a giant planet, and planets ordinarily have moons that are only about one-thousandth their sizes. Selene, on the contrary, is approximately one-fourth the diameter of its planet, and as much as 1/81 of Earth's mass. The seasonal changes that we have detected on Earth have not been seen on Selene.

During the last few years, we have sent a series of rockets to Earth. The Terra 1 and Terra 2 did not succeed in traveling the long distance to Earth, but Terra 3 flew by at a distance of 8 billion centirams (remember that 1 ram is the length of the left antenna of Queen Schrip, who reigned from the year 15,363 to 16,437) and succeeded in getting a series of photographs from close up. They showed a planet that is mostly covered with maria, which correspond to the darker areas. Radar reflectivities indicate that they may be covered with dihydrogen oxide (H_2O). There are mountain areas and very few craters. The largest peaks (Fig. 16-1) are smaller than our own volcanoes, and the largest canyon, in a raised land mass that extends from slightly above the equator far into the southern hemisphere, is about the size of our own Great Canyon. Since Earth is a larger planet than ours, these canyons and mountains are smaller than ours with respect to the size of the planet.

Terra 4 went into orbit around Earth, and took a series of photographs over its 10 phobon lifetime (1 phobon, of course, is the length of time it takes for Phobos to orbit the Mars, and corresponds to about 1/100 of an Earth selenth). Four relatively smooth areas were chosen as prime landing sites.

Terra 5 attempted a crash landing on Earth last year, and succeeded in slowing its velocity as it passed through Earth's considerable atmosphere. But contact with it was lost a few seconds after landing; perhaps it was covered over by whatever material makes up the greenish areas (Fig. 16–2). Those of our scientists who say that this material is dihydrogen oxide claim that this mysterious disappearance supports their theory.

We have not been able to establish the presence of any intelligent life on the planet. Indeed, the presence of life would seem to depend on establishing that for Earth, as on the Mars, the seasonal blowing of dust that takes place provides shelter from solar radiation for individual Earthians, assuming that they, as we, cannot stand exposure to sunlight. We look for signs of Earthian work. Some of our intelligence analysts think that they have detected a long serpentine streak traversing one of the continents and some signs of a checkerboard pattern on a large scale, which could indicate the presence of agriculture. But these detections are marginal, and must be checked further. More recent observations seem to show traces of cities (Fig. 16–3).

A somewhat later look in the Martian archives might turn up the following ideas, as reported by Paul A. Weiss of the Rockefeller Institute.

A summary of our report is that we have discovered life on Earth! It is the discovery of the millennium for us Martians. From a height of 5 million rams, we could see streaks of light moving like waves across the landscape, often in two channels right next to each other and flowing in opposite directions. When we came closer, we saw that each luminous knot had an independent existence and had two white lights in front and two red lights behind. They were phototactic, attracted by a flickering light source. But rather than rush into the source, they stopped just short of it and formed a crystalline array. Then they remained immobile, perhaps sleeping. Their luminous activity diminished (Fig. 16–4).

Inside these Earthians, we eventually discovered even smaller objects, probably parasites (Fig. 16–5). They are lodged, for the most part, in the interior of Earthians. They never stray far off from the Earthians, even when they are disgorged, so they must be dependent on the Earthians for sustenance. The parasites are more numerous in larger hosts (Fig. 16–6); since host size reflects host age, the accessory bodies obviously multiply inside the hosts as the latter grow.

We can show that the Earthians are alive, for they metabolize by taking in sustenance sucked in through tubes inserted in their rears (Fig. 16–7). They give off wastes mostly as gas and smoke. Some areas of Earth are positively fouled with the waste products. We also noted signs of grooming, with a wiping motion, but only in front (Fig. 16–8). Astonishingly, this was started and stopped in synchrony by all the members of the population, which may imply that the Earthians have brains.

16.2 THE EARTH'S INTERIOR

The study of the Earth's interior and surface is called *geology*. Much of our knowledge of the structure of the Earth's interior comes from *seismology*. Geologists study how the Earth vibrates in various places as a result of large shocks, such as earthquakes. These vibrations travel through different types of material at different speeds. Also, when they strike the boundary between different types of material, these *seismic waves* are reflected and

Figure 16–5 If cars are considered to be the dominant life form on Earth, then the people would often be considered to live inside them.

Figure 16–6 Larger vehicles have more people inside them.

Figure 16–7 Cars have many of the characteristics that we consider to imply that something is alive.

Figure 16–8 A car grooming itself, or at least grooming its windshield.

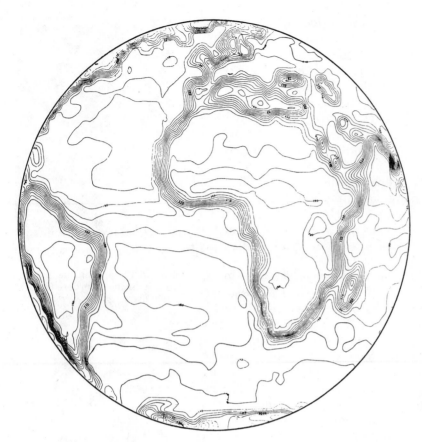

Figure 16–9 If we are to understand our observations of other planets, we must have a good idea of what astronomers on other planets, with similar equipment to ours, could observe about the Earth. This map shows how the Earth would appear from Venus, as viewed at a wavelength of 1 cm, with a telescope with a resolution of 1 arc sec. The contours represent lines of equal intensity of radio signal. The map is based on Nimbus 5 satellite observations that have been degraded in resolution (that is, reduced in quality) to simulate the effect of the distance from Venus to the Earth. Such observations might indicate that the Earth is a very interesting place for further study.

refracted, just as light is when it strikes glass. From piecing together seismological and other geological evidence, geologists have been able to develop a picture of the Earth's interior (Fig. 16–10).

The innermost region is called the *core*. It consists primarily of iron and nickel. The central part of the core may be solid, but the outer part is probably a very dense liquid. Outside the core is the *mantle,* and on top of the mantle is the thin outer layer called the *crust*. The upper mantle and crust together are called the *lithosphere*. The lithosphere is a rigid layer, and surrounds a more plastic zone that is partially melted.

How did such a layered structure develop? As the Earth formed there was a lot of debris around it, in the form of dust and rocks. It is likely that the young Earth was subject to constant bombardment from this debris. This heated up the surface to the point where it began to melt, producing lava. However, this process could not have been responsible for the heating of the interior of the Earth, because the rock out of which the Earth was made conducts heat very poorly. (We say its *thermal conductivity* is low.) Much of the heat for the Earth's interior came from energy released as the Earth formed from an accretion of particles.

Another source of heat for the interior is the *natural radioactivity* within the Earth. Certain isotopes are unstable, even though they are formed naturally. That is, they spontaneously undergo nuclear reactions and eventually form more stable isotopes. In these reactions, high energy particles such as electrons or alpha particles (helium nuclei) are given off. These

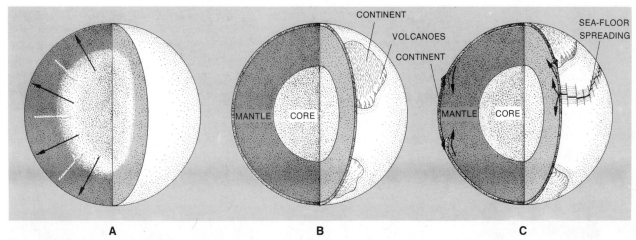

Figure 16–10 The structure of the Earth and stages in its evolution. *(A)* Tens of millions of years after its formation, radio-active elements, along with gravitational compression and impact of debris, produced melting and differentiation. Heavy materials sank inward and light materials floated outward. During this time, the original atmosphere (consisting mostly of hydrogen) was blown away by the solar wind. The atmosphere that replaced it contained methane, ammonia and water. *(B)* The heaviest materials form the core, and the lightest materials form the crust. *(C)* The crust has broken into rigid plates that carry the continents and move very slowly away from areas of sea-floor spreading. From "The Earth" by Raymond Siever. Copyright © 1975 by Scientific American, Inc. All rights reserved.

energy to these atoms. The rock becomes hotter. There was a sufficient amount of radioactive material in the Earth to cause a great deal of heating. Also, since the thermal conductivity of the material was low, the heat generated was trapped. As a result the interior got hotter and hotter.

After about a billion years, the Earth's interior had become so hot that the iron melted, and sank to the center, forming the core. Eventually other materials also melted. As the Earth eventually began to cool, various materials, because of their different densities and *freezing points* (the temperature at which they change from liquid to solid), solidified at different distances from the center. This process is called *differentiation*, and is responsible for the present layered structure of the Earth.

16.3 CONTINENTAL DRIFT

In the process of differentiation, most of the radioactive elements ended up in the outer layers of the Earth. Thus, there is, effectively, a heat source not far below the ground. This leads to a general *heat flow* outward through the upper mantle and crust. The power flowing through each square centimeter is 10 million times less than that necessary to light an average light bulb. The power reaching the Earth's surface from the interior is thousands of times less than that reaching it from the Sun. However, the terrestrial heat flow does have important geological consequences.

In some geologically active areas (Fig. 16–11), the heat flow rate is much higher than average, which indicates that the source of heat is close to the surface. Some scientists feel that the outflowing *geothermal energy* in these regions can be tapped as an energy source.

An important consequence of the heating just below the crust is that the crust sits on top of a hot layer. The rock in this zone is not hot enough to melt completely, but the material becomes soft, and behaves in a plastic fashion.

Figure 16–11 Thermal activity beneath the Earth's surface results in geysers. The thermal area shown here is in Rotorua, New Zealand.

The lithosphere, a cooler, more rigid layer, made up of crust and uppermost mantle, actually floats on top of this plastic layer. This rigid layer is segmented into *plates*, thousands of kilometers in extent but only about 50 km thick. The hot material beneath the rigid plates is being churned very slowly in convective motions, so the plates are carried around over the surface of the Earth. This theory is called *continental drift* or *plate tectonics* because the continents sit on top of the plates.

The boundaries between the plates are geologically active areas. Therefore, these boundaries are traced out by the regions where earthquakes

Figure 16–12 This plot of all earthquakes from 1963 through 1973 greater than 4.5 on the Richter scale shows that earthquakes occur preferentially at plate boundaries. The six principal tectonic plates are labelled, and arrows show whether the plates are converging or diverging.

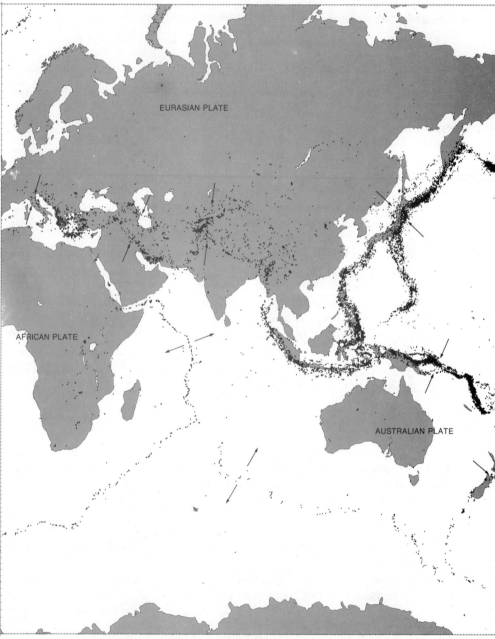

occur, as illustrated in Figure 16–12. The boundaries where two plates are moving apart mark regions where molten material is being pushed up from the hotter interior to the surface. The most famous example of this is the *mid-Atlantic ridge* (Fig. 16–13). Molten material is being forced up and is being deposited as lava flows on either side, producing new sea floor. The motion of the plates is also responsible for the formation of the great mountain ranges. For example, the Indian subcontinent is thought to have collided with the Asian continent, producing the Himalayas.

The theory of continental drift, once in disrepute, is now generally

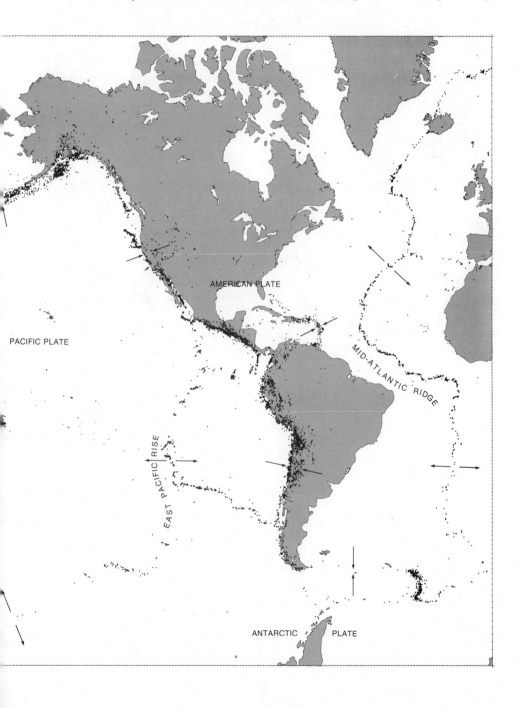

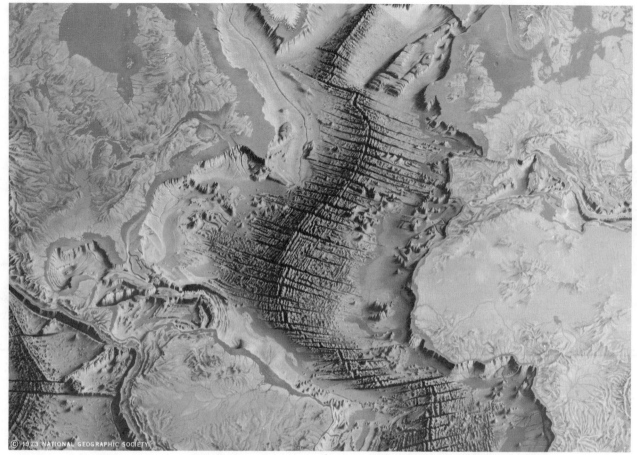

Figure 16–13 This map of the Atlantic ocean floor shows how the continents sit on the plates. North and South America are at the left; Europe and Africa are at the right. The feature running from north to south in the middle of the ocean is the *Mid-Atlantic Ridge*. It marks the boundary between plates that are moving apart.

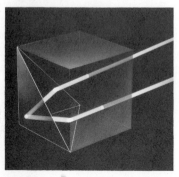

Figure 16–14 A cube, coated on the inside with reflecting material, has the property of reflecting an incoming light ray back in exactly the same direction from which it came, no matter what that direction was.

accepted. The continents, in this theory, were once connected as two supercontinents, one called Gondwanaland (after a province of India of geologic interest) and a northern supercontinent called Laurasia. (These may, in turn, have once separated from a single supercontinent called Pangaea, which means "all lands.") Over the millennia, they have moved apart as plates have separated. We can see from the shapes how they originally fit together, and verify this by finding similar fossils and rock types along two opposite coastlines, once adjacent, but now widely separated. In the future, we expect California to separate from the rest of the United States, Australia to be linked to Asia, and the Italian "boot" to disappear.

We can now measure continental drift directly by shooting lasers from positions on different plates at special corner reflectors (Fig. 16–14) left on the moon by the astronauts or carried in Earth orbit by satellites (Fig. 16–15). These corner reflectors return light beams that hit them directly back in the direction from which the beams came. Thus by timing the interval between sending the laser pulse from Earth and receiving the echo, we can accurately tell the distance to the corner reflectors. One of the satellites carries a plaque showing continental drift (Fig. 16–16).

A

B

Figure 16–15 *(A)* The corner reflectors left on the Moon by the Apollo astronauts. *(B)* A satellite carrying corner reflectors into Earth orbit to aid in accurate positioning of locations on Earth.

16.4 TIDES

It has long been accepted that the phenomenon of tides is most directly associated with the Moon. This is because the tides each day are about an hour later than the same tides the previous day, in keeping with the fact that the Moon transits about 53 minutes later each day. Tides are a result of the fact that the force of gravity exerted by the Moon (or any other body) gets weaker as you get farther away from it.

Tides depend on the difference between the gravitational attraction at different points. The forces that cause tides are often called *differential forces*, since it is the difference that counts.

To explain the tides in Earth's oceans, we will assume, for simplicity, that the Earth is completely covered by a layer of water. In the most naive approach, we might say that the water closest to the Moon is attracted toward the Moon with the greatest force and is thus raised the greatest height above the Earth's surface. Thus high tides would occur at a given location on Earth as this location passed closest to the Moon as the Earth rotated. If this picture were all there were to the case, high tides would occur about once a day. However, we know that there are two high tides a day, separated by roughly 12½ hours.

To see how we get two high tides, we must look at the situation in a little

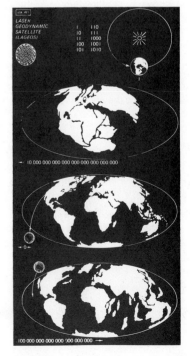

Figure 16–16 A NASA satellite bearing 426 retroreflectors was launched into a circular Earth orbit in 1976. It is being used to provide information on the Earth's rotation and crustal movement. The satellite, called LAGEOS (Laser Geodynamic Satellite), is expected to survive in orbit for 8 million years. In case the satellite is discovered millions of years hence, it bears a series of 3 views *(bottom)* of the continents in their locations 200 million years in the past, at present, and 200 million years in the future, according to the theory of continental drift.

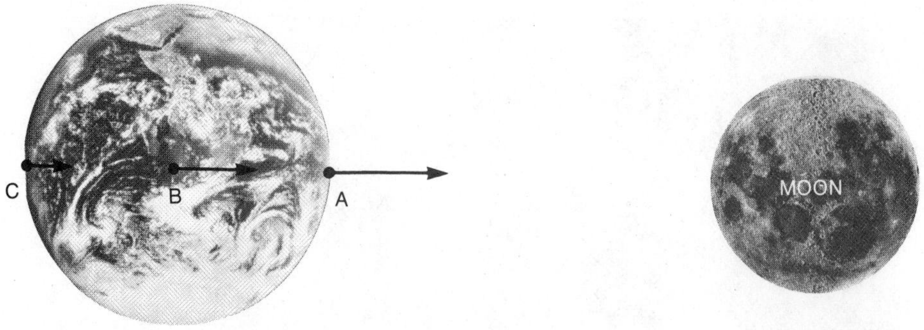

Figure 16–17 A schematic representation of the tidal effects caused by the Moon. The arrows represent the acceleration of each point that results from the gravitational pull of the Moon. The nearest point has the greatest acceleration. (Actually, the differences in acceleration are much smaller than indicated here.)

We have discussed such differential forces in more detail in Section 9.8 in the context of the evolution of binary stars and will discuss them again in Section 19.1, where we will see that tidal forces cause the rings of Saturn, Jupiter, and Uranus.

Figure 16–18 Mont St. Michel is shown here at low tide, when it is connected to the mainland by dry land, and the fields shown here are available for grazing. At high tide these fields are covered by water, and Mont St. Michel is an island. The abbey there was founded in 708, and the Gothic buildings date from the thirteenth century.

As the result of the friction of the tides, the Earth's rotation is slowing down slightly; that is, the day is growing slightly longer. Because angular momentum is conserved, the Moon, as a result, is receding slowly from the Earth.

more detail, as depicted in Figure 16–17. We consider three points, A, B, and C, where B represents the solid Earth, and A and C represent parts of the ocean on opposite sides of the Earth. We will take point A to be closest to the Moon. Since the Moon's gravity weakens as we get farther away from it, the pull produced by the Moon's gravity is greatest for point A and least for point C. If the Earth and Moon were not in orbit about each other, all these points would fall toward the Moon. However, since the closest one has the greatest acceleration, the three points would tend to move apart as they fall.

Now we must recall that the Earth and Moon are in orbit about each other. (We can think, loosely, that the Moon is in orbit around the Earth.) For orbital motion, the orbiting object actually is falling. However, as it falls, it also moves over by an amount great enough so that the distance from the center of the orbit doesn't change (if the orbit is circular). Essentially, the Moon is falling all the time, but the Earth curves away at the same rate that the Moon falls, so the Moon turns out to be orbiting.

Now let us return to our three points. They will still experience different accelerations, and the tendency to move apart will still be there, even though we are taking into account the fact that the Moon is orbiting the Earth. Thus we see that the high tide on the side of the Earth that is near the Moon is a result of the water being pulled away from the Earth. However, the high tide on the opposite side of the Earth results from the Earth being pulled away from the water.

The Sun also has an effect on the Earth's tides, but it is only about half as much as that of the Moon. Though the Sun exerts a greater force on the Earth than does the Moon, the Sun is so far away that its force does not change very much from one side of the Earth to the other. At a time of new or full Moon, the tidal effects of the Sun and Moon are in the same direction, and we have very high tides, called *spring tides*. At the time of the first and last quarter Moon, the effects do not add, which gives rise to the smaller *neap tides*.

Though this explanation works for the overall nature of tides, the details of tides at individual locations on the shore also depend on such factors as the variation of the depth of the ocean floor or the shape of the channel. The difference between high and low tide is some locations is spectacular (Fig. 16–18).

16.5 THE EARTH'S ATMOSPHERE

The pressure in the atmosphere (Fig. 16–19) falls off sharply with height above the Earth's surface. We define the pressure at the surface of the Earth to be *one atmosphere*. By expressing the pressure in atmospheres, we are really taking the ratio of the pressure at any level to that at the surface.

The temperature (Fig. 16–20) varies less regularly with altitude.

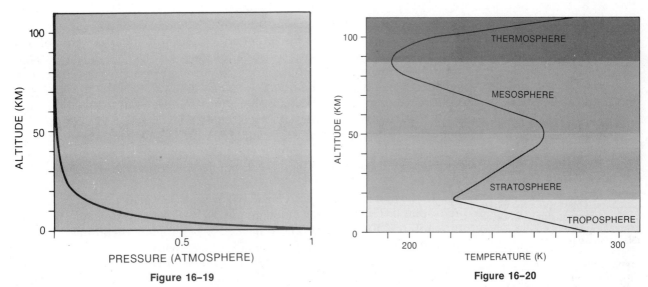

Figure 16–19

Figure 16–20

It is convenient to divide the atmosphere into layers (Fig. 16–21), according to the composition and the physical processes that determine the temperature.

Our common experience, and all of the Earth's weather, is confined to the very thin *troposphere*. At the top of the troposphere, the pressure is only about 10 per cent of its value at the ground. A major source of heat for the troposphere is infrared radiation from the ground, which is, in turn, heated by visible radiation from the Sun. Very close to the ground, most of the energy transport is by convection, the rising of warm air. In any case, the source of heat for the troposphere is the ground.

Above the troposphere are the *stratosphere* and the *mesosphere*. The upper stratosphere and lower mesosphere contain the *ozone layer*. (Ozone is a molecule, consisting of three oxygen atoms, that absorbs ultraviolet radiation from the Sun.) The absorption of ultraviolet radiation by the ozone means that these layers get much of their energy directly from the Sun. The temperature is thus higher than that at the top of the troposphere.

Above the mesosphere is the *ionosphere*, where many of the atoms are ionized. This is where the most energetic photons from the Sun (such as

Figure 16–19 Pressure in the Earth's atmosphere as a function of altitude. Note how rapidly it decreases.

Figure 16–20 Temperature in the atmosphere as a function of altitude. In the troposphere, the energy source is the ground, so the temperature falls off with altitude. Higher temperatures in other layers result from direct absorption of solar ultraviolet and shorter wavelength radiation.

We will talk more about ozone and its importance for life on Earth in the next section.

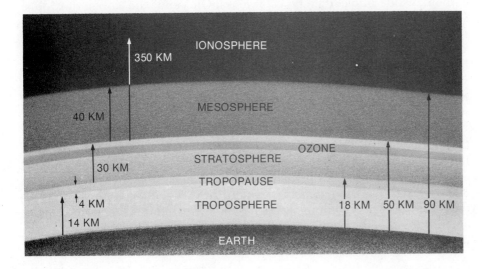

Figure 16–21 The layers of the Earth's atmosphere showing the location of the part of the stratosphere where ozone is formed. Once formed, the ozone is transported to lower stratospheric layers.

Figure 16–22 Much of our information about the atmosphere comes from satellites; we even see such pictures on our nightly television weather programs. This cloud cover photograph was made from a satellite of the National Oceanic and Atmospheric Administration (NOAA); outlines of the states and coastlines have been added. Hurricane Belle is visible in this photograph, which was taken on August 9, 1976.

ionized. This is where the most energetic photons from the Sun (such as x-rays) are absorbed. Thus the temperature gets quite high. Because of this rising temperature, this layer is also called the *thermosphere*. Because of the ionization, the ionosphere contains many free electrons. These electrons reflect very long wavelength radio signals. When the conditions are right, radio waves bounce off the ionosphere, and this allows us to tune in distant radio stations. When solar activity is high, there may be so much of this bouncing, called "skip," that CB radios may pick up so many distant stations as to make local callers inaudible. There have been proposals to open additional CB bands in regions of the radio spectrum where "skip" would occur less.

In the past several years, considerable effort has gone into studies of our atmosphere. This has been aided greatly by observations from above, made by satellites (Fig. 16–22). There have also been many theoretical studies. The equations that tell us how the atmosphere will behave are essentially the same as the equations of structure for stars, except that the sources of energy are different. These equations are solved in large computers, often the same ones used to solve the stellar structure problems, in an effort to make a model of how the atmosphere changes over a period of weeks and thus predict the weather. The rotation of the Earth also has a very important effect in determining how the winds blow. Comparison of the circulation of winds on the Earth, which rotates in 1 Earth day, on the slowly-rotating Venus, which rotates in 243 Earth days, and on rapidly-rotating Jupiter, which rotates in 11 Earth hours, helps us understand the weather on Earth.

16.6 OF WATER AND LIFE

Why is it that there is so much water on Earth, in our atmosphere, and in our oceans? The water seems to be necessary for life. How have features that have made Earth a hospitable planet for life as we know it made the other planets so unattractive? Can we make sure we are not changing the Earth's atmosphere so that we will wind up with an atmosphere like that of, say, Venus, which has clouds of sulfuric acid drops?

On the Earth, the origin of life changed the future course of our planet. Life forms produced much of the oxygen that we now have.

The water on the Earth has acted as a catalyst to spur the fixation of carbon dioxide into certain types of rocks. On Venus the carbon dioxide is still in the atmosphere—in fact, it makes up over 95 per cent of the atmosphere. The carbon dioxide is the major cause of trapping solar energy in Venus' atmosphere, thus keeping the surface of Venus very hot, too hot for life. Apparently water and plant life have saved the Earth from suffering the same fate as Venus. Perhaps if life had arisen on Venus at the same time that it arose on Earth (Fig. 16–23), it would have modified Venus' atmosphere as it did the terrestrial one.

Ozone (O_3) plays a very vital role on Earth: it keeps out ultraviolet radiation from the Sun. The part of the ultraviolet that has wavelengths slightly longer than 3000 Å, where the Earth's atmosphere becomes transparent, causes suntanning at the beach. None of the wavelengths shorter than 3000 Å pass through the ozone. If they did, they would kill us. Most scientists think that amounts of ultraviolet radiation even slightly increased over the amount we already receive, whether longer or shorter than 3000 Å, would increase the incidence of skin cancer.

Scientists have recently been concerned about the way that our technology may threaten the ozone layer (Fig. 16–21) of our Earth. The problem surfaced in an analysis of the effect of supersonic airplanes, which fly in the stratosphere. Airplanes exhaust nitrous oxides, and if the nitrous oxides could migrate to the upper stratosphere, they would destroy some of the ozone. This would allow more ultraviolet radiation to reach the Earth's surface. Ordinary jets don't fly as high as the stratosphere, but supersonic transports (SST's) do. There was concern that a fleet of SST's in commercial service, putting out nitrous oxides for half-a-dozen hours a day, could seriously deplete the ozone. This possibility was one of the major reasons why the American SST program was terminated. (An interesting sidelight is that more recent work shows that SST's might even increase the amount of ozone.)

It was later realized that a more serious problem was the gases that are used as refrigerants in refrigerators and air conditioners and were used as propellants in some aerosol cans. There is evidence these gases are accumulating in the stratosphere and depleting the ozone. These gases were originally chosen for aerosol use because they were thought to be inert, that is, non-interacting. However, it now seems that they are not as inert as they originally had been thought to be. Once they are released into the atmosphere, the propellant gases (such as some types of Freon, a commonly used trade name for fluorocarbons) eventually break down, and the individual chlorine atoms can transform the ozone into other molecules. The chlorine returns to its atomic state after breaking down the ozone (chemically, it acts as a catalyst), and so each chlorine atom can wreak havoc on the ozone for years. We may already have put enough gases into the atmosphere to make a permanent change in the ozone layer. Spacecraft are measuring the chlorine content of the upper atmosphere in order to improve our assessment of the situation, radio astronomers are measuring molecules in the atmosphere, and other astronomical methods are also being applied. Chemists are calculating the cycle of reactions in more detail. The effect seems to be even more serious than we had first realized. The use of Freons in aerosol cans seems non-essential and has been limited. "Essential uses," such as refrigeration and industrial processing of metals, are still permitted. But even though the United States has limited its use of these gases, the rest of the world hasn't very much. The U.S. now makes less than half the worldwide supply.

One lesson to learn from all these considerations is that we can be glad that the basic scientific research is being done, because it exposes the possible side effects of ordinary things we otherwise take for granted: fast airplanes and cans of deodorant. It's easier to say "go out and design an airplane that can go 2000 miles per hour" than to be sure you have thought of all the possible side effects that the decision might have on everything else on Earth. The dangers from refrigerants and aerosol propellants were understood in part because analyses were going on to understand chlorine compounds in Venus' atmosphere. Studies of the atmospheres of other planets make us appreciate the fundamental benefits of our own atmosphere. With the new technology that is continuing to be developed, we must make certain that basic scientific research goes forward on all fronts so that we will be prepared in the future for whatever problems might arise. Research limited to applied problems won't be enough.

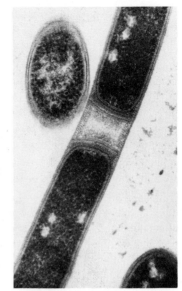

Figure 16–23 Microscopic views of two methanospirillium cells, a type of methane-consuming cell that may be modern examples of some of the most primitive life that arose on Earth billions of years ago.

"Basic scientific research" is research that is fundamental in nature; where it might be applied may not be obvious when the research is done. Whatever applications may be made of basic research often follow many years later. Work in astronomy usually qualifies as "basic research."

16.7   RADIATION BELTS

In January 1958, the first American space satellite carried aloft, among other things, a device to search for charged particles that might be orbiting the Earth. This device, which was under the direction of James A. Van Allen of Iowa State University, detected a region filled with charged particles of high energies. (We now know that there are actually two such regions surrounding the Earth, like a small and a large doughnut, as is shown in Figure 16–24.) The boundaries of the *Van Allen belts* are not sharp, but the inner one extends from about 1000 km above the surface of the Earth upward to a distance of about 1 Earth radius. The outer belt extends from about 3 to 5 Earth radii.

These particles are trapped by the Earth's magnetic field. Charged particles can only move in the direction of magnetic field lines, and not across the field lines. Usually a charged particle will follow a path that spirals around a magnetic field line. As a particle moves from above the equator, toward one of the magnetic poles, the field gets stronger, so the field lines get closer together. Eventually as the field lines converge on the poles, they make considerable angles with each other (though they never cross). When this happens the particle can travel no farther, since otherwise it would cross a neighboring field line. The magnetic force stops the particle and makes it go back in the other direction. This is known as a *magnetic mirror*. It is one of the leading methods that scientists are trying to use in terrestrial laboratories to keep a hot plasma (gas of charged particles) trapped in one place long enough to start fusion. Controlled fusion may one day provide much of our energy production, though major technical problems still remain to be solved.

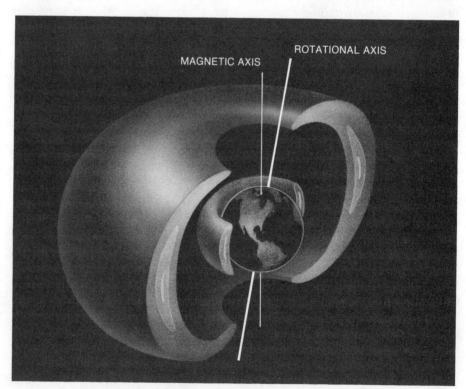

Figure 16–24   The Van Allen radiation belts. The smaller and larger belts are actually doughnut-shaped regions of charged particles trapped by the Earth's magnetic field.

OUTLINE AND SUMMARY

The Earth as seen from another planet (Section 16.1)
   Terrain and atmosphere
   Signs of life difficult to detect from afar
   We must be careful about interpreting extraterrestrial
      observations in terms of our own terrestrial expe-
      rience.
Structure of Earth's interior (Section 16.2)
   Core: iron and nickel most likely in a dense liquid
      state
   Mantle and crust: together called lithosphere
   Natural radioactivity within Earth is source of heat for
      Earth's interior
   Layered structure of Earth due to differentiation
Continental drift (Section 16.3)
   Geothermal energy as a result of heat flow from active
      areas
   Continents sit on top of slowly moving plates; thus
      the study of continental drift is also called plate
      tectonics
Tides (Section 16.4)

   Differential forces are those that cause tides
   Spring tides and neap tides
Earth's atmosphere (Section 16.5)
   All weather confined to thin troposphere
   Ozone layer between stratosphere and mesosphere
   Ionosphere (also called thermosphere) reflects radio
      waves
Water and life (Section 16.6)
   Water on Earth spurred fixation of CO_2 into certain
      types of rocks
   Life on Earth altered our atmosphere, creating O_2 and
      O_3 (ozone), which protects us from harmful ul-
      traviolet radiation
   Detrimental effect of chloro-fluorocarbons on ozone
      layer
Radiation belts (Section 16.7)
   Discovered by James Van Allen from data obtained
      from first American satellite
   Composed of charged particles trapped by Earth's
      magnetic field

QUESTIONS

†1. (a) Imagine that you are observing the Earth from an altitude of 100 kilometers. Estimate the angular size of a house, a football field, and an average size city. What features do you think would point to the existence of intelligent life? (b) What about from 1000 km up?

2. Consult an atlas and compare the size of the Grand Canyon in Arizona with the Rift Valley in Africa. How do they compare in size with the giant canyon on Mars, which is about the diameter of the United States?

3. Plan a set of experiments or observations that you, as a Martian scientist, would have an unmanned spacecraft carry out on Earth. What data would your spacecraft radio back if it landed in a corn field? In the Sahara? In the Antarctic? In Times Square?

4. Of the following in the Earth, which is the densest? (a) crust; (b) mantle; (c) lithosphere; (d) core.

5. (a) Explain the origins of tides. (b) Discuss the tides if the Moon were twice as far away from the Earth as it actually is.

6. What is the source of most of the radiation that heats the troposphere?

7. Look at a globe and make a list of which pieces of the various continents probably lined up with each other before the continents drifted apart.

8. Draw a diagram showing the positions of the Earth, Moon, and Sun at a time when there is the least difference between high and low tides.

†9. At what height above the Earth's surface is the pressure 1 per cent of its value at the ground?

10. How does the temperature vary with height in the troposphere? In the stratosphere?

11. Why are we worried about fleets of SST's? How do they differ from the existing fleets of normal jets as far as ozone is concerned?

---

†This indicates a question requiring a numerical solution.

CHAPTER 17

═══ MARS ═══

AIMS:
To discuss the atmosphere and other
features of Mars, to see how spacecraft
in orbit and on the surface have given
new results, and to assess the chances
of finding life there

Mars has long been the planet of greatest interest to scientists and non-scientists alike. Its interesting appearance as a reddish object in the night sky and some of the past scientific studies that have been carried out have made Mars the prime object of speculation as to whether or not extra-terrestrial life exists there.

In 1877, the Italian astronomer Giovanni Schiaparelli published the results of a long series of visual telescopic observations he had made of Mars. He reported that he had seen *canali* on the surface. When this Italian word for "channels" was improperly translated into "canals," which seemed to connote that they were dug by intelligent life, public interest in Mars increased.

We now know that the channels or canals Schiaparelli and other observers reported are not present on Mars—the positions of the *canali* do not even always overlap the spots and markings that are actually on the Martian surface (Fig. 17–1). But hope of finding life in the solar system springs eternal, and the latest studies have indicated the presence of considerable quantities of liquid water in Mars' past, a fact that leads many astronomers to hope that life could have formed during those periods.

Figure 17–1 A drawing of Mars and a photograph, both made at the close approach of 1926. We now realize that the "canals" seen in the past do not usually correspond to real surface features.

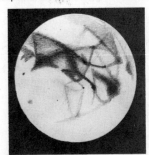

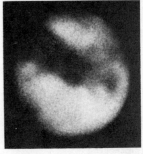

Taken by Viking Orbiter 2 as it approached the dawn side of Mars, in early August 1976. At the top, with water ice cloud plumes on its western flank, is Ascreaus Mons, one of the giant Martian volcanoes. In the middle is the great rift canyon called Valles Marineris, and near the bottom is the large, frosty crater basin called Argyre. The south pole is at the bottom. The symbol for Mars appears at the top of this page.

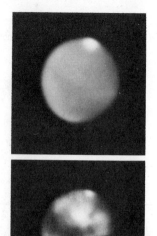

Figure 17–2 Photographs of Mars in violet *(top)* and in infrared *(bottom)* light. Mars has much less surface contrast in the violet.

Mars' orbit, about 1.5 A.U. in radius, has an eccentricity of 9 per cent, so at some of its oppositions it is closer to the Earth than at others. Obviously, these more favorable oppositions are better for observing, because at these times the disk of Mars appears greater in angular size. It varies from 14 to 25 seconds of arc at opposition, from about 1/130th to 1/60th the size of the full Moon. At the opposition of September 1988, Mars will be especially close to Earth, only 59 million km away.

17.1 CHARACTERISTICS OF MARS

Mars is a small planet, 6800 km across, which is only about half the diameter of Earth or Venus, although one-and-a-half times that of Mercury. Mars' atmosphere is thin—at the surface its pressure is only 1 per cent of the surface pressure of Earth's atmosphere—but it might be sufficient for certain kinds of life.

From the Earth, we have relatively little trouble in seeing through the Martian atmosphere to inspect the planet's surface (Fig. 17–2), except when a Martian dust storm is raging. At other times, we are limited mainly by the turbulence in our own thicker atmosphere, which causes unsteadiness in the images and limits our resolution to 60 km at best (in other words, we cannot see detail smaller than 60 km even under the best observing conditions).

Unlike the orbits of Mercury or Venus, the orbit of Mars is outside the Earth's, so it is much easier to observe in the night sky (Fig. 17–3). We can simply observe it in the direction opposite from that of the Sun. We are then looking at the lighted face. When three celestial bodies are in a line, the alignment is called a *syzygy*, and the particular case when the planet is on the opposite side of the Earth from the Sun is called an *opposition*. These oppositions occur at intervals of 26 Earth months on the average.

Mars revolves around the Sun in 23 Earth months. The axis of its rotation is tipped at a 25° angle from the plane of its orbit, nearly the same as the Earth's 23½° tilt. Because the tilt causes the seasons, we know that Mars goes through a year with four seasons just as the Earth does.

From the Earth, we have long watched the effect of the seasons on Mars. In the Martian winter, in a given hemisphere, there is a polar cap. As the Martian spring comes to the northern hemisphere, for example, the north polar cap shrinks and material at more temperate zones darkens. The surface of Mars is always mainly reddish, with darker gray areas that appear blue-green for physiological reasons of color contrast (see Color Plate 29). In the spring, the darker regions spread. Half a Martian year later, the same thing happens in the southern hemisphere.

One possible explanation that previously seemed obvious for these changes is biological; Martian vegetation could be blooming or spreading in the spring. But there are other explanations, too. The theory that presently seems most reasonable is that each year at the end of southern-hemisphere springtime, a global dust storm starts, covering the entire surface with

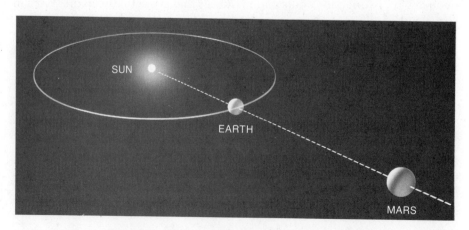

Figure 17–3 When Mars is at opposition, it is high in our night-time sky.

light dust. Then the winds get as high as hundreds of kilometers per hour and blow fine, light-colored dust off the slopes. This exposes the dark areas underneath. Gradually over the next Martian year, dust is also stripped away in certain places, such as near craters and other obstacles. Finally, a global dust storm starts up again to renew the cycle. The reddish color probably comes from iron oxide (rust) mixed in with the dust.

17.2 SPACE OBSERVATIONS FROM MARINER 9

Mars has been the target of a series of spacecraft launched by both the United States and the Soviet Union. The earliest of these spacecraft were flybys, which attempted to photograph and otherwise study the Martian surface and atmosphere in the few hours they were in good position as they passed by (Table 17–1).

In 1971, the United States sent out a spacecraft not just to fly by Mars for only a few hours, but actually to orbit the planet and send back data for a year or more. This spacecraft, Mariner 9, went into orbit around Mars in November 1971, after a 5-month voyage from Earth.

In the next section, we will discuss the Vikings, the most recent spacecraft to visit Mars.

But when the spacecraft successfully reached Mars, our first reaction was as much disappointment as it was elation. While Mariner 9 was en route, a tremendous dust storm had come up, almost completely obscuring the entire surface of the planet. Only the south polar cap and 4 dark spots were visible.

The storm began to settle after a few weeks, with the polar caps best visible through the thinning dust. Finally, three months after the spacecraft arrived, the surface of Mars was completely visible, and Mariner 9 could proceed with its mapping mission. It mapped the entire surface at a resolution of 1 km, and further photographed selected areas at a resolution 10 times better than this limit. The spacecraft was to last another year, and so more than completed its assigned tasks—in fact, we even got the scientific bonus of being able to see how the dust acted at the end of the storm. These observations gave us information both about the winds aloft and about the heights and shapes of the surface features.

TABLE 17–1 UNMANNED PROBES LAUNCHED TO MARS

Spacecraft	Launched by	Arrival Date	Comment
Mariner 4	USA	July 1965	Flyby
Mariner 6	USA	July 1969	Flyby
Mariner 7	USA	August 1969	Flyby
Mariner 9	USA	November 1971	Orbiter
Mars 2	USSR	November 1971	Orbiter; lander lost
Mars 3	USSR	December 1971	Orbiter; lander worked 20 s
Mars 4	USSR	February 1974	Flew by; no observations
Mars 5	USSR	February 1974	Orbiter
Mars 6	USSR	March 1974	Orbiter; lander failed
Mars 7	USSR	March 1974	Orbiter; lander failed
Viking 1	USA	June 1976	Orbiter and lander
Viking 2	USA	August 1976	Orbiter and lander

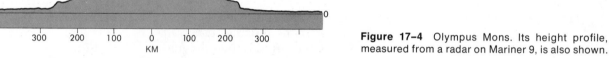

Figure 17–4 Olympus Mons. Its height profile, measured from a radar on Mariner 9, is also shown.

"Geo-" is from the Greek for Earth. Strictly speaking, "geology" applies only to the Earth, but the term is now more widely used, especially for the application of geological principles to the Moon and planets.

The tallest volcano on Earth is Mauna Kea (Fig. 16–1), in the Hawaiian islands, if we measure its height from its base deep below the ocean. Mauna Kea is only 9 km high, taller than Everest. And remember, it is on Earth, a much bigger planet than Mars.

The data taken after the dust storm had ended showed four major "geological" areas on Mars: volcanic regions, canyon areas, expanses of craters, and terraced areas near the poles.

A chief surprise of the Mariner 9 mission was the discovery of extensive areas of vulcanism on Mars. The four dark spots on the Martian surface proved to be volcanoes. The largest of the volcanoes, which corresponds in position to the surface marking long known as Nix Olympica, "the snow of Olympus," is named Olympus Mons, "Mount Olympus." It is a huge volcano—600 kilometers at its base and about 25 kilometers high (Fig. 17–4). Olympus Mons is crowned with a crater 65 kilometers wide; Manhattan Island could be easily dropped inside the crater.

Another surprise on Mars was the discovery of systems of canyons. One tremendous canyon (Fig. 17–5)—about 5000 kilometers long—is as big as the United States and comparable in size to the Rift Valley in Africa, the longest geological fault on Earth. Here again the size of the canyon with respect to Mars is proportionately larger than any geologic formations on the Earth. Venus, the Earth, and Mars each seems to have a large rift.

Perhaps the most amazing discovery on Mars was the presence of sinuous channels. These are on a smaller scale than the *canali* that Schiaparelli had seen, and are entirely different phenomena. Some of the channels show tributaries (Fig. 17–6), and the bottoms of some of the channels show the characteristic features that stream beds on Earth have. Even though water cannot exist on the surface of Mars under today's conditions, it is

Figure 17–5 The tremendous canyon, now named Valles Marineris—Mariner Valley—is about 5000 km long, nearly the diameter of the United States. The canyon is 120 km wide and 6 km deep.

difficult to think of other ways to explain them satisfactorily than to say that the channels were cut by running water in the past.

The indication that water once flowed on Mars is particularly interesting because biologists feel that water is necessary for the formation and evolution of life. The presence of water on Mars, therefore, even in the past, may indicate that life could have formed and may even have survived. If we could discover life on another planet that arose independently of life on Earth, comparison of the life forms would tell us what is important to cells and to life, and what things may be of peripheral importance. For us on the Earth, the implications of these discoveries for advances in medicine, not to mention theology, are obvious.

If there had been water on Mars in the past, and in quantities great enough to cut the river beds, where has it all gone? Most of the water would probably be in a permafrost layer beneath middle latitudes and polar regions.

Some of the water is bound in the polar caps that extend down to latitude

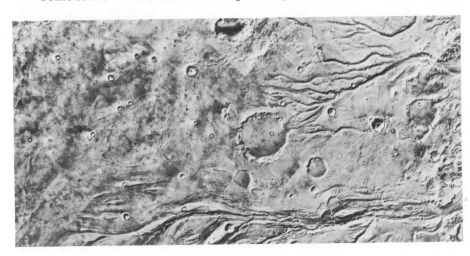

Figure 17–6 Stream channels with tributaries on Mars indicate to most scientists that water flowed there in the past.

A

B

Figure 17–7 *(A)* This mosaic of Mariner 9 views shows the north polar cap, which is shrinking as a result of Martian springtime. The volcano Olympus Mons is visible at the bottom. *(B)* Over 1500 Mariner 9 photographs were pasted onto a globe to make the mosaic shown here. We are looking down on the residual north polar cap.

The composition of the atmosphere is measured by looking at the spectrum. This can even be done from the Earth, but problems arise because the spectrum of the Earth's own atmosphere interferes, particularly at the wavelengths of the water vapor lines.

50° in the winter (Fig. 17–7). Up to the time of Mariner 9, we thought that the polar caps were mostly frozen carbon dioxide—"dry ice"; the presence of water was controversial. We now know that the large polar caps visible during the winter are carbon dioxide. But when a cap shrinks during its hemisphere's summer, a residual polar cap of water ice remains.

If the water trapped in the polar caps or under the surface of Mars was released some time in the past, as liquid water on the surface or as water vapor in the atmosphere, it may have been sufficient to make Mars verdant. Perhaps the polar caps melted, releasing quantities of water because dust that settled on the ice caused more sunlight to be absorbed or a widespread episode of vulcanism caused a general heating. Climatic change could also have resulted from a change in the orientation of Mars' axis of rotation, from a change in the solar luminosity, or from a variation in atmospheric composition (which would affect the greenhouse effect).

We found that the Martian atmosphere is composed of 90 per cent carbon dioxide with small amounts of carbon monoxide, oxygen, and water.

We measured the density of the atmosphere by monitoring the changes of radio signals from spacecraft as they went behind Mars. As the spacecraft were occulted by the atmosphere of the planet, we could tell the rate at which the density drops. From the composition and the way the density varies with height, we derived the variations of pressure and temperature. The surface pressure is 1 per cent of that which exists on the Earth's surface. The pressure decreases with altitude slightly more slowly than it does on Earth.

17.3   VIKING!

In the summer of 1976, two U.S. spacecraft named Viking reached Mars after flights of about 10 months. Each spacecraft contained two parts: an orbiter and a lander. The orbiter served two roles; it not only used its cameras and other instruments to map and analyze the surface but also served as a relay station for the radio signals from the lander to Earth. The lander studied the rocks and weather near the surface of Mars. Its other role was surely no less significant: its major task was to sample the surface and decide whether there was life on Mars!

Viking 1 reached Mars in June of 1976 after a flight that led to spectacular large-scale views of the Martian surface (Color Plate 30). Viking 1 orbited Mars for a month while it radioed back pictures to scientists on Earth who were trying to determine a relatively safe spot to land.

The orbiter part of the mission alone was extremely successful. Observations were obtained at higher resolution than that provided by Mariner 9 of the giant volcanoes (Fig. 17–8) and of the huge 5000-km-long valley (Fig. 17–9).

These detailed views of Mars allow us to better interpret the similarities and differences that this planet of extremes—such as huge canyons and gigantic volcanoes—has with respect to the Earth. For example, Mars has exceedingly large, gently sloping volcanoes but no signs of the long mountain ranges or the equivalent of the deep mid-ocean ridges that on Earth tell us that plate tectonics has been and is taking place. The large volcanoes on Mars are of a type called "shield volcanoes" that have gently sloping sides (Figure 17–10). The slopes are gradual because the lava spread rapidly. On Earth, we also have steep-sided volcanoes, which occur where the continental plates are overlapping, as in the Aleutian islands or for Mount Fujiyama.

Perhaps the volcanic features on Mars are so huge because continental drift is absent there. If molten rock flowing upward causes volcanoes to form, as is thought to be the case, on Mars the features just get bigger and bigger rather than move away from the underlying source.

Sterilization procedures have been worked out for spacecraft sent to Mars and the other planets so they do not bring along microorganisms that would contaminate the planet's atmosphere, leaving life forms from the contamination to be "discovered" by subsequent spacecraft. The spacecraft are baked under high temperatures for some time.

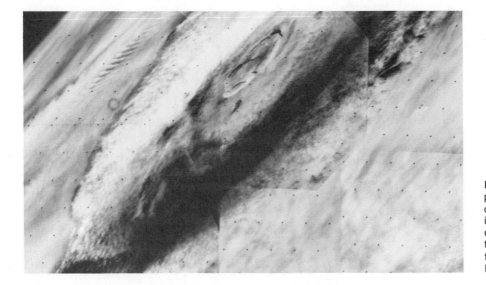

Figure 17–8 Olympus Mons photographed by the Viking 1 orbiter. Clouds extend up most of its 25-km height. The volcanic crater, 80 km across, extends into the stratosphere and was cloud free. Mars limb shows several haze layers.

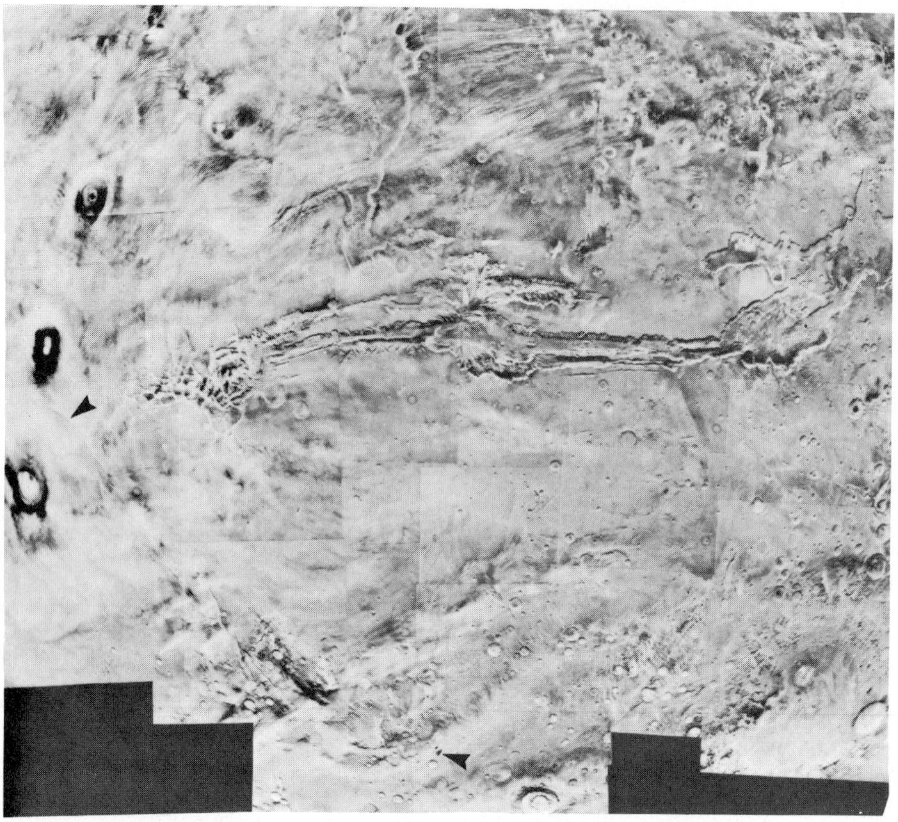

Figure 17–9 A mosaic of 102 photographs of Mars from Orbiter 1 in 1980, near the end of the spacecraft's lifetime. Valles Marineris, as long as the United States is wide, stretches across the center. Three huge volcanoes are at left. The sharp line near them, marked with an arrow, is an unusual feature that is either a weather front or an atmospheric shock wave. The lower arrow shows the shadows of a group of clouds that are at an elevation of 28 km (91,000 ft). The largest cloud is nearly 32 km long. (NASA photo)

The first choices for a landing site, made long in advance based on Mariner 9 pictures, had to be rejected when Viking's pictures revealed dangerous boulders or craters that could overturn the spacecraft. But finally a site was selected, and on July 20, 1976, exactly seven years after the first manned landing on the Moon, Viking 1's lander descended safely onto a plain called Chryse.

Figure 17–10 Shield volcanoes on Earth include Haleakala (shown here) and Mauna Kea in Hawaii and Kilimanjaro in Africa. Martian volcanoes are all of this type, with gently sloping sides.

Figure 17–11   This 100° view of the Martian surface was taken from the Viking 1 lander, looking northeast at left and southeast at right. It shows a dune field with features similar to many seen in the deserts of Earth. From the shape of the peaks, it seems that the dunes move from upper left to lower right. The large boulder at the left is about 8 meters from the lander and is 1 × 3 meters in size. The boom that supports Viking's weather station cuts through the center of the picture. ("Chance of precipitation," the local newscaster would say, "0 per cent.")

The views showed rocks of several kinds (Color Plates 32–34), covered with reddish-brown material that is probably an iron oxide compound. Sand dunes were also visible (Fig. 17–11). The sky on Mars turns out to be pink (Color Plates 32–34), from reddish dust suspended in the air as a result of one of Mars' frequent dust storms.

A series of experiments aboard the lander was designed to search for signs of life. A long arm was deployed (Color Plate 32) and a shovel at its end dug up a bit of the Martian surface (Fig. 17–12). The soil was dumped into three experiments that searched for such signs of life as respiration and metabolism. The results were astonishing at first. The biological experiments sent back signals that seemed similar to those that would be caused on Earth by biological rather than by mere chemical processes. But later results were less spectacular and non-biological explanations seem more likely. It is probable that some strange chemical process mimicked life in these experiments.

One important experiment gave much more negative results for the chance that there is life on Mars. It analyzed the soil and looked for traces of organic compounds. On Earth, many organic compounds left over from dead forms of life remain in the soil; the life forms themselves are only a tiny fraction of the organic material. On Mars, who knows? Perhaps life forms evolved that efficiently used up their predecessors. Still, the absence of organic material from the Martian soil is a strong argument against the presence of life on Mars.

The Viking 2 lander descended on September 3, 1976, on a site that was thought to be much more favorable to possible life forms than was the Viking 1 site. Again, Viking 2 found a variety of types of rocks (Fig. 17–13 and Color Plate 33) and a pink sky. The atmosphere at this second site, Utopia Planitia, contains three or four times more water vapor than has been observed near Chryse. Viking 2's experiments on the search for life sent back data similar to Viking 1's.

Anyway, we mustn't always ask for a yes or no answer. Even if the life signs detected by Viking come from chemical rather than biological processes, we have still learned of some fascinating new chemistry going on.

Figure 17–12   The first trench, 7 cm wide × 5 cm deep × 15 cm long, excavated by Viking 1 to collect Martian soil for the life-detection experiments. From the fact that the walls have not slumped, we can conclude that the soil is as cohesive as wet sand.

Figure 17-13 The first photograph of the surface of Mars taken from Viking 2 on September 3, 1976. A wide variety of rocks are seen to lie on a surface of fine-grained material. One of the lander's footpads can be seen at the lower right.

When life arose on Earth, it probably took up chemical processes that had been previously in existence. Similarly, if life arose on Mars in the past or would normally arise there in the future (assuming our visiting there doesn't contaminate Mars and ruin the chances for the beginning of indigenous life), we might expect the life forms to use the chemical processes that already existed. So even if we haven't detected life itself, we may well have learned important things about its origin.

The presence of two orbiters circling Mars allowed a division of duties. One continued to relay data from the two Viking landers to Earth, while the orbit of the other was changed so that it passed over Mars' north pole. The main polar cap, thought to be made of dry ice, had retreated as a result of summertime and only a residual ice cap was left. From the amount of water

Figure 17-14 This oblique view from the Viking orbiter shows Argyre, the smooth plain at left center. The Martian atmosphere was unusually clear when this photograph was taken, and craters can be seen nearly to the horizon. The brightness of the horizon results mainly from a thin haze. Detached layers of haze can be seen to extend from 25 to 40 km above the horizon and may be crystals of carbon dioxide.

Box 17.1 Viking Experiments that Searched for Life

Each Viking lander carried three biological experiments: (1) a gas-exchange experiment; (2) a labeled-release experiment; and (3) a pyrolytic-release experiment. Each experiment examined Martian soil. The results were essentially the same at both landing sites.

(1) The gas-exchange experiment looked for changes in the atmosphere caused by metabolism of organisms in the soil. Soil was mixed in the lander with a small amount of a nutrient medium. (The medium consisted of organic compounds dissolved in water.) Enough medium was added to humidify the chamber without wetting the soil. At first, carbon dioxide and oxygen—gases whose release on Earth is connected with life—were given off rapidly at both sites. This seemed exciting, but then the release of gases stopped. Two and a half weeks later, enough medium was added to saturate the soil, and incubated for seven months. The production of carbon dioxide rose at first and then leveled off, while the production of oxygen continued to decline. All this action can be accounted for with chemical reactions; it seems that the original reactions, which seemed exciting at first, were the result of the chemical interaction of the soil with water vapor.

(2) The labeled-release experiment used a medium containing simpler but cosmically common organic compounds that included some atoms of the radioactive isotope of carbon, carbon 14. When the nutrient was first added to the soil in the lander, carbon dioxide was released and its radioactive carbon 14 atoms were counted. But the CO_2 probably came from chemical decomposition of the soil, because when more nutrient was added no additional gas was released.

(3) The pyrolytic-release experiment used a furnace to pyrolyze—cause a chemical change by heat—a series of samples of Martian soil. The soil was sealed in a chamber along with Martian atmosphere and illuminated with simulated Martian sunlight from a xenon lamp. Small amounts of radioactive carbon monoxide and carbon dioxide were included, and after five days the soil was analyzed to see if any of the radioactive carbon had found its way into the organic material in the soil. Most of these tests indicated the presence of radioactive carbon dioxide, which was the expected indication that life was present in the soil. But when a sample of soil was exposed to high temperature for three hours before being tested, the production of carbon dioxide was reduced to 10 per cent of its previous value rather than being eliminated. It seems unlikely that any life would be able to survive this treatment, so it makes a chemical explanation seem possible here too.

The fourth experiment relevant to the search for life was the chemical experiment: the mass spectrometer, which measures the mass of the molecules it samples and thus identifies them. It found no evidence for organic compounds in the soil, though it was able to make very sensitive measurements.

In sum, though the first results seemed to show that life may be present, the later results did not back this idea. Still, the pyrolysis experiment is difficult to interpret conclusively.

vapor detected with the orbiter's spectrograph, and from the fact that the temperature was too high for this residual ice cap to be frozen carbon dioxide, the conclusion has been reached that it is made of water ice. This important result indicates the probable presence of lots of water on the Martian surface in the past—in accordance with the existence of the stream beds—and seems to raise the probability that life has existed or does exist on Mars.

Figure 17-15

Figure 17-16

Figure 17-15 This Martian storm on August 9, 1978, resembles satellite pictures of storms on Earth, showing the counterclockwise circulation typical at northern latitudes. The temperature is too warm for the clouds to be carbon dioxide; they therefore must be made of water ice. The front-filled crater at top right is 92 km in diameter. The white patches at top are outlying regions of the north polar remnant cap.

Figure 17-16 This view from May, 1979, shows a thin coating of water ice on the rocks and soil, an occurrence that happens every Martian year at this time and lasts about 100 days. The ice coating is very thin, perhaps less than 0.01 mm thick.

Observations from Orbiter showed Mars' atmosphere (Fig. 17–14). The lengthy period of observation allowed the discovery of weather patterns on Mars. With its rotation period similar to that of Earth, many features of Mars' weather are similar to our own (Fig. 17–15).

Both Viking landers and orbiters lasted long enough to show seasonal changes on Mars. The Viking landers showed the seasonal buildup of frost (Fig. 17–16). One of the Viking landers may survive another ten years, sending back data on a limited basis. The orbiters showed dust storms (Fig. 17–17), which generally form in the early summer in the southern hemisphere and in winter in the northern hemisphere. The storms are seasonal partly because Mars is closest to the Sun at this time and the increased solar heating puts the atmosphere in circulation.

What's the next step? There are no definite plans, but the Viking missions have been so spectacularly successful that it seems reasonable that further exploration of Mars would be valuable. Carl Sagan, a Cornell astronomer who not only has worked with the Viking observations but also has become the foremost spokesman to the public about the project and its

Figure 17-17 At left *(A)* we see a region of the Martian surface partly covered with clouds of water ice; at right *(B)* we see the same part of the surface 9 months later at an oblique angle. It includes a dust storm 600 km across, about the size of Colorado. The storm is moving eastward, from left to right; the leading edge is sharp and the trailing portion is diffuse.

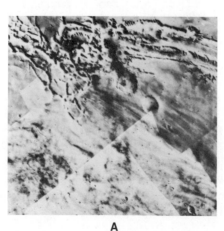

A

B

results, points out the frustration at not being able to see over the horizon. He and others ask for a roving vehicle that could travel over the Martian surface. Enough spare parts exist from Viking that this could be done at a cost that is reasonable for such complex space experiments. But current budgetary limitations make it unlikely that we can hope in the 1980's for a Mars rover or a mission to bring Mars' soil back to Earth. Manned space flights to Mars would be much more expensive and so are even farther off.

17.4 SATELLITES OF MARS

Mars has two moons. In a sense, these are the first moons we have met in our study of the solar system, for Mercury and Venus have no moons at all, and the Earth's Moon is so large relative to its parent that we may consider the Earth and Moon as a double planet system. The moons of Mars, Phobos (from the Greek for Fear) and Deimos (from the Greek for Terror), are mere chunks of rock, only 27 and 15 kilometers across, respectively.

Phobos and Deimos are very minor satellites. They revolve very rapidly, Phobos only 6000 and Deimos only 20,000 km above the surface. Phobos completes an orbit of Mars in only 7 hours and 40 minutes, more rapidly than a Martian day, which is, as we have said, about the same length as an Earth day. Deimos orbits in about 30 hours, slower than Mars' rotation period, just as our Moon revolves more slowly than the Earth's rotation period. Because Phobos orbits more rapidly than Mars rotates, an observer on the surface of Mars would see Phobos moving conspicuously backwards compared to Deimos, the other planets, and the stars, which all move from east to west across the sky.

It is amusing to note that in 1727 Jonathan Swift invented two moons of Mars for *Gulliver's Travels*, long before the moons were discovered. This is widely considered to be a lucky guess, but it was a reasonable guess at that time, for numerological reasons. It may even have dated back to Kepler, who had reasoned in 1610 that the planets interior to the Earth, Mercury and Venus, had no known moons, the Earth had one, while Jupiter, the next

In Greek mythology, Phobos and Deimos were companions of Ares, the equivalent of the Roman war god, Mars.

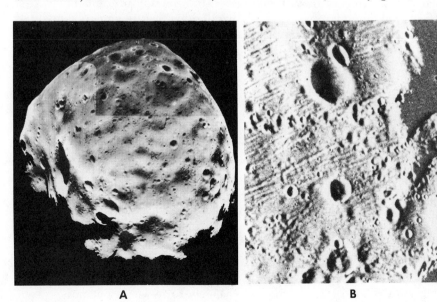

A   **B**

Figure 17–18 *(A)* A Viking 1 mosaic of photographs of Phobos. The largest two craters on Phobos have been named after Asaph Hall and after Angelina Stickney, his wife, who encouraged the search. *(B)* A close-up view of Phobos from Viking Orbiter 2. Phobos' north pole is near upper left. It is heavily cratered as was expected but, surprisingly, shows striations and chains of small craters. Similar chains of small craters on Earth's Moon, on Mars, and on Mercury were formed by secondary cratering from a larger impact.

Figure 17–19 This photograph of Deimos was obtained by Viking Orbiter 1 from a range of 3300 km. About half of the side facing the camera is illuminated, and that lighted portion measures about 12 × 8 km. While Mariner 9 pictures taken nearly five years ago from greater distances showed only a few large craters on Deimos, the improved resolution here reveals a heavily-cratered and presumably very old surface. At least a dozen craters are prominent—the two largest measuring 1.3 km and 1 km. Craters as small as 100 meters pockmark the surface. Still more recent photographs, taken when the Viking 2 orbiter went within 65 km of Deimos, show a surprisingly smooth surface. Thus it, and Phobos as well, may be covered with dust.

planet out from Mars, had 4 known moons. Two is an intermediate value. Both Swift's moons and the real ones are very small and revolve very quickly. The two non-fictional moons were discovered in 1877 by Asaph Hall of the U.S. Naval Observatory in Washington, D.C.

Mariner 9 and the Vikings made closeup photographs and studies of Phobos and Deimos (Figs. 17–18 and 17–19). Each turns out to be not at all like our Moon, which is a spherical, planet-like body. Phobos and Deimos are just cratered chunks of rock that may have been broken off on impact with meteoroids. They are too small to have enough gravity to have made them round.

Phobos, which is only about 27 km in its longest dimension and 19 km in its shortest, has a crater on it that is 8 kilometers across, a large fraction of Phobos' circumference. Deimos is even smaller than Phobos, only around 15 km in its longest dimension and 11 km in its shortest. Now that we know their sizes, we can calculate their albedos; Phobos and Deimos turn out to be about as dark as the darkest maria on the Moon. From more detailed but similar comparisons of how the albedos—the fraction of light reflected—vary from wavelength to wavelength, we deduce that they may be made of dark carbon-rich ("carbonaceous") rock, similar to some of the asteroids.

We can conclude that the moons are fairly old because they have been around long enough for the surface to be extensively cratered. The rotation of both moons is gravitationally linked to Mars with the same side always facing the planet, just as our Moon is linked to Earth.

SUMMARY AND OUTLINE

Observations from the Earth (Section 17.1)
   Rotation period
   Surface markings
   Dust storms
   Seasonal changes
Space observations (Section 17.2)
   Many space probes sent from the U.S. and U.S.S.R.

Mariner 9 orbiter mapped all of Mars (Section 17.2)
   Volcanic regions indicating geological activity
   Canyons larger than Earth counterparts
   Craters and terraced areas near the poles
   Signs of water evident
   Underlying layer of polar caps
   Sinuous channels, some with tributaries

Channel bottoms show markings similar to terrestrial stream beds
Viking (Section 17.3)
    Orbiters and landers
    Closeup photographs of surface rocks
    Analysis of soil

Biological explanation seemed possible at first
Chemical explanation now seems most likely
Residual polar cap is made of water ice
Satellites of Mars (Section 17.4)
    Phobos and Deimos
    Both irregular chunks of rock

QUESTIONS

1. Outline the features of Mars that make scientists think that it is a good place to search for life.

2. If Mars were closer to the Sun, would you expect its atmosphere to be more or less dense than it is now? Explain.

†3. From the relative masses and radii (see Appendix 5), verify that the density of Mars is about 70 per cent that of Earth.

4. Approximately how much more solar radiation strikes a square meter of the Earth than a square meter of Mars?

5. Compare the tallest volcanoes on Earth and Mars relative to the diameters of the planets.

6. What evidence is there that there is, or has been, water on Mars?

7. Why is Mars' sky pink?

8. Describe the composition of Mars' polar caps. Explain the evidence.

9. List the various techniques for determining the composition of the atmosphere of Mars.

10. List the evidence from Viking for and against the existence of life on Mars.

11. (a) Aside from the biology experiments, list three types of observations made from the Viking landers.
    (b) What are two types of observations made from the Viking orbiters?

12. Why aren't Phobos and Deimos regular, round objects like the Earth's Moon?

TOPICS FOR DISCUSSION

1. Discuss the problems of sterilizing spacecraft, and whether we should risk contaminating Mars by landing spacecraft.

2. Discuss the additional problems of contamination that manned exploration would bring. Analyze whether, in this context, we should proceed with manned exploration in the next decade or two.

---

†This indicates that the question requires a numerical solution.

CHAPTER 18

JUPITER

AIMS:
To describe Jupiter, the dominant
planet of the solar system, and to
discuss an example of a class of planets
very different from the inner,
terrestrial planets

The planets beyond the asteroid belt—Jupiter, Saturn, Uranus, and Neptune—are very different from the four "terrestrial" planets. These *giant planets*, or *Jovian planets*, not only are much bigger and more massive, but are also less dense. This suggests that the internal structure of these giant planets is entirely different from that of the four terrestrial planets.

Jupiter, also called Jove, was the chief Roman deity.

Jovian is the adjectival form of Jupiter.

18.1 FUNDAMENTAL PROPERTIES

The largest planet, Jupiter, dominates the Sun's planetary system. Jupiter is 5 A.U. from the Sun and revolves once every 12 years. It alone contains two thirds of the mass in the solar system outside of the Sun, 318 times as much mass as the Earth.

Jupiter has at least 14 moons of its own and so is a miniature planetary system in itself. It is often seen as a bright object in our night sky, and observations with even a small telescope reveal bands of clouds across its surface and show four of its moons. Jupiter is more than 11 times greater in diameter than the Earth. From its volume and mass, we calculate its density to be 1.3 gcm⁻³, not much greater than the 1 gcm⁻³ density of water. This tells us that any core of heavy elements (such as iron) that Jupiter may have does not make up as substantial a fraction of Jupiter's mass as the cores of the inner planets make up of their planets' masses. Jupiter, rather, is mainly composed of the lighter elements hydrogen and helium. Jupiter's chemical composition is closer to that of the sun and stars than it is to that of the Earth (Fig. 18–1).

Jupiter's mass can be readily measured by following the orbits of its satellites and using Kepler's third law. Jupiter has only 1/1000 the mass of the Sun, too little for nuclear fusion to have begun in its interior.

Figure 18–1 The composition of Jupiter.

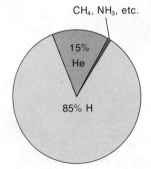

A mosaic of Jupiter and its four Galilean satellites made from the Voyager 1 spacecraft. Callisto with its impact bull's-eye is at lower right, Ganymede is at lower left, relatively bright Europa is shown against Jupiter, and Io appears in the background at left. The symbol for Jupiter appears at the top of the page.

Figure 18-2 Jupiter, photographed from the Earth, shows belts and zones of different shades and colors. By convention (that is, the set of terms we arbitrarily choose to use) the bright horizontal bands are called *zones* and the dark horizontal bands are called *belts*. Adjacent belts and zones rotate at different speeds.

Jupiter rotates very rapidly, once every 10 hours. Undoubtedly, this rapid spin rate is a major reason for the colorful bands, which are clouds spread out parallel to the equator. The bands appear in subtle shades of orange, brown, gray, yellow, cream, and light blue, and are beautiful to see (Fig. 18–2). They are in constant turmoil; the shapes and distribution of bands continually change in a matter of days.

Interestingly, the clouds do not rotate as would the surface of a solid body (Fig. 18–3): the clouds at the equator rotate slightly faster than the clouds nearer the poles, completing their rotation about five minutes earlier each time around. Thus every fifty days, the clouds at the equator have made an additional rotation. Not only the uppermost layers of Jupiter's clouds but also much deeper levels probably rotate at differing velocities. Jupiter isn't solid; it has no crustal surface at all. At deeper and deeper levels its gas just gets denser and denser, eventually liquefying.

The rapid rotation makes the planet bulge at the equator. The distance from pole to pole is substantially less—7 per cent—than the equatorial diameter; we say that Jupiter is *oblate* (Fig. 18–4). Jupiter's axis of rotation is only 3° from the axis of Jupiter's orbital revolution around the Sun, so that, unlike the Earth and Mars, Jupiter has no seasons.

Figure 18-3 Jupiter's differential rotation, which causes the points that were aligned at time T_1 to be separated at time T_2, spreads out the clouds into bands. The rate at which this happens is exaggerated in this drawing.

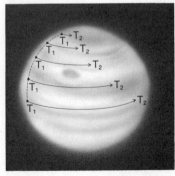

The most prominent feature of the visible cloud surface of Jupiter is a large reddish oval known as the *Great Red Spot* (see Fig. 18–2 and Color Plate 38). It is about 14,000 km × 30,000 km, many times larger than the Earth, and drifts about slowly with respect to the clouds as the planet rotates. The Great Red Spot is a relatively stable feature, for it has been visible for at least 150 years, and maybe 300 years. Sometimes it is relatively prominent, and at other times the color may even disappear for a few years. The Great Red Spot also changes in shape.

In 1955, intense bursts of radio radiation were discovered to be coming from Jupiter. Though it was quite a surprise to detect any radio signals at all, several kinds of signals were detected. Some bursts seemed to come from the atmosphere. It has been discovered that the bursts of radio emission are correlated with passages of Jupiter's innermost major moon, Io, above the regions from which the bursts are coming. At shorter radio wavelengths, Jupiter emits continuous radiation.

The fact that Jupiter emits radio waves indicated that Jupiter, even more so than the Earth, has a strong magnetic field and strong *radiation belts* (actually, belts filled with magnetic fields in which particles are trapped, large-scale versions of the Van Allen belts of Earth). We reach this conclusion because we can account for the presence of the radio radiation only with mechanisms that involve such magnetic fields and belts. Some of the radio emission comes from interactions of high-energy particles passing through space with Jupiter's magnetic field.

18.2 JUPITER'S MOONS: EARLY VIEWS

Jupiter has at least 15 satellites. Four of the innermost satellites are very nicely named: Io, Europa, Ganymede, and Callisto. At 5216 km in diameter, Ganymede is one of the biggest satellites in the solar system (even larger than Titan). These four moons are called the *Galilean satellites* (Fig. 18–5). They were discovered by Galileo in 1610 when he first looked at Jupiter with his telescope. These satellites were named Io, Europa, Ganymede, and Callisto by Simon Marius, a German astronomer who independently discovered them.

The Galilean satellites have played a very important role in the history of astronomy. It was the fact that these particular satellites were noticed to be going around another planet (Fig. 18–6), like a solar system in miniature, that supported Copernicus' heliocentric model of our solar system. Not everything revolved around the Earth!

The dozen moons of Jupiter that were discovered before the last decade fall into three groups. The first group includes the five innermost satellites then known (Amalthea and the Galilean satellites), which are also the largest bodies. Their orbits are all less than 2 million km in radius. The orbits of the four moons in the second group are all about 12 million km in radius. The

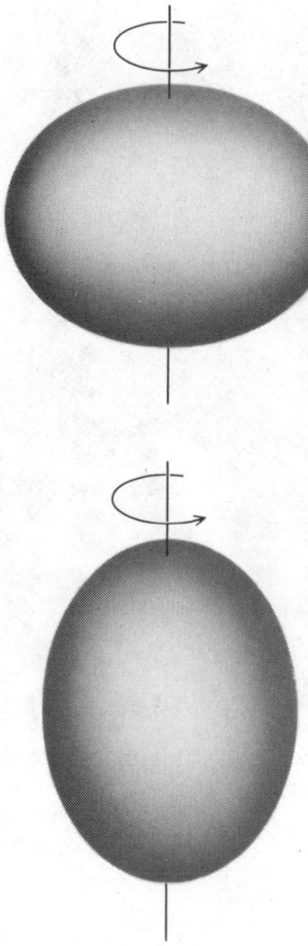

Figure 18–4 The top ellipsoid is *oblate*; the bottom ellipsoid is *prolate*.

Figure 18–6 Galileo's sketches of the changes in the position of the satellites.

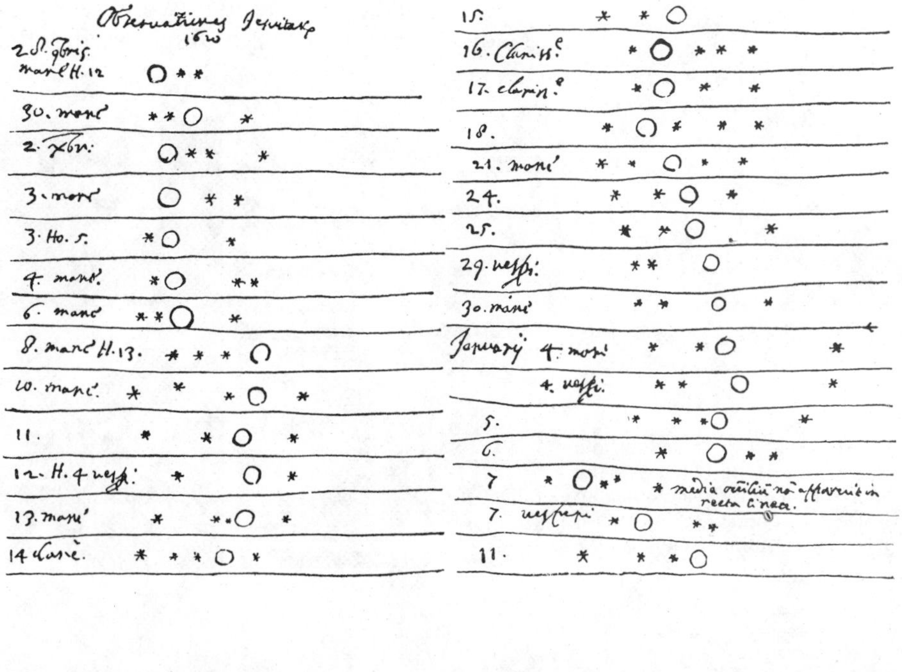

Figure 18–5 Jupiter with the four Galilean satellites. These satellites were named Io, Europa, Ganymede, and Callisto by Simon Marius, a German astronomer who independently discovered them.

Figure 18–7 The photograph on which the thirteenth moon of Jupiter, shown with an arrow, was discovered. The moon is named Leda. The telescope was set to track across the sky along with Jupiter; thus stars show as trails. The trails here correspond to anonymous—unnamed and un-numbered—faint stars.

orbits of the four outermost moons, which are in the third group, are 21 to 24 million km in radius. These four outermost moons revolve in the direction opposite from that of Jupiter's rotation (that is, in retrograde direction), while the other moons revolve prograde. Moons in the outer two groups may be bodies captured by Jupiter's gravity after their formation, for they have high inclinations and/or eccentricities.

In the last half-dozen years, additional moons of Jupiter have been discovered by ground-based observations. Charles Kowal has been using the Palomar Schmidt telescope to detect moons all the way down to about 22nd magnitude, a level of intensity fainter than had been previously studied (Fig. 18–7). Spacecraft to Jupiter have also discovered at least two other satellites.

In Section 18.4 we shall see how spacecraft observations from close up have changed our understanding of the moons. Formerly pinpoints of light, they are now known worlds.

18.3 SPACECRAFT OBSERVATIONS

Each of the planets we have discussed so far has recently been visited by spacecraft that have greatly advanced our knowledge about them. Jupiter is another spectacular example. Two spacecraft, Pioneer 10 and Pioneer 11, gave us our first close-up views of the colossal planet in 1973 and 1974. A second revolution in our understanding of Jupiter occurred in 1979, when Voyager 1 and Voyager 2 also flew by Jupiter (Fig. 18–8).

Each of the spacecraft carried many types of instruments to measure properties of Jupiter, its satellites, and the space around them, but the experiment of most popular interest was always the imaging equipment. The Pioneers did not carry any equipment specifically designed for sending back pictures, but did provide a valuable set of observations with a device whose special purpose was to take polarization measurements. The resolution of the images was improved over the best images we can obtain from Earth by a factor of five.

The Voyagers provided a quantum leap in quality (Fig. 18–9) for several reasons. First, the Voyagers were Mariner-type spacecraft, which held steady in three dimensions in space rather than rotating on one axis like the

Figure 18–8 (A) This duplicate of Pioneers 10 and 11 hangs in the National Air and Space Museum in Washington, D.C. (B) An artist's conception of the Pioneer space-craft during a flyby.

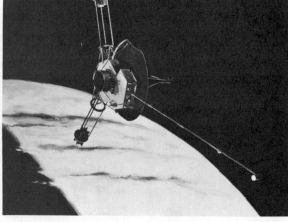

A   B

Box 18.1 Jupiter's Satellites in Mythology

All the moons except Amalthea are named after lovers of Zeus, the Greek equivalent of Jupiter. Amalthea, a goat-nymph, was Zeus's nurse and in gratitude was made into the constellation Capricorn. Zeus changed Io into a heifer to hide her from Hera's jealousy; hence the horns of the crescent moon. Io was tied up anyway and even after she escaped was chased by a stinging gadfly. She swam the sea we now call "Ionian."

Ganymede was a Trojan youth carried off by an eagle to be Jupiter's cup bearer (the constellation Aquarius). Callisto was punished for her affair with Zeus by being changed into a bear. She was then slain by mistake, and rescued by Zeus by being transformed into the Great Bear in the sky. Jealous Hera persuaded the sea god to forbid Callisto to ever bathe in the sea, which is why Ursa Major never sinks below the horizon.

Europa was carried off to Crete on the back of Zeus, who took the form of a white bull. She became Minos' mother. Pasiphae (also the name of one of Jupiter's moons) was the wife of Minos and the mother of the Minotaur. One of her daughters was Phaedra.

In the 17th century, the moons were used to measure the speed of light. In 1675, Olaus Roemer in Denmark noticed that the moons reached their predicted positions later than expected when Jupiter was on the far side of the Sun and earlier when it was on the near side. He attributed this difference to light having a finite speed and derived a value that is close to the currently accepted value.

The name Voyager was chosen instead of the routine number in the Mariner series to heighten public interest.

Pioneers. Second, in the years between the launch of the Pioneers and the 1977 launches of the Voyagers, miniaturization of electronics had improved so much that the Voyagers could send back data at 100 times the rate of the Pioneers.

In all cases, the images were telemetered back to Earth in the form of raster scans. (When an image is scanned from side to side and top to bottom

Figure 18-9 Io traversing the surface of Jupiter, as seen from Voyager 2 during its approach in June 1979. The Voyagers revealed that the motions of gas in the bands and zones of Jupiter's atmosphere are much more complicated than had been suspected.

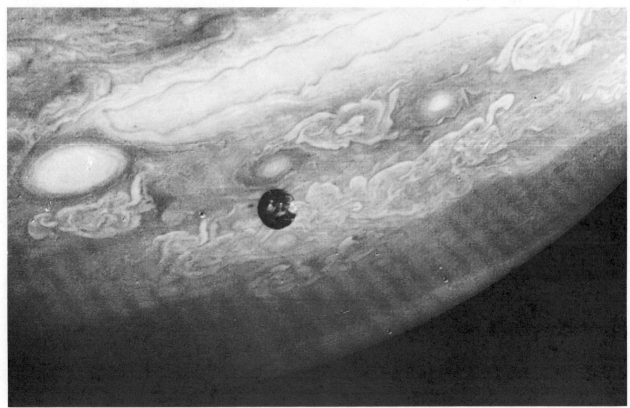

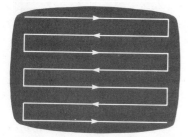

Figure 18–10 Ordinary televisions and many imaging devices scan in a *raster* mode, in which the picture is built up line by line. A sample raster scan with 7 lines is shown. A television scans similarly, though it has 525 lines. Each full Voyager image contained 800 individual lines of data each containing 800 individual picture elements (called "pixels"), making a total of 640,000 pixels per image.

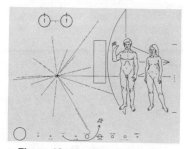

Figure 18–11 The plaques borne by Pioneers 10 and 11. A man and a woman are shown standing in front of an outline of the spacecraft, for scale. The spin-flip of the hydrogen atom, which will be described in Chapter 23, is shown at top left. It also provides a scale, because even travelers from another solar system would know that the wavelength of the transition is a certain length, which we call 21 cm. The sun and planets, including distinctively ringed Saturn, and the spacecraft's trajectory from Earth, are shown at bottom. (Will the visitors realize that we had not yet discovered rings around Jupiter and Uranus?) The directions and periods of several pulsars are shown at left. Numbers are given in the binary system.

Figure 18–12 A mosaic of the Great Red Spot, made from Voyager 2 at a range of 2 million km from Jupiter. The white oval south of the Great Red Spot is similar in structure. Both rotate in the anticyclonic—counterclockwise—sense.

(Fig. 18–10), like a TV picture, the result is called a *raster pattern.*) In as little as 48 seconds, the Voyagers could send a full picture back to Earth, where detailed computer work improved the images. The computers also assembled the two-color images from the Pioneers and the multi-color images from the Voyagers (Color Plate 37) to make full-color pictures. (The Voyagers carry several color filters covering various ultraviolet and visible spectral regions.)

The Pioneer and Voyager spacecraft are travelling fast enough to escape the solar system. As they depart, following encounters in some cases with Saturn as well as Jupiter, the spacecraft are radioing back valuable information on interplanetary conditions. In about 80,000 years the spacecraft will be about one parsec (about three light years) away from the solar system. The Pioneers bear plaques that give some information about their origin and about life on Earth, in case some interstellar traveler from another solar system might pick it up (Fig. 18–11). The Voyagers bear more elaborate mementos, as we shall describe in Section 21.4.

18.3a The Great Red Spot

The Great Red Spot shows very clearly in many of the images (Fig. 18–12 and Color Plate 38). It is a gaseous island many times larger across than the Earth, whose top is about 8 km above the neighboring cloud tops. Now that we have close-up observations first from the Pioneers and then from the Voyagers, we know that the Spot is the vortex of a violent, long-lasting

anticyclonic storm, similar in many ways to large storm areas of Earth's weather.

Although we had not been able to see much structure in the Great Red Spot prior to 1979, the Voyager observations have such fine resolution that we can actually use time-lapse observations to study the rotation period of the Spot. We also see how it interacts with surrounding material (Fig. 18–13), which includes clouds and smaller spots. Many other similar storms (that is, spots) were observed on Jupiter from the Voyager spacecraft, but none was as large or as colorful as the Great Red Spot.

Why has the Great Red Spot lasted this long? Perhaps heat energy flows into the storm from below it, maintaining its energy supply. The storm certainly contains more mass than hurricanes or other cyclonic storms on Earth, which makes it more stable. Further, Jupiter has no continents or other structure to break up the storm, unlike the case for Earth. Still another possible contributing factor may be that Jupiter's clouds radiate less efficiently than Earth's.

There have been several suggestions as to the causes of the color of the Spot. Some of the possibilities are predicated on the idea that the Spot is a giant storm in which material from lower atmospheric layers may be rapidly brought upward, changing color in the process because solar radiation acts on it or because it interacts with the water vapor in Jupiter's atmosphere. The water vapor was discovered not from space, but rather from infrared observations made in 1974 from a jet aircraft flying above most of the Earth's own water vapor. Two hydrocarbons, ethane and acetylene, were also discovered in Jupiter's atmosphere in 1974, as was phosphine, one phosphorus plus three hydrogen atoms, which may change to P_4 to give Jupiter's Great Red Spot its color. Since acetylene is not a stable compound, it must be produced constantly in order to have the abundance we observe today. Theoreticians suggested that it may be produced by ultraviolet light acting on methane or by lightning in Jupiter's atmospheric storms. Extensive lightning storms, including giant-size lightning strikes called *superbolts*, were discovered from the Voyagers, as were giant aurorae (Fig. 18–14).

18.3b Jupiter's Atmosphere

Heat emanating from the interior of Jupiter produces huge convection currents. The bright bands on Jupiter, the zones, are rising currents of gas

Figure 18–13 A time-lapse sequence of the Great Red Spot, showing the flow of gas. The pictures are taken every other rotation of Jupiter, making the interval 22 Earth hours. Note how a white cloud enters the spot's circulation and begins to be swept around. The Great Red Spot rotates with a period of about 6 Earth hours.

Storms on Earth always rotate in the same direction in a given hemisphere (north or south). In the northern hemisphere, they rotate counterclockwise around low pressure regions; we call such storms "cyclonic." Around highs, we have "anticyclonic" storms, which rotate the other way. The directions are reversed in the southern hemisphere. The Great Red Spot is an anticyclonic storm.

Figure 18–14 This long exposure of the dark side of Jupiter, photographed from Voyager 1, shows an aurora on the limb near the Jovian north pole. The bright spots on the disk are probably lightning, comparable to the brightness of superbolts seen at the tops of terrestrial tropical thunderstorms.

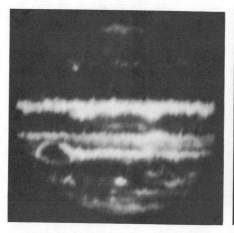 

Figure 18–15 *(Left)* Ground-based observations with the 5-m Palomar telescope at an infrared wavelength show bright regions where the temperature is highest, presumably where we can see deepest into the clouds. The Great Red Spot appears on the left limb as a dark area encircled by a bright ring, indicating that the Spot is cooler than the region that surrounds it. *(Right)* This is a view in the visible taken one hour later from Voyager.

Figure 18–16 Pioneer 11 gave us this first view of the north pole of Jupiter. The bands visible at lower latitudes break up into small eddies. The Great Red Spot is visible.

Figure 18–17 A computer reconstruction of Voyager images of Jupiter's south pole to show the view that would be seen from directly above the pole. An irregularly shaped region right at the pole for which no photographs were available at this 600-km resolution appears as black. The photograph shows the Great Red Spot and a disturbance trailing from it that extends half-way around the planet. The equal spacing of the smaller white spots suggests that some wave action may be taking place. The Jet Propulsion Laboratory, which made this image, has also made a similar image of the northern hemisphere.

driven by this convection (Fig. 18–15). The dark regions, the belts, are falling gas. The tops of these dark belts are somewhat lower (about 20 km) than the tops of the zones and so are about 10 K warmer. The cloud bands themselves are cyclonic patterns, pulled out to surround the planet by Jupiter's rotation.

Pioneer 11 gave us the first opportunity to get a look at Jupiter from a different point of view than we have on Earth. It came in over the polar region. It took this path for two reasons: first, it wanted to avoid possible damage by the intense Jovian radiation belts. When Pioneer 10 swept by Jupiter in late 1973, its instruments had narrowly escaped crippling damage during its passage through the radiation belts. Pioneer 11 came three times closer to Jupiter, passing 40,000 km (0.6 Jupiter radii) above the cloud tops. Kepler's second law shows that Pioneer 11 was travelling much faster than Pioneer 10, and so spent less time in the radiation zone. Also, it was farther from the belts, which are concentrated toward Jupiter's magnetic equator.

Figure 18–16

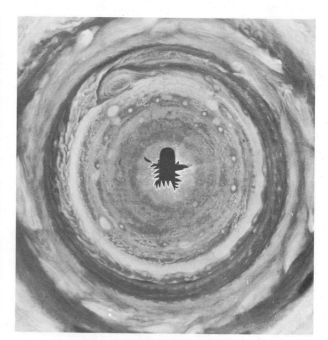

Figure 18–17

Another reason it took this path was scientific. We can never see the polar regions from Earth, as Jupiter's axis is nearly perpendicular to our orbit. So scientists were anxious to get their first look at Jupiter's poles (Fig. 18–16). The results were noteworthy; at high latitudes the circulation pattern of cloud bands that we see at the equator is destroyed and the bands break up into eddies. The Voyagers went a step further, and allowed detailed mapping of both poles (Fig. 18–17). With information like this, we can now study "comparative atmospheres": those of the Earth, Venus, Mars, and Jupiter.

18.3c   Jupiter's Interior

Data from the Pioneers and Voyagers increased our understanding not only of the atmosphere but also of the interior of Jupiter (Fig. 18–18). Most of the interior is in liquid form. Jupiter's central temperature may be between 13,000 and 35,000 K. The central pressure is 100 million times the pressure of the Earth's atmosphere measured at our sea level because of all the mass of Jupiter pressing in. Because of this high pressure, Jupiter might even have an interior composed of ultracompressed hydrogen surrounding a rocky core consisting of 20 Earth masses of iron and silicates.

The liquid hydrogen is probably in a state called "metallic" because it would conduct heat and electricity; these properties are basic to our normal terrestrial definition of "metal." This metallic region is probably where Jupiter's magnetic field is generated by dynamo action. (The interaction of mass, motion, and an electric field can create magnetism. This process also probably takes place in the sun to make sunspots. It also takes place in "dynamos" on Earth, which convert mechanical energy into electricity.) The metallic region makes up 75 per cent of Jupiter's mass.

Jupiter radiates twice as much heat as it receives from the Sun, leading to Jupiter's complex and beautiful cloud circulation pattern. There must be

Comparative studies of the planets help us understand the circulation of clouds and winds, for example. Rapidly rotating Jupiter differs from slowly rotating Venus, and from the Earth and Mars, which rotate in about 24 hours. Also, Jupiter has heat inputs below (internal) and above (the Sun), while the Earth's only major source is the Sun. Further, the Earth's atmosphere is bounded below by a solid surface while Jupiter's is not. Differences in composition are also extreme.

The unusual use of the word "metal" by stellar astronomers to mean all elements heavier than helium is not relevant here.

Figure 18–18 *(A)* The temperature of Jupiter increases toward its center. When we observe in the infrared, even from Earth, most of Jupiter's atmospheric gases don't absorb the radiation; only clouds limit our view. Since the belts are brighter in the infrared, we are seeing hotter gas, which comes from deeper down. The zones are cloudy at higher levels. Thus the zones are rising atmosphere while the belts are sinking atmosphere. The colors, found in the belts, may have to do with sulfur compounds in the middle clouds or perhaps with organic molecules. *(B)* The current model of the interior of Jupiter.

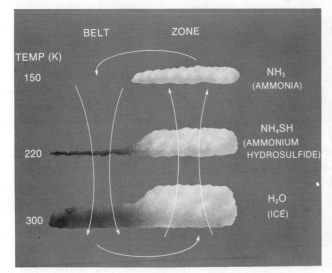

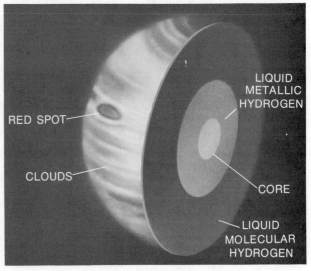

A   B

some internal energy source—perhaps the energy remaining from Jupiter's collapse from a primordial gas cloud 20 million km across to a protoplanet 700,000 km across, 5 times the present size of Jupiter. This catastrophic phase of collapse started, it has been suggested, when the temperature grew sufficiently high to break up the hydrogen atoms. The rapid phase may have taken only 3 months to occur, following the 70,000 years it had previously taken to shrink from a more diffuse cloud. Jupiter is undoubtedly still contracting. Jupiter lacks the mass necessary by a factor of about 75, however, to have heated up enough for nuclear reactions to begin.

Jupiter is truly a mini-solar-system. Its early luminous state no doubt caused it to have satellites of higher density closer in and lower density ones farther out (see Appendix 6), similar to the distribution of the Sun's planets.

18.3d Jupiter's Magnetic Field

The Pioneer missions showed that Jupiter's tremendous magnetic field is even more intense than many scientists had expected, a result confirmed by the Voyagers (Fig. 18–19). (The existence of a magnetic field was known because of Jupiter's radio emission.) At the height of Jupiter's clouds the magnetic field is 10 times that of the Earth, which itself has a strong field.

The inner field is toroidal, shaped like a doughnut, containing several shells like giant versions of the Earth's Van Allen belts. High-energy protons and electrons are trapped there. The satellites Amalthea, Io, Europa, and Ganymede travel through this region. The middle region, charged particles being whirled around rapidly by the rotation of Jupiter's magnetic field, does not have a terrestrial counterpart. The outer region is the interaction with the solar wind. Sometimes high-energy particles are ejected from the outer region and travel across interplanetary space. Such particles have been detected by spacecraft far across the solar system from Jupiter.

When the solar wind is strong, Jupiter's outer magnetic field (shaped like a flattened pancake) is pushed in.

The magnetic field of Jupiter interacts with the solar wind as far as 7,000,000 km from the planet and forms a shock wave, as does the bow of a ship plowing through the ocean. Pioneer 11 observed a 10-hour fluctuation

Figure 18–19 An artist's view of Jupiter's magnetosphere, the region of space occupied by the planet's magnetic field. It rotates at hundreds of thousands of kilometers per hour along the planet, like a big wheel with Jupiter at the hub. The inner magnetosphere is shaped like a doughnut with the planet in the hole. The highly unstable outer magnetosphere is shaped as though the outer part of the doughnut had been squashed. The outer magnetosphere is "spongy" in that it pulses in the solar wind like a huge jellyfish. It often shrinks to one-third of its largest size.

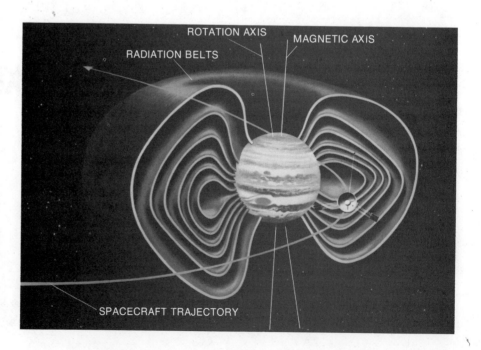

ROTATION AXIS
MAGNETIC AXIS
RADIATION BELTS
SPACECRAFT TRAJECTORY

in the magnetic field, matching Jupiter's period of rotation. The spacecraft discovered that this fluctuation occurs because the magnetic axis is tilted 10 degrees from the axis of rotation and the center of the field is not exactly at the center of the planet. The magnetic field thus resembles a wobbly disk. Also, Jupiter's magnetic poles are reversed compared with the Earth's: our north pole is on the same side of the plane of the planetary orbits as is Jupiter's south pole.

At the time that Pioneer 10 reached the distance of Saturn's orbit in 1976, it passed into a zone where the solar wind was absent. This is undoubtedly the effect of a long magnetic tail that Jupiter has. The tail may be a billion kilometers long and results from the stretching of some of Jupiter's magnetic field lines by the solar wind. These field lines prevent the solar wind from entering the tail region, which extends over 635 million km from Jupiter away from the Sun, 100 times longer than the Earth's magnetic tail. Saturn enters Jupiter's magnetic tail once every 20 years, including April 1981.

Voyager 1 found a region of interplanetary space between Jupiter and Mars that has a temperature of 300 billion to 400 billion degrees Celsius. These temperatures, the highest yet measured in the solar system, were located in a layer of charged particles (hydrogen, sulphur, and oxygen ions) that are believed to come from the satellite Io. The temperatures were measured about 5 million kilometers from the Jovian surface at Jupiter's magnetopause, the location where Jupiter's magnetic field is just strong enough to maintain the particles in one place. The high temperatures were not hazardous to the spacecraft because the density of ions is extremely low there.

High temperature merely means that the individual ions are moving around with high velocity. Thus, though each ion may have a relatively large energy, there are few ions so the total energy is low.

*18.3e Spacecraft to Jupiter

All these discoveries, and more, were made with spacecraft that were subject to the most severe environmental conditions and design limitations. Because they must function so far from the Sun, the Pioneers and Voyagers carry nuclear power generators instead of the solar cells that we can use on spacecraft sent to the terrestrial planets.

The spacecraft had to be especially reliable because of the long time duration of their journeys. And they were on their own to a large extent during the most crucial moments of their Jupiter flyby because they were then so far away that radio signals took about 52 minutes to travel from the spacecraft to Earth. Thus even if something had gone wrong with Voyager, it would have taken 52 minutes for this information to reach the flight controllers at the Jet Propulsion Laboratory in Pasadena, and an additional 52 minutes for any new instructions to travel back to the spacecraft, at least an hour and a half in all.

Even detecting the signal from the spacecraft is a good trick. The amount of energy collected from a Voyager by JPL's 63-meter radio telescope is so small that it would have to be collected for billions of years in order to light a Christmas tree bulb for just one second!

Figure 18–20 The volcanoes of Io can be seen erupting on this photograph. Linda Morabito, a JPL engineer, discovered the first volcano. She saw the large volcanic cloud barely visible on the right limb while studying images that had been purposely overexposed to bring out the stars so that they could be used for navigation. The plume extends upward for about 250 km and is scattering sunlight toward us. The bright spot just leftward of Io's terminator is a second volcanic plume projecting above the dark surface into the sunlight.

18.4 JUPITER'S MOONS CLOSE UP

Through Voyager closeups, the satellites of Jupiter have become known to us as worlds with personalities of their own. The four Galilean satellites,

Figure 18–21 Two views, taken two hours apart, as an erupting volcano rotates onto Io's limb.

Densities derived from Voyager observations (g cm⁻¹):

Io	*3.5*
Europa	*3.0*
Ganymede	*1.9*
Callisto	*1.8*

Io's surface, orange in color and covered with strange formations, led Bradford Smith of the University of Arizona, the head of the Voyager imaging team, to remark that "It's better looking than a lot of pizzas I've seen."

in particular, were formerly known only as dots of light. Even the Pioneers had detected only a hint of surface structure. Since these satellites range between 0.9 and 1.5 times the size of our own Moon, they are substantial enough to have interesting surfaces and histories (Color Plate 39 and the picture opening this Chapter).

Io provided the biggest surprises. It had been known that particles had been given off from Io as it went around Jupiter, but Voyager 1 discovered that these were the results of active volcanoes on the satellite (Figs. 18–20 and 18–21). Eight volcanoes were seen actually erupting, many more than erupt on the Earth at any one time. When Voyager 2 went by a few months later, seven of the same volcanoes were still erupting.

Io's surface (Figs. 18–22 and 18–23 and Color Plates 40 and 41) has been transformed by the volcanoes, and overall is the youngest surface we have observed in our solar system. Gravitational forces from the other Galilean satellites pull Io slightly inward and outward from its orbit around Jupiter, flexing the satellite. Its surface moves in and out by as much as 100 meters. This creates heat from friction which presumably heats the interior and leads to the vulcanism. The surface of Io is covered with the sulfur compounds given off. Io certainly wouldn't be a pleasant place to visit.

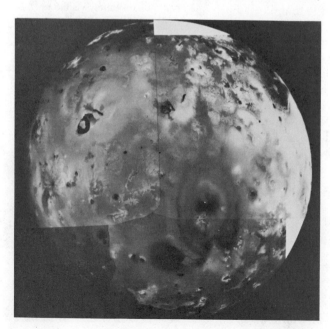

Figure 18–22 A mosaic of Io with a resolution of 8 km, showing volcanic features and no craters. The heart-shaped region at lower center surrounds an erupting volcano named for Pele, the Hawaiian volcano goddess. Its shape changed substantially between this Voyager 1 image and the Voyager 2 images.

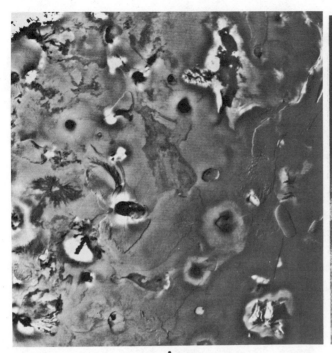

A   B

Figure 18–23 *(A)* A closeup of Io's surface, showing a variety of volcanic flows and calderas, mountains, plains, and scarps. *(B)* A volcanic caldera 50 km in diameter on Io. Dark flows over 100 km long extend from it.

Sulfur and other material given off by Io fill a doughnut-shaped region that surrounds Io's orbit. Further, as Io moves in its orbit through the Jovian magnetic field, a huge electric current is set up between Io and Jupiter itself in a tube of magnetic field called a *flux tube* (Fig. 18–24).

Europa (Color Plate 42), the brightest of Jupiter's Galilean satellites, has a very flat surface and is covered with narrow dark stripes (Fig. 18–25). This suggests that the surface we see is ice. The markings may be fracture systems in the ice. Few craters are visible, suggesting that the ice was soft enough below the crust to close in the craters. Either internal radioactivity or a gravitational heating like that inside Io may have provided enough heat to allow this.

The largest satellite, Ganymede (Color Plate 43), shows many craters (Fig. 18–26) alongside grooved terrain (Fig. 18–27). Ganymede is larger than Mercury but is less dense. It may contain large amounts of water and ice surrounding a rocky core. Ganymede's rotation is linked to its orbital period. The appearance of the side that always faces away from Jupiter is dominated by a dark cratered region (Fig. 18–28). Few craters are visible near its center. It has been tentatively concluded that the formation of the younger grooved terrains may have erased the evidence of the ancient impacts there.

Callisto (Color Plate 44) has so many craters (Fig. 18–29) that its surface is the oldest of Jupiter's Galilean satellites. A huge bull's-eye formation (Fig. 18–30) contains about 10 concentric rings, no doubt resulting from a huge impact. Spectra indicate that Callisto is probably also covered with ice.

Voyager also observed Amalthea, Jupiter's innermost satellite. This small chunk of rock is irregular and oblong in shape. It is comparable in size to many asteroids, ten times larger than the moons of Mars and ten times smaller than the Galilean satellites.

Voyager 2 discovered two previously unknown satellites of Jupiter,

All 4 Galilean satellites always keep their same face toward Jupiter.

Figure 18–24 As Io orbits Jupiter, a tube of magnetic field, a *flux tube,* extends up out of the plane of its orbit and carries a huge current.

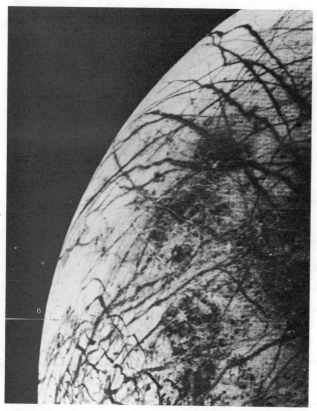

Figure 18–25 The surface of Europa, which is very smooth, is covered by this complex array of streaks. Few impact craters are visible.

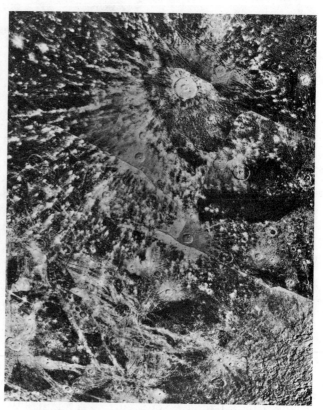

Figure 18–26 This mosaic of Ganymede, Jupiter's largest satellite, shows numerous impact craters, many with bright ray systems. The large crater at upper center is about 150 km across. The mountainous terrain at lower right is a younger region, as is the grooved terrain at bottom center.

Figure 18–27 Large areas of Ganymede are covered with terrain that is covered with many grooves tens of kilometers wide and, in some cases, thousands of kilometers long. Younger grooves cover older grooves.

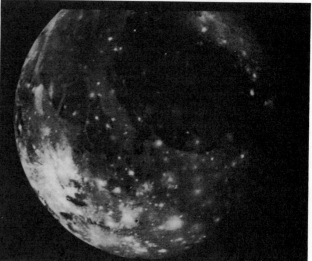

Figure 18–28 The side of Ganymede facing away from Jupiter has a dark feature about 3200 km across, probably an impact basin. Recent impact craters appear as bright spots.

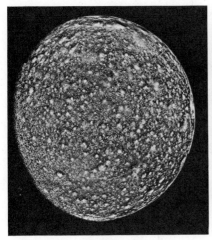

Figure 18-29

Figure 18-30

Figure 18-29 Callisto is cratered all over, very uniformly. Several have rays or concentric rings. No craters larger than 50 km are visible, from which we deduce that Callisto's crust cannot be very firm. The fact that the limb is so smooth indicates that there is no high relief. Because there are so many craters, the surface must be very old, probably over 4 billion years.

Figure 18-30 The other side of Callisto contains this remarkable impact basin, similar to Mare Oriental on the Moon and the Caloris Basin on Mercury. The bright spot is about 600 km across and the outer ring is about 2600 km in diameter. The rings may be ripples that were caused by a meteoritic impact and then quickly froze. We now see the rings as variations in albedo; they have lost their relief.

making the total of known moons 14, with an additional moon (one of those discovered with the Schmidt telescope at Palomar) as yet unconfirmed.

18.5 JUPITER'S RING

Voyager 1 discovered a wispy ring of material around Jupiter (Fig. 18–31). As a result, Voyager 2 was targeted to take a series of photographs of the ring. From the far side looking back, the ring appeared unexpectedly bright (Fig. 18–32) even to those who had programmed the cameras to make the exposures on the basis of the Voyager 1 observations. The probable explanation for this brightness is that small particles in the ring scatter the light in the forward direction, that is, toward us. Within the main ring, fainter material appears to extend down to Jupiter's surface (Fig. 18–33).

The ring particles may come from Io, or else they may come from comet and meteor debris or from material knocked off the innermost moons by meteorites. The individual particles are probably in the ring only temporarily. Perhaps Jupiter's newly discovered innermost moon causes the rings to end where they do. We can now start to compare Jupiter's rings with rings of other planets.

In the next chapter, we shall discuss the fundamental ideas about how gravity leads to the formation of planetary rings, in the context of the most famous ring of all: that of Saturn.

Only a few years ago, we thought that Saturn was the only ringed planet in the solar system, and constructed theories to explain why that is so. Now we know that Jupiter and Uranus have rings as well, and we therefore suspect that we may find one around Neptune too when we are able to look carefully enough, or even around the Sun.

18.6 FUTURE EXPLORATION OF JUPITER

Voyager 1 and Voyager 2 were spectacular successes. The data they provided about Jupiter itself, its satellites, and its ring, dazzled the eye and revolutionized our understanding.

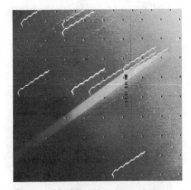

Figure 18-31 The image from Voyager 1 on which Jupiter's ring was discovered. The trails and wavy motion of the star images result from spacecraft motion during the 11-minute exposure. The ring, similarly, appears as a multiply-exposed streak.

Figure 18-32 Jupiter's ring extends outward from Jupiter's bright limb in this photograph taken by Voyager 2 looking back from the dark side. The lower ring image is cut short by Jupiter's shadow.

Figure 18–33 A mosaic of Jupiter's ring from Voyager 2. It is slightly blurred by spacecraft motion, especially in the left-most image.

"And David put his hand in his bag, and took thence a stone, and slung it, and smote the Philistine [Goliath] in his forehead."
First Samuel 17:49

Figure 18–34 An artist's drawing of the probe portion of the Jupiter Orbiter Probe, also known as Project Galileo, as it descends into the Jovian atmosphere. The Jupiter Orbiter Probe is to be the first planetary mission launched from the Space Shuttle.

Jupiter's large mass makes it a handy source of energy to use in order to send probes to more distant planets. Just as a sling, like David's, transfers energy to a stone, some of Jupiter's energy can be transferred to a spacecraft through a gravitational interaction. This *gravity assist* method has been used to send the Pioneer 11 and the Voyagers beyond Jupiter to Saturn. Voyager 2 may even go on to Uranus and Neptune.

NASA is planning Project Galileo, which will provide one spacecraft to orbit Jupiter and another to drop a probe into Jupiter's clouds. The project may use a gravity assist from Mars to get to Jupiter. The Orbiter will orbit Jupiter a dozen times in a 20-month period, coming so close to several of Jupiter's moons that pictures will have a resolution 10 to 100 times greater even than those from the Voyagers. The Probe will transmit data for an hour as it falls through the Jovian atmosphere (Fig. 18–34). We expect to lose contact with it after it penetrates the clouds for 130 km or so. It should give us accurate measurements of Jupiter's composition. The comparison to the Sun's composition should help us understand Jupiter's origin.

The two parts of Galileo had been scheduled for a joint launch aboard Space Shuttle in 1982, but the continued delays in the Space Shuttle program and the unfortunate fact that the Shuttle's lifting capabilities are not as great as had been planned has forced a delay in the Galileo launches until 1984. Further, the two parts of Galileo will be launched separately, adding substantially to the cost of the mission (and increasing the chances of future cuts in its funding).

SUMMARY AND OUTLINE

Fundamental properties (Section 18.1)
  Highest mass and largest diameter of any planet
  Low density: 1.3 g cm^{-3}
  Composition primarily of hydrogen and helium; some ammonia and methane present
  Rapid rotation, causing oblateness of disk
  Colored bands on surface; Great Red Spot
  Intense bursts of radio radiation
  Strong magnetic field and radiation belt
Jupiter's moons (Section 18.2)
  At least 15 satellites
  Galilean satellites are largest
  Three groups of satellites
    Innermost moons, including Galilean satellites: orbits less than 2 million km in radius

  4 next moons: orbits about 12 million km in radius
  Outermost moons: orbits 21 to 24 million km in radius
Spacecraft observations (Section 18.3)
  Pioneer 10 and 11 flybys in 1973 and 1974
  Voyager 1 and 2 flybys in 1979, with greatly increased resolution
  Many photographs taken of cloud structure
  Great Red Spot is a giant anticyclonic storm
  No solid surface: gaseous atmosphere and liquid interior
  Jupiter radiates 2 to 3 times the heat it receives from the Sun (confirmed previous observations)
  Colored bands the result of huge convection currents and the rapid rotation

Pioneer 11 and Voyagers photographed the Jovian polar regions, a task impossible from Earth
Tremendous magnetic field detected
Spacecraft observations of Jupiter's moons (Section 18.4)
Io is transformed by volcanoes, 8 of which were seen erupting

Europa has a flat surface, criss-crossed with dark markings
Ganymede has many craters and grooved terrain
Callisto is covered with craters, including a bull's-eye
Amalthea is irregular and oblong
Jupiter's ring (Section 18.5)
A narrow ring, composed of small particles

QUESTIONS

1. Why does Jupiter appear brighter than Mars despite its greater distance from the Earth?

2. Assume that the Jovians put a dye in their atmosphere such that there is a green line running from the north pole to the south pole. Sketch the appearance of this line a few days later.

3. Even though Jupiter's atmosphere is very active, the Great Red Spot has persisted for a long time. How is this possible?

4. (a) How did we first know that Jupiter has a magnetic field? (b) What did the recent studies show?

5. In Roemer's measurement of the speed of light, compare how "late" the moons of Jupiter appeared to arrive in their predicted positions when Jupiter was at its farthest point from the Earth compared to when Jupiter was closest to the Earth.

6. It has been said that Jupiter is more like a star than a planet. What facts support this statement?

7. What advantages over the 5-m Palomar telescope on Earth did Voyagers 1 and 2 have for making images of Jupiter?

8. What are two other types of observations other than photography made from the Voyagers to Jupiter?

9. How does the interior of Jupiter differ from the interior of the Earth?

10. From measurements made on Figure 18–13, calculate—in km hr^{-1}—the speed of rotation of the white blob caught in the Great Red Spot.

11. Which moons of Jupiter are icy? Why?

12. Contrast the volcanoes of Io with those of Earth.

13. Compare the surfaces of Callisto, Io, and the Earth's Moon. Show what this comparison tells us about the ages of features on the surfaces.

14. Using the information given in the text and in Appendices 5 and 6, sketch the Jupiter system including all the moons and the ring. Mark the groups of moons.

15. (a) Describe the "gravity assist" method. (b) What does it help us do?

PROJECTS

1. Describe in some detail the origins of the names of each of Jupiter's moons in Greek mythology.

2. Over a four-hour interval one night, plot the positions of the Galilean satellites at half-hour intervals.

3. Over a one-week interval, plot the positions of the Galilean satellites from night to night. Using the observations in 2 and 3 together, deduce the projection on the sky of the orbits of the individual satellites.

TOPICS FOR DISCUSSION

1. Although many of the most recent results about Jupiter came from spacecraft, prior ground-based studies had told us many things. Discuss the status of our pre-1973 knowledge of Jupiter, and specify both some things about which space research did not add appreciably to our knowledge and some things about which space research led to a major revision of our knowledge.

2. If we could somehow suspend a space station in Jupiter's clouds, the astronauts would find a huge gravitational force on them. What are some of the effects that this would have? (The novel *Slapstick* by Kurt Vonnegut deals with some of the problems that would arise if gravity were different in strength from what we are used to.)

CHAPTER 19

SATURN, URANUS, NEPTUNE, AND PLUTO

♄ ♅ ♆ ♇

AIMS:

To study the three other giant planets: Saturn, Uranus, and Neptune; to marvel at Saturn's rings; and to see how little we know about Pluto

19.1 SATURN

Saturn is the most beautiful object in our solar system, and possibly even the most beautiful object we can see in the sky. The glory of its system of rings makes it stand out even in small telescopes.

Saturn, like Jupiter, Uranus, and Neptune, is a giant planet. Saturn is 9.5 A.U. from the Sun, and has a lengthy year, equivalent to 30 Earth years.

The giant planets are characterized by low densities. Saturn has the lowest density of any planet in our solar system: only 0.7 g cm^{-3}, which is 70 per cent the density of water. Thus, if we could find a big enough bathtub, Saturn, like Ivory Soap, would float (Fig. 19–1). The bulk of Saturn is made of hydrogen molecules and of helium. Deep enough in the interior where the pressure is sufficiently high, the hydrogen molecules are converted into metallic hydrogen. Saturn could have a core of heavy elements making up 20 per cent of its interior.

The rings extend far out in Saturn's equatorial plane, and they are inclined to the planet's orbit by 27° (Fig. 19–2). Over a 30-year period, we sometimes see them from a vantage point of 27° above their northern side, sometimes from 27° below their southern side, and at intermediate angles at intermediate times. When seen edge on, they are all but invisible (Fig. 19–3).

Saturn's mass is 95 times that of Earth, and its diameter, without the rings, is 9 times greater.

Saturn is shown in Color Plates 45 and 46.

Figure 19–1 Saturn's density is lower than that of water.

Saturn and some of its moons, photographed from the Voyager 1 spacecraft. Dione is in the foreground, Tethys and Mimas are above it and to Saturn's right, Enceladus and Rhea are to the left, and Titan in its distant orbit is at the top. The symbols for Saturn, Uranus, Neptune, and Pluto appear from left to right at the top of this page.

355

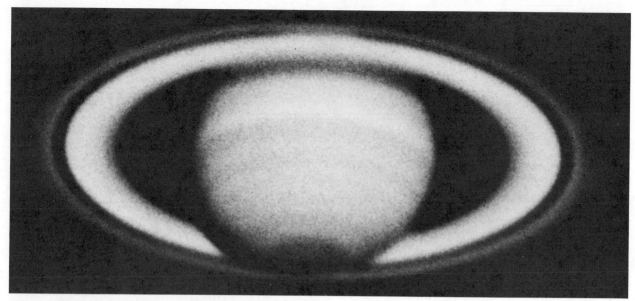

Figure 19–2 The rings of Saturn, photographed on January 19, 1973. (New Mexico State University Observatory photograph)

Artificial satellites that we send up to orbit around the Earth are constructed of sufficiently rigid materials that they do not break up even though they are within the Earth's Roche limit.

Figure 19–3 The rings of Saturn photographed at various times. The rings extend far out in the equatorial plane of Saturn, and range from about 70,000 to over 135,000 km from the planet's center. They are inclined to Saturn's orbit by 27°. Over a 30-year period, we sometimes see them from a vantage point of 27° above their northern side, sometimes from 27° below their southern side, and at intermediate angles at intermediate times. When seen edge on, they are all but invisible. (Lowell Observatory photographs)

19.1a Saturn's Rings

The rings of Saturn are either material that was torn apart by Saturn's gravity or material that failed to accrete into a moon at the time when the planet and its moons were forming. These bits of matter spread out in concentric rings around Saturn. There is a sphere for each planet, called *Roche's limit*, or the *Roche limit*, inside of which blobs of matter cannot be held together by their mutual gravity. The forces that tend to tear the blobs apart from each other are called *tidal forces* (which are also discussed in Section 9.8 on binary stars and in Section 16.4 on the Earth); they arise because some blobs are closer to the planet than others and are thus subject to higher gravity. (The differing gravitational forces of the Moon on different places

---

°Box 19.1   The Roche Limit

E. A. Roche defined the limit in 1849 as the smallest distance within which a liquid body does not break up because the pull of the body's own gravitation holding itself together is stronger than the tidal forces pulling it apart. Similar limits can be defined for solid bodies and non-spherical bodies, and for the case where the orbiting body is taking on individual molecules. The limits depend on both the mass of the planet and the density of the ring particles. Thus the existence of rings at a given radius can help distinguish whether the rings are made of ice or of stony material. Saturn's rings, to be within the Roche limit, must be made primarily of ice.

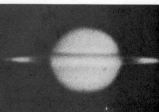

on the Earth, and to a lesser extent the differing gravitational forces of the Sun, cause the ocean tides on Earth.) The radius of the Roche limit varies with the amount of mass in the parent body. The Sun also has a Roche limit, but all the planets lie outside it. All the moons of the various planets lie outside the respective Roche limits. Saturn's rings lie inside Saturn's Roche limit, so it is not surprising that the material in the rings is not collected into a single orbiting body.

Box 19.2   Cassini's Division

In 1610, Galileo had used his new invention, the telescope, to discover that Saturn was not round; it seemed to have "ears." The Dutch astronomer Christian Huygens published an anagram in 1656 explaining that Saturn has a ring; Huygens' more conventional publication of his result in a book followed three years later. Cassini, who worked in Paris some years later, observed the largest gap in the ring in 1675, and this gap is named after him (the Cassini division). The idea that the ring is composed of many bodies orbiting independently was proposed by James Clerk Maxwell in England in 1856. The existence of the rings of Saturn had an important role in the development of the nebular hypothesis by Kant and Laplace.

There are several concentric rings around Saturn. The brightest ring is separated from a fainter outer ring by what is called *Cassini's division* (Fig. 19–2). Another ring is inside the brightest ring. The gaps in the ring structure result from gravitational effects from Saturn's satellites. Finer structure in the rings has been studied from spacecraft, as we shall see.

The rotation of Saturn's rings can be measured directly from Earth with a spectrograph, for different parts of the rings have different Doppler shifts (Fig. 19–4). We know that the rings are not solid objects, because of their differential rotation. Also, on at least one occasion, stars occulted by the rings could be seen shining dimly through. Even though the rings are 275,000 km across, they are very thin from top to bottom, relatively much flatter than a phonograph record. Radar waves were first bounced off the rings in 1973. The result of the radar experiments show that the particles in the ring are probably rough chunks of ice at least a few centimeters and possibly a meter across. Infrared spectra show that they are covered with ice.

19.1b   Ground-Based Observations

Like Jupiter, Saturn rotates very quickly on its axis, also in about 10 hours. As a result, it is oblate by 10 per cent. Saturn has delicately colored bands of clouds. The clouds rotate about 10 per cent more slowly at high

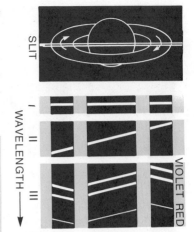

Figure 19–4   If a slit is laid across Saturn, as shown at top, then the spectrum appears as shown at III on the bottom. If Saturn were not rotating at all, then no Doppler shift would be seen, as in I. If Saturn and its rings rotated as a solid body, then the spectrum would appear as in II. The fact that the spectrum looks like III indicates that Saturn itself rotates as a solid body but that each rock in the rings has an orbit that obeys Kepler's laws.

A similar spectrographic method tells us about the rotation of spiral galaxies.

The powerful radar beam from the 305-m (1000-ft) radio telescope at Arecibo, Puerto Rico, travelled 2 hours and 14 minutes at the speed of light to Saturn and back to the 64-m (210-ft) JPL antenna at Goldstone, California.

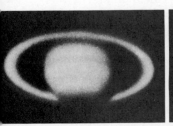

Figure 19–5 A 1980 image of Saturn showing additional satellites and a wide, faint extra ring. The image of Saturn is blocked out so that the images of these other faint objects can be seen.

latitudes than they do at the poles. Methane, ammonia, and molecular hydrogen have been detected spectrographically in the clouds.

Although we know very little about what is under Saturn's clouds, we calculate that Jupiter and Saturn should turn liquid under the tremendous pressures present at great depths, but there is no evidence that there is a solid surface at any level above the core. If you fell into Jupiter or Saturn, you would descend into denser and denser slush.

Saturn gives off radio signals, as does Jupiter, an indication to earthbound astronomers that Saturn also has a magnetic field. Just as for Jupiter, Saturn has a source of internal heating.

Until recent discoveries from the Earth and from space, we knew of 9 moons of Saturn. All but the most recently discovered have beautiful names taken from Greek mythology.

Additional moons of Saturn have been reported from time to time. The best time to look from the Earth is at the times when Saturn's rings are viewed edge on and thus fade from view, making it easier to see any moons. The discovery of at least two moons was reported from the 1966 passage of the Earth through the plane of the rings, but the observations were difficult to confirm. The 1980 passage is revealing several more possible moons (Fig. 19–5), and it is not yet possible to tell whether some of the images seen are the moons previously reported. The spacecraft observations that we shall soon discuss are also discovering new moons of Saturn.

All the moons, except Titan, range from about 130 km to 1600 km across. Planet-size Titan, however, is a different kind of body. An atmosphere has been detected on Titan, and methane has been found in it. Because of this atmosphere, a greenhouse effect may have warmed the surface of Titan, and so Titan has become one of the more interesting places in the solar system on which we can search for life—some astronomers think that there may even be more chance of finding life there than on Mars.

Saturn's moons are named after the Titans, the children and grandchildren of Gaea, the goddess of the Earth, who had been fertilized by drops of Uranus' blood. The 9 outermost moons are called Mimas, Enceladus, Tethys, Dione (mother of Aphrodite), Rhea, Titan, Hyperion (the father of Helios), Iapetus, and Phoebe.

All the moons except Phoebe and Iapetus are in orbits inclined no more than 1° or 2° to the plane of Saturn's equator. Iapetus' orbit is inclined 15°. Phoebe's orbit is inclined 150° and is very eccentric, so Phoebe may well be an asteroid captured at some time in the past.

19.1c Space Observations of Saturn

Pioneer 11, which had passed Jupiter in 1974, reached Saturn in 1979. But Pioneer 11 had the handicap of having to travel across the diameter of the solar system from Jupiter to reach Saturn, while the Voyagers had a much shorter route. Voyager 1 reached Saturn in 1980, only a year after Pioneer 11. Between the time of the launch of Pioneer 11 and the launch of the Voyagers, experimental and electronic capabilities had improved so much that the Voyagers could transmit data at 100 times the rate of Pioneer II. Voyager 2 is scheduled to reach Saturn in 1981.

The most obvious observations to describe are those of the rings themselves, which were visible for the first time from a substantially different vantage point from the one we have on Earth. The resolution of Voyager 1's cameras was much higher than the resolution of Pioneer 11's instrument, which had been specifically designed for making measurements of polarization rather than for photography. The backlit view obtained by Pioneer 11 (Fig. 19–6) showed that Cassini's division, visible as a dark (and thus apparently empty) band from Earth, appeared bright, with a dark line of material running through it. The outer "A" ring showed structure as well. The brightest ring observed from Earth, the "B" ring, appeared dark on the Pioneer 11 view, presumably because it is too opaque to allow light to pass through it. The rings appear to have low mass and therefore low density, which suggests that they are made up largely of ice. This would also explain the atomic hydrogen that was found around the rings, probably resulting from dissociation of water ice.

Analysis of gravity field and temperature measurements suggest that Saturn's core extends about 13,800 km from the center, making it about twice the size of the Earth. It is so compressed by Saturn's mass that the core contains about 11 Earth masses of matter. The inner core is mostly iron and rock, and the outer core consists of ammonia, methane, and water. The next 21,000 km out seem to consist of liquid metallic hydrogen, which is consistent with the discovery of Saturn's magnetic field.

Saturn radiates into space about 2.5 times more heat than it absorbs from the Sun. One interpretation is that only 1/3 of Saturn's heat is left over from its formation and from continuing gravitational contraction, with the rest of the heat being generated by helium sinking through the liquid hydrogen in Saturn's interior.

The passage of Voyager 1 by Saturn was one of the most glorious events of the space program. Though during the spacecraft's approach, the subtleness of colors in Saturn made photographs of the surface less spectacular than those of Jupiter, the structure and beauty of the rings during the closest approach dazzled all. (See the Cover, Color Plate 46, and Fig. 19–7.)

A new "F" ring was discovered outside the already known rings, separated by a 3000-km gap known as the "Pioneer division."

Figure 19–6 Though Pioneer 11, which passed Saturn in 1979, gave images far inferior to those from the Voyagers, it gave the first view of Saturn's rings when illuminated by the Sun from the other side. The rings appeared in many ways as inverses of the rings as seen from the front. Saturn's moon Rhea is below the planet. (NASA/ Ames; U. of Arizona image processing) Compare the Voyager image on page 545.

Voyager encounters:
Voyager 1   November 1980
Voyager 2   August 1981

Figure 19–7 Saturn and its satellites Tethys *(outer left),* Enceladus *(inner left)* and Mimas *(right top of rings)* in a mosaic taken by Voyager 1 from a distance of 18 million km, 2 weeks before closest approach. (All Voyager photos courtesy of Jet Propulsion Laboratory/NASA)

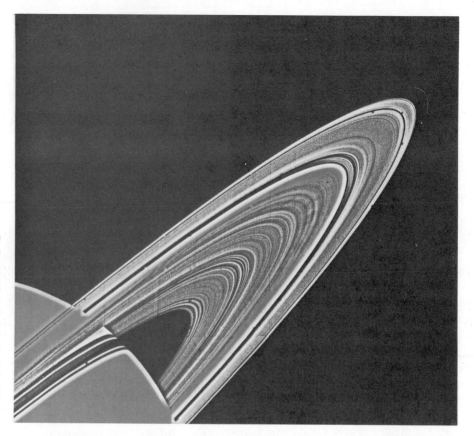

Figure 19–8 The complexity of Saturn's rings. About 100 can be seen on this enhanced image, taken from a range of 8 million km. Several narrow rings can even be seen inside the dark Cassini division. A moon discovered by Voyager 1, Saturn's 14th, appears at upper right, just inside the narrow F-ring. It apparently shepherds particles, like a sheep dog, keeping them from falling inward. Another newly discovered moon, just outside the ring, keeps the particles from escaping.

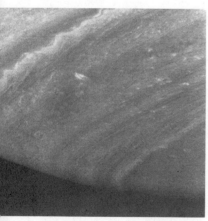

Figure 19–9 Waves and eddies in Saturn's clouds at mid-southern and polar latitudes, where the large-scale bands and zones break down into such small-scale features.

The closer the spacecraft got to the rings, the more rings became apparent (Fig. 19–8). Before Voyager, scientists discussed whether there were three, four, five, or six rings. But as Voyager approached, the photographs increasingly showed that each of the known rings was actually divided into many thinner rings. By the time Voyager 1 had passed Saturn, we knew of hundreds of rings or even a thousand.

Before the flyby, it had been thought that the structure of the rings and the gaps between them were determined by gravitational effects of Saturn's moons on the orbiting ring particles. But this explanation had been worked out with only a half-dozen rings known. The ring structure Voyager discovered is too complex to be explained in this way alone. Consideration of the sixteen or so satellites cannot explain a thousand rings. The rings may have objects tens of kilometers across in them in addition to the icy snowballs that probably make up the bulk of the ring material. These larger objects may help determine the distribution of matter in the rings.

The structure in Saturn's clouds is of much lower contrast than that in Jupiter's clouds. Even so, cloud structure was revealed to Voyager at its closest approach (Fig. 19–9). A few circulating ovals similar to Jupiter's Great Red Spot and ovals were detected, as was turbulence in the belts and zones. Study of these features also gave the first direct view of the rotation of Saturn's clouds and allowed detailed measurements of the velocities of rotation to be made.

Pioneer 11 had revealed the magnetic field at Saturn's equator to be somewhat weaker than had been anticipated, only ⅔ of the field present at the Earth's equator. Remember, though, that Saturn is much larger than the Earth and so its equator is much farther from its center. Saturn's magnetic

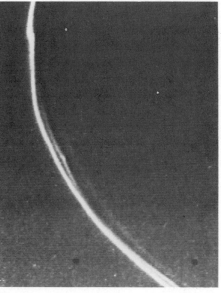

A   **B**

Figure 19–10 (*A*) This composite shows detail in one of Saturn's inner rings. The horizontal line marks the border between the two photos, each of which shows a different edge of the rings. In the dark gap in the center of each picture we see a bright ring that is narrowed in the lower picture and slightly broadened and displaced in the upper. (*B*) Complex structure in the outer, F, ring shown in Figure 19–8. Two narrow, braided bright rings are visible, as is a broader diffuse component about 35 km wide.

axis appears to be aligned with its rotational axis, unlike Jupiter's, the Earth's, and Mercury's. Perhaps the source of Saturn's magnetic field lies close to Saturn's center. The total strength of the magnetic field is 1000 times stronger than Earth's and 20 times weaker than Jupiter's. Saturn's radiation belts are disrupted by the rings.

Everyone had expected that collisions between particles in Saturn's rings would make the rings be perfectly round and uniform. Thus, the discovery from Voyager 1 that there was changing radial structure in the rings came as a big surprise. "Spokes" can be seen in the rings when they are seen at the proper angle (look back at Fig. 19–7). Differential rotation must make the spokes dissipate in hours after they form. The spokes look dark from the front side, but look bright from behind, the view from Voyager when it passed Saturn and looked back. This information gives the distribution of particles in the spokes, since light is bounced around differently by particles of different sizes.

It also came as a surprise that some of the rings are not round. Figure 19–10A compares the rings seen on opposite sides of Saturn. At least one of the rings is brighter and displaced on one side compared with its appearance on the other. And scientists were astonished to find that the outer ring, the narrow F-ring discovered by Pioneer 11, seems to be made of three braids (Fig. 19–10*B*).

Farther out from Saturn than the rings, we find the satellites of Saturn. Like those of Jupiter, they now seem more like independent worlds than like mere moons. From points of light they have been transformed into bodies with features visible in detail. The largest of Saturn's satellites, Titan, is larger than the planet Mercury and has an atmosphere (Fig. 19–11). Voyager has shown that Titan, under its atmosphere, is slightly smaller than Jupiter's Ganymede.

Ground-based spectra had detected methane, revealing that Titan has a thin atmosphere. Studies of how the radio signals faded when Voyager went behind Titan showed that Titan's atmosphere is denser than Earth's. And Voyager's ultraviolet spectrometer detected nitrogen, which makes up the bulk of Titan's atmosphere. The methane is only a minor constituent,

Figure 19–11 (*A*) Titan was disappointingly featureless even to Voyager's cameras because of its thick smoggy atmosphere. Its northern polar region was relatively dark. (*B*) Haze layers can be seen at the limb, with divisions at altitudes of 200, 375, and 500 km.

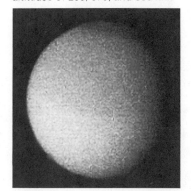

A

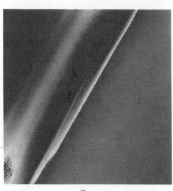

B

A, Mimas

B, Mimas

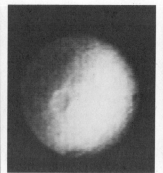

C, Tethys

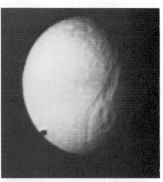

D, Tethys

Figure 19–12 (*A*) The impact feature on Mimas is about 130 km in diameter. (*B*) The other side of Mimas, showing the trough crossing the center that may be the result of the impact. (*C*) A bright circular feature on Tethys. (*D*) Craters and a 750-km-long valley on Tethys.

perhaps 1 per cent. Several layers of haze have been detected. Titan is surrounded by a huge doughnut-shaped hydrogen cloud that presumably results from the breakdown of its methane (CH_4).

The atmosphere completely shrouds Titan so that no surface can be seen. Titan's northern polar region appears darker than its equator or southern hemisphere. It seems that Titan's atmosphere is opaque because of the action of sunlight on chemicals in it, forming a sort of "smog." Smog on Earth forms in a similar way. The hydrocarbons formed may give Titan its reddish tint. Some of the color may result from the bombardment of Titan's atmosphere by protons and electrons trapped in Saturn's magnetic field. Titan, like Jupiter's Io, generates an electric current as it moves.

It may well be that chemicals akin to gasoline fall out of Titan's sky as rain. Lakes or oceans of liquid nitrogen may cover some part of Titan's surface, though it may be too warm for these to form away from the poles. These chemicals on Titan would be similar to those from which we think life evolved on the primitive Earth. But it would probably be too cold on Titan for life to begin. The temperature near the surface, deduced from measurements made with Voyager's infrared radiometer, is only about −183°C (−300°F), somewhat warmed by the greenhouse effect but still extremely cold.

Saturn's other moons, all icy, reveal a variety of surfaces. They are so cold that the ice acts as rigid as rock and can retain craters. The moon's mean densities are all 1.0 to 1.5, which means that they are probably mostly ice throughout with some rocky material included. Four of them are over 1000 km across, so we are dealing with major bodies (see Appendix 6), about ⅓ the size of Earth's Moon. For Saturn as for Jupiter, most satellites perpetually have the same side facing their planet. Let us consider the major ones in order from the inside out.

Moon	Radii (km) from Voyager
Mimas	*195*
Enceladus	*245*
Tethys	*525*
Dione	*555*
Rhea	*765*
Iapetus	*725*

E, Dione

F, Dione

G, Rhea

Figure 19–12 (*E* and *F*) Dione, showing impact craters, debris, and ridges or valleys. (*G*) Rhea, with craters as large as 300 km across, many with central peaks.

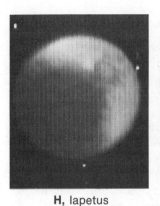

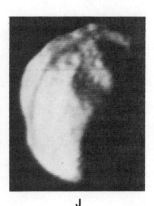

H, Iapetus   **I**   **J**

Figure 19–12 (*H*) Iapetus, which has its trailing side 5 times brighter than its leading side. The dark side may have accumulated dust spiraling in toward Saturn. A large circular feature is also visible. (*I*) Saturn's tooth-shaped 11th moon, only 135 by 70 km. (*J*) The thin line that can be seen to move across this moon in the 13 minutes between views is the shadow of an otherwise unseen ring.

Mimas boasts a huge impact structure that is over one-quarter the diameter of the entire moon (Fig. 19–12*A*). The crater has a raised rim and a central peak, typical of large impact craters on the Earth's Moon and on the terrestrial planets. The whole satellite is saturated with craters. The canyon visible half-way around Mimas (Fig. 19–12*B*) may be a result of the impact. The energy of the impact may have shattered the satellite and been focused half-way around. Many craters a few kilometers across are visible. Enceladus' surface, not shown here, is smoother than that of the other moons.

Tethys also has a large circular feature (Fig. 19–12*C*), about 180 km across. The material to its left on the picture is darker than the material to its right. The side of Tethys that faces Saturn (Fig. 19–12*D*) shows not only many craters but also a large valley about 750 km long and 60 km wide.

Dione too shows many impact craters, some of them with rays of debris (Fig. 19–12*E*). Many valleys are visible in Dione's icy crust (Fig. 19–12*F*).

We see impact craters on Rhea in Fig. 19–12*G*. Some, with sharp rims, must be fresh. Others, with subdued rims, must be ancient. Next out is Titan, which we have already discussed, and then Hyperion.

Strangely, the side of Iapetus (Fig. 19–12*H*) that precedes in its orbit is about 5 times darker than the side that trails. The large circular feature is probably an impact feature outlined by dark material. Saturn's outermost satellite is Phoebe.

In addition to these 9 moons long known, several others have been detected from the ground in the last decades, and several have been confirmed or discovered by the Pioneer and Voyager spacecraft. Voyager even provided closeups of one of the recent ground-based discoveries (Fig. 19–12*I*). It is too small (only 135 km × 70 km) for gravity to have pulled it into a spherical shape. This satellite, oddly, apparently all but shares an orbit with another newly discovered satellite. The orbits are not absolutely identical, so the two approach each other every few years. Their gravitational fields are strong enough to affect the other only when they are very close to each other. When this happens, they twirl around each other, interchange orbits, and are off on their merry ways again separating from each other. No other case of such a gravitational interaction is known in the solar system. This pair of satellites probably controls the outer edge of the ring system.

It is difficult to say how many moons Saturn has. Indeed there may be many orbiting in the rings that are 50 or 100 km across.

Yet another superlative for Voyager 1 is the view it provided of the crescent Saturn after it passed the planet and looked back (Fig. 19–13). The back side of Saturn is slightly illuminated by sunlight reflected off the rings.

Figure 19–13 The crescent of Saturn and the planet's rings, taken from 1,500,000 km from the far side of Saturn as Voyager 1 departed. The rings' shadows are visible cutting across the over-exposed crescent.

19.2 URANUS

Figure 19–14 Uranus, photographed from the ground in blue light, shows no surface markings.

Uranus and Neptune both have greenish casts when observed from the Earth, which led in times past to many diffuse objects in the sky (some of which also appear green but which are not planets at all) being named "planetary nebulae."

Figure 19–15 Its axis of rotation lies in the plane of Uranus' orbit.

There are two other giant planets beyond Saturn: Uranus (U'ranus) and Neptune. Both Uranus and Neptune are large planets, about 50,000 km across and about 15 times more massive than the Earth. The densities of Uranus and Neptune are low (1.2 and 1.7 gcm^{-3}, respectively). Their albedoes are high, which indicates that they are covered with clouds.

Uranus was discovered to be a planet by William Herschel in England in 1781, but it had been plotted on sky maps for about a hundred years prior to that with the thought that it was just another star. It revolves around the Sun in 84 years, at an average distance of more than 19 A.U. from the Sun. Uranus never gets any larger in the sky than 3.6 seconds of arc, and studying its surface structure from the surface of the Earth is very difficult (Fig. 19–14).

Even the photographs from the balloon Stratoscope, which in 1970 carried a 90-cm telescope up to an altitude of 24 km, above most of the Earth's atmosphere, showed no detail on Uranus' surface. Stratoscope's resolution was 1/6 arc sec, an improvement of about 6 times. Since Stratoscope would have seen belts on Uranus like those of Jupiter and Saturn, we know that Uranus does not have them. Molecules in its atmosphere may simply be scattering the incoming sunlight; this would also account for the high albedo. The Stratoscope observations suggest that Uranus is surrounded by a thick layer of methane clouds, with a semi-transparent atmosphere of molecular hydrogen above the methane. Both methane and molecular hydrogen have been observed with ground-based spectrographs.

The other planets rotate such that their axes of rotation are roughly parallel to their axes of revolution around the Sun. Uranus is different, for its axis of rotation is roughly perpendicular to the other planetary axes. Uranus' axis of rotation lies in the plane of its orbit (Fig. 19–15). Sometimes, one of Uranus' poles faces the Earth, 21 years later its equator crosses our field of view, and then another 21 years later we face the other pole. Thus there are

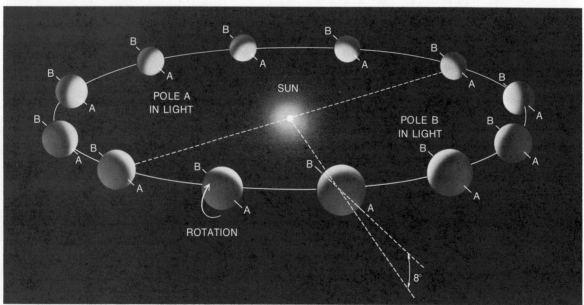

strange seasonal effects on Uranus. When we understand just how the heating affects the clouds, we will be closer to understanding our own Earth's weather systems. Uranus is always very cold. Studies of its infrared radiation give a temperature of 58 K. Thus there is no evidence for a heat source, unlike the case for Jupiter, Saturn, and Neptune.

It is said that Uranus rotates in the retrograde direction, and so it does—but barely. Its rotation is approximately perpendicular to the normal direction of rotation of the other planets. Mainly, from our point of view, we see it rotating sometimes as do the hands of a clock or in the reverse sense (with the pole facing us) and sometimes from top to bottom (with the poles at the sides). All the other planets always appear to rotate from side to side (with the poles at the top and bottom). Over the last few years different studies of Uranus' rotation have given different values for the period. The recent values vary widely, from 12 to 24 hours, which indicates how inaccurate our knowledge is of the distant members of our solar system.

One method of finding the period is to observe how the Doppler effect broadens a sharp spectral absorption line. The broadening results because the light from the entire planet enters the spectrograph simultaneously. This radiation includes Doppler-shifted contributions from each side of the planet.

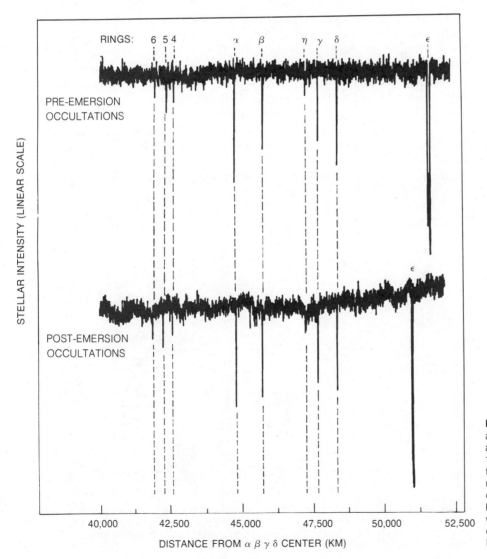

Figure 19–16   As Uranus drifted across the sky on March 10, 1977, and passed in front of a star (SAO 158687), dips in the intensity of the star showed the presence of rings occulting the starlight. The observations were made from the Kuiper Airborne Laboratory and were confirmed by ground-based observations. (Observations by J. L. Elliot, E. Dunham, and D. Mink)

Radio astronomers at the Jet Propulsion Laboratory discovered that radio emissions that originate in Uranus' atmosphere had become 30 per cent stronger over a 10-year period. The results can be explained if Uranus' atmosphere is either warming up or is becoming clearer to the passage of radio waves. The change may result from the changing orientation of Uranus' axis of rotation with respect to the Sun. The north pole is now turning toward the Sun as part of the 84-year period of changing orientation. It will point only 15° from the Earth and Sun in 1987.

In 1977, Uranus occulted a faint star. The occultation was predicted to be visible only from the Indian Ocean southwest of Australia, so a team of Cornell scientists flew in NASA's Kuiper Airborne Laboratory, an instrumented airplane, to observe it. Surprisingly, about half an hour before the predicted time of occultation, a few slight dips (Fig. 19–16) were detected in the brightness of the star. (The equipment had been turned on early because the exact time of the occultation was uncertain.) Similar dips were recorded, in the reverse order, about half an hour after the occultation. Furthermore, ground-based observers detected similar dips, even though the star was never completely occulted for them. These dips indicate that Uranus is surrounded by at least five rings (Fig. 19–17). They are called, from the innermost outward, by the first five Greek letters. Three more occultations in 1978 confirmed the discovery of these rings and suggested the possibility of additional ones. All the rings have radii between 42,000 and 52,000 km, 1.7 to 2.1 times the radius of the planet.

The rings of Uranus have been further observed, and more have been discovered. Nine are now known. Three were discovered by further analysis of the original discovery data while one was discovered at a subsequent occultation. Reexamination of the old photographs from the Stratoscope balloon revealed that the shadow of the rings could be detected on Uranus' disk.

Remember that the size of the Roche limit depends not only on the planet's mass but also on the density of the ring material. Unlike Saturn's rings, which are within Saturn's Roche limit only if they are made of ice, the rings of Uranus and Jupiter are within those planets' Roche limits for either ice or rock.

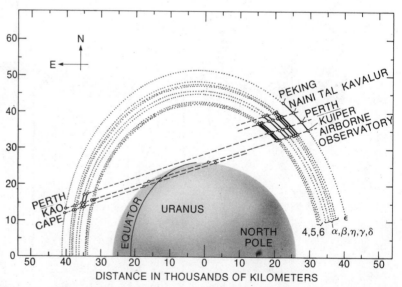

Figure 19–17   The positions of the dips in the observations by which the rings of Uranus were discovered are marked in the top diagram. Some of the rings have been assigned Greek letters and other have been called by numbers. The epsilon ring seems to be particularly out of round, and is precessing about 2° per day.

Uranian rings are very narrow (some are only a few km wide). How can narrow rings exist, when the tendency of colliding particles would be to spread out? One possibility is that a small satellite in each ring keeps the particles together.

The Caltech infrared group managed to observe the rings directly in reflected sunlight at a wavelength of 2.2 microns. At that infrared wavelength, Uranus is very dark because of absorption from atmospheric methane, while the reflectivity of the rings (always relatively poor) remains the same as at other wavelengths. The rings completely encircle the planet, but are not resolved individually. The low reflectivity of the rings probably indicates that, unlike Saturn's rings, they do not have icy surfaces.

Uranus has five moons (Fig. 19–19), each with a beautiful name. From the innermost to the outermost, they are Miranda, Ariel, Umbriel, Titania, and Oberon. The innermost, Miranda, was discovered fairly recently, in 1948, and is the smallest. The other four moons range from 400 to 1000 km across, but the radii are very poorly known. Very little is known about them, but indications that water ice exists on Umbriel, Titania, and Oberon have been found.

Voyager 2 will reach Uranus in 1986.

19.3 NEPTUNE

Neptune is even farther away than Uranus, 30 A.U. compared to about 19 A.U. Its orbital period, by Kepler's third law, is thus 165 years. Its discovery was a triumph of the modern era of Newtonian astronomy. Neptune had not been known until mathematicians analyzed the deviations from an elliptical orbit shown by Uranus. These deviations are small, but were detectable, and could have been caused by gravitational interaction with another, as yet undetected, planet.

The denouement is one of astronomy's most famous anecdotes. John C. Adams in England (Fig. 19–20) predicted positions for the new planet in

All the good reasons that astronomers had thought of over the years to explain why Saturn was the only planet with rings have been proved wrong first by the rings of Uranus and more recently by the ring of Jupiter.

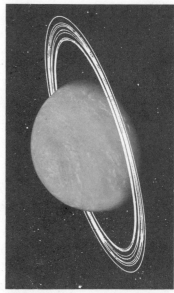

Figure 19–18 An artist's sketch of Uranus and its rings. (Drawn from data of J. L. Elliot, E. Dunham, L. H. Wasserman, R. L. Millis, and J. Churms)

OBERON

TITANIA

ARIEL

URANUS

MIRANDA

UMBRIEL

Figure 19–19 Uranus and its moons. A few stars are also in the field of view. (Photo by William Liller with the 4-m telescope at Cerro Tololo)

Figure 19–20 "1841. July 3. Formed a design in the beginning of this week, of investigating, as soon as possible after taking my degree, the irregularities in the motion of Uranus which are yet unaccounted for; . . ."

The value adopted by the International Astronomical Union in 1976 (see Appendix 5) gives heavy weight to the occultation value but also considers other results.

1845, but the astronomy professor at Cambridge did not bother to try to observe this prediction of a recent college graduate. Adams was prevented (by the butler) from interrupting the Astronomer Royal's dinner, so the two did not meet. Though Adams left a copy of his calculations, when the Astronomer Royal requested further information, partly to test Adams' abilities, Adams did not take the request seriously and did not respond at first. The very "proper" Astronomer Royal took offense and did not choose to have further dealings with Adams. The story then continues in France, where a year later Urbain Leverrier was independently working on predicting the position of the undetected planet.

When the Astronomer Royal saw in the scientific journals that Leverrier's work was progressing well (Adams' work had not been made public), for nationalistic reasons he began to be more responsive to Adams' calculations. But the search for the new planet, though begun in Cambridge, was carried out half-heartedly. Neither did French observers take up the search. By this time it was 1846. Leverrier sent his predictions to an acquaintance at Berlin, J. Galle, who enthusiastically began observing and discovered Neptune within hours.

Years of acrimonious debate followed over who (and which country) should receive the credit for the prediction. We now tend to credit both Adams and Leverrier (and mock that particular Astronomer Royal and Cambridge professor).

We now have the benefit of hindsight, and can examine the calculations of Adams and Leverrier more carefully. One assumption was necessary for them to make their calculations—they had to guess a radius for the orbit of Neptune in order to calculate its expected direction in the sky. They used the value from Bode's law (Section 12.10). This value was 39 A.U., substantially larger than the value we have since measured for Neptune. The error happened not to be of importance for the configuration of the planets at the time they were working, but at other times their calculations would not have given the right position. So there certainly was an element of luck in their successful prediction.

Neptune's orbit is so large that it takes a very long time to travel around the Sun, and it has not yet made a full orbit since it was discovered. Its angular size in our sky is so small that it is always very difficult to study. Even measuring its diameter accurately is hard. In 1968, Neptune occulted (passed in front of) an eighth magnitude star, a star about as bright as Neptune itself. This occultation was visible only from Japan, Australia, and New Zealand. From the fact that the star is known only by its catalogue number (BD −17° 4388, pronounced BD minus seventeen degrees, four, three, eight, eight), you can tell that it was a humdrum and unspectacular star. However, when it was occulted by Neptune it became very important. From observations of the rate at which the star dimmed, astronomers could deduce information about Neptune's upper atmosphere and eventually the atmosphere's temperature and pressure structure. Further, from the length of time that the star was hidden by the disk of Neptune, they deduced that Neptune's diameter was 2.3 seconds of arc, equivalent to 50,450 km at that distance, accurate to ±200 km. The astronomers could calculate the diameter from the occultation because they knew how fast Neptune moves across the sky with respect to the fixed stars. The longer the star was hidden from view, the greater the diameter of Neptune must be.

The diameter had been known to be about this value, similar to Uranus' diameter, but it was interesting to get a more accurate value. Accurate values are important for calculating the planet's density, thus leading to deductions about its composition. (Since the volume of a sphere is proportional to the cube of its radius, and the density is the mass divided by the volume, any inaccuracy in the radius leads to a much larger inaccuracy in the density.)

Structure on Neptune has finally been definitely detected from the ground (Fig. 19–21) on images taken electronically in 1979 with charged-coupled devices (CCD's), which were described in Section 2.11. The several images obtained over a 100-minute interval show discrete clouds in the northern and southern hemisphere separated by a dark equatorial band. Motion of the clouds caused by Neptune's rotation can also be seen.

Neptune, like Uranus, appears greenish in a telescope. Only hydrogen, methane, and ethane have thus far been detected with spectrographs. Like Uranus, its rotational period is difficult to determine; the periods derived differed by factors of 3 or 4. The latest values are in the 18- to 22-hour range.

Neptune has two moons (Fig. 19–22). Triton (a sea god, son of Poseidon) is large, probably a little larger than our Moon. A second moon, Nereid (a sea nymph), is small. Nereid is perhaps only 600 kilometers across, and is in a very eccentric orbit with an average radius 15 times greater than Triton's. Nereid never gets brighter than 20th magnitude, which is near the limit of our observational capabilities even with the largest telescopes.

19.4 PLUTO

Pluto, the outermost known planet, is a deviant. It has the most eccentric orbit—an orbit that is most different from a circle. Its orbit also has the greatest inclination with respect to the ecliptic plane, near which the other

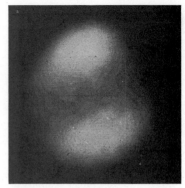

Figure 19–21 Neptune, photographed with a charged-coupled device in the infrared light (8900 Å) that is strongly absorbed by methane gas in Neptune's atmosphere. The bright regions are where non-absorbing clouds of methane ice crystals overlie the methane gas. The image was taken on a night of exceptionally good seeing; Neptune was only 2.5 arc sec in diameter. The CCD provided the infrared sensitivity necessary to take the image with a short exposure. (Image by H. Reitsema, with the 1.5-cm reflector of the Catalina Observatory)

Figure 19–22 Neptune and its satellites. Triton is close to Neptune, and Nereid is at upper right.

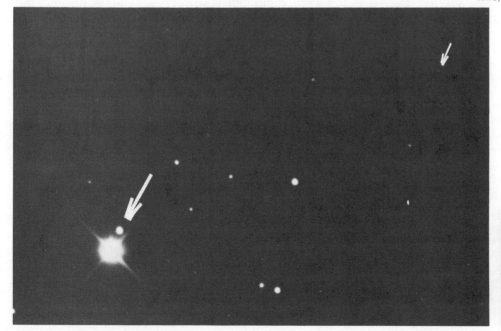

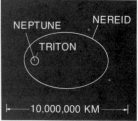

Figure 19–23 Small sections of the plates from which Tombaugh discovered Pluto. On February 18, 1930, Tombaugh noticed that one dot among many had moved between January 23, 1930 *(left)* and January 29, 1930 *(right)*.

Since Pluto is within 10 years of perihelion out of a 248-year period, it is about as bright at opposition as it ever gets from Earth. It hasn't been as bright—about magnitude 13.5—for over 200 years. It should be barely visible through a 20-cm (8-inch) telescope under dark sky conditions.

planets revolve. Pluto's orbit is inclined by 17°, while Mercury's is inclined by only 7° and the other planets are all inclined by 4° or less. In a drawing of planetary orbits, Pluto sticks out.

Pluto will reach perihelion, its closest possible distance from the Sun, in 1989. Its orbit is so eccentric that part of its orbit lies inside the orbit of Neptune. It is now on that part of its orbit, and will remain there until 1999. Thus, in a sense, Pluto is the eighth planet for a while, though the significant point is really that the semi-major axis of Pluto's orbit, 40 A.U., is greater than that of Neptune.

The discovery of Pluto was a result of a long search for an additional planet which, together with Neptune, was causing perturbations in the orbit of Uranus. Finally, in 1930, Clyde Tombaugh found the dot of light that is Pluto (Fig. 19–23) after a year of diligent study of photographic plates at the Lowell Observatory. From its slow motion with respect to the stars from night to night (Fig. 19–24), it was identified as a new planet. Its period of revolution is almost 250 years.

19.4a   Pluto's Mass and Size

Figure 19–24 Pluto's motion can be seen in these photographs taken on successive nights.

Even such basics as the mass and diameter of Pluto are very difficult to determine. It has been hard to deduce the mass of Pluto because the procedure required measuring Pluto's effect on Uranus, a more massive body.

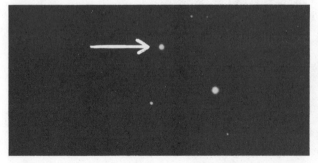 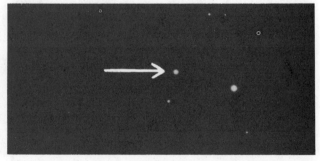

(The orbit of Neptune is too poorly known to be of much use.) Moreover, Pluto has made less than one revolution around the Sun since its discovery, thus providing little of its path for detailed study. As recently as 1968, Pluto was thought to have a mass 91 per cent that of the Earth. Later studies of the orbit of Uranus indicated that Pluto had a mass only 11 per cent that of the Earth, but these observations were very uncertain. On the basis of these data, we were not able to reliably determine Pluto's mass.

The situation changed drastically in 1978 with the surprise discovery that Pluto has a satellite. The presence of a satellite allows us to deduce the mass of the planet by applying Newton's form of Kepler's third law. In this age of space exploration, it is refreshing to see that important discoveries can be made with ground-based telescopes.

A U.S. Naval Observatory scientist, James W. Christy, was studying plates taken with the Observatory's 1.5-m telescope to refine our knowledge of Pluto's orbit. He noticed that some of the photographs seemed to show a bump on the side of Pluto's image (Fig. 19–25). The bump was much too large to be a mountain, and turns out to be a moon orbiting with the period that we had previously measured for the variation of Pluto's brightness—6 days 9 hours 17 minutes. The elongated image of Pluto has since been seen both on older photographs of Pluto and on ones taken since in the positions that have been predicted on the basis of past observations. As a result, the existence of the moon is being accepted. It has been named Charon, the name of the boatman who rowed passengers across the River Styx to Pluto's realm in Greek mythology (and pronounced "Shar on," similar to the name of the discoverer's wife, Charlene). We may have to wait for the Space Telescope to see Pluto and Charon resolved from each other.

From the photograph one can get an approximate idea of the distance between Pluto and Charon. This allows us to calculate the sum of the masses of Pluto and Charon. If we assume that Pluto and its moon have the same albedos and densities, we can compute the masses of each. Charon is 5 or 10 per cent of the mass of Pluto, and Pluto is only 1/500 the mass of the Earth, ten times less than had been suspected even recently.

Until about the same time as the discovery of Charon, the best method for determining the radius of Pluto, as it is for Neptune, had been to observe a stellar occultation. Pluto passed near a 15th magnitude star in 1965, and this unique passage was observed very closely to see if the star would be occulted. But the star was never hidden from view. From this fact, astronomers knew that Pluto appeared smaller in the sky than the minimum angular separation of the position of the center of Pluto and the position of the star, both of which were known accurately (Fig. 19–26). Since we know the distance to Pluto, simple trigonometry gave a limit to the radius of Pluto. This observation showed that Pluto had to be smaller than 6800 km across, which confirmed that Pluto was closer in size to the terrestrial than to the Jovian planets.

We recently learned that Pluto is even smaller than this limit. Infrared spectral studies show the presence of methane ice on Pluto's surface. Since ice has a high albedo, the planet can be relatively small and still reflect the amount of light that we measure. And some of the light comes from Pluto's moon, leaving less of the light to come from Pluto itself.

Only in 1979 was the diameter of Pluto measured directly. The tech-

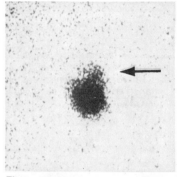

Figure 19–25 An image of Pluto taken on July 2, 1978, with the 1.55-m reflector of the U.S. Naval Observatory in Flagstaff, Arizona, shows a bump that has been interpreted as a satellite. The bump has been seen on one side or the other of the planet, fitting the predictions of an orbit that has been calculated. The satellite has been named Charon.

On the basis of photographs like this one, the separation of Charon from Pluto has been estimated to be about 17,000 km. From the separation and the period, a mass has been derived for Pluto, which turns out to have only 0.2 per cent the mass of the Earth.

Charon may be as large as a third the size of Pluto. Further, it is separated from Pluto by only seven or eight Pluto diameters (compared to the 30 Earth diameters that separate the Earth and the Moon). So Pluto/Charon make a double planet system.

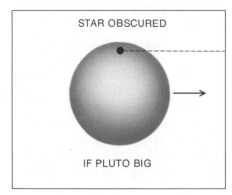
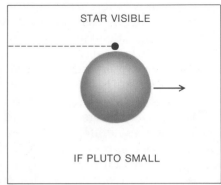

Figure 19–26 The fact that a 15th magnitude star was not occulted by Pluto gives us a limit for how large the diameter of Pluto could be.

Speckle interferometry has since detected the separation of Pluto and Charon and given values of distance and mass agreeing with those given above.

nique of speckle interferometry (Section 6.3) was used with the 5-m Hale telescope. Pluto's diameter measured in this way is between 3000 and 3600 km. This is much smaller than Mars (6800) or Mercury (4800) and is approximately the same size as our Moon (3500), adding to the evidence that we should perhaps not consider Pluto as one of the major planets.

19.4b What Is Pluto?

The new values of the mass and radius of Pluto can be used to derive Pluto's density, the mass divided by the volume. The average density turns out to be very low, between 0.5 and 1 times the density of water, depending on which end of the possible range of radii is more nearly correct. Thus Pluto must be made of frozen materials. Its composition is more similar to that of the satellites of the giant planets than to that of the terrestrial planets. Since spectral evidence indicates that silicates are probably present in addition to methane ice, the higher end of the density range and thus the lower end of the diameter range (3000 km) seems most reasonable.

Now that we know Pluto's mass, we calculate that it is far too small to cause the perturbations in Uranus' orbit that originally led to Pluto's discovery. Thus the prediscovery prediction was really wrong, and the discovery of Pluto was purely the reward of hard work in conducting a thorough search in a zone of the sky near the ecliptic.

If we were standing on Pluto, the Sun would be over a thousand times fainter than it is to us on Earth. We would need a telescope to see the solar disk. Pluto is not massive enough to retain much of an atmosphere. Because the methane frost on its surface is continually changing to methane gas, there is probably a tenuous atmosphere of methane. It may be mixed in with heavy gases like argon to keep it from escaping.

Pluto does not have the kind of hydrogen/helium atmosphere that the giant planets have.

No longer does Pluto seem so different from the other outer planets. Pluto remains strange in that it is so small next to the giants, and that its orbit is so eccentric and so highly inclined to the ecliptic. The evidence revives the thinking that Pluto may be a former moon of one of the giant planets, probably Neptune, and escaped because of a gravitational encounter with another planet.

The orbits of Neptune and Pluto are affected at present by their mutual gravities in such a way that their orbital positions relative to each other repeat in a cycle every 20,000 years. The two planets can never come closer to each other than 18 A.U. But Pluto may still have broken away from Neptune when another planet passed nearby. The same event could have broken off a piece of Pluto to become Charon. The original orbit that Pluto and Charon went into, in this model, has since been modified to the current orbit.

Pluto is so far away that there are no current plans to send spacecraft there. Our recent breakthrough discoveries of its mass and radius, however, have come from ground-based observations. With the continued progress in ground-based astronomy, the pending launch of the Space Telescope, and a little serendipity, we can still hope for further major jumps in our understanding of Pluto in the near future.

19.5 OTHER PLANETS?

Every few years one sees newspaper headlines reporting the possible discovery of "Planet X," a tenth planet. So far, none of these reports has turned out to be true.

If a tenth planet had a sufficiently small mass, or was sufficiently far away, then we could not rule out its existence on the basis of current observations of planetary orbits. We could discover it only by a very lucky accident of happening to look in the right place for it. Of course, if an object we might discover is much smaller than Pluto we might not even want to call it a full-fledged planet.

One recent prediction of "Planet X" was based on the study of orbits of comets, even though ejections of matter from comets sometimes modify their orbits unpredictably. In any case, a mass and orbit for a perturbing planet were deduced. But it was soon realized that a planet of this mass and with an orbit at the inferred inclination would have long since disrupted the orbits of the outer planets that we do see. And besides, a search made with the 1.2-meter (48-inch) Schmidt telescope at Palomar didn't turn up the object. It would seem that we must treat all predictions of "Planet X" with great skepticism until and unless an actual photographic discovery is made.

If we limit our search to the ecliptic, then Tombaugh has set a limit on the presence of planets exterior to Pluto. He extended his planetary search after he discovered Pluto, and his observational material rules out the presence of a Neptune-sized body in the ecliptic within 270 A.U. from the Sun.

But what about a tenth planet very close to the Sun, inside the orbit of Mercury? For hundreds of years people have searched for such a planet, and its "discovery" has been reported on several occasions. This nonexistent planet even has a name: Vulcan, after the Roman god of fire, for it would be extremely hot because of its closeness to the Sun.

One ingenious attempt to explain why Vulcan has never been seen is to say that it is always directly on the other side of the Sun from the Earth. But if it is at an orbital distance less than Mercury's, Kepler's laws show that it could not remain directly opposite the Earth. And if Vulcan is of appreciable size, it would have revealed itself by gravitationally perturbing Mercury's orbit. As we have seen (in Section 7.11), it took Einstein's general theory of relativity to account for the details of Mercury's orbit, but the orbit is now satisfactorily explained.

The best time to search for an inner planet is during a total solar eclipse, when one can see stars in the sky near the Sun. No new planet has ever been seen during eclipses, although several experiments have been carried out to look for one. Such experiments have only turned up an occasional new comet. Based on these observations, we know that any additional body in orbit around the Sun could not be bigger than about ten kilometers across, far from real planetary size.

For the present, it seems fair to conclude that the Sun has nine, and only nine, planets.

SUMMARY AND OUTLINE

Saturn (Section 19.1)
  Giant planet with system of rings
  Lowest density of planets: 0.7 g cm^{-3}
  Ring system
    27° inclination to orbit
    Orbiting chunks of rock and ice
    Inside Roche's limit for Saturn
    Cassini's division and other divisions caused by gravitational effects of moons
    Very thin from top to bottom
    Rocks studied with radar
  Rapid rotation period, causing oblateness
  Methane, ethane, and molecular hydrogen detected in atmosphere
  No evidence of solid surface
  Magnetic field present
  Internal heating
  At least 9 moons, including Titan, largest moon in solar system
  Atmosphere detected on Titan giving rise to speculation as to the chance of finding life there
  Pioneer 11 and Voyager observations from close up
Uranus (Section 19.2)
  Large planet with low density and high albedo
  Discovered by Herschel in 1781
  Atmosphere of methane and molecular hydrogen
  No surface structure can be seen
  Axis of rotation is in the plane of its orbit and retrograde
  Short rotation period
  Low or no internal heating
  Rings discovered at an occultation

  5 known moons
  Diameter more accurately known as a result of a stellar occultation
Neptune (Section 19.3)
  Large planet with low density and high albedo
  Discovery by Adams and Leverrier part science and part luck
  Atmosphere of methane and molecular hydrogen
  Diameter measured more accurately at a stellar occultation
  2 known moons
  Internal heating
Pluto (Section 19.4)
  Most eccentric orbit in solar system
  Greatest inclination to ecliptic
  Rotation accurately determined due to differing albedo
  Discovered in 1930 after diligent search along the ecliptic
  Radius, mass (and thus density) difficult to determine
  Discovery of a moon (Charon) allows mass to be deduced
    Mass of Pluto is only 0.2 per cent that of Earth
  Near occultation of star indicates that diameter is less than 6800 km
    Other considerations indicate that diameter is much smaller still
  Density thus is 0.7 that of water, similar to satellites of outer planets
Search for additional planets (Section 19.5)
  Search by Tombaugh failed to turn up any new planets within 270 A.U. from Sun
  Existence of planet inside Mercury's orbit ruled out

QUESTIONS

1. What are the similarities between Jupiter and Saturn? What are the differences?

†2. What is the angular size of the Sun as viewed from Saturn? How many times smaller is this than the angular size of the Sun we see from the Earth?

†3. When Jupiter and Saturn are closest to each other, what is the angular size of Jupiter as viewed from

Saturn? How does this compare with the angular size of Jupiter as viewed from the Earth?

4. Describe the major developments in our understanding of Saturn, from ground-based observations to Pioneer 11 to Voyager.

5. Why are the moons of the giant planets more appealing for exploration than the planets themselves?

6. What is strange about the direction of rotation of Uranus, and how might that affect Uranus' weather?

7. Why do we know so little about Uranus and Neptune compared with Jupiter and Saturn?

8. Explain how the occultation of a star can help us learn the diameter of a planet.

9. The inaccuracies in measuring the diameter of Neptune mean that we know only that Neptune's diameter is probably between 50,250 km and 50,650 km. What percentage accuracy is this? To what percentage of accuracy is the density known?

†10. What fraction of its orbit has Neptune traversed since it was discovered?

†11. What fraction of its orbit has Pluto transversed since it was discovered?

12. What evidence suggests that Pluto is not a "normal" planet?

13. The position of Pluto in the sky was accurately known, so why were astronomers unsure whether it would occult a particular star in 1965?

14. Summarize the evidence that suggests that Pluto is not a giant planet.

15. From the separation of Pluto and Charon, show how to calculate the mass of Pluto.

---

†This indicates a problem requiring a numerical solution.

REVIEW QUESTIONS ON THE PLANETS

1. List the planets in order of
   (a) increasing size,
   (b) increasing density,
   (c) increasing distance from the Sun.

2. Which planets probably have significant internal energy sources?

3. Mars and Venus are the planets potentially "most like" the Earth, but there are substantial differences. To what degree do you think the various differences can be ascribed to their different distances from the Sun, and to what extent must other explanations be invoked?

4. Which two planets come closest together?

5. If you were an astronomer and could set up an observatory on any planet, which would you choose? To what degree would your decision be influenced by the part of the spectrum in which you are interested?

6. Of all the moons in the solar system, the one whose existence is hardest to understand is the Earth's. Why?

7. Which planets have radiation belts like the Earth's? What property of the planet determines whether or not it will have radiation belts?

8. Classify the planets according to those whose atmosphere is thicker or thinner than that of the Earth.

9. Compare and contrast the moons of the planets. Which moons are the largest? Which have atmospheres? Which are bigger than some planets? Which have volcanoes? Which are icy?

10. Which planets have rings? Would you expect that rings could exist around Neptune? Mercury?

CHAPTER 20

COMETS, METEOROIDS, AND ASTEROIDS

AIMS:

To describe the non-planetary members of our solar system, and to see how their history may provide us with our best information about the origin of the solar system

Besides the planets and their moons, there are many other objects in the family of the Sun. The most spectacular, as seen from Earth, are some of the comets. Bright comets have been noted throughout recorded history, instilling in observers great awe of the heavens.

It has been realized since the time of Tycho Brahe, who studied the comet of 1577, that comets are phenomena of the solar system rather than of the Earth's atmosphere. From the fact that the comet did not show a parallax when observed from different locations on Earth, Tycho deduced that the comet was at least three times farther away from the Earth than the Moon.

Asteroids and meteoroids are other residents of our solar system. We shall see how, along with the comets, they may prove to be storehouses of information about the solar system's origin.

Comets have long been seen as omens. "When beggars die, there are no comets seen; The heavens themselves blaze forth the death of princes."

Shakespeare,
Julius Caesar

20.1 COMETS

Every few years, a bright comet fills our sky with its tremendous tail (Fig. 20–1). From a small, bright area called the *head,* a *tail* may extend gracefully over one-sixth (30°) or more of the sky. Although the tail may give an impression of motion, because it extends out to only one side, the comet

Comet West, a bright comet that was visible to pre-dawn observers in the northern hemisphere in 1976.

Figure 20–1 Comet Ikeya-Seki over Los Angeles in 1965, observed from one of the solar towers at the Mount Wilson Observatory. Photographs like this are taken with ordinary 35-mm cameras; this one was a 32-sec exposure on Tri-X film at f/1.6.

The "long hair" that is the tail led to the name comet, *which comes from the Greek for "long-haired star,"* aster kometes.

does not move visibly across the sky as we watch. With binoculars or a telescope, however, an observer can accurately note the position of the head with respect to nearby stars and detect that the comet is moving at a slightly different rate from the stars as comet and stars rise and set together. A photograph of even a few seconds duration will show the relative motion. By the next night, the comet may have moved 2°, four times the diameter of the Moon. (Both its right ascension and its declination change.)

Within days or weeks this bright comet will have faded below naked-eye brightness, though it can be followed for additional weeks with binoculars and then for additional months with telescopes.

The tail of a comet is directed away from the Sun. Thus, if we see the comet setting in the western sky after sunset, its tail would extend easterly, directed upward toward the zenith and away from the Sun. Conversely, if we see the comet rising in the eastern sky before dawn, its tail would extend westerly toward the zenith.

Most comets are much fainter than the one described above. Half a dozen to a dozen new comets are discovered every year, and most become known only to astronomers. An additional few are "rediscovered"—that is, from the orbits derived from past occurrences it can be predicted when and approximately where in the sky a comet will again become visible. Up to the present time, over 600 comets have been discovered. In addition to the several new comets discovered each year, the returns of a similar number of already known comets are also detected each year.

20.1a The Composition of Comets

At the center of the head of a comet is the *nucleus*, which is at most a few kilometers or so across. It is composed of chunks of matter. The most widely

Box 20.1 Discovering a Comet

A comet is generally named after its discoverer, the first person to see it (or the first two or three people if they independently find it without too much time separating the first observations). Comets are also assigned letters and numbers. First the letters are assigned in order of discovery in a given year. Then, a year or two later when all the comets that passed near the Sun are likely to be known, Roman numerals are assigned in order of their perihelion passage. The two comets, for example, discovered by Lubos Kohoutek in March 1973, were both named Comet Kohoutek. The one that eventually went close to the Sun was first known as 1973f. It was assigned the number 1973XII at the end of 1974, a year after it passed perihelion.

Many discoverers of comets are amateur astronomers, including some amateurs who examine the sky each night in hope of finding a comet. To do this, one must know the sky very well, so that one can tell if a faint, fuzzy object is a new comet or a well-known nebula. It is for that reason that Messier made his famous eighteenth century list of nebulae (Appendix 9).

If you find a comet, telegraph the International Astronomical Union Central Bureau for Astronomical Telegrams, at the Smithsonian Astrophysical Observatory in Cambridge, Massachusetts (TWX 7103206842 ASTROGRAM CAM). (Telegraph address—RAPID SATELLITE CAMBMASS). You may alternatively telephone (617) 864-5758, though this is a less desirable method. In any case, you should identify the direction of the comet's motion, its position, and its brightness. Don't forget to identify yourself with your name, address, and telephone number. If you are the first (or maybe even the second or third) to find the comet, it will be named after you.

The record for discovering comets belongs to the former caretaker of the Marseilles Observatory, Jean Louis Pons, who discovered 37 of them between 1801 and 1827. Several Japanese amateur astronomers have found many comets: Minoru Honda has found a dozen. After work, he spends many hours each night scanning the sky. But the comet may become sufficiently bright for discovery only while it is daylight or cloudy in Japan. So there is hope for less dedicated observers.

Many comets, particularly the ones discovered by professional astronomers, are discovered not by eye but only by examination of photographs taken with telescopes. The photographs are usually those taken for other purposes, and so the discovery of a comet is serendipity—a fortuitous extra discovery—at work.

accepted theory of the composition of comets, advanced in 1950 by Fred L. Whipple of the Harvard and Smithsonian Observatories, is that the nucleus is like a *dirty snowball*. The nucleus may be ices of such molecules as water (H_2O), carbon dioxide (CO_2), ammonia (NH_3), and methane (CH_4), with dust mixed in.

This model explains many observed features of comets, including the fact that the orbits of comets do not appear to accurately follow the laws of gravity. When sunlight evaporates the ices, the molecules are expelled from the nucleus. This action generates an equal and opposite reaction, the same force that runs jet planes. Since the comet nucleus is rotating, the force is not always directly away from the Sun, even though the evaporation is triggered on the sunny side. In this way, comets show the effects of non-gravitational forces in addition to the effect of the force of the solar gravity.

The nucleus itself is so small that it is impossible to observe directly from Earth. It is surrounded by the *coma* (pronounced cō′ ma), which may

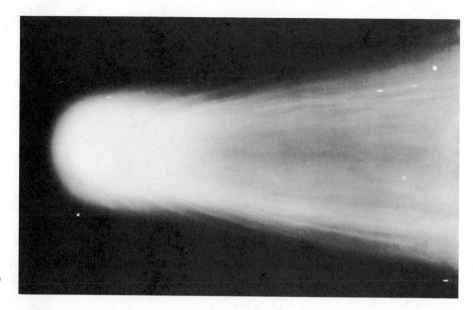

Figure 20-2 The head of Halley's Comet in 1910.

The nucleus and coma together are the head of a comet (Fig. 20-2).

grow to be as large as 100,000 km or so across (Fig. 20-2). The coma shines partly because its gas and dust are reflecting sunlight toward us and partly because gases liberated from the nucleus are excited enough by sunlight that they radiate. (Since they are excited by ultraviolet radiation and radiate in the visible, this is an example of *fluorescent* processes, which are described in Box 23-1.) The spectrum of a comet head shows sets of lines from simple molecules (Fig. 20-3).

With the second Orbiting Astronomical Observatory, we became able to observe in comets the Lyman alpha line of hydrogen, which is in the ultraviolet. In this way, it was discovered in 1970 that a huge hydrogen cloud (Fig. 20-4) surrounds the head. This hydrogen cloud may be a million miles in diameter! It probably results from the breakup of water molecules by ultraviolet light from the Sun.

The tail can extend as much as 1 A.U. (150,000,000 km), and so comets can become the largest objects in the solar system. But the amount of matter in the tail is very small—the tail is a much better vacuum than we can make in laboratories on Earth.

Many comets actually have two tails (Fig. 20-5). Both extend generally in the direction opposite to that of the Sun, but are different in appearance. The *dust tail* is caused by dust particles that had been impurities in the ices of the nucleus, released when the ice was vaporized. The dust particles are left behind in the comet's orbit; they are blown slightly away from the Sun by the pressure caused by photons of solar light hitting the particles (this is

Figure 20-3 Part of the spectrum of Comet Tago-Sato-Kosaka, photographed in 1969 at the European Southern Observatory in Chile. Molecules give not just single lines but rather bands (groups) of lines; one of the bands of CN is visible at the left. Bands of CH and C_3 are seen in the rest of the photograph.

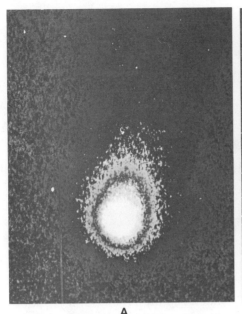

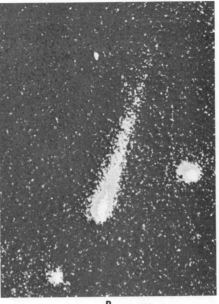

A

B

Figure 20–4 *(A)* These contours of intensity result from a photograph of Comet Kohoutek taken by the Skylab astronauts outside the Earth's atmosphere. The photograph was taken in the Lyman alpha line of hydrogen at 1216 Å in the ultraviolet, and shows the hydrogen halo. This halo was about 1° across, about 2,500,000 km in diameter, at the time of this observation. *(B)* These contours result from a photograph that was taken immediately after Fig. 20–4A, but through a filter that did not pass the Lyman alpha line. Thus the hydrogen halo does not show, though the tail does. The tail was about 2° long, about 5,000,000 km in length. Hot stars from the background constellation, Sagittarius, also show.

known as "radiation pressure"). As a result of the comet's continued orbital motion, the dust tail usually curves smoothly behind the comet. The *gas tail* (also called the *ion tail*) is composed of ions (such as CO^+, N_2^+, CO_2^+, and CH^+) blown out more or less straight behind the comet by the solar wind. As puffs of ionized gas are blown out and as the solar wind varies, the ion tail takes on a structured rather than a smooth appearance. Each puff of matter can be seen. The magnetic fields in the solar wind carry only the ionized matter along with them; the neutral atoms are left behind in the coma.

AUGUST 22, 1957

AUGUST 24, 1957

Figure 20–5 In Comet Mrkos, the straight *ion tail,* extending toward the top, and the *dust tail,* gently curving toward the right, were clearly distinguished.

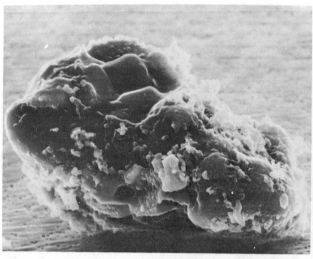 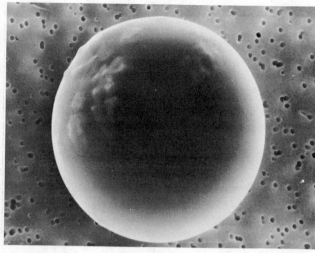

A

B

Figure 20-6 These tiny extra-terrestrial dust particles, magnified about 5000 times by a scanning electron microscope, were captured at an altitude of 20 km, as the particles drifted downward. Most of the particles are 0.01 mm across and weigh a billionth of a gram. *(A)* Most of the particles were fluffy like this one. Each contains perhaps a million grains of different minerals. *(B)* Hard spheres like this one were rarer. They must have resulted from some form of melting.

Dust that may be from comets has been captured by sending up sticky plates on a NASA U-2 aircraft. The plane was sent high above the layer of terrestrial pollution. The abundances of the elements are similar to those of meteorites rather than to terrestrial material (Fig. 20–6). The particles probably date back to the origin of the solar system. One reason to think so is the evidence of small anomalies in the abundances of certain elements that would have been caused by the supernova that may have triggered the formation of our solar system.

A comet—head and tail together—contains less than a billionth of the mass of the Earth. It has been said that comets are as close as something can come to being nothing.

20.1b The Origin and Evolution of Comets

Let us trace the life of a comet from birth onward.

It is now generally accepted that there are hundreds of millions of incipient comets surrounding the solar system in a sphere perhaps 50,000 A.U. (almost 1 light year) in radius. This sphere is known as the *Oort comet cloud* after Jan H. Oort, the Dutch astronomer who advanced the theory in 1950. The total mass of matter in the cloud is only 10 to 100 times the mass of the Earth. Occasionally one of the incipient comets leaves the Oort cloud, perhaps because gravity of a nearby star has tugged it out of place, and the comet approaches the Sun (Fig. 20–7). Its orbit is a long ellipse (that is, an ellipse of high eccentricity). If the comet passes near the Jovian planets, its orbit is altered by the gravity of these massive objects. Because the Oort cloud is spherical, comets are not limited to the plane of the ecliptic and come in randomly from all angles.

As the comet gets closer to the Sun, the solar radiation that reaches it begins to vaporize the molecules in the nucleus. We have not generally been able to detect the molecules in the nucleus directly—the *parent molecules*—though we would clearly love to do so, but we have mostly detected in the head the simpler molecules into which the parent molecules break down—the *daughter molecules*.

Examples of daughter molecules are H, OH, and O, which might be results of the breakdown of the parent H_2O. Similarly, daughters NH and NH_2 can result from parent NH_3.

The tail begins to form, and as more and more of the nucleus is va-

Figure 20–7 Halley's Comet has a very elliptical orbit so it goes out far beyond the outer planets even though its perihelion distance is only 0.6 A.U.

porized, the tail grows longer and longer. Still, even though the tail can be millions of kilometers long, it is so tenuous that only 1/500 of the mass of the nucleus may be lost. Thus a comet may last for many passages around the Sun.

The comet is brightest and its tail is generally longest at about the time it passes perihelion (the closest point in its orbit to the Sun). However, because of the angle at which we view the tail from the Earth, it may not appear the longest at this stage.

Following perihelion, as the comet recedes from the Sun, its tail fades; the head and nucleus receive less solar energy and fade as well. The comet may be lost until its next return, which could be as short an interval as 3.3 years (as it is for Encke's Comet) or as long as 80,000 years (as it is for Comet Kohoutek) or more. Periodic comets lose a little mass reappearance after reappearance, and the comet eventually disappears. We shall see in Section 20.2 that some of the meteoroids are left in its orbit. Some of the asteroids, particularly those that cross the Earth's orbit, may be dead comet nuclei.

Because new comets come from the places in the solar system that are farthest from the Sun and thus coldest, they probably contain matter that is unchanged since the formation of the solar system 4.6 billion years ago. So the study of the constituents of comets is important for understanding the early stages of solar system formation.

20.1c Halley's Comet

Figure 20–8 Edmond Halley.

In 1705, the English astronomer Edmond Halley (Fig. 20–8) applied a new method developed by his friend Isaac Newton to determine the orbits of comets from observations of the positions of the comets in the sky. He reported that the orbits of the bright comets that had appeared in 1531, 1607, and 1682 were about the same. Because of this, and because the intervals between appearances were approximately equal, Halley suggested that we were observing a single comet orbiting the Sun, and predicted that it would again return in 1758. The recovery of this bright comet on Christmas night of that year, 16 years after Halley's death, was the proof of Halley's hypothesis (and Newton's method); the comet has since been known as Halley's Comet. It seems probable that the bright comets reported every 74 to 79 years since 87 B.C. (and possibly even before then, in 240 B.C.) were earlier appearances. The fact that Halley's Comet has been observed at least 27 times

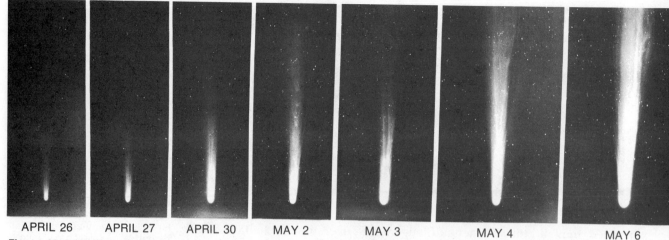

| APRIL 26 | APRIL 27 | APRIL 30 | MAY 2 | MAY 3 | MAY 4 | MAY 6 |

Figure 20–9 Halley's Comet in 1910.

Mark Twain was born during the appearance of Halley's Comet in 1835, and often said that he came in with the comet and would go out with it. And so he did, dying during the 1910 return of Halley's Comet.

The European mission is named Giotto after the 14th century Italian painter who included Halley's Comet in a painting.

endorses the calculations that have been made that show that less than 1 per cent of a cometary nucleus' mass is lost at each perihelion passage.

Halley's Comet went especially close to the Earth during its 1910 return (Fig. 20–9), and the Earth actually passed through its tail. Many people had been frightened that the tail would somehow damage the Earth or its atmosphere, but the tail had no noticeable effect. It was known to most scientists even then that the gas and dust in the tail were too tenuous to harm our environment.

Halley's Comet reappears every 74 to 79 years; Jupiter and Saturn perturb its orbit enough to cause the variation. Since we can count on its reappearance soon, prior to its February 9, 1986 perihelion (Fig. 20–10) we can make plans long in advance to observe it. NASA's plans to launch a spacecraft to fly near it have been curtailed for financial reasons. A probe would even have been sent into the nucleus of Halley's Comet. The probe would have been able to sample parent molecules directly. This spacecraft would have been designed to go on to rendezvous with Comet Temple 2 in mid-1988; that is, the spacecraft would match the comet's orbit so as to get a closeup view of the nucleus over an extended period of time. Alternative plans may yet come to pass. The European and Japanese space agencies are considering their own missions. A scaled-down NASA effort remains possible.

The 1986 reappearance of Halley's Comet will not be as spectacular as its 1910 passage, however, for the Earth and comet will be on opposite sides of the sun when the comet is brightest. Thus the comet with its faint tail is not expected to appear as spectacular as, for example, 1975n, Comet West (which is shown in the photograph opening this chapter and is discussed in the following section).

20.1d Comet Kohoutek and Comet West

In March of 1973, Lubos Kohoutek was studying faint asteroids on photos he had taken at the Hamburg Observatory in Germany where he works. On two of these plates he discovered faint comets—about 16th magnitude—which were named for him when he sent word of the discoveries to the IAU Central Bureau for Astronomical Telegrams.

Further analysis of his plates of the second comet, including a further set of plates he took later in the month and a two-month-old plate that he found

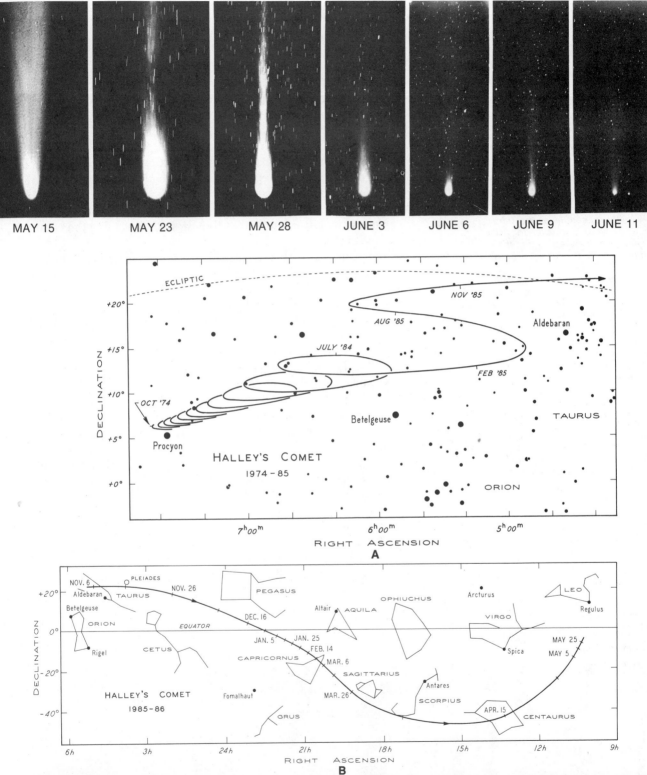

MAY 15   MAY 23   MAY 28   JUNE 3   JUNE 6   JUNE 9   JUNE 11

A

B

Figure 20–10. *(A)* For some time before its perihelion, Halley's Comet will follow an orbit that seems to circle in the sky as a result of the Earth's revolution around the Sun. When Halley's Comet draws close to us *(B)* its apparent motion across the sky will grow more rapid. Observers can use these charts to search for Halley's Comet. Its tail will appear broadside in December 1985 and March 1986.

also showed the comet, revealed that he had made a most unusual discovery: the comet would not reach perihelion for about nine months and would pass very close to the Sun—within only 21 million km (0.14 A.U.)—thus becoming very bright. Because of these predictions the second comet, 1973f and the second Comet Kohoutek of the year, became known the world over.

Never before had a potentially bright comet been discovered this far from the Sun—at the distance of Jupiter's orbit—or this long before perihelion. Thus it was impossible to make an accurate prediction of the amount of brightening, but it seemed that it might brighten from 16th to −5th magnitude, a factor of 250,000,000. On the basis of this prediction scientists readied all kinds of experiments that required months to prepare.

But the comet did not brighten quite as much as had been expected, and the astronomers revised toward fainter levels their predictions of the comet's eventual maximum brightness. Somehow these revised and updated predictions never became as widely circulated as the original hopes that the comet would be spectacular.

Comet Kohoutek was at its brightest within a few days of its perihelion passage. During that time it was too close to the Sun to be seen by anyone except the astronauts in Skylab, who photographed (see Color Plate 17) and sketched (Fig. 20–11) the comet. The comet was actually almost as bright as Venus. The shape and size of the comet and the tail changed rapidly. A day after perihelion, the astronauts discovered a brilliant spike—an anti-tail—pointing toward the Sun, the opposite direction from that in which comet tails point (Fig. 20–11).

A week after perihelion the comet drew far enough away in the sky from the Sun that its tail could be seen by observers on Earth. Unfortunately, it was just barely bright enough to be seen with the naked eye, and was not the spectacular object to which many had been looking forward. In particular, it could not be seen by observers in cities, where light pollution made the sky brighter than the comet's tail. Nevertheless, the tail stretched gracefully over many degrees of sky (Fig. 20–12). Observers with optical telescopes and radio telescopes on the ground were able to get good spectra and discover molecules that had never before been seen in comets.

Why did Kohoutek appear so bright so early on? It has been hypothesized that Kohoutek may have been a "virgin comet," a comet making its first passage around the Sun. Thus, an especially great amount of material was vaporizing when the comet was first discovered, making it appear brighter than average for that distance from the Sun and from the Earth. Once that layer of material escaped, lower layers were more ordinary.

The long advance notice even allowed the schedule of the third crew of astronauts due to visit Skylab to be delayed so that they would be aloft during Comet Kohoutek's perihelion passage. Anyone who has ever worked with a space experiment will especially realize how major an event it is to change such a schedule.

Figure 20–11 Sketches of Comet Kohoutek and its sunward spike by the astronauts aboard the third Skylab mission. This rare phenomenon had previously been seen to such an extent in only one other modern comet, and to any extent in only a dozen comets. The Kohoutek observations showed that the spike contained meteoroid particles as large as 1 mm across, far larger than the dust in a comet tail. These particles are affected by the pressure of the Sun's radiation. The nucleus and the particles have moved differently over a period of months, long enough to allow them to be seen separately. Adding the effect of perspective, because we are observing from the Earth, it seems probable that the particles in the spike were actually left behind the comet and only appear projected onto the sunward direction.

DEC 29, 1973 PERIHELION +1 DAYS   DEC 30, 1973 PERIHELION +2 DAYS   DEC 31, 1973 PERIHELION +3 DAYS   JAN 2, 1974 PERIHELION +5 DAYS

Figure 20–12 A photograph of Comet Kohoutek with the 1.2-meter Schmidt telescope on Palomar Mountain. Although the comet was too faint to be seen well with the naked eye, it was a fine photographic object and was bright enough to enable unique scientific observations to be made.

We can think of the Sun vaporizing layer after layer of the material of a comet's nucleus, similar to the way in which we can peel off layer after layer from an onion. The idea that the comet was shedding dust was borne out by infrared observations, since dust radiates in the infrared. The fact that a subsequent comet's infrared brightness dropped by a factor of a hundred within a few weeks lends support to the layered-onion theory.

Comet Kohoutek is now receding into space, and will go out as far as 4000 A.U. While it is far from the Sun, we can see from Kepler's second law that it will move very slowly. It won't be back in our vicinity for another 80,000 years.

Astronomers often say that comets are unpredictable. Comet Kohoutek, which was a popular bust and an astronomer's delight, bore out that idea.

Bright comets ordinarily appear with much less notice than Kohoutek had. In late 1975, Richard West, of the European Southern Observatory, discovered on a plate taken in Chile a comet that within a couple of months became an object very easy to see with the naked eye. After perihelion Comet West was visible in the predawn sky to northern hemisphere observers (see Color Plate 23). For a few days its tail extended over 30°. Bits of dust in the nucleus were ejected at an irregular rate, which led to the delicate structure in the tail that is visible in the photograph opening this chapter. Indeed, the nucleus itself broke into at least four pieces. On the basis of the orbit, it might not be back for a million years; it might even have been set on a course of ejection from the solar system, in which case it will never return.

20.2 METEOROIDS

There are many small chunks of matter in interplanetary space, ranging up to tens of meters across. When these chunks are in space, they are called *meteoroids*. When one hits the Earth's atmosphere, friction slows it down and heats it up—usually at a height of about 100 km—until all or most of it is vaporized. Such events result in streaks of light in the sky, which we call *meteors*. (Meteors are popularly known as *shooting stars*.) The brightest meteors can reach magnitude −15 or even −20, which is brighter than the full moon. We call such bright objects *fireballs* (Fig. 20–13). We can sometimes even hear the sounds of their passage and of their breaking up into smaller bits. When a fragment of a meteoroid survives its passage through the Earth's atmosphere, the remnant that we find on Earth is called a *meteorite*. We now also refer to objects that hit the surfaces of the Moon or of other planets as meteorites.

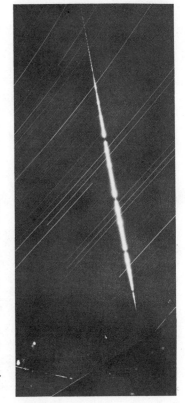

Figure 20–13 A fireball observed in 1970 by the "Prairie Network" of the Smithsonian Astrophysical Observatory, a network of wide-field cameras spaced around the mid-west of the United States in hope of pinpointing a meteor's path and thus permitting the meteorite to be found. A shutter in the camera rotates so that breaks occur in the image of the meteor trail at intervals of 1/20 sec. This permits timing of the motion. The fireball pictured here was photographed by two of the cameras of the Network, which led to the discovery of a meteorite at Lost City, Oklahoma. The trail persisted for about 8 seconds, though the short breaks in the trail from which this could normally be measured do not show because of film over-exposure except at the top.

20.2a Types of Meteorites

We have seen in Chapter 13 that meteorites have formed almost all of the craters on the Moon. Tiny meteorites less than a millimeter across, called *micrometeorites,* are the major cause of erosion on the Moon. Micrometeorites also hit the Earth's upper atmosphere all the time, and remnants of the material they deposit can be collected for analysis from balloons or airplanes. The dust particles shown in Figure 20–6 are examples. The micrometeorites, which may have been only the size of a grain of sand when they first hit the Earth's atmosphere, are often sufficiently slowed down before they are vaporized that they can reach the ground. Dust resulting from micrometeorites can be sampled by collecting bits of ice from the Arctic or the Antarctic, or from mountain tops.

Space is full of meteoroids of all sizes, with the smallest particles being the most abundant. Most of the small particles, less than about 1 mm across, come from comets. Most of the large particles, more than about 1 cm across, come from collisions of asteroids in the asteroid belt. It is generally these larger meteoroids that become meteorites.

There are several major kinds of meteorites. Most of the meteorites that are found have a very high content—about 90 per cent—of iron; the rest is nickel. These *iron meteorites* (or, for short, *irons*) are thus very dense—that is, they weigh quite a lot for a given volume (Fig. 20–14).

Most meteorites that hit the Earth are stony in nature and are often referred to simply as *stones.* Most of these stony meteorites are of a type called *chondrites,* because they contain rounded particles called "chondrules." The stony meteorites without chondrites are called *achondrites;* they may have once been chondrites, and melted and cooled. Because stony meteorites resemble ordinary rocks, they are not usually discovered unless their fall is observed. That explains why most meteorites discovered at random are irons. But when a fall is observed, most meteorites recovered are

Figure 20–14 The Ahnighito meteorite, a 34-ton iron-nickel meteorite, was discovered by Eskimos in Greenland in the early 1800's. The photographs show the meteorite being moved into the new Hall of Meteorites at the American Museum of Natural History in New York. (Photos by John Pazmino)

Figure 20–15 The Barringer meteor crater in Arizona. It is 1.2 km in diameter. Dozens of other terrestrial craters are now known, many from aerial or space photography. Most are not as clearly marked, since erosion has taken a toll.

Tektites, small, rounded glassy objects that are found at several locations in the Earth's southern hemisphere, may have splashed out from a single meteorite impact. The origin of tektites, including whether they came from the Earth or from the Moon, has long been controversial, though a terrestrial origin seems most probable.

stones. The stony meteorites have a high content of silicates; only about 10 per cent of their mass is iron and nickel.

The largest meteorite crater on Earth may be a depression over 400 km across deep under the Antarctic ice pack. This is comparable with the size of lunar craters. Another very large crater, in Hudson's Bay in Canada, is filled with water. Most meteorite craters on Earth are either disguised in such ways or have eroded away. A large crater that is obviously meteoritic in origin is the Barringer Crater in Arizona (Fig. 20–15). It is the result of what was perhaps the most recent large meteor to hit the Earth, for it was formed only 25,000 years ago.

Every few years a meteorite is discovered on Earth immediately after its fall (Fig. 20–16). The chance of a meteorite's landing on someone's house is very small, but it has happened! Often the positions of fireballs in the sky are tracked in the hope of finding fresh meteorite falls. The newly discovered meteorites are rushed to laboratories for chemical analysis of their con-

Figure 20–16 (A) A bright fireball was observed and photographed on February 5, 1977. Analysis indicated that it probably landed near Innisfree, Alberta, Canada. Scientists flew out to search the snow-covered wheat fields and found this 2 kg meteorite. It is only the third meteorite recovered whose previous orbit around the Sun is known. (B) A meteor crater 2 meters wide and 5.5 meters deep, one of a hundred from the fall in Kirin Province of China on March 8, 1976. A fireball was seen before the meteorites landed. The largest meteorite is the largest stony meteorite ever recovered; it weighs 1770 kg (3894 lbs).

A   **B**

TABLE 20–1 METEORITES

Composition	Seen Falling	Finds
Irons	6%	66%
Stony-irons	2%	8%
Stones	92%	26%

stituents before the Earth's atmosphere or human handlers can contaminate them. Many meteorites have recently been found in the Antarctic, where they have been well-preserved as they accumulated over the years.

*20.2b   Carbonaceous Chondrites

Chondrites are stony meteorites that usually contain small spherical grains known as chondrules. Objects that contain large quantities of carbon are carbonaceous.

One *carbonaceous chondrite* (a rare kind of stony meteorite with a high carbon content) was found to contain simple amino acids. The object was the Murchison meteorite, which fell near Murchison, Victoria, Australia in 1969. The formation of such complicated building blocks of life in cold, isolated places like meteoroids is one of several indications that the precursors of life develop naturally. We shall be developing more of this evidence in the following chapter.

The question was raised whether these amino acids were truly extraterrestrial or had rather contaminated the samples on Earth. Analysis showed that the amino acids contained equal quantities of types that cause the polarization of light to be rotated in right-handed and left-handed senses; as a result, no rotation of the plane of polarization is seen. Since all amino acids in living beings on Earth cause polarized light to be rotated in the left-handed sense, this indicates that the amino acids in the Murchison meteorite are extraterrestrial.

We can measure the ages of the meteorites through study of the ratios of radioactive and non-radioactive isotopes contained in them. The measurements show that the meteoroids were formed at times up to 4.6 billion years ago, the time of the beginning of the solar system. Furthermore, the chondrites show no sign of ever having melted, and may date from the formation itself; some may be the planetesimals. The analysis of the abundances of the elements in meteorites thus tells us about the solar nebula from which the solar system formed. In fact, up to the time of the first landing on the Moon, meteorites were the only extraterrestrial material we could get our hands on.

The largest carbonaceous chondrite available for scientific study is the Allende meteorite, which fell in Mexico in 1969. Sufficient quantities have been available for analysis that our knowledge of the abundances in meteorites has been greatly improved. For example, we can demonstrate that enough radioactive aluminum (Al^{26}) existed long ago to provide enough heat to melt those meteorites that show signs of melting.

Some of the carbonaceous chrondrites may be cometary debris.

Most of our knowledge of the abundances of the elements in the solar system has come either from analysis of the solar spectrum or from analysis of meteorites. Until recently it was thought that all the analyses were consistent with uniform abundances of each element throughout the solar nebula. But evidence is growing, including in particular the high quality data from the Allende meteorite, that abundances varied from place to place in the solar

Figure 20–17 A meteor crossing the field of view while the Palomar Schmidt was taking a 15-minute exposure of Comet Kobayashi-Berger-Milon on August 11, 1975. The cluster of galaxies in Ursa Major and Canes Venatici is also in the field; M106 is the most prominent galaxy visible. This photograph shows objects that are at four entirely different scales of distance. The meteor *(arrow)* is in the Earth's atmosphere; the comet is in interplanetary space; stars are in our galaxy, and the cluster of galaxies is extremely far away.

nebula. The abundances of common elements like oxygen may vary by 5 per cent; the abundances of rare earths can vary by greater amounts.

20.2c *Meteor Showers*

Meteors often occur in *showers*, that is, times when the rate at which meteors are seen is far above average. On any clear night a naked-eye observer may see a few *sporadic* meteors an hour, that is, meteors that are not part of a shower (Fig. 20–17). (Just try going out to a field in the country and watching the sky for an hour.) During a shower several meteors may be visible to the naked eye each minute, though this is rare. Meteor showers occur at the same time each year (Table 20–2), and probably represent the passage of the Earth through the orbits of defunct comets. The meteoroids in the orbit of a former comet may be the products of the decay of the comet.

TABLE 20–2 METEOR SHOWERS°

Name	Date of Maximum	Duration Above 25% of Maximum	Approximate Limits	Number per Hour at Maximum
Quadrantids	Jan 4	1 day	Jan 1–6	110
Lyrids	April 22	2 days	April 19–24	12
Eta Aquarids	May 5	3 days	May 1–8	20
Delta Aquarids	July 27–28	–	July 15–Aug 15	35
Perseids	Aug 12	5 days	July 25–Aug 18	68
Orionids	Oct 21	2 days	Oct 16–26	30
Taurids	Nov 8	–	Oct 20–Nov 30	12
Leonids	Nov 17	–	Nov 15–19	10
Geminids	Dec 14	3 days	Dec 7–15	58

°The number of sporadic meteors per hour is 7 under perfect conditions. The visibility of showers depends mostly on how bright the Moon is on the date of the shower, which depends on its phase. Meteors are best seen with the naked eye; using a telescope or binoculars merely restricts your field of view.

Meteorites usually do not result from showers, so presumably the meteoroids that cause a shower are very small.

The rate at which meteors are seen usually increases after midnight on the night of a shower, because that side of the Earth is then facing and plowing through the oncoming interplanetary debris. If the Moon is gibbous or full, then the sky is too bright to see the shower well. The meteors in a shower are seen in all parts of the sky, but their trajectories all seem to emanate from a single point in the sky called the *radiant*. A shower is usually named after the constellation that contains its radiant. The existence of a radiant is just an optical illusion because perspective makes a set of parallel paths approaching you appear to emanate from a point.

Many theories have been advanced for the cause of an explosion near the Tunguska River in Siberia in 1908, including odd ones like black holes and antimatter. Most scientists think it was a fragment of material entering our atmosphere, so let us discuss it here in the section on meteors and meteorites. Since no remnant has been found, even though trees were blown outward for kilometers around, it may rather have been a small fragment of a comet, and vaporized completely.

Apollo asteroids are discussed on page 395.

It is becoming credible that an Apollo asteroid's impact caused the extinction of dinosaurs on Earth, an event that is known to have happened over a short interval of time.

20.3 ASTEROIDS

The nine known planets were not the only bodies to result from the agglomeration of planetesimals 4.6 billion years ago. Thousands of *minor planets*, called *asteroids*, also resulted. They are detected by their small motions in the sky relative to the stars (Fig. 20–18).

Most of the asteroids are found in elliptical orbits whose average size is between the sizes of the orbits of Mars and Jupiter. The zone where most asteroids are found is called the *asteroid belt*. Indeed, it was not a surprise when the first asteroid was discovered, on January 1, 1801, the first day of the nineteenth century, because Bode's law (described in Section 12.10) predicted the existence of a new planet in approximately that orbit.

The first asteroid was discovered by a Sicilian clergyman/astronomer, Giuseppe Piazzi, and was named Ceres after the patron saint of Sicily, the Roman goddess of harvests. Though Piazzi followed Ceres' motion in the sky for over a month, he became ill and the object moved into evening twilight. Ceres was lost. But the problem led to a great advance in mathematics: the young mathematician Carl Friedrich Gauss developed an important way of plotting the orbits of objects given only a few observations. The orbit Gauss worked out led to the rediscovery of Ceres. Modern versions of Gauss' method are still in use today, though calculations are now done by computer rather than by hand.

Though the discovery of Ceres was a welcome surprise, even more surprising was the subsequent discovery in the next few years of three more asteroids. Bode's law hadn't called for **them!** The new asteroids were named Pallas, Juno, and Vesta, also after goddesses, which began the generally observed tradition of assigning female names to asteroids.

1 Ceres has a mass of 5.9×10^{-10} that of the Sun; 2 Pallas and 4 Vesta each have about one-fifth as much mass.

The next asteroids weren't discovered until 1845, and over 2000 are now named and numbered. They are assigned numbers when their orbits be-

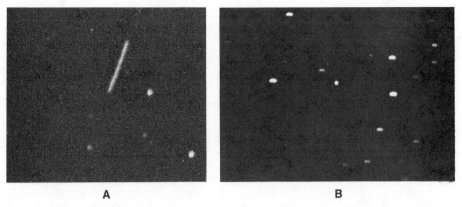

A   **B**

Figure 20–18 *(A)* Asteroids leave a trail on a photographic plate when the telescope is tracking at a sidereal rate, that is, with the stars. *(B)* In this photograph of the asteroid Menzel, named after the late Harvard astrophysicist, the telescope was tracking with the asteroid. Thus the asteroid appears as a dot and the stars are trails.

come well determined, leading to such names as 1 Ceres, 16 Psyche, and 433 Eros. Only half a dozen asteroids discovered are known to be larger than 300 km across, and over 200 more are larger than 100 km across. Small asteroids may only be 1 km or less across, but the largest asteroids known (Fig. 20–19) are of the size of some of the moons of the planets. All the asteroids together contain less mass than the Moon. Perhaps 100,000 asteroids could be detected with Earth-based telescopes if we wanted to work on it.

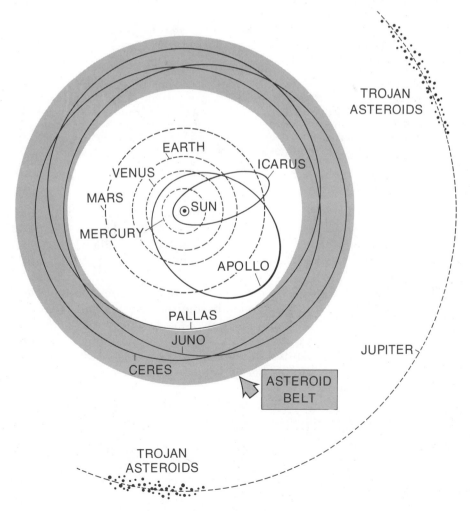

Figure 20–19 Some asteroid orbits. The asteroid belt, in which most asteroids lie, is shaded.

200 KM

CERES

PALLAS

VESTA

HYGIEA

INTERAMNIA

DAVIDA

PSYCHE

Figure 20–20 The sizes of the larger asteroids and their relative albedos.

Box 20.2   Measuring the Sizes of Asteroids

The most accurate way to measure the size of an asteroid is to follow its passage in front of a star. By timing when the star disappears and reappears, we can measure its diameter along one line across the asteroid. If several observers time the occultation and reappearance from different positions on the Earth, we get enough information to plot an accurate shape. The stars are so far away that the shadow of the asteroid they project on the Earth is the full size of the asteroid.

The technique was first used in 1975 for the asteroid Eros, which occulted a star of magnitude 3.6. The predictions of where the shadow would fall were revised only four hours before the occultation, and groups of observers rushed to take up positions along a line between Amherst, Massachusetts, and Waterbury, Connecticut. Three of the groups saw the star abruptly disappear for 2.6 to 3.4 seconds. Analysis of the observations, with knowledge of the speed that Eros was moving in the sky with respect to the stars, directly gives the dimension of Eros at the orientation it then had. Its longest diameter was 22 km. Radar observations could also be made because Eros was only 23 million km from the Earth. All the observations together show that Eros is 13 × 15 × 36 km.

A larger asteroid, Juno, occulted a 9th magnitude star in 1979. Four of the observing stations were within the band of the occultation, which crossed the southwestern United States. Juno proved to be elliptical, with major and minor axes of 292 and 243 km.

An occultation of an 11th magnitude star in 1978 gave us data for Pallas. The information, coupled with an earlier determination of the orientation of Pallas' axis, indicates the asteroid is 558 × 526 × 532 km. A determination of the radius by speckle interferometry gave a value 25 per cent larger for this object.

Accurate measurements of the diameter of an asteroid are important for computing its albedo and its density. Since albedo depends on area, which depends on the square of the diameter, and density depends on volume, which depends on the cube of the diameter, the percentage uncertainty in diameter makes a larger percentage error in albedo or density. Only the occultation method leads to sufficiently accurate knowledge of the albedo and the density to distinguish among carbonaceous, stony, and nickel-iron materials.

Though most asteroids are located between the orbits of Mars and Jupiter, from 2.2 to 3.2 A.U. from the Sun (Fig. 20–20), there is plenty of empty space between them. Asteroids rarely come within a million kilometers of each other. Still, asteroids may collide with each other occasionally, forming meteoroids.

The Pioneers and Voyagers en route to Jupiter and beyond travelled through the asteroid belt for many months and showed that the amount of dust among the asteroids is not increased over the amount of interplanetary dust in the vicinity of the Earth. Particles the size of dust grains are only about three times more plentiful in the asteroid belt, and smaller particles are less plentiful. So the asteroid belt will not be a hazard for space travel to the outer parts of the solar system.

In some cases, we can see interplanetary dust directly from the Earth. From locations with exceptionally clear skies, dust in interplanetary space appears as the *zodiacal light,* the reflection of sunlight off the dust grains. The F-corona seen at solar eclipses is also light reflecting off interplanetary dust.

*20.3a Special Groups of Asteroids

The *Apollo asteroids* have orbits whose close points—the perihelia (the plural of perihelion)—are inside the orbit of the Earth. The orbits thus cross the Earth's orbit. We know of about two dozen asteroids of this type, beside the prototype, Apollo. There may be 750 in all. One asteroid even goes inside the orbit of Mercury, and has been named Icarus, after the inventor in Greek mythology who flew too near the Sun. (Asteroids outside the asteroid belt sometimes have names other than feminine.) On June 14, 1968, Icarus came within 6 million km of the Earth.

Most Apollo asteroids will probably collide with the Earth eventually, because their orbits may intersect the Earth's. Luckily, there are only a few dozen Apollos greater than 1 km in diameter. Apollo asteroids may well be the source of most meteorites. In this model, the meteorites are the debris from Apollos broken up during their visit to the asteroid belt.

Another unusual type of asteroid is the Trojan asteroids, which always remain in orbit around the Sun oscillating about the point 60° ahead of or behind Jupiter. (These locations are known as the Lagrangian points, after the French scientist Lagrange.) At the Lagrangian points the gravitational pulls of the Sun and of Jupiter balance out so that material remains in space without being pulled toward one or the other.

The first Trojan asteroid to be discovered was named Achilles. Later it was decided to name the asteroids in the Lagrangian point ahead of Jupiter after Greek heroes and those in the point trailing Jupiter after Trojan heroes; but 617 Patroclus, a Greek, is in the Trojan camp. Similarly, there is one Trojan, 624 Hector, in the Greek camp.

*20.3b The Compositions of Asteroids

Since 1970, new methods have enabled us to better measure the sizes and albedos of asteroids. One method is based on the amount of infrared radiation emitted by asteroids at wavelengths of 10 or 20 microns. A second important method is based on studies of the polarization of visible light, since the polarization of darker asteroids seems to change more as the asteroid's position changes than it does for lighter ones. Some asteroids are as dark as coal, with albedos of only 3 per cent. Indeed, the surface may be covered with carbon or carbon compounds. Other asteroids have albedos of 50 per cent.

Such considerations, and also asteroid spectra, lead us to the conclusion that asteroids are made of different materials from each other, and represent the chemical compositions of different regions of space. The asteroids at the inner edge of the asteroid belt are mostly stony in nature, while the ones at the outer edge are darker (as a result of being more carbonaceous). Most of the small asteroids that pass near the Earth—Icarus, Eros, Toros, Geographos, and Alinda—are among the stony group. Three of the largest asteroids—Ceres, Hygeia, and Davida—are among the carbonaceous group. A third group may be mostly composed of iron and nickel. The differences may be a direct result of the variation of the solar nebula with distance from the protosun. Differences in chemical composition also show that we must discard the old theory that the asteroids represent the breakup of a planet that once existed between Mars and Jupiter.

Asteroids have also had different types of histories since their formation. 4 Vesta, for example, is covered with basalt, which probably resulted from lava. Thus some asteroids are differentiated, that is, layered. We think that Vesta was heated to the 1000 K necessary to cause the lava by radioactive decay inside. However, if radioactive materials were distributed uniformly in the asteroid belt, we do not understand why even bigger asteroids like Pallas and Ceres do not have basalt on their surfaces as well. Perhaps the distribution of short-lived radioactive isotopes like Al^{26} was not uniform. The asteroids and their meteoritic offspring may provide the data to show us the details of how the solar system was formed.

When the solar system was forming, proto-Jupiter probably became so massive that it disrupted a large region of space around it. The planetesimals in that region may have been sent into eccentric orbits, and may have collided at high velocity. In this way, they broke each other apart and sent the fragments off in orbits of high eccentricity and inclined to the ecliptic plane to become the asteroids.

Figure 20–21   Chiron. This is the plate on which it was discovered. It was quite a surprise that a new type of object should be dis-covered within our solar system.

Chiron, in Greek mythology, was the wisest of the centaurs and the teacher of Achilles; if similar ob-jects are discovered, they could be called centaurian and named after other centaurs.

20.4   CHIRON

Chiron, a strange object in our outer solar system, was discovered in 1977 as a very faint object on a photograph taken with the Palomar Schmidt camera (Fig. 20–21). Preliminary determinations of its orbit permitted ob-servations to be found as far back as 1895, so its orbit is now well-determined. It has a 50-year period, an orbital eccentricity of 0.38, and an orbital inclina-tion of 7°. It ranges in distance from the Sun from 8.5 A.U., within the orbit of Saturn, to almost as far out as the orbit of Uranus. No other objects are known with similar orbits and brightness.

From its brightness and distance, we can calculate that Chiron is only about 300 km across. (It appears as only a point of light.) Whether it is an asteroid (perhaps the first discovered of a trans-Saturnian belt of asteroids) or an exceptionally large comet that is too far from the sun for ices to evaporate is unknown.

Chiron's last perihelion was in 1945. It was not observed then, perhaps because of wartime cuts in observing. We have already seen that it was observed in 1895, its first perihelion after the discovery of photography. Chiron's next perihelion will occur in 1995. As it approaches the Earth and Sun, it will certainly be the subject of scrutiny.

SUMMARY AND OUTLINE

Comets (Section 20.1)
    The head (the nucleus + the coma together); the tail
    Dust tail is sunlight reflecting off particles blown back by radiation pressure; ion tail (gas tail) is sunlight re-emitted by ions blown back by the solar wind
    Dirty snowball theory: nucleus composed of rock covered with ices
    Coma is gases vaporized from nucleus; molecules are present
    Ultraviolet space observations revealed huge hydro-gen cloud

Origin of comets: Oort comet cloud; comets detached by gravity
Comet orbits are ellipses
Halley's Comet followed for last 27 returns, but 1986 return will not be as spectacular as some past passages
Comet Kohoutek did not become the popular display that had been hoped for, but nonetheless gave unique data because of its relatively high bright-ness and because it was discovered early
Comet West became very bright with more typical notice of 2 months

Meteoroids (Section 20.2)

Meteoroids: in space; *meteors:* in the air; *meteorites:* on the ground

Most meteorites that hit the Earth are chondrites, which resemble stones and are therefore often undetected. Iron meteorites are more often found

Meteorites may be chips of asteroids and thus give us material with which to investigate the origin of the solar system

Meteorites provide us with much of our information about abundances in the solar system. We could use them to map out inhomogeneities in the solar nebula if we only knew what parts of the solar nebula the different meteorites came from

Some meteors arrive in showers, others are sporadic

Asteroids, also called minor planets (Section 20.3)

Most in asteroid belt, 2.2 to 3.2 solar radii (between orbits of Mars and Jupiter)

Apollo asteroids come within orbit of the Earth

Trojan asteroids in the Lagrangian points (Jupiter ±60°)

Asteroids have different compositions, albedos, and histories

Chiron (Section 20.4)

8.5 to 20 A.U. orbit; 50-year period

First in a trans-Saturnian belt?

An asteroid-like object or a comet head?

QUESTIONS

1. In what part of its orbit does a comet travel head first?

2. Why is the Messier catalogue important for comet hunters?

3. Would you expect comets to follow the ecliptic? Explain.

4. How far is the Oort comet cloud from the Sun, relative to the distance from the Sun to Pluto?

5. The energy that we see as light from a comet comes from what source or sources?

6. Many comets are brighter after they pass close to the Sun than they are on their approach. Why should that be?

7. Which part of a comet has the most mass?

8. Explain why Comet West, seen in the photograph opening this chapter, showed delicate structure in its tail. Are we observing mainly the dust tail or the ion tail?

9. Why was the large cloud of hydrogen that surrounds the head of a comet not detected until 1970?

10. What is the relation of meteorites and asteroids?

11. Why do most meteoroids not reach the surface of the Earth?

12. Why do some meteor showers last only a day while others can last several weeks?

13. Why are meteorites important in our study of the solar system?

14. Does the asteroid belt fit Bode's law? How might this be interpreted in terms of the history of the solar system?

15. How would the occultation of a star by Eros tell us about the possibility that Eros has an atmosphere?

OBSERVING PROJECT

Observe the next meteor shower. No instruments are necessary.

CHAPTER 21

LIFE IN THE UNIVERSE

AIMS:

To discuss the possibility of intelligent life existing elsewhere in the Universe besides the Earth, and to assess our chances of communicating with such life; to give an astronomer's view of UFO's; and to give scientific criteria for "truth" and definitions of "hypotheses" and "theories"

We have discussed the nine planets and the dozens of moons in the solar system, and have found most of them to be places that seem very hostile to terrestrial life forms. Yet some locations besides the Earth—Mars with its signs of ancient running water and perhaps Titan—have characteristics that allow us to convince ourselves that life may have existed there in the past or might even be present now or develop in the future.

In our first real attempt to search for life on another planet, the Viking landers (described in Section 17.3) carried out biological and chemical experiments with Martian soil. The proper interpretation is probably that there is no life on Mars. Rather, chemical processes explain the Viking results. Later in this chapter, we will discuss the criteria that we use to decide whether or not to accept such radical new conclusions as the existence of extraterrestrial life.

Since life probably would preferentially arise on planets, this chapter is included along with the discussion of the solar system.

Recently the number of scientists studying the possibility that life exists in other locations in the Universe has grown. The money spent to search for signs of life and the efforts expended to use modern radio telescopes to search for extraterrestrial signals are indications of the growing acceptance of such studies.

Since it seems reasonable that life, as we know it, anywhere in the Universe would be on planetary bodies, let us first discuss the chances of life arising elsewhere in our solar system. Then we will consider the chances that life has arisen in some more distant parts of the galaxy in which we live or elsewhere in the Universe.

21.1 · THE ORIGIN OF LIFE

It would be very helpful if we could state a clear, concise definition of life, but unfortunately that is not possible. Biologists state several criteria that are ordinarily satisfied by life forms, reproduction, for example. Still, there exist forms on the fringes of life—viruses, for example, which need a host organism in order to reproduce—and scientists cannot always agree whether some of these things are "alive" or not.

In writing science fiction, authors sometimes conceive of beings that show such signs of life as the capability for intelligent thought, even though the beings may share few of the criteria that we ordinarily recognize. In Fred Hoyle's novel *The Black Cloud*, for example, an interstellar cloud of gas and dust is as alive as—and smarter than—you or I. But we can make no concrete deductions if we allow such wild possibilities, and exobiologists prefer to limit the definition of life to forms that are more like "life as we know it."

This rationale implies, for example, that extraterrestrial life is based on complicated chains of molecules that involve carbon atoms. Life on Earth is governed by deoxyribonucleic acid (DNA) and ribonucleic acid (RNA), two carbon-containing molecules that control the mechanisms of heredity. Chemically, carbon is able to form "bonds" with several other atoms simultaneously, which makes these long carbon-bearing chains possible. In fact, we speak of compounds that contain carbon atoms as *organic*.

Carbon is one of the more abundant elements on a cosmic scale (see Table 7–1), although its abundance is far below those of hydrogen or helium. Another reasonably abundant element that can make several chemical bonds simultaneously is silicon, and some scientists have theorized that silicon-based life may exist somewhere. Serious consideration of life forms seems limited to life based on chains of carbon or silicon atoms. No evidence of silicon chemistry has ever been found, though, in meteorites or in interstellar space. This strengthens the probability that we need consider only carbon chemistry.

If we accept this limitation, then we ask how hard it is to build up the long organic chains. To the surprise of many, an experiment was performed twenty-five years ago that showed that the construction of organic molecules was much easier than had been supposed.

At the University of Chicago, at the suggestion of Harold Urey, Stanley Miller put several simple molecules in a glass jar. Water vapor (H_2O),

Figure 21–1   It is all in the point of view. © 1974 United Feature Syndicate, Inc.

methane (CH_4), and ammonia (NH_3) were included along with hydrogen gas. Miller exposed the mixture to electric sparks, simulating the lightning that may exist in the early stages of the formation of a planetary atmosphere. After a few days, he found that long chains of atoms had formed in the jar, and that these organic molecules were even complex enough to include simple amino acids, the building blocks of life.

At the University of Maryland, Cyril Ponnamperuma has carried out modern versions of these experiments, and created organic molecules from simple actions on simpler molecules. Further, he has been synthesizing Jupiter's atmosphere by filling a container with similar gases, exposing them to ultraviolet light, and sending electric sparks through them. Brown and yellow organic compounds, resembling the colors of Jupiter's clouds, form. This suggests that organic compounds also exist on Jupiter.

Ponnamperuma and his colleagues have also examined material from meteorites to search for organic compounds. The Murchison meteorite may have been contaminated by organic compounds after it landed on Earth, but the Maryland group also found amino acids in two meteorites that had been frozen in ice in the Antarctic for hundreds of thousands of years. Half of the amino acids, when examined with polarized light, affect the light in the opposite manner than do all 20 of the Earth's normal amino acids, which indicates that the organic material is extraterrestrial in origin.

It seems clear, however, that mere amino acids or even DNA molecules are not life itself. A jar containing a mixture of all the atoms that are in a human being is not the same as the human being in person. This is the vital gap in the chain; astronomers certainly are not qualified to say what supplies the "spark" of life.

Still, it is reasonable to many astronomers that since it is not very difficult to form complex molecules, life may well have arisen not only on the Earth but also in other locations in our solar system. Even if life is not found in our solar system, there are so many other stars in space that it would seem that some of them could have planets around them and that life could have arisen on some of the planets independently of the origin of life on the Earth.

21.2 OTHER SOLAR SYSTEMS?

We have seen how difficult it was to detect even the outermost planets in our own solar system. At the four-light-year distance of Proxima Centauri, the nearest star to the Sun, any planets would be too faint for us to observe directly.

But even though we cannot hope to see the light reflected by even a giant planet, there is a faint hope of detecting such a planet by observing its gravitational effect on the star itself. One such method is the same that has been used to discover white dwarfs (see Section 9.4). We study the proper motion of a star—its apparent motion across the sky with respect to the other stars (as discussed in Section 5.6)—and see if the star follows a straight line or appears to wobble slightly. The part of astronomy that involves the measurement of the positions of objects in the sky is called **astrometry.**

The most likely stars to observe in a search for planets are the stars with the largest proper motion, for these are likely to be the closest to us and thus any wobble would cover the greatest angular extent we could hope for. The star with the largest proper motion, Barnard's star, is only 1.8 parsecs (6 light

Figure 21–2 The Sproul 61-cm visual refractor is used for photographic and visual studies of our stellar neighborhood. A long-range search for very faint stellar and planet-like companions to nearby stars has been made since 1937 from the ever-increasing plate collection, the largest of its kind to come from one telescope. Several possible candidates for sub-stellar objects have been discovered along with a number of unseen faint low-mass stellar companions to nearby stars.

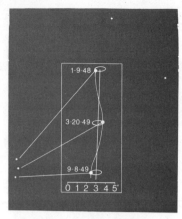

Figure 21–3 The photograph is a composite of three photos of Barnard's star, taken at the Sproul Observatory at intervals of approximately 6 months. The stars at the top and right have small proper motions and small parallaxes, and appear in the same position on each of the three negatives. Barnard's star, on the other hand, shows its proper motion of over 10 arc sec per year. It also shows a lateral displacement caused by parallax, the result of the Earth's orbit around the Sun. The displacement caused by the possible planets does not show at this scale.

Astrometry has been undergoing a revolutionary change in methods recently, with the advent of new machines that automatically measure positions on photographic plates. Photocells can be used to "center" an automatic measuring engine on the center of a star image, and the position can be recorded directly by a computer, or at least on punched cards. These new procedures greatly increase not only the accuracy but also the speed of studying the astrometric plates that are taken.

Soon astrometric observations from the Hipparchos spacecraft and the Space Telescope will revolutionize the field.

years) away from the Sun. It is the fourth nearest star to the Sun; only the members of the Alpha Centauri triple system are nearer.

At the Sproul Observatory (Fig. 21–2), a major astrometric center, Peter van de Kamp has studied observations of Barnard's star taken from 1937 on, and in 1962 reported that he had discovered a wobble in its proper motion (Fig. 21–3). He interpreted this wobble to indicate the presence of a giant planet, larger even than Jupiter, in a very elliptical orbit. After further observations and analysis, he reported that the data were even better explained if there were two giant planets present; then the orbits could be more nearly circular. Van de Kamp also found evidence, though less confidently, for a planet around another of the nearest stars, ε Eridani.

Within the last few years, however, George Gatewood of the Allegheny Observatory in Pittsburgh, and Heinrich Eichhorn of the University of South Florida, have reported that their study of other long-term observations of Barnard's star does not show the wobble in its proper motion. A restudy of the Sproul astrometric plates by van de Kamp shows that for a period of a few years the measured values were different from the trend of preceding and following years; the difference may have started when the telescope's lens was remounted in 1949. A restudy of the data with this in mind significantly reduced the amount of wobble that may be present.

At present, the matter is unresolved. Gatewood now finds that his own data contain the suggestion of a deviation from straight-line proper motion for Barnard's star, but not conclusively so. He admits the possibility that the current data could imply the presence of planets, though this current indication of a wobble does not seem sufficiently convincing that he would have reported it in the absence of the historical context. Van de Kamp's new analysis of his own data still indicates a wobble that can best be interpreted in terms of planets orbiting Barnard's star, but this is a different wobble from the one he had reported previously. Note that if these planets exist, they would have the mass of Jupiter or Saturn. Any Earth-sized planets would not have enough mass to produce a perceptible wobble in their star's proper motion.

At present, most astronomers feel less certain about the results for Barnard's star than they did a few years ago. Further, the extrapolation from the nearest stars that almost all stars have solar systems is less certain than it had been. Observations from the Space Telescope may resolve the situation fairly soon.

A spectrographic survey carried out at Kitt Peak seems to endorse the idea that many stars have planets. One tenth of the 123 solar-type stars and one sixth of the hotter stars under study showed variations over time in their radial velocities. This would correspond to the star's going back and forth as a planet swings around it, in order to keep the gravitational center of the system moving straight. Some of the invisible companions have masses that are too small to allow them to be stars. Some of these may be planets rather than black dwarfs; just how many, it is difficult to say.

21.3 THE STATISTICAL CHANCES FOR EXTRATERRESTRIAL INTELLIGENT LIFE

Instead of phrasing one all-or-nothing question about life in the Universe, we can break down the problem into a chain of simpler questions. This procedure was developed by Frank Drake, a Cornell astronomer, and extended by, among others, Carl Sagan of Cornell and Joseph Shklovskii, a Soviet astronomer.

21.3a The Probability of Finding Life

First we consider the probability that stars at the centers of solar systems are suitable to allow intelligent life to evolve. For example, an O star would probably stay on the main sequence for too short a time to allow intelligent life to evolve.

Second, we ask what the chances are of a suitable star having planets. The import of the previous section is that the chances are probably pretty high, but that we are less certain of this conclusion than we had thought.

Third, we need planets with suitable conditions for the origin of life. A planet like Jupiter might be ruled out, for example, because it lacks a solid surface and because its surface gravity is high. (Alternatively, though, one could consider a liquid region, if it were at a suitable temperature, to be as advantageous as were the oceans on Earth to the development of life here.)

Fourth, we have to consider the fraction of the suitable systems on which life actually begins. This is the biggest uncertainty, for if this fraction is zero (with the Earth being a unique exception), then we can get nowhere with this entire line of reasoning. Still, the discovery that amino acids can be formed in laboratory simulations of primitive atmospheres, and the discovery of complex molecules containing half a dozen or more atoms even in interstellar space (see Section 24.3), indicate to many astronomers that it is not as difficult as had been thought to form complicated molecules. Amino acids, much less complicated than DNA but also basic to life as we know it, have even been found in meteorites. The discovery from radio observations

Most stars have spectral types F, G, K, and M; these might be suitable sites for planets on which life can evolve, though M stars, like Barnard's star, might be too cool. Many multiple stars may be ruled out because the temperatures at their planets might vary rapidly with time as the planets orbited (Fig. 21–4).

Analysis of the fossils of 3.5-billion-year-old single-celled organisms, archaebacteria (Fig. 21–5), indicate that life may have formed very early on in the Earth's history. One class of these fossils that take in carbon dioxide and hydrogen and give off methane has RNA (genetic material) that is different from that of either plants or animals and supports the idea that life evolved before oxygen appeared in the Earth's primeval atmosphere.

Figure 21–4 Some stable planetary orbits around multiple stars include ones close to one star *(A)* or far from both *(B)*. There are no stable figure-8 orbits around both stars.

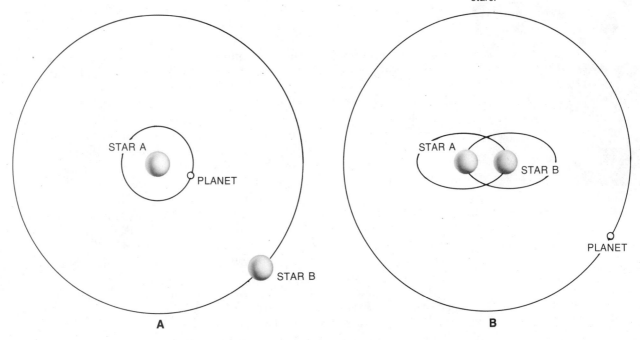

STAR A

PLANET

STAR B

A

STAR A

STAR B

PLANET

B

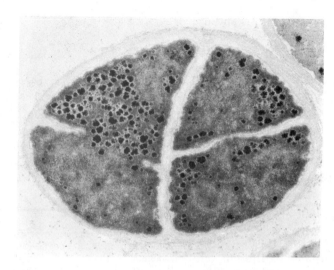

Figure 21–5 An archaebacterium through a microscope. (Courtesy of J. G. Zeikus, University of Wisconsin)

But how long does the phase last when the radio radiation of technological civilizations escapes into space? Already signals that once were broadcast throughout the atmosphere are being put underground in cables and so no longer are radiated into space. Perhaps the period over which a planet is noisy in the radio spectrum lasts only the few decades between the discovery of radio and the development of cable TV and video disks.

Figure 21–6A From *The Day the Earth Stood Still.* (© 1951 Twentieth Century-Fox Film Corporation. All rights reserved.) The movie described a mission to Earth sent in order to restrain our warlike nature.

that molecules (such as carbon monoxide) are associated with sites of star formation strengthens the link. Many astronomers choose to think that the fraction of planetary systems on which life begins may be high.

If we want to have meaningful conversations with the aliens, however, we must have a situation where not just life but intelligent life has evolved. We cannot converse with algae or paramecia. Furthermore, the life must have developed a technological civilization capable of interstellar communication. These considerations reduce the probabilities somewhat, but it has still been calculated that there are likely to be technologically advanced civilizations of intelligent life within a few hundred light years of the Sun.

Now one comes to the important question of the lifetime of the technological civilization itself. We now have the capability of destroying our civilization either dramatically in a flurry of hydrogen bombs or more slowly by, for example, altering our climate or increasing the level of atmospheric pollution. It is a sobering question to ask whether the lifetime of a technological civilization is measured in decades, or whether all the problems that we have, political, environmental, and otherwise, can be overcome, leaving our civilization to last for millions or billions of years.

21.3b Evaluating the Probability of Life

At this point we have reached an interesting result, for one can try to assess (really, to guess, for the most part) fractions for each of the above steps in the chain of reasoning and multiply the fractions together. A quite reasonable set of assumptions leads to the conclusion that there may be billions of planets in this galaxy on which life may have evolved. The nearest one may be within dozens of light years of the Sun. On this basis, many if not most professional astronomers have come to feel that intelligent life probably exists in many places in the universe. Carl Sagan's latest estimate is that there may be a million stars supporting technological civilizations in the Milky Way Galaxy.

But though this point of view has been gaining increasing favor over the last decade, a reaction has recently started. Michael Hart of Trinity University and others have evaluated Drake's equations with a more pessimistic set of numbers and concluded that we earthlings may be alone in our galaxy or in the Universe. Hart suggested that earlier estimates may have been 100 to 1000 times too high.

In particular, Hart calculated the width of the zone around each type of star in which temperatures would be suitable for life to arise. He found that if the Earth were even only 5 per cent closer to the Sun than we are, that a runaway greenhouse effect would have occurred, and that a runaway glaciation would have occurred if we were slightly farther away. Thus the size of the habitable zone around a star is smaller than had been believed, and for stars only 20 per cent smaller than the Sun disappears entirely. And stars more than 20 per cent more massive than the Sun don't stay stable long enough—about 4 billion years—for life to evolve to our stage.

One reason for doubting that the universe is teeming with life is the fact that they have not established contact with us. Where are they all? To complicate things further, is it necessarily true that if intelligent life evolved, they would choose to explore space or to send out messages?

Philip Morrison of MIT, one of the originators of the first proposal to search for radio signals from extraterrestrial life, summarized the current situation to the 1979 General Assembly of the International Astronomical Union.

Morrison pointed out that it was then the 20th anniversary of the original proposal that he and Guiseppe Cocconi made to search for radio signals, and that we still knew three facts: (1) no radio signals from afar have been detected; (2) no extraterrestrials are known on Earth; and (3) we are here. He pointed out that we now probably have a fourth fact: that there is no life on Mars (which would make the current evidence of the probability of life arising equal to ½—Earth yes and Mars no).

The organizer of the IAU symposium, Michael Papagiannis of Boston University, pointed out that even if colonization of space took place at only 1

Figure 21–6B Luke on a Taun-Taun. © Lucasfilm, Ltd. (LFL) 1980. All rights reserved. From the motion picture: *The Empire Strikes Back,* courtesy of Lucasfilm, Ltd.

The large attendance at special sessions about the search for life indicates the widespread interest among professionals in the subject.

The lack of agreement among astronomers is indicated by the four talks that were given to assess the value of N, the number of advanced civilizations in our galaxy. The positions taken were that N is very small, that N is very large, that N is neither very small nor very large, and that N is either very small or very large.

Box 21.1   The Probability of Life in the Universe

Our discussion follows the lines of an equation written out to estimate the number of civilizations in our galaxy that would be able to contact each other. In 1961, Frank Drake of Cornell wrote

$$N = R_* f_p n_e f_l f_i f_c L,$$

where R_* is the rate at which stars form in our galaxy, f_p is the fraction of these stars that have planets, n_e is the number of planets per solar system that are suitable for life to survive (for example, those that have Earth-like atmospheres), f_l is the fraction of these planets on which life actually arises, f_i is the fraction of these life forms that develop intelligence, f_c is the fraction of the intelligent species that choose to communicate with other civilizations and develop adequate technology, and L is the lifetime of such a civilization. Now that we have seen that Jupiter's moons are worlds with personalities of their own, perhaps we should replace f_p with a new factor f_{mp} (moons and planets) that is f_p multiplied by the average number of suitable moons per planet in a solar system.

The largest uncertainty is f_l, which could be essentially zero or could be close to 1. The result is also very sensitive to the value one chooses for L—is it 1 century or a billion years? Do civilizations destroy themselves? Or perhaps they just go off the airwaves. Depending on the alternatives one chooses, one can predict that there are dozens of communicating civilizations within 100 light years or that the Earth is unique in the galaxy in having one.

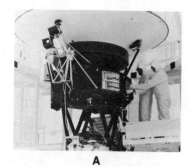

A

B

Figure 21–7 *(A)* Voyager 2 being prepared for launch. The workers are loading a goldplated copper record with two hours' worth of Earth sounds and a coded form of photographs. The sounds include a car, a steamboat, a train, a rainstorm, a rocket blastoff, a baby crying, animals in the jungle, and greetings in various languages. Musical selections include Bach, Beethoven, rock, jazz, and folk music. *(B)* 116 slides are included on the record, and one was taken by the author of this book. It is in the geology sequence, and shows Heron Island on the Great Barrier Reef in Australia, in order to illustrate an island, an ocean, waves, beach, and signs of life.

light year per century, that would still mean that the entire galaxy would have been colonized in less than 1 per cent of its lifetime, equivalent to 1 week of a human's life.

21.4 INTERSTELLAR COMMUNICATION

What are the chances of our visiting or being visited by representatives of these civilizations? Pioneers 10 and 11 and Voyagers 1 and 2 are even now carrying messages out of the solar system in case an alien interstellar traveler should happen to encounter these spacecraft (Fig. 21–7). Still, it seems unlikely that humans can travel the great distances to the stars, unless some day we develop spaceships to carry whole families and cities on indefinitely long voyages into space.

We may note that the Sun has another 4 or 5 billion years to go in its current stable state, while recorded history on Earth has existed for only about 5000 years, one millionth of the age of the solar system. There is plenty of time in the future for interstellar space travel to develop.

We can hope even now to communicate over interstellar distances by means of radio signals. Much thought has been given to what signals to send out, and to what is the best frequency to use, as will be discussed in the following section. What is the chance of hearing another civilization or having our own messages heard? Note that we have known the basic principles of radio for only a hundred years, and that powerful radio, television, and radar transmitters have existed for less than 50 years. Even now we are sending out signals into space on the normal broadcast channels. A wave bearing the voice of Caruso is expanding into space, and at present is 50 light years from Earth. And once a week a new episode of *Mork and Mindy* is carried into the depths of the Universe. Military radars are even stronger.

The Sun would appear to be a variable radio source as seen from most directions in space, because the Earth's signals would come from the vicinity of the Sun. The signal would vary with a period of 24 hours since it would peak periodically (Fig. 21–8) each day as specific concentrations of transmitters rotated to the side of the Earth facing our listener (Fig. 21–9).

In 1974 the giant radio telescope at Arecibo, Puerto Rico (Fig. 21–10),

Figure 21–8 The distribution of television transmitters on the Earth. As the Earth rotates, the irregular distribution would lead to a periodic signal. (Courtesy Woodruff Sullivan, University of Washington.)

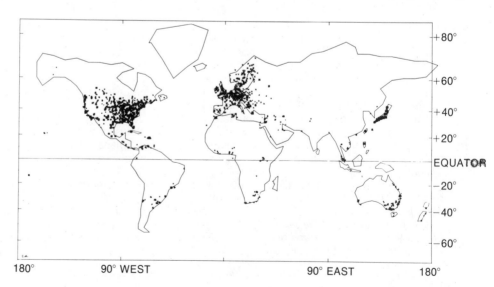

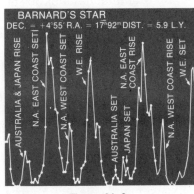

Figure 21–9

Figure 21–10

Figure 21–9 The strength of the signal in the television broadcast band as it would be received at Barnard's star in the course of a day shows variations that would indicate the presence of life on Earth.

Figure 21–10 The Arecibo telescope in Puerto Rico, used to send a message into space.

was upgraded, in that the precision and quality of the mesh of the bowl-like surface of the dish was improved. The telescope is 305 meters across, the largest telescope on Earth. At the rededication ceremony, a powerful signal was sent out into space at the fundamental radio frequency of hydrogen radiation, bearing a message from the people on Earth (Fig. 21–11). This is the only major signal that we have purposely sent out as our contribution to the possible interstellar dialogue.

The signal from Arecibo was directed at the globular cluster M13 in the constellation Hercules on the theory that the presence of 300,000 closely packed stars in that location would increase the chances of our signal being received by a civilization on one of them. But the travel time of the message (at the speed of light) is 24,000 years to M13, so we certainly could not expect to have an answer before twice 24,000, or 48,000 years, have passed. If anybody (or any**thing**) is observing our Sun when the signal arrives, the radio brightness of the Sun will increase by 10 million times for a 3-minute period. A similar signal, if received from a distant star, could be the giveaway that there is intelligent life there.

21.5 THE SEARCH FOR LIFE

For planets in our solar system, we can search for life by direct exploration (as we did with Viking on Mars), but electromagnetic waves, which travel at the speed of light, seem a much more sensible way to search for extraterrestrial life outside our solar system. How would you go about trying to find out if there was life on a distant planet? If we could observe the presence of abundant oxygen molecules or a discovery from the spectrum of a distant star that the molecules there are out of their normal balance, that would tell us that there must be life there. The Space Telescope, for example, should be able to detect the infrared band of radiation from molecular oxygen on a planet orbiting Alpha Centauri.

The most promising way to detect the presence of life at great distances appears to be a search in the radio part of the spectrum. But it would be too

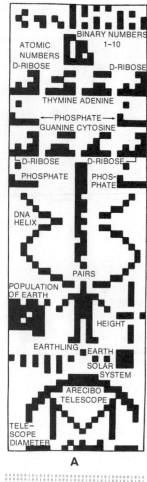

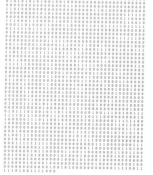

Figure 21–11 *(A)* The message sent to M13, with a translation into English added. *(B)* The message was sent as a string of 1679 consecutive characters, in 73 groups of 23 characters each. There were two kinds of characters, each represented by a frequency; the two kinds of characters are reproduced here as 0's and 1's.

overwhelming a task to listen for signals at all frequencies in all directions at all times. One must make some reasonable guesses on how to proceed.

There are a few frequencies in the radio spectrum that seem especially fundamental. Neutral hydrogen in space, for example, has a basic spectral line at a wavelength of 21 cm. This corresponds to 1420 MHz, a frequency over ten times higher than stations at the high end of the normal FM band. (This hydrogen radiation is of great importance in studying our galaxy, and will be discussed in Section 23.4.) We might conclude that creatures on a far-off planet would decide that we would be most likely to listen near this frequency because it is so fundamental (or because water is important for their lives).

On the other hand, perhaps the great abundance of radiation from hydrogen itself would clog this frequency, and one or more of the other frequencies that correspond to strong natural radiation should be preferred. The "water hole," the wavelength range between the radio lines of H and OH, is at a minimum of radio noise from background celestial sources, the telescope's receiver, and the Earth's atmosphere, and so is another favored possibility. If we ever detect radiation at any frequency, I am sure that we would immediately start to wonder why we had not realized all along that this frequency was the obvious choice.

In 1960, Frank Drake used a telescope at the National Radio Astronomy Observatory to listen for signals from two of the nearest stars. He was searching for any abnormal kind of signal, a sharp burst of energy, for example, such as pulsars emit. He was able to devote to this investigation, known as Project Ozma (after the princess of the land of Oz in L. Frank Baum's stories), only 200 hours of observing time, distributed over a few months.

The observations were later extended in a more methodical search called Ozma II. Starting in 1973, Ben Zuckerman of the University of Maryland and Patrick Palmer of the University of Chicago systematically monitored over 600 of the nearest stars of supposedly suitable spectral type. They used computers to search the data they recorded for any sign of special signals; Ozma II obtained data at 10 million times the rate of the original Ozma. Needless to say, nothing significant turned up; the world will know it rapidly if anything ever does.

Over a dozen separate searches have been undertaken in the radio spectrum, some concentrating on all-sky coverage and others on coverage of individual stars over time. Radio telescopes at Berkeley's Hat Creek Observatory, Ohio State, the National Radio Astronomy Observatory in West Virginia and the Algonquin Observatory in Ontario have been used. Even a search for strong ultraviolet radiation from a laser-like source has been made from a telescope in space.

A grandiose proposal has been made to build a huge array of over 1000 radio telescopes, each 100 meters in diameter, covering an area about 10 kilometers on a side. The total area collecting radiation would be almost that of a single telescope with a 10-kilometer diameter! The array would be fantastically sensitive. It could detect the Earth's leakage signals at a distance of perhaps 100 light years. This plan, called Project Cyclops (Fig. 21–12), would cost 5 billion dollars. Its construction would depend on a decision that the search for interstellar life in this way was an important national priority, a decision that seems unlikely to be made in the foreseeable future.

Of course, valuable astronomical research could be carried out with the Cyclops array when it was not being used for interstellar monitoring, but nobody is about to spend 5 billion dollars on straight astronomy either. Still, the amount is not out of line with the cost of space research. In any case, lesser proposals to use existing telescopes to listen for interstellar radio signals are more likely to be funded. Much of the emphasis now is on obtaining electronic equipment to analyze the incoming signals very carefully. In particular, signals of very narrow frequency range would be considered suspicious.

More and more astronomers in both the United States and in the Soviet Union are interested in "communication with extraterrestrial intelligence," which is becoming known by the acronym CETI, and in the simpler and less expensive "search for extraterrestrial intelligence," SETI. The programs include scanning a wide range of frequencies in all directions in the sky for suitable strange signals.

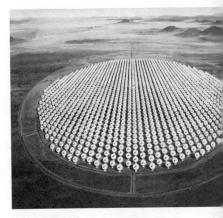

Figure 21–12 An artist's conception of the Project Cyclops array.

*21.6 UFO's

But why, you may ask, if most astronomers accept the probability that life exists elsewhere in the Universe, do they not accept the idea that unidentified flying objects (UFO's) represent visitation from these other civilizations (Fig. 21–13)? The answer to this question leads us not only to explore the nature of UFO's but also to consider the nature of knowledge and truth. The discussion that follows is a personal view, but one that is shared by many scientists.

First of all, most of the sightings of unidentified flying objects that are

Scientific spinoffs of SETI include improved radio spectral-line surveys of stars leading to better knowledge of interstellar matter and the discovery of new radio sources. The information on the effects of terrestrial radiation and the atmosphere on radio signals will also help improve our deep-space tracking of spacecraft and communication links.

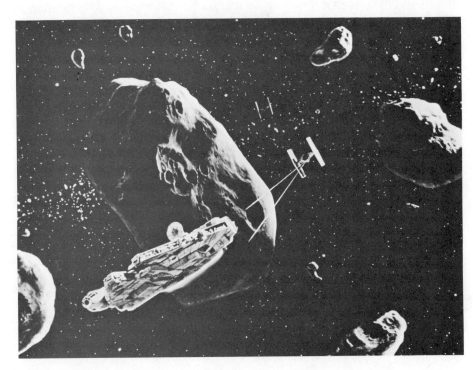

Figure 21–13 Chase through the asteroids. © Lucasfilm, Ltd. (LFL) 1980. All rights reserved. From the motion picture: *The Empire Strikes Back,* courtesy of Lucasfilm, Ltd.

Figure 21–14 A UFO fabrication: an aluminum plate topped by a cottage cheese container on which black dots have been placed to simulate portholes. The landing gear is constructed from hemispheres of ping-pong balls.

Figure 21–15 The two UFO's are really lights shining through the hole in a phonograph record viewed obliquely. The background was added later.

It is mainly non-scientists who feel that UFO's are worth investigating. And it should be pointed out that there is a lot of money involved in reporting sightings of UFO's—since the topic is so popular there are books and articles to be written, and interviews to be given, all of which have a financial return. One must be skeptical of the motives of those who push reports of UFO sightings or want their organizations to continue UFO investigations.

reported can be explained in terms of natural phenomena. Astronomers are experts on strange effects that the Earth's atmosphere can display, and many UFO's can be explained by such effects. For example, I know that every time Venus shines brightly on the horizon, I will get a lot of telephone calls in my capacity as local astronomer asking me about the UFO. A planet or star low on the horizon can seem to flash red and green because of atmospheric refraction, a phenomenon not generally realized by non-scientists. Atmospheric effects can affect radar waves as well as visible light.

Sometimes other natural phenomena—flocks of birds, for example—are reported as UFO's. One should not accept explanations that UFO's are flying saucers from other planets before more mundane explanations—including hoaxes, exaggeration, and fraud—are exhausted (Figs. 21–14 and 21–15).

The topic is of tremendous public interest. As the result of public pressure, a committee was appointed under governmental auspices a few years ago to investigate UFO's. A distinguished physicist, the late E. U. Condon, headed the group, which worked from 1966 to 1969. They asked the leading flying saucer groups to send them the information about the cases that seemed to prove most convincingly that UFO's really have an extraterrestrial identification. After exhaustive analysis, the Condon group was able to account for these cases on the basis of more commonplace effects. Only when too little information is available—as when one has only a report that "something bright was seen in the northern sky for ten minutes and it moved around"—do the UFO reports become impossible to explain without invoking flying saucers. (UFO enthusiasts would not agree with this evaluation.)

Further, it has been shown that for many of the effects that have been reported, the UFO's would have been defying well-established laws of physics if their reported motions actually took place. Where are the sonic booms, for example, from rapidly moving UFO's? Scientists treat challenges to laws of physics very seriously, as much of our science and technology is based on them. If the laws of physics are seriously challenged by UFO's, then we would have to start questioning other applications we make of these laws, such as the safety of skyscrapers or the lift of airplanes.

Almost every professional astronomer feels that UFO's can be so obviously and completely explained by natural phenomena that they are not worthy of more of our time. Astronomers just know too much about the properties of the atmosphere to give credence to most of the UFO reports or to think that further investigation is a worthwhile investment of time or money that could be put to better pursuits.

Some individuals may ask why we reject the identification of UFO's with flying saucers, when that explanation is "just as good an explanation as any other." Let us go on to discover what we mean by "truth" and how that applies to the above question.

*21.7 OF TRUTH AND THEORIES

At every instant, what is happening can be explained in a variety of ways. When we flip a light switch, for example, we assume that the switch closes an electric circuit in the wall and allows the electricity to flow. But it is certainly possible, although not very likely, that the switch activates a relay that turns on a radio that broadcasts a message to an alien on Mars. The Martian then sends back a telepathic message to the electricity to flow, and the light goes on. The latter explanation sounds unlikely to the point that we don't seriously consider it. We would even call the former explanation *true*, without qualification. Even the fact that the signal sent in 1976 to cut the ribbon to open the National Air and Space Museum went to the Viking spacecraft orbiting Mars and back before activating a robot arm a few feet away from the signal's starting point, doesn't lead us to doubt what happens when we flip a wall switch.

We regard as *true* the simplest explanation that satisfies all the data we have about any given thing. This principle is known as *Occam's Razor;* it is named after a fourteenth-century British philosopher. Without it, we would always be subject to such complicated doubts that we would accept nothing as known.

Science is based on Occam's Razor, though we don't usually bother to think about it. For example, we accept the notions of Copernicus and Kepler that the planets orbit around the Sun largely because they provide a relatively simple explanation for all the planetary motions. It is possible to conceive of ways to explain the solar system with everything orbiting around the Earth, or around even Pluto, but the equations would be much more complicated. We say without hesitation that "the Earth revolves around the Sun," or that "it is a fact that the Earth revolves around the Sun," or that "the fact that the Earth revolves around the Sun is true," even though we cannot absolutely, completely, and forevermore rule out all other descriptions.

Sometimes we call something true that might be more accurately described as a *theory.* The scientific method is based on hypotheses and theories. A *hypothesis* is an explanation that is advanced to explain certain facts. When it can be shown that the hypothesis actually explains most or all of the facts known, then we call it a *theory.* We usually test a theory by seeing whether it can predict things that were not previously observed, and then by trying to confirm whether the predictions are valid. We might call the theory, if it passes these tests, a *well-established theory.* Others might prefer to reserve the word *theory* for this stage and call the previous stages *hypotheses* or *conjectures.* The dividing line between hypotheses and theories is not always clear.

An example of a theory is the Newtonian theory of gravitation, which succeeded for many years in explaining almost all the planetary motions. Only a discrepancy of 43 seconds of arc per century in the advance of the perihelion of Mercury, as was described in Section 7.11, remained unexplained. Now, since approximately 1919, we accept Einstein's general theory of relativity as a better explanation of gravitation, and say that Newtonian theory is only approximately valid in regions of space that are limited in size and in which the force of gravitation present is not very large. Is Newton's theory "true"? Yes, in most regions of space. Is Einstein's theory "true"? We say so, although we may also think that one day a new theory will come along that is more general than Einstein's in the same way that Einstein's is more general than Newton's.

How does this view of truth tie in with the previous discussion of

The distance record for round-about signalling is held by a signal from JPL in California that was sent to Voyager 1 near Jupiter, returned to a tracking station in Australia, sent to JPL in Pasadena, and relayed to a science museum in Hutchinson, Kansas. There it went to a laser that set off a photocell that ignited an explosive charge for the groundbreaking.

Occam's Razor, sometimes called the Principle of Simplicity, *is a razor in the sense that it is a cutting edge that allows a distinction to be made among theories.*

UFO's? Every moment we must make decisions about what to do, based on what we think will happen next. We wouldn't even take a step if we thought that a yawning chasm would open up before us in the midst of our living rooms. Scientists have assessed the probability of UFO's being flying saucers from other worlds, and have decided that the probability is so low that the possibility is not even worth considering. We have better things to do with our time and with our national resources. We have so many other, simpler explanations of the phenomena that are reported as UFO's that we apply Occam's Razor and decide to call the identification of UFO's with extraterrestrial visitation *false*. UFO's may be unidentified, but they are probably not flying, nor for the most part are they objects.

SUMMARY AND OUTLINE

The origin of life (Section 21.1)
    Earth life based on complex chains of carbon-bearing (organic) molecules
    Carbon, as well as silicon, forms such chains easily
    Organic molecules are easily formed under laboratory conditions that simulate primitive atmospheres
Evidence for other solar systems (Section 21.2)
    Search for wobbles in proper motions of stars
    Evidence for a planet around Barnard's star once was positive, but is currently probably negative
Statistical chances for extraterrestrial intelligent life (Section 21.3)
    Problem broken down into stages
    Major uncertainties include what the chance is that life will form given the component parts, and what the lifetime of a technological civilization is likely to be

Signals we are sending from Earth (Section 21.4)
    Leakage of radio, television, and radar signals
    Signal beamed toward globular cluster M13 from Arecibo
The search for life (Section 21.5)
    Experiments on Viking spacecraft to Mars
    Projects Ozma, Ozma II, and other searches
    Soviet and American CETI and SETI plans
UFO's explainable as natural phenomena (Section 21.6)
    Application of Occam's Razor—the Principle of Simplicity—rules out the existence of UFO's
Definitions of truth, hypothesis, and theory (Section 21.7)

QUESTIONS

1. What is the significance of the Miller-Urey experiments?

2. Assume that a star has a planet with a mass one tenth that of the star. Sketch the path of the star and the planet in the sky.

3. Does a star with a large proper motion necessarily have planets? Explain.

4. Discuss the evidence for and against the idea that Barnard's star has planets around it.

5. Describe how the Space Telescope will aid in the search for planets around distant stars.

6. If one tenth of all stars are of suitable type for life to develop in the case that planets exist around them, and 1 per cent of all stars have planets, and 10 per cent of all planetary systems have a planet at a suitable distance from the star for a comfortable environment, what fraction of stars have a planet with conditions suitable for life? How many such stars would there be in our galaxy?

7. Describe the limitations for the evolution of life in solar systems much more massive than the sun.

8. List the means by which we might detect extraterrestrial intelligence.

9. Describe how Occam's Razor might have been applied during the debate over the geocentric and heliocentric pictures of the solar system.

10. Give a non-scientific example of a hypothesis, and suggest how it might be tested. At what stage would the hypothesis become a theory?

TOPICS FOR DISCUSSION

1. Comment on the possibility of scientists in another solar system observing the Sun, suspecting it to be a likely star to have populated planets, and beaming a signal in our direction.

2. Discuss the extent of the beliefs of the members of your class in UFO's.

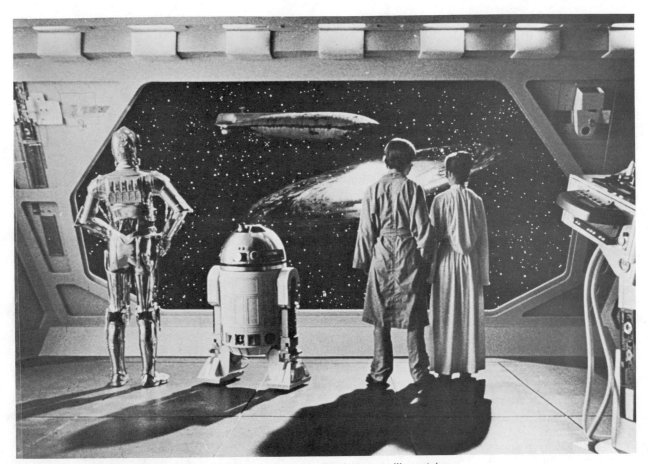

Figure 21–16 As C3PO, R2D2, Luke, and Princess Leia gaze outward, we too will now take up the study of our galaxy, other galaxies, and beyond. © Lucasfilm, Ltd. (LFL) 1980. All rights reserved. From the motion picture: *The Empire Strikes Back,* courtesy of Lucasfilm, Ltd.

PART V

On the clearest nights, when we are far from city lights, we can see a hazy band of light stretched across the sky. This is the *Milky Way*—the aggregation of dust, gas, and stars that makes up the galaxy in which the sun is located.

Don't be confused by the terminology: the Milky Way itself is the band of light that we can see from the earth, and the Milky Way Galaxy is the whole galaxy in which we live. Like other galaxies, our Milky Way Galaxy is composed of a hundred billion stars plus many different types of gas, dust, planets, etc. The Milky Way is that part of the Milky Way Galaxy that we can see with the naked eye in our nighttime sky.

The Milky Way appears very irregular in form when we see it stretched across the sky—there are spurs of luminous material that stick out in one direction or another, and there are dark lanes or patches in which nothing can be seen. This is just the manifestation of the splotchy distribution of the dust, stars, and gas.

Here on earth, we are inside our galaxy together with all of the matter we see as the Milky Way. Because of our position, we see a lot of matter when we look in the plane of our galaxy. On the other hand, when we look "upwards" or "downwards" out of this plane, our view is not obscured by matter, and we can see past the confines of our galaxy. We are able to see distant galaxies only by looking in parts of our sky that are away from the Milky Way, that is, by looking out of the plane of the galaxy.

The gas in our galaxy is more or less transparent to visible light, but the small solid particles that we call "dust" are opaque. So the distance we can see through our galaxy depends mainly on the amount of dust that is present. This is not surprising: we can see great distances through our

The Origin of the Milky Way by Tintoretto, circa 1578. (The National Gallery, London)

414

THE MILKY
WAY GALAXY

gaseous air on earth, but let a small amount of particulate material be introduced in the form of smoke or dust thrown up from a road and we find that we can no longer see very far. Similarly, the dust between the stars in our galaxy dims the starlight by scattering it in different directions besides absorbing some of it.

The presence of so much dust in the plane of the Milky Way Galaxy actually prevents us from seeing very far toward its center—we can see only 1/10 of the way toward the galactic center itself. This explains the dark lanes across the Milky Way, which are just areas of dust, obscuring any emitting gas or stars. The net effect is that we can see just about the same distance in any direction we look in the plane of the Milky Way. These direct optical observations fooled scientists at the turn of the century into thinking that the earth was near the center of the universe.

It was not until the 1920's that the American astronomer Harlow Shapley realized that we were not in the center of the galaxy. He was studying the distribution of globular clusters and noticed that they were all in the same general area of the sky as seen from the earth. They mostly appear above or below the galactic plane and thus are not obscured by the dust. When he plotted their distances and directions (using the methods described in Chapter 6), he saw that they formed a spherical halo around a point thousands of light years away from us (see Fig. 22–1). Shapley's touch of genius was to realize that this point must be the center of the galaxy.

In recent years astronomers have also been able to use wavelengths other than optical ones, especially radio wavelengths, to study the Milky Way Galaxy. We shall first discuss the types of objects that we find in the Milky Way Galaxy, and then go on to describe how radio astronomy and other methods of observation have given us new information about our galaxy.

CHAPTER 22

STRUCTURE OF THE MILKY WAY GALAXY

AIMS:
To understand and review the basic
components of the Milky Way Galaxy
and to understand why spiral
structure forms

We have now described the stars, which are important constituents of any galaxy, and the planets that accompany at least one of them. But the stars do not exist alone; they exist in pairs and clusters, and are accompanied in a galaxy by gas and dust. In this chapter, we describe the gas and dust that we see in the optical part of the spectrum as nebulae. We also describe the overall structure of the Milky Way Galaxy and how, from our location inside it, we detect this structure.

In the next chapter, we will discuss the gas and dust between the stars more fully, including observations in other parts of the spectrum besides the visible.

The current picture that we have of our own galaxy (Fig. 22–1) is that it has the general shape of a pancake with a bulge at its center. This *nuclear bulge* is about 2000 parsecs in radius, with the galactic *nucleus* at its midst. The nucleus itself is only about 5 parsecs across. The nuclear bulge has the shape of a flattened sphere and does not show spiral structure.

Outside the bulge, *spiral arms* unwind in the shape of a pinwheel. The region of the spiral arms, called the galactic *disk*, extends out to about 15,000 parsecs (15 kiloparsecs) or so from the center of the galaxy. The values now being measured for the sun's distance from the center are mostly about 8.5 kpc. Some evidence exists, though, that the distance might be closer to the former, "traditional," value of 10 kpc, two thirds of the outer radius of the disk. The disk is very thin, 2 per cent of its width, like a phonograph record. It

Later in this chapter we shall see that the nucleus is best studied in the x-ray, radio, and infrared parts of the spectrum.

The Milky Way in Sagittarius.

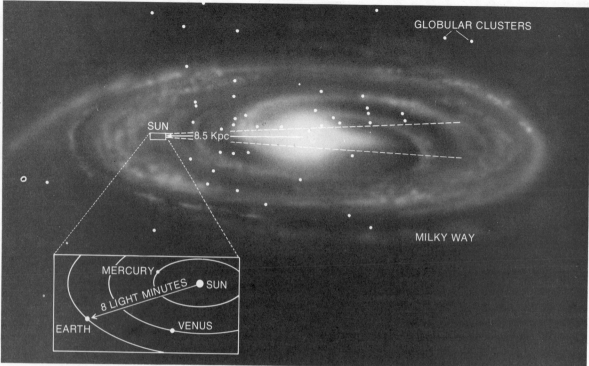

Figure 22–1 An artist's conception of the Milky Way Galaxy, with the actual positions of the globular clusters for which distances are known. Absorption by interstellar matter blocks our view of almost all objects between the dotted lines. From the fact that most of the clusters appear in less than half of our sky, Shapley deduced that the galactic center is in the direction indicated. The drawing shows the *nuclear bulge* surrounded by the *disk*, which contains the spiral arms. The globular clusters are part of this *halo*.

"Nebula" is Latin for fog or mist. The plural is usually nebulae *rather than "nebulas."*

contains all the young stars. It is slightly warped at its ends, perhaps by interaction with our satellite galaxies, the Magellanic Clouds. Older stars, including the globular clusters, form a galactic *halo* around the disk. This halo is at least as large in diameter as the disk, but extends far above and below the plane of our galaxy.)

22.1 NEBULAE

(Not all the gas and dust in our galaxy has coalesced into stars. A *nebula* is a cloud of gas and dust that we see in visible light. When we see the gas actually glowing in the visible part of the spectrum, we call it an *emission nebula*. When we see the dust appear as a dark silhouette against other glowing material, we call the object an *absorption nebula*. When we see dust reflecting light in our direction from stars nearby it, then we have a *reflection nebula*.)

Another class of objects was once known as "spiral nebulae," since they looked like glowing gas with "arms" spiraling away from their centers. (We still speak of the Great Nebula in Andromeda, Color Plate 68.) However, these spiral nebulae are now known to be galaxies in their own right, and indeed contain nebulae of their own. Historically, the debate was whether they were part of our own galaxy or "island universes," as they turn out to be (Chapter 25). The use of the term "nebula" for these particular objects is of historical interest only.

The Great Nebula in Orion (Color Plate 56) is an emission nebula. In the winter sky, it can readily be observed with the eye looking through even a small telescope, but only with long photographic exposures or large telescopes can we study its structure in detail. In the center of the nebula are four closely grouped bright stars called the Trapezium, which provide the energy to make the nebula glow. In the region of the Orion Nebula we think stars are being born this very minute.

Sometimes a cloud of dust obscures our vision in some direction in the sky (Fig. 22–2). This cloud is called an absorption nebula (or, sometimes, a dark nebula).

The Horsehead Nebula (Fig. 22–3 and Color Plate 59) is an example of an object that is both an emission and an absorption nebula simultaneously. A bit of absorbing dust intrudes onto emitting gas, outlining the shape of a horse's head for us. We can see that the horsehead is a continuation of a dark area in which very few stars are visible.

The North America Nebula (Color Plate 60), gas and dust that has a shape in the sky similar to the shape of North America on the earth's surface, is another good example of both an emission and an absorption nebula. The red emission comes from glowing gas spread across the sky. This gas is not completely opaque to visible light; of the stars that are visible in the nebula, some are between the sun and nebula but others are behind the nebula.

The boundary of the North America Nebula as we see it, however, is not just the point where the glowing gas stops. For the North America Nebula, dust absorbs light in the area that corresponds to the Gulf of Mexico. We can tell this because we see fewer stars there.

Figure 22–2 Across the Milky Way we see many absorption nebulae, regions where dark absorbing matter prevents us from seeing many stars. The most prominent on this photograph is marked with an arrow.

Figure 22–3 The Horsehead Nebula, IC 434, in Orion.

Figure 22-4 Reflection nebulae in the Pleiades, M45.

The clouds of dust surrounding some of the stars in the Pleiades (Fig. 22-4 and Color Plate 48) are examples of reflection nebulae—they merely reflect the starlight toward us without emitting much radiation of their own. Many other reflection nebulae are known (Fig. 22-5).

The nebulae are particularly beautiful objects because of their interesting configurations, and because the different processes of emission and absorption cause color effects that, although they are too faint to be visible to the naked eye, can be captured on photographs. Color Plates 48 to 64 all show nebulae of one type or another.

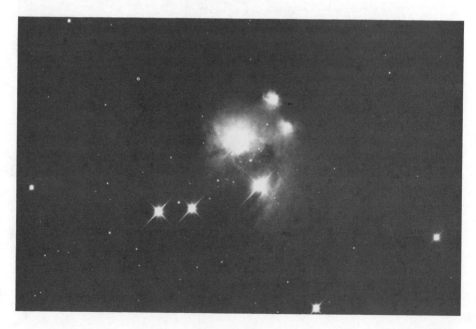

Figure 22-5 A reflection nebula, NGC 7129, in Cepheus.

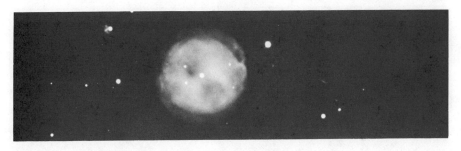

Figure 22–6 The Owl Nebula, M97, a planetary nebula in Ursa Major.

22.1a Stellar Remnants

Some of the most beautiful nebulae in the sky are composed of gas thrown off in the late stages of evolution of stars of various masses. We normally speak of these objects differently than we do for other nebulae.

The shells of gas thrown off by dying stars of approximately the mass of the sun are known as *planetary nebulae* (Figs. 22–6 and 22–7 and Color Plates 49 to 51).

The Ring Nebula (Fig. 22–8 and Color Plate 49) in the constellation Lyra is the best known planetary nebula, for it can be seen even with small telescopes. The colors can be seen on long-exposure photographs but not with the eye directly, even through a telescope. New capabilities of making high resolution images in the radio part of the spectrum have provided additional information about the structure and evolution of planetary nebulae.

The gaseous shells of planetary nebulae can actually be seen to expand over a 10- or 20-year time scale. We can extrapolate back to the time when the shells were blown off, assuming that their motion away from their central stars has not slowed down or accelerated very much. Such calculations indicate that planetary nebulae can only be 20,000 to 50,000 years old—not very old on an astronomical time scale.

Shells of novae also carry matter into interstellar space.

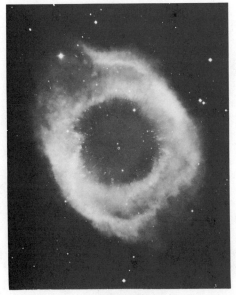

Figure 22–7

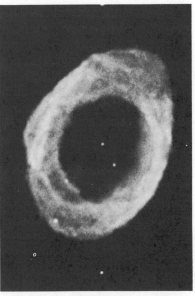

Figure 22–8

Figure 22–7 NGC 7293, a planetary nebula in the constellation Aquarius, the Water Bearer.

Figure 22–8 The Ring Nebula, M57, a planetary nebula in Lyra.

Figure 22-9 The Veil Nebula, NGC 6960, a supernova remnant.

Supernovae remnants (Fig. 22-9 and Color Plates 52 to 55) are described in Section 10.2.

More catastrophic stellar explosions, which result from the deaths of more massive stars, also throw matter into interstellar space. These supernova remnants can now be studied especially well in both the radio and x-ray parts of the spectrum.

22.2 THE CENTER OF OUR GALAXY

Optical and radio astronomical methods have become standard for studying our galaxy. As we have seen, our observations of our galaxy in the visible part of the spectrum are restricted because of the presence of interstellar dust. We did not even know that the center of our galaxy (Fig. 22-10) lies in the direction of the constellation Sagittarius as seen from earth until Shapley deduced the fact from his study of the distribution of globular clusters.

In recent years, observations in other parts of the electromagnetic spectrum have become increasingly important for the study of the Milky Way Galaxy. The center of our galaxy is a prominent source in each of these spectral regions. In the radio region of the spectrum, for example, Sagittarius A (Fig. 22-11) has long been known as one of the strongest sources in the sky.

22.2a Infrared Techniques

Astronomers working in the infrared usually use the unit of microns for wavelength. 1 micron (1 μ) is 1/1,000,000 m, and is 10,000 Å.

For many years the sky has been intensively studied at optical wavelengths up to about 8500 Å (0.85 micron) and to a lesser extent up to 1.1 microns, which is in the near infrared. The sky has also been studied for decades at radio wavelengths down to one or two centimeters (10,000 or 20,000 microns). But until the last few years, the sky has been studied very little at the wavelengths in between, which are the millimeter and the infrared wavelengths.

We saw in Section 2.13a that the lack of film sensitivity to the infrared, whose photons contain relatively low energy, is a major reason for this lack. Neither have especially sensitive electronic detection devices been available in the infrared. Atmospheric limitations, caused by the presence of only a few windows of transparency, have been another major factor. Since one can

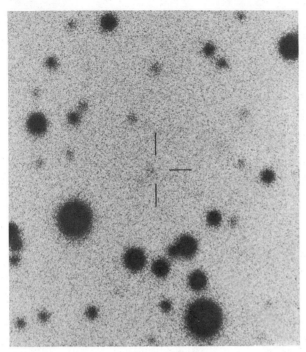

Figure 22–10

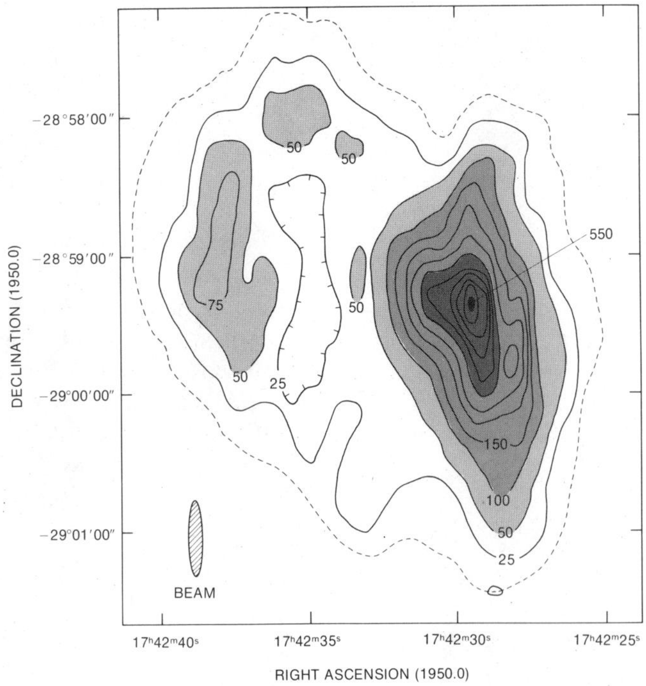

Figure 22–11

We shall consider star formation in Section 24.5.

Figure 22–10 The galactic center region shown in a negative print. The position of the center of our galaxy is marked with straight lines.

Figure 22–11 A radio map of the galactic center. The intense source at right is called Sgr A.

observe better in the infrared from locations where there is little water vapor overhead, the new emphasis on infrared telescopes at sites like Mauna Kea in Hawaii should improve the situation immeasurably.

Yet another difficulty comes from the fact that ordinary heat radiation is in the infrared. The earth's atmosphere radiates conspicuously in the infrared, so the radiation coming into a telescope includes infrared radiation from the atmosphere, from the source being observed, and from the telescope itself. To limit the telescopic contribution of radiation, the detection equipment is usually bathed in liquid nitrogen, or even in liquid helium, which is colder and more expensive.

Robert Leighton and Gerry Neugebauer of Caltech mapped the sky in the 2.2-micron window during the mid 1960's. They surveyed the whole part of the sky that can be observed from the top of Mount Wilson, and detected some 20,000 infrared sources! Remember that there are only about 6000 stars that can be seen with the naked eye.

Leighton and his group have now installed telescopes in the Owens Valley in California to study much longer wavelengths. They also have plans for a telescope at a high altitude site, perhaps Mauna Kea. They are studying wavelengths close to the infrared/radio boundary at 1 mm; we thus say that they are working in the *submillimeter* and *millimeter* regions.

Many of the infrared sources are cool stars but many others do not coincide in space with known optical sources. The map of the radio sky doesn't look like the map of the optical sky, and the map of the infrared sky doesn't resemble either of the others. Most of these infrared-emitting objects, however, unlike the radio objects, are in our galaxy. Following the survey program, some of the more interesting infrared objects have been individually studied much more carefully. Many turn out to be intimately connected with stars in formation.

22.2b The Galactic Nucleus

One of the brightest infrared sources in our sky is located at the position of the radio source Sagittarius A (Fig. 22–12), which marks the center of our galaxy. Since the amount of scattering by dust varies with wavelength, we can see farther through interstellar space in the infrared than we can in the visible. In particular, in the infrared we can see 8.5 kiloparsecs to the center of our galaxy.

This infrared source subtends 1 arc min, and so is about 3 parsecs (about 10 light years) across. This makes it a very small source for the prodigious amount of energy it emits: as much energy as if there were 80 million suns radiating. It is also a strong and variable x-ray source.

We don't know just what it is in the center of our galaxy that causes this radiation, but across the spectrum this small source does seem to give off as much as 0.1 per cent of the total radiation from our galaxy. It is interesting to know that the center of our galaxy is qualitatively different from the outer parts. (Other galaxies have even more prominent infrared, radio, and x-ray sources in their nuclei.)

Several models have been advanced to account for the radiation from the galactic nucleus. In one model, bright sources of ultraviolet or visible radiation, such as a dense group of stars, are present there. These sources would be surrounded by shells of dust that heat up to about 50 K, thus providing the infrared luminosity. The equivalent of about 10 solar masses of dust, not an inconceivable amount, would have to be present to make this model work. Similar processes, on a smaller scale, are known for shells surrounding hot stars.

Another model, much more speculative, says that both matter and antimatter exist in the center of our galaxy. They would annihilate each other, as matter and antimatter do, and a tremendous amount of energy would be liberated in this annihilation. The energy could tend to push the matter and antimatter away from each other, preventing the situation from "running away," that is, continuously escalating at an increasing rate. This model is certainly not proved at present, for we do not definitely know of any large-scale existence of antimatter at all, but neither can it be ruled out. The

Figure 22–12 *(A)* A high resolution map of the center of the Milky Way at a wavelength of 2.2 microns. The resolution is about 1.2 arc min, and the height of the entire image is about 1.1°. At this wavelength, we see stars rather than dust. Dust becomes apparent at longer wavelengths. *(B)* A photograph in visible light taken in the direction of the center of our galaxy, showing exactly the same region observed in the infrared in *A*. In visible light, interstellar dust completely hides the galactic center, and we are seeing only relatively nearby stars and gas. (Infrared image by Eric Becklin, Institute for Astronomy, University of Hawaii)

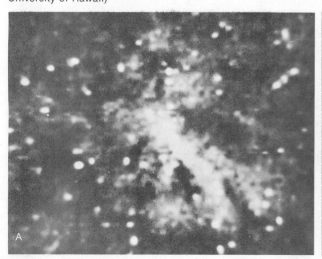
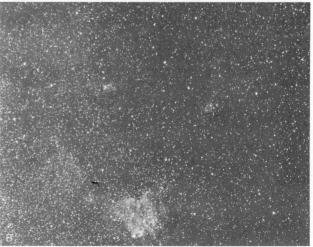

detection in 1979 of a gamma-ray spectral line from the direction of the galactic center is being interpreted in terms of the production of these particular gamma rays in a matter-antimatter annihilation. In particular, an electron is meeting its anti-particle, a *positron,* and the two are annihilating each other completely, releasing energy in the amount of $E = mc^2$, where m is the sum of the masses of the two particles.

In the very center of the nucleus, radio astronomers have discovered an extremely narrow source. It is only about 10 astronomical units across, smaller than the orbit of Jupiter. It is giving off a great deal of energy (though not as much as the nuclei of some distant galaxies). The leading model for the galactic center at the moment, with black holes in fashion, is that a high-mass black hole is present there. This galactic center model, with a black hole at its core, explains particularly well why the radio source at the galactic center is so small in extent. The black hole would contain 10 thousand to a million solar masses. Matter spiralling in would heat up and give off the large amount of energy that we detect. Some of the matter would be in the form of interstellar dust and gas, and would be giving off the observed radio and infrared radiation. Radio studies of the motions of gas near the center endorse the idea that a very massive source is present there.

22.3 HIGH-ENERGY SOURCES IN OUR GALAXY

The study of our galaxy provides us with a wide range of types of sources. Many of these have been known for many years from optical studies (Fig. 22–13). We have just seen how the infrared sky looks quite different. The radio sky provides still a different picture. Technological advances in the last few years have enabled us to study the distribution of sources in our galaxy in the x-ray and gamma-ray region of the spectrum as well.

Figure 22–13 A drawing of the Milky Way, made under the supervision of Knut Lundmark at the Lund Observatory in Sweden. 7000 stars plus the Milky Way are shown in this panorama, which is in coordinates such that the Milky Way falls along the equator.

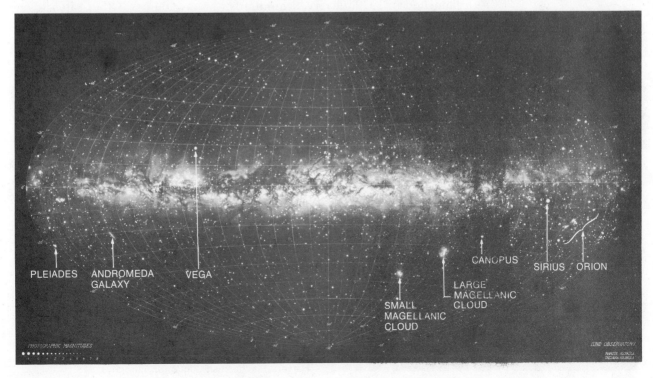

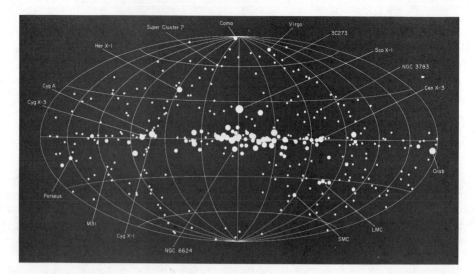

Figure 22–14 An x-ray map of the Milky Way showing the objects observed by the Uhuru spacecraft as of mid-1978, the Fourth Uhuru Catalogue.

We have already mentioned (in Section 11.5 on black holes) some of the results provided by the Uhuru satellite.

Non-solar x-ray astronomy began in 1962, when Riccardo Giacconi and colleagues discovered x-rays from a source in the constellation Scorpius (and named it Scorpius X-1). Their group, from American Science and Engineering, Inc., and MIT in Cambridge, Mass., was soon joined in studying non-solar x-rays by Herbert Friedman and colleagues at the Naval Research Laboratory in Washington, D.C. Their work throughout the 1960's was carried out with rockets rather than with satellites. A few dozen sources, including Scorpius X-1, the Crab Nebula, and the Virgo cluster of galaxies, were found.

Most of the x-ray sources that we detect are located in our galaxy (Fig. 22–14). The first reasonable map of the x-ray sky was made with a NASA satellite in its Small Astronomy Satellite series (SAS-1). This Uhuru satellite, launched in 1970, observed hundreds of x-ray sources. Other satellites to study x-rays have been launched by the U.S., by the Soviet Union, by Britain from a launch pad in Kenya, and by the Netherlands from the United States.

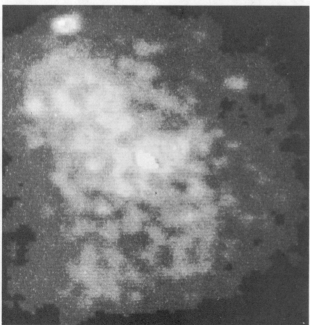

Figure 22–15 This computer interpretation of an x-ray image shows a bright point at the galactic center. This view is identical to that shown in Figure 22–12, except that the scale of this image should be reduced by a factor of 1.5 relative to that of the infrared and optical images. (Einstein Observatory x-ray image by Michael Watson, Paul Hertz, and colleagues, Harvard-Smithsonian Center for Astrophysics)

Most of our knowledge of x-ray sources now comes from the series of U.S. High-Energy Astronomy Observatories. HEAO-1 mapped 1500 x-ray sources. HEAO-1 and HEAO-2 (Einstein) together have studied many of them over periods of time. Although many of the x-ray sources are in our galaxy, others are extragalactic. The Einstein Observatory has made it possible to study not only the x-ray source at the center of our own galaxy (Fig. 22–15) but also the distribution of small x-ray sources in nearby galaxies. The Andromeda Galaxy previously had been only barely detected as a single x-ray source, but the Einstein Observatory has found many sources distributed through its central region (Fig. 22–16). Since it is easier to understand galaxies seen from outside than it is from the perspective of our interior view of the Milky Way Galaxy, this can help us in understanding our own galaxy.

Some of the observations of x-ray bursts from globular clusters detected by these satellites have been discussed in Section 6.6.

Box 22–1 Superbubble in Cygnus

A huge ring or shell of gas glowing in x-rays has been discovered. This superbubble in the interstellar gas, visible in x-rays from HEAO-1 but invisible to optical studies, is in the constellation Cygnus, high in our summer sky. It is 1200 light years in diameter, about 5 per cent of the radius of our galaxy. Though pieces had been detected before, it was too large for its structure to be obvious until HEAO-1 made a large-scale sweep of the sky. It contains the most energy of any single feature of our galaxy, more than 10 times the total output of the sun since its birth.

The gas in the superbubble is at 2 million K. Ten thousand solar masses of material is present. Although supernovae are the most powerful explosions we know of, 30 supernovae would have been necessary to provide enough energy to form the superbubble. Thus its discoverers, Webster Cash of the University of Colorado and Philip Charles of the University of California at Berkeley, suggested that a series of supernovae exploded in the region. In their model, a first supernova 3 million years ago sent a shock wave into the dark cloud known as the Great Rift in Cygnus, forming other massive stars. These stars in turn eventually became supernovae, and so on. The force of the explosions swept out the bubble. Searches are under way for similar bubbles in our own and in nearby galaxies. The sun may well have been formed in a similar chain of star formation.

The satellite experiment was built by Elihu Bolt of NASA's Goddard Space Flight Center and Gordon Garmire of Caltech.

Figure 22–16 An x-ray image of the central region of M31, the Andromeda galaxy, taken with the Einstein Observatory. Twenty x-ray sources can be seen in this image. Though most x-ray sources in our own Milky Way galaxy are in the galactic plane, that is not true for this group of sources in Andromeda. (Einstein Observatory x-ray image by Leon Van Speybroeck and colleagues, Harvard-Smithsonian Center for Astrophysics)

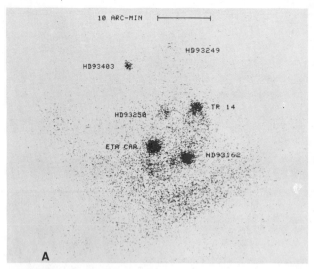

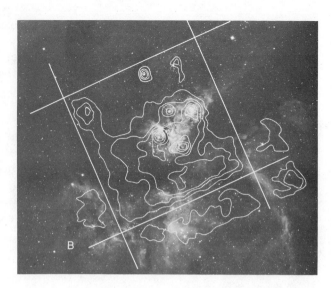

Figure 22–17 *(A)* Einstein Observatory x-ray image of the Eta Carinae Nebula. Several strong x-ray sources are visible in this region. *(B)* X-ray contours overlaid on an ultraviolet photograph of the Eta Carinae Nebula. (Frederick D. Seward and colleagues, Harvard-Smithsonian Center for Astrophysics)

NASA is planning a Gamma Ray Observatory to go into earth orbit in 1985.

Figure 22–18 The gamma-ray map made by the European COS-B spacecraft shows the plane of the Milky Way and several individual gamma-ray sources. The map includes the region between 25° above and 25° below the plane of the Milky Way.

One of the more interesting locations in the sky is the region of the Eta Carinae Nebula (Color Plates 57 and 58). This object, which is located in our own galaxy, is not visible at the northern latitudes of the United States. Eta Carinae itself may be an old supernova remnant, but there is extensive nebulosity in the area and a variety of other sources. The Einstein Observatory has detected x-rays from the stars and nebulae in the vicinity (Fig. 22–17).

Gamma-ray astronomy is even harder to carry out than x-ray astronomy, and only a few small satellites had been launched to study gamma rays prior to HEAO-3. Although an occasional strange object, like the Crab Pulsar, can be detected at every wavelength of the spectrum, including the x-ray and gamma-ray regions, astronomers were hard put to find optical counterparts of other gamma-ray sources. Also, many gamma-ray observations have been made from Vela satellites, U.S. satellites whose prime purpose is to detect nuclear explosions in space.

One interesting gamma-ray observation is the discovery of a general background of gamma-rays concentrated along the plane of our galaxy (Fig. 22–18). This diffuse background is not concentrated in particular objects, not even in the direction of the galactic center. Presumably, these gamma-rays are caused by the interaction of *cosmic rays* with interstellar matter. Even though the cosmic rays are already formed with high energies at such sites as supernovae, many of them are brought up to even higher energies by processes taking place throughout the galaxy.

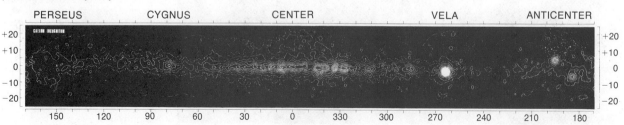

Another gamma-ray observation is the occasional occurrence of mysterious bursts, which come from seemingly random locations in the sky. One of the problems in the past has been limited observational ability to pinpoint the location of the bursts. The presence of gamma-ray detectors in space not only aboard HEAO-3 but also aboard such spacecraft as Pioneer Venus should help this situation. Piling surprises upon surprises, an international set of sensors put in space by observers from the U.S., Europe, and the U.S.S.R. has discovered a new type of gamma-ray burst ten times more powerful than any previous bursts. The burst seemed to come from a supernova remnant associated with the Large Magellanic Cloud, an object that is hundreds of times farther away from us than we had assumed the sources of gamma-ray bursts would be.

A powerful burst was observed on March 5, 1979. The initial burst, only 1/5000 s in duration, was followed for a few minutes by pulses with an 8-s period. The periodicity may indicate the presence of a neutron star. During the burst, the object gave off energy at a rate greater than the entire Milky Way Galaxy.

Studies of electromagnetic radiation like x-rays and gamma-rays and of rapidly moving cosmic ray particles are part of the new field of *high-energy astrophysics*. High-energy astrophysics is an especially active field for theorists, as well as observers. The theorists study, among other things, how radiation of high energies can be generated in interactions between matter and other matter or between matter and radiation.

22.4 THE SPIRAL STRUCTURE OF THE GALAXY

22.4a Bright Tracers of the Spiral Structure

When we look out past the boundaries of the Milky Way Galaxy, usually by observing in directions above or below the plane of the Milky Way, we can see a number of galaxies with arms that appear to spiral outward from near their centers. In this section we shall discuss some of the evidence that our own Milky Way Galaxy also has spiral structure.

We shall discuss spiral galaxies beyond the Milky Way (Color Plates 66, 68, 69, 71, 72, and 73) in Chapter 25.

It is always difficult to tell the shape of a system from a position inside it. Think, for example, of being somewhere inside a maze made of tall hedges. We might be able to see through some of the foliage and be reasonably certain that there were layers and layers of hedges surrounding us, but we would find it difficult to trace out the pattern. If we could fly overhead in a helicopter, though, the pattern would become very easy to see.

Similarly, we have difficulty tracing out the spiral pattern in our own galaxy, even though the pattern would presumably be apparent to someone located outside our own galaxy. Still, by meticulously noting the distances and directions to objects of various types, we can tell something about the Milky Way's spiral structure.

Galactic clusters are good objects to use for this purpose, for they are known to always be in the spiral arms. We can find the distance to a galactic cluster from its color-magnitude diagram. We have now determined the distances to over 200 of these clusters.

We think that spiral arms are regions where young stars are to be found. Other signs of young stars besides galactic clusters are regions of ionized hydrogen known as H II regions (pronounced "H two regions"). We know from studies of other galaxies that H II regions are preferentially located in spiral arms. In studying the locations of the H II regions, we are really studying the locations of the O stars and the hotter B stars, since it is

Figure 22–19 The positions of young galactic clusters and H II regions in our own galaxy are projected on a photograph of the spiral galaxy NGC 1232, which has the same linear diameter as our own galaxy.

ultraviolet radiation from these hot stars that provides the energy for the H II regions to glow. These O and B stars, which must be relatively young because their lifetimes are short, must therefore tend to be formed in the spiral arms; in the next section we shall discuss a theory that explains why this should be the case.

We can find the distances to H II regions, in some cases, by studying the intensities of their hydrogen spectral lines (H β, for example), and comparing the observed intensities with the intensities of similar lines in regions whose distances are known in some other fashion. From the comparison we can deduce how distant the H II regions must be in order for the hydrogen lines to have their observed intensities. Only a few dozen distances have been calculated by this method, however.

When the positions of the galactic clusters and H II regions of known distances are studied, by plotting their distances and directions as seen from earth, they appear, to trace out three spiral arms (Fig. 22–19). The spacings between the arms and the widths of the arms appear consistent with spacings and arm widths that we can observe in other galaxies.

These observations are carried out in the visible part of the spectrum. Note that even when we are observing the O and B stars, which are very bright and therefore can be seen at distances that are relatively great compared to other stars, the optical map shows only regions of our galaxy close to the sun. Interstellar dust prevents us from studying parts of the spiral arms farther away from the sun. Another very valuable method of studying the spiral structure in our own galaxy involves a spectral line of hydrogen in the radio part of the spectrum. Radio waves penetrate the interstellar dust, and we are no longer limited to studying the local spiral arms. This radio method will be discussed in Section 23.4.

*22.4b Differential Rotation

The latest calculations indicate that the sun is approximately 8.5 kiloparsecs from the center of our galaxy. From spectroscopic observations of the Doppler shifts of globular clusters, which do not participate in the galactic rotation, or of distant galaxies, we can tell that the sun is revolving around the center of our galaxy at a speed of approximately 250 kilometers per second. At this velocity, it would take the sun about 250 million years to travel once around the center; this period is called the *galactic year*. But not all stars revolve around the galactic center in the same period of time. The central part of the galaxy rotates like a solid body. Beyond the central part, the stars that are farther out have longer galactic years than stars closer in.

If this system of *differential rotation*, with differing rotation speeds at different distances from the center, has persisted since the origin of the galaxy, we may wonder why there are still only a few spiral arms in our galaxy and in the other galaxies we observe. The sun could have made fifty revolutions during the lifetime of the galaxy, but points closer to the center would have made many more revolutions. Thus the question arises: why haven't the arms wound up very tightly?

*22.4c Density-Wave Theory

The leading current solution to this conundrum is a theory first suggested by the Swedish astronomer B. Lindblad and elaborated mathematically by the American astronomers C. C. Lin at M.I.T. and Frank Shu, now at Berkeley. They say, in effect, that the spiral arms we now see are not the same spiral arms that were previously visible. In their model, the spiral arm pattern is caused by a spiral *density wave*, a wave that moves through the stars and gas in the galaxy. This density wave is a wave of compression, not of matter being transported. It rotates more slowly than the actual material, and causes the density of material to build up as it passes by. A shock wave—a moving surface at which a compression takes place, as in a sonic boom from an airplane—is also built up. The shock wave heats the gas.

So, in the density-wave model, the spiral arm we see at any given time does not represent the actual motion of individual stars in orbit around the galactic center. We can think of the analogy of a crew of workers painting a white line down the center of a busy highway. A bottleneck occurs at the location of the painters, and an observer in an airplane would see an increase in the number of cars at that place. As the line painters continued slowly down the road at, say, 8 km/hr (5 mph), the airborne observers would seem to see the place of increased density move down the road at that slow speed. But they would not be seeing the individual cars, which could still speed down the highway at 80 km/hr (50 mph), slow down briefly as they cross the region of the bottleneck, and then resume their high speed.

Similarly, we might only be viewing some galactic bottleneck at the spiral arms. The gas is compressed there, both from the increased density itself and, to a greater extent, from the shock wave formed by other gas piling into this slowed-up gas from behind. The compression causes heating, and also leads to the formation of protostars that collapse to become stars. Thus the spiral density wave leads to star formation, though the individual stars themselves would move across the arms as they revolve around the center of the galaxy.

It may help to think of the analogy of a sound wave, which is also a compression wave. The sound wave travels through the air at a rapid rate, but the air itself doesn't move at that rate in the direction of the sound wave. (You certainly don't feel a wind every time a friend talks to you.) The mass of air has no net velocity at all.

Similarly, the density wave can pass through the galactic matter even though the galactic matter does not share in its motion. As the density wave passes a given point, stars begin to form there because of the increased density. The distribution of new stars can take a spiral form (Fig. 22–20). The new stars heat the interstellar gas so that it becomes visible. In fact, we do see young, hot stars and glowing gas outlining the spiral arms, which are checks of this prediction of the density-wave theory.

Some astronomers do not accept the density-wave theory; the problem of explaining why the density wave persists is a major objection. One alternative theory takes a view quite different from that of the density-wave theory, saying that stars are produced by a chain reaction and are then spread out into spiral arms by the differential rotation of the galaxy. The chain reaction begins when high mass stars become supernovae. The expanding shells from the supernovae trigger the formation of stars in nearby regions. Some of these new stars become massive stars, which become supernovae, and so on.

Computer modeling of the chain-reaction/differential-rotation process gives values that seem to agree with the observed features of spiral galaxies (Fig. 22–21). Each galaxy has a differential velocity, with regions at different

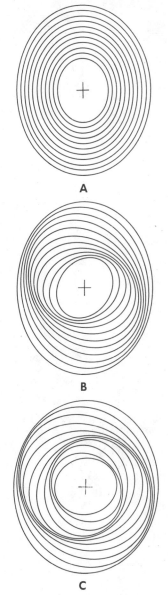

Figure 22–20 Each part of the figure includes the same set of ellipses; the only difference is the relative alignment of their axes. Consider that the axes are rotating slowly and at different rates. The compression of their orbits takes a spiral form, even though no actual spiral exists. The spiral structure of a galaxy may arise from an analogous effect.

A

B

C

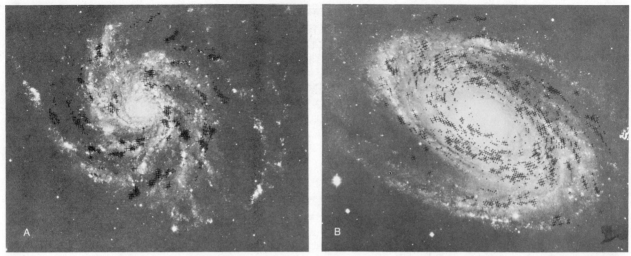

Figure 22-21 The model galaxy plotted as +'s superimposed upon a photograph of the galaxy. The models were computed using observed values of rotational velocities for each distance from the center. *(A)* M101. *(B)* M81, including the effect of projecting the model at a 58° angle. (Courtesy of Humberto Gerola and Philip E. Seiden, IBM Watson Research Center)

distances from the center rotating at different speeds. At some distance from the center, the speed is a maximum. The degree of winding depends on the value for this maximum velocity (Fig. 22–22).

Other alternative explanations involve gravitational interactions. Spiral arms might be pulled out of the cores during a collision or near passage of two galaxies in space. It seems unlikely, however, for enough of these gravitational interactions of galaxies to have taken place to account for the large number of spiral galaxies that we observe. We shall discuss the effects of gravitational interactions in Section 25.2b.

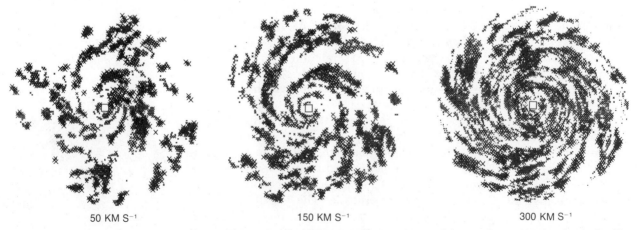

50 KM S⁻¹   150 KM S⁻¹   300 KM S⁻¹

Figure 22-22 Model galaxies for different values of the maximum rotational velocity. The leftmost corresponds to Hubble type Sc, while the rightmost corresponds to Hubble type Sa.

Color Plate 47 (top): 47 Tucanae, a prominent globular cluster visible from the southern hemisphere. (Cerro Tololo Inter-American Observatory photo with the 4-m telescope)

Color Plate 48 (bottom): The Pleiades. M45, is a galactic cluster in the constellation Taurus, the bull. Reflection nebulae are visible around the brightest stars. (Palomar Observatory, California Institute of Technology photo with the 1.2-m Schmidt camera)

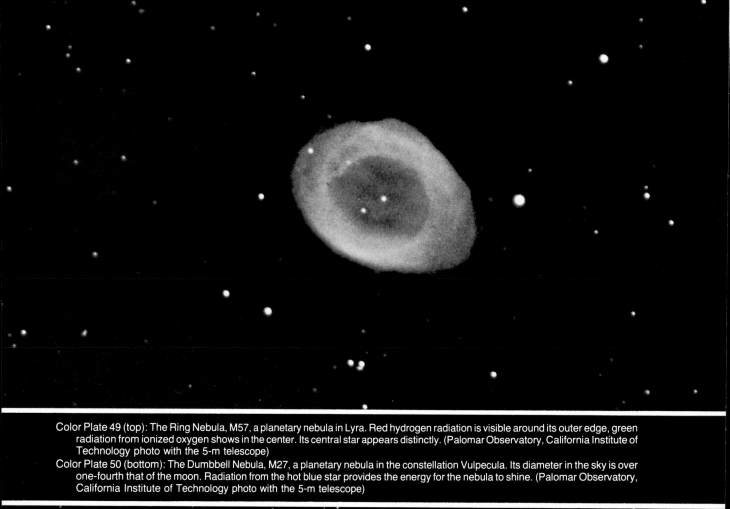

Color Plate 49 (top): The Ring Nebula, M57, a planetary nebula in Lyra. Red hydrogen radiation is visible around its outer edge, green radiation from ionized oxygen shows in the center. Its central star appears distinctly. (Palomar Observatory, California Institute of Technology photo with the 5-m telescope)

Color Plate 50 (bottom): The Dumbbell Nebula, M27, a planetary nebula in the constellation Vulpecula. Its diameter in the sky is over one-fourth that of the moon. Radiation from the hot blue star provides the energy for the nebula to shine. (Palomar Observatory, California Institute of Technology photo with the 5-m telescope)

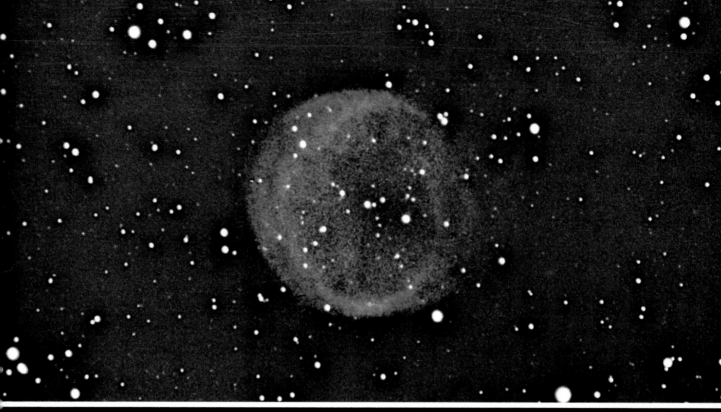

Color Plate 51 (top): A planetary nebula, NGC 6781, in the constellation Aquila. (Palomar Observatory, California Institute of Technology photo with the 1 2-m Schmidt camera)

Color Plate 52 (bottom, left): A radio image of the Cassiopeia A supernova remnant, observed at a wavelength of 20 cm with the VLA. (Courtesy of Philip E Angerhofer, Richard A. Perley, Bruce Balick, and Douglas Milne with the VLA or NRAO)

Color Plate 53: (bottom, right): An x-ray image of the Cassiopeia A supernova remnant, observed with the Einstein Observatory. The bright ring is thought to be the region associated with the expanding shock front No pulsar has been detected (Courtesy of S. S. Murray and colleagues at the Harvard-Smithsonian Center for Astrophysics)

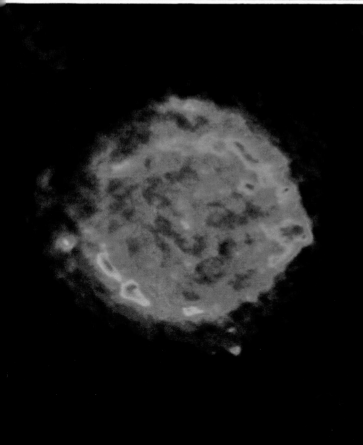

CAS A SUPERNOVA REMNANT
EINSTEIN OBSERVATORY

60 ARC-SECS

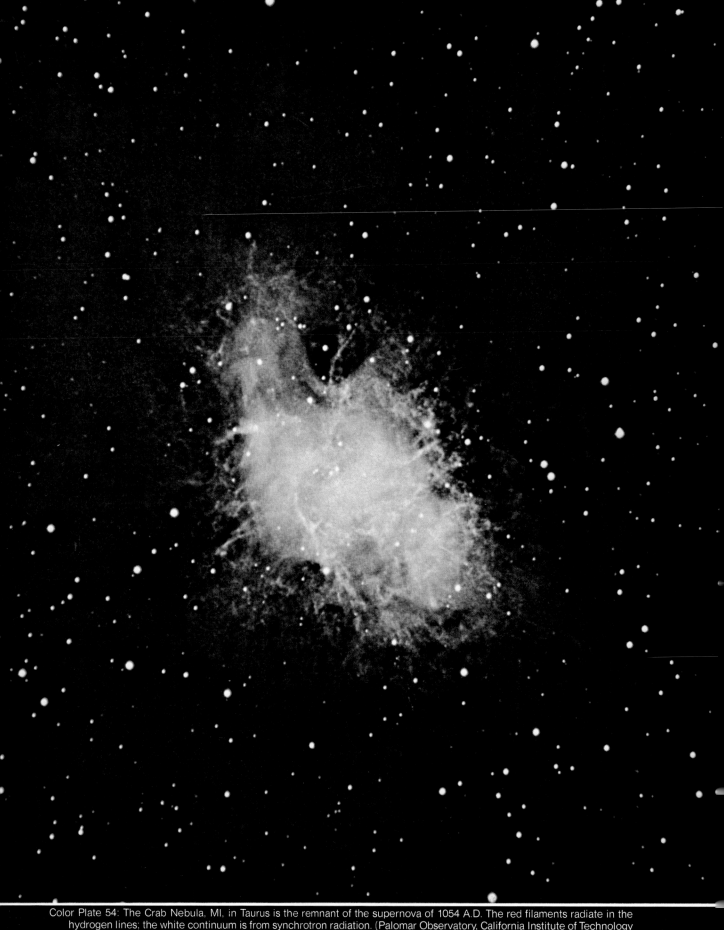

Color Plate 54: The Crab Nebula, M1, in Taurus is the remnant of the supernova of 1054 A.D. The red filaments radiate in the hydrogen lines; the white continuum is from synchrotron radiation. (Palomar Observatory, California Institute of Technology photo with the 5-m telescope)

Color Plate 55 (top): The Veil Nebula, NGC 6992, part of the Cygnus loop, a supernova remnant. (Palomar Observatory, California Institute of Technology photo with the 1.2-m Schmidt camera)

Color Plate 56 (bottom): The Orion Nebula, M42, is glowing gas excited by the Trapezium, four hot stars. The nebula contains stars in formation: it is 25 ly across and 400 parsecs away. (Palomar Observatory, California Institute of Technology photo with the 5-m telescope)

Color Plate 57: The Eta Carinae Nebula, NGC 3372, in the southern constellation Carina. Visually, this is the brightest part of the Milky Way. The central dark cloud superimposed on the brightest gas is the Keyhole Nebula.(GK. "eta") Carinae is the brightest star left of the Keyhole. (Cerro Tololo Inter-American Observatory photo)

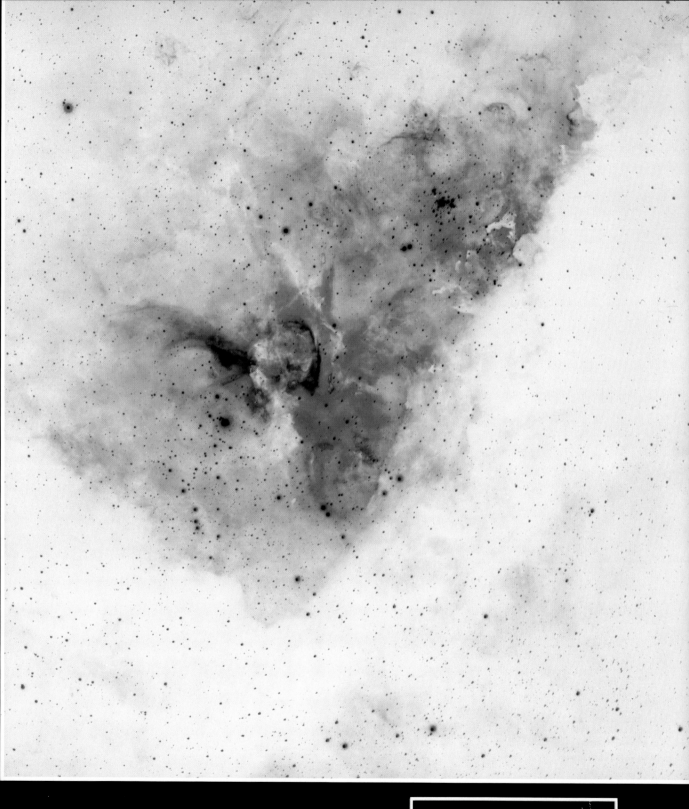

Color Plate 58: A false-color view of the Eta Carinae Nebula (compare the direct view in Color Plate 57). Green represents the forbidden line of doubly ionized oxygen, which requires a high level of excitation, and pink represents the forbidden line of singly ionized sulfur, which requires lower amounts of energy to excite it. The Keyhole Nebula shows clearly. Spectroscopic and proper motion studies of the Eta Carinae Nebula show that there are internal chaotic motions super-imposed on a general expansion.

RING

ETA
CARINAE

KEYHOLE

5 arc min

Color Plate 59: The Horsehead Nebulae, NGC 2024, in Orion, is an absorbing region superimposed on a region emitting red radiation characteristic of hydrogen. (Palomar Observatory, California Institute of Technology photo with the 1.2-m Schmidt camera)

SUMMARY AND OUTLINE

Our galaxy has a nuclear bulge, a nucleus, a disk, and a halo
Nebulae (Section 22.1)
  Emission
  Absorption (dark)
  Reflection
  "Spiral"
  Planetary
  Supernova remnants
Infrared observations (Section 22.2)
  Windows in the spectrum
  Technological difficulties
  Radiation from the galactic nucleus indicates a lot of energy is generated in a small volume there
High-energy astronomy (Section 22.3)
  The Uhuru and HEAO-1 satellites

Gamma-ray background radiation is the result of cosmic ray interaction with matter
Spiral structure of the galaxy (Section 22.4)
  Bright tracers of the spiral structure (Section 22.4a)
    Difficult to determine shape of our galaxy because of obscuring dust and gas
    H II regions are used to determine spiral structure
  Differential rotation (Section 22.4b)
    Sun is approximately 8.5 kpc from center of galaxy
    Central part of galaxy rotates as a solid body
    Outer part, including the spiral arms, is in differential rotation.
  Density-wave theory (Section 22.4c)
    Leading current solution to why spiral arms have not wound up very tightly
    We see the effect of a wave of compression; the distribution of mass itself is not spiral in structure

QUESTIONS

1. Why do we think that our galaxy is a spiral?

2. How would the Milky Way appear if the sun were close to the edge of the galaxy?

3. Sketch (a) a side view and (b) a top view of the Milky Way Galaxy, showing the shapes and relative sizes of the nuclear bulge, the disk, and the halo. Mark the position of the sun.

4. Compare (a) absorption (dark) nebulae, (b) reflection nebulae, and (c) emission nebulae.

5. How can something be both an emission and an absorption nebula? Explain and give an example.

6. If you see a blue star surrounded by a red nebula, is this an emission or a reflection nebula? Explain.

7. Why may some infrared observations be made from mountain observatories while all x-ray observations must be made from space?

8. Illustrate, by sketching the appropriate Planck radiation curves, why cooling an infrared tele-

scope from room temperature to 4 K greatly reduces the background noise.

9. What are two tracers that we use for the spiral structure of our galaxy? What are two reasons why we expect them to trace spiral structure?

10. Why will x-rays expose a photographic plate while infrared radiation will not?

11. What types of objects give off strong infrared radiation?

12. What types of objects give off x-rays?

13. Discuss how observations from space have added to our knowledge of our galaxy.

14. Why does the density-wave theory lead to the formation of stars?

15. What is another system, besides the Milky Way Galaxy, that exhibits differential rotation?

16. Comment on our understanding of the gamma-ray bursts.

CHAPTER 23

THE INTERSTELLAR MEDIUM

AIMS:
To discuss the interstellar medium, and to understand the techniques (especially radio astronomy and ultraviolet astronomy) used to study it

The gas and the dust between the stars is known as the *interstellar medium.* The nebulae represent regions of the interstellar medium in which the density of gas and dust is higher than average. The interstellar medium contains the elements in what we call their *cosmic abundances*, that is, the overall abundances they have in the cosmos. This cosmic abundance is roughly 90 per cent hydrogen atoms, 9 per cent helium atoms, and less than 1 per cent of heavier atoms, though it is important to note that essentially all the heavier atoms are represented.

23.1 H I AND H II REGIONS

For many purposes, we may consider interstellar space as filled with hydrogen at an average density of about 1 atom per cubic centimeter, although individual regions may have densities departing greatly from this average. Regions in which the atoms of hydrogen are predominantly neutral are called *H I regions* (the Roman numeral "I" referring to the first, or basic,

---

The 100-meter radio telescope at Effelsburg, near Bonn, Germany.

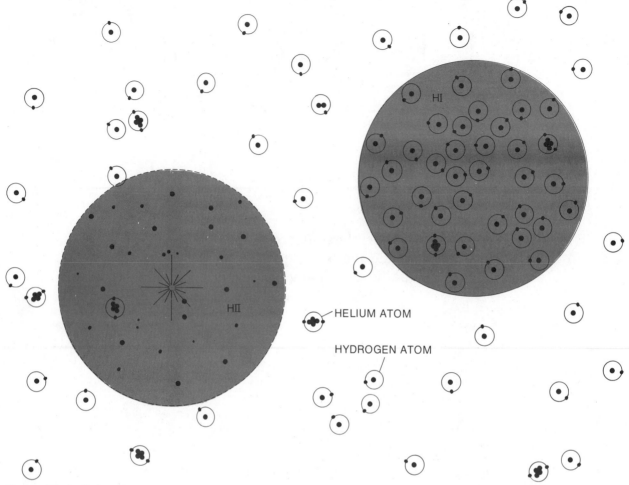

HELIUM ATOM

HYDROGEN ATOM

Figure 23–1 *H I regions* are regions of neutral hydrogen of higher density than average and *H II regions* are regions of ionized hydrogen. The protons and electrons that result from the ionization of hydrogen by a hot star, and the neutral hydrogen atoms, are shown schematically. The outlines of the regions are shown for illustrative purposes only; in space, of course, the regions are not outlined. The larger dots represent protons or neutrons and the smaller dots represent electrons. The central star that causes the H II region is shown.

state). Where the density of an H I region is high enough, pairs of hydrogen atoms combine to form molecules (H_2). The densest part of the gas associated with the Orion Nebula might have a million or more hydrogen molecules per cubic centimeter.

Ionized hydrogen, with one electron missing, is known as H II (the "second state"). Since hydrogen, which makes up the overwhelming proportion of interstellar gas, contains only one proton and one electron, a gas of ionized hydrogen contains individual protons and electrons. Wherever a hot star provides enough energy to ionize hydrogen, an H II region (Fig. 23–1) results. Emission nebulae are such H II regions. They glow because the gas is heated. Several mechanisms of heating are known, most of which result in emission lines appearing.

But even though the nebulae are comparatively dense, they are still considerably less dense than is the gas in, for example, the air we breathe on earth. In fact, even the nebulae with the highest densities are still many orders of magnitude less dense than the best vacuums that can be made in laboratories on earth, so that the nebulae provide scientists with a way of studying the basic properties of gases under conditions that are unobtainable on earth.

˚Box 23.1   Fluorescence

Bright spectral lines are emitted at certain wavelengths in H II regions by a *fluorescent* process. A *fluorescent* process is one in which photons of a high energy are transformed, through interactions with atoms, to photons of lower energy. (Equivalently, we can think of fluorescence as a process that transforms photons of short wavelength to photons of longer wavelength.)

In particular, very hot stars give off many photons at wavelengths shorter than 912 Å, which is the deep ultraviolet wavelength that is the longest at which photons can ionize hydrogen. These short-wavelength photons ionize the hydrogen near the hot stars, which makes H II regions. We cannot observe these high-energy photons directly, for even those that get out of the H II region without giving up their energy to a hydrogen atom cannot penetrate the earth's atmosphere. But we can observe their effect with telescopes on earth through the following fluorescent process:

The electrons continually recombine with the hydrogen ions, and the resultant neutral hydrogen atoms are usually on their higher energy levels. They then jump down through a series of middle energy levels until they reach the ground state. This means that emission lines in the Balmer series appear in the visible (and other hydrogen emission lines appear in other parts of the spectrum). The atoms are quickly ionized again, and the fluorescent process takes place all over again.

In addition, in some planetaries, a specific ultraviolet transition excites doubly ionized nitrogen and oxygen ions (N III and O III) and emission lines of these ions also show. This process in nitrogen and oxygen is thus also a fluorescent one, for it begins with ultraviolet photons and winds up with visible photons.

Note that "fluorescent" is spelled with a "uo," opposite to the "ou" that appears in the word "flour."

23.2 INTERSTELLAR REDDENING AND EXTINCTION

Distant stars in the plane of our galaxy are obscured from our vision. In addition, many stars that are still close enough to be visible are partially obscured. The amount of obscuration varies with the wavelength at which we observe. The blue light is *scattered* by dust in space more efficiently than the redder light is; that is, for a given distance through the dust more of the blue light has been bounced around in every direction. Thus less of the blue light comes through to us than the red light, and the stars look redder—we say that the stars are *reddened*. This reddening is thus a consequence of the scattering properties of the dust. It has nothing to do with "redshifts," since in the present case the spectral lines are not shifted in wavelength.

Over the years many studies have been made of the *interstellar reddening*. These studies have shown that in the visible, the amount of reddening varies approximately inversely with the wavelength; that is, the amount of reddening is proportional to one divided by the wavelength. For example, red light of 7000 Å is dimmed about half as much as ultraviolet light of 3500 Å. This law of reddening is an empirical result (that is, a result based on observation rather than on theory).

The amount of scattering together with the amount of actual absorption of visible radiation is known as the *extinction*. In the blue part of the spectrum, the total extinction in the 8.5 kpc between the sun and the center of the galaxy is about 25 magnitudes. Most of this takes place far from the sun, in regions with a high dust content. Even a tremendously bright object located near the center of our galaxy would be dimmed too much—25 magnitudes—to be seen from the earth in this part of the spectrum. Since the amount of reddening decreases with wavelength, we have more hope of seeing to the galactic center in the infrared.

The extinction of visible light relatively near the sun in the plane of our galaxy is about 1 magnitude per thousand parsecs (1 magnitude per kiloparsec). Thus a star 1000 parsecs away would appear about 1 magnitude fainter than it would if there were no interstellar dust, a star 2000 parsecs away would appear 2 magnitudes dimmer, and so on. The amount of extinction is actually very irregular, in that it varies from this average value as one looks in different directions.

The amount of extinction has been found empirically to be proportional to the amount of reddening. So by measuring the colors of stars, which gives the reddening, we can derive the extinction.

Figure 12–7 shows the formation of blue skies and red sunsets.

> **Box 23.2   Why Is the Sky Blue?**
>
> A similar scattering by the electrons in air molecules in the terrestrial atmosphere makes the sky blue. Further, when the sun is near the horizon, we have to look diagonally through the earth's layer of air. Our line of sight through the air is then longer than a line of sight straight up. Thus when the sun is low in the sky, most of the blue is scattered out before it reaches us. Relatively more red light reaches us. This accounts for the reddish color of sunsets. The light that reaches us from the part of the sky away from the sun is scattered sunlight, and is therefore predominantly blue. Electrons scatter visible wavelengths of light much more efficiently than dust does. Electrons scatter light based on the fourth power of the wavelength. Thus ultraviolet light at 3500 Å is scattered $2^4 = 16$ times more efficiently than light at 7000 Å in the infrared.

Traditionally, studies of reddening and extinction have been used to find the distances to stars. Let us consider two stars of identical spectral type, which we can find by looking at their spectral lines. The stars thus have the same distribution of energy with wavelength. If the first star is located out of the plane of the galaxy as seen from earth, that is, away from the Milky Way, it is not reddened because there isn't much interstellar material between it and us. If the second star is located near the galactic plane, it is reddened. The relative amount of reddening gives us the amount of extinction. The extinction, in turn, tells us how much interstellar material the light has passed through. From the average density of interstellar material, we can then calculate the approximate distance to the second star.

Or we can apply the above rule of thumb linking extinction with distance: 1 mag/kpc.

Unfortunately, the reddening is splotchy in nature, and we cannot be certain that there is no volume of dust of unusually high or unusually low density in our line of sight. The edges of dark nebulae are good examples of how the amount of reddening of two stars almost but not quite in the same direction could be very different. Also, reddening can be caused by shells of dust surrounding the star in addition to arising in interstellar space. We must use this method of determining distances with caution or limit its use to statistical studies.

We have just been discussing how to observe the dust by studying how it affects starlight that passes through it. It is difficult to observe the dust directly, but the dust is heated a bit by radiation, thus causing the dust to radiate. The dust never gets very hot, so its radiation peaks in the infrared. The radiation from dust scattered among the stars is too faint to be detected, but the radiation coming from clouds of dust surrounding stars has been observed.

Similarly, as the interstellar gas is "invisible" in the visible part of the spectrum (except at the wavelengths of certain weak spectral lines), special techniques are needed to observe the gas in addition to observing the dust. Radio astronomy is the most widely used technique for studying the interstellar gas. We shall discuss it at length in this chapter and the next.

In Chapter 24, we shall discuss how infrared studies have revealed to us how stars are forming in the midst of dust.

23.3 RADIO OBSERVATIONS

From antiquity until 1930, observations in a tiny fraction of the electromagnetic spectrum, the visible part, were the only way that observational astronomers could study the universe. Most of the images that we have in our minds of objects in our galaxy (e.g., nebulae) are based on optical studies, since most of us depend on our eyes to discover what is around us.

Optical astronomy has shown us how to sort the objects in the universe into stars and galaxies. In our own galaxy we can tell the nebulae apart from the stars, distinguish the spectral types of the stars, and, from another point of view, recognize different types of clusters. We have used the Doppler effect and the period-luminosity relation to measure the distances to stars and clusters of stars. Astronomy has made good use of the limited amount of data available to it, and most of the descriptions in this text up to this point have been based on optical studies.

If they had to make a choice, however, between observing only optical radiation or only everything else, many astronomers might choose "everything else." The rest of the electromagnetic spectrum carries more information in it than do the few thousand angstroms that we call visible light. We will first discuss the basic techniques of radio astronomy, and then go on to see how radio astronomy joins with other observing methods to investigate interstellar space.

23.3a *Continuum Radio Astronomy*

All radio astronomy research in the early days was of the continuum (Fig. 23–2). In radio astronomy, as in optical astronomy, that means that we

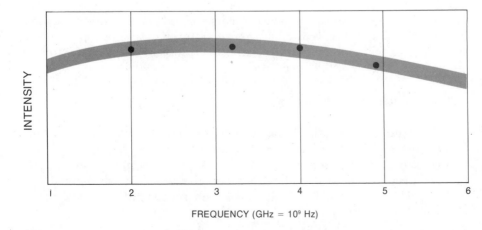

FREQUENCY (GHz = 10⁹ Hz)

Figure 23–2 When we speak of a *radio spectrum*, we often mean the way the intensity of radiation changes over a wide range of frequencies. In the graph shown, the intensity was measured only at four widely separated frequencies, and the shaded curve was fit to the observed points.

consider the average intensity of radiation at a given frequency without regard for variations in intensity over small frequency ranges, that is, we ignore any spectral lines. When radio "continuum spectra" were measured, the continuum levels were measured at widely separated frequencies. These continuum values, and the spectra derived from them by joining the values by straight lines or other simple smooth curves, provided information about the mechanism that causes the continuum emission.

Since radio waves cannot be focused into an image on a photographic plate, there is an important difference in the way radio and optical observations must be made. At a radio telescope, the observer simply measures the total amount of radiation that is coming from the direction in which the telescope is pointing. To make a radio map with a single dish, the pointing of the telescope must be physically changed; the telescope may be scanned from point to point across the image.

In one sense, an analogue in optical astronomy to continuum radio astronomy is not photography, but rather the use of *photocells* to record data. There too one can only measure the total intensity of light that is coming into the telescope from the direction in which the telescope is pointed; we do not get a "picture" without rastering—sweeping across, down a bit, back, down a bit, across, etc. New methods in both optical and radio astronomy are finally allowing astronomers to make images without using photographic plates.

To take a radio continuum spectrum of an object, one must keep the telescope pointing at the object as it passes across the sky, and measure the intensity of radiation at different frequencies one after the other. To make these frequency changes one is sometimes required to physically climb up to the focal point of the telescope where the antenna (called the "feed") is located, and change the piece of receiving equipment that is installed there. In order to take a radio spectrum that covers a particularly large range in frequency, one may even have to use several different telescopes, each of which works efficiently at a particular part of the frequency range.

As radio astronomical observations got under way, it was immediately apparent that the brightest objects in the radio sky, that is, the objects that give off the most intense radio waves, are not identical with the brightest objects in the optical sky. The radio objects were named with letters and with the names of the constellations in which they are located. Thus Taurus A is the brightest radio object in the constellation Taurus; we now know it to be the Crab Nebula. Sagittarius A is the center of our galaxy; Sagittarius B is another radio source nearby, whose emission is caused by clouds of gas near the galactic center.

Extragalactic means outside our galaxy.

Many of the radio objects have been discovered to lie outside our galaxy, and we shall discuss them in subsequent chapters. Quasars (Chapter 26), for example, are an important group of extragalactic objects that radiate strongly in the radio region of the spectrum.

*23.3b Synchrotron Radiation

Continuum radio radiation can be generated by several processes. One of the most important is *synchrotron emission* (Fig. 23–3), the process that produces the radiation from Taurus A. Taurus A is a supernova remnant, and lines of magnetic field extend throughout the visible Crab Nebula and beyond. Electrons, which are electrically charged, tend to spiral around

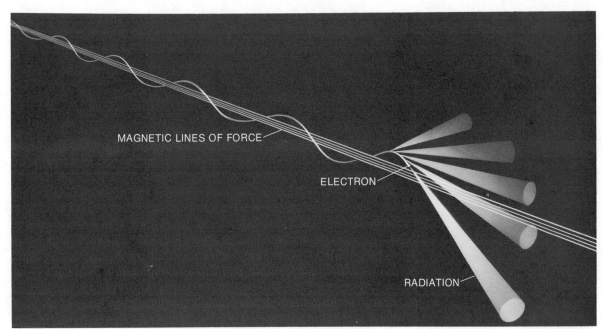

Figure 23–3 Electrons spiralling around magnetic lines of force at velocities near the speed of light (we say "at relativistic velocities") emit radiation in a narrow cone. This radiation, which is continuous and highly polarized, is called synchrotron radiation. Synchrotron radiation has been observed in both optical and radio regions of the spectrum.

magnetic lines of force. Electrons of high energy spiral very rapidly, at speeds close to the speed of light. We say that they move at "relativistic speeds," since the theory of relativity must be used for calculations when the electrons are going that fast. Under these conditions, the electrons radiate very efficiently. (This is the same process that generates the light seen in electron synchrotrons in laboratories on earth, hence the name synchrotron radiation.) The suggestion that the synchrotron mechanism causes radiation from various astronomical sources was first made by several Soviet theoreticians about 1950. (Synchrotron radiation is highly polarized, and the discovery a few years later that the optical radiation from the Crab Nebula is highly polarized (Fig. 23–4) was an important confirmation of this suggestion. The radio radiation from the Crab and many other sources is also highly polarized.)

The intensity of synchrotron radiation is related not to the temperature of the astronomical body that is emitting the radiation, but rather to the strength of its magnetic field and to the number and energy distribution of the electrons caught in that field. Since the temperature of the object cannot

Figure 23–4 Photographs of the Crab Nebula taken through filters that pass visible light polarized at the angles shown with the arrows. A non-polarized source would appear the same when viewed at any angle of polarization. The pictures at different polarization angles look very different from each other. This shows that the light from the Crab is highly polarized, which implies that it is caused by the synchrotron mechanism.

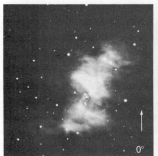
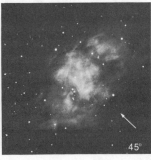

be derived from knowledge of the intensity of the radiation, we call this radiation *non-thermal* radiation. Synchrotron radiation is but one example of such non-thermal processes. The synchrotron process can work so efficiently that a relatively cool astronomical body can give off a tremendous amount of such radiation, at a given frequency, perhaps so much that it would have to be heated to a few million degrees before it would radiate as much *thermal* radiation at that frequency.)

By thermal radiation we mean continuous radiation whose spectrum is directly related to the temperature of the gas.

23.4 THE RADIO SPECTRAL LINE FROM INTERSTELLAR HYDROGEN

(In about 1950, though radio astronomers were very busy with continuum work, there was still a hope that a radio spectral line might be discovered. This discovery would be important for many reasons, but especially because it would allow Doppler shift measurements to be made.)

(What is a radio spectral line? Just as an optical spectral line corresponds to a wavelength (frequency) in the optical spectrum that is more (for an emission line) or less (for an absorption line) intense than neighboring wavelengths or frequencies, a radio spectral line corresponds to a frequency (or wavelength) at which the radio noise is slightly more, or slightly less, intense.\

If a radio station broadcasted just a hum, it would appear as an emission line on our home radios; terrestrial transmissions are normally "modulated" (the voice or music is carried as a modulation on a steady "carrier" signal).

23.4a The Hydrogen Spin-Flip

(The most likely candidate for a radio spectral line that might be discovered was a line from the lowest energy levels of interstellar hydrogen atoms. This line was predicted to be at a frequency of 1420 MHz, equivalent to a wavelength of 21 cm. Since hydrogen is by far the most abundant element in the universe, it seems reasonable that it should produce a strong spectral line. Furthermore, since most of the interstellar hydrogen has not been heated by stars or by any other strong mechanism, it is most likely that this hydrogen is in its state of lowest possible energy.\

The formation of the hydrogen spectrum is discussed in Section 3.4.

(We have seen how the Balmer series of hydrogen is visible in the optical spectrum of the sun and other stars. This series of lines comes from transitions of electrons that cause the hydrogen atom to change between its second lowest principal energy state and other, higher states. Thus the transition from level 3 to level 2 is called Balmer alpha. Actually, since for many years the Balmer series was the only one that could be observed, it is usually called hydrogen alpha, or H α (H alpha). The transition from level 4 to level 2 is H β (H beta), from level 5 to level 2 is H γ (H gamma), and so on. (Astronomers all know the Greek alphabet or at least the first few letters.))

(The hydrogen lines that result from transitions to or from the lowest principal energy state (the *ground state*) to the higher states are called the Lyman series. Ly α (Lyman alpha), which falls at a wavelength of 1216 Å in the ultraviolet, does not come through the earth's atmosphere, so Ly α from celestial sources must be observed from satellites. In particular, the Copernicus satellite has made valuable studies of the Lyman series of hydrogen.)

One speaks interchangeably of ground level *or* ground state.

(As we discussed, spectral lines come not from the energy levels themselves but from changes between one energy level and another. The energy that corresponds to a transition of an electron in an atom from one energy state to another, appears as a photon of one and only one wavelength. The

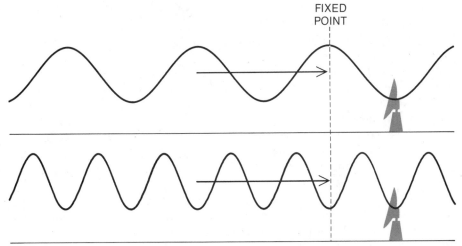

FIXED
POINT

Figure 23–5 Since all electro-magnetic radiation travels at the same speed (the speed of light) in a vacuum, fewer waves of longer wavelength *(top)* pass an observer in a given time interval than do waves of a shorter wave-length *(bottom)*. If the wavelength is half as long, twice as many waves pass, i.e., the frequency is twice as high. The wavelength times the frequency is constant, with the constant being the speed of light; $\lambda\nu = c$.

Box 23.3 Conversion from Frequency to Wavelength

Although optical astronomers usually use wavelength units, radio astronomers usually use *frequency* units. If a wave travels at a constant speed—in this case at the speed of light—fewer peaks will cross a given point in a given time when the wavelength is longer. Thus the frequency at which peaks in the wave pass a given point is decreased. The converse is also true. The simple equation that links wavelength and frequency is wavelength × frequency = c. Wavelength is usually denoted with the Greek letter λ (lambda) and frequency is usually denoted with the Greek letter ν (nu), so the equation is written $\lambda\nu = c$ (Fig. 23–5). The energy E is simply expressed as $h\nu$, where h is Planck's constant.

Students often find this dual terminology—sometimes in wavelengths and sometimes in frequencies—confusing. So do astronomers. After doing research in a given field, one actually thinks in the appropriate units; radio astronomers think in frequency units. So at meetings at which both optical and radio astronomers are present, there is considerable translating of units back and forth.

The one basic number that must be remembered is the speed of light: 3×10^{10} cm s^{-1}. From that, frequency-wavelength conversions can be derived easily. For example, if a radio wave has a wavelength of 1 cm, its frequency is 3×10^{10} cm s^{-1} ÷ 1 cm = 3×10^{10} hertz (hertz is the name for cycles per second, and is abbreviated Hz). Since M is the symbol for mega (10^6), and G is the symbol for giga (10^9), the frequency of this line would be 3×10^{10} Hz = 30×10^9 Hz = 30 GHz.

Figure 23–6 21-cm radiation re-sults from an energy difference between two sub-levels in the lowest principal energy state of hydrogen. The energy difference is much smaller than the energy difference that leads to Lyman α. So (because $E = h\nu = hc/\lambda$) the wavelength is much longer.

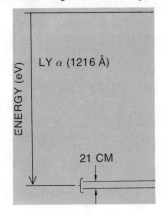

wavelength of this radiation is determined, as we have seen in Section 3.3, by the formula E = constant ÷ wavelength. The constant is Planck's constant, h, times the speed of light, c.

Let us now return to consideration of the hydrogen line at 1420 MHz = 1420×10^6 Hz. To find its wavelength, we divide the frequency into the speed of light, 3×10^{10} cm s^{-1} ÷ 1.42×10^9 Hz, and get 21 cm.

This line at 21 cm comes not from a transition to the ground state from one of the higher states, not even from level 2 to level 1 as does Lyman alpha, but rather from a transition between the two sublevels into which the ground state of hydrogen is divided (Fig. 23–6).

For this astronomical discussion, it is sufficient to think of a hydrogen atom as an electron orbiting around a proton. Both the electron and the proton have the property of spin; each one has angular momentum (see Section 12.9) as if it were spinning on its axis.

With this terminology, the spin of the electron can be either in the same direction as the spin of the proton or in the opposite direction. The rules of quantum mechanics prohibit intermediate orientations. If the spins are in opposite directions, the energy state of the atom is very slightly lower than the energy state occurring if the spins are in the same direction. The energy difference between the two states is equal to $h\nu = h \times 1420 \times 10^6$ Hz, which is the same as saying that a transition from the upper to the lower state gives rise to a 1420 MHz photon.

If an atom is sitting alone in space in the upper of these two energy states, with its electron and proton spins aligned in the same direction, it has a certain small probability of having the spinning electron spontaneously flip over to the lower energy state and emit a photon. We thus call this a *spin-flip* transition (Fig. 23–7). The photon of hydrogen's spin-flip corresponds to radiation at a wavelength of 21 cm (Fig. 23–8). If we were to watch any particular group of hydrogen atoms, we would find that it would take 11 million years before half of the electrons had undergone spin-flips; we say that the *half-life* is 11 million years for this transition. But even though the probability of a transition taking place in a given interval of time is very low, there are so many hydrogen atoms in space that enough 21-cm radiation is given off to be detected.

We have described how an *emission line* can arise at 21 cm. But what happens when continuous radiation passes through neutral hydrogen gas? In this case, some of the electrons in atoms in the lower state will absorb a 21-cm photon and flip over, putting the atom into the higher state. Then the

Figure 23-7 When the electron in a hydrogen atom flips over so that it is spinning in the opposite direction from the spin of the proton *(top)*, an emission line at a wavelength of 21 cm results. When an electron takes energy from a passing beam of radiation to flip from spinning in the opposite direction from the proton to spinning in the same direction *(bottom)*, then a 21-cm line in absorption results.

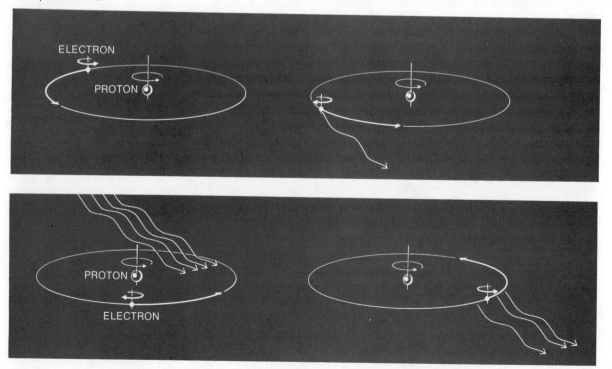

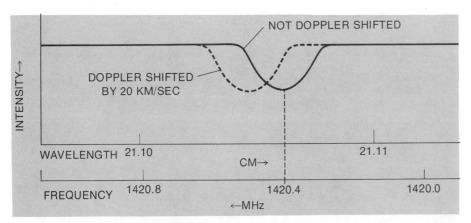

NOT DOPPLER SHIFTED

DOPPLER SHIFTED
BY 20 KM/SEC

INTENSITY →

WAVELENGTH 21.10   CM→   21.11

FREQUENCY 1420.8   1420.4   1420.0
←MHz

Figure 23–8 The position of the hydrogen line, shown here in absorption, can be given in either frequency or wavelength units. These numbers are commonly rounded to 21 cm or 1420 MHz in conversation. Any Doppler shift, of course, shifts the absorption in wavelength and frequency.

radiation that emerges from the gas will have a deficiency of such photons and will show the 21-cm line in *absorption* (Fig. 23–7).

In 1944, H. C. van de Hulst, a Dutch astronomer, predicted that the 21-cm emission would be strong enough to be observable as soon as the proper equipment was developed. It took seven more years for the instrumental capability to be built up. In 1951, scientists both in Holland at Leiden and in the United States at Harvard were building equipment to detect the 21-cm radiation. An unfortunate fire set back the Dutch effort many months. In the meantime, the work at Harvard went on. Electronic equipment was built in a physics lab to observe in the direction of the galactic center through a small antenna stuck out a window (into which passing undergraduates occasionally lobbed snowballs). Finally, the Harvard team, consisting of a graduate student named Harold Ewen and his advisor, Edward M. Purcell, went "on the air" and succeeded in observing the 21-cm line in emission (Fig. 23–9). Soon the Dutch group and then a group in Australia confirmed the detection of the 21-cm line. Spectral-line radio astronomy had been born.

23.4b Mapping Our Galaxy

21-cm hydrogen has proved to be a very important tool for studying our galaxy because it passes unimpeded through the dust that prevents optical observations very far into the plane of the galaxy. Using 21-cm observations, astronomers can study the distribution of gas in the spiral arms. We can detect this radiation from gas located anywhere in our galaxy, even on the far side, whereas light waves penetrate the dust clouds in the galactic plane only about 10 per cent of the way to the galactic center.

But here again we come to the question that bedevils much of astronomy: how do we measure the distances? Given that we detect the 21-cm radiation from a gas cloud (since the cloud contains neutral hydrogen, it is an H I region), how do we know how far away the cloud is from us?

The answer can be found by using a model of rotation for the galaxy, that is, a description of how each part of the galaxy rotates. As we have already learned, the outer regions of galaxies rotate differentially; that is, the gas nearer the center rotates faster than the gas farther away from the center.

Figure 23–10 shows a simplified version of differential rotation. Because of the differential rotation, the distance between us and point A is decreasing. Therefore, from our vantage point at the sun, point A has a net

Figure 23–9 Harold Ewen with the horn radio telescope he and Edward Purcell used to discover the 21-cm line.

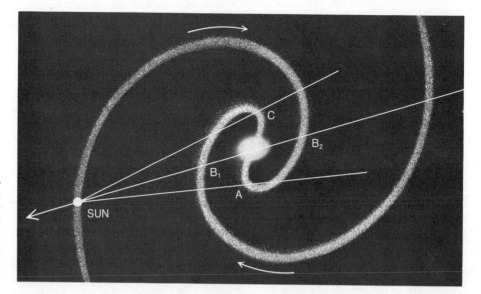

Figure 23–10 Because of the differential rotation, the cloud of gas at point A appears to be approaching the sun, and the cloud of gas at point C appears to be receding. Objects at points B_1 or B_2 have no net velocity with respect to the sun, and therefore show no Doppler shifts in their spectra.

velocity toward us. Thus its 21-cm line is Doppler shifted toward shorter wavelengths. If we were talking about light, this shift would be in the blue direction; even though we are discussing radio waves we say "blueshifted" anyway. If we look from our vantage point at gas cloud C, we see a redshifted 21-cm line, because its higher speed of rotation is carrying C away from us. But if we look straight toward the center, clouds B_1 and B_2 are both passing across our line of sight in a path parallel to that of our own orbit. They have no net velocity toward or away from us. Thus this method of distance determination does not work when we look in the direction of the center, nor indeed in the direction of the anti-center.

Once we measure the Doppler shift, we can deduce the net velocity. In the inner part of the galaxy, we can deduce the law of differential rotation with distance from the center since the highest velocity cloud along each line of sight must be the cloud that is closest to the center. We can get the distance of that cloud from the center by simple geometry. Once we work out the law of differential rotation, then we can tell how far each cloud is from the center of the galaxy by figuring out where along our line of sight in a given direction the cloud would have the proper velocity to match the observations. Clouds farther from the galactic center take longer to revolve than do clouds closer to the center, in a manner similar to Kepler's third law (Fig. 23–11). By observing in different directions, we can build up a picture of the spiral arms.

Unfortunately, gas clouds have not only a velocity of revolution around the center of the galaxy but also random velocities to and fro. When we look at the center of the galaxy, we see that there are some clouds of gas moving outward with an average velocity of 50 km s^{-1}. These motions, both systematic and random, place a fundamental uncertainty on the conclusions from this method. But this is the best that we can do.

Note that each line of sight in which we look crosses a circle of gas at a given distance from the center at two points. Thus an ambiguity arises, because a cloud at either of these two points would have the same Doppler velocity with respect to the sun. We have to resort to methods other than 21-cm studies to resolve this ambiguity. We know, for example, the average

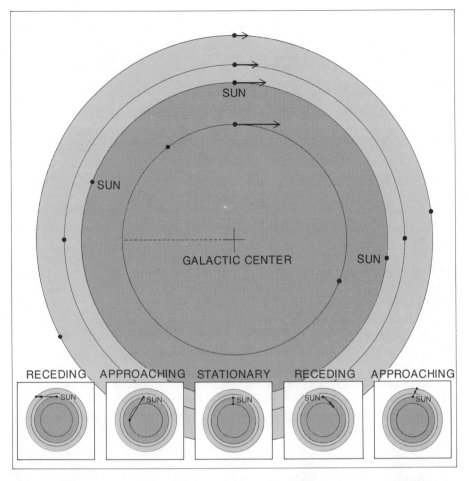

Figure 23–11 Because of differential rotation in our galaxy, stars or gas clouds that were lined up at one time (upper dots on top part of diagram) are spread out by the time they have gone about 1/4 of the way *(right)* or about 3/4 of the way *(left)* around the galaxy. The inserts show the net velocity that stars or gas clouds at different distances from the galactic center would have with respect to the sun. Objects interior to the sun's orbit around the galactic center (darker shaded area) are orbiting faster than does the sun, and objects farther out than the sun's orbit (lighter shaded area) are orbiting more slowly than does the sun. Thus measurements of the Doppler shift of 21-cm radiation can be used to construct a map of the hydrogen clouds, which in turn map out spiral structure.

size of H I regions, and so can often tell whether the emitting cloud is at the farther or nearer point by its angular size in the sky. Alternatively, we can sometimes tell by noting whether the 21-cm radiation seems relatively weak or strong.)

Of course, the whole sky cannot be seen from either the northern hemisphere or the southern hemisphere of the earth. Therefore maps must be made from observations in both hemispheres. Also, the 21-cm maps show many narrow arms (Fig. 23–12) but no clear pattern of a few broad spiral

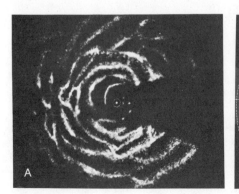
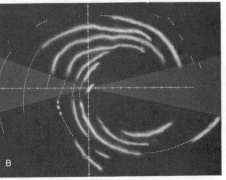

Figure 23–12 Two artists' impressions of the structure of our galaxy based on 21-cm data. Differences between the two maps are probably not real and give an idea of the limitations of the method. Because hydrogen clouds located in the directions either toward or away from the galactic center have no radial velocity with respect to us, we cannot find their distances.

arms like those we see in other galaxies. Is our galaxy really a spiral at all?

Although questions remain, it is clear that the 21-cm radiation is at the base of our mapping efforts for our own galaxy, and has allowed us to make major advances in our understanding of the Milky Way Galaxy.

23.5 MEASURING THE MASS OF OUR GALAXY

We can measure the mass of our galaxy by studying the velocity of rotation of gas clouds. All the mass inside the radius of the cloud's orbit acts as though it were concentrated at one point, while the mass outside the cloud's orbit has no effect on its velocity. (Outside the cloud's orbit, the gravitational pull in one direction exactly balances out the gravitational pull in the opposite direction.) The velocity of rotation tells us what the period would be, so from Kepler's third law we can calculate the mass inside the orbit of the gas cloud.

It is thus important to measure the velocity of rotation of our galaxy as far out from the center as possible. Velocities have been measured inside the sun's orbit from measurements of the 21-cm line, using the method discussed in the previous section. Leo Blitz of the University of California at Berkeley has now extended these measurements out twice as far using the same method but with radiation from carbon monoxide instead of hydrogen. (We shall discuss the carbon monoxide radiation in more detail in the next chapter.)

A graph of the velocity of rotation vs. distance from the center is called a *rotation curve* (Fig. 23–13). It had been expected that our galaxy's rotation curve would stop increasing and begin to decrease when measurements were made far enough out that the galaxy ended. Then no more mass would be included inside as we went to larger radii. Kepler's law would begin to give the effect we are familiar with for the planets, with longer periods farther out. But the curve does not turn over and begin to decrease, which indicates that the galaxy is larger and contains more mass than we had thought. It now seems that our galaxy is twice as massive as the Andromeda Galaxy and contains about 8×10^{11} solar masses.

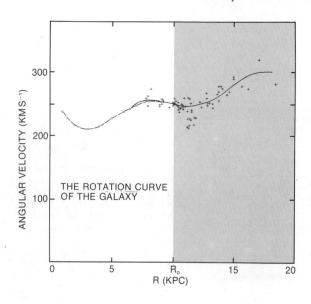

Figure 23–13 The rotation curve of our galaxy. The inner part of the curve (dotted line) is based on 21-cm hydrogen observations. It had been anticipated that beyond about 10 parsecs the velocity would decline following Kepler's third law, because there would be no additional mass added as we go further out. But by observing carbon monoxide we are newly able to make direct measurements of the outer regions (shaded). The observations (crosses) show that the velocity of rotation doesn't decline. This must mean that there is more mass in the outer regions of the galaxy than we had anticipated. (Courtesy of Leo Blitz, University of California at Berkeley)

SUMMARY AND OUTLINE

H I and H II regions (Section 23.1)
Interstellar reddening and extinction (Section 23.2)
Radio observations (Section 23.3)
   Continuum radio astronomy
      Spectra are measured over a broad frequency range
      Radiation generated by synchrotron emission,
         electrons spiralling rapidly in a magnetic field,
         is found in many non-thermal sources
   Synchrotron radiation
Radio spectral line from interstellar hydrogen (Section
   23.4)

21-cm spectral line from neutral hydrogen was discovered in 1951
Line occurs at 21 cm through a spin-flip transition,
   corresponding to a change between energy sub-
   divisions of the ground level of hydrogen
Both emission and absorption have been detected
21-cm radiation is used to map the galaxy
Distances are measured using differential rotation
   and the Doppler effect
Measuring mass of our galaxy (Section 23.5)
   Revised measurements have doubled its previously
   accepted size

QUESTIONS

1. Explain how a fluorescent process can transform a 900 Å photon into several photons in the visible region of the spectrum.

2. List two relative advantages and disadvantages of radio astronomy and optical astronomy.

†3. The angular resolution of the 100-meter telescope at a wavelength of 21 cm is approximately 9 arc minutes. How large a telescope would you need to get the same angular resolution at a wavelength of 2 mm? You will want to refer to the discussion of resolution in Section 2.3.

4. How does the procedure of making a radio map of a region differ from taking a photograph?

5. Why is emission from cool regions of space most likely to be detectable at radio wavelengths?

†6. What is the frequency of 1-mm waves?

†7. What is the wavelength of 108 MHz waves, those you tune in at the right end of your FM radio dial?

8. What determines whether the 21-cm lines will be observed in emission or absorption?

9. If our galaxy rotated like a rigid body, with each point rotating with the same period no matter what the distance from the center, would we be able to use the 21-cm line to determine distances to H I regions? Explain.

10. Describe how a spin-flip transition can lead to a spectral line, using hydrogen as an example. Could deuterium also have a spin-flip line?

---

†This indicates a question requiring a numerical solution.

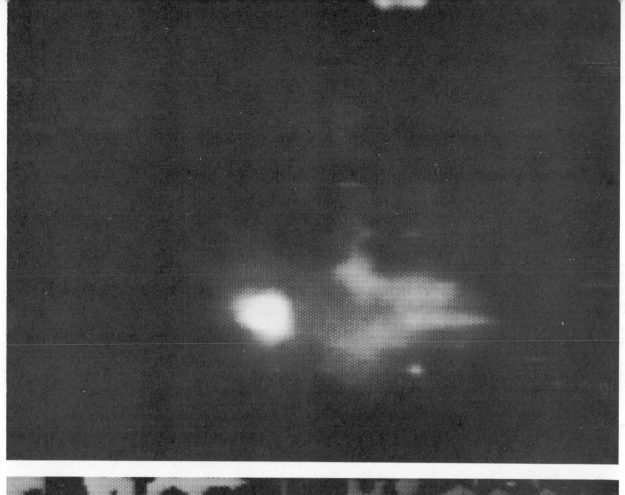

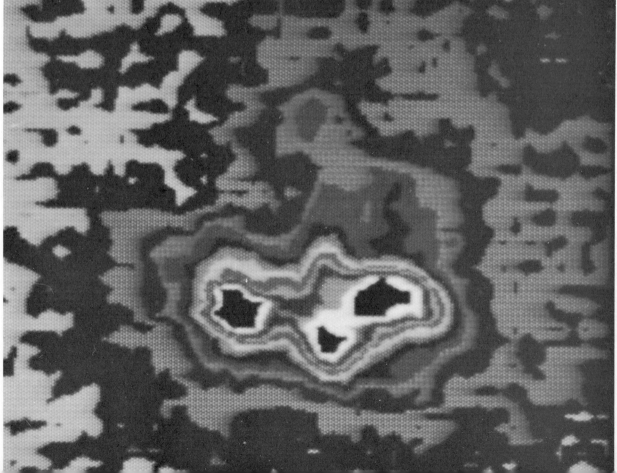

CHAPTER 24

INTERSTELLAR MOLECULES AND STAR FORMATION

AIMS:
To see how studies of the interstellar medium made in the radio and infrared regions of the spectrum are leading us to an understanding of how stars are born

In many regions of the sky, sometimes individually and sometimes in clusters, we see young stars. They must be forming even now. But how do we find their nurseries and look into their cradles?

Many of the young stars are associated with nebulae of various types and with dark clouds of interstellar matter. The new stars are presumably forming out of that gas and dust. However, our view in visible light is blocked by the very gas and dust out of which the stars can form. Fortunately, the new techniques available through spectral studies in the radio part of the spectrum and through observations in the infrared penetrate the gas and dust. Many of the studies involve the discovery and observation of molecules in space. We have become able to study the birth processes of stars, including stars in formation and infant stars. In this chapter, we shall describe the research that has led to a new understanding of the formation of stars.

24.1 MOLECULAR HYDROGEN

While individual atoms of hydrogen, whether neutral or ionized, have been extensively observed in the interstellar gas, hydrogen molecules (H_2)

"How do stars form? In the early days when observational astronomy was entirely optical, it was hoped that large new telescopes such as the 200'' would lead us to the answer. Astronomers such as Bok and Herbig struggled valiantly amid the muck. But, alas, the interstellar dust could not be pushed aside.... The spectacular rise of molecular radio astronomy and infrared astronomy has allowed us to pull back the dusty veil and view the grand spectacle as it unfolds. Whether we shall be clever enough to understand the message remains to be seen." From a review article by Ben Zuckerman and Patrick Palmer in Annual Reviews of Astronomy and Astrophysics.

Infrared images of the Kleinman-Low Nebula in Orion, a region of active star formation. The brightest part of the upper image is the Becklin-Neugebauer object, probably one of the youngest stars known. A map of equal-intensity lines appears at bottom. (Courtesy of Robert D. Gehrz, J. A. Hackwell, and Gary Grasdalen, Wyoming Infrared Obs.)

have been observed only recently. Even though astronomers have long thought that molecular hydrogen could be a major constituent of the interstellar medium, it was simply not possible to observe hydrogen in molecular form. This is because at the low temperature of interstellar space, only the lowest energy levels of molecular hydrogen are excited, and no lines linking these levels fall in the optical or radio regions of the spectrum. We had to wait to observe from space in order to observe lines from H_2 because these lines occur in the far ultraviolet. Our atmosphere prevents these lines from reaching us on earth.

In 1970, George Carruthers of the Naval Research Laboratory in Washington succeeded in using a rocket to observe interstellar ultraviolet absorption lines from H_2. However, the rocket flight lasted only a few minutes, so the cloud in front of only one star could be studied.

In 1972, a 90-cm (36-inch) telescope was carried into orbit aboard NASA's third Orbiting Astronomical Observatory, named "Copernicus" in honor of that astronomer's 500th birthday (which occurred a year later). The telescope, operated by a group of scientists from the Princeton University Observatory, was largely devoted to observing interstellar material. The observers could point the telescope at a star and look for absorption lines caused by gas in interstellar space as the light from the star passed through the gas en route to us. Since it is easiest to pick up interstellar absorption lines if the star itself has no lines of its own, the scientists observed in the direction of B stars, which have few lines and are very bright (Fig. 24–1). (O stars would do as well, but there are many fewer of them.) The scientists preferred B stars that were rapidly rotating, so that even the few lines they do have would be smeared out by the Doppler effect. Further, because of their brightness, B stars are visible at greater distances than are stars of most spectral types.

Since the hydrogen molecule is easily torn apart by ultraviolet radiation, the observers did not find much molecular hydrogen in most directions. But whenever they looked in the directions of highly reddened stars, they found a very high fraction of hydrogen in molecular form: more than 50 per cent. Presumably, in these regions the dust shields the hydrogen from being torn apart by ultraviolet radiation. There are also theoretical grounds for believing that the molecular hydrogen is formed on dust grains, so it seems reasonable that the high fraction of H_2 is found in the regions where there are more dust grains.

Half the gas in the disk of our galaxy may be in the form of molecular hydrogen.

Figure 24–1 From above the atmosphere of the earth, the Copernicus satellite looks toward a star that provides a more-or-less continuous spectrum. (A rapidly-rotating B star is usually chosen, because B stars have few lines and those lines are washed out by Doppler shifts resulting from the velocity of rotation.) When a cloud of gas is in the path, then absorption or emission from molecules or atoms in the cloud results.

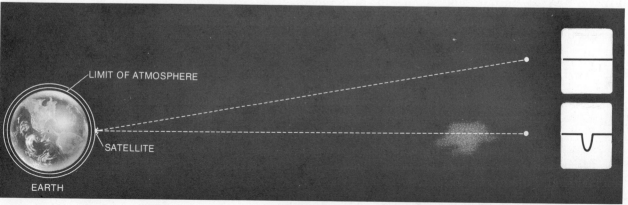

LIMIT OF ATMOSPHERE

SATELLITE

EARTH

24.1a Interstellar Deuterium

The Copernicus satellite also made observations of deuterium, a rare isotope of hydrogen. In the same shielded areas in which they observed molecular hydrogen, the astronomers also found a small fraction (10^{-6}) of H_2 in the HD form, that is, a molecule consisting of one atom of the ordinary isotope of hydrogen tied to one atom of deuterium ($_1H^2$).

The amount of deuterium in interstellar space gives us information about the origin of the universe, so the observations of HD caused some excitement in the astronomical community. But the results proved difficult to interpret, because D and H can combine at a different rate from the rate at which H combines with other H's. Thus the ratio of D to H in the HD form could be different from the ratio of the total amount of D to the total amount of H. Thus the relative abundances of HD and H_2 do not necessarily indicate the overall ratio of D to H.

Later on, however, the Copernicus satellite was able to detect deuterium in the form of individual atoms by studying the Lyman lines from hydrogen and deuterium, which also lie in the ultraviolet. We will analyze the importance of these observations in the chapter on cosmology, in Section 27.7.

24.2 INTERSTELLAR OPTICAL MOLECULAR LINES

Over the last few decades, several optical spectral lines were discovered that originated from interstellar space rather than from the stars themselves. These lines can be observed through ordinary optical telescopes, and until 1963, they were the only interstellar lines known besides the line at 21 cm (Fig. 24–2). They come from atoms, molecules, or radicals (other multi-atomic units) in space, including CN, CH, and CH^+. Besides these lines, there are an additional three dozen unidentified very diffuse (i.e., fuzzy) features in the optical spectrum, all of which lie between 4400 Å and 6800 Å. It may be that this set of diffuse features, which has been primarily studied by George Herbig of the Lick Observatory, results from absorption by interstellar dust particles. These particles may be only 300 Å across, about 1/10th the size of the dust particles that cause the interstellar reddening.

24.3 RADIO SPECTRAL LINES FROM MOLECULES

Even though optical studies of the interstellar medium have been going on for decades, recently much more information has come from ultraviolet observations made from space and radio studies made from the earth. For several years after the 1951 discovery of 21-cm radiation, spectral-line radio astronomy continued with just the one spectral line. Astronomers tried to find others. One prime candidate was OH, hydroxyl, which should be rela-

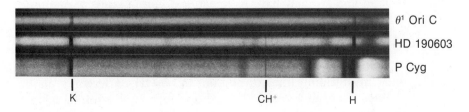

θ^1 Ori C

HD 190603

P Cyg

K CH^+ H

Figure 24–2 Absorption lines in the visible part of the spectrum caused by ionized calcium (H and K lines) and the molecular ion CH^+ in clouds of interstellar gas lying in the direction of 3 stars, whose names are given to the right of the individual spectra.

tively abundant, as molecules go, because it is a combination of the most abundant element, hydrogen, with one of the most abundant of the remaining elements, oxygen. OH has four lines close together at about 18 cm in wavelength, and the relative intensities expected for the four lines had been calculated.

It wasn't until 1963 that other radio spectral lines were discovered, and four new lines were indeed found at 18 cm. But the intensity ratios were all wrong to be OH, according to the predicted values, and for a time we spoke of the "mysterium" lines. The idea of discovering a new element was not unprecedented—after all, unknown lines at the solar eclipse in 1868 had been assigned to an unknown element, "helium," because they occurred only (as far as was known at that time) on the sun. Of course, the periodic table of elements has been filled in during the last 100 years, so a new element wasn't really expected. Mysterium turned out to be OH after all, but with strange processes affecting the excitation of the energy levels of OH and thus amplifying certain of the lines at the expense of others, which are weakened. This in itself is very interesting. The process is that of masering.

24.3a Masers

Masers (maser is an acronym for *m*icrowave *a*mplification by *s*timulated *e*mission of *r*adiation) and lasers, their analogue using *l*ight instead of *r*adio waves, are of great practical use on the earth. For example, masers are used as sensitive amplifiers. Masers were "invented" on earth not long before they were found in space.

For maser action to take place, the numbers of molecules or atoms in the various possible energy states must be different from the numbers that would normally be there. By "normally" we mean that under most conditions we can readily calculate the numbers of molecules or atoms in the various possible energy states according to standard methods that depend mainly on the temperature of the gas. For a maser, we must build up an exceedingly large number of molecules at a given energy level higher than the ground (lowest) level. Under certain conditions, the excited electrons can be triggered to all jump down together to a lower energy state, generating an intense emission line.

Figure 24–3 The 43-m (140-ft) telescope at Green Bank. Note the astronomer standing under the telescope.

To build up this number, there must be some method of *pumping* enough energy into the system to excite the molecules to the given excited level. Various mechanisms come into play in particular situations for particular molecules. For example, in some cases, ultraviolet radiation from a nearby star can pump molecules from the ground state to a given excited state. If infrared instead of ultraviolet radiation does the pumping, we say that we have an "infrared pump" to excite the maser. In any case, by allowing for pumping to specific levels we can see why the ratios of intensities of the lines can be different from the intensities that would be expected in the absence of pumping. We may have to consider different kinds of pumping mechanisms for OH in different regions.

24.3b Molecular Lines

The four "mysterium" lines turned out to be OH undergoing maser action. OH lines have now been detected not only in absorption against background emission, but also in emission. They are quite widespread and

Figure 24–4 An overall view of the National Radio Astronomy Observatory's field site at Green Bank, West Virginia. Three dishes that are used together as an interferometer are at right. The 91-m (300-ft) dish is in front of the 43-m (140-ft) dish at left.

are used to tell us about the physical conditions in the clouds from which the OH radiation comes.

The interstellar abundance of OH was soon calculated from the measurement of OH radiation. It was very much less than that of isolated hydrogen or oxygen atoms, which seemed reasonable since in the virtual vacuum of interstellar space there is little chance for oxygen and hydrogen atoms to interact. Still, radio astronomers were disappointed with the low OH densities: only one OH molecule for every billion H atoms. It seemed quite unlikely that the quantities of any molecules composed of three or more atoms would be great enough to be detected. The chance of three atoms getting together in the same place, it seemed, would be very small.

In 1968, however, Berkeley's Charles Townes and colleagues used their Hat Creek Observatory to observe at the radio frequencies in the centimeter range that were predicted to be the frequencies of water (H_2O) and ammonia (NH_3). The spectral lines of these molecules proved surprisingly strong, and were easily detected. In fact, we had had the capability for some years to make equipment sensitive enough to detect them, but nobody had tried to make the observations.

Soon afterwards, a group of radio astronomers used the 43-m telescope (Figs. 24–3 and 24–4) at the National Radio Astronomy Observatory in Green Bank, West Virginia, to search for interstellar formaldehyde (H_2CO) at a wavelength of 6 cm. They succeeded in detecting the formaldehyde molecule, the first molecule that contained two "heavy" atoms, that is, two atoms other than hydrogen.

By this time, it was apparent that the earlier notion that it would be difficult to form molecules in space was wrong. There has been much research on this topic, but the mechanism by which molecules are formed has not yet been satisfactorily determined. For some molecules, including molecular hydrogen, it seems that the presence of dust grains is necessary. In this scenario, one atom would hit a dust grain and stick to it (Fig. 24–5). It may be thousands of years before a second atom hits the same dust grain, and even longer before still more atoms hit. But these atoms may stick to the dust

Figure 24–5 Hydrogen molecules are formed in space with the aid of dust grains at an intermediate stage.

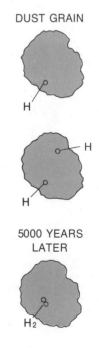

DUST GRAIN

H

H

H

5000 YEARS LATER

H_2

The discoverers of interstellar ammonia and water vapor included Albert C. Cheung (now at the University of California at Davis), David M. Rank (now at Lick), Charles H. Townes and William J. Welch (of the University of California at Berkeley), and Donald Thornton (now at George Washington University). The group who discovered interstellar formaldehyde soon thereafter included Ben Zuckerman (University of Maryland), Patrick Palmer (University of Chicago), David Buhl (now at NASA's Goddard Space Flight Center), and Lewis E. Snyder (now at the University of Illinois). These pioneers in spectral-line radio astronomy, joined by the many others now in the field, at first spent much of their time searching for new molecules, while now all the scientists are spending most of their time learning about the physical processes by studying the molecules.

The formulas are examples for illustration and are not for memorization. Note that they include only H, C, O, and N.

grain rather than bouncing off, which gives them time to join together. Complex reactions may take place on the surface of the dust grain. Then, somehow, the molecule must get off the dust grain back into space as a gas. Perhaps incident ultraviolet radiation or the energy released in the formation of the molecule allows the molecule to escape from the grain surface.

Though hydrogen molecules form on dust grains, there is a strong body of opinion that holds that most of the other molecules are formed in the interstellar gas without need for grains. Recent theoretical and laboratory studies indicate that reactions between neutral molecules and ionized molecules may be particularly important. Many of these chains start with molecular hydrogen, formed on grains, being ionized by cosmic rays. There are still many gaps in the theory. It is likely that no single process forms all the interstellar molecules.

After the discoveries of water, ammonia, and formaldehyde in interstellar space, further discoveries came one after another. Many different radio astronomers looked up the wavelengths of likely spectral lines from molecules containing abundant elements and were able to observe the radiation. Sometimes the lines were in absorption and sometimes in emission. Eventually, astronomers and chemists started measuring frequencies especially for use in searches at the telescope.

The list of molecules discovered expanded gradually from three-atom molecules like ammonia and water, and four-atom molecules like formaldehyde, to even more complex molecules. Formic acid ($HCOOH$) with five atoms, methyl cyanide (CH_3CN) with six, and cyanohexatriyne (HC_7N) are examples of such heavier molecules (see Appendix 12). The discovery of methyl alcohol (CH_3OH) in 1970 was greeted jocularly, and cases of liquor were bet on the discovery of ethyl alcohol (CH_3CH_2OH), the drinking kind. In 1974, the ethyl alcohol molecule was discovered too. (There are 10^{28} fifths, calculated at 200 proof, in the molecular cloud in which it was observed.)

At first most discoveries were made in the centimeter range of the spectrum, and attempts to work at the longer wavelengths closer to a meter

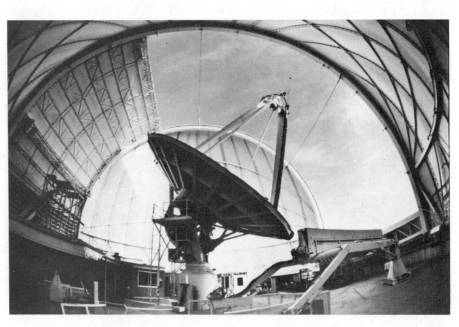

Figure 24–6 The 11-m (36-ft) telescope of the National Radio Astronomy Observatory, seen here through a fish-eye lens, is on Kitt Peak in Arizona (see Fig. 2–26). It is used for observations at millimeter wavelengths, and has been used to discover many interstellar molecules.

proved fruitless. Then the technology at the shorter wavelengths—from 2 to 10 mm—improved, largely because of work done in developing sensitive instrumentation at the Bell Laboratories, which was also where Jansky had worked. Many more molecules were found in these short *millimeter radio wavelengths*. With the new technology, the leadership in discovering molecules passed from the 43-m (140-ft) radio telescope of NRAO at Green Bank to an 11-m (36-ft) radio telescope (Fig. 24–6) that NRAO operates at Kitt Peak in Arizona. The 11-m telescope is able to observe at millimeter wavelengths. The 43-m telescope cannot observe at millimeter wavelengths. There is currently emphasis on millimeter radio astronomy, and new millimeter telescopes are under construction or have just been completed. The University of Massachusetts at Amherst's 14-m (45-ft) telescope is a recent example (Fig. 24–7).

Also, there have been further advances in the technology of designing and building sensitive radio receivers. Maser action in the receiver itself has been used to amplify very faint signals observed at centimeter wavelengths. Some receivers are cooled by surrounding them with liquid nitrogen or liquid helium—the technology of ultracooling is called *cryogenics*. The advances in receiver technology continue to be linked to research on more mundane communications problems.

Over four dozen molecules have now been discovered in interstellar space. Over 240 spectral lines remain unidentified, some of which are undoubtedly from still additional molecules.

24.4   ANALYZING INTERSTELLAR SPACE

There is much more to spectral-line radio astronomy than simply discovering new spectral lines. For one thing, studying the lines provides information about physical conditions—temperatures, densities, and motions, for example—in the gas clouds that emit the lines. For example, formaldehyde radiates only when the density is roughly ten times that of the gas that radiates carbon monoxide. The clouds that emit molecular lines are usually so dense that hydrogen atoms have combined into hydrogen molecules and very little 21-cm radiation is emitted. Some of the lines, for instance, those of carbon monoxide (CO), have been detected in many directions in space, and maps have been made of the distribution of these molecules (Fig. 24–8). Sensitive radio receivers can do this, even though the densities of these molecules in the densest clouds may be only one in every cubic meter.

We can also study the physical conditions in H II regions, which are less dense. In this case, we observe a set of radio spectral lines coming from atoms rather than from molecules. These are called *recombination lines* because they result when an electron that has separated from an atom when the atom was ionized recombines with, that is, rejoins, the rest of the atom. The electron usually recombines on a high energy level, and then jumps down to lower and lower levels (Fig. 24–9), giving off a series of spectral lines in the radio spectrum as it does so. The strongest recombination lines are from hydrogen, but weaker lines from helium and still heavier elements have also been detected.

An additional contribution from observations of interstellar molecules has come from studies of the relative abundances of different isotopes. For example, not only can formaldehyde (H_2CO) be studied in its usual form,

Recently, new devices that make infrared spectra have been used to discover molecules in solar-system size shells around sources that radiate continuous spectra in the infrared. Acetylene (C_2H_2) was discovered in this way.

Note that the 43-m telescope operating at a wavelength of 10 cm is 430 wavelengths across, whereas the 11-m telescope operating at 2 mm is 5500 wavelengths across. Thus the 11-m telescope is effectively larger than the 43-m telescope, relative to the wavelengths at which observations are being made. This has the practical effect of providing higher angular resolution.

Figure 24–7 The radome enclosing the 14-m millimeter-wave radio telescope of the University of Massachusetts—Amherst. The telescope is located at the Quabbin Reservoir in central Massachusetts.

H_2CO normally consists of 2 atoms of hydrogen, 1 atom of C^{12}, and 1 atom of O. It can, more rarely, have another isotope of carbon, such as C^{13}, instead of C^{12}. It could also contain other hydrogen or oxygen isotopes.

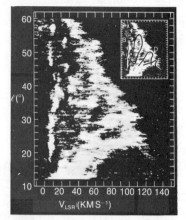

Figure 24–8 The distribution of radiation from the carbon monoxide in our galaxy, made at low resolution with a small radio telescope at Columbia University. This resolution matches that of 21-cm maps. The vertical axis shows galactic longitude, roughly the angle from the center of the galaxy around the Milky Way as seen from the earth. (No information is shown in the perpendicular direction, galactic latitude). The horizontal axis shows the velocity of the clouds that emit the CO. Note that in each direction, we see a wide range of velocities, indicating that we are seeing several clouds. The inset shows how this map corresponds to several spiral arms observed at 21-cm. (Courtesy of Richard Cohen)

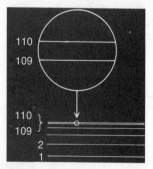

Figure 24–9 The recombination lines that we observe in the radio part of the spectrum are transitions among very high energy states of hydrogen (and other light atoms). H 109 α, for example, the transition between hydrogen's levels 110 and 109, is at a frequency of 5 GHz, which is equivalent to a wavelength of 6 cm. The notation is similar to that of Lyman α, a transition between the lowest two levels.

which contains the most common isotope of carbon, C^{12}, but also it can be studied in the form $H_2C^{13}O$. Knowledge of the relative abundances of the different isotopes is very important in developing theories of the formation of the elements. The study of these abundances had been the subject of much previous work with other methods. Who would have thought fifteen years ago that these isotopic abundances would best be measured with this—then unknown—method?

It even turns out that basic parameters that describe the structure of some of these molecules may best be measured through astronomical spectral lines studied in space rather than in the laboratory. Some molecules don't have spectral lines that are amenable to the measurement processes that can be used on laboratory samples on earth, and others can't even be formed on earth.

Another important conclusion reached by spectral-line studies is that complex molecules can be readily formed by natural processes. Some of the molecules that have been discovered in space are "organic" in nature, in that they contain carbon. We are even approaching the level of complexity at which we can find simple amino acids, the building blocks of life. And we know (see Section 21.1), even though we have not found amino acids themselves, that amino acids can be readily formed in a laboratory. All we need do is pass electricity through mixtures containing some of the molecules that have been discovered in space, or illuminate such a mixture with ultraviolet light. So it seems that it is much easier than we had originally thought to form the molecules from which life is made. These considerations add support to the beliefs of many astronomers who feel that life has probably arisen spontaneously at many different locations in the universe.

Spectral-line molecular investigations have brought a new type of immigrant into the astronomy profession—chemists. Sometimes these scientists have used their expertise in the lab to measure the frequencies of the lines, something that must be known in order to observe them in radio sources. Sometimes they use their knowledge and training to help interpret the formation of these complex molecules. In any case, this is an example of how astronomy can incorporate another field of science.

24.5 THE FORMATION OF STARS

Most radio spectral lines seem to come only from a very limited number of places in the sky; carbon monoxide is the major exception, for it is widely distributed across the sky.

The study of various lines can tell us a lot about the nature of a region of space, even deep inside a dark cloud. Infrared radiation also does much better in penetrating the dust that protects molecular clouds. So infrared and radio observations have been walking hand in hand over the last years toward providing us with an understanding of how stars are formed from dense regions of gas and dust.

Many molecular clouds are about 50 parsecs across, fragmented into many denser bits of 1 parsec in size. These bits must become much smaller and denser yet to form a star or, in many cases, several stars. One of the important questions in this field is related to the formation of the brightest stars, those of spectral types O and B, which are stars of large mass. These are usually found in loose groupings called OB associations. These associations are often divided into subgroups that appear to have been formed at different

Figure 24–10 The globule Barnard 335. The small cloud of gas and dust appears as a "hole" in the distribution of stars. We assume that the stars are roughly uniformly distributed and that such holes result because of the extinction caused by dust in the line of sight. Emission from carbon monoxide has been observed from this region, signifying that something is really present. The angular size of this globule is about 4 arc minutes.

Radio studies of interstellar molecules and infrared studies have recently led to a major improvement in our understanding of stars in formation. The discussion in this section thus completes the cycle we began in the chapters on stellar evolution, for the older stars have been born, died, and spread out their material through interstellar space.

times. Since O and B stars don't live very long, we usually find them very close to the clouds from which they have formed. Astronomers have studied the stars and clouds for clues on how star formation started and continued in these regions. The explanations under investigation include the passage of a spiral arm, a nearby supernova explosion, and a nearby H II region expanding into the cloud. These massive stars may form within 10,000 or 100,000 years; less massive stars would take much longer.

Sometimes, we see smaller objects, called globules, or *Bok globules*, after the astronomer Bart Bok, who has made an extensive study of these objects. Some of the smaller globules are visible in silhouette against H II regions, as in Color Plate 61 (top right of picture, for example). Others are larger, and appear isolated against the stellar background, as in Figure 24–10. It has been suggested that the globules may be on their way to becoming stars. There is some doubt as to whether this is true for the smaller globules, but molecular observations of the larger globules indicate that they may contain sufficient mass for gravitational collapse to be taking place.

24.5a A Case Study: The Orion Molecular Cloud

Many radio spectral lines have been detected only in a particular cloud of gas located in the constellation Orion, not very far from the main Orion Nebula. This *Orion Molecular Cloud* (Fig. 24–11), which contains about 500 solar masses of material, is itself buried deep in nebulosity. It is relatively accessible to our study because it is only about 500 parsecs from us. Even though less than 1 per cent of the Cloud's mass is dust, that is still a sufficient amount of dust to prevent ultraviolet light from nearby stars from entering

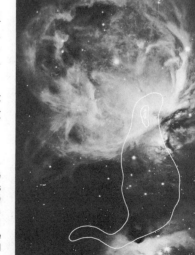

Figure 24–11 The contours show the molecular cloud associated with the Orion Nebula. The molecular cloud is actually on the far side of the glowing gas of the nebula, but the radio waves from the molecules penetrate the nebula and are observed with radio telescopes like the NRAO 11-m (36-ft) dish on Kitt Peak. The contours of radio emission correspond roughly to regions of different density. The densest part (the smallest region) also includes a region of infrared radiation called the Kleinmann-Low Nebula and a strong point infrared emitter called the Becklin-Neugebauer object. Weaker emission from the molecules actually extends beyond the last contour and even far beyond the range of this photograph.

The bright region at the lower part of the photo, NGC 1977, appears to the naked eye as the northern star in the sword of Orion but actually consists of a number of stars and a small H II region. The shape of the outermost contour shown here, indicates that the H II region may be expanding against the molecular cloud.

and breaking the molecules apart. Thus the molecules can accumulate in number.

We know that young stars are found in this region—the Trapezium, a group of four hot stars readily visible in a small telescope are the source of ionization and of energy for the Orion Nebula. The Trapezium stars are relatively young, about 100,000 years old. The Orion Nebula (Color Plate 56), prominent as it is in the visible, is a gas located along the side of the molecular cloud nearest to us, as shown in the model put forward by Ben Zuckerman of the University of Maryland (Fig. 24–12). The nebula contains much less mass than does the molecular cloud.

The properties of the molecular cloud can be deduced by comparing the radiation from various molecules and by studying the radiation from each molecule individually. The density, 1000 particles per cubic centimeter in the outer limits at which the cloud is visible to us, increases toward the center, and it may actually be as dense as 10^6 particles/cm³ at its center. This is still billions of times less dense than our earth's atmosphere, and 10^{16} times less dense than the star that may eventually form, though it is substantially denser than the average interstellar density of about 1 particle per cm³.

One of the brightest of all the infrared sources in the sky, an object in the Orion Nebula that was discovered by Eric Becklin, now at the University of Hawaii, and Gerry Neugebauer of Caltech, is right in the midst of the Orion Molecular Cloud. This *Becklin-Neugebauer object*, also known for short as the *B-N object*, is about 200 astronomical units across. Its temperature is 600 K. The B-N object is the prime candidate for a star about to be born.

The B-N object is behind so much dust as seen from earth that it appears very faint. But a new device at the Kitt Peak National Observatory now enables us to take high-quality infrared spectra of this object. In 1978, this spectrometer succeeded in making observations of two of the hydrogen infrared series of spectral lines. From comparing certain of the spectral lines,

Another example of star formation in a molecular cloud is M17, shown in Color Plate 62. The relatively sharp edge at right marks the left edge of the molecular cloud, which contains both infrared objects and maser sources.

Figure 24–12 The structure of the Orion Nebula and the Orion Molecular Cloud, proposed by Ben Zuckerman of the University of Maryland.

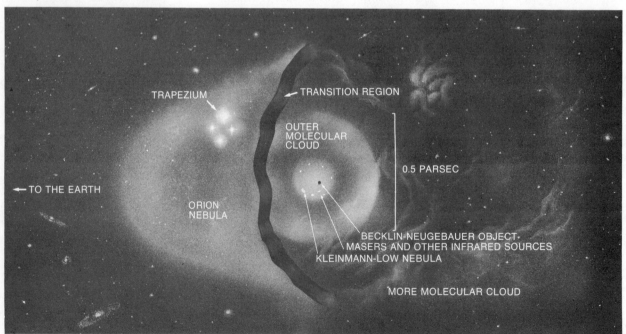

the scientists could deduce the amount by which the B-N object is reddened. They also concluded that the object is a very young star of spectral type B, a massive, hot star. It is already, but just barely, operating on the basis of nuclear fusion.

Other infrared sources are also present nearby, and probably also contain young stars or stars in formation. Small intense sources of maser radiation from various molecules also exist nearby. Such interstellar masers can only exist for small sources, about the size of our solar system, and are also signs of stars in formation.

If the Becklin-Neugebauer object is indeed a star at the distance of the Orion Nebula, we can calculate how much its light would have to be reddened to appear to us so bright in the infrared and faint in the visible. The extinction in the visible region of the spectrum would be at least 50 magnitudes (a factor of 10^{20}).

24.5b Future Infrared Observations

Observations in the infrared are proceeding at a rapid pace, particularly with the existence of new facilities such as those on Mauna Kea. A variety of infrared sources is under study.

IRAS, the NASA/Netherlands/UK Infrared Astronomy Satellite, with an 0.6-m telescope, will be launched in 1982. It will map the sky with 2-arc-min resolution. Cooled by liquid helium, it will provide a gain in sensitivity of 1000 at 10 and 20 microns and will make the first full survey at 50 and 100 microns. It may discover a million new sources.

In 1986 or 1987, NASA plans to launch an 0.85-m infrared telescope with the Space Shuttle. This Shuttle Infrared Telescope Facility, like IRAS, will be cooled to a temperature of only 2 K in order to allow sensitive observations to be made. The telescope will feed several instruments to allow both photometry and spectroscopy. It will be operated aboard the European Spacelab.

An instrumented airplane known as the Kuiper Airborne Observatory provides excellent data.

24.5c The Youngest Stars

How does a star appear just after it is born? It will probably still be associated with dust, and the radiation from some of the dust should give it an excess amount of radiation in the infrared over what is normal for stars of the same visible color. It may well have emission lines visible in its spectrum. And its intensity may not have settled down to a steady condition. T Tauri stars (see Section 8.1) have all these characteristics. They are thus considered by many to be exceedingly young stars, just about to reach the main sequence.

Four T Tauri stars have been observed to erupt. One did so in Orion in 1936, and has been named FU Orionis (following long after ZZ and AA in the standard system of nomenclature for variable stars). Three other examples of this class of *FU Orionis stars* are now known, enough objects to make astronomers think this may be a general phenomenon rather than an isolated instance. The variable star V1057 Cygni in the Swan, which brightened in 1969, was the second example. Both FU Orionis and V1057 Cygni brightened over 100 times, partly because they changed in surface temperature and partly because they increased in size. V1057 Cygni grew from 4 to 16 times the radius of the sun over a period of a year. A third FU Orionis star, V1515 Cygni, has had a more gradual rise in brightness between the turn of the century and 1958. The fourth FU Orionis star is much fainter so its properties are not well known.

All four are associated with arcs of dust, signs that the stars had not been

T Tauri stars are generally thought to be pre-main–sequence objects containing about 1 solar mass. They would thus be expected to become stars of spectral types G, F, or A.

around very long or else the dust would be vaporized. Evidence for dust closer in to the stars is the fact that their infrared radiation is brighter than would be expected on the basis of their temperatures. The dust is probably from the remainder of the protostellar nebulae. The velocities of the stars in space are the same as the velocities of the molecular clouds near which they appear, clinching the association of the two.

The eruptions are probably repetitive; FU Orionis can already be seen to be fading, though much more slowly than it brightened. From the meager statistics of these four examples, it seems that such an eruption takes place in a T Tauri star about every 10,000 years.

Spectral lines that result from water vapor have recently been discovered in the spectra of two of the stars. By contrast, the T Tauri star HL Tau shows absorption by ice—frozen water—in its infrared spectrum near a wavelength of 3 microns. It may represent an even earlier stage in stellar evolution. The brightening of an FU Orionis star may melt the ice that once surrounded it.

*24.6 AT A RADIO OBSERVATORY

What is it like to go observing at a radio telescope? First, you decide just what you want to observe, and why. You have probably been working in the field before, and your reasons might tie in with other investigations under way. Then you need to know the frequencies of the spectral lines; they may be available in books or tables, or they may have to be measured specially for you in chemical laboratories. Perhaps it was a newly available set of radio frequencies received in the mail from a colleague that made you decide to observe a particular molecule.

Then you decide at which telescope you want to observe; let us say it is one of the telescopes of the National Radio Astronomy Observatory. You first send in a written proposal to the NRAO headquarters, where it is read and evaluated. If the proposal is approved, it is placed in a queue waiting for observing time. You might be scheduled to observe for a five-day period to begin six months to a year after you have submitted your proposal.

At the same time you might be applying for support to carry out the research, usually to the National Science Foundation. Your proposal would possibly contain requests for some support for yourself, perhaps for a summer, and support for a student or students to work on the project with you. It might also contain requests for funds for computer time at your home institution, some travel support, and funds to support the eventual study of the data and the publication of the research. The time spent at the telescope is usually the shortest part of the overall effort.

When your observing time comes, you pack your bags and go to the airport. It doesn't really matter whether you are at a big university or at a small college, or even on the staff of the national observatory itself—you still have to pack your bags and go off to the telescope for a few days. With a multimillion dollar telescope, it is important that it be located at a good observing site and that the observing time be used as efficiently as possible. The cost of a few additional plane trips is small compared to the other costs. You are not charged directly for telescope time—that is covered in the overall budget of the observatory itself.

The National Radio Astronomy Observatory (NRAO) has its headquarters at Charlottesville, Virginia. It has several telescopes at Green Bank in

Figure 24–13A The 11-meter telescope of the National Radio Astronomy Observatory.

Figure 24–13B   The VLA, west of Socorro, New Mexico.

West Virginia, for study of wavelengths longer than about 1 cm, at Kitt Peak in Arizona, for study of millimeter wavelengths (Fig. 24–13), and west of Socorro in New Mexico for use of an array of telescopes (the Very Large Array) that permits high resolution images to be made. Travel to the telescopes is usually time consuming. For example, to observe at the Very Large Array, you not only have to fly to Albuquerque, and drive to the post office address of Socorro, but then you have to drive another hour west out onto a broad plain. All the observing locations have dormitories and dining halls so that astronomers can stay near the telescopes.

At the observatory, you will meet the other members of your team if they have come from different places. If this is your first time at this particular telescope, care will have been taken to see that an astronomer experienced at observing there is present to help you get started.

The astronomers sit at a computer console monitoring the data as they come in from the telescope or array of telescopes (Fig. 24–14). The electronics that are used to treat the signal incoming from the feed (Fig. 24–15) are as important a part of the system as the dish itself. A trained observing assistant actually runs the mechanical aspects of the telescope. These observing assistants are regular employees of the observatory and work in shifts. The astronomers arrange their own time schedules so that they can

Figure 24–14   The taking of spectral line data is often controlled by a scientist or observing assistant at a keyboard linked to a computer. A screen on which the spectra can be continually seen is at top left. The view here is inside the control room of the 67-m telescope at Parkes, Australia.

Unlike the optical sky, which is blue in the daytime, the radio sky background remains dark even when the sun is up. As long as they don't point their telescopes to within a few degrees of the sun, radio astronomers can observe anywhere in the sky at any hour of the day or night.

Figure 24–15 Radio astronomy depends on electronics as much as optical astronomy has depended on optics. Shown here is the inside of the autocorrelator—a device used for observing radio spectral lines—at the 64-m telescope at Parkes. A failure of any of the thousands of electronic parts shown can mean a breakdown of the whole procedure of collecting data.

observe around the clock—one doesn't want to waste any observing time. This is unlike optical astronomers who can work all night and sleep for part of the daytime.

You give the observing assistant the coordinates of the point in the sky that you want to observe, and the telescope is pointed for you. The electronic systems are particularly advanced, and a computer can display the incoming data on a video screen. You can even use the computer to manipulate the data a bit, perhaps adding together the results from different five-minute chunks of "integration time," that is, exposures, that you have made. At the Very Large Array, huge computers combine the output from two dozen telescopes and show you a color-coded image, with the colors referring to the brightness. The data are stored on magnetic computer tape. You would take home with you both copies made on paper or film and the computer tapes themselves for further study.

Depending on just what you are observing, your results may or may not be immediately apparent to you on the computer screen. Some spectral lines are so intense in certain sources that you can see the emission line on the screen in just a few minutes. Some lines are so faint that you may have to integrate for many hours, or even days or months, to reach an acceptable level of sensitivity.

When you have finished observing one source, perhaps because it has set below the horizon, you ask the operator to point the telescope to another source, and off you go again.

At the end of your observing run, you take the data back to your home institution to complete the analysis. You are expected to publish the results as soon as possible in one of the standard scientific journals, probably after you have given a paper about the results at a professional meeting, such as that of the American Astronomical Society. The discussions you have with other astronomers in your field are a professionally valuable and personally interesting part of your job. Astronomers have colleagues to talk to about research of mutual interest wherever they go in the world.

Often a research paper deals with some special aspect of an object or set of objects. To generalize and to draw basic conclusions, many sets of observations are often necessary. The conclusions now being reached about how stars are formed, for example, make use of observations made by many individuals and groups. Radio spectra from telescopes such as the 11-m on Kitt Peak, from the newly-available high-resolution images from the Very Large Array, and from observations made in other parts of the spectrum are used.

SUMMARY AND OUTLINE

Molecular hydrogen (Section 24.1)
Interstellar ultraviolet absorption lines from H_2, with abundances up to 50 per cent, are observed from above the earth's atmosphere
Copernicus satellite detected not only normal H_2 but also deuterium, which has cosmological importance
Interstellar optical molecular lines (Section 24.2)
CN, CH, CH^+

Radio spectral lines from molecules (Section 24.3)
"Mysterium" lines discovered at 18 cm—later identified as OH affected by masering process
Masers first developed artificially on earth, but later discovered to exist in space
Maser action occurs after atoms are excited by energy-pumping mechanisms so that many are in the same excited state
Three dozen molecules have been discovered in space, including some as massive as amino acids

Many molecules are found at short millimeter radio wavelengths

Analyzing interstellar space (Section 24.4)

Radio recombination lines result after an electron recombines with an ion in an H II region

Analysis of molecular lines and recombination lines tells us physical conditions (e.g., temperature, density, and motions)

The formation of stars (Section 24.5)

Distribution of CO, which emits at low density, is widespread, while emission from other molecules, which emit at higher densities, is more localized

Molecular clouds contain gas that is collapsing to become stars

Molecules are associated with dark clouds, such as the Orion Molecular Cloud, where the molecules are shielded from being torn apart by ultraviolet radiation

The Becklin-Neugebauer object, a compact infrared source deep inside the Orion Molecular Cloud, has been revealed through infrared spectral observations to be a young B star

At a radio observatory (Section 24.6)

Stages of conceiving the project, applying for time, observing, reducing data, publishing results

QUESTIONS

1. Why did it take so long to discover interstellar hydrogen molecules?

2. Why do astronomers look at rapidly rotating stars of spectral class B when they want to study interstellar molecules?

3. Why might the fraction of deuterium in the form of HD molecules be different from the overall fraction of deuterium compared to hydrogen in the universe?

4. In what forms and which spectral lines did the Copernicus satellite observe interstellar deuterium?

5. What forms the diffuse interstellar absorption lines?

6. What was "mysterium"? Why was it thought to be strange?

7. How does a maser work?

8. Why did the abundance of heavy molecules predicted from observations of hydroxyl give too low a value?

9. Why are dust grains important for the formation of interstellar molecules?

10. (a) How many interstellar molecules have 2 atoms, 3 atoms, etc.? (b) How many interstellar molecules contain one heavy (i.e., non-H) atom, 2 heavy atoms, etc.? (c) What fraction of the known interstellar molecules are organic? (d) How many interstellar molecules are chemicals of which you have heard previously?

11. What is cryogenics?

12. Which molecule is found in the most locations in interstellar space?

13. What are recombination lines? Can Lyman α be a recombination line? Can Hα?

14. Is it better to study molecular clouds in the visible or in the infrared? Why?

15. Why might we expect to find O and B stars near molecular clouds?

16. Describe the relation of the Orion Nebula and the Orion Molecular Cloud.

17. Describe the Becklin-Neugebauer object. Why do scientists find it interesting?

18. What future plans do we have for infrared studies from space?

19. What is an FU Orionis star? What stage of stellar evolution are we observing?

20. Optical stellar astronomers can observe only at night. In what time period can radio astronomers observe?

TOPICS FOR DISCUSSION

1. What does the discovery of fairly complex molecules in space imply to you about the existence of extraterrestrial life?

2. Discuss which other sciences contribute to the study of interstellar space and how they do so.

PART VI

Though the individual stars that we see are all part of the Milky Way Galaxy, discussed in the preceding three chapters, we cannot be so categorical about the locations of the conglomerations of gas and stars that can be seen through telescopes. Once they were all called "nebulae," but we now restrict the meaning of this word to gas and dust in our own galaxy. Some of the objects that were originally classed as nebulae turned out to be huge collections of gas, dust, and stars located far from our Milky Way Galaxy and of a scale comparable to that of our galaxy. These objects are galaxies in their own right, and are both fundamental units of the universe and the stepping stones that we use to extend our knowledge to tremendous distances.

In the 1770's, a French astronomer named Charles Messier was interested in discovering comets. To do so, he had to be able to recognize whenever a new fuzzy object appeared in the sky. He thus compiled a list of about 100 diffuse objects that could always be seen. To this day, these objects are commonly known by their *Messier numbers*. Messier's list contains the majority of the most beautiful objects in the sky, including nebulae, star clusters, and galaxies. The list, compiled to search for comets, turns out to have this much more general importance.

Soon after, William Herschel, in England, compiled a list of 1000 nebulae and clusters, which he expanded in subsequent years to include 2500 objects. Herschel's son John continued the work, incorporating observations made in the southern hemisphere. In 1864, he published the *General Catalogue of Nebulae*. In 1888, J. L. E. Dreyer published a still more extensive catalogue, *A New General Catalogue of Nebulae and Clusters of Stars,* the *NGC,* and later published two supplementary *Index Catalogues, IC's.* The 100-odd non-stellar objects that have Messier numbers are known by them, or else by their numbers in Dreyer's catalogue.

466

The large reflector, with a mirror 6 feet (1.8 meters) across, built by the Earl of Rosse in Ireland in 1845. Problems with maintaining

GALAXIES
AND BEYOND

Thus the Great Nebula in Andromeda = M31 = NGC 224. The Crab Nebula = M1 = NGC 1952.

an accurate shape for the mirror, which was made of metal, led to the telescope's abandonment.

When larger telescopes were turned to the Messier objects in later years, especially by Lord Rosse in Ireland in about 1850, some of the objects showed traces of spiral structure, like pinwheels. They were called "spiral nebulae." But where were they located? Were they close by or relatively far away?

When such telescopes as the 0.9-m Crossley reflector at Lick in 1898, and later the 1.5-m and 2.5-m reflectors on Mount Wilson, began to photograph the "spiral nebulae," they revealed many more of them. The shapes and motions of these "nebulae" were carefully studied. Some scientists thought that they were merely in our own galaxy, while others thought that they were very far away, "island universes" in their own right, so far away that the individual stars appeared blurred together. (The name "island universes" originated with the philosopher Immanuel Kant in 1755.)

The debate raged until 1924, when observations made at the Mount Wilson Observatory by Edwin Hubble proved that there were indeed other galaxies in the universe besides our own. In fact, we think of galaxies and clusters of galaxies as fundamental units in the universe. The galaxies are among the most distant objects we can study; quasars are objects on a galactic scale that are, for the most part, even farther away.

Galaxies and quasars can be studied in most parts of the spectrum. Radio astronomy, in particular, has long proved a fruitful method of study. The study across the spectrum of galaxies and quasars provides tests of physical laws at the extremes of their applications and links us to cosmological consideration of the universe on the largest scale.

467

CHAPTER 25

════ GALAXIES ════

AIMS:

To discuss the different types of galaxies that are observed in various parts of the spectrum, to see that galaxies are fundamental units of the universe, to study the expansion of the universe, to consider how interferometry is now allowing radio astronomy to make significant advances in the study of galaxies, and to consider how and why galaxies may have formed

On April 26, 1920, Harlow Shapley (pronounced to rhyme with "map lee") (Fig. 25–1) and Heber D. Curtis were brought together at the National Academy of Sciences in Washington to discuss the scale of our own galaxy and the nature of the "spiral nebulae," matters on which they had become known in the preceding years as the major protagonists. The *Shapley-Curtis debate* is an interesting example of the scientific process at work, though it did not settle the question of the nature of the "spiral nebulae."

The arguments used by Shapley and Curtis were several in number, and involved many of the concepts that we have dealt with in earlier chapters. Shapley and Curtis mainly debated our own galaxy's size. We have seen in the Introduction to Part V how Shapley's earlier research had led him to realize that the Milky Way was ten times larger than had been thought, though Curtis did not accept this result. The nature of the "spiral nebulae" was treated in the final paragraphs of the published versions of their statements. Curtis' conclusion that the "spiral nebulae" were external to our galaxy was based in large part on the notion that our galaxy is much smaller

Figure 25–1 Harlow Shapley.

An edge-on view of a galaxy of type Sb, NGC 4565, in the constellation Coma Berenices. Several other galaxies also appear on this picture, though they are much farther away and thus much smaller. The irregular objects about 2 cm upward from the center of NGC 4565, for example, are distant galaxies.

than we now know it to be. Curtis had several reasons for the view that the "spiral nebulae" were actually far-off galaxies. These included his analysis of what were then called "novae" that in 1917 were discovered to be going off from time to time in the "spiral nebula" in Andromeda. He reasoned that if the "spiral nebulae" were external to our galaxy, then the absolute magnitudes of the novae would be consistent with the absolute magnitude of novae in our own galaxy.

Shapley, on the other hand, argued that the "spiral nebulae" were close by because proper motion for points, probably stars, in several of them had been detected by his colleague Adriaan van Maanen. This proper motion was presumably caused by the rotation of these "nebulae." Also, Shapley had earlier noted that one of the "novae"—S Andromedae—that had erupted in the Andromeda "spiral nebula" in 1885, would have had to be much brighter than the other novae observed in that "nebula." This made Shapley feel that the evidence from "novae" was not internally consistent, that is, some pieces of evidence were inconsistent with other pieces.

In fact, both astronomers were using incomplete or fallacious evidence. Nobody at that time knew about interstellar absorption (Section 23.2). This dimmed distant stars and thus made them seem farther away, which led to a general overestimate of the distance scale. Van Maanen's observations of proper motions in "spiral nebulae" were incorrect; later observations were to show that these objects do not show proper motions. Some of Shapley's feelings were based on incomplete knowledge, for S Andromedae was actually a supernova rather than a nova. (At that time one spoke of "Tycho's nova," which we now call Tycho's supernova.) The distinction between novae and supernovae was not realized until the work of the Swedish astronomer Knut Lundmark, published in 1920. Even Lundmark's research left unanswered questions, and Curtis did not agree that the "novae" fell into two such well-defined classes.

Curtis was correct that the "spiral nebulae" were comparable objects to our own galaxy, but for the wrong reasons. Shapley, on the other hand, came to the wrong conclusion but followed a proper line of argument that was unfortunately based on incorrect and inadequate data.

The question of the distance to the "spiral nebulae" was settled only in 1924 by Edwin Hubble (who was shown in Fig. 2–34). He used the Mount Wilson telescopes to observe Cepheid variables in three of the "spiral nebulae." His definitive conclusion (following the line of argument we described in Section 6.4b) was that the "spiral nebulae" were outside our own galaxy and of dimensions not overwhelmingly different from it. Since Hubble's work there has been no doubt that the spiral forms we see in the sky are galaxies like our own. For the rest of the book we shall strictly use the term *spiral galaxies;* the incorrect, historical term, "spiral nebula," often hangs on in certain contexts, chiefly when we discuss the "Great Nebula in Andromeda," which is actually a spiral galaxy.

25.1 TYPES OF GALAXIES

Hubble went on to use the Mount Wilson telescopes to study the different types of galaxies. Actually, spiral galaxies are in the minority; there are many galaxies that have elliptical shapes and others that are irregular or abnormal in appearance. In 1925, Hubble set up a system of classification of

All novae in our galaxy reach the same maximum absolute magnitude.

Even as late as 1929, Hubble wrote that the fact that the visual magnitude at maximum of S Andromedae was 8.0 "places it at once in that mysterious class of exceptional novae which attain luminosities that are respectable fractions of the total luminosities of the systems in which they appear."

Figure 25–2 A galaxy of Hubble type E0, M87, NGC 4486, in the constellation Virgo. Globular clusters can be seen in the outer regions. This is actually a peculiar elliptical galaxy, and is a strong source of radio radiation.

Figure 25–3 M31, the Great Galaxy in Andromeda, a type Sb spiral, with its accompanying elliptical galaxies NGC 205 *(top left)* and M32 *(bottom right).*

galaxies that we shall discuss below; we normally describe a galaxy by its *Hubble type.*)

25.1a Elliptical Galaxies

Most of the galaxies of which we know are elliptical in shape (Fig. 25–2). The largest of these *elliptical galaxies* may contain 10^{13} solar masses and may be 10^5 parsecs across (approximately the diameter of our own galaxy); these *giant ellipticals* are rare. Much more common are *dwarf ellipticals*, which may contain "only" a few million solar masses and be only 2000 parsecs across.

Elliptical galaxies range from nearly circular in shape, which Hubble called *type E0*, to very elongated, which Hubble called *type E7*. Galaxies of various amounts of apparent oblateness (a measure of the difference between the longest and shortest diameters) are in between. The spiral Andromeda Galaxy, M31, shown in Figure 25–3 and in Color Plate 68, is accompanied by two elliptical companions. It is obvious on the photographs that the companions are much smaller than M31 itself.)

25.1b Spiral Galaxies

Although spiral galaxies, with arms unwinding smoothly from the central regions, are a minority of all the galaxies in the universe, they form a

Figure 25–4 Normal spiral galaxies.

NGC 1201   Type S0   NGC 2811   Type Sa   NGC 2841   Type Sb   NGC 628 M74   Type Sc

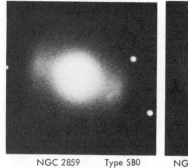
NGC 2859   Type SB0

NGC 175   Type SBab

NGC 1300   Type SBb(s)

NGC 1073   Type SBc

Figure 25–5 Barred spiral galaxies.

Figure 25–6 The Large Magellanic Cloud *(left)* and the Small Magellanic Cloud *(right)*, photographed with the new British 1.2-m Schmidt telescope at Siding Spring, Australia, as part of the current project to extend the National Geographic Society–Palomar Observatory Sky Survey to the southern hemisphere. The Royal Observatory, Edinburgh, Scotland, operates the telescope as a national facility, and the exposed plates from Siding Spring are sent to Edinburgh for analysis.

majority in certain particular groups of galaxies. Also, since they are brighter than the more abundant small ellipticals, we tend to see the spirals as dominant in a given volume of space, while really the fainter ellipticals may make up the majority.

Sometimes the arms are tightly wound around the nucleus; Hubble called this type S*a*, the S standing for "spiral" (Fig. 25–4). Categories of spirals with arms less and less tightly wound (that is, looser and looser) are called *types Sb* (Plates 71 and 72) and *Sc* (Plates 66 and 69). The nuclear bulge as seen from edge-on is less and less prominent as we go from Sa to Sc galaxies, as can be seen from the difference between the figures opening this chapter and those in this section. It has been found spectroscopically that galaxies rotate in the sense that the arms trail.

Spiral galaxies can be 10,000 to 30,000 parsecs across. They contain 10^9 to over 10^{11} solar masses; since most stars are of less than 1 solar mass this means that spirals contain over 10^9 to over 10^{11} stars—we think our own galaxy has 10^{11} (100 billion).

In about one third of the spirals, the arms unwind not from the nucleus but rather from a straight *bar* of stars, gas, and dust that extends to both sides of the nucleus (Fig. 25–5). These are again classified in the Hubble scheme from *a* to *c* in order of increasing openness of the arms, but with a *B* for "barred" inserted: *SBa*, *SBb*, and *SBc*. There is actually a complete range in the size of the bar from not visible to dominant in the appearance of a galaxy, so non-barred ("normal") spirals and barred spirals may not really be distinct types from each other.

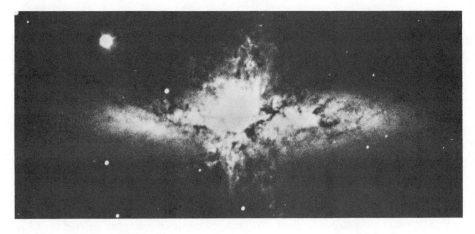

Figure 25–7 M82, a most unusual galaxy that is a powerful source of radio radiation. It was once thought to be exploding, but then gentler processes were thought to cause its form and nonthermal radiation. Now there is new evidence that it may be exploding after all.

25.1c Irregular Galaxies

A few per cent of the galaxies show no regularity, neither spiral nor elliptical. The Magellanic Clouds, for example, are irregular galaxies; they were shown in Figure 6–16 and enlargements appear in Figure 25–6. Sometimes traces of regularity—perhaps a bar—can be seen. Irregular galaxies are classified as *Irr.*

Irregular galaxies that can be resolved into nebulae, stars, and clusters are called *type Irr I.* Other galaxies appear amorphous and cannot be resolved into nebulae, stars, and clusters. They are called *type Irr II.*

25.1d Peculiar Galaxies

In some cases, as in M82 (Fig. 25–7), it appears at first as though an explosion has taken place in what might have been a regular galaxy. Another possibility, though, is that we are seeing light from the galaxy's nucleus scattered toward us by dust in the filaments. The ring galaxy shown in Figure 25–8A probably resulted from the passage of one galaxy through another. For other peculiar galaxies, such as those shown in Figure 25–8B, we have little idea of what might have gone on.

Peculiar galaxies are classified as the corresponding Hubble type followed by *(pec):* for example, *Sa (pec).*

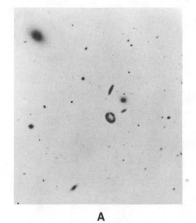

A   B

Figure 25–8 *(A)* A group of galaxies photographed at the prime focus of the Soviet 6-m telescope. One component of this group is a ring-like galaxy without a nucleus. It contains much gas and many hot stars. The nearby galaxy in the upper left corner, NGC 4513, has a ring of gas around it. The print is a negative to bring out the faint structure. *(B)* NGC 2685, a peculiar galaxy, type S0 (pec), in Ursa Major. It seems to be wrapped with helical filaments around a second axis of symmetry; the reasons for this are not understood.

25.1e The Hubble Classification

Hubble drew out his scheme of classification in a *tuning-fork diagram* (Fig. 25–9). The transition from ellipticals to spirals is represented by *type S0*, which resembles spirals in having the shape of a disk but does not have spiral arms.

It has since been shown, from optical observation and from studies of the 21-cm hydrogen line, that the amount of gas between the stars in galaxies depends on the type of galaxy. Elliptical galaxies have essentially no gas or dust between their stars, while spiral galaxies have interstellar gas and dust. The relative amount of gas increases from types Sa (or SBa) to Sc (or SBc). Irregular galaxies usually have even denser interstellar media than do spirals.

At first it was thought that the arrangement of galaxies in the Hubble classification, and the differing amounts of gas in different types, might indicate that one type of galaxy evolves into another, but that is no longer thought to be the case. We believe that the differences in gas content result from differing conditions at the time of formation of the galaxies, and may be the result of processes that led some of the galaxies to become spiral and others elliptical.

There is in any case a correlation between the gas content (and thus the Hubble type) and the spectral types of stars that we see in different galaxies. Only in the galaxies with a substantial gas and dust content—mostly the Sc, SBc, and Irr galaxies—are the O and B stars to be found. Since these stars have short lifetimes on a stellar scale, they must have been formed comparatively recently, within the last several million years. Elliptical galaxies contain only older, cooler stars; this difference from spirals can be detected even with systems of filters, as the elliptical galaxies appear redder than spirals. Nor do we find H II regions, which surround hot stars, in elliptical galaxies.

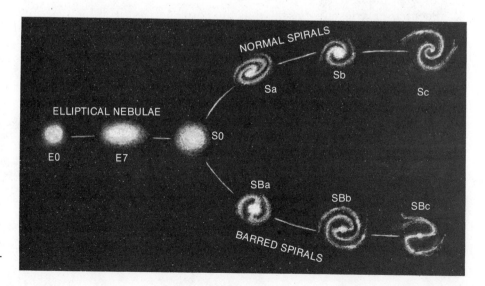

Figure 25–9 The Hubble classification of galaxies.

25.1f Seyfert and N Galaxies

In the years since Hubble's work, other types of galaxies have been found that, although they can be fit into the Hubble classification scheme, have certain characteristics that make them suitable for special mention.

For example, in 1943, Carl Seyfert discovered a type of galaxy with a nucleus that is small and especially bright with respect to the arms (Fig. 25–10). *Seyfert galaxies* are defined both by this trait and by the appearance of broad emission lines in their spectra. The presence of emission lines indicates that hot gas is present in the nucleus, perhaps heated by violent activity there. The fact that the lines are exceptionally broad in wavelength is interpreted to mean that there are rapid motions of mass in the cores of Seyfert galaxies, causing Doppler shifts. The discovery of x-rays and strong infrared emission from many Seyfert nuclei is consistent with the idea that the nuclei are sites of activity.

Two to five per cent of galaxies are Seyferts. We do not know whether this means that all galaxies spend 2 to 5 per cent of their lives in this stage, or whether a small percentage of galaxies spend most of their lives in this stage.

Another type of galaxy with a bright nucleus is called an *N galaxy*. The intensity of optical radiation from the nucleus of an N galaxy overwhelms the radiation from the rest of the galaxy.

The Seyfert galaxies and the idea that M82 appeared to be exploding were the first indications of the existence of *active galaxies*, galaxies with internal activity on a powerful scale. Now the existence of radio, infrared, and x-ray emission from a variety of galaxies has changed our preconceived idea that galaxies were quiet places. We shall see that Seyfert galaxies and N galaxies are coming under increasing consideration because of their possible relation to quasars.

Figure 25–10 The Seyfert galaxy NGC 4151, showing its bright nucleus.

The difference between Seyfert and N galaxies is partly based on their appearance and partly based on spectroscopic observations. The distinction is partly historical and is not always clear cut.

*25.2 THE ORIGIN OF GALACTIC STRUCTURE

The overall shapes of galaxies, ranging from spherical (as in an E0 elliptical) to disk-shaped with only a slight central bulge (as in an Sc spiral) can be understood as a consequence of the conservation of angular momentum. Presumably the elliptical galaxies were formed out of masses of gas that had only small overall angular momentum. For the spiral galaxies, on the other hand, the higher angular momenta provided a contribution from what we often call "centrifugal force" that tended to balance the inward force of gravity in one plane. Perpendicular to this plane, there was no such force acting to oppose gravity, and the gas collapsed into a disk.

25.2a Density Waves

We have discussed the density-wave explanation for spiral structure in Section 22.4c. We showed there that the pattern that we see as a spiral may be the result of increased density in a spiral form, resulting in increased compression of the gas and thus star formation. But the individual stars, once formed, usually orbit the galactic center at rates very different from that of the overall pattern.

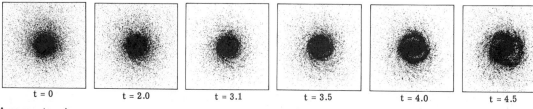

t = 0	t = 2.0	t = 3.1	t = 3.5	t = 4.0	t = 4.5

Figure 25–11 A computer simulation by Frank Hohl of NASA's Langley Research Center in Virginia, showing the evolution of spiral structure.

25.2b Gravitational Effects and Interacting Galaxies

Most astronomers believe that most spiral arms result from density waves. But other explanations must still be considered. For example, we must consider gravitational interactions, especially in certain cases.

In the three hundred years since Newton presented the basic laws of gravity, many solutions have been found to equations that show the gravitational workings of a system containing one large mass near its center and another, lesser mass in orbit around the larger mass. To describe this system, we need only study the gravitational interaction of two bodies; this is thus called the *two-body problem*. In our solar system, we understand the orbit of a planet as basically a two-body problem involving the planet and the sun. The effects that the planets have on each other are much less than the effect that the sun has on each, and the mutual interactions of the planets are treated as small deviations (called *perturbations*) from the situation that would be present if only the sun-planet two-body problem had to be solved.

Even a situation that is as easy to state as a three-body problem, where the mutual gravitational interactions of three bodies must be studied, has no known general solution. Only certain limited cases of the three-body problem can be solved exactly.

One way of treating complicated gravitational problems is to use a computer to simulate the evolution of a system over time. One can consider a system of any number of masses at an instant in time, have the computer calculate all the independent gravitational interactions between each pair of masses, and use this information to see what the system would look like a short time later. In this way, one can generate a series of "snapshots" of the system, and thus follow its evolution (Fig. 25–11).

For a few of the peculiar galaxies, evidence exists that gravity is the dominant force in forming "arms" (often called "tails"). The situation depends on the interaction of two galaxies, and current calculations indicate that such interactions are too rare to have caused the many spirals that we see. Nonetheless, in particular cases, the effects of gravitational interactions can probably be dominant (Fig. 25–12).

Two American astronomers, Juri Toomre and Alar Toomre, have used computer methods to follow the gravitational interaction between two galaxies that come very close together. In order to simplify the calculations, they made the assumption that the mass of each of the galaxies is concentrated at its center, and that the interactions of the individual pairs of particles they considered did not have to be taken into account. Figure 25–13 shows a time series they have calculated for the interaction of two massive galaxies. Long arms are drawn out by tidal forces; the original angular momentum of the galaxies contributes to the graceful curvature. At bottom

Figure 25–12 An interacting pair of galaxies, known as VV 33, whose components are linked with a thin bar. The photograph was obtained with the 6-m telescope. The print is a negative to bring out the faint structure.

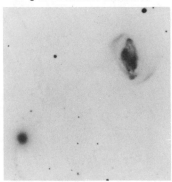

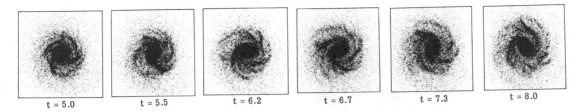

| t = 5.0 | t = 5.5 | t = 6.2 | t = 6.7 | t = 7.3 | t = 8.0 |

we see the appearance of the galaxies at the time when they have moved sufficiently far away that they no longer affect each other to further change their form. The result of the computer simulation is very similar in appearance to that of a pair of galaxies known as "the Antennae," shown in Figure 25–14.

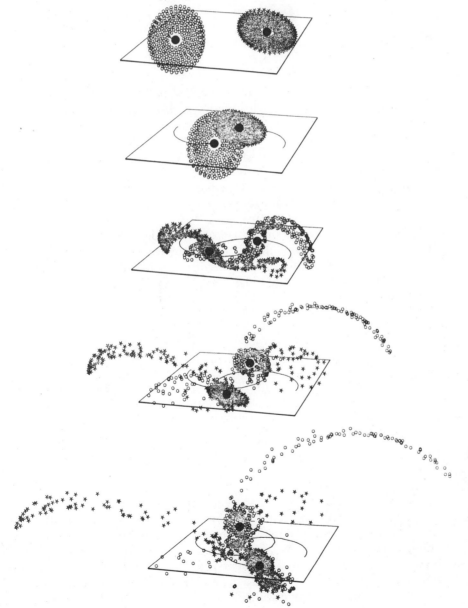

Figure 25–13 Drawings made on a computer by Alar and Juri Toomre, showing a time sequence of the very close encounter of two identical model galaxies. The large dot represents the center of each galaxy; the mass and central force of each galaxy is concentrated at its center for the purposes of this calculation. A disk of 350 particles is associated with each galaxy. The drawings are separated by an interval of 200 million years in the evolution of the galaxies.

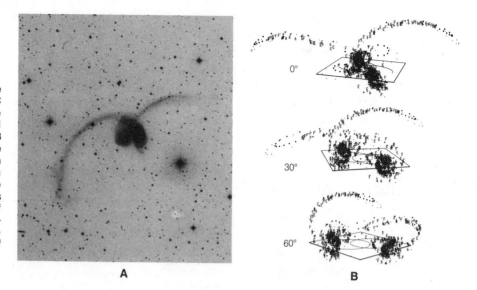

A   B

Figure 25–14 (A) A negative print of the pair of galaxies NGC 4038 and NGC 4039, known as the Antennae. (B) The numerical simulation shown in Fig. 25–13 closely resembles the Antennae when observed from a certain angle, as shown at top. From other angles, however, the resemblance is not obvious; this makes us realize that problems of the third-dimension and perspective limit our ability to recognize the true structure of galaxies.

Color Plate 67 for "The Mice," another example of a gravitational interaction of galaxies

Figure 25–14B shows the result we saw in Figure 25–13 from three different points of view. We are fortunate in being able to see this beautiful pair of galaxies broadside, for it is the prettiest view.

The above example involved two massive galaxies. But the interaction of two galaxies of very unequal mass can lead to the more massive one developing forms, at least in its outer parts, that resemble spiral structure. Note that the Milky Way has close galactic companions—the Magellanic Clouds—so it is possible that there was a gravitational contribution to the origin of the spiral arms of our own galaxy.

An alternate class of theories to explain the persistence of spiral structure involves the magnetic fields that are known to be present in galaxies. The field—which for our own galaxy is only one millionth the strength of the earth's magnetic field—is aligned along the spiral arms, as is known from observations of polarization of starlight. Magnetic and other theories cannot yet be ruled out as possible contributors to the spiral structure.

25.3 CLUSTERS OF GALAXIES

Clusters of galaxies are on a completely different scale from that of the galactic cluster, a type of star cluster.

Careful study of the positions of galaxies and their distances from us has revealed that most galaxies are part of groups or clusters. Groups have just a handful of members, while *clusters of galaxies* may have hundreds of members.

25.3a The Local Group

The two dozen or so galaxies nearest us form the *Local Group*. The Local Group contains a typical distribution of types of galaxies and extends over a volume 1 megaparsec in diameter. It contains three spiral galaxies, each 15 to 50 kiloparsecs across—the Milky Way, Andromeda, and M33. There are four irregular galaxies, each 3 to 10 kiloparsecs across, including the Large and Small Magellanic Clouds. At least a dozen other dwarf irregulars are known. The rest of the galaxies are ellipticals, including 4 regular ellipticals, each 2

Figure 25–15 The fuzzy, non-stellar objects Maffei I and Maffei II, photographed in the infrared by Hyron Spinrad of Berkeley. A bright star, with a ring around it because it is so overexposed, also appears. The objects are severely reddened because so much dust is present in their direction.

to 5 kiloparsecs across, two of which are the companions to the Andromeda Galaxy. The others are dwarf ellipticals, mostly less than 2 kiloparsecs across.

Previously unknown candidates for membership in the Local Group are occasionally found. Some of these newly discovered galaxies have been difficult to discover even though they are so close because they lie in the plane of our galaxy and are thus hidden from our view by dust (Fig. 25–15).

25.3b *Farther Clusters of Galaxies*

There are apparently other small groups of galaxies in the vicinity of the Local Group, each containing only a dozen or so members. The nearest cluster of many galaxies (a rich cluster as opposed to a poor cluster) can be observed in the constellation Virgo (the Virgin) and is called the Virgo Cluster. It covers a region in the sky over 6° in radius, 12 times greater than the angular diameter of the moon. The Virgo Cluster contains hundreds of galaxies of all types. It is about 2 million parsecs across, and is located about 20 million parsecs away from us.

Other rich clusters are known at greater distances, including the Coma Cluster in the constellation Coma Berenices (Berenice's Hair). The Coma Cluster has spherical symmetry and a concentration of galaxies toward its center, not unlike the distribution of stars in a globular cluster. Thus it is a *regular cluster* as opposed to an *irregular cluster*.

Rich clusters of galaxies are generally x-ray sources. There are at least two possibilities for the origin of the observed x-rays. First, it could be the combination of emission from a few discrete x-ray souces.* Second, it could be radiation from a very hot (10^8 K) gas filling intergalactic space.

There may be as many as 10,000 galaxies in a cluster, and the density of galaxies near the center of a rich cluster may be higher than that near the Milky Way by a factor of one thousand to one million. Thousands of clusters of galaxies are known.

One interesting question to ask is whether these clusters of galaxies are in turn grouped into clusters of clusters—*superclusters*. Current evidence indicates that this next higher order of clumping also seems to take place

Figure 25–16A The rich cluster of galaxies in the constellation of Hercules.

---

*Each source radiates by a mechanism technically known as "inverse Compton scattering," which can add energy to low-energy photons, converting them to x-rays. It does this through the interaction of the photons with electrons that are moving at relativistic velocities. This process appears to be very important to the understanding of x-ray astronomy.

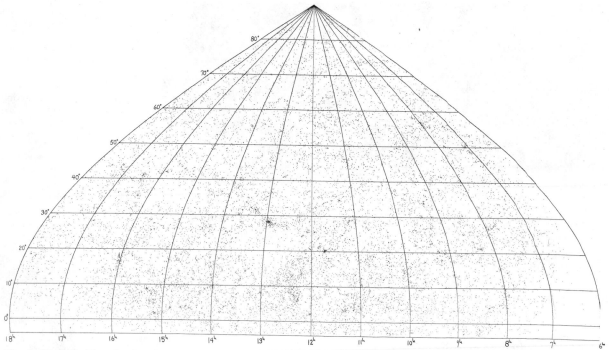

Figure 25–16B The distribution of the 20,000 galaxies brighter than magnitude 15.7 in the northern celestial hemisphere from the catalogue of Fritz Zwicky. The galaxies seem quite uniformly distributed aside from the obvious clusters. The Coma cluster appears near the center of the diagram, at 13ʰ and 30°. The cluster Abell 1367 appears to its lower right. These two clusters, along with four groups and several isolated galaxies, make up a supercluster. The Hercules supercluster appears at lower left, near 16ʰ and 18°. The Virgo cluster is so near that it is spread out over too much of the sky to be apparent. (Diagram by John Tonry, Marc Davis, and John Huchra, Center for Astrophysics)

(Fig. 25–16A). Stephen A. Gregory of Bowling Green State University and Laird Thompson of the University of Hawaii have concluded that every nearby very rich cluster is located in a supercluster (Fig. 25–16B). The Local Group, the several similar groupings nearby, and the Virgo Cluster form one such cluster of clusters, the *Local Supercluster*. This cluster of clusters contains 100 member clusters and is on the order of 100 megaparsecs across.

Does the clustering continue in scope? Are there clusters of clusters of clusters, and clusters of clusters of clusters of clusters, and so on in a hierarchical sequence? The evidence at present is that this is not so. It seems that clusters of clusters, such as the Local Supercluster, are the largest scale of inhomogeneity in the universe.

25.4 THE EXPANSION OF THE UNIVERSE

In the decade before the problem of the location of the "spiral nebulae" was settled, Vesto M. Slipher of the Lowell Observatory took many spectra that indicated that the spirals had large redshifts. This work was to lead to a profound generalization. In 1929 Hubble announced that galaxies in all directions are moving away from us and that there is a direct proportionality between the distance of a galaxy and its redshift. Hubble, in collaboration at Mt. Wilson with Milton L. Humason, went on during the 1930's to establish the relation more fully. It is known as *Hubble's law*. The redshift is presumably caused by the Doppler effect (which was described in Section 5.5), and the law is usually stated in terms of the velocity that corresponds to the measured wavelength rather than in terms of the redshift itself.

Hubble's law states that the velocity of recession of a galaxy is proportional to its distance. It is written

$$v = H_o d,$$

where v is the velocity, d is the distance, and H_o is the present-day value of the constant of proportionality, which is known as *Hubble's constant* (Figs. 25–17, 25–18, and 25–19).

The most generally accepted value of Hubble's constant, measured by Allan Sandage and Gustav Tammann at the Hale Observatories, is 50 km s^{-1} Mpc^{-1}. This is almost a factor of 10 lower than the value that Hubble originally announced, but Sandage and Tammann have been able to make use of new developments in finding the distance to distant galaxies to derive the improved value, incorporating as well earlier corrections to the distance scale. Some scientists, based on other observations, think that the value is somewhat larger. Values of 75 and 100 have their proponents. Let us use 50 for the rest of this book for convenience, as most astromers do.

Hubble's constant is given in units that appear strange, but it merely states that for each megaparsec (3.3 × 10^6 light years) of distance from the sun, the velocity increases by 50 km s^{-1} (thus the units are 50 km s^{-1} per Mpc). From Hubble's law, we see that a galaxy at 10 Mpc would have a redshift corresponding to 500 km s^{-1}; at 20 Mpc the redshift of a galaxy would correspond to 1000 km s^{-1}; and so on.

We use a subscript "o" on H_o in order to retain the letter H for describing how Hubble's constant might vary over time. Obviously, if Hubble's constant varies over time, it isn't really a constant.

The major import of Hubble's law is that objects in the universe in all directions are moving away from us; the universe is expanding. Since the time when Copernicus moved the earth out of the center of the universe (and the time when Shapley moved the earth and sun out of even the center of the Milky Way Galaxy), we have not liked to think that we could be at the center of the universe. Fortunately, Hubble's law can be accounted for without our having to be at any such favored location, as we see below.

Imagine a raisin cake (Fig. 25–20) about to go into the oven. The raisins are spaced a certain distance away from each other. Then, as the cake rises, the raisins spread apart from each other. If we were able to sit on one of those raisins, we would see our neighboring raisins move away from us at a certain speed. It is important to realize that raisins farther away from us would be moving away faster: not only would the distance from us to the neighboring raisin have increased but also the additional distance beyond the neighbor to the farther raisin would have increased. Thus no matter in what direction we looked, the raisins would be receding from us, with the velocity of recession proportional to the distance.

The next important point to realize is that it doesn't matter which raisin we sit on; all the other raisins would always seem to be receding. Of course, the raisin cake is finite in size and the universe may have no limit, but other than that, the analogy is exact. The fact that all the galaxies appear to be receding from us does not put us in a unique spot in the universe; there is no center to the universe. Each observer at each location would observe the same picture.

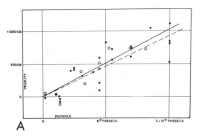

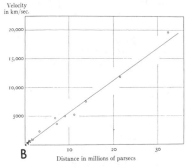

Figure 25–17 (*A*) Hubble's original diagram from 1929. Dots are individual galaxies; open circles are from groups of galaxies. The scatter to one side of the line or the other is substantial. (*B*), By 1931, Hubble and Humason had extended the measurements to greater distances, and Hubble's law was well established. All the points shown in the 1929 work appear bunched near the origin of this graph. v = H$_o$d represents a straight line of slope H$_o$. These graphs use older distance measurements than we now use, and so give different values for H$_o$ than we now derive.

Figure 25–18 Spectra are shown at right for the galaxies at left all reproduced to the same scale. Distances are based on Hubble's constant = 50 kms⁻¹ Mpc⁻¹. Notice how the farther away a galaxy is, the smaller it looks. The arrow below each horizontal streak of spectrum shows how far the H and k lines of ionized calcium are redshifted. The spectrum of an emission-line source located inside the telescope building appears as vertical lines above and below each galactic spectrum to provide a comparison with redshift known to be zero.

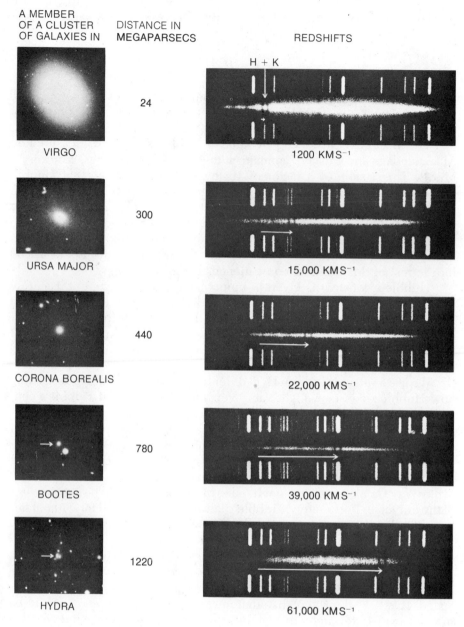

A MEMBER OF A CLUSTER OF GALAXIES IN

DISTANCE IN **MEGAPARSECS**

REDSHIFTS

H + K

VIRGO — 24 — 1200 KMS⁻¹

URSA MAJOR — 300 — 15,000 KMS⁻¹

CORONA BOREALIS — 440 — 22,000 KMS⁻¹

BOOTES — 780 — 39,000 KMS⁻¹

HYDRA — 1220 — 61,000 KMS⁻¹

Figure 25–19 The Hubble diagram for the galaxies shown in Fig. 25–18.

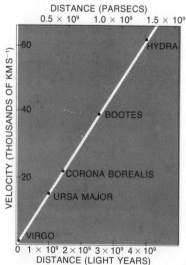

Note also that the raisin cake—and the universe—must be expanding with respect to something. We are assuming that we are still able to measure lengths on an unchanging scale. In our analogy, the size of the raisins themselves is not changing; only the separations are changing. (In the universe, the galaxies themselves and the clusters of galaxies are not expanding; only the distances between the clusters are increasing.)

Note also that individual stars in our galaxy can appear to have small redshifts or blueshifts, caused either by their peculiar velocities or by the differential galactic rotation. Also, even some of the nearer galaxies have random velocities of sufficient size, or velocities less than our rotational

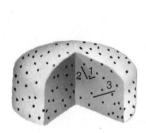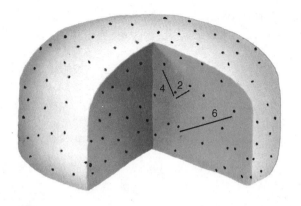

Figure 25–20 From every raisin in a raisin cake, every other raisin seems to be moving away from you at a speed that depends on its distance from you. This leads to a relation like the Hubble law between the velocity and the distance. Note also that each raisin would be at the center of the expansion measured from its own position, yet the cake is expanding uniformly. For a better analogy with the universe, consider an infinite cake; clearly, there is no center to its expansion.

velocity in our galaxy, so that they are approaching us. But except for these few nearby cases, all the galaxies are receding. The redshift we measure for a given galaxy is the same no matter in which part of the spectrum we observe. Optical and radio observations of redshifts are shown in Figure 25–21.

The major problem for setting the Hubble law on the firmest footing is finding the distances to the galaxies for which redshifts are measured. Only for the nearest galaxies can we detect Cepheid or RR Lyrae variable stars. Beyond those galaxies we use such methods as assuming that the magnitudes of supergiant stars or sizes of H II regions are more or less the same as they are in our own galaxy, and calculating, respectively, spectroscopic parallaxes or the distances that would make the observed angular dimensions of the H II regions correspond with the linear dimensions we know for those objects in the Milky Way. At still greater distances, we assume that the brightest member in a cluster of galaxies has the same absolute magnitude as all other brightest members of other clusters; sometimes to limit the difficulty that one or two members might be exceptionally bright, we rather consider the third-brightest member of a cluster. (The distribution of magnitudes of galaxies in clusters has also been considered, though this is much more difficult to do.)

Certainly the methods we have to use grow less and less precise as we get farther and farther away from the sun. In particular, we are seeing galaxies that emitted their light very long ago. A galaxy seen in that youthful stage may well have had a very different brightness than the nearby galaxies that we see. For example, there is some evidence that the brightest galaxies in clusters may be incorporating other galaxies and so growing especially bright; this is called *galactic cannibalism.*

Another uncertainty in observing very distant objects comes from the fact that we observe radiation that was emitted in the ultraviolet even though it is now redshifted into the visible. We do not yet know much about ultraviolet spectra of galaxies. We may thus be making a systematic error in assessing their distances. In any case, all the current evidence indicates that when we correct for such effects as best we can, the Hubble relation continues to be a straight-line proportionality deep into space.

Beyond a certain range, we can no longer independently measure distances, and our only method of assessing distance is application of the Hubble law to the observed redshifts. A handful of galaxies have thus far been detected with redshifts larger than 0.75. One of these, 3C 343.1, is

One radio galaxy that is not rotating contains a rapidly rotating cloud of ionized hydrogen, something that seems easiest to explain if the cloud were the remnant of another galaxy that was "eaten."

We discuss the redshift formula and its application in Section 26.2, since more quasars than galaxies have such high redshifts. The redshift of 3C 343.1 is so great, 75 per cent of the speed of light, that the formula used for the Doppler shift must be the version that takes the special theory of relativity into account.

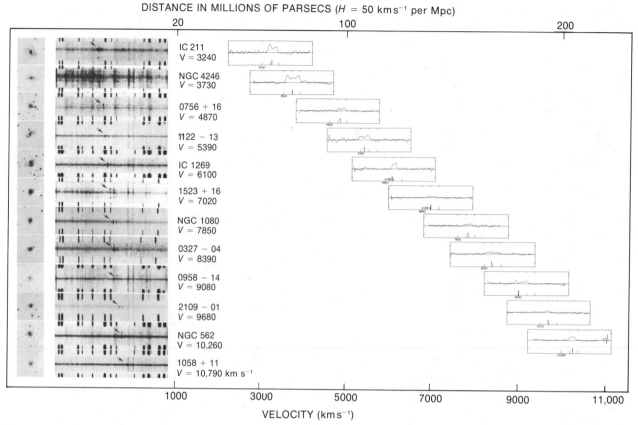

DISTANCE IN MILLIONS OF PARSECS ($H = 50$ km s^{-1} per Mpc)

VELOCITY (km s^{-1})

Figure 25–21 In addition to the optical spectra of galaxies of various distances, this figure also shows the 21-cm line observed in these galaxies. In each of these 21-cm spectra, the radial velocity of the galaxy is found from the Doppler shift of the center of the emission feature.

The time that it took the light to reach us is called the look-back time.

shown in Figure 25–22. Its redshift indicates that the object is over 8 billion light years away. Since the light has been travelling for 8 billion years to reach us, we are seeing the galaxy as it was 8 billion years ago. We say that the *look-back time* for this galaxy is 8 billion years. Studies of such distant galaxies might tell us how the overall properties of galaxies have evolved over time.

25.5 ACTIVE GALAXIES

We have seen how a radio eye has been opened on the universe, and how observations made in the radio part of the spectrum have helped us understand the structure of our own Milky Way and of objects such as supernova remnants in it. However, most of the objects that we detect in the radio sky (that is, locations from which more than the background level of radio radiation can be detected) turn out not to be located in our galaxy. The study of these *extragalactic radio sources* is a major subject of this section.

The core of our galaxy, the radio source we call Sagittarius A, is one of the strongest radio sources that we can observe in our galaxy. But if the Milky Way Galaxy were at the distance of other galaxies, its radio emission would be very weak.

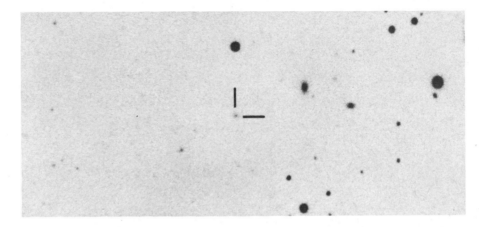

Figure 25–22 One of the farthest known galaxies, 3C 343.1, is barely observable even with the 4-m Kitt Peak telescope, as shown on this negative print. It has a redshift of 0.75, which means that the spectral lines that are observed are shifted to the red by 75 per cent of their original values. (The line of ionized oxygen that is observed in the laboratory at the ultraviolet wavelength 3727 Å, for example, appears in this object at a wavelength in the red of 6522 Å.) The size of the redshift indicates that the object is over 8 billion light years away.

We are able to detect a small amount of radio emission from many spiral galaxies. Among ellipticals, some seem to have quite strong radio emission while others at equal distances seem radio quiet. The radio radiation is presumably created by such processes as synchrotron radiation (which was described in Section 23.3), but the strength of the radiation is not exceptional and we shall not further deal with such "normal" radio galaxies.

Since the earliest work on radio astronomy, it has been clear that there is a class of galaxies whose members emit quite a lot of radio radiation, many orders of magnitude (that is, many powers of ten) more than "normal" radio galaxies. These have often been called "peculiar" radio galaxies; we shall limit the use of the term *radio galaxy* to mean these relatively powerful radio sources. They often appear optically as peculiar giant elliptical galaxies.

The strength of the emission from a radio galaxy does not follow Planck's law in having a smooth distribution of radiation diminishing in intensity on both sides of some wavelength at which there is a maximum of intensity. Since radiation that follows Planck's law is called *thermal*, the emission from a radio galaxy is *non-thermal*. Radio galaxies are usually also strong emitters of x-rays and of infrared radiation. By analogy with the term "radio galaxy," we speak of an *x-ray galaxy* for such a strong x-ray emitter.

25.5a Some Radio Galaxies of Renown

The first radio galaxy to be detected, Cygnus A, radiates about a million times more energy in the radio region of the spectrum than does the Milky Way Galaxy. A map of the source is shown in Figure 25–23, with an optical image superimposed. As has been known for decades, Cygnus A, and dozens of other radio galaxies, emit radio radiation mostly from two zones, called *lobes*, located far to either side of the optical object that is visible. The optical object—two fuzzy blobs—has been the subject of much analysis, but its makeup is still not understood. At first, it was thought to be two galaxies in collision, but that idea was discarded when it was realized that there would not be enough collisions of galaxies to explain the many similar radio sources that had by then been discovered. Perhaps Cygnus A is a single galactic nucleus in the process of splitting, or perhaps opaque gas is blocking our view of part of a single object.

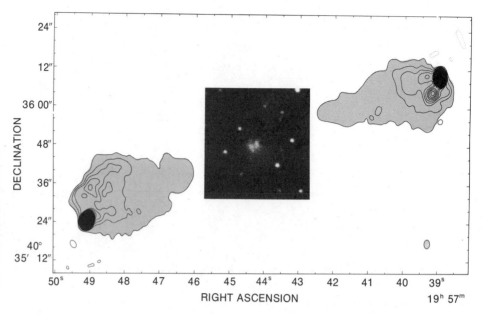

Figure 25–23 A radio map of Cygnus A, with shading and contours indicating the intensity of the radio emission. A photograph of the faint optical object or objects observable is superimposed at the proper scale. The beam size is shown by the ellipse at the lower right. An enlargement of the central region appears above. The radio map was made not with a single dish but with an interferometer.

Such *double-lobed structure* is typical of many radio galaxies. Some have a third, central location of radio emission as well. Others did not seem to have such emission from the center as well as emission from the two lobes. But recent work at higher frequencies than had been customarily used for observing indicates that galactic nuclei probably always or almost always join the lobes in being radio sources, though the amount of emission from the nuclei may be relatively faint at the lower frequencies at which most observations have been made. It has long been thought that perhaps the gas that is emitting radiation in the lobes was ejected from the nucleus. Perhaps the different relative strength of the nucleus compared with the strength of the lobes for different sources means that the sources are in different states of evolution.

Still other radio galaxies seem to emit radiation only from a single central volume, although that volume is usually greater in extent than the size of the optical image.

Often there are peculiarities in the optical images that correspond to radio sources of all kinds. For example, a small jet of gas can be seen on short exposures of M87 (see Fig. 25–30), which corresponds to the powerful radio galaxy Virgo A. Light from the jet is polarized, which confirms that the synchrotron process is at work here. M82, the galaxy that has the appearance of being in a state of explosion (Fig. 25–7), is also a radio galaxy.

Another strange optical image is shown in Figure 25–24 and Color Plate 70. This is the radio source Centaurus A, whose optical notation is NGC 5128. In this case, there are two different pairs of radio sources, a pair of lobes located outside the optical image and on each side of it, covering 5° to 8° of sky as seen from the earth, and a second pair of radio emitters located symmetrically to either side of the central dark band but close enough in to fall on the optical image, covering 7 arc min of sky. A weak radio source at the center of this galaxy varies over time by a factor of at least 4, and appears to be originating in this central "point source."

Observations of this galaxy with the 4-m reflector at Cerro Tololo in the southern hemisphere have revealed the presence of a faint jet of gas extend-

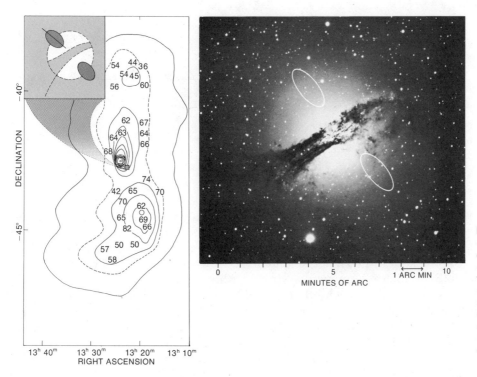

Figure 25–24 A radio map *(left)* of Centaurus A, with shading and contours indicating the intensity of the radio emission. In addition to the two large lobes, two smaller regions of radio emission are superimposed on the optical galaxy, NGC 5128, which is shown at right. The optical image is very peculiar, and resembles an elliptical galaxy with thick dust lanes wrapped around it.

ing over 40,000 parsecs from the center of the galaxy. This endorses the picture that an explosion took place, both expelling the jet and providing energy for the radio radiation.

*25.6 RADIO INTERFEROMETRY

25.6a Radio Interferometers

The resolution of single radio telescopes is very low, because of the long wavelength of radio radiation. Single radio telescopes may only be able to resolve structure a few minutes of arc or even a degree or so across, depending both on the wavelength of observation and on the size of the dish. For the past few years, the techniques of interferometry have been applied to radio astronomy, with the result that arrays of radio telescopes now in existence can map the sky with resolutions higher than the 1 arc sec or so that we can get with optical telescopes (Fig. 25–25). Let us first describe how these radio

Figure 25–25 The Andromeda Galaxy as it would appear at different resolutions.

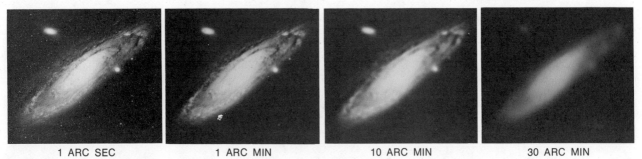

1 ARC SEC   1 ARC MIN   10 ARC MIN   30 ARC MIN

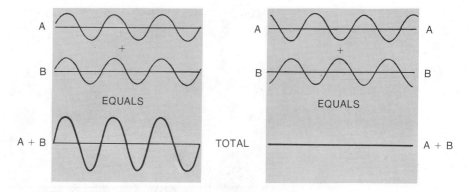

Figure 25–26 If electromagnetic waves—light or radio waves, for example—are *in phase* (i.e., oscillating in step), as at left, then the waves add in strength. If the waves are out of phase, as at right, then they can *interfere* with each other and cancel out. The phenomenon of interference cannot be understood if we think only in terms of particles (namely, photons).

interferometers work. Then we will discuss some of the high resolution results that have been obtained.

The wave structure of electromagnetic radiation manifests itself in the existence of *interference* phenomena, in which two beams of radiation—both of which are non-zero in strength (that is, have some intensity)—can be superimposed on each other to yield alternating bands, called *fringes,* of zero and non-zero radiation. For visible light, these fringes correspond to light and dark bands. The existence of interference cannot be understood when light is treated as a collection of particles (the photons), but can easily be understood on the basis of the wave theory (Fig. 25–26). (According to the laws of quantum mechanics, electromagnetic radiation has a dual nature in that it sometimes has properties of waves and sometimes has properties of photons.)

When we have a telescope—say the single dish of a radio telescope or the single mirror of an optical telescope—it receives radiation from a point in space as wavefront after wavefront of parallel radiation. The radiation is *coherent,* which means that all the waves are in step, with peaks arriving together, then troughs, then peaks, and so on.

The resolution of a single dish radio telescope depends on the diameter of the telescope (let us consider, for the moment, radiation of one frequency so that we do not have to discuss the variation of resolution with frequency). If, as we see in Figure 25–27, we could somehow retain only the outer zone of the dish, the resolution would remain the same (though the collecting area would be decreased, so we would have to collect the signal for a longer time to get the same intensity).

If we can maintain the coherence of the incoming radiation at two different dishes—that is, if we can maintain our knowledge of the relative arrival times of the wavefront at each of two dishes—we can retain the same

Figure 25–27 A large single mirror (*A*) can be thought of as a set of smaller mirrors (*B*). Since the resolution of a telescope for light of a certain wavelength depends only on the telescope's aperture, retaining only the outermost segments (*C*) matches the resolution of a full-aperture mirror. We can use a property of light called *interference* to analyze the incoming radiation. The device is then called an *interferometer.*

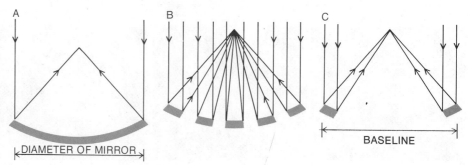

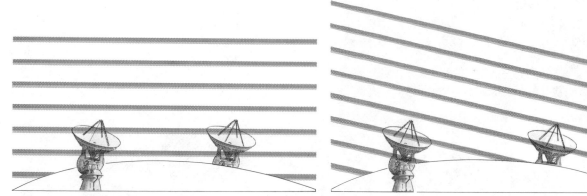

resolution as though we had one whole dish of the spacing of the two small dishes. Whereas for a single dish, we would call this maximum spacing of the two most distant points from which we can detect radiation by the name "diameter," for a two-dish interferometer, we call it the *baseline*.

The main problem in building an interferometer is to maintain knowledge of the coherence of radiation from one dish to the other, that is, the distance that a given wavefront has to travel after it hits one dish until it hits the other (Fig. 25–28). Since wavefronts always travel at the same speed—the speed of light—this is equivalent to measuring the time of arrival of the waves of radiation to a high degree of accuracy.

Since the delay in arrival time between the waves depends on the angular position of an object in the sky with respect to the baseline, by studying the time delay one can figure out angular information about the object.

Figure 25–28 In the left half of the figure, a given wave peak reaches both dishes simultaneously, so the amplitudes of the waves add. In the right half, the wave peak reaches one dish while a minimum of the wave reaches the other; the amplitudes subtract and zero total intensity results. Thus the interference phenomenon is set up, as long as we maintain a constant difference in the travel time of the electrical signals from the dishes where they are received to the control room where they are put together.

25.6b Very-Long-Baseline Interferometry

The breakthrough in timekeeping came with the invention of atomic clocks, which can keep time to an accuracy of one part in 10^{15} for the latest hydrogen masers (which is equivalent to a drift of three hundred-billionths of a second in a year). A time signal from an atomic clock can be recorded on a tape recorder on one tape band while the radio signal is put on an adjacent tape band. This signal can be compared at any later time with the signal from the other dish, synchronized accurately through comparison of the clock signals. The ability to have the time recorded so accurately freed radio astronomers of the need to have the dishes in direct contact with each other during the period of observation. Now all that is necessary is that the two telescopes observe the same object at the same period of time; the comparison of the signals can take place in a computer weeks later. With this ability, astronomers can make up an interferometer of two or more dishes very far apart, even thousands of kilometers. This technique is called *very-long-baseline interferometry (VLBI)*.

The maximum baseline that can be used in VLBI research is approximately that of the diameter of the earth (Fig. 25–29), 12,700 km. This corresponds at wavelengths of a few centimeters to resolutions as small as 0.0001 arc sec, far better than resolutions that can be gotten with optical telescopes. So radio astronomy has moved in recent years from a situation of providing inferior resolution to a situation of providing superior resolution.

Figure 25–29 VLBI techniques with a baseline that is the diameter of the earth allow radio sources to be studied with extremely high resolution.

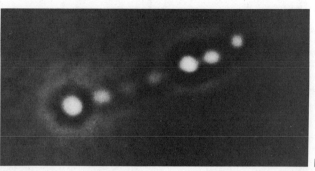

Figure 25–30 *(A)* The galaxy M87, which corresponds to the radio source Virgo A, as photographed with the Kitt Peak 4-m telescope. The inset shows a shorter exposure, on which a small jet of gas is visible. *(B)* Computer enhancement has been used on a photograph taken with the Palomar 5-m telescope to bring out details of the jet. North is at top and East is at left in all photographs.

Box 25.1 The Peculiar Galaxy M87

Also made possible by this extremely high resolution were observations of the nucleus of the radio galaxy M87 (Fig. 25–2), one of the brighter members of the Virgo Cluster. The galaxy turns out to have an odd optical appearance on short exposures (Fig. 25–30), for a jet of gas can be seen. The galaxy corresponds to the powerful radio galaxy Virgo A. Light from the jet is polarized, which confirms that the synchrotron process is at work here.

High resolution observations of the nucleus of M87 made with VLBI techniques have shown that it is only 0.001 arc sec across, which at even its distance of 50 million light years makes it a very small source in which to generate so much energy, which is emitted across the spectrum from x-rays to radio waves. Indeed, the galaxy gives off much more energy than do other galaxies of its type; the central region is as bright as 10^8 suns.

Optical studies of the motion of matter circling M87's nucleus have allowed estimates of the mass of the nucleus to be made, and indicate that 5 billion solar masses of matter are there. The stars are moving so fast that a huge mass must be present to hold them in. Other optical studies disclosed the presence of an extremely bright point of light in the center of the galaxy. The observations were made possible by new electronic detection equipment at the telescopes at Kitt Peak and Palomar. The higher resolution observations that will soon become possible from the Space Telescope should tell us more about this exotic source. At present, the observations are best explained by the presence of a giant black hole at M87's nucleus.

A further surprise from M87 was the discovery from radio observations that the galaxy is emitting rapid-fire radio pulses, each less than a thousandth of a second long. Each pulse contains 10,000 times as much energy as the sun radiates in a second and a billion times more energy than a pulse from an ordinary pulsar. This is consistent with the black-hole model, with a burst being emitted each time a hot spot on a source orbiting the central black hole came around. The observations were made possible by new electronic and computer techniques that were used with radio telescopes at the Dudley Observatory in Schenectady, New York, and at Arecibo.

VLBI techniques are difficult and time-consuming to apply. Also, when we work at high resolution, we sample only a small area of the sky at any time, and it therefore takes longer to study a region at high resolution than it does at low resolution. Therefore, VLBI techniques can be applied only to very small areas of sky. But for those few areas, chosen for their special interest, our knowledge of the structure of radio sources has been fantastically improved.

We have already mentioned (in Section 7.11) that VLBI techniques have provided an accurate measurement for the deflection of electromagnetic waves by the mass of the sun, providing a confirmation of Einstein's general theory of relativity. The discovery of the extremely small source at the nucleus of our galaxy (Section 22.2b) is also a VLBI result.

Other VLBI observations have not only revealed the presence of a few small components in far-off radio sources like quasars, but also have shown that in some cases these components are separating at angular velocities that seem to correspond (at the distances of these objects based on Hubble's law) to velocities greater than the speed of light. Since this result is impossible, other theoretical explanations of the data have been sought. One model is the *Christmas tree model*, or *marquee model* (as in a theater marquee), in which there are really several components flashing on and off. Thus when we see two components apparently separated by a very much greater distance than they were earlier, we would really be seeing two different components from the previous set of observations. The object need not be actually expanding at all.

Alternatively, when one applies the special theory of relativity, one can find situations where projection effects of rapidly moving objects can give the appearance of moving at speeds greater than the speed of light. If two objects are moving rapidly apart from each other, one toward and the other away from us, the light from the farther object takes longer to reach us and thus originated earlier than the light we receive at the same time from the nearer object. Under these circumstances, the movement at speeds greater than light could be only apparent and not actual. Evidence is accumulating that this latter model based on projection effects is the proper explanation.

25.6c   Aperture Synthesis Techniques

By suitably arranging a set of radio telescopes across a landscape, one can simultaneously make measurements over a variety of baselines, because each two telescopes in the set can be considered to be a pair with a different baseline from each other pair (Fig. 25–31). Thus with such an arrangement one can more rapidly put together a map of a radio source than one can from two-dish interferometers. Also, the existence of several dishes instead of just two means that there is that much more collecting area.

Note that the resolution of an interferometer depends on the baseline, so that at any one time the resolution is quite good along the line in which the telescopes lie but is only the resolution of a single dish in the perpendicular direction. Fortunately, we can take advantage of the fact that the earth's rotation changes the orientation of the telescopes with respect to the stars (Fig. 25–32). Thus by observing over a 12-hour period of time, one can improve the resolution in all directions.

In effect, we have synthesized the existence of a large telescope covering an elliptical area whose longest diameter is the same as the maximum

We speak of superluminal velocities, velocities faster than that of light.

Figure 25–31   The lines joining the several dishes making up an interferometer represent pairs of dishes that give results equivalent to having several baselines simultaneously. The same results could be gathered with only two dishes, one of which was moved around, but it would take longer. With the three dishes shown, one can simultaneously make measurements with three different baselines.

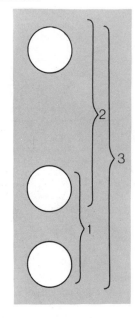

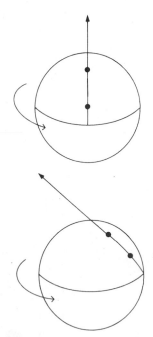

Figure 25-32 The earth's rotation carries around a straight line of dishes so that it maps out an ellipse in the sky. The figure shows the view of a line of dishes on the earth that we would see from the direction of the source being observed.

We have already seen pictures made with the VLA at various places in the book.

Figure 25-33 One of the first completed antennas of the VLA, being transported along the railroad track at the Socorro site.

Figure 25-34 The VLA, near Socorro, New Mexico.

separation of the outermost telescopes. This type of interferometric technique is known as *aperture synthesis,* and was pioneered by Martin Ryle in England. The interferometers that he built at Cambridge have been responsible for many high-resolution maps.

The most fantastic aperture synthesis radio telescope has been constructed in New Mexico by the National Radio Astronomy Observatory. It is composed of 27 dishes, each 26 m (85 ft) in diameter (Fig. 25–33), arranged in the shape of a "Y" over a flat area 27 km (17 mi) in diameter (Fig. 25–34). Because the array is in the shape of a "Y" rather than a straight line, it does not need to wait for the earth to rotate in order to synthesize the aperture. The "Y" is delineated by railroad tracks, on which the telescopes can be transported to 72 possible observing sites. The control room at the center of the "Y" contains a large computer to analyze all the signals. The system is prosaically called the *Very Large Array (VLA).*

The VLA can make pictures of a field of view a few minutes of arc across, with resolutions comparable to the 1 arc sec of optical observations from large telescopes, in about 10 hours. Though this resolution is lower than the resolution obtainable for a limited number of sources with VLBI techniques, the ability to make pictures in such a short time is invaluable. And we have a lot to learn at a resolution of 1 arc sec (actually 0.6 arc sec at an observing wavelength of 6 cm and 2.1 arc sec at 21 cm).

25.6d Aperture Synthesis Observations

At Westerbork in the Netherlands, twelve telescopes, each 25 m (82 ft) in diameter, are spaced over a 1.6 km (1 mi) baseline. Westerbork has discovered several giant double radio sources (Fig. 25–35), much larger than any of the double-lobed sources previously known. Some are hundreds of times larger than our own galaxy (Fig. 25–36). These are the largest single objects currently known in the universe.

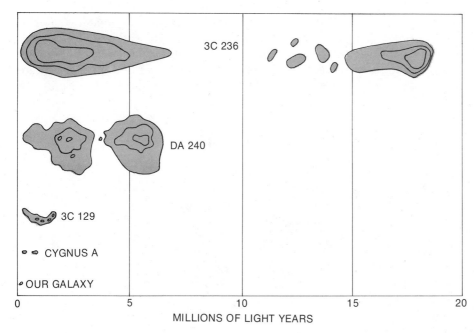

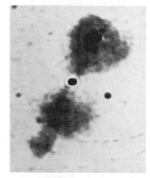

Figure 25–35 A Westerbork synthesis map of DA 240, a giant radio galaxy with two lobes of emission. It is 34 arc min across in our sky, larger than the full moon. Its central radio component coincides with a distant galaxy observed in visible light.

Interferometer observations have revealed the existence of a class of galaxies with tails extending out from one side. They are called *head-tail galaxies*, and resemble tadpoles in appearance (Figs. 25–37 and 25–38). As these galaxies move through intergalactic space, it is thought that they expel the clouds of gas that we see as the tails. High resolution observations of one such galaxy with the VLA (Fig. 25–37) shows that the source at the nucleus is less than 0.1 arc sec across, corresponding to a diameter at the distance of this galaxy to only 0.01 parsec. A narrow continuous stream of emission leads away from the nucleus and into the tail. Such observations are being used to understand both the galaxy itself and the intergalactic medium through which it is moving.

Figure 25–37 The head-tail galaxy NGC 1265 examined with the high resolution of the VLA. The resolution of the close-up at the right is about 1 arc sec. The size of the beam is shown by the ellipse at lower right. (Courtesy of F. N. Owen, J. O. Burns, and L. Rudnick, National Radio Astronomy Observatory)

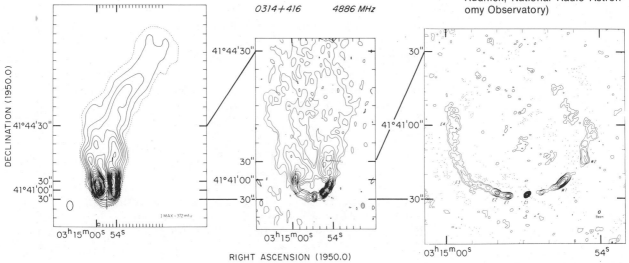

Figure 25-38 A Westerbork radio map of the head-tail radio source 3C 129, converted to a radio-photograph in which the brightness of the image corresponds to the intensity of radio emission at a wavelength of 21 cm (near but not at the hydrogen wavelength). The resolution is about 30 arc sec. The front end of the head of the radio galaxy corresponds to the position of a faint optical galaxy, a member of a rich cluster of galaxies.

The discrete blobs that we can see in the tails indicate that the galaxies give off puffs of ionized gas every few million years as they chug through intergalactic space. Perhaps by studying these puffs, we can learn about the main galaxies themselves as they were at earlier stages in their lives. Head-tail galaxies seem to be a common although hitherto unknown type; most are found in rich clusters of galaxies.

*25.7 THE FORMATION OF GALAXIES

The idea that galaxies are important building blocks of the universe was not proved until 1924, so the separation of the problem of how the galaxies were formed from the cosmogonical problem we discussed in Section 12.11 did not become necessary until then. Since that time and to the present, most astronomers have believed that galaxies originated through some sort of *gravitational instability*. In this theory, a fluctuation in density either developed or pre-existed in the gas from which the galaxy was to form. This fluctuation grew in mass and then collapsed and cooled until the galaxy was formed. A fluctuation containing a large amount of mass would form a cluster of galaxies. After the galaxy or galaxies were formed, stars and indeed planets could form in a similar manner.

When Hubble's law was discovered, the theoretical problem of galaxy formation became more difficult, for one had to predict what would happen to gravitational instabilities in an expanding universe. When the primeval black body radiation (which we discuss in Section 27.5) was discovered in 1965, its existence became one of the major considerations, for it was so strong long ago that it was the dominating factor.

One problem all along with the gravitational instability theories has been the fact that calculations seem to indicate that fluctuations would not grow in size sufficiently rapidly to have given rise to objects on the scale of galaxies or clusters of galaxies. And where did the fluctuations come from? Were they present during the initial instants after the formation of the universe, or did they arise as chance statistical variations in density? Further, why do galaxies spin? Where does their angular momentum come from? Though the gravitational instability theory remains dominant, the above difficulties have not been entirely overcome.

One idea gaining currency is that shock waves are important for starting the gravitational instability. We have already seen that there is evidence that a shock wave from a supernova explosion started the collapse of the nebula that became our solar system. Now Jeremiah Ostriker of Princeton has suggested that more extreme shock waves from massive supernovae started the collapse of the masses of gas that became galaxies.

He proposes that the core of a galaxy is formed in a rapid chain reaction of very large supernovae. The chain reaction would take place in only a million years or so, a brief time on an astronomical scale. Where several shock waves met, the effect would be enhanced. This might account for the fact that galaxies formed in clusters (Fig. 25–39). The discovery of metal atoms in intergalactic gas clouds provides backing for the supernova theory, because these heavy atoms were presumably formed in the supernovae.

In one sense, the gravitational instability theories hark back to the Aristotelian view that the universe was fundamentally simple, and that order

Figure 25–39 A cluster of galaxies in the constellation Coma Berenices. A close look will reveal that many of the small dots are not round and sharp, and are thus galaxies rather than stars.

cannot follow from basic disorder. The contrary view is that the universe was very complex at first and has evolved to its current relatively simple stage (assuming that the regularities we see when we study galaxies indicate a basic simplicity). The latest version of these alternative theories involves a fundamental *cosmic turbulence* that existed since the origin of the universe. At a certain stage in the expansion of the universe, amounts of mass suitable to become galaxies or clusters of galaxies would tend to separate out from the overall distribution of matter, carrying an intrinsic spin with them from the turbulence.

Such theories had fallen into disrepute until the last ten years because theoretical calculations had indicated that pre-existing turbulence would have disappeared in the early stages of the universe. But Leonid Ozernoi and his collaborators at the Lebedev Physics Institute in Moscow have found reasons why this need not have been so, and have elaborated on cosmic turbulence theories. With pre-existing turbulence, it is easier to understand why galaxies spin than it is on the basis of theories of gravitational instability. But it can be objected that we have merely changed the question from "where does the spin of the galaxies come from?" to "where does the turbulence come from?" without providing a fundamental answer.

Turbulence *in a material is a disturbed state with swirls and eddies in motion.*

Much theoretical work is currently being devoted to increasing our understanding of the dynamics of gases and to how the galaxies may have formed.

SUMMARY AND OUTLINE

Observations and catalogues of non-stellar objects
  Lord Rosse's early observations of spiral forms
  Messier's catalogue, General Catalogues by the Herschels, New General Catalogue (NGC) and Index Catalogues (IC) by Dreyer
  Galaxies as "island universes"
  Shapley-Curtis debate, 1920, dwelt on size of our own galaxy
  Incorrect measurements of proper motion and ignorance of the distinction between novae and supernovae led to the lack of knowledge of what galaxies are
  Matter was settled when Hubble observed Cepheids in galaxies
  Hubble classification (Section 25.1)
    Elliptical galaxies (E0–E7)
    Spiral galaxies (Sa–Sc) and barred spiral galaxies (SBa–SBc)
    Irregular galaxies (Irr I, Irr II)
    Peculiar galaxies (pec)
    Seyfert and N galaxies—bright cores
Origin of galactic structure (Section 25.2)
  Density waves lead to spiral appearance, though the actual distribution of matter may not be spiral
  Tidal effects create tails in interacting galaxies
Clusters of galaxies (Section 25.3)
  Local Group
  Rich clusters, such as Virgo and Coma, are x-ray sources
  Local Supercluster exists

The universe is expanding (Section 25.4)
  Hubble's law, $v = H_o d$, expresses how the velocity of expansion increases with increasing distance
  Best current value for Hubble's constant is $H_o = 50$ km s^{-1} Mpc^{-1}
  The expansion is universal, and has no center
Active galaxies (Section 25.5)
  Some objects are powerful sources of non-thermal radiation in the radio, x-ray, and infrared spectral regions
  Double-lobed shape is typical of radio galaxies; sometimes a peculiar optical object is present at the center
Radio interferometry (Section 25.6)
  Single-dish radio telescopes have low resolution; interferometers give resolution as high or higher than optical observations
  VLBI (very-long-baseline interferometry) uses widely separated dishes to provide the highest resolution now possible
  Aperture synthesis arrays provide quicker maps, still retain high resolution
  Current interferometers at Cambridge in England, Westerbork in the Netherlands, and VLA (Very Large Array) in U.S.
  Giant radio galaxies and head-tail galaxies studied
Galaxy formation (Section 25.7)
  Gravitational instability is leading model
  Some astronomers believe in pre-existing cosmic turbulence

QUESTIONS

1. Shapley and Curtis both argued on the basis of incomplete knowledge. In view of what we now know, evaluate the points that they made.

2. The sense of rotation of galaxies is determined spectroscopically. How is this done?

3. Which classes of galaxies are the most likely to have new stars forming? What evidence supports this?

4. What makes us think that the nuclei of Seyfert galaxies are very active?

5. How is it possible that galaxies could exist close to our own and not have been discovered before?

6. To measure the Hubble constant, you must have a means (other than the redshift) to determine the distances to galaxies. What are two methods that are used?

Color Plate 60: The North America Nebula, NGC 7000, in Cygnus. Absorbing dust is silhouetted against glowing gas to give the shape of the North American continent. (Palomar Observatory, California Institute of Technology photo with the 1.2-m Schmidt camera)

Color Plate 61 (top): M16, the Eagle Nebula in the constellation Serpens. Hydrogen radiation makes it appear red. The bright stars at the upper right are hot and young, and are part of a galactic cluster. The small, dark regions may be protostars. (Palomar Observatory, California Institute of Technology photo with the 5-m telescope)

Color Plate 62 (bottom): M17, the Omega Nebula in Sagittarius. (Palomar Observatory, California Institute of Technology photo with the 1.2-m Schmidt camera)

Color Plate 63: The Trifid Nebula, M20, in Sagittarius is glowing gas divided into three visible parts by absorbing lanes. The blue nebula at the top is unconnected to the Trifid. (Palomar Observatory, California Institute of Technology photo with the 5-m telescope)

Color Plate 64 (top): The Lagoon Nebula, M8, in Sagittarius. Red hydrogen light is emitted by gas that is excited by the radiation of very hot stars buried within the nebula; dark filaments of material within the cloud emit strong infrared radiation. Also, several peculiar variable stars in this nebula occasionally flare up. The Lagoon Nebula is about 60 light years across and is located about 6500 light years away from us. (Kitt Peak National Observatory photo with the 4-m telescope)

Color Plate 65 (bottom): The Large Magellanic Cloud, a satellite galaxy of our own Milky Way Galaxy. It is best visible from the southern hemisphere. (Photo by Hans Vehrenberg)

Color Plate 66 (top): The Whirlpool Galaxy, M51, in Canes Venatici, a type Sc spiral galaxy. At the end of one of its arms, a companion galaxy, NGC 5195, appears. (U.S. Naval Observatory photo)

Color Plate 67 (bottom): "The Mice," two interacting galaxies (NGC 4676 A and B). The original photograph was computer-enhanced with a special system at the Kitt Peak National Observatory. The computer system provided the colors in order to bring out the details. Tidal forces, caused by gravity, have distorted the shapes of the galaxies.

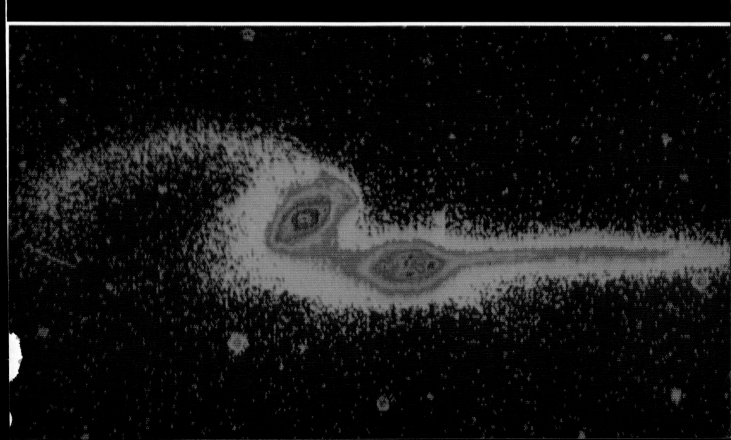

Color Plate 68: The Andromeda Galaxy, also known as M31 and NGC 224, the nearest spiral galaxy to the Milky Way. It is accompanied by two elliptical galaxies, NGC 205 (below) and M32 (above). M31 is Hubble type Sb. (Palomar Observatory, California Institute of Technology photo with the 1.2-m Schmidt camera)

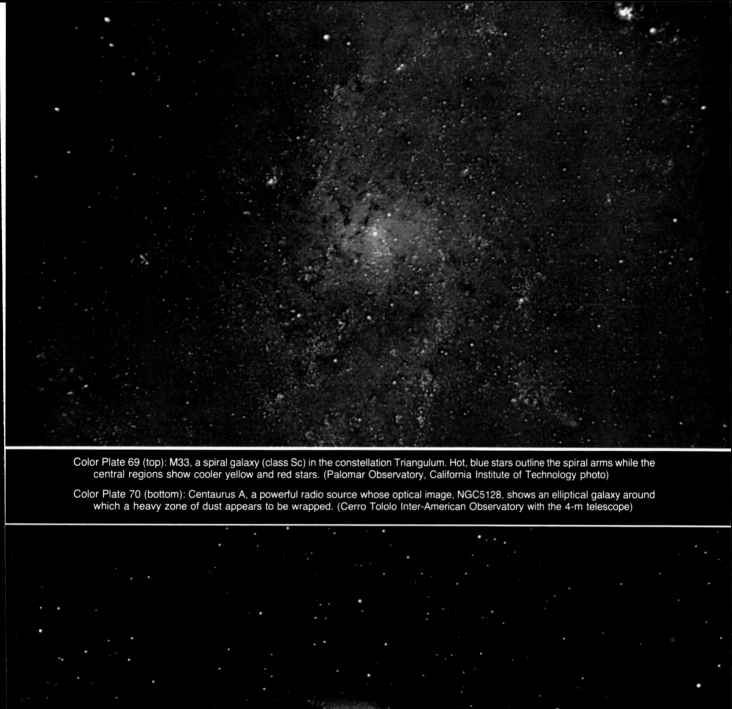

Color Plate 69 (top): M33, a spiral galaxy (class Sc) in the constellation Triangulum. Hot, blue stars outline the spiral arms while the central regions show cooler yellow and red stars. (Palomar Observatory, California Institute of Technology photo)

Color Plate 70 (bottom): Centaurus A, a powerful radio source whose optical image, NGC5128, shows an elliptical galaxy around which a heavy zone of dust appears to be wrapped. (Cerro Tololo Inter-American Observatory with the 4-m telescope)

Color Plate 71: NGC 7331, an Sb spiral galaxy in Pegasus. This galaxy may be linked with the group of galaxies known as Stephan's Quintet. (Palomar Observatory, California Institute of Technoloty photo with the 5-m telescope)

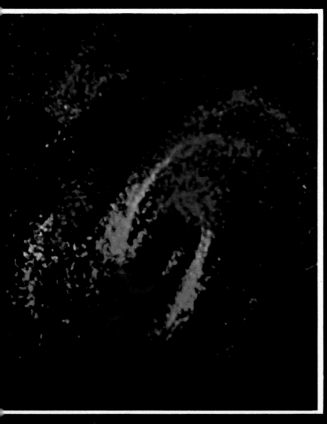

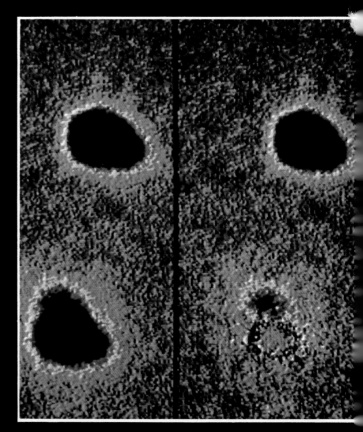

Color Plate 72: An image of 21-cm radiation from the Sb galaxy M81 in Ursa Major. Brightness represents intensity, color represents the Doppler shift: red = recession, violet = approach, green = no shift. (Westerbork data from A. H. Rots and W.W. Shane; imaged at NRAO)

Color Plate 73: The double quasar. At left we see a false-color view of the two images, which are separated by only 6 seconds of arc. At right bottom, the top image has been subtracted from the bottom image. The remainder is the galaxy that is acting as a gravitational lens. (Courtesy of Alan Stockton, Institute for Astronomy, University of Hawaii)

7. Does Alpha Centauri, the nearest set of stars to us, show a redshift that follows Hubble's law? Explain.

8. What are three pieces of evidence that indicate that M87 may have a giant black hole in it?

†9. (a) At what velocity in km s^{-1} is a galaxy 5 million parsecs away receding from us? (b) Express this velocity in mi s^{-1} and mi hr^{-1}.

†10. How far away (in km) is a galaxy with a redshift of 0.2?

---

†This indicates a question requiring a numerical solution.

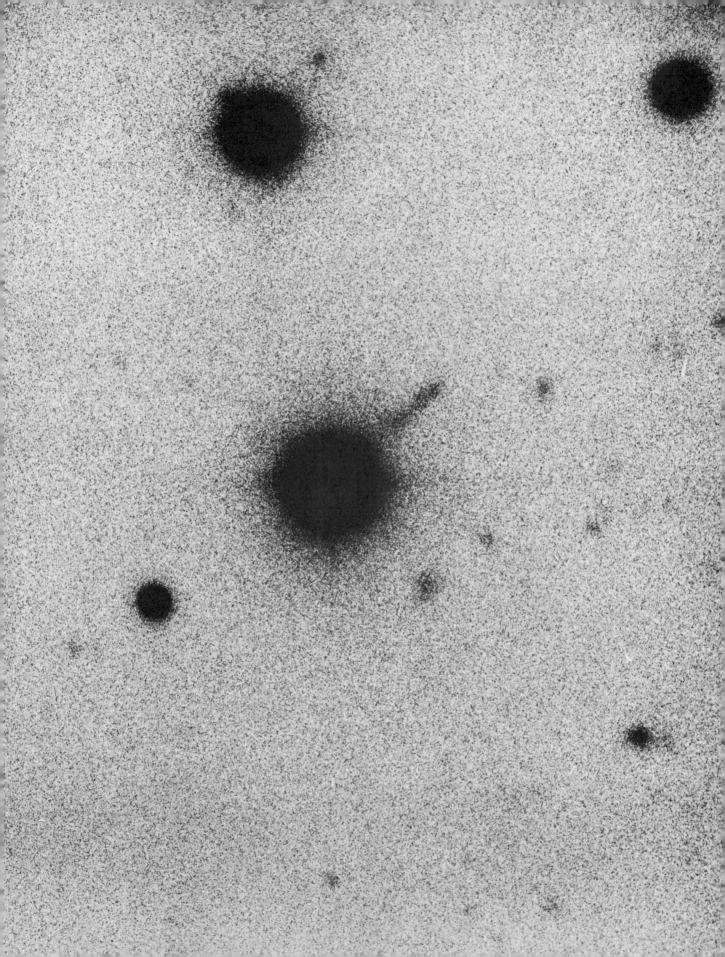

CHAPTER 26

══════════QUASARS══

AIMS:
To describe the discovery of quasars
and their significance as probes of the
early stages of the universe

(Quasars are enigmatic objects that, like stars, appear as points of light in the sky. But unlike the stars that we see, the quasars occur in the farthest reaches of the universe and thus may be the key to our understanding the history and structure of space. We study quasars in both the optical and radio parts of the spectrum. Although they are relatively faint in visible light, they are among the strongest radio sources in our sky, and therefore must be prodigious radiators of energy. Ordinary methods of generating energy are not sufficient. Consequently, the quasars raise fundamental questions about our basic physical laws and how energy is generated in the universe.)

(The word *quasar* originated from QSR, a contraction of "quasi-stellar radio source." However, the most important characteristic of quasars is not that they are emitting radio radiation but that they are travelling away from earth at tremendous speeds. After quasars were named, radio-quiet "quasi-stellar objects" were discovered and are also called quasars. Astronomers can deduce distances by observing the quasar spectra, measuring the Doppler shifts, and applying Hubble's law (Section 25.4). Since the quasars have the largest redshifts known, Hubble's law tells us that they are the most distant objects that we see.)

Most astronomers accept the idea that quasars are the objects most distant from us, but a small minority contends that they are in fact relatively close to us. The reasoning that leads to these opposite conclusions will be discussed later in this chapter. Even if the quasars are relatively near, which

An enlarged portion of a negative of the quasi-stellar radio source 3C 273, taken with the 5-m Hale telescope. The object, the largest of the black disks, looks like any 13th magnitude star, except for the faint narrow jet that is visible out to about 20 seconds of arc, equivalent to 50,000 parsecs, from the quasi-stellar object.

seems more and more unlikely as new evidence is discovered, they would still be important to astronomy. They would then show that our present methods of measuring distances to faraway objects are not always valid. Whether nearby or fantastically distant, quasars have strange properties that raise basic questions.

26.1 THE DISCOVERY OF QUASARS

Accurate positions can be measured for radio sources either by interferometry or by lunar occultation.

Quasars are a discovery resulting from the interaction of optical astronomy and radio astronomy. When maps of the radio sky turned out to be very different in appearance from maps of the optical sky, many astronomers set out to correlate the radio objects with visible ones. The sun was easy to identify, and some of the brighter galaxies also proved to be radio sources (as discussed in Section 25.5).

Single-dish radio telescopes did not give sufficiently accurate positions of objects in the sky to allow identifications to be made, so interferometers (Section 25.6a) had to be used. Most of the radio objects catalogued could be identified with optical objects, but a few had no clear identifications. Allan Sandage of the Mt. Wilson and Palomar Observatories and Thomas Matthews of Caltech's Owens Valley Radio Observatory (Matthews is now at the University of Maryland) used the Owens Valley interferometer to improve the accuracy of the positions and the 5-meter telescope to photograph the suspected regions. At least one of the strong radio sources, 3C 48 (the 48th source in the 3rd Cambridge catalogue), seemed to be suspiciously near a faint (16th magnitude) bluish star (Fig. 26–1), and Sandage and Matthews reported this identification of a "radio star." At that time, 1960, no stars had been found to emit radio waves, with the sole exception of the sun, whose weak radio radiation we can detect only because of the sun's proximity to the earth.

Figure 26–1 The first quasar, 3C 48, photographed with the 5-m Hale telescope.

In Australia, a large radio telescope was used to observe the passage of the moon across the position in the sky of another bright radio source, 3C 273. We can take advantage of the fact that we know the position of the moon in the sky very accurately. When it occults—hides—a radio souce, then we know that at the moment the signal strength decreases, the source must have passed behind the advancing limb of the moon. We don't know at which point along the limb, though. Later on, the instant the radio source emerges, the position of the lunar limb marks another set of possible positions. Where these two sets of possible positions overlap is the actual position of the source. Three lunar occultations of 3C 273 occurred within a few months, and from the data, a very accurate position for the source and a map of its structure was derived. The optical object, which is 13th magnitude, is not completely starlike in appearance, for a luminous jet appears to be connected to the point nucleus (as shown in the photograph opening this chapter).

There seem to be two components of the radio emission from 3C 273. One coincides with the jet, and the other coincides with the bluish stellar object, clinching the identification as a quasar. Maarten Schmidt of Mt. Wilson and Palomar photographed the spectrum of this "quasi-stellar radio source" with the 5-m Hale telescope. The spectra of 3C 273 and 3C 48, both bluish quasi-stellar objects, showed emission lines, but the lines did not agree in wavelength with the spectral lines of any of the elements. The lines

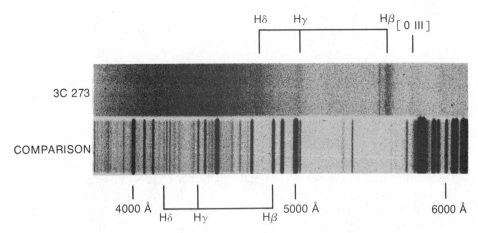

Figure 26–2 Spectrum of the quasar 3C 273. The lower spectrum consists of hydrogen and helium lines; it serves to establish the scale of wavelength. The upper part is the spectrum of the quasar, an object of 13th magnitude. The Balmer lines Hβ, Hγ, and Hδ in the quasar spectrum are at longer wavelengths than in the comparison spectrum. The redshift of 16 per cent corresponds, according to Hubble's law, to a distance of three billion light years. Note that the comparison spectrum represents hydrogen and helium sources on earth (more particularly, located inside the Palomar dome).

had the general appearance of spectral lines emitted by a gas of medium temperature, though.

The breakthrough came in 1963. At that time, Schmidt noted that the spectral lines of 3C 273 (Fig. 26–2) seemed to have the same pattern as lines of hydrogen under normal terrestrial conditions. Schmidt then made a major scientific discovery: he asked himself whether he could simply be observing a hydrogen spectrum that had been greatly shifted in wavelength by the Doppler effect. The Doppler shift required would be huge: each wavelength would have to be shifted by 16 per cent toward the red to account for the spectrum of 3C 273. This would mean that 3C 273 is receding from us at approximately 16 per cent of the speed of light. Immediately, Schmidt's colleague Jesse Greenstein recognized that the spectrum of 3C 48, another of the bluish stellar objects associated with radio sources, could be similarly explained. All the lines in the spectrum of 3C 48 were shifted by 37 per cent, a still more astounding redshift.

(Later, absorption lines were discovered in quasars as well as the emission lines just discussed. Many quasars have several systems of emission and absorption lines of differing redshifts. Some of these lines may be formed in clouds of gas of differing velocities surrounding the quasars.)

(Over 1500 quasars have been discovered, with redshifts ranging up to over 350 per cent. Candidate objects that have a high probability of being quasars can be found by looking for star-like objects that seem unusually strong in the blue and ultraviolet (Fig. 26–3). But one must take spectra to prove that they are quasars, a time-consuming procedure on such faint objects. Astronomers estimate that there have been perhaps a million quasars in the universe. By now, however, most of them have probably lived out their lifetimes; that is, they have given off so much energy that they are no longer quasars. Perhaps about 35,000 quasars now exist.)

E. Margaret Burbidge of the University of California at La Jolla and Roger Lynds of Kitt Peak, along with Schmidt, have measured many redshifts, turning suspected quasar identifications into known quasars.

26.2 THE REDSHIFT IN QUASARS

(It seems most reasonable that the redshift arises from the Doppler effect. Most objects in the universe show some Doppler shift with respect to the earth. For velocities that are small compared to the velocity of light, the amount of shift in the spectrum is written simply, for rest wavelength λ, velocity v, and speed of light c, as

Figure 26–3 This object appears stellar but is known to be not a star but a quasar.

$$\frac{\Delta\lambda}{\lambda} = \frac{v}{c} \qquad \text{(as discussed in Section 5.5).}$$

Astronomers often use the symbol z to stand for $\Delta\lambda/\lambda$, the amount of the redshift.

26.2a Doppler Shifts of Stars and Quasars

The stars in our galaxy have redshifts and blueshifts; these Doppler shifts are small (they occur from random motions of stars or from organized motions around the center of our galaxy) but are large enough for us to measure. The spectrum of the star that has the largest radial velocity with respect to the earth has a Doppler shift of two tenths of one per cent, that is z = 0.2 per cent = 0.002. Thus this star is moving at two tenths of one per cent of the speed of light. Unlike stars, which have slight redshifts and blueshifts, all galaxies except the nearest are redshifted with respect to the sun.

The Doppler shifts of quasars are much greater than those of most galaxies. Even for 3C 273, the brightest quasar, z = 0.16 (read "a redshift of 16 per cent"). Thus Hubble's law implies that 3C 273 is as far away from us as are distant galaxies. Only a handful of galaxies are known to have greater redshifts.

The only object in our galaxy with a Doppler shift of this magnitude is SS 433 (Section 10.14).

*26.2b Computing a Doppler Shift

Let us consider a redshift of 0.2, to pick a round number, and calculate its effect on a spectrum (Fig. 26–4). z = 0.2 implies that v/c = 0.2, and therefore

$$v = 0.2c = 0.2 \times 3 \times 10^{10} \text{ km s}^{-1} = 6 \times 10^4 \text{ km s}^{-1}.$$

If a spectral line were emitted at 4000 Å in the quasar, at what wavelength would we record it on our photographic plate on earth? (Let us use round numbers.)

$\Delta\lambda/\lambda = \Delta\lambda/4000 = 0.2$. Therefore $\Delta\lambda = 0.2 \times 4000$ Å = 800 Å.

Note that we have only calculated $\Delta\lambda$, the shift in wavelength. The new wavelength is equal to the old wavelength plus the shift in wavelength,

$$\lambda_{\text{new}} = \lambda_{\text{original}} + \Delta\lambda = 4000 \text{ Å} + 800 \text{ Å} = 4800 \text{ Å}.$$

Thus the line that was emitted at 4000 Å in the quasar was recorded at 4800 Å on earth.

Similarly, a line that was emitted at 5000 Å is also shifted by 0.2, and 0.2 of 5000 Å is 1000 Å. $\lambda + \Delta\lambda = 5000 + 1000 = 6000$ Å. The spectrum is thus not merely displaced by a constant number of angstroms, but is stretched more and more toward the higher wavelengths.

Figure 26–4 The wavelengths of the visible part of the spectrum are shown at top, and the wavelengths at which spectral lines would appear after undergoing a redshift of 0.2 ($\Delta\lambda/\lambda$ = 0.2) is at bottom. The lines are shifted by 0.2 times their original wavelengths, and wind up at 1.2 times their original wavelengths. The non-relativistic formula $\Delta\lambda/\lambda = v/c$ can be applied to give the approximate velocity; at still smaller redshifts this formula is even more accurate.

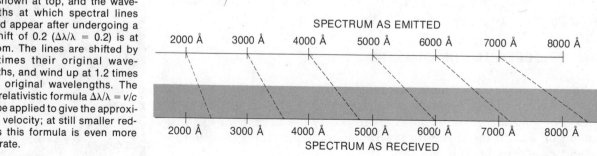

SPECTRUM AS EMITTED
2000 Å   3000 Å   4000 Å   5000 Å   6000 Å   7000 Å   8000 Å

2000 Å   3000 Å   4000 Å   5000 Å   6000 Å   7000 Å   8000 Å
SPECTRUM AS RECEIVED

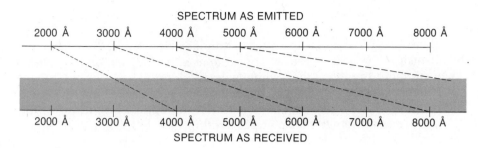

SPECTRUM AS EMITTED

SPECTRUM AS RECEIVED

Figure 26–5 For a redshift of 1 ($\Delta\lambda/\lambda = 1$), as shown here, the lines are shifted *(bottom)* by an amount equal to their original wavelengths, and wind up at twice their original wavelengths. The velocity is a significant fraction of the speed of light, and the simple non-relativistic formula for the Doppler shift cannot be applied.

*26.2c Relativistic Doppler Shifts

This simple Doppler formula above is valid only for velocities much less than c, the speed of light. For speeds closer to the speed of light, we must use a formula from the special theory of relativity:

$$\frac{\Delta\lambda}{\lambda} = \sqrt{\frac{1 + v/c}{1 - v/c}} - 1.$$

Positive values of v correspond to receding objects. When v is much less than c, then the relativistic formula approximates the non-relativistic formula. But when v gets very close to c, $\Delta\lambda/\lambda$ gets greater than 1 even though v is still less than c. We still use the letter z to stand for $\Delta\lambda/\lambda$.

For $z = 1$, we merely have the shift, $\Delta\lambda$, equaling the original wavelength, λ (Fig. 26–5). The quasar with the largest known redshift, $z = 3.53$, has the wavelengths of its lines shifted by 353 per cent, as was mentioned previously. This makes the new wavelengths $3.53 + 1 = 4.53$ times the original wavelengths. The velocities of recession of all quasars are still less than the speed of light, as the special theory of relativity tells us they must always be.

26.3 THE IMPORTANCE OF QUASARS

If we accept that the redshifts of quasars are caused by the Doppler effect, and we apply Hubble's law, we realize that the quasars are the farthest known objects in the universe. At the moment, we have no other method of measuring distances to objects so far away, and so it is important to test Hubble's law. Any deviations from the law would no doubt show up in the farthest objects, namely the quasars.

If the quasars were found not to satisfy Hubble's law, moreover, then doubt would be cast on all distances derived by Hubble's law. In that case, we would never know whether we were observing an object that satisfied the law or one that did not.

If we accept the quasars as the farthest objects, then they are billions of light years away from us and their light has taken billions of years to reach us. Thus we are looking back in time when we observe the quasars, and we hope that they will help us understand the early phases of our universe. For example, a detailed survey of the entire sky to look for quasars has turned up many new examples, and has shown that the number of quasars per volume of space increases as you go outward. Thus the number of quasars in the universe at earlier times was greater. Among other things, this shows that the universe has evolved.

26.4 THE ENERGY PROBLEM

If the quasars are as far away as this orthodox view holds, then they must be intrinsically very luminous to appear to us at their observed intensities. They are more luminous than whole galaxies.

When quasars were discovered, scientists went back to the historical files of photographic plates. They measured the optical brightness of the objects on many photographic plates, extending back to 1887 for 3C 273, and found that the brightness varied on a time scale of weeks or months.

If you want to pack as much energy into a quasar as they seem to radiate, you must have a certain amount of space in which to pack it. A quasar cannot be too small and still contain the 100 million solar masses that many astronomers think is necessary to generate that much energy. Also, if something is, say, a tenth of a light year across, you would expect that it could not vary in brightness in less than a tenth of a year. After all, one side cannot signal to the other side, so to speak, to join in the variation (Fig. 26–6). Somehow the whole object has to know at the same time to do something, and that ability is limited by the speed of light. So the variations in intensity mean that the quasars are probably fairly small.

This can be verified to a certain extent with studies using very-long-baseline-interferometry techniques, which can observe on a scale of 1/1000 arc sec, and now with the VLA, the Very Large Array. Many quasars, though their optical images appear quasi-stellar, are typical double-lobed radio sources. Some, like 3C 273, are asymmetric. Though the lobes of quasars can extend out quite far, the nuclei are below the limit of resolution.

One possible model for the energy generation in a quasar is nuclear fusion in an extremely massive body (containing thousands of solar masses) known as a *supermassive star*. There is no confirming observation that such supermassive stars exist.

A model of many exploding supernovae seemed attractive for a time because one could explain the observed sporadic variations in the intensity as variations in the number going off from one moment to another. But the notion does not seem acceptable in terms of the amount of mass required. Almost a dozen supernovae a day would have to be going off in some cases.

Other unconvincing suggestions have also been made to explain the source of energy in quasars. For example, perhaps matter and antimatter are annihilating each other.

The absolute magnitude of 3C 279, another quasar, has been as high as −31, 10 magnitudes (and therefore 10,000 times) brighter than the Andromeda Galaxy.

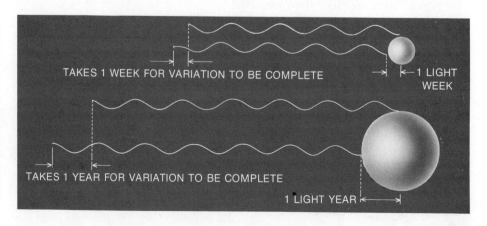

Figure 26–6 The figure illustrates why a large object can't fluctuate in brightness as rapidly as a smaller object. Say that each object abruptly brightens at one instant. The wave emitted from the top of the object takes somewhat longer to reach us than the wave emitted from the side of the object nearest us, just because of the additional distance it has to travel. We don't see the full effect of the variation in brightness until we have the waves from all parts of the object. This simply takes longer for larger objects than it does for smaller ones.

TAKES 1 WEEK FOR VARIATION TO BE COMPLETE · ←1 LIGHT WEEK

TAKES 1 YEAR FOR VARIATION TO BE COMPLETE · 1 LIGHT YEAR

At present, the general consensus is that quasars have giant black holes in their centers, larger scale versions of the maxi black holes that may well exist in the centers of our own and other galaxies. As mass falls into the black hole, energy is given off, in much the same way but on a much larger scale than energy is given off from black holes resulting from collapsed stars. The details have not been worked out, and certainly no proof has been found that this is the case, but nonetheless this picture has become the majority view.

It may be a long time before we know definitively, or agree on, whether black holes or some other method provides the energy source. Perhaps the answer will prove to be a new method, yet unknown, that can be applied to help our energy problems here on earth.

26.5 THE ORIGIN OF THE REDSHIFTS

26.5a Non-Doppler Methods

The most obvious explanation of quasar redshifts is the Doppler effect. But the distances implied are so large and the energy problem so difficult, that other explanations have been investigated very thoroughly.

Some scientists feel that the problem of providing enough energy is so serious that quasars must be close to us. If the quasars are close, we must think of some other way of accounting for their great redshifts. There are at least three other ways. They are all unattractive to many astronomers because they challenge Hubble's law and thus cast doubt on our knowledge of the whole scale of the universe.

The first way also relies on velocity and assumes that quasars are close but are going away from us very rapidly. Quasars could be relatively close to us yet show high redshifts if they had been ejected from the center of a galaxy at very high velocities (Fig. 26–7). If such quasars occurred in another galaxy, we would expect to see some of them going away from us and some coming toward us. But all the quasars have redshifts; none of them has a blueshift, so astronomers think that this idea is unlikely. But if we ever find a blueshifted quasar, much of astronomy will have to be rewritten.

One explanation that allows the quasars to have exploded from the center of our galaxy or a nearby galaxy, yet still explains why there are no blueshifted quasars, is that the explosion took place so long ago that all the quasars have expanded past us and are now receding. So much energy would be required for such tremendous and perhaps repeated explosions that the energy problem would not be resolved.

A second kind of redshift is "gravitational." According to Einstein's theory of relativity, spectral lines are redshifted very slightly when they leave a star or any object that has a large mass and hence a large gravitational pull. We have seen (Section 9.5) that the sun slightly redshifts the lines in the spectrum that leave it and that a white dwarf star, which is much denser than the sun, redshifts the lines much more. But scientists were not able to work out such a system that would be free of other, unobserved consequences. And eventually, measurements were made of redshifts from the faint material that surrounds some quasars. The gravitational redshift theory predicts that the redshift would vary with distance from the quasar, but the observations contradict this. The theory has thus been discarded.

A third possibility is that the quasars are acting on principles that we do

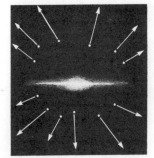

Figure 26–7 If quasars were local, and were ejected from our galaxy, that could explain why they all have high redshifts. But a tremendous amount of energy is necessary to make the ejection. So we do not find ourselves any closer to a solution of the energy problem—where all this energy comes from.

not yet understand—some new kind of physics. Some specific modifications of physical laws have even been suggested. But most scientists feel (as discussed in Section 21.7) that the basic philosophy of science forbids us from inventing new physical laws as long as we can satisfactorily use the existing ones to explain all our data.

26.5b Evidence for the Doppler Effect

Some of the strongest evidence that the quasars are at their *cosmological distances* (their distances according to Hubble's law) instead of *local* comes from the survey of quasar numbers. The quasars increase with distance from us at a steady rate, which is difficult to account for on a local model but occurs naturally if quasars are at their distances according to Hubble's law.

Now there is even stronger evidence that the redshift is a valid estimator of distance for quasars. A strange object named BL Lacertae has been known for years. At first, it was thought to be merely a variable star, but in recent years it was suspected to be stranger than that, partly because of its rapidly varying radio emission. Its spectrum showed only a featureless continuum, with no absorption or emission lines, so little could be learned about it.

Two Caltech astronomers had the idea of blocking out the bright central part of BL Lacertae with a disk held up in the focus of their telescope, thereby obtaining the spectrum of the haze of gas that seems to surround the star. On one of the several nights that they observed, the central source was relatively faint and the spectrum of the surrounding haze of gas could be observed with an electronic device. The spectrum turns out to be a faint continuum with absorption lines that are redshifted by 7 per cent. Thus BL Lacertae is an extremely nearby quasar! Other similar objects (now called Lacertids or BL Lacs) have been found to also have redshifts of this order.

If we assume that quasars are at the distances assigned by Hubble's law, then they have properties similar to those of Lacertids. Since few doubt that the Lacertids are at the distances that correspond to their redshifts by Hubble's law, this strengthens the notion that the quasars are indeed at their own Hubble law distances.

Other probable quasars with similarly small redshifts have since been found (Fig. 26–8). We had not noticed them before because they are located near the galactic plane and their light is thus severely dimmed. The quasar shown was discovered as the optical identification of an x-ray source, which may well be the best way of discovering radio-quiet quasars. It is only 800 million light years away. This source is dimmed by almost 100 times by dust in the plane of our galaxy. If it had been located outside the galactic plane, it would appear brighter than 12th magnitude and would undoubtedly have been discovered long ago. The discovery of this and other objects intermediate in properties between ordinary galaxies and traditional quasars endorses the interpretation of quasars as distant objects.

26.6 ASSOCIATIONS OF GALAXIES WITH QUASARS

A principal attack on the theory that quasars are at great distances has come from Halton Arp of the Hale Observatories. His observations have led him to conclude that quasars might be linked to galaxies, both the peculiar and the ordinary types. Arp contends that he has found many examples in

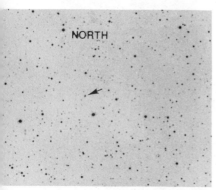

Figure 26–8 QSO 0241 + 622, studied by Bruce Margon, now of the University of Washington, and Karen Kwitter, now of Williams College, appears stellar and has a luminosity far exceeding that of normal galaxies, just as quasars do. Its spectrum resembles those of many quasars and Seyferts. Its redshift is only 4 per cent.

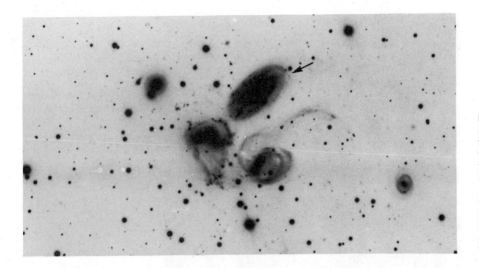

Figure 26–9 Stephan's Quintet, in a negative print of a photograph taken by Halton Arp with the 5-m telescope at the Hale Observatories. The galaxy with the discordant redshift is marked with an arrow. Less exposed photographs show that the dark area to the lower left of this galaxy has two nuclei and so is two interacting galaxies. Thus there are indeed five members of this "quintet."

which the quasars are located on opposite sides of a galaxy from each other. He argues that the quasars and the galaxy must therefore be linked. We know the distances to the galaxies from Hubble's law; if the quasars and galaxies are indeed physically linked, then the quasars are at the same distance from us as the galaxies. These distances are much smaller than the quasars' velocities of recession indicate.

A similar argument that Doppler distances cannot be trusted is made with Stephan's Quintet (Fig. 26–9), five apparently linked galaxies, one of which has a redshift very different from the others. Such cases can be explained without need to reject the Hubble law if the object with a discordant redshift only happens to appear in almost the same line of sight as do the other objects. Associations of a quasar with a distant group of galaxies with redshifts that do agree with each other have been found.

In other examples, it seems that a galaxy and a quasar with different redshifts are actually linked by a bridge of material. But this too can be a projection effect, and there is no agreement that the two are physically connected.

Arp's data are subject to several objections, the main one being statistical. Arp feels that the associations he finds of objects of differing redshifts are genuine and not the result of chance. He points out that the probability of finding these objects so close together in the sky is very small. But most astronomers feel that the fact that a quasar and a galaxy are seen in the same direction from the earth does not prove that they are physically linked; they could be at different distances (Fig. 26–10). The associations would be harder to explain in this way if we found a group of galaxies in which not just one but two discrepant redshifts were found. No such group is known. So far, most astronomers do not accept Arp's arguments and feel that quasars are in fact very far away.

Alan Stockton of the University of Hawaii carried out an important statistical test of the association of quasars with galaxies. He selected 27 relatively nearby quasars that were so luminous that the energy problem would be particularly severe. He carefully examined the regions around the quasars, and for 17 of the quasars found galaxies apparently associated with them. The number of associated galaxies was 29.

"Any coincidence," said Miss Marple to herself, "is always worth noticing. You can throw it away later if it is only a coincidence."
Agatha Christie
NEMESIS
(London: Collins, 1971, p. 58)

Figure 26–10 One example of how even improbable things can happen. The galaxies shown at center, called 145-IG-03, appear to be interacting on this Cerro Tololo plate: But are two galaxies interacting or three? Surprisingly, the stellar-appearing object at the end of the luminous bridge—the round dot at the lower right end of the apparently connected objects—turns out (from its spectrum) to be a normal star of spectral type K and with a redshift very close to 0. This star is superimposed on the more distant galaxies to an accuracy of less than 0.1 arc sec. The example shown here points up the difficulty of concluding on the basis of probability arguments that objects are physically associated.

For 8 of the quasars, at least one of the galaxies nearby had a redshift identical to that of the quasar (Fig. 26–11). Since he had chosen his sample of quasars before knowing what he would find, he was able to apply standard statistical tests to assess the probability that he could randomly find galaxies so apparently close to quasars in the sky. The probability of an effect of projection making galaxies appear so close to such a high fraction of quasars is less than one in a million. This indicates that the quasars and associated galaxies are physically close together in space. The results strongly endorse the idea that the distances to quasars are those we derive from Hubble's law.

Figure 26–11 In these photographs by Alan Stockton, the central objects are quasars, whose catalogue numbers are given below the photos, and the numbered objects are galaxies. The redshifts of the galaxies and the quasars agree. Note how the quasars are brighter than galaxies at the same distances.

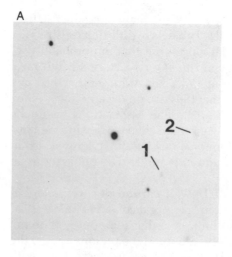

QSO 1004 + 130

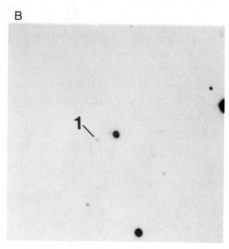

QSO 1512 + 370

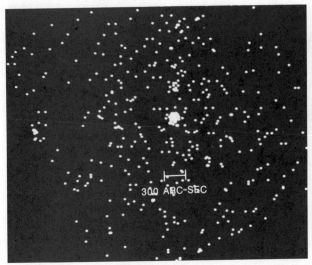

Figure 26–12

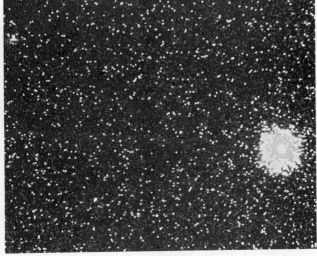

Figure 26–13

26.7 QUASARS OBSERVED FROM SPACE

The earth's atmosphere prevents the ultraviolet and x-ray region of quasar spectra from reaching us, but our ability to launch rockets and spacecraft has eliminated this handicap. Since many of the spectral lines we observe in the visible spectra of quasars were emitted by the quasar in the ultraviolet and redshifted into the visible, it is particularly important to study the ultraviolet spectra of relatively bright nearby quasars to find out more about that part of a quasar's spectrum.

The first ultraviolet quasar spectrum was made by a Johns Hopkins University group, who launched a 40-cm telescope with a high-resolution spectrograph on a rocket. The strongest line they observed was Lyman alpha, and other spectral lines showed up as well. More recently, the International Ultraviolet Explorer is allowing the observation of many more of the brightest quasars.

The real breakthrough in observing quasars has come with the launch of the Einstein Observatory. Only a few quasars had been observed previously in the x-ray region of the spectrum. 3C 273, which is reasonably close for a quasar and which gives off millions of times more energy than our galaxy in both the visible and the x-ray spectral regions, had been the first. The sensitivity of Einstein brought us from a situation in which very few quasars were known to have x-ray emission to a situation in which dozens of x-ray emitting quasars are known. Some of the quasars observed are among the farthest known in that they have redshifts greater than 2 (Fig. 26–12). Even an observation of 3C 273 turned up a new distant quasar in the background (Fig. 26–13). The objects are confirmed as quasars only when optical spectra are taken and large redshifts are found.

Surprisingly, more of the quasars being discovered with the Einstein Observatory have low redshifts rather than high redshifts. Combined x-ray and optical studies of these newly discovered quasars may change our ideas of how quasars are distributed in space and how they evolved with time.

Figure 26–12 The quasar B2 1225+31 at a redshift of 2.2 is visible in this 100-minute Einstein Observatory exposure. Quasar numbers like this refer to their celestial coordinates. (Observations were made by Harvey Tananbaum and colleagues at the Harvard-Smithsonian Center for Astrophysics.)

Figure 26–13 The bright object at lower right is the prominent quasar 3C 273, but a newly-discovered quasar is visible at upper right as well in this x-ray image taken with the Einstein Observatory. The objects are too small to be resolved, though the image of 3C 273 has spread considerably because of overexposure. (Observations were made by Harvey Tananbaum and colleagues at the Harvard-Smithsonian Center for Astrophysics.)

Figure 26–14 The gravity of a massive object, perhaps a galaxy, in the line of sight can form multiple images of an object. If the alignment of the earth, the intermediate object, and the distant quasar is perfect, we would see a ring. A slight misalignment would make crescents or individual images.

26.8 THE DOUBLE QUASAR

The astronomical world was agog in 1979 at the discovery of a pair of quasars so close to each other that they might be two images of a single object. The story began when inspection of the Palomar Sky Survey charts, taken years ago with the Schmidt telescope there, revealed a close pair of 17th magnitude objects at the position corresponding to a radio source.

The two images are separated by only 6 seconds of arc. Their redshifts are identical to the third decimal place: z = 1.4136 ± 0.0015.

When the spectra of the objects were taken at Kitt Peak, both objects turned out to be quasars, and their spectra looked identical. Even stranger, their redshifts were essentially identical. The brand new Multiple Mirror Telescope took some of its first spectra of these objects. The MMT measurements for the redshifts of the quasars showed that they were even closer to each other than the previous observations had shown.

Dennis Walsh of the University of Manchester, Robert F. Carswell of Cambridge University, and Ray J. Weymann of the University of Arizona first identified the double quasar and suggested that it resulted from a gravitational lens.

It seems improbable that two independent quasars should be so similar in both spectral lines and redshift, so the scientists who took the first spectra suggested that both images showed the same object! They suggested that a gravitational bending of the quasar's radiation was taking place. We have discussed such gravitational bending as a consequence of Einstein's general theory of relativity and seen that it has been verified for the sun (Section 7.11). For the quasars, a massive object between the quasar and us is acting as a gravitational lens, bending the radiation from the quasar one way on one side and the other way on the other side (Fig. 26–14). As a result, we see the image of the quasar in at least two places.

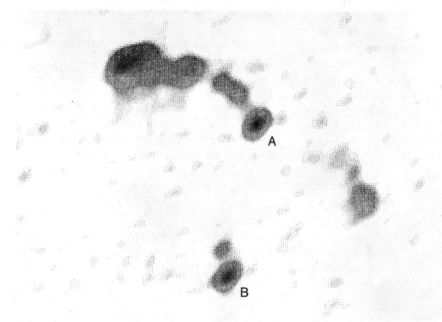

Figure 26–15 A radio map of the double quasar, Q 0957+561 A and B, made with the VLA. The sources marked A and B correspond to the optical objects. Additional images are also seen, which could result from an off-center gravitational lens. (Courtesy of B. F. Burke, P. E. Greenfield, and D. H. Roberts, MIT)

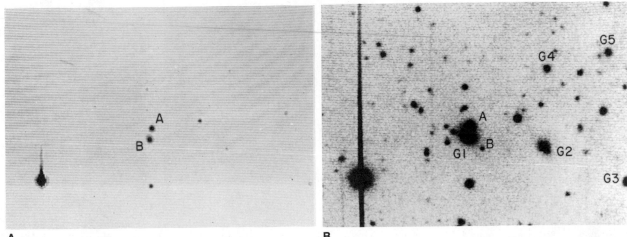

A   **B**

Detailed radio maps of the region have been made with several interferometers, most notably with the VLA. The VLA's map (Fig. 26–15) shows the two sources and others as well.

One test of whether a gravitational lens is present is to see if the brightness fluctuations of the two optical images are similar. Since the light from the two images travels different paths through space, though, the fluctuations could be displaced in time from each other by months or years.

Our knowledge of the double quasar was greatly increased by a picture taken with an electronic camera (using CCD's, devices described in Section 2.11a). The observation shows (Fig. 26–16) that a cluster of galaxies lies between us and the quasar. Quasar image B is seen through the brightest member of the cluster. The gravitational effect of the cluster adds to that of this galaxy in making the gravitational lens.

The gravitational lens picture was strongly endorsed by a photograph that seems actually to show the intervening galaxy (Fig. 26–17). It seems to be a giant elliptical galaxy with a redshift about one-fourth that of the quasar. The galaxy is best seen in Color Plate 73.

The double quasar seems to be the first known example of a gravitational lens. Another multiple quasar, this one with three (and perhaps even five) identical images close together, has since been discovered. It presumably also shows a gravitational lens effect.

26.9 QUASARS, LACERTIDS, AND GALAXIES

In recent years, the important question has become whether quasars are truly a new phenomenon or just previously unfamiliar types of galaxies.

Figure 26–16 Two views of a photograph of the double quasar taken with a CCD camera at the 5-m Hale telescope. *(A)* The frame is reproduced so that the twin images, marked A and B, stand out. *(B)* The frame is reproduced so that the fainter parts of the image show. We then see a cluster of galaxies, some of which are marked (G1–G5). 90 per cent of the objects on the frame are galaxies. A and B are now overexposed and blur out. G1 provides most of the gravity for the lens effect. Though its image does not show clearly, its redshift can be measured spectroscopically. (Courtesy of Jerome Kristian, James A. Westphal, and Peter Young, Palomar Observatory)

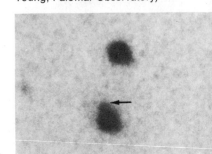

Figure 26–17 This optical photograph of the double quasar, 0957+61A and B, shows the intervening galaxy. The 17th magnitude quasar images are bluish and are separated by only 5.7 seconds of arc. Their redshifts are identical to the third decimal place: $z = 1.4136 \pm 0.0015$. This photograph by Alan Stockton of the University of Hawaii shows the intervening galaxy only 1 second of arc from the "B" quasar image. Its redshift is about 0.37. The objects are in the constellation Ursa Major.

TABLE 26–1   ENERGIES OF GALAXIES AND QUASARS

| | **Relative Luminosity** | | |
	X-Ray	*Optical*	*Radio*
Milky Way	1	1	1
Radio galaxy	100–5,000	2	2,000–2,000,000
Seyfert/N galaxy	300–70,000	2	20–2,000,000
Quasar: 3C 273	2,500,000	250	6,000,000

A₁   A₂   A₃   B

Figure 26–18 *(A)* Increasing exposure time reveals more and more galactic structure around the Seyfert galaxy NGC 4151, which appears almost stellar on the least exposed photograph. *(B)* The nucleus of the Seyfert galaxy photographed from an altitude of 25,000 m (80,000 ft) with the 90-cm (36-in) telescope carried aloft in 1970 by the Stratoscope II balloon.

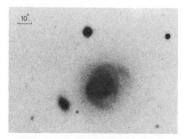

Figure 26–19 A Seyfert galaxy, ESO 113-IG 45, with an exceptionally luminous nucleus. This is a case of a quasar (or almost quasar) in the center of a spiral galaxy.

Some spiral galaxies have especially bright nuclei while others have fainter centers. One type of galaxy, discovered by Carl Seyfert of the Mount Wilson Observatory in 1943 and named after him, has a very bright nucleus indeed (Fig. 26–18). Can the quasars that we see be only the bright nuclei of galaxies? After all, although we can see both the nuclei and the spiral arms of close Seyferts, only the nuclei would be visible if such galaxies were very far away (Fig. 26–19). Another type of galaxy, an N galaxy, also has an especially bright core.

Jerome Kristian of the Hale Observatories has studied a sample of about 24 quasars to see whether they could be galaxies with bright cores. He attempted to predict which would be close enough to us to reveal some structure around the core, which would appear as points like quasars, and which would be in between and thus might or might not show structure. He then studied the best photographs of these objects, and found that almost all objects in which one would expect to see structure did in fact show structure, almost all objects that would not show structure because they were too far away did not show structure, and that the middle group was mixed. It thus seems reasonable to believe that quasars are a class of objects somehow related to the cores of galaxies. Perhaps quasars and galaxies are the same, but are simply in different stages of evolution.

Further, properties of quasars like strength of radio, optical, and x-ray emission and the types of lines in their spectra are now known to be similar to those of certain types of galaxies. Also, as we saw, certain properties of quasars are similar to those of Lacertids. And the Lacertids seem to be special "events" in the center of a mass of gas, just as quasars may be similar to bright "events" in the centers of Seyfert galaxies or N galaxies. Observations of nearby Seyferts and N galaxies indicate that such "events" can happen late in the lifetime of a galaxy. Since most quasars are very far away, the light we see now was emitted so long ago that it had been thought that such events might only be linked to young galaxies.

In sum, there seems to be a strong relation between quasars and the cores of galaxies. This linkage may make the quasars seem somewhat less strange, but at the same times causes galaxies to seem more exotic. Lacertids, Seyfert galaxies, and quasars are probably the same basic phenomenon, though in different stages of evolution or activity. If we had happened to know of BL Lacertae earlier on, quasars would not have looked as strange to us when they were discovered.

SUMMARY AND OUTLINE

Discovery of quasars (Section 26.1)
  QSR—quasi-stellar radio sources
  Interaction of optical and radio astronomy
  Position from lunar occultations or from interferometry
  Huge Doppler shifts in spectra
  Radio quiet objects also included as quasars
Doppler shift in quasars (Section 26.2)
  Large redshifts mean large velocities of recession and, by Hubble's law, great distances
Importance of quasars (Section 26.3)
  Test of Hubble's law, which is thus a test of the accuracy of the distance scale
  Great distance, which means we view them as they were in early stages of the universe
Energy problem (Section 26.4)
  Too much energy required to be accounted for by ordinary processes
  Possible explanations
    Supermassive star
    Immense number of supernovae
    Annihilation of matter and anti-matter
    Presently unknown method
    Generally accepted explanation: giant black holes
Non-Doppler explanations of redshifts (Section 26.5)
  Quasars ejected from cores of galaxies at very high velocities; no evidence of blueshifted quasars, however
  Gravitational redshifts; conflict with observations

Quasars operating under new physical laws?
All these alternative explanations challenge Hubble's law
Peculiar galaxies and quasars (Section 26.6)
  Question of physical link between galaxies and quasars
  Stephan's Quintet and other sources with discrepant redshifts might be explained as chance alignments
Space observations of quasars (Section 26.7)
  Rocket and IUE studies in ultraviolet
  Einstein Observatory observations of quasars with large redshifts and of numerous faint quasars
The double quasar (Section 26.8)
  Two quasars with identical properties located very close together
  Could be two images of one quasar formed by a gravitational lens
  Recent optical evidence supports gravitational lens theory
Quasars and galaxies (Section 26.9)
  Are quasars truly a new phenomenon?
  Seyfert and N galaxies also have bright nuclei
  BL Lacertae and other nearby objects known to have distances corresponding to Hubble's law have properties in common with quasars
  Quasars seem to be events in cores of galaxies at some evolutionary stage

QUESTIONS

1. Why is it important to find the optical objects that correspond in position with radio sources?

†2. (a) You are heading toward a red traffic light so fast that it appears green. How fast are you going?
(b) At \$1 per mph over the speed limit of 55 mph, what would your fine be in court?

†3. A quasar is receding at 1/10 the speed of light.
(a) If its distance is given by the Hubble relation, how far away is it?
(b) At what wavelength would the 21-cm line appear?

†4. A quasar has $z = 0.3$. What is its velocity of recession in km s^{-1}?

†5. A quasar is observed and it is found that a line whose rest wavelength is 3000 Å is observed at 4000 Å.
(a) How fast is the quasar receding?
(b) How far away is it if its distance is given by the Hubble relation?

6. What do quasi-stellar radio sources and radio-quiet quasars have in common?

7. Why does the rapid time variation in some quasars make the "energy problem" even more difficult to solve?

8. What are three differences between quasars and pulsars?

9. Briefly list the objections to each of the following "local" explanations of quasars:
(a) They are local objects flying around at large velocities.
(b) The redshift is gravitational.

10. If quasars were proved to be local objects, how would this help solve the "energy problem"?

11. Describe the implications of the observations of quasars in the x-ray spectrum. What new ability allowed these observations to be made?

12. What features of some quasars suggest that quasars may be closely related to galaxies?

---

†This indicates a question requiring a numerical solution.

CHAPTER 27

COSMOLOGY

AIMS:
To study the origin and evolution of
the universe, to discuss the origin of
the elements, and to consider the
future course the universe will take

In Armagh, Ireland, in the mid-seventeenth century, Bishop Ussher declared that the universe was created at 9 A.M. on Sunday, October 23rd, in the year 4004 B.C. Nowadays we are less certain of the details of our origin.

We study the origin of the universe as part of the study of the universe as a whole. This larger field is called *cosmology*. Our discussion of astronomy has brought us to consider many different types of celestial objects and many different ways of observing them. All this observational knowledge and all these techniques, in conjunction with theoretical calculation, must be brought to bear on cosmological problems in order to understand the most fundamental questions about our universe.

The study of where we have come from and of what the universe is like leads us to consider where we are going. Is the universe now in its infancy, in its prime of life, or in its old age? Will it die? It is difficult for us, who, after all, spend a lot of time thinking of topics like "what shall I watch on TV tonight" or "what's for dinner," to realize that we can think seriously about the structure of space around us. It is awesome to realize that we can conclude what the future of the universe will be. One must take a little time every day, as Alice was told when she was in Wonderland, to think of impossible things. By and by we become accustomed to concepts that may seem overwhelming at first. You must sit back and ponder when studying cosmology; only in time will many of the ideas that we shall discuss take shape and form in your mind.

Albert Einstein in the Swiss Patent Office in Berne in 1905.

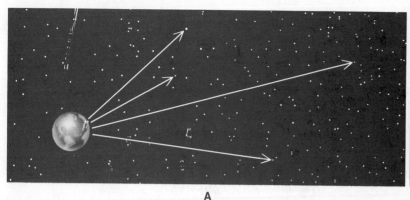

A

B

Figure 27-1 *(A)* If we look far enough in any direction in an infinite universe, our line of sight will hit the surface of a star. This leads to Olbers' paradox. *(B)* This painting by the Belgian surrealist René Magritte shows a situation that is the equivalent to the opposite of Olbers' paradox.

We have phrased the paradox in terms of stars, though we know that the stars are actually grouped into galaxies. But we can carry on the same argument with galaxies, and deduce that we must see the average surface brightness of galaxies everywhere. This is patently not what we see.

*27.1 OLBERS' PARADOX

Many of the deepest questions of cosmology can be very simply phrased. Why is the sky dark at night? Analysis of this simple observation leads to profound conclusions about the universe. We can easily see that the night sky is basically dark, with light from stars and planets scattered about on a dark background. But a bit of analysis will show that if there is a uniform distribution of stars in space, it shouldn't be dark anywhere in the sky. If we look in any direction at all we will eventually see a star (Fig. 27–1), so the sky should appear uniformly bright. We can make the analogy to our standing in a forest. There, we would see some trees that are closer to us and some trees that are farther away, but if the forest is big enough our line of vision will always eventually stop at the surface of a tree. If all the trees were painted white, we would see a white expanse all around us (Fig. 27–2). Similarly, when looking up at the night sky we would expect the sky to have the uniform brightness of the surface of a star.

The fact that this argument doesn't work, and that the sky is dark at night, is called *Olbers' paradox*. Wilhelm Olbers phrased it in 1823, although the question had been discussed at least a hundred years earlier. The solution of Olbers' paradox leads us into basic considerations of the structure of the universe. This solution could not have been advanced in Olbers' time because it depends on astronomical discoveries that have been made more recently.

*27.1a Surface Brightness and Olbers' Paradox

Olbers' paradox arises because in whatever direction we look, our line of sight winds up on the surface of a star. It is tempting to explain the paradox merely by claiming that the farthest stars are so distant that they are too faint to see, but this potential explanation is not valid. The brightness of each part of the surface of a star does not decrease no matter how far the star is away. In this case we are interested in the *surface brightness* of the star, that is, how much energy is given off by a unit area (such as 1 cm²). This surface brightness rather than the *flux*, the amount of light emitted by an object that passes through 1 square centimeter at our location, is the significant quantity for Olbers' paradox.

You may be familiar with the concept of surface brightness from an example in photography. Suppose that we adjust our camera by taking a light meter reading when we are close to our subject. When we step back to take our picture, these settings don't change. They stay constant because the surface brightness of the objects remains constant even though the object is farther away.

We can understand why by realizing that the surface brightness is the flux divided by the total angular area. For objects farther and farther away, the flux decreases but the apparent area of the object decreases at the same rate, namely with the inverse square of the distance (Section 5.2). Thus the surface brightness remains the same.

27.1b   Solutions to Olbers' Paradox

One might think that the easiest way out of this paradox, as was realized in Olbers' time, is simply to say that interstellar dust is absorbing the light from the distant stars and galaxies. But this doesn't solve the problem, because the dust would have to be everywhere because the sky is generally dark in all directions. This widely distributed dust would soon absorb so much energy that it would heat up and begin glowing. Given a long enough time, all the matter in the universe would begin glowing with the same brightness, and we would have our paradox all over again.

(One solution to Olbers' paradox lies in part in the existence of the redshift, and thus in the expansion of the universe (Section 25.4). This does not mean that the answer to the paradox is just that visible light from distant galaxies is redshifted out of the visible, for at the same time ultraviolet light is continually being redshifted into the visible. The point is rather that each quantum of light undergoes a real diminution of energy as it is redshifted. The energy emitted at the surface of a faraway star or galaxy is diminished by this redshift effect before it reaches us.)

(But the redshift is not the only part to the solution of Olbers' paradox. As we look out in space, we are looking back in time, because the light has taken a finite amount of time to travel. If we could see out far enough, we would possibly see back to a time before the stars were formed. E. R. Harrison of the University of Massachusetts has pointed out that this is an equally valid way out of the dilemma. In fact, Harrison calculates that this explanation is a more important contribution to the solution of the paradox than is the existence of the redshift. In most directions, we would not expect to see the surface of a star for 10^{24} years. We would thus have to be able to see stars 10^{24} years back in time for the sky to appear uniformly bright. The stars don't burn that long and the universe is simply not that old. On the basis of Hubble's law, an age of somewhat over 10^{10} years seems to be the maximum.)

The fact that we have to know about the expansion and the age of the universe to answer Olbers' question—why is the sky dark at night?—shows how the most straightforward questions in astronomy can lead to important conclusions. In this case, we find out about the expansion of the universe or about the lifetimes of the universe and the stars in it.

27.2   THE BIG BANG THEORY

Astronomers looking out into space have noticed that on the large scale the universe looks about the same in all directions. That is, ignoring the presence of local effects such as our being in the plane of a particular galaxy and thus seeing a Milky Way across the sky, the universe has no direction that is special. Further, it seems that there is no change with distance either, except insofar as time and distance are linked.

(This notion has been codified as the *cosmological principle*: **the universe is homogeneous and isotropic throughout space.** The assumption of

Figure 27–2   The lower halves of trees are often painted in Mexican parks, and provide an example of a similar phenomenon to seeing a uniform expanse of starlight.

Since the energy, E, of a quantum at a given wavelength, λ, corresponding to a certain frequency, ν, is E = hν = hc/λ, the quantum has at first an energy corresponding to its original wavelength. As its wavelength gets longer because of the Doppler effect and Hubble's law, the formula E = hc/λ shows that its energy diminishes. This energy is actually lost from the quantum.

Cosmologists often assume that all the matter in the universe is spread out uniformly in space. Of course, that isn't exactly true, but it is a great convenience in many theoretical calculations not to have to worry about "details" like galaxies.

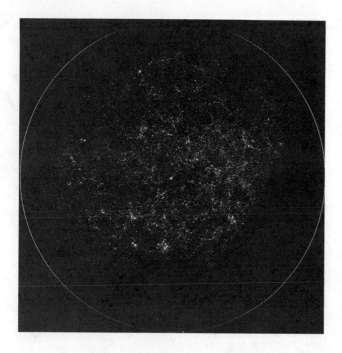

Figure 27–3 The angular distribution of the 1,000,000 northern hemisphere galaxies brighter than magnitude 19. The north galactic pole is at the center and the galactic equator is at the edge. The gray tone in each of 2,000,000 rectangles corresponds to the number of galaxies in it. The Coma cluster appears near the center though the Virgo cluster is lost in the background. The distribution seems isotropic.

"Big bang" is the technical as well as the popular name for these cosmologies; professionals write about the big bang in the scientific journals. The theories are based on Einstein's general theory of relativity.

Remember that the galaxies themselves are not expanding; the stars in a galaxy don't tend to move away from each other.

homogeneity says that the distribution of matter doesn't vary with distance from the sun, and the assumption of *isotropy* (Fig. 27–3) says that it is about the same no matter in which direction one looks. Actually, of course, we know that we have to look in certain directions to see out of our galaxy without having our vision ended by the interstellar dust, but remember that we are ignoring inhomogeneities or lack of isotropy on this small scale.

The explanations of the universe that astronomers now accept are known as *big bang* theories. Basically, these cosmologies say that once upon a time there was a great big bang that began the universe. From that moment on, the universe expanded, and as the galaxies formed they shared in the expansion. The big bang theories satisfy the cosmological principle.

Many students ask whether the fact that there was a big bang means that there was a center of the universe from which everything expanded. The answer is no; first of all, the big bang may have been the creation of space itself. Furthermore, the matter of this primordial cosmic egg was everywhere at once. There may be an infinite amount of matter in the universe, so it is possible that at the big bang an infinite amount of matter was compressed to an infinite density while taking up all space.

Consider a two-dimensional analogy to a universal expansion: the surface of a rubber balloon covered with polka dots. Let us consider the view from one dot. As the balloon is blown up, all the dots would seem to recede from us. No matter which dot we were on, all the other dots would seem to recede. The surface of the balloon is two-dimensional yet has no edge. Even if we go infinitely far in any direction, we never reach a boundary. There is no center to the surface from which all dots are actually expanding. It is much more difficult for most of us to visualize a three-dimensional situation (or, including time as a fourth dimension, a four-dimensional situation of space-time). Yet the above analogy is a good one, because our universe can expand uniformly yet have no center to the expansion.

What will happen in the future? One possibility is that the universe will

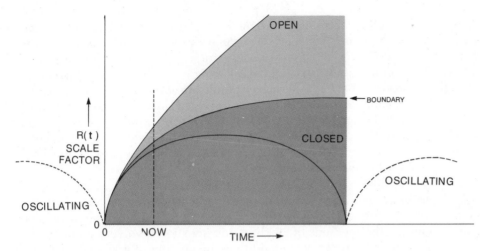

Figure 27-4 To trace back the growth of our universe, we would like to know the rate at which its rate of expansion is changing. Big bang models of the universe are shown; the vertical axis represents a scale factor, *R*, that represents some measure of distances and how they change as a function of time, t. The universe could be open and expand forever or closed and begin to contract again. If it is closed, we do not know whether it would oscillate or whether we are in the only cycle of expansion or contraction it will ever undergo. We will see later that current evidence favors the open model.

Astronomers use a quantity called the *deceleration parameter*, which is given the symbol q_0, to describe how fast the expansion is slowing down, $q_0 = 1/2$ marks the dividing line between an open universe (q_0 less than 1/2, including 0) and a closed universe (q_0 greater than 1/2, including 1).

continue to expand forever. This case is called an *open universe*. It corresponds to the case where the universe is infinite. The other possibility is that at some time in the future the universe will stop expanding and will begin to contract. This case is called a *closed universe* (Fig. 27–4), and corresponds to the case where the universe is finite. In a closed universe, we would eventually reach a situation that we might call a "big crunch."

What was present before the big bang? There is no real way to answer this question. For one thing, we can say that time began at the big bang, and that it is meaningless to talk about "before" the big bang because time didn't exist. We do not now think that the universe can remain in a static condition, so it seems unlikely that the universe was always just sitting there in an infinitely compressed state, whatever "always" means.

Of course, these possibilities don't answer the question of why the big bang happened. Another possibility is that there had been a prior big bang and then a recollapse, and that this most recent big bang was one in an infinite series of bangs. This version of a closed universe theory is called the *oscillating universe*. But the oscillating universe theory doesn't really tell us anything about the origin of the universe because in this case there was no origin. The concepts discussed in these paragraphs will not be easy to digest. They may take hours, years, or a lifetime to come to terms with.

If we consider Hubble's law with the current value for Hubble's

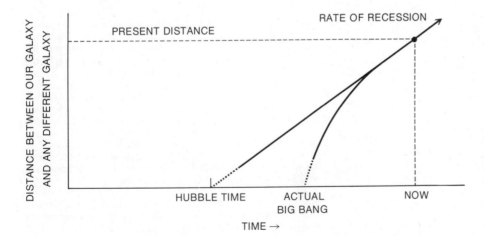

Figure 27-5 If we could ignore the effect of gravity, then we could trace back in time very simply; the Hubble time corresponds to the inverse of the Hubble constant ($1/H_0$). Actually, gravity has been slowing down the expansion. The vertical axis again represents some scale factor.

constant—and the cosmologists Allan Sandage and Gustav Tammann have evidence that Hubble's constant is the same for relatively nearby galaxies as it is for very distant galaxies—we can extrapolate backward in time (Fig. 27–5). We simply calculate when the big bang would have had to take place for the universe to have reached its current state at its current rate of expansion. This calculation indicates that the universe is somewhere between 13 billion and 20 billion years old. The range of values indicates in part the uncertainty in Hubble's constant and in part the uncertainty in the calculation of how much the effect of gravity would have by now slowed down the expansion of the universe.

Open or closed, the universe will last at least another 40 billion years, so we have nothing to worry about for the immediate future. Nevertheless, the study of the future of the universe is an exceptionally interesting investigation. Some of the methods of tackling this question involve measuring the amount of mass in the universe and thus the amount of gravity (as we will discuss in Section 27.7). Other methods of determining our destiny involve looking at objects as distant as possible to see if any deviation from the Hubble law can be determined. This investigation of distant bodies has been carried out many times over the years, but a deviation from the straight-line relation between velocity and distance known as Hubble's law has not been found conclusively. The deviations have always been within the uncertainty. One major uncertainty comes from the fact that the best independent measure of the distances to the farthest galaxies comes from their brightnesses (Fig. 27–6).

We now realize that the farthest galaxies, which would provide the best determination of a deviation from Hubble's law, may well have been very different long ago when they emitted the light we are now receiving. Thus we cannot assume that even such basic properties as their size and brightness were similar then to their size and brightness now. Since determining the distance to these distant galaxies depends on understanding their properties, it now seems that such considerations of galaxy evolution provide too much uncertainty to allow us to determine whether the universe is open or closed in this way.

How galaxies evolve with time is not known. Were they brighter or fainter long ago? One possibility is that galaxies grow fainter with age. There is also evidence that some giant galaxies in clusters may be consuming other galaxies in those clusters. This galactic cannibalism *would make the remaining galaxies become brighter.*

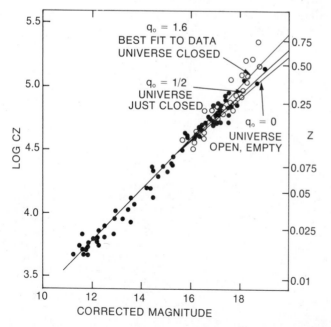

Figure 27–6 This Hubble diagram, plotted in terms of redshift and magnitude, shows faint clusters of galaxies. The latest measurements of J. Kristian, A. Sandage, and J. Gunn of the Hale Observatories are shown as open circles, added to the dots of prior observations. The horizontal axis shows the magnitude, which is a measure of distance. The magnitude has been corrected for various subtle effects, such as interstellar absorption and the appearance of a galaxy to observing equipment. In principle, we can determine the future of the universe from the slight deviations at upper right of the curves from a straight line. If the curve appears to the upper left of a straight line, the expansion is slowing down. If the curve falls far enough to the left, then the universe is closed, finite, and will eventually begin to contract. If it is not so far to the left, then the universe is open, infinite, and will expand forever. Even though the data curve slightly upward from a straight line, the data are currently not sufficient to definitely conclude between these possibilities. A special difficulty that prevents us from doing so is the effect of the galaxies' evolution, which is very uncertain.

*27.3 GENERAL RELATIVITY AND COSMOLOGY

Although we have discussed the big bang theories in words rather than in equations, these theories actually arise as solutions to a set of equations that Einstein advanced as part of his general theory of relativity. Einstein tried to find solutions to his equations and the solutions he found in 1917 corresponded to a universe that was either expanding or contracting. But at that time the observational evidence for the expansion of the universe was not known, and it was thought that the universe was static. So Einstein deliberately and artificially modified his equations to make his model universe static. He did this by arbitrarily adding a constant to one side of the equations. This constant is called the *cosmological constant.*

Another solution to Einstein's original equations was worked out by the Dutch scientist Willem de Sitter, also in 1917 (Fig. 27-7). But his solution had a deficiency too—it corresponded to a universe without any matter in it! So de Sitter's solution is also unsatisfactory, though the overall density of matter in the universe is fairly low. De Sitter himself found a dozen years later that the density of the universe was too high for his solution to be valid.

Figure 27–7 Einstein and de Sitter at the Mount Wilson Observatory in 1931.

Note that even though Einstein's and de Sitter's solutions are not in agreement with current observational evidence, the investigations were still worth carrying out. Work on them developed methods that were later useful, and these extreme solutions helped us to understand the more realistic solutions once they were found. Much of the research done in science does not lead to ultimate answers at the first try.

Sets of solutions that correspond to more physically reasonable universes were worked out in the early 1920's by the Soviet mathematician Alexander Friedmann. Friedmann's solutions are the mathematical formulations of the big bang cosmologies that we now accept. The Belgian abbé Georges Lemaître independently discovered similar solutions in 1927. He suggested that the original condition, the "cosmic egg," from which the universe is expanding was hot and dense. He thus deserves credit for the current view that the universe began with a *hot big bang.*

Unfortunately, astronomers did not come to know of the solutions of Friedmann and Lemaître until 1930, after Hubble had found the velocity-distance relationship we know as Hubble's law.

*27.4 THE STEADY STATE THEORY

The cosmological principle, that the universe is homogeneous and isotropic, is very general in scope, but starting in the late 1940's, three British scientists began investigating a principle that is even more general. Hermann Bondi, Thomas Gold, and Fred Hoyle considered what they called the *perfect cosmological principle,* that the universe is not only homogeneous and isotropic in space but also **unchanging in time.** According to Occam's razor (Section 21.6), we must accept the simplest theory that is in accord with all the observations, but it is a matter of personal preference whether the cosmological principle or the perfect cosmological principle is more simple.

The theory that follows from the perfect cosmological principle is called the *steady state theory* (Fig. 27–8); it has certain philosophical differences from the big bang cosmologies. For one thing, according to the steady state theory, the universe never had a beginning and will never have an end. It always looked just about the way it does now and always will look that way. Some scientists are glad to see the question of what happened before the big bang, or the need for finding a cause for the big bang, eliminated and thus find the steady state theory attractive.

But the steady state must be squared with the fact that the universe is expanding. How can the universe expand continually but not change in its overall appearance? For the density of matter to remain constant, new matter

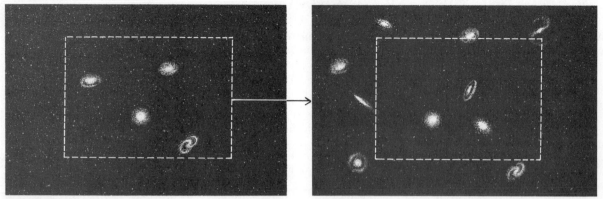

Figure 27–8 In the steady state theory, as the dotted box at left expands to fill the full box at right, new matter is created to keep the density constant. In the picture, the four galaxies shown at left can all still be seen at right, but new galaxies have been added so that the number of galaxies inside the dotted box is about the same as it was before.

must be created at the same rate that the expansion would decrease the density. Only in this way can the density remain the same.

The matter created in the steady state theory is not simply matter that is being converted from energy by $E = mc^2$. No, this is **matter that is appearing out of nothing,** and is thus equivalent to energy appearing out of nothing.

Many scientists objected very strenuously to the idea that matter could appear out of nothing. After all, the "law" that matter is conserved (i.e., cannot be created or destroyed), generalized by Einstein into a law of conservation of matter and energy, seems very basic. Nevertheless, since we know the rate at which the universe is expanding, we can calculate how much new matter would have to be created. It works out to be only one hydrogen atom per cubic centimeter of space every 10^{15} years, which is equivalent to one thousand atoms of hydrogen per year in a volume the size of the Astrodome in Houston. This value is far too small for us to be able to measure or even for us to hope to measure. So the deviation that the steady state theory would require from the "law" of conservation of matter and energy is too small to be ruled out on observational grounds.

For many years a debate raged between proponents of the big bang theories and proponents of the steady state theory. The evidence that came in—usually seeming to indicate that distant objects were somehow different than closer ones, which would show that the universe was evolving—seemed to favor the big bang cosmologies over the steady state theory. But none of this evidence was conclusive. Usually either alternative explanations for the data were conceivable or the steady state theory could be modified, sometimes extensively, to be consistent with the discoveries.

The discovery of quasars provided some of the strongest evidence against the steady state theory. The quasars are for the most part located far away in space, which corresponds to their having been more numerous at an earlier time. Thus something has been changing in the universe, which is not acceptable in the steady state theory. As the evidence that the quasars were indeed at these distances grew, the status of the steady state theory diminished. A detailed survey carried out at the Hale Observatories establishes that the number of quasars increases steadily with increasing distances from us.

In the next section we shall discuss still stronger evidence that provided the crushing blow against the steady state theory.

27.5 THE PRIMORDIAL BACKGROUND RADIATION

In 1965, a discovery of the greatest importance was made: radiation was detected that is most readily explained as a remnant of the big bang itself.

That much is easy to state, and if we have indeed discovered radiation from the big bang itself then clearly the steady state theory is discredited. But the explanation of just how the observed radiation is interpreted to come from the big bang is fairly technical and requires an understanding of abstract concepts. However, the sheer importance of the discovery warrants our taking the time to consider this matter thoroughly.

The discovery was made by Arno A. Penzias and Robert W. Wilson of the Bell Telephone Laboratories in New Jersey (Fig. 27–9). They were testing a radio telescope and receiver system (Fig. 27–10) to try to track down all possible sources of static. The discovery, in this way, thus parallels Jansky's discovery of radio emission from space, which marked the beginning of radio astronomy.

Penzias and Wilson were observing at a wavelength in the radio spectrum. After they had subtracted the contributions of all known sources to the static, they were left with a residual signal that they could not explain. The signal was independent of the direction they looked, and did not vary with time of day or season of the year. The signal corresponded to the very small amount of radiation that would be put out at that frequency by a black body at a temperature of only 3 K, 3° above absolute zero.

At the same time, Robert Dicke, P. J. E. Peebles, David Roll, and David Wilkinson at Princeton University had, coincidentally, predicted that radiation from the big bang should be detectable by radio telescopes. First, they concluded that radiation from the big bang permeated the entire universe, so that its present-day remnant should be coming equally from all directions. Second, they concluded that this radiation would have the spectrum of a black body, which simply means that the amount of energy coming out at different wavelengths can be described by giving a temperature. Third, they predicted that though the temperature was high at some time in the distant

Figure 27–9 Arno Penzias *(left)* and Robert W. Wilson *(right)* with their antenna in the background. Penzias and Wilson shared in the 1978 Nobel Prize in Physics for their discovery.

Figure 27–10 The large horn-shaped antenna at the Bell Telephone Laboratories' space communication center in Holmdel, New Jersey. Penzias and Wilson found more radio noise than they expected at the wavelength of 7 cm at which they were observing. After removing all possible sources of noise (faulty connections, loose antenna joints, contributions from nesting pigeons), a certain amount of radiation remained. It was the 3° background radiation.

Actually, the Princeton group was not the first to predict that such radiation might be present. In the late 1940's and early 1950's, Ralph Alpher, Robert Herman, and George Gamow had made similar predictions, but this earlier work was at first overlooked.

Figure 27–9

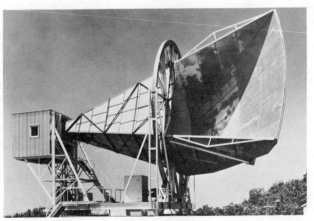

Figure 27–10

Most other mechanisms to account for the black-body radiation predict that the intensity of the radiation would vary slightly from direction to direction in the sky. For example, if the radiation originated in discrete sources, then we would expect the radiation to be most intense in the direction of the strongest sources.

We have discussed black bodies in Section 3.2. Basically, the emission from a black body follows Planck's law of radiation (Figs. 3–4 and 3–8) in that for a given temperature there is an equation that tells us the intensity of radiation at each wavelength. The key fact to remember about Planck's law is that specifying just one number—the temperature—is enough to define the whole Planck curve.

It is tempting to think of black bodies as physical rather than conceptualized objects, and that is a temptation that must be fought here in order to understand the radiation from the big bang. It is quite normal for us to try to visualize abstract concepts in concrete terms. Horatio, for example, speaks to Hamlet of "the morn, in russet mantle clad. . . ." Of course, the morn wasn't wearing any such thing; it has no particular shape to be clad.

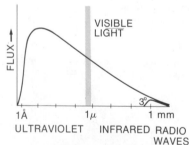

Figure 27–11 Planck curves for black bodies at different temperatures. Radiation from a 3° black body (the inner curve at lower right) peaks at very long wavelengths. On the other hand, radiation from a very hot black body (the upper curve) peaks at very short wavelengths.

past, the radiation would now correspond to a black body at a particular very low temperature, only a few degrees above absolute zero (Fig. 27–11). If radiation could be found that came equally from all directions and had the low-temperature black-body spectrum that matched the prediction, then the theory would be confirmed.

Though the Bell Labs and Princeton groups were only a few kilometers apart from each other in New Jersey, the two groups found out about each other only in a conversation with someone at MIT in Massachusetts. Once they compared notes, they agreed to a simultaneous publication of results, with the Princeton's group theoretical paper coming first and the Bell Labs' group observational paper following. The latter reported the data but did not mention the cosmological interpretation, in part because Penzias and Wilson had not worked on the interpretation and in part because they thought that their observational result might stand the test of time better than the theory.

The Princeton group continued the process of building their own receiver to observe at a different radio wavelength. They were shortly able to measure the intensity at this wavelength, and found that it also corresponded to that of a black body at 3°. This tended to confirm the idea that the radiation was indeed from a black body, and thus that it resulted from the big bang.

Because the big bang took place simultaneously everywhere in the universe, radiation from the big bang filled the whole universe. The radiation thus has the property of being isotropic to a very high degree—that is, it is the same in any direction that we observe. It was generated all through the universe at the same time, so its remnant must seem to come from all around us now. The fact that the observed radiation was highly isotropic was thus strong evidence that it came from the big bang.

27.5a The Origin of the Background Radiation

The leading models of the big bang consider a hot big bang. The temperature was billions upon billions of degrees in the fractions of a second following the beginning of time.

In the millennia right after the big bang, the universe was opaque. Photons did not travel very far before they were absorbed by hydrogen ions. After the photons were absorbed, they would soon be reradiated. This recycling process gives black body radiation corresponding to the temperature of the matter.

Gradually the universe cooled. After about a million years, when the temperature of the universe reached 3000 K, the temperature and density were sufficiently low for the hydrogen ions to combine with electrons to become hydrogen atoms. This *recombination* took place suddenly. Since hydrogen has mainly a spectrum of lines rather than a continuous spectrum and no electrons remained free, the gas suddenly began to absorb photons at only a few wavelengths instead of at all wavelengths. Thus from this time on, most photons could travel all across the universe without being absorbed by matter. The universe had become transparent.

Since matter rarely interacted with the radiation from that time to the present, it no longer continually recycled the radiation. We were thus left with the radiation at the temperature it had at the instant when the universe became transparent. This radiation is travelling through space forever. Observing it now is like studying a fossil. As the universe continues to expand, the radiation retains the shape of a Planck curve but the curve corresponds to cooler and cooler temperatures.

Since the radiation is present in all directions, no matter what we observe in the foreground, it is known as the *background radiation*. In reference to its origin, it is also called the *primordial background radiation*.

In the above scenario, the universe has changed from a hot opaque place to its current transparent state. Such a change is completely inconsistent with the steady state theory.

27.5b   The Temperature of the Background Radiation

Planck curves corresponding to cooler temperatures have lower intensities of radiation and have the peaks of their radiation shifted toward longer wavelengths. We have seen that radiation from the sun, which is 6000 K, peaks in the yellow-green and that radiation from a cool star of 3000 K peaks in the infrared. The universe, much cooler yet, peaks at still longer wavelengths: the peak of the black-body spectrum is at the dividing wavelength between infrared and radio waves. In a moment, we shall see just how cool the universe is.

Now, we recall that Penzias and Wilson measured a particular flux at the wavelength measured by their equipment. This flux corresponds to what a black body at a temperature of only 3 K would emit. Thus we speak of the universe's *3° background radiation**. Following the original measurement at Bell Labs and at Princeton, other groups soon measured values at other radio wavelengths (Fig. 27–12).

Unfortunately, though, for many years we were able to observe only at wavelengths longer than that of the peak, so only the right-hand half of the curve was defined. All the points corresponded to a temperature of approximately 3°, but still, a black-body curve does have a peak. It would have been more satisfying if some of the points measured lay on the left side of the peak.

---

*In the Système International, this would be called 3 K radiation, but we shall join the astronomical community in continuing to call it 3° radiation.

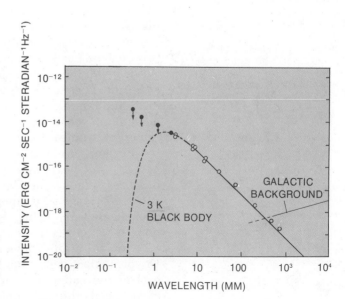

Figure 27–12 Ground-based measurements of the background intensity. The direct measurements are shown as open circles and upper limits set by observations of interstellar molecules are shown as solid dots. The atmosphere prevents ground-based observations from being made at wavelengths shorter than the peak of the curve. The presence of a background of radiation from sources in our galaxy prevents observations to the right of the points on the graph.

The importance of the discovery of the background radiation cannot be overstressed. Dennis Sciama, the British cosmologist, put it succinctly by saying that up to 1965 we carried out all our calculations knowing just one fact: that the universe expanded according to Hubble's law. After 1965, he said, we had a second fact: the existence of the background radiation. That may be a bit oversimplified, but it is essentially true.

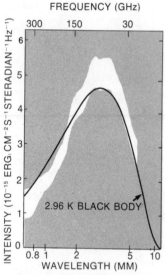

Figure 27–13 The balloon observations of D. Woody and P. L. Richards showing the agreement of the data *(shaded)* with the theoretical curve for a black body *(solid line)*. The radio observations we have discussed previously are all located in the right hand lower portion of the graph.

Points lying on the left side of the peak would prove conclusively that the radiation followed a black-body curve and was not caused by some other mechanism that could produce a straight line, or some other form that happened to mimic a 3° black-body curve in the centimeter region of the spectrum.

Though it would have been desirable to measure a few points on the short wavelength side of the black-body curve, astronomers were faced with a formidable opponent that frustrated their attempts to measure these points: the earth's atmosphere. Our atmosphere absorbs most radiation from the long infrared wavelengths up to about 6 mm. This is a severe limitation as the turnover in the 3° black-body curve occurs at about 2 mm. For some years, we had only an indirect method of showing that the black-body curve indeed peaked at about 2 mm.

More direct observations of the region beyond the peak of the Planck curve were attempted, including observations from rockets, but the attempts failed because of observational difficulties.

In 1975, infrared observations were finally made that seem to have proved unequivocally that the radiation follows a black-body curve. Groups in the United States and England launched instruments in balloons that travelled high enough to get above most of the earth's atmosphere. The astronomers were able to measure the short wavelength side of the black-body curve down to 0.6 mm, and found that their measurements agreed with a black-body curve of 3°. More recent measurements appear in Figure 27–13.

So at present astronomers consider it settled that radiation has been detected that could only have been produced in a big bang. Accepting this clearly rules out the steady state theory. It means that some version of the big bang theories must hold, though it doesn't settle the question of whether the universe will expand forever, or eventually contract.

27.5c The Background Radiation as a Tool

Now that the spectrum has been measured so precisely on both sides of the peak of intensity, we have moved beyond the point of confirming that the radiation is black-body. We can now study its deviations from the general black-body curve and from isotropy.

In particular, it is certainly striking on Figure 27–13 that the observations match a black-body curve very well. However, the spectrum does seem a little below the theoretical curve on the left side and a little above it on the right side. Obtaining this observational curve from the actual data that are measured requires subtracting out a contribution from the atmosphere remaining above the balloon. It is yet to be determined whether the deviation comes from some error in this process or whether it is of cosmological origin.

Further, careful studies have been made from airplanes (Fig. 27–14) of how constant the background radiation is from direction to direction. For this purpose, studies at only one wavelength need be made. It seems that a very slight difference in temperature—about one part in three thousand—has been measured from one particular direction in space to the opposite direction.

This **anisotropy** (deviation from isotropy) is what would result from the Doppler effect if our sun was moving at 350 km s⁻¹ with respect to the background radiation. Since we know the velocity with which our sun is moving as it orbits the center of our galaxy, we can remove the effect of our

A   B

Figure 27–14 *(A)* The U-2 jet used to carry out observations of the anisotropy of the background radiation from an altitude of 20 km. *(B)* Radiometers to measure anisotropy at two different frequencies in the modified upper hatch of the U-2. The larger pair of horn antennas measures radiation at a wavelength of 0.9 mm; the smaller pair of horns measures radiation at a wavelength of 0.5 mm. Two horns are used for each wavelength because comparative measurements can be made more accurately than absolute measurements of the intensity. The observations have been carried out in the northern hemisphere by Berkeley scientists George F. Smoot, Marc V. Gorenstein, Richard A. Muller, and Phil M. Lubin. Similar results have been obtained from balloon flights of Edward S. Cheng, Peter R. Saulson, and David T. Wilkinson of Princeton and Brian E. Corey of MIT.

Observations that relate infrared brightnesses of galaxies with distances measured from radio studies give a result similar to that deduced from the anisotropy in the background radiation. M. Aaronson, J. Huchra, J. Mould, W. T. Sullivan, R. A. Schommer, and G. D. Bothun conclude that our local region of space is moving rapidly with respect to distant parts of the universe. They deduce a Hubble constant of 90 and so an age for the universe of only 10 billion years. Since this is less than the age of some globular clusters, a discrepancy remains to be resolved.

sun's orbital motion. We deduce that our galaxy is moving 520 km s^{-1} relative to the background. Our galaxy is moving in roughly the direction of the constellation Hydra. Since some measurements indicate that the Virgo Cluster of galaxies is moving at a much smaller velocity with respect to us, it may mean that the whole Virgo Cluster is moving in the direction that we derive from the measurements of anisotropy. We do not know any particular reason why our galaxy and the Virgo Cluster should have such a large velocity.

Once the effect of this motion-caused anisotropy is subtracted, the remainder of the radiation seems very isotropic. We can conclude, for example, that the density of the universe does not vary on a large scale by more than one part in a thousand, or else the original density fluctuations in the universe would show up as an anisotropy. Original density fluctuations would also show up as a polarization of the background radiation. No polarization has been detected so far.

27.5d Future Studies of the Background Radiation

COBE (Fig. 27–15), **CO**smic **B**ackground **E**xplorer, is a NASA spacecraft scheduled for 1985 launch to study the background radiation. It should be able to study the spectrum over a wide wavelength range and measure the anisotropy at four wavelengths, both 10 times as precisely as the current measurements.

Figure 27–15 An artist's conception of the COBE (Cosmic Background Explorer) spacecraft, scheduled for launch in 1983.

*27.6 THE CREATION OF THE ELEMENTS

How did the tremendous explosion we call the big bang result in the universe we now know, with galaxies and stars and planets and people and flowers? Obviously, many complex stages of formation have taken place, and what was torn asunder at the beginning of time has now taken the form of an organized system.

We can trace the expansion backward in time (see Fig. 27–4) and can calculate how long ago all the matter we see would have been collapsed together. This calculation leads to the idea that the big bang took place some 13 to 20 billion years ago. It is difficult to comprehend that we can meaningfully talk about the first few **seconds** of that time so long ago. But we can indeed set up sets of equations that satisfy the physical laws we have derived, and make computer simulations and calculations that we think tell us a lot about what happened right after the origin of the universe. In this section we will go back in time beyond the point where the background radiation was set free to travel through the universe.

According to these calculations, the universe at first had a fantastically high density and temperature. After about 5 seconds, the universe had cooled to a few billion degrees. Only simple kinds of matter—protons, neutrons, electrons, neutrinos, and photons—were present at this time. The number of protons and neutrons was relatively small, and they grew relatively more important as the temperature continued to drop.

Antiparticles to the particles mentioned above—antiprotons, antineutrons, positrons, and antineutrinos—were present as well. We are not certain why the present-day universe seems to be almost entirely made of matter instead of antimatter. Perhaps it resulted from an imbalance of matter and antimatter that can be traced all the way back in time.

At earlier times, the exotic particles now being observed by physicists in giant accelerators had been present as well. Even isolated quarks may have been present, though they would have disappeared by the time 1 microsecond went by. The mini black holes described in Section 11.6 may have been formed in this era.

James Cronin of the University of Chicago and Val Fitch of Princeton won the 1980 Nobel Prize in Physics for a discovery about the weak interaction (see Box 8.2) that may explain why more matter than antimatter survived the first instants after the big bang. The results of their experiment imply that the rate at which matter and antimatter were produced in early nuclear reactions could have been different.

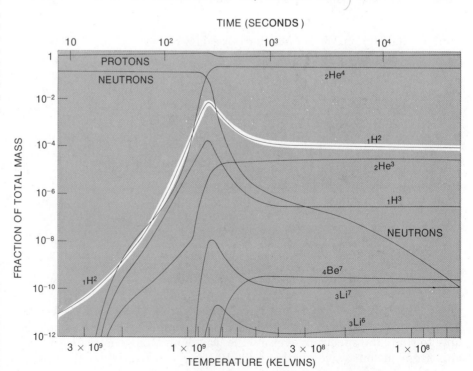

Figure 27–16 A theoretical model, worked out by Robert V. Wagoner of Stanford, that shows the changing relative abundances of the light elements in the first minutes after the big bang. Time is shown on the top axis and the corresponding temperature is shown on the bottom axis.

After about a hundred seconds, the temperature dropped to a billion degrees (which is low enough for a deuterium nucleus to hold together), and the protons and neutrons began to combine into heavier assemblages, the nuclei of the heavier isotopes of hydrogen and elements like helium and lithium (Fig. 27–16). The study of the formation of the elements is called *nucleosynthesis.*

The first nuclear amalgam to form was simply a proton and neutron together. We call this a *deuteron;* it is the nucleus of deuterium, an isotope of hydrogen (see Fig. 8–6). Then two protons and a neutron could combine to form the nucleus of a helium isotope, and then another neutron could join to form ordinary helium. Within minutes, the temperature dropped to 100 million degrees, too low for nuclear reactions to continue. Nucleosynthesis stopped, with about 25 per cent of the mass of the universe in the form of helium. Nearly all the rest was and is hydrogen.

To test this theory, we would like to study the distribution of helium around the universe, to see if it tends to be approximately this percentage everywhere. But helium is very difficult to observe; it has few convenient spectral lines for us to study, and those are usually observable only in hot stars and gaseous nebulae, where a lot of energy is available to excite them. Furthermore, even when we can observe helium in stars we are observing only the surface layers, which do not necessarily have the same abundances of the elements as the interiors of the stars.

Even though "the helium problem" has been extensively studied, one cannot yet conclusively say that the helium is uniformly distributed around the universe, though the evidence tends to support that conclusion.

Only the lightest elements were formed in the early stages of the universe. The nuclear state of mass 5 and again the state of mass 8 are not stable and don't hold together long enough to form still heavier nuclei by the addition of another proton. Only hydrogen (and deuterium) and helium (and perhaps traces of lithium) would be formed in the big bang.

The heavier elements must, then, have been formed at times after the big bang. E. Margaret Burbidge, Geoffrey Burbidge, William A. Fowler, and Fred Hoyle showed in 1957 how both heavier elements and additional amounts of lighter elements can be synthesized in stars. In a stellar interior, processes such as the triple-alpha process (Section 8.4) can get past these mass gaps. Elements are also synthesized in supernova explosions.

At present we believe that element formation took place in two stages. First, light elements were formed soon after the big bang. Later, the heavier elements and additional amounts of most of the lighter elements were formed in stars or in stellar explosions.

Some of the earliest quantitative work on nucleosynthesis in the big bang was described in an article published under the names of Ralph Alpher, Hans Bethe, and George Gamow in 1948. Actually, Alpher and Gamow did the work, and just for fun included Bethe's name in the list of authors so that the names would sound like the first three letters of the Greek alphabet. These letters seem particularly appropriate for an article about the beginning of the universe.

Thus heavy elements in our bodies do not date back to the earliest part of time. They were formed more recently, in the interiors of stars or in supernovae in our own galaxy.

27.7 THE FUTURE OF THE UNIVERSE

How can we predict how the universe will evolve in the distant future? We know that for the present it is expanding, but will that always continue? We have seen that the steady state theory now seems discredited, so let us discuss the alternatives that are predicted by different versions of big bang cosmologies.

Basically, the question is whether there is enough gravity in the universe to overcome the expansion. If gravity is strong enough, then the

Physicists and astronomers were excited in 1980 by a report of an experiment that indicated indirectly that neutrinos have mass. Even if verified, though, we still do not know whether or not this would provide enough mass to close the universe.

expansion will gradually stop, and a contraction will begin. If gravity is not strong enough, then the rate of expansion might slow, but the universe would continue to expand forever) just as a rocket sent up from Cape Canaveral will never fall back to earth if it is launched with a high enough velocity.

To assess the amount of gravity, we must determine the average mass in a given volume of space, that is, the density. It might seem that to find the mass in a given volume we need only count up the objects in that volume: one hundred billion stars plus umpteen billion atoms of hydrogen plus so much interstellar dust, etc. But there are severe limitations to this method, for there are many kinds of mass that are invisible to us. How much matter is in black holes or in the form of neutrinos, for example? Until a few years ago, we couldn't measure the molecular hydrogen in space, and until a few decades ago we couldn't measure the atomic hydrogen in space either, although their abundances are now known to us.

Still another way mass could be hidden is connected to the fact that an intergalactic medium would be hard to detect if it were hot enough. It would thus be almost entirely ionized, and no 21-cm radiation would be emitted. A hot gas, however, would radiate x-rays, so through x-ray observations we can set limits on the amount of hot intergalactic gas that can be present.

We must turn to methods that assess the amount of mass by effects that don't depend on the mass being visible. All mass has gravity, for example, so we are led to study the gravitational attraction on a large scale.

27.7a The Missing Mass Problem

One place to investigate large-scale gravitational attractions is in clusters of galaxies. Clusters of galaxies appear to be large-scale stable configurations that have lasted for a long time. But since galaxies have random velocities in various directions with respect to the center of the mass of the cluster, why don't the galaxies escape?

The *missing mass problem is the discrepancy found when the mass derived from consideration of the motions of galaxies in clusters is compared with the mass that we can observe.*

If we assume that the clusters of galaxies are stable in that the individual galaxies don't disperse, we can calculate the amount of gravity that must be present to keep the galaxies bound. Knowing the amount of gravity in turn allows us to calculate the mass that is causing this gravity. When this supposedly simple calculation is carried out for the Virgo Cluster, it turns out there should be 50 times more mass present than is observed; 98 per cent of the mass expected is not found. This is called the *missing mass problem.*

For any given cluster of galaxies, we can find an *ad hoc* way out of the missing mass problem. For example, we can assume that the Virgo Cluster is exploding rather than being stable, which would mean that we could no longer assume that the outward forces are in balance against gravity. But similar calculations carried out for other clusters also indicate that mass is missing. For the Coma Cluster, a relatively large and nearby cluster of galaxies in the constellation Coma, 90 per cent of the mass seems to be missing.

The missing mass problem seems to be widespread. Still it may simply be the case that when clusters of galaxies form, their members generally have higher velocities than they will have when the galaxy settles down (we say, technically, *relaxes*) to its final state. But we have assumed that we can say that there is a balance between the outward motion we measure and gravity pulling the galaxies inward. On this assumption, we deduce how much mass there must be to provide enough gravity to prevent the galaxies from escaping. But this assumption may not be valid. The rapid motion may be a sign of youth rather than of the presence of a lot of mass. Thus we cannot use this method to calculate how much mass there should be if clusters of galaxies are too young to have "relaxed."

Some clusters seem so old that it is hard to avoid the conclusion that they have "relaxed," leaving us with the missing mass problem very strongly. In Section 27.7b, we shall examine another method of measuring the density. It also does not depend on whether mass is visible or invisible.

27.7b Deuterium and Cosmology

Perhaps the major method now in use to assess the density of the universe concerns itself with the abundance of the light elements. Since these light elements were formed soon after the big bang (as we saw in Section 27.6), they tell us about conditions at the time of their formation. If we can find what the density was then, we can use our knowledge of the rate that the universe has been expanding to determine what the density is now.

The light elements include hydrogen, helium, lithium, beryllium, and boron (see Appendix 4).

But somehow we must distinguish between the amount of these elements that was formed in the big bang, and the amount that has been subsequently formed in stars. This consideration complicates the calculations for helium, for example, because helium formed in stellar interiors and then spewed out in supernovae has been added to the primordial helium formed in the big bang.

Fortunately, the deuterium isotope of hydrogen is free of this complication. We do not think that any is formed in stars, so all the deuterium now in existence was formed at the time of the big bang.

Deuterium has a second property that makes it an important probe of the conditions that existed in the first fifteen minutes after the big bang, when the deuterium was formed. The amount of deuterium that is formed is particularly sensitive to the density of matter at the time of formation. In other words, a slight variation in the primordial density makes a larger change in the deuterium abundance than it does in the abundance of other isotopes or elements. We measure the amount of deuterium relative to the amount of hydrogen.

Deuterium, "heavy hydrogen," contains one proton and one neutron in its nucleus instead of just the single proton of ordinary hydrogen. We are most familiar with deuterium as a constituent of "heavy water," HDO, which is used in atomic reactors. Deuterium is potentially an important constituent of the fusion process we are trying to harness to provide energy on earth.

Why is the ratio of deuterium to (ordinary) hydrogen sensitive to density? Deuterium very easily combines with an additional neutron. This combination of one proton and two neutrons is another hydrogen isotope (tritium). The new neutron quickly decays into a proton, leaving us with a combination of two protons and one neutron, which is an isotope of helium. If the universe was very dense in its first few minutes, then it was easy for the deuterium to meet up with neutrons, and almost all the deuterium "cooked" into helium. If, on the other hand, the density of the universe was low, then most of the deuterium that was formed still survives.

The result of theoretical calculations showing the relation of the amount of surviving deuterium and the density of the universe is shown in Figure 27–17. This graph can be used to find the density that would have been present soon after the big bang, if the ratio of deuterium to hydrogen can be determined observationally.

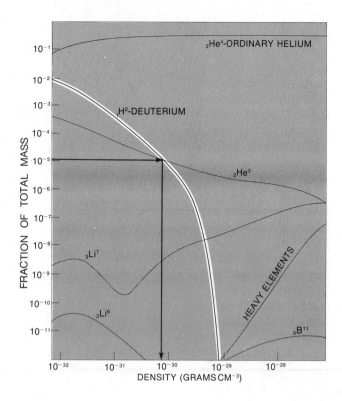

Figure 27–17 The horizontal axis shows the current day cosmic density of matter. From our knowledge of the approximate rate of expansion of the universe, we can deduce what the density was long ago. The abundance of deuterium, outlined in white, is particularly sensitive to the time when the deuterium was formed. Thus present-day observations of the deuterium abundance tell us what the cosmic density is, by following the arrow on the graph.

Unfortunately, deuterium is very difficult to observe. It has no spectral lines accessible to optical observation. Deuterium makes up one part in 6600 of the hydrogen in ordinary seawater, but it was not known how this related to the cosmic abundance of deuterium.

It was not until 1972 that anybody succeeded in detecting deuterium in interstellar space. Diego A. Cesarsky, Alan T. Moffett, and I observed a very faint absorption feature that, if real, is the absorption line caused by the spin-flip transition of deuterium (see Section 23.4). We used the 40-meter (130-foot) radio telescope at Caltech's Owens Valley Radio Observatory and pointed it in the direction of the center of our galaxy, toward the radio source Sagittarius A. Sagittarius A is a strong continuous background source, and the absorption line we probably observed was caused by deuterium located in space between the galactic center, about 10 kiloparsecs away, and the earth.

At the same time, Penzias and Wilson were discovering deuterium as one of the constituents of a "deuterated" molecule, DCN, in the cloud of gas and dust in the constellation Orion. It is difficult to find a deuterium abundance from such molecular observations because the processes of interstellar molecular formation are not well understood. They have since mapped the strength of this DCN emission line in several molecular clouds in our galaxy. The amount of DCN seems to be less in the direction of the center of the galaxy than it is in other directions, probably because there are more stars in that direction consuming the deuterium.

By now, the Lyman lines of deuterium have been detected in absorption by the Princeton experiment aboard the Copernicus Orbiting Astronomical Observatory. They looked in the direction of a nearby star in order to try to detect an absorption line caused by deuterium between the star and the earth. The Copernicus satellite can study, however, only the nearest few hundred parsecs to us, 5 per cent of the distance to the galactic center. The Copernicus results (Fig. 27–18) establish a fraction of deuterium to hydrogen of 14 parts per million.

Further studies of the Copernicus data, which seemed simple to interpret at first, indicate that there may be variations in the deuterium abundance from place to place even relatively nearby. We expect, though, that primordial nucleosynthesis would have led to a homogeneous and isotropic distribution of deuterium, and that our nearby region would be fairly homogeneous. This is being further studied. Nothing in astronomy ever seems to wind up as simple as it seems at first.

Notwithstanding these uncertainties, all the studies of deuterium thus far agree that the amount of deuterium is such that there is not, and was not, enough mass to reverse the expansion of the universe. The universe is apparently open. The results indicate that the universe will not fall back on itself in a big crunch.

27.7c Galaxies, Quasars, and Cosmology

Even though over the years optical searches for deviations from Hubble's law have tended to favor a closed universe, new optical studies by the same group of people give a different result. Allan Sandage of the Hale Observatories and Gustav Tammann of the University of Basel in Switzerland have long studied the distances to galaxies. Joined by Amos Yahil of SUNY at Stony Brook, they have now analyzed the motion of nearby galaxies (Fig. 27–19). They tried to see if there is enough mass in concentrations like the Virgo Cluster to distort the uniform outward flow of galaxies that Hubble's law predicts. The results indicate that there is not enough mass present, and so that the universe is open.

One of the latest contributions comes from space observations of x-rays. A diffuse background of x-ray emission had been discovered and HEAO-1 had measured its spectrum. The shape of the spectrum had suggested that a lot of hot previously undiscovered intergalactic material was present. There might have been enough to "close the universe," that is, to provide enough gravity to make the universe closed.

But when the Einstein Observatory was launched, and was able to examine the x-ray background in detail, this idea changed. Riccardo Giacconi and his colleagues showed that much of this background really resulted from discrete quasars that could not be resolved by the earlier satellites' instruments. Long exposures on arbitrarily chosen fields that were apparently blank on shorter exposures revealed that faint quasars are present (Fig. 27–20). The x-ray emission from these faint, distant quasars seems to be enough to provide much or all of the total x-ray background. So the need for otherwise invisible matter to provide the x-ray background has been reduced, and without this matter the universe again seems open.

27.7d The Long Run

It is interesting that all the above measurements suggest the same result, that the universe will not stop its expansion, and that the galaxies will

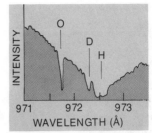

Figure 27–18 Observations from the Copernicus satellite of the Lyman γ (gamma) lines of interstellar deuterium and normal hydrogen. The broad absorption that takes up the whole graph is from hydrogen; its center is marked with an *H*. The corresponding deuterium line is the narrow dip marked with a *D*. It is narrower than the overall hydrogen line because the abundance of deuterium is less than that of hydrogen. A narrow oxygen line (marked with an *O*) also appears.

The contributions of optical astronomy will be revolutionized in a few years with the launch of Space Telescope.

Figure 27–19 The group of nearby galaxies under study. The box is at 75 Mpc. Note the large complex of galaxies in the upper part of the diagram. They surround the Virgo Cluster, and extend all the way down to the Local Group, which is at the center. Yahil, Sandage, and Tamman analyzed the rate at which the nearby galaxies are slowing down because of the gravity from the Virgo complex of galaxies, independent of whether the mass was in the galaxies themselves or in intergalactic matter.

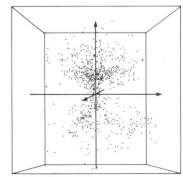

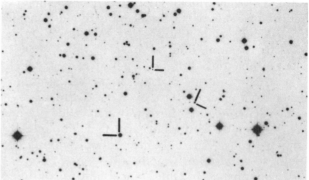

Figure 27–20 A long exposure with the Einstein Observatory, 12 hours in duration, revealed new sources where none had been known before. The x-ray image *(left)* can be compared with an optical plate obtained with the Palomar Schmidt telescope *(right).* Three sources are apparent. The upper source corresponds to a previously unknown quasar of magnitude 19.8 and $z = 0.5$. The source to the right corresponds to a previously unknown quasar of magnitude 17.8 and $z = 1.96$. The source at the left corresponds to an ordinary star of spectral type G0 and magnitude 13. The optical image includes objects down to 22nd magnitude. (X-ray image from Riccardo Giacconi and colleagues at the Harvard-Smithsonian Center for Astrophysics. Optical image from Wallace Sargent and Charles Kowal at the Hale Observatories.)

continue to fly away from each other. Of course, new theoretical ideas or new observations may yet change this picture. In any case, it is a good example of modern astronomical research to note that these results were obtained from large telescopes on earth, radio observations using a variety of techniques, and ultraviolet and x-ray observations using telescopes in space. We are making use of all the methods we know to tell us the future of the universe.

SUMMARY AND OUTLINE

Olbers' paradox and its solutions (Section 27.1)
   The expansion of the universe diminishes energy in quanta
   We are looking back before the formation of the stars
Big bang theory (Section 27.2)
   Cosmological principle: universe is homogeneous and isotropic
   Hubble's constant and the "age" of the universe—13 to 20 billion years
The solutions to Einstein's equations and the cosmological constant (Section 27.3)
Steady state theory (Section 27.4)
   Perfect cosmological principle: universe is homogeneous, isotropic, and unchanging in time
   Continuous creation of matter is implied
   Demise of the steady state theory with discovery of quasars and 3° black-body radiation
Primordial background radiation (Section 27.5)
   Provides strong support for the big bang theory
   Evidence for 3° black-body radiation
   Direct radio measurements were made but only on one side of the peak of the black-body curve

New infrared observations provide convincing evidence for the radiation corresponding to a 3° black body because they confirm the peaked nature of the radiation
   Anisotropy indicates motion of the sun and the Virgo Cluster
Creation of the elements (Section 27.6)
   Origin of the light elements in the first minutes after the big bang
   Heavier elements formed in the interiors of stars and in supernovae
Future of the universe (Section 27.7)
   The missing mass problem: discrepancy between mass derived by studying motions of galaxies in clusters of galaxies and the visible mass
   Deuterium-to-hydrogen ratio depends sensitively on cosmic density
      All deuterium observed was formed in the first minutes after the big bang
      Current evidence from deuterium is that universe is open, i.e., expansion will not stop
If neutrinos have mass, the universe must be closed

Optical evidence from motions of nearby galaxies agrees that universe is open

Discovery from Einstein Observatory that x-ray background is really from faint quasars indicates that little hot interstellar gas is present and so agrees with open universe

QUESTIONS

1. Hindsight has allowed solutions to be found to Olbers' paradox. For example, knowing that the universe is expanding, we can come up with a solution. However, based on the reasoning in this chapter, do you think it is possible that scientists might have used Olbers' paradox to reach the conclusion that the universe is expanding before it was determined observationally? Explain.

2. We say in the chapter that the universe is homogeneous on a large scale. What is the largest scale on which the universe does not seem to appear homogeneous?

3. List observational evidence in favor of and against each of the following theories: (a) the big bang theory, and (b) the steady state theory.

4. What is "perfect" about the perfect cosmological principle?

5. If galaxies and radio sources increase in luminosity as they age, when we look to great distances then we are seeing them when they were less intense than they are now. We thus cannot compare them directly to similar nearby galaxies or radio sources. If this increase in luminosity with time is a valid assumption, does it tend to make the universe seem to decelerate at a greater or lesser rate than the actual rate of deceleration? Explain.

†6. For a Hubble constant of 50 kms^{-1} MPC^{-1} show how you calculate the Hubble time, the age of the universe ignoring the effect of gravity. (Hint: Take $1/H_0$, and simplify units so that only units of time are left.)

7. In actuality, if current interpretations are correct, the 3° background radiation is only indirectly the remnant of the big bang, but is directly the remnant of an "event" in the early universe. What event was that?

†8. Apply Wien's displacement law to the solar photospheric temperature and spectral peak in order to show where the spectrum of a 3° blackbody peaks.

9. What does the anisotropy of the background radiation tell us?

10. Why must we resort to indirect methods to find out if the universe is open or closed?

11. Why is determining the abundance of deuterium so important?

12. What are two pieces of evidence that suggest that the universe is open?

---

†This indicates a question requiring a numerical answer.

TOPICS FOR DISCUSSION

1. Which is more appealing to you: the cosmological principle or the perfect cosmological principle? Discuss why we should adopt one or the other as the basis of our cosmological theory.

2. What effect would finding out whether the universe will expand forever, will begin to contract into a final black hole, or will contract and then oscillate with cycles extending into infinite time, have on the conduct of one's life or thoughts?

CHAPTER 28

ASTRONOMY NOW

AIMS:
To summarize some of the future
developments that we may expect in
astronomical research, and to discuss
the mechanics of carrying out
astronomical research

We have seen in the preceding chapters that astronomy encompasses
not only a wide range of subjects but also a wide range of techniques. No
longer can a single astronomer hope to be expert in all phases of the science.
We have consistently seen how new discoveries often open more questions
than they answer. The advance of science seems to leave us farther and
farther behind!

Up to this point, this book has been organized by type and size of object,
from stars (and their planets) at the beginning up to the universe as a whole
at the end. Let us take a brief tour of topics that are now under active consideration, organized, for a change, by wavelength in the electromagnetic
spectrum. The discussion should give us an idea of new discoveries that are
expected in the next few years.

28.1 CURRENT PLANS

High-energy processes in the universe give off radiation in the
shortest wavelengths of the electromagnetic spectrum, namely gamma-rays
and x-rays. These high-energy processes give off or accelerate cosmic rays
as well. Of course, we are dependent on satellites above the earth's at-

This Voyager view from the far side of Saturn shows the backlighted rings as the inverse in
brightness of the rings as seen from above. The C-ring, the inner faint ring as seen from earth,
and the Cassini division, entirely dark in the view from earth, are bright.

mosphere to study these electromagnetic waves and cosmic ray particles efficiently. The continuing study of data from the three High-Energy Astronomy Observatories will certainly continue to reveal significant new results. In the meantime, theoreticians continue to make progress in understanding the processes that make the high-energy radiation and the cosmic rays.

Studies from the Einstein Observatory should also lead to a better understanding of black holes. Much data has been recorded from Cygnus X-1 and other black hole candidates, and we look forward to hearing the discoveries. After all, we want to know whether the current tendency to invoke the presence of black holes to explain puzzling phenomena is a panacea or a fad. At present, prospective black holes are being found all over on many scales of size and mass.

In the next longer wavelength region, the International Ultraviolet Explorer satellite continues to study the stars, nebulae, galaxies, interstellar space, quasars, and just about everything else. The launch of the Space Telescope in 1983, with its ultraviolet and visible observing ability, its increased resolution, and its ability to observe fainter objects, will make the next edition of this book differ from this one even more than this one deviates from its predecessor. In solar research, the data from the Solar Maximum Mission continue to be studied and together with other new data from the ground should improve our understanding of solar activity. We should also come to understand in better fashion the interplay between the sun and the earth, including the solar effect on terrestrial climate.

In visible light, a number of new giant ground-based telescopes are just beginning their operation. Our ability to observe the optical sky has just increased manyfold. Electronic devices to receive and record the photons that are focused by the huge mirrors increase the capability of even the older telescopes. Among the many topics of current interest investigated in this region of the spectrum are the mechanisms by which mass is lost by stars and nebulae to the interstellar medium and the distribution of the most distant

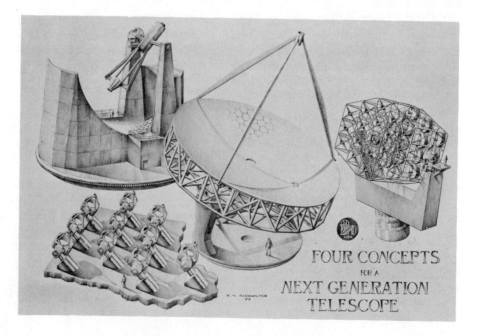

Figure 28–1 A study group at the Kitt Peak National Observatory is working on the Next Generation Telescope (NGT). Whether it will be a mosaic of small mirrors, a collection of medium-sized telescopes, or something else entirely must be decided. The equivalent aperture will be about 25 meters (1000 inches).

FOUR CONCEPTS
FOR A
NEXT GENERATION
TELESCOPE

Figure 28-2

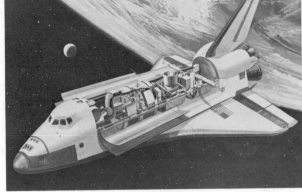

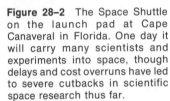

Figure 28-3

quasars and galaxies. Even though many new facilities are available, we continue to plan for telescopes of the future (Fig. 28-1).

The infrared is a burgeoning area of research. We are seeing stars in formation and so are getting to understand better the process of stellar birth. We can study dust wherever it is in space—in the solar system or between the stars. New telescopes devoted largely to infrared studies have recently become operational. The availability of new sites like Mauna Kea and new techniques like that used for the Multiple Mirror Telescope make us able to observe the infrared sky much better. Also, electronic instrumentation used to study this part of the spectrum continues to improve rapidly.

Plans for future telescopes in space are highly dependent on NASA's Space Shuttle (Fig. 28-2). It will launch the Space Telescope and many other telescopes as well. At times it will carry the European Spacelab (Fig. 28-3), which will provide a shirt-sleeve environment for scientists in space. The non-reusable European Ariane rocket will also launch spacecraft.

In radio astronomy, there will continue to be tremendous strides in making high-resolution pictures of astronomical objects. The VLA (Very Large Array) has recently been completed. It enables large-scale mapping of the sky and detailed study of individual objects to be carried out in a manner and quantity never before possible (Fig. 28-4). In spectral-line radio astronomy, scientists have shifted their interest away from mere discovery

Figure 28-2 The Space Shuttle on the launch pad at Cape Canaveral in Florida. One day it will carry many scientists and experiments into space, though delays and cost overruns have led to severe cutbacks in scientific space research thus far.

Figure 28-3 An artist's conception of Spacelab being carried into orbit by the Space Shuttle. Astronauts are working in shirt-sleeves in a pressurized module. Telescopes and antennas are in the vacuum of space behind them.

Figure 28-4 Several telescopes of the VLA in New Mexico.

of new molecules toward using the molecular studies to analyze physical conditions. The National Science Foundation is considering the construction of a 25-m radio telescope for use in the millimeter range, which would aid in such studies. Mauna Kea is a possible site. The molecular studies overlap considerably with studies in the infrared, particularly for understanding star formation.

Direct sampling is limited to our solar system, but whenever possible is of the greatest value. Measurements taken on the surface of the moon and of Mars and from probes descending through the atmosphere of Venus have enabled us to undertake detailed analyses that were otherwise impossible. The Galileo probe through Jupiter's clouds and perhaps a comet probe should similarly expand our knowledge.

This is but the beginning of a long list of topics of active research. Different astronomers put high priority on very different areas. I suppose that we are fortunate that we don't all want to work on the same thing at the same time! There is surely no shortage of research topics available for astronomers.

Perhaps the most significant gains will come when observations from several different parts of the spectrum are put together. The sun, for example, is under careful scrutiny from wavelengths down below 0.01 angstroms up to wavelengths above 3 meters, a factor of more than a trillion (10^{12}).

In a textbook of astronomy, it is relatively easy to get across a sense of progress in observational results compared to transmitting a sense of the importance of theoreticians in the whole enterprise. The computer, especially, has enhanced the ability of theoreticians to make detailed models of what is occurring in a stellar atmosphere, in a stellar interior, or in a disk of material around a black hole.

Even without the computer, our theoretical understanding of certain basic processes has increased greatly. For example, the study of properties of the interaction of matter and a magnetic field is a field that is tongue-twistingly named *magnetohydrodynamics* and often abbreviated MHD. Our improved understanding of MHD is proving useful not only for explaining astrophysical situations but also for immediate application on earth in generating power. Theoreticians study MHD both with and without computers. The ability of a scientist to sit quietly, confidently calculating on a computer or with pencil and paper how bodies 10^{26} (100,000,000,000,000,000,000,000,000) times the astronomer's size will act, remains at the heart of modern research.

28.2 THE SCIENTIFIC ENTERPRISE

Science of all kinds, including astronomy, grows increasingly dependent on large instrumentation. It becomes more and more difficult—and is really nigh unto impossible—to carry out significant research as a lone individual. Even theoreticians are often dependent on the availability of large computers, and their calculations may be limited in accuracy mainly by the amount of computer time that is available to them.

All this points to the importance of funding to provide the equipment and the financial backing for the scientists' time and effort. Although many smaller sources of funds exist, the role of government funding has increased to the point where it is clearly the dominant source (Fig. 28–5). A govern-

"No one has yet programmed a computer to be of two minds about a hard problem, or to burst out laughing. . . ."

Lewis Thomas,
LIVES OF A CELL
(New York: Viking Press, 1974)

Figure 28–5 The source of funding for most scientific research in astronomy.

ment supports scientific research for a variety of reasons. Its interest in supporting science for the sake of knowledge itself is one of them, just as governments often support museums of art, or opera companies.

But this is surely not the major reason why governments support science. Usually there is a desire for practical results. Astronomers do not ordinarily undertake their investigations for practical ends, but over the past centuries a number of basic ideas and processes of practical importance have been discovered in an astronomical context. We can confidently expect that today's astronomical research will pay off similarly in the long run.

In the United States, the National Science Foundation is the organization that is the primary source of support for basic research in astronomy (Fig. 28–6). The astronomy section of the NSF has been organized into two divisions. One, the Astronomy Centers Section, has responsibility for the national observatories that are wholly supported by the NSF: the Kitt Peak National Observatory, the National Radio Astronomy Observatory, the Arecibo Ionospheric Observatory, and the National Center for Atmospheric Research (where the atmospheres of stars are studied as well as the atmosphere of the earth). Second, the NSF gives out a number of grants to scientists at colleges and universities, sometimes to allow them to maintain facilities or projects at their home institutions and sometimes to allow them to use the shared national facilities.

In Canada, in most European countries, in Australia, in the Soviet Union, and elsewhere, government organizations also have important influences on the state of astronomical research.

Astronomy has also certainly benefited by the interest of NASA in astronomical problems. The cost of space experiments is many times the cost of ground-based observing, so one must be prepared to think on another scale of expenses when one considers using a satellite to make observations. Of course, most of the costs of experimenting in space are not really from the astronomical experiments but are rather engineering costs. And the tremendous amounts of money that are spent do not just go up into outer space; the money is spent here on earth, largely to pay contractors and subcontractors to build, test, and maintain the equipment.

Even NASA's yearly budget is only enough to run the Department of Health and Human Services for about one week.

Astronomers often benefit from apparatus or projects that were set up mainly for other purposes. The whole Apollo program to land astronauts on

Figure 28–6 From the 1980 eclipse expedition to India.

the moon by the end of the decade of the 1960's had important political implications and implications for national pride and goals; these led the United States government to spend on the project sums previously unparalleled in the history of scientific support. But for whatever reasons the astronauts were sent to the moon, or for whatever reasons Skylab was sent into orbit, the astronomers were at the ready to make whatever observations they could with the new techniques available.

Taking advantage of the largess of NASA brings risks with it too. NASA has invested so much effort and money in developing the Space Shuttle, a project that has been fraught with difficulties and overruns, that funds available for astronomical research itself have recently been severely cut.

Non-governmental organizations also sometimes fund astronomical research. For example, the National Geographic Society has supported a series of eclipse expeditions and financed the fundamentally important National Geographic Society-Palomar Observatory Sky Survey with the Schmidt camera on Palomar. The Research Corporation, a private corporation that derives its income from patents assigned to it, sometimes gives grants for astronomical projects.

28.3 GRANTS

How does a scientist, say a professor at a university, go about getting a grant from the National Science Foundation? The first stage is to have a clear idea of the project under consideration, which would normally require some preliminary research. This preliminary research must be supported in some other way, perhaps by funds from the university. Eventually the project would be developed to a point where a proposal for a grant could be written (Fig. 28–7).

The proposal might be some two dozen pages in length. It would contain a clear exposition of the proposed project, its importance, and why it should be undertaken in this specific manner. The scientist's ability to undertake the project would have to be justified; research ability demonstrated by prior published papers would be an important factor in assessing this. The budget is also an important part of the proposal.

At the astronomy section of the NSF, the program managers would send out copies of the proposal to perhaps half a dozen scientists across the country who are knowledgeable in the specific field of research under discussion. These scientists are asked to act as anonymous *referees* of the proposal in a process known as *peer review*. The referees donate their time to this purpose as part of every scientist's role in keeping the national scientific enterprise running.

The program managers try to assess the referees' reports and assign ratings and priorities to each proposal. Then they try to divide their limited funds appropriately. There are always many more acceptable proposals than can be funded.

Figure 28–7 A mock proposal, including most of the kinds of descriptions and budgeting stages that a real proposal would have.

HOPKINS OBSERVATORY
WILLIAMS COLLEGE
WILLIAMSTOWN, MASS. 01267
To: Universal Creation Foundation
REQUEST FOR SUPPLEMENT TO U.C.F. GRANT
#000-00-00000-001
"CREATION OF THE UNIVERSE"
This report is intended only for the
internal uses of the contractor.
Period: Present to Last Judgment
Principal Creator: Creator
Proposal Writers and Contract Monitors:
Jay M. Pasachoff and Spencer R. Weart

BACKGROUND

Under a previous grant (U.C.F. Grant #000-00-00000-001), the Universe was created. It was expected that this project would have lasting benefits and considerable spinoffs, and this has indeed been the case. Darkness and light, good and evil, and Swiss Army knives were only a few of the useful concepts developed in the course of the Creation. It was estimated that the project would be completed within four days (not including a mandated Day of Rest, with full pay), and the 50% overrun on this estimate is entirely reasonable, given the unusual difficulties encountered. Infinite funding for this project was requested from the Foundation and granted. Unfortunately, this has not proved sufficient. Certain faults in the original creation have become apparent, which it will be necessary to correct by means of miracles. Let it not be said, however, that we are merely correcting past errors; the final state of the Universe, if this supplemental request is granted, will have many useful features not included in the original proposal.

PROGRESS TO DATE

Interim progress reports have already been submitted ("The Bible," "The Koran," "The Handbook of Chemistry and Physics," etc.). The millennial report is currently in preparation, and a variety of publishers for the text (tentatively entitled, "Oh, Genesis!") will be created. The Gideon Society has applied for the distribution rights. Full credit will be given to the Foundation.

Materials for the Universe and for the Creation of Man were created out of the Void at no charge to the grant. A substantial savings was generated when it was found that materials for Woman could be created out of Man, since the establishment of Anti-Vivisection Societies was held until Phase Three. Given the limitations of current eschatological technology, it can scarcely be denied that the Contractor has done His work at a most reasonable price.

SUPPLEMENT

We cannot overlook a certain tone of dissatisfaction with the Creation which has been expressed by the Foundation, not to mention by certain of the Created. Let us state outright that this was to be expected, in view of the completely unprecedented nature of the project. The need for a supplement is to be ascribed solely to inflation (not to be confused with expansion of the Universe, which was anticipated). Union requests for the accrual of Days of Rest at the rate of one additional Day per week per millennium ($Dw^{-1}m^{-1}$) must also be met. Concerning the problem of Sin, we can assure you that extensive experimentation is under way. Considerable experience is being accumulated and we expect a breakthrough before long. When we are satiated with Sin, we shall go on to consider Universal Peace.

We cannot deny—in view of the cleverness of the Foundation's auditors—that the bulk of the supplemental funds will go to pay off old bills. Nevertheless we do not anticipate the need for future budget requests, barring unforeseen circumstances. If this project is continued successfully, additional Universe—anti-Universe pairs can be created without increasing the baryon number, and we would keep them out of the light cone of the Original. By the simple grant of an additional Infinity of funds (and note that this proposal is merely for Aleph Null), the officers of the Foundation will be able to present their Board of Directors with the accomplished Creation of one or more successful Universes, instead of the current incomplete one.

We will attempt to minimize the additional delays that may temporarily exist during the changeover from fossil fuels to fusion for some minor locations in the Universe. For the time being, elements with odd masses (hydrogen, lithium, etc.) will be created only on odd days of the month and those with even masses (helium, beryllium, etc.) on the even days, except, of course, for Sundays. The 31st of each month will be devoted to creation of trans-uranic elements. The Universe has been depleted of deuterium; new creation of this will take place only on February 29th in leap years.

PROSPECTIVE BUDGET

Remedial miracles on fish of the sea	∞
Remedial miracles on fowl of the air	∞
Creeping things that creep upon the earth, etc.	∞
Hydrogen	n/c (created)
Heavier elements	n/c (nucleosynthesis)
(Note: The carbon will be reclaimed and ecologically recycled)	
Mountains (Sinai, Ararat, Palomar)	∞
Extra quasars, neutron stars	∞
Black holes (no-return containers)	∞
Miscellaneous, secretarial, office supplies, etc.	∞
Telephone installation (Princess model, white, one-time charge, tax included)	$16.50

SALARIES

Creator (1/4 time)	at His own expense
Archangels	
Gabriel	1 trumpet (Phase 5)
Beelzebub	misc. extra brimstone (low sulfur)
Others	assorted halos
Prophets	
Moses	stone tablets (to replace breakage)
Geniuses	finite

N.B. Due to the Foundation's regulations and changes in Exchange Rates, we have not yet been able to reimburse Euclid (drachmae), Leonardo (lire), Newton (pounds sterling), Descartes (francs), or Reggie Jackson (dollars). Future geniuses will be remunerated indirectly via the Alfred Nobel Foundation.

Graduate students (2 at 2/5 time)	reflected glory

MONITORING EQUIPMENT & MISC.

1 5-meter telescope (maintenance)	finite
Misc. other instruments, particle accelerators, etc.	large but finite
Travel to meetings	∞
Pollution control equipment	+40%
Total	$\infty + 40\% = \infty$
Overhead (51.97%)	∞
Total funds requested	∞

Starting date requested:

Immediate. Pending receipt of supplemental funds, layoffs are anticipated to reach the 19.5% level in the ranks of angels this quarter.

Figure 28–7 *See opposite page for legend*

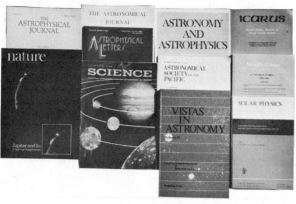

Figure 28–8

Figure 28–9

Figure 28–8 Some of the professional journals from around the world.

Figure 28–9 Some of the magazines and journals that regularly publish non-technical articles about astronomy.

The process of review can take six months or more, so prospective proposers have to decide long in advance what they want to do. One day, successful proposers receive in the mail copies of letters announcing the awarding of grants. The projects can then get under way.

When the research is complete and a paper is prepared for publication, peer review plays an important part as well. The papers that astronomers submit to most of the major journals are sent to anonymous reviewers, who not only pass on the suitability of work for publication but also sometimes make suggestions for improvement.

There is a wide variety of journals for scientific articles (Fig. 28–8). Many other magazines and journals publish more general articles about astronomy that can be read by both professionals and the general public (Fig. 28–9).

28.4 THE NEW ASTRONOMY

In the decade of the eighties, astronomy is maintaining its momentum as one of the most active sciences. It is exploding with new results.

The discussions in this book have necessarily been limited to a fraction of the topics with which contemporary astronomers concern themselves. Yet we have mentioned topics and ongoing investigations that will provide a basis for the astronomy of the future.

Close to earth, we still have much data to analyze from recent visits to most of the planets. Some planetary exploration will continue (Fig. 28–10), depending on budgetary limitations. We are better prepared now to study comets than we have ever been, and the reappearance of Halley's comet in 1985 or of another bright comet will lead to much excellent data.

The sun remains the center of active investigation, not only for its fundamental properties as the nearest example to earth of a star but also because of its effect on our environment in space (Fig. 28–11). From neutrinos and particles in the solar wind to the entire electromagnetic spectrum, we can study the sun in more ways that we can study any other object in the universe.

Figure 28–10 This artist's concept depicts the Galileo spacecraft releasing a probe that will penetrate deep into the atmosphere of the planet Jupiter. The event will occur when the spacecraft is 100 days from Jupiter encounter.

The stars and galaxies are also under scrutiny across the electromagnetic spectrum. They are all much fainter than the sun, of course, and so are correspondingly more difficult to study. But we seem to be approaching the point where we begin to understand the fundamental processes that govern stars and galaxies (although I would hate to read these words from the perspective of a hundred years hence, when it will be seen that we really now know nothing at all).

We are also making progress on our understanding of cosmology, a subject that has been at the center of human thought for over two thousand years but about which we have a lot to learn. The detection of the primordial radiation has brought us tidings of the big bang itself, and we hope that careful measurements of this radiation in different directions will give us more information about the inhomogeneities in the first stages of the universe.

Astronomy has come a long way since early humans first noticed the objects in the sky and their motions. The pace of progress is dramatically increasing, and it is exciting to see the discoveries made in such quick succession. Still, we have at least as far to go.

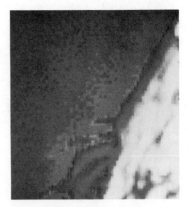

Figure 28–11 Observations of a solar prominence in the ultraviolet, taken with the Solar Maximum Mission satellite in 1980.

SUMMARY AND OUTLINE

Astronomy, from short wavelengths to long, plus direct sampling (Section 28.1)
The scientific enterprise (Section 28.2)
  Importance of funding
  National Science Foundation (NSF) oversees most funding for astronomy in the U.S.

NASA space missions
Grants (Section 28.3)
  How a proposal is prepared and submitted
  Peer review
Astronomy, from the earth to the outer universe (Section 28.4)

TOPICS FOR DISCUSSION

1. What are the astronomical projects you think are the most important for the next decade?
2. What is the value of astronomy to you?

APPENDICES

Appendix 1

Greek Alphabet

Upper Case	Lower Case	Pronunciation
A	α	alpha
B	β	beta
Γ	γ	gamma
Δ	δ	delta
E	ϵ	epsilon
Z	ζ	zeta
H	η	eta
Θ	θ	theta
I	ι	iota
K	κ	kappa
Λ	λ	lambda
M	μ	mu
N	ν	nu
Ξ	ξ	xi
O	o	omicron
Π	π	pi
P	ρ	rho
Σ	σ	sigma
T	τ	tau
Υ	υ	upsilon
Φ	ϕ	phi
X	χ	chi
Ψ	ψ	psi
Ω	ω	omega

Appendix 2

Measurement Systems

Metric units

Basic units

length	meter (m)
volume	liter (l)
mass	gram (gm or g)
time	second (sec or s)

Other metric units

$$1 \text{ micron } (\mu) = 10^{-6} \text{ meter}$$
$$1 \text{ Ångstrom } (\text{Å or A}) = 10^{-10} \text{ meter}$$
$$= 10^{-8} \text{ cm}$$

Prefixes for use with basic units of metric system

Prefix	Symbol	Power		Equivalent
tera	T	10^{12} =	1,000,000,000,000	Trillion
giga	G	10^{9} =	1,000,000,000	Billion
mega	M	10^{6} =	1,000,000	Million
kilo	k	10^{3} =	1,000	Thousand
hecto	h	10^{2} =	100	Hundred
deca	da	10^{1} =	10	Ten
– – –	–	10^{0} =	1	One
deci	d	10^{-1} =	.1	Tenth
centi	c	10^{-2} =	.01	Hundredth
milli	m	10^{-3} =	.001	Thousandth
micro	μ	10^{-6} =	.000001	Millionth
nano	n	10^{-9} =	.000000001	Billionth
pico	p	10^{-12} =	.000000000001	Trillionth
femto	f	10^{-15} =	.000000000000001	
atto	a	10^{-18} =	.000000000000000001	

Examples: 1000 meters = 1 kilometer = 1 km
10^6 hertz = 1 megahertz = 1 MHz
10^{-3} s = 1 millisecond = 1 ms

Some other units used in astronomy

Energy:	1 joule (J) = 10^7 ergs
	1 electron volt (eV) = 1.60207×10^{-12} ergs
Power:	1 watt (W) = 1 joule per s $(\text{J} \cdot \text{s}^{-1})$
Frequency:	hertz (Hz) = cycles per s (s^{-1})
	1 megahertz (MHz) = 10^6 Hz

Conversion factors

1 cm = 0.3937 in		1 in = 25.400 mm
1 m = 1.0936 yd		= 2.54 cm
1 km = 0.6214 mi		1 ft = 0.3048 m
= 5/8 mi		1 yd = 0.9144 m
1 kg = 2.2046 lb		1 mi = 1.6093 km
1 gm = 0.0353 oz		= 8/5 km
		1 lb = 0.4536 kg

Appendix 3

Basic Constants

Physical constants

Speed of light*	c	$= 299\ 792\ 458\ \text{m} \cdot \text{s}^{-1}$
Constant of gravitation*	G	$= 6.672 \times 10^{-11}\ \text{m}^3 \cdot \text{kg}^{-1} \cdot \text{s}^{-2}$
Planck's constant	h	$= 6.6262 \times 10^{-27}\ \text{erg} \cdot \text{s}$
Boltzmann's constant	k	$= 1.3806 \times 10^{-16}\ \text{erg} \cdot \text{kelvin}^{-1}$
Stefan-Boltzmann constant	σ	$= 5.66956 \times 10^{-5}\ \text{erg} \cdot \text{cm}^{-2} \cdot \text{deg}^{-4} \cdot \text{s}^{-1}$
Wien displacement constant	$\lambda_{max}T$	$= 0.289789\ \text{cm} \cdot \text{K} = 28.9789 \times 10^6\ \text{Å} \cdot \text{K}$
Mass of hydrogen atom	m_H	$= 1.6735 \times 10^{-24}\ \text{gm}$
Mass of neutron	m_n	$= 1.6749 \times 10^{-24}\ \text{gm}$
Mass of proton	m_p	$= 1.6726 \times 10^{-24}\ \text{gm}$
Mass of electron	m_e	$= 9.1096 \times 10^{-28}\ \text{gm}$
Rydberg's constant	R	$= 1.09677 \times 10^5\ \text{cm}^{-1}$

Mathematical constants

$\pi = 3.1415926536$
$e = 2.7182818285$

Astronomical constants

Astronomical unit*	1 A.U.	$= 1.495\ 978\ 70 \times 10^{11}\ \text{m}$
Solar parallax*	$\pi_\odot$	$= 8.794148\ \text{arc sec}$
Parsec	pc	$= 206\ 264.806\ \text{A.U.}$
		$= 3.261633\ \text{light years}$
		$= 3.085678 \times 10^{18}\ \text{cm}$
Light year	ly	$= 9.460530 \times 10^{17}\ \text{cm}$
		$= 6.324 \times 10^4\ \text{A.U.}$
Tropical year (1900)* — equinox to equinox		$= 365.24219878\ \text{ephemeris days}$
1 Julian century*		$= 36525\ \text{days}$
1 day*		$= 86400\ \text{s}$
Sidereal year		$= 365.256366\ \text{ephemeris days}$
		$= 3.155815 \times 10^7\ \text{s}$
Mass of sun*	$M_\odot$	$= 1.9891 \times 10^{33}\ \text{gm}$
Radius of sun*	$R_\odot$	$= 696000\ \text{km}$
Luminosity of sun	$L_\odot$	$= 3.827 \times 10^{33}\ \text{erg} \cdot \text{s}^{-1}$
Mass of earth*	$M_\oplus$	$= 5.9742 \times 10^{27}\ \text{gm}$
Equatorial radius of earth*	$R_\oplus$	$= 6\ 378.140\ \text{km}$
Mean distance center of earth to center of moon		$= 384\ 403\ \text{km}$
Radius of moon*	R_M	$= 1\ 738\ \text{km}$
Mass of moon*	M_M	$= 7.35 \times 10^{25}\ \text{gm}$
Solar constant	S	$= 135.3\ \text{mW} \cdot \text{cm}^{-2}$
Direction of galactic center	α	$= 17^h42.4^m$
(precessed for 1950)	δ	$= -28° 55'$

---

*Adopted as "IAU (1976) system of astronomical constants" at the General Assembly of the International Astronomical Union.

Appendix 4

Elements and Solar Abundances*

Atomic Number	Element	Name	Atomic Weight	Solar Abundance (Relative to Hydrogen = 10^{12})
1	H	hydrogen	1.01	10^{12}
2	He	helium	4.00	6.3×10^{10}
3	Li	lithium	6.94	10×10^1
4	Be	beryllium	9.01	1.4×10^1
5	B	boron	10.81	1.3×10^2
6	C	carbon	12.01	4.2×10^8
7	N	nitrogen	14.01	8.7×10^7
8	O	oxygen	16.00	6.9×10^8
9	F	fluorine	19.00	3.6×10^4
10	Ne	neon	20.18	3.7×10^7
11	Na	sodium	22.99	1.9×10^6
12	Mg	magnesium	24.31	4.0×10^7
13	Al	aluminum	26.98	3.3×10^6
14	Si	silicon	28.09	4.5×10^7
15	P	phosphorus	30.97	3.2×10^5
16	S	sulphur	32.06	1.6×10^7
17	Cl	chlorine	35.45	3.2×10^5
18	Ar	argon	39.95	1.0×10^6
19	K	potassium	39.10	1.4×10^5
20	Ca	calcium	40.08	2.2×10^6
21	Sc	scandium	44.96	1.1×10^3
22	Ti	titanium	47.90	1.1×10^5
23	V	vanadium	50.94	1.0×10^4
24	Cr	chromium	52.00	5.1×10^5
25	Mn	manganese	54.94	2.6×10^5
26	Fe	iron	55.85	3.2×10^7
27	Co	cobalt	58.93	7.9×10^4
28	Ni	nickel	58.71	1.9×10^6
29	Cu	copper	63.55	1.1×10^4
30	Zn	zinc	65.37	2.8×10^4
31	Ga	gallium	69.72	6.3×10^2
32	Ge	germanium	72.59	3.2×10^3
33	As	arsenic	74.92	
34	Se	selenium	78.96	
35	Br	bromine	79.90	
36	Kr	krypton	83.80	
37	Rb	rubidium	85.47	4.0×10^2
38	Sr	strontium	87.62	7.9×10^2
39	Y	yttrium	88.91	1.3×10^2
40	Zr	zirconium	91.22	5.6×10^2
41	Nb	niobium	92.91	7.9×10^1
42	Mo	molybdenum	95.94	1.4×10^2
43	Tc	technetium	98.91	
44	Ru	ruthenium	101.07	6.8×10^1
45	Rh	rhodium	102.91	2.5×10^1
46	Pd	palladium	106.4	3.2×10^1
47	Ag	silver	107.87	7.1
48	Cd	cadmium	112.40	7.1×10^1
49	In	indium	114.82	4.5×10^1
50	Sn	tin	118.69	1.0×10^2
51	Sb	antimony	121.75	1.0×10^1
52	Te	tellurium	127.60	
53	I	iodine	126.90	
54	Xe	xenon	131.30	
55	Cs	caesium	132.91	$<7.9 \times 10^1$

Appendix 4 continued on the opposite page

Atomic Number	Element	Name	Atomic Weight	Solar Abundance (Relative to Hydrogen = 10^{12})
56	Ba	barium	137.34	1.2×10^2
57	La	lanthanum	138.91	1.3×10^1
58	Ce	cerium	140.12	3.5×10^1
59	Pr	praseodymium	140.91	4.6
60	Nd	neodymium	144.24	1.7×10^1
61	Pm	promethium	146	
62	Sm	samarium	150.4	5.2
63	Eu	europium	151.96	5.0
64	Gd	gadolinium	157.25	1.3×10^1
65	Tb	terbium	158.93	
66	Dy	dysprosium	162.50	1.1×10^1
67	Ho	holmium	164.93	
68	Er	erbium	167.26	5.8
69	Tm	thulium	168.93	1.8
70	Yb	ytterbium	170.04	7.9
71	Lu	lutetium	174.97	5.8
72	Hf	hafnium	178.49	6.3
73	Ta	tantalum	180.95	
74	W	tungsten	183.85	5.0×10^1
75	Re	rhenium	186.2	≤ 0.5
76	Os	osmium	190.2	5.0
77	Ir	iridium	192.2	7.1
78	Pt	platinum	195.09	5.6×10^1
79	Au	gold	196.97	5.6
80	Hg	mercury	200.59	$< 1.3 \times 10^2$
81	Tl	thallium	204.37	7.9
82	Pb	lead	207.19	8.5×10^1
83	Bi	bismuth	208.98	$< 7.9 \times 10^1$
84	Po	polonium	210	
85	At	astatine	210	
86	Rn	radon	222	
87	Fr	francium	223	
88	Ra	radium	226.03	
89	Ac	actinium	227	
90	Th	thorium	232.04	1.6
91	Pa	protactinium	230.04	
92	U	uranium	238.03	< 4.0
93	Np	neptunium	237.05	
94	Pu	plutonium	242	
95	Am	americium	242	
96	Cm	curium	245	
97	Bk	berkelium	248	
98	Cf	californium	252	
99	Es	einsteinium	253	
100	Fm	fermium	257	
101	Md	mendelevium	257	
102	No	nobelium	255	
103	Lr	lawrencium	256	
104	Rf	rutherfordium	261	
105	Ha	hahnium	262	
106		(Reported 1974)	263	

°The solar abundance for hydrogen has been arbitrarily set at 10^{12}.

Atomic weights are averages for terrestrial abundances.

Solar abundances from John E. Ross and Lawrence H. Aller, *Science, 191*, 1223, 1976. Gaps indicate that the element has not been observed on the sun.

Appendix 5

The Planets

Appendix 5a. *Intrinsic and Rotational Properties*

Name	Equatorial Radius km	÷Earth's	Mass ÷Earth's	Mean Density (gm/cm³)	Oblate- ness	Surface Gravity (Earth = 1)	Sidereal Rotation Period	Inclination of Equator to Orbit	Apparent Magnitude at Mean Opposition
Mercury	2,439	0.3824	0.0553	5.44	0.0	0.378	59^{d}	$<28°$	−1.8
Venus	6,052	0.9489	0.8150	5.24	0.0	0.894	244.3^{d} R	177°	−4.3
Earth	6,378.140	1	1	5.497	0.0034	1	$23^{h}56^{m}04.1^{s}$	23°27′	−
Mars	3,397.2	0.5326	0.1074	3.9	0.009	0.379	$24^{h}37^{m}22.662^{s}$	23°59′	−2.01
Jupiter	71,398	11.194	317.89	1.3	0.063	2.54	$9^{h}50^{m}$ to $>9^{h}55^{m}$	3°05′	−2.55
Saturn	60,000	9.41	95.17	0.7	0.098	1.07	$10^{h}39.9^{m}$	26°44′	+0.67
Uranus	26,230	4.4	14.56	1.0	0.01	0.8	12^{h} to $24^{h} \pm 4^{h}$	97°55′	+5.52
Neptune	24,300	3.8	17.24	1.7	0.02	1.2	15^{h} to 20^{h}	28°48′	+7.84
Pluto	1,500–1,800	0.4	0.02	0.5–0.8	?	?	$6^{d}9^{h}17^{m}$	?	+14.9

R signifies retrograde rotation.

The masses and diameters are the values recommended by the International Astronomical Union in 1976, except for Uranus and new values for Pluto. Densities and surface gravities were calculated from these values. The length of the Martian day is from G. de Vaucouleurs (1979). Uranus values from Elliot, J. L., E. Dunham, D. J. Mink, and J. Churms (*Ap. J. 236*, 1026, 1980). Ranges are given for recent measurements of the rotation periods of Uranus and Neptune. Pluto is now magnitude 13.8; it will be much closer than its mean opposition distance for decades.

Appendix 5b. *Orbital Properties*

Name	Semimajor Axis A.U.	10^{6} km	Sidereal Period Years	Days	Synodic Period (days)	Eccentricity	Inclination to Ecliptic
Mercury	0.3871	57.9	0.24084	87.96	115.9	0.2056	7°00′26″
Venus	0.7233	108.2	0.61515	224.68	584.0	0.0068	3°23′40″
Earth	1	149.6	1.00004	365.26	−	0.0167	0°00′14″
Mars	1.5237	227.9	1.8808	686.95	779.9	0.0934	1°51′09″
Jupiter	5.2028	778.3	11.862	4337	398.9	0.0483	1°18′29″
Saturn	9.5388	1427.0	29.456	10,760	378.1	0.0560	2°29′17″
Uranus	19.1914	2871.0	84.07	30,700	369.7	0.0461	0°48′26″
Neptune	30.0611	4497.1	164.81	60,200	367.5	0.0100	1°46′27″
Pluto	39.5294	5913.5	248.53	90,780	366.7	0.2484	17°09′03″

Mean elements of planetary orbits for 1980, referred to the mean ecliptic and equinox of 1950 (Seidelmann, P. K., L. E. Doggett, and M. R. DeLuccia, *Astronomical Journal 79*, 57, 1974). Periods are calculated from them.

Appendix 6

Planetary Satellites

Satellite	Semimajor Axis of Orbit (km)	Sidereal Revolution Period (d)	(h)	(m)	Orbital Eccentricity	Orbital Inclination (°)	Diameter (km)	Mass ÷ Mass of Planet	Mean Density (gm · cm⁻³)	Discoverer	Visible Magnitude at Mean Opposition Distance
SATELLITE OF THE EARTH											
The Moon	384,500	27	07	43	0.055	18–29	3476	0.01230002*	3.34	—	−12.7
SATELLITES OF MARS											
Phobos	9,370	0	07	39	0.015	1.1	14 × 11 × 10	3.0 × 10⁻⁹	2	Hall (1877)	11.6
Deimos	23,520	1	06	18	0.0008	0.9–2.7v	8 × 6 × 6	6.3 × 10⁻¹⁰	2	Hall (1877)	12.7
SATELLITES OF JUPITER											
1979 J1	129,000	0.297			0		(20)			Voyager 2 (1979)	
XVI 1979 J3	127,000	0	07	04			(40)			Synnott/Voyager 1 (1980)	
XIV Adrastea	134,000	0	07	09			(20)			Jewett/Voyager 2 (1979)	
V Amalthea	180,000	0	11	57	0.003	0.4	270 × 165 × 153	18 × 10⁻¹⁰		Barnard (1892)	14.1
XV 1979 J2	222,000	0	16	11			(80)			Synnott/Voyager 1 (1980)	
I Io	422,000	1	18	28	0.000	0	3632 ± 60	4.70 × 10⁻⁵*	3.5	Galileo (1610)	5.0
II Europa	671,000	3	13	14	0.000	0.5	3126 ± 60	2.56 × 10⁻⁵*	3.0	Galileo (1610)	5.3
III Ganymede	1,070,000	7	03	43	0.001	0.2	5276 ± 60	7.84 × 10⁻⁵*	1.9	Galileo (1610)	4.6
IV Callisto	1,885,000	16	16	32	0.01	0.2	4820 ± 60	5.6 × 10⁻⁵*	1.8	Galileo (1610)	5.6
XIII Leda	11,110,000	240			0.147	26.7	(10)	5 × 10⁻¹³		Kowal (1974)	20
VI Himalia	11,470,000	251			0.158	27.6	170	8.5 × 10⁻¹⁰		Perrine (1904)	14.7
X Lysithea	11,710,000	260			0.130	29.0	(20)	0.010 × 10⁻¹⁰		Nicholson (1938)	18.4
VII Elara	11,740,000	260			0.207	24.8	80	0.35 × 10⁻¹⁰		Perrine (1905)	16.4
XII Ananke	20,700,000	617 R			0.169	147	(20)	0.007 × 10⁻¹⁰		Nicholson (1951)	18.9
XI Carme	22,350,000	629 R			0.207	164	(30)	0.020 × 10⁻¹⁰		Nicholson (1938)	18.0
VIII Pasiphae	23,330,000	735 R			0.378	145	(40)	0.077 × 10⁻¹⁰		Melotte (1908)	17.7
IX Sinope	23,370,000	758 R			0.275	153	(30)	0.015 × 10⁻¹⁰		Nicholson (1914)	18.3
—										Kowal (1975)	20

SATELLITES OF SATURN

This situation with the inner satellites of Saturn is unclear. We must await analysis of ground-based observations from the 1979–80 passage through the plane of Saturn's rings and Pioneer 11 and Voyager results.

Satellite	Semimajor Axis of Orbit (km)	Sidereal Revolution Period (d)	(h)	(m)	Orbital Eccentricity	Orbital Inclination (°)	Diameter (km)	Mass ÷ Mass of Planet	Mean Density (gm · cm⁻³)	Discoverer	Visible Magnitude at Mean Opposition Distance
1979 S5	140,500				—	—	(50)			Pioneer 11 (1979)	
11	151,000	0	16	40	0.0	0.0	(200)			Foundation and Larson (= Pioneer 11 Sighting)	14
10 Janus	160,000	0	17	59	0.0	0.0	(200)	3 × 10⁻⁸		Dollus (1966) (unconfirmed)	14
1979 S3	169,000									Pioneer 11 (1979)	
1 Mimas	187,000	0	22	37	0.020	1.5	(400)	6.59 × 10⁻⁸	1.5†	W. Herschel (1789)	12.1
2 Enceladus	238,000	1	08	53	0.004	0.0	(500)	1.48 × 10⁻⁷	1†	W. Herschel (1789)	11.8
3 Tethys	295,000	1	21	18	0.000	1.1	1040 ± 120†	1.10 × 10⁻⁶	1.0†	Cassini (1684)	10.3
4 Dione	378,000	2	17	41	0.002	0.0	1000 ± 240†	2.04 × 10⁻⁶	1.5†	Cassini (1684)	10.4
5 Rhea	526,000	4	12	25	0.001	0.4	1600 ± 200†	3.2 × 10⁻⁶	1	Cassini (1672)	9.7
6 Titan	1,221,000	15	22	41	0.029	0.3	5800	2.41 × 10⁻⁴*	1.4†	Huygens (1665)	8.4
7 Hyperion	1,481,000	21	06	38	0.104	0.4	225 ± 30†	2 × 10⁻⁷		Bond (1848)	14.2
8 Iapetus	3,561,000	79	07	56	0.028	14.7v	1450 ± 200†	3.94 × 10⁻⁶		Cassini (1671)	11.0
9 Phoebe	12,960,000	550	11	R	0.163	150	(240)	5.2 × 10⁻⁸		W. Pickering (1898)	16.5

SATELLITES OF URANUS

Satellite	Semimajor Axis of Orbit (km)	Sidereal Revolution Period (d)	(h)	(m)	Orbital Eccentricity	Orbital Inclination (°)	Diameter (km)	Mass ÷ Mass of Planet	Mean Density (gm · cm⁻³)	Discoverer	Visible Magnitude at Mean Opposition Distance
5 Miranda	130,000	1	09	56	0.017	3.4	(300)	1 × 10⁻⁶		Kuiper (1948)	16.5
1 Ariel	191,000	2	12	29R	0.003	0	(800)	15 × 10⁻⁶		Lassell (1851)	14.4
2 Umbriel	260,000	4	03	27R	0.004	0	(550)	6 × 10⁻⁶		Lassell (1851)	15.3
3 Titania	436,000	8	16	56R	0.002	0	(1000)	50 × 10⁻⁶		W. Herschel (1787)	14.0
4 Oberon	583,000	13	11	07R	0.001	0	(900)	29 × 10⁻⁶		W. Herschel (1787)	14.2

SATELLITES OF NEPTUNE

Satellite	Semimajor Axis of Orbit (km)	Sidereal Revolution Period (d)	(h)	(m)	Orbital Eccentricity	Orbital Inclination (°)	Diameter (km)	Mass ÷ Mass of Planet	Mean Density (gm · cm⁻³)	Discoverer	Visible Magnitude at Mean Opposition Distance
Triton	354,000	5	21	03R	0.000	160.0	3640–5280†	2 × 10⁻³*		Lassell (1846)	13.6
Nereid	5,570,000	365	5		0.76	27.4	(300)	10⁻⁶		Kuiper (1949)	18.7

SATELLITE OF PLUTO

Satellite	Semimajor Axis of Orbit (km)	Sidereal Revolution Period (d)	(h)	(m)	Orbital Eccentricity	Orbital Inclination (°)	Diameter (km)	Mass ÷ Mass of Planet	Mean Density (gm · cm⁻³)	Discoverer	Visible Magnitude at Mean Opposition Distance
Charon	20,000?	6	9	17	0?	105?	(1000)	0.05–0.1?		Christy (1978)	16–17

Based on a table by Joseph Veverka in the *Observer's Handbook 1980* of the Royal Astronomical Society of Canada, updated for satellites of Mars, Saturn, Jupiter, and Pluto. Many of Veverka's values are from J. Burns.

*Values recommended by the International Union in 1976.

Diameters and densities marked with a † from Cruikshank, D. (*Reviews of Geophysics and Space Physics 17*, 165, 1979). Values for the Galilean satellites are from Voyager (Smith, B. A. et al., *Science 204*, 951, 1979). Amalthea diameter from Voyager. Values for Phobos and Demos from Veverka, J. (private communication, 1980; and *Vistas in Astronomy 22*, 163, 1978). Charon data from R. S. Harrington and J. W. Christy, (*A. J. 85*, 168, 1980). Inclinations greater than 90° are retrograde rotation, and R signifies retrograde revolution. Apparent magnitudes at discovery given for Jupiter XIII and XIV. Apparent magnitude at discovery given for Pluto's satellite.

Speckle interferometry studies with the Canada-France-Hawaii-telescope (L.A.U. Circular 3509, September 1980) give 4000 and 2000 km, respectively, for the diameter of Pluto and Charon, a total mass of 1/300 that of Earth, and densities of 0.5 gcm⁻³.

Appendix 7

Brightest Stars

Star	Name	Position 1980.0 R.A.	Dec.	Apparent Magnitude (V)	Spectral Type		Absolute Magnitude	Distance (ly)	Proper Motion (s · yr^{-1})	Radial Velocity (km · s^{-1})
1. α CMa A	Sirius	06 44.2	−16 42	−1.46	Al	V	+1.42	8.7	1.324	−7.6
2. α Car	Canopus	06 23.5	−52 41	−0.72	F0	Ib-II	−3.1	98	0.025	+20.5
3. α Boo	Arcturus	14 14.8	+19 17	−0.06	K2	IIIp	−0.3	36	2.284	−5.2
4. α Cen A	Rigil Kentaurus	14 38.4	−60 46	0.01	G2	V	+4.37	4.2	3.676	−24.6
5. α Lyr	Vega	18 36.2	+38 46	0.04	A0	V	+0.5	26.5	0.345	−13.9
6. α Aur	Capella	05 15.2	+45 59	0.05	G8	III?+F	−0.6	45	0.435	+30.2
7. β Ori A	Rigel	05 13.6	−08 13	0.14v	B8	Ia	−7.1	900	0.001	+20.7
8. α CMi A	Procyon	07 38.2	+05 17	0.37	F5	IV-V	+2.6	11.4	1.250	−3.2
9. α Ori	Betelgeuse	05 54.0	+07 24	0.41v	M2	Iab	−5.6	520	0.028	+21.0
10. α Eri	Achernar	01 37.0	−57 20	0.51	B3	Vp	−2.3	118	0.098	+19
11. β Cen AB	Hadar	14 02.4	−60 16	0.63v	B1	III	−5.2	490	0.035	−12
12. α Aql	Altair	19 49.8	+08 49	0.76	A7	IV-V	+2.2	16.5	0.658	−26.3
13. α Tau A	Aldebaran	04 34.8	+16 28	0.86v	K5	III	−0.7	68	0.202	+54.1
14. α Vir	Spica	13 24.1	−11 03	0.91v	B1	V	−3.3	220	0.054	+1.0
15. α Sco A	Antares	16 28.2	−26 23	0.92v	M1	Ib+B	−5.1	520	0.029	−3.2
16. α PsA	Fomalhaut	22 56.5	−29 44	1.15	A3	V	+2.0	22.6	0.367	+6.5
17. β Gem	Pollux	07 44.1	+28 05	1.16	K0	III	+1.0	35	0.625	+3.3
18. α Cyg	Deneb	20 40.7	+45 12	1.26	A2	Ia	−7.1	1600	0.003	−4.6
19. β Cru	Beta Crucis	12 46.6	−59 35	1.28v	B0.5	III	−4.6	490	0.049	+20.0
20. α Leo A	Regulus	10 07.3	+12 04	1.36	B7	V	−0.7	84	0.248	+3.5
21. α Cru A	Acrux	12 25.4	−62 59	1.39	B0.5	IV	−3.9	370	0.042	−11.2
22. ε CMa A	Adhara	06 57.8	−28 57	1.48	B2	II	−5.1	680	0.004	+27.4
23. λ Sco	Shaula	17 32.3	−37 05	1.60v	B1	V	−3.3	310	0.031	0
24. γ Ori	Bellatrix	05 24.0	+06 20	1.64	B2	III	−4.2	470	0.015	+18.2
25. β Tau	Elnath	05 25.0	+28 36	1.65	B7	III	−3.2	300	0.178	+8.0

Based on a table compiled by Donald A. MacRae in the *Observer's Handbook 1980* of the Royal Astronomical Society of Canada, amended with information from W. Gliese (private communication, 1979).

Italics indicate an average value of a variable radial velocity.

Appendix 8

The Nearest Stars*

No.	Name	R.A. α h	R.A. m	Dec. δ °	Dec. δ '	Parallax π ''	Distance (pc)	Proper Motion μ (''yr⁻¹)		Radial Velocity Vr (km s⁻¹)	Spectral Type	V	B-V	Mv	Luminosity (L☉ = 1)
1	Sun										G2 V	−26.72	0.65	4.85	1.0
2	Proxima Cen	14	28	−62	36	0.772	1.30	3.85	282°	−16	dM5e	11.05	1.97	15.49	0.00006
	α Cen A	14	38	−60	46	.750	1.33	3.68	281	−22	G2 V	−0.01	0.68	4.37	1.6
	α Cen B										K0 V	1.33	0.88	5.71	0.45
3	Barnard's star	17	56	+04	36	.545	1.83	10.31	356	−108	M5 V	9.54	1.74	13.22	0.00045
4	Wolf 359	10	56	+07	10	.421	2.38	4.70	235	+13	dM8e	13.53	2.01	16.65	0.00002
5	BD +36°2147 (Lalande 21185)	11	03	+36	07	.397	2.52	4.78	187	−84	M2 V	7.50	1.51	10.50	0.0055
6	L 726−8 = A	1	37	−18	04	.387	2.58	3.36	80	+29	dM6e	12.52	1.85	15.46	0.00006
	UV Cet = B									+32	dM6e	13.02		15.96	0.00004
7	Sirius A	6	44	−16	42	.377	2.65	1.33	204	−8	A1 V	−1.46	0.00	1.42	23.5
	Sirius B										DA	8.3	−0.12	11.2	0.003
8	Ross 154 (V 1216 Sgr)	18	49	−23	50	.345	2.90	0.72	104	−4	dM5e	10.45	1.70	13.14	0.00048
9	Ross 248 (HH And)	23	40	+44	04	.314	3.18	1.60	176	−81	dM6e	12.29	1.91	14.78	0.00011
10	ε Eri	3	32	−09	32	.303	3.30	0.98	271	+16	K2 V	3.73	0.88	6.14	0.30
11	Ross 128 (FI Vir)	11	47	+00	58	.298	3.36	1.38	152	−13	dM5	11.10	1.76	13.47	0.00036
12	61 Cyg A	21	06	+38	38	.294	3.40	5.22	52	−64	K5 V	5.22	1.17	7.56	0.082
	61 Cyg B										K7 V	6.03	1.37	8.37	0.039
13	ε Ind	22	03	−56	52	.291	3.44	4.70	123	−40	K5 V	4.68	1.05	7.00	0.14
14	BD +43° 44 A (GX And)	10	18	+43	54	.290	3.45	2.90	82	+13	M1 V	8.08	1.56	10.39	0.0061
	+43 44 B (GQ And) (Groombridge 39 AB)									+20	M6 V e	11.06	1.80	13.37	0.00039
15	L 789−6	22	38	−15	28	.290	3.45	3.26	46	−60	dM7e	12.18	1.96	14.49	0.00014
16	Procyon A	7	39	+05	17	.285	3.51	1.25	214	−3	F5IV-V	0.37	0.42	2.64	7.65
	Procyon B										DF	10.7		13.0	0.00055
17	BD +59°1915 A	18	42	+59	36	.282	3.55	2.29	325	0	dM4	8.90	1.54	11.15	0.0030
	+59 1915 B							2.27	323	+10	dM5	9.69	1.59	11.94	0.0015
18	CD −36°15693 (Lacaille 9352)	23	05	−35	59	.279	3.58	6.90	79	+10	M2 V	7.35	1.48	9.58	0.013
19	G 51−15	8	29	+26	51	.278	3.60	1.27	242			14.81	2.06	17.03	0.00001
20	τ Ceti	1	43	−16	03	.277	3.61	1.92	297	−16	G8 V	3.50	0.72	5.72	0.45
21	BD +5°1668	7	27	+05	27	.266	3.76	3.77	171	+26	dM5	9.82	1.56	11.94	0.0015
22	L 725−32 (YZ Cet)	1	11	−17	06	.261	3.83	1.32	62	+28	dM5e	12.04	1.83	14.12	0.00020
23	CD −39°14192	21	16	−38	58	.260	3.85	3.46	251	+21	M0 V	6.66	1.40	8.74	0.028
24	Kapteyn's star	5	11	−44	59	.256	3.91	8.72	131	+245	M0 V	8.84	1.56	10.88	0.0039
25	Krüger 60 A	22	27	+57	36	.253	3.95	0.86	246	−26	dM3	9.85	1.62	11.87	0.0016
	Krüger 60 B (DO Cep)										dM5e	11.3	1.8	13.3	0.0004
26	BD −12°4523	16	30	−12	36	.247	4.05	1.18	183	−13	dM5	10.11	1.60	12.07	0.0013
27	Ross 614 A	6	28	−02	48	.246	4.07	1.00	133	+24	dM7e	11.10	1.71	13.12	0.00049
	Ross 614 B (V 577 Mon)											14		16	0.00004
28	van Maanen's star	0	48	+05	19	.232	4.31	2.99	155	+54	DG	12.37	0.56	14.20	0.00018
29	Wolf 424 A	12	33	+09	09	.230	4.35	1.76	279	−5	dM6e	13.16	1.80	14.97	0.00009
	Wolf 424 B (FL Vir)										dM6e	13.4		15.2	0.00007
30	CD −37°15492	0	04	−37	27	.225	4.44	6.11	112	+23	M4 V	8.56	1.46	10.32	0.0065
31	L 1159−16 (TZ Ari)	1	59	+13	00	.224	4.46	2.09	149		dM8e	12.26	1.82	14.01	0.00022
32	BD +50°1725 (Groombridge 1618)	10	10	+49	33	.222	4.50	1.45	250	−26	K7 V	6.59	1.36	8.32	0.041
33	CD −46°11540	17	28	−46	53	.216	4.63	1.06	147		dM4	9.37	1.53	11.04	0.0033
34	G 158−27	0	06	−07	38	.214	4.67	2.04	204		dM	13.74	1.95	15.39	0.00006
35	CD −49°13515	21	32	−49	11	.214	4.67	0.81	184	+8	M1 V	8.67	1.46	10.32	0.0065
36	CD −44°11909	17	37	−44	17	.213	4.69	1.16	217		M5	10.96	1.65	12.60	0.00079
37	BD +68°946	17	37	+68	22	.213	4.69	1.31	196	−22	M3.5V	9.15	1.50	10.79	0.0042
38	G 208−44 = A	19				.211	4.74	0.74	143			13.41	1.90	15.03	0.00008
	G 208−45 = B											13.99	1.98	15.61	0.00005
39	BD −15°6290	22	52	−14	22	.209	4.78	1.14	124	+9	dM5	10.17	1.60	11.77	0.0017
40	o² (40) Eri A	4	14	−07	41	.207	4.83	4.08	213	−42	K1 V	4.43	0.82	6.01	0.34
	40 Eri B	4	13	−7	44			4.07	212	−21	DA	9.52	0.03	11.10	0.0032
	40 Eri C (DY Eri)									−45	dM4e	11.17	1.66	12.75	0.00069
41	BD +20°2465 (AD Leo)	10	19	+19	58	.206	4.85	0.49	264	+11	M4.5Ve	9.43	1.54	11.00	0.0035
42	L 145−141	11	44	−64	42	.206	4.85	2.68	97		DC	11.50	0.19	13.07	0.00052
43	70 Oph A	18	04	+2	31	.203	4.93	1.12	167	−7	K0 V	4.22	0.86	5.76	0.43
	70 Oph B										K5 V	6.00		7.54	0.084
44	BD + 43°4305 (EV Lac)	22	46	+44	14	.200	5.00	0.83	236	−2	dM5e	10.2	1.6	11.7	0.0018
45	Altair	19	50	+08	49	.198	5.05	0.66	54	−26	A7IV,V	0.76	0.22	2.24	11.1
46	AC +79°3888	11	44	+78	48	.193	5.18	0.89	57	−119	sdM4	10.80	1.60	12.23	0.0011
47	G 9−38 = A	8	57	+19	50	.192	5.21	0.89	267		m	14.06	1.84	15.48	0.00006
	LP426−40 = B										m	14.92	1.93	16.34	0.000025
48	BD +15°2620 (Lalande 25372)	13	45	+15	01	.192	5.21	2.30	129	+15	M4 V	8.49	1.44	9.91	0.0095

*Courtesy of W. Gliese (private communication, 1979).

Appendix 9 Messier Catalogue

M NGC	α 1980 h m	δ ° ′	m_v	Description
1 1952	5 33.3	+22 01	11.3	Crab Nebula in Taurus
2 7089	21 32.4	−00 54	6.3	Globular cluster in Aquarius
3 5272	13 41.3	+28 29	6.2	Globular cluster in Canes Venatici
4 6121	16 22.4	−26 27	6.1	Globular cluster in Scorpio
5 5904	15 17.5	+02 11	6	Globular cluster in Serpens
6 6405	17 38.9	−32 11	6	Open cluster in Scorpio
7 6475	17 52.6	−34 48	5	Open cluster in Scorpio
8 6523	18 02.4	−24 23		Lagoon Nebula in Sagittarius
9 6333	17 18.1	−18 30	7.6	Globular cluster in Ophiuchus
10 6254	16 56.0	−04 05	6.4	Globular cluster in Ophiuchus
11 6705	18 50.0	−06 18	7	Open cluster in Scutum
12 6218	16 46.1	−01 55	6.7	Globular cluster in Ophiuchus
13 6205	16 41.0	+36 30	5.8	Globular cluster in Hercules
14 6402	17 36.5	−03 14	7.8	Globular cluster in Ophiuchus
15 7078	21 29.1	+12 05	6.3	Globular cluster in Pegasus
16 6611	18 17.8	−13 48	7	Open cluster in Serpens
17 6618	18 19.7	−16 12	7	Omega Nebula in Sagittarius
18 6613	18 18.8	−17 09	7	Open cluster in Sagittarius
19 6273	17 01.3	−26 14	6.9	Globular cluster in Ophiuchus
20 6514	18 01.2	−23 02		Trifid Nebula in Sagittarius
21 6531	18 03.4	−22 30	7	Open cluster in Sagittarius
22 6656	18 35.2	−23 55	5.2	Globular cluster in Sagittarius
23 6494	17 55.7	−19 00	6	Open cluster in Sagittarius
24 6603	18 17.3	−18 27	6	Open cluster in Sagittarius
25 IC4725	18 30.5	−19 16	6	Open cluster in Sagittarius
26 6694	18 44.1	−09 25	9	Open cluster in Scutum
27 6853	19 58.8	+22 40	8.2	Dumbbell Nebula; planetary nebula in Vulpecula
28 6626	18 23.2	−24 52	7.1	Globular cluster in Sagittarius
29 6913	20 23.3	+38 27	8	Open cluster in Cygnus
30 7099	21 39.2	−23 15	7.6	Globular cluster in Capricornus
31 224	0 41.6	+41 09	3.7	Andromeda Galaxy (Sb)
32 221	0 41.6	+40 45	8.5	Elliptical galaxy in Andromeda; companion to M31
33 598	1 32.8	+30 33	5.9	Spiral galaxy (Sc) in Triangulum
34 1039	2 40.7	+42 43	6	Open cluster in Perseus
35 2168	6 07.6	+24 21	6	Open cluster in Gemini
36 1960	5 35.0	+34 05	6	Open cluster in Auriga
37 2099	5 51.5	+32 33	6	Open cluster in Auriga
38 1912	5 27.3	+35 48	6	Open cluster in Auriga
39 7092	21 31.5	+48 21	6	Open cluster in Cygnus
40 −	−	−		Double star in Ursa Major
41 2287	6 46.2	−20 43	6	Open cluster in Canis Major
42 1976	5 34.4	−05 24		Orion Nebula
43 1982	5 34.6	−05 18		Orion Nebula; smaller part
44 2632	8 38.8	+20 04	4	Praesepe; open cluster in Cancer
45 −	3 46.3	+24 03	2	The Pleiades; open cluster in Taurus
46 2437	7 40.9	−14 46	7	Open cluster in Puppis
47 2422	7 35.6	−14 27	5	Open cluster in Puppis
48 2548	8 12.5	−05 43	6	Open cluster in Hydra
49 4472	12 28.8	+08 07	8.9	Elliptical galaxy in Virgo
50 2323	7 02.0	−08 19	7	Open cluster in Monoceros

Appendix 9 continued on the opposite page

M NGC	α 1980 δ		m_v	Description
	h m	° '		
51 5194	13 29.0	+47 18	8.4	Whirlpool Galaxy; spiral galaxy (Sc) in Canes Venatici
52 7654	23 23.3	+61 29	7	Open cluster in Cassiopeia
53 5024	13 12.0	+18 17	7.7	Globular cluster in Coma Berenices
54 6715	18 53.8	−30 30	7.7	Globular cluster in Sagittarius
55 6809	19 38.7	−31 00	6.1	Globular cluster in Sagittarius
56 6779	19 15.8	+30 08	8.3	Globular cluster in Lyra
57 6720	18 52.9	+33 01	9.0	Ring Nebula; planetary nebula in Lyra
58 4579	12 36.7	+11 56	9.9	Spiral galaxy (SBb) in Virgo
59 4621	12 41.0	+11 47	10.3	Elliptical galaxy in Virgo
60 4649	12 42.6	+11 41	9.3	Elliptical galaxy in Virgo
61 4303	12 20.8	+04 36	9.7	Spiral galaxy (Sc) in Virgo
62 6266	16 59.9	−30 05	7.2	Globular cluster in Scorpio
63 5055	13 14.8	+42 08	8.8	Spiral galaxy (Sb) in Canes Venatici
64 4826	12 55.7	+21 48	8.7	Spiral galaxy (Sb) in Coma Verenices
65 3623	11 17.8	+13 13	9.6	Spiral galaxy (Sa) in Leo
66 3627	11 19.1	+13 07	9.2	Spiral galaxy (Sb) in Leo; companion to M65
67 2682	8 50.0	+11 54	7	Open cluster in Cancer
68 4590	12 38.3	−26 38	8	Globular cluster in Hydra
69 6637	18 30.1	−32 23	7.7	Globular cluster in Sagittarius
70 6681	18 42.0	−32 18	8.2	Globular cluster in Sagittarius
71 6838	19 52.8	+18 44	6.9	Globular cluster in Sagitta
72 6981	20 52.3	−12 39	9.2	Globular cluster in Aquarius
73 6994	20 57.8	−12 44		Open cluster in Aquarius
74 628	1 35.6	+15 41	9.5	Spiral galaxy (Sc) in Pisces
75 6864	20 04.9	−21 59	8.3	Globular cluster in Sagittarius
76 650	1 40.9	+51 28	11.4	Planetary nebula in Perseus
77 1068	2 41.6	−00 04	9.1	Spiral galaxy (Sb) in Cetus
78 2068	5 45.8	+00 02		Small emission nebula in Orion
79 1904	5 23.3	−24 32	7.3	Globular cluster in Lepus
80 6093	16 15.8	−22 56	7.2	Globular cluster in Scorpio
81 3031	9 54.2	+69 09	6.9	Spiral galaxy (Sb) in Ursa Major
82 3034	9 54.4	+69 47	8.7	Irregular galaxy (Irr) in Ursa Major
83 5236	13 35.9	−29 46	7.5	Spiral galaxy (Sc) in Hydra
84 4374	12 24.1	+13 00	9.8	Elliptical galaxy in Virgo
85 4382	12 24.3	+18 18	9.5	Elliptical galaxy (SO) in Coma Berenices
86 4406	12 25.1	+13 03	9.8	Elliptical galaxy in Virgo
87 4486	12 29.7	+12 30	9.3	Elliptical galaxy (Ep) in Virgo
88 4501	12 30.9	+14 32	9.7	Spiral galaxy (Sb) in Coma Berenices
89 4552	12 34.6	+12 40	10.3	Elliptical galaxy in Virgo
90 4569	12 35.8	+13 16	9.7	Spiral galaxy (Sb) in Virgo
91 −	−	−		M58 ?
92 6341	17 16.5	+43 10	6.3	Globular cluster in Hercules
93 2447	7 43.6	−23 49	6	Open cluster in Puppis
94 4736	12 50.1	+41 14	8.1	Spiral galaxy (Sb) in Canes Venatici
95 3351	10 42.8	+11 49	9.9	Barred spiral galaxy (SBb) in Leo
96 3368	10 45.6	+11 56	9.4	Spiral galaxy (Sa) in Leo
97 3587	11 13.7	+55 08	11.1	Owl Nebula; planetary nebula in Ursa Major
98 4192	12 12.7	+15 01	10.4	Spiral galaxy (Sb) in Coma Berenices
99 4254	12 17.8	+14 32	9.9	Spiral galaxy (Sc) in Coma Berenices
100 4321	12 21.9	+15 56	9.6	Spiral galaxy (Sc) in Coma Berenices

Appendix 9 continued on the following page

Appendix 9 *Continued*

M NGC	α 1980	δ	m_v	Description
	h m	° ′		
101 5457	14 02.5	+54 27	8.1	Spiral galaxy (Sc) in Ursa Major
102 –	–	–		M101 ?
103 581	1 31.9	+60 35	7	Open cluster in Cassiopeia
104 4594	12 39.0	−11 35	8	Sombrero Nebula; spiral galaxy (Sa) in Virgo
105 3379	10 46.8	+12 51	9.5	Elliptical galaxy in Leo
106 4258	12 18.0	+47 25	9	Spiral galaxy in (Sb) Canes Venatici
107 6171	16 31.8	−13 01	9	Globular cluster in Ophiuchus
108 3556	11 10.5	+55 47	10.5	Spiral galaxy (Sb) in Ursa Major
109 3992	11 56.6	+53 29	10.6	Barred spiral galaxy (SBc) in Ursa Major

Positions and magnitudes based on a table in the *Observer's Handbook 1980* of the Royal Astronomical Society of Canada.

Appendix 10

The Constellations

Latin Name	Genitive	Abbreviation	Translation
Andromeda	Andromedae	And	Andromeda*
Antlia	Antliae	Ant	Pump
Apus	Apodis	Aps	Bird of Paradise
Aquarius	Aquarii	Aqr	Water Bearer
Aquila	Aquilae	Aql	Eagle
Ara	Arae	Ara	Altar
Aries	Arietis	Ari	Ram
Auriga	Aurigae	Aur	Charioteer
Boötes	Boötis	Boo	Herdsman
Caelum	Caeli	Cae	Chisel
Camelopardalis	Camelopardalis	Cam	Giraffe
Cancer	Cancri	Cnc	Crab
Canes Venatici	Canum Venaticorum	CVn	Hunting Dogs
Canis Major	Canis Majoris	CMa	Big Dog
Canis Minor	Canis Minoris	CMi	Little Dog
Capricornus	Capricorni	Cap	Goat
Carina	Carinae	Car	Ship's Keel**
Cassiopeia	Cassiopeiae	Cas	Cassiopeia*
Centaurus	Centauri	Cen	Centaur*
Cepheus	Cephei	Cep	Cepheus*
Cetus	Ceti	Cet	Whale
Chamaeleon	Chamaeleonis	Cha	Chameleon
Circinus	Circini	Cir	Compass
Columba	Columbae	Col	Dove
Coma Berenices	Comae Berenices	Com	Berenice's Hair*
Corona Australis	Coronae Australis	CrA	Southern Crown
Corona Borealis	Coronae Borealis	CrB	Northern Crown
Corvus	Corvi	Crv	Crow
Crater	Crateris	Crt	Cup
Crux	Crucis	Cru	Southern Cross
Cygnus	Cygni	Cyg	Swan

Appendix 10 continued on the opposite page

Appendix 10 *Continued*

Latin Name	Genitive	Abbreviation	Translation
Delphinus	Delphini	Del	Dolphin
Dorado	Doradus	Dor	Swordfish
Draco	Draconis	Dra	Dragon
Equuleus	Equulei	Equ	Little Horse
Eridanus	Eridani	Eri	River Eridanus*
Fornax	Fornacis	For	Furnace
Gemini	Geminorum	Gem	Twins
Grus	Gruis	Gru	Crane
Hercules	Herculis	Her	Hercules*
Horologium	Horologii	Hor	Clock
Hydra	Hydrae	Hya	Hydra* (water monster)
Hydrus	Hydri	Hyi	Sea serpent
Indus	Indi	Ind	Indian
Lacerta	Lacertae	Lac	Lizard
Leo	Leonis	Leo	Lion
Leo Minor	Leonis Minoris	LMi	Little Lion
Lepus	Leporis	Lep	Hare
Libra	Librae	Lib	Scales
Lupus	Lupi	Lup	Wolf
Lynx	Lyncis	Lyn	Lynx
Lyra	Lyrae	Lyr	Harp
Mensa	Mensae	Men	Table (mountain)
Microscopium	Microscopii	Mic	Microscope
Monoceros	Monocerotis	Mon	Unicorn
Musca	Muscae	Mus	Fly
Norma	Normae	Nor	Level (square)
Octans	Octantis	Oct	Octant
Ophiuchus	Ophiuchi	Oph	Ophiuchus* (serpent bearer)
Orion	Orionis	Ori	Orion*
Pavo	Pavonis	Pav	Peacock
Pegasus	Pegasi	Peg	Pegasus* (winged horse)
Perseus	Persei	Per	Perseus*
Phoenix	Phoenicis	Phe	Phoenix
Pictor	Pictoris	Pic	Easel
Pisces	Piscium	Psc	Fish
Piscis Austrinus	Piscis Austrini	PsA	Southern Fish
Puppis	Puppis	Pup	Ship's Stern**
Pyxis	Pyxidis	Pyx	Ship's Compass**
Reticulum	Reticuli	Ret	Net
Sagitta	Sagittae	Sge	Arrow
Sagittarius	Sagittarii	Sgr	Archer
Scorpius	Scorpii	Sco	Scorpion
Sculptor	Sculptoris	Scl	Sculptor
Scutum	Scuti	Sct	Shield
Serpens	Serpentis	Ser	Serpent
Sextans	Sextantis	Sex	Sextant
Taurus	Tauri	Tau	Bull
Telescopium	Telescopii	Tel	Telescope
Triangulum	Trianguli	Tri	Triangle
Triangulum Australe	Trianguli Australis	TrA	Southern Triangle
Tucana	Tucanae	Tuc	Toucan
Ursa Major	Ursae Majoris	UMa	Big Bear
Ursa Minor	Ursae Minoris	UMi	Little Bear
Vela	Velorum	Vel	Ship's Sails**
Virgo	Virginis	Vir	Virgin
Volans	Volantis	Vol	Flying Fish
Vulpecula	Vulpeculae	Vul	Little Fox

*Proper names.
**Formerly formed the constellation Argo Navis, the Argonaut's ship.

Appendix 11

A Survey of the Sky

Go out at dusk, to a place where you have a good view of the sky. The farther from city lights you are, the better. The moon may well be visible already; if you know where to look you can even often see it in the daytime. As darkness comes, the first other objects you can see are usually the planets. The brightest planet visible is usually Venus if it is low in the western sky. (Venus never can get too far from the sun in the sky, so it must be in the part of the sky where the sun is setting.) When Venus is visible at this time, following the sun into the western horizon, we call it "the evening star." (When it is, on the other hand, ahead of the sun, it appears as "the morning star" instead.)

The positions of the planets are given monthly in the popular magazines listed in the back of this book.

The next brightest object in the sky is often Jupiter. Though the sun, the moon, and the planets will follow approximately the same path across the sky, the *ecliptic*, the outer planets such as Jupiter are not restricted to being close to the sun. So Jupiter may be anywhere along the ecliptic at nightfall.

Jupiter and Venus are outstanding even to the naked eye. And a small telescope will show you the four brightest satellites of Jupiter as well as horizontal belts on the planet, and the phases of Venus. A big telescope isn't necessary; Galileo discovered these things in the early 17th century with a telescope only a few centimeters across.

Mars is a planet that orbits outside the earth's orbit, so it can appear anywhere along the ecliptic. Its reddish tinge singles it out to the naked eye.

Mercury doesn't get nearly as bright as Venus or Jupiter, but you may see it close to the western horizon at sunset. Since Mercury orbits the sun in only 88 days, it is in the evening sky for about a month at a time, the morning sky for about a month at a time, and too close to the sun to see at all for the rest of the time.

Saturn, likewise, is not so bright in the sky as to stand out to the naked eye, but a view with a telescope is breathtaking. It really has rings! They aren't hard to see at all. And my excitement at looking at Saturn—a feeling that I never lose—is increased by realizing that even now spacecraft are on their way to take closeup photographs and make other measurements, just as the other planets we have mentioned have been visited in recent years.

When the sky is truly dark, the stars come out. In the northern part of the sky, you can find the Big Dipper, a group of stars with four stars making the bowl and three more making the handle. You can find the Big Dipper on any of the four star maps that are bound into this book. The Big Dipper appears right side up in the evening sky in the fall and upside down in the evening sky in the spring.

The zenith is the point in the sky directly overhead.

The two front stars of the bowl of the Big Dipper act as the "pointers" to the north star. Follow the line determined by these stars about 30 degrees, one third of the way across the sky between the horizon and the zenith (the direction straight up), and you will come to Polaris, the north star. Polaris is at the end of the handle of the Little Dipper, although the Little Dipper is not as easy to pick out in the sky as the Big Dipper. Polaris always remains in the same position in the sky; that is, always in the same direction and always at the same height above your horizon. Thus it is of great importance in navigation. Stars and groups of stars that always stay above the horizon, including the Big Dipper and the Little Dipper, are called *circumpolar*. They appear on all of the star maps.

From *The Odyssey of Homer*

Glorious Odysseus [Ulysses], happy with the wind, spread sails and taking his seat artfully with the steering oar he held her on her course, nor did sleep ever descend on his eyelids as he kept his eye on the Pleiades and late-setting Boötes, and the Bear, to whom men give also the name of the Wagon, who turns about in a fixed place and looks at Orion, and she alone is never plunged in the wash of the Ocean.

The Autumn Sky

As it grows dark on an autumn evening, the pointers in the Big Dipper will point upward toward Polaris. Almost an equal distance on the other side of Polaris as the distance from the pointers to Polaris is a W-shaped constellation named Cassiopeia. Cassiopeia, in Greek mythology, was married to Cepheus, the king of Ethiopia (and the subject of the constellation that neighbors Cassiopeia to the west). Cassiopeia appears sitting on a chair, as shown in the opening illustration of Chapter 3.

Continuing across the sky away from the pointers, we next come to the constellation Andromeda, who in Greek mythology was Cassiopeia's daughter. In Andromeda, you might see a faint hazy patch of light; this is actually the center of the nearest galaxy to our own and is known as the Great Galaxy in Andromeda. It takes a telescope or a long-exposure photograph to reveal that it is a galaxy. Though this is one of the nearest galaxies, it is much farther away than any of the individual stars that we see.

Turning southwest in the sky from Andromeda, but still high overhead, are four stars that appear to make a square known as the Great Square of Pegasus. Pegasus was the flying horse best known to us as the symbol of the Mobil Oil Company.

If it is really dark out (which probably means that you are far from a city and also that the moon is not full or almost full), you will see the Milky Way crossing the sky high overhead. It will appear as a hazy band across the sky, with ragged edges and dark patches and rifts in it. The Milky Way passes right through Cassiopeia.

Moving northeast from Cassiopeia, along the Milky Way, we come to the constellation Perseus; he was the Greek hero who slew the Medusa. (He flew off on Pegasus, who is conveniently nearby in the sky and, from his winged mount, saw Andromeda, whom he saved.) On the edge of Perseus nearest to Cassiopeia, with a small telescope we can see two hazy patches of stars that are really clusters of hundreds of stars called *galactic clusters* (also called open clusters). This *double cluster in Perseus,* also known as h and χ (chi) Persei, provides two of the galactic clusters that are easiest to see with small telescopes.

In the other direction from Cassiopeia (whose W is relatively easy to find), we find a cross of bright stars directly overhead. These are in the constellation Cygnus, the Swan. In the direction marked by Cygnus, in a location from which we receive a stream of wildly varying x-rays, we think a black hole is located. Also in Cygnus is a particularly dark region of the Milky Way, called the Northern Coal Sack. Dust in space in that direction prevents us from seeing as many stars as we do in other directions in the Milky Way. Cygnus contains a prominent grouping of stars known as the Northern Cross. Slightly to the west is another bright star, Vega, in the constellation Lyra (the Lyre). And farther westward we come to the constellation Hercules, named for the Greek hero who performed twelve great labors, including bringing back the golden apples. In Hercules is an older, larger type of star cluster called a globular cluster. It is known as M13, the *globular cluster in Hercules* (see the figure opening Chapter 6).

The Winter Sky

As the autumn proceeds and the winter approaches, the constellations we have discussed appear closer and closer to the western horizon for the same hour of the night. By early evening on January 1, Cygnus is setting in the western sky, while Cassiopeia and Perseus are overhead.

To the south of the Milky Way, near Perseus, we can now see a group of six stars close together in the sky. The grouping can catch your attention as you scan the sky. These are the Pleiades, traditionally the Seven Sisters of Greek mythology, the daughters of Atlas. This is another example of a galactic cluster. Binoculars or a small telescope reveal dozens of stars there; a large telescope will ordinarily show too small a region of sky for you to see the Pleiades well at all. So a bigger telescope isn't always better.

Further toward the east, rising earlier every evening, is the constellation Orion, the Hunter. Orion is perhaps the easiest constellation of all to pick out in the sky, for three bright stars close together in a line make up its belt. Orion is warding off Taurus, the Bull. A reddish star, Betelgeuse (beetle-juice would not be far wrong for pronunciation, though some say "beh-tel-jooz"), marks Orion's shoulder, and symmetrically on the other side of his belt, the bright bluish star Rigel (rī′gel or rī′jel) marks his heel. Betelgeuse is an example of a red supergiant star; it is hundreds of millions of kilometers across, bigger itself than the earth's orbit. Extending down from Orion's belt is his sword. A telescope, or a long exposure with a camera set to track the stars, reveals a beautiful region known as the Great Nebula in Orion, or the Orion Nebula. Its shape can be seen in even a smallish telescope; however, only photographs reveal its color (see Color Plate 56). It is the site of formation of new stars.

Rising after Orion is Sirius, the brightest star in the sky. Orion's belt points right to it. Sirius appears blue-white, which indicates that it is very hot. Sirius is sufficiently bright to stand out to the naked eye. It is part of the constellation Canis Major, the Great Dog. (You can remember its proximity to Orion by thinking of it as Orion's dog.)

Back toward the top of the sky, between the Pleiades and Orion's belt, is a V-shaped group of stars known as the Hyades. They mark the face of Taurus, the Bull, while the Pleiades are riding on Taurus' shoulder. The Hyades are an open cluster, as are the Pleiades. In Greek mythology, Jupiter turned himself into a bull to carry Europa over the sea.

The Spring Sky

We can tell that spring is approaching when the Hyades and Orion get closer and closer to the western horizon each night, and finally are no longer visible when the sun sets. Now a pair of stars, the twins (Castor and Pollux), are nicely placed for viewing in the western sky at sunset. Pollux is slightly reddish, while Castor is not, and they are about the same brightness. Castor and Pollux were the Roman military gods. The constellation is called Gemini, the Twins.

On spring evenings, the Big Bear is overhead, and anything in the dipper would spill out. Leo, the Lion, is just to the south of the zenith (follow the pointers backwards). Leo looks like a backward question mark, with the bright star Regulus at its base. Regulus marks the lion's heart. The rest of Leo, to the east of Regulus, is marked by a triangle of stars. Many people visualize a sickle-shaped head and a triangular tail.

If we follow an arc begun by the stars in the handle of the Big Dipper, we come to a bright reddish star, Arcturus, another supergiant. It is in the kite-shaped constellation Boötes, the Herdsman.

Sirius sets right after sunset in the springtime; however, another bright star, Spica, is rising in the southeast in the constellation Virgo, the Virgin. It is farther along the arc from the Big Dipper through Arcturus. Vega, a star that is almost as bright, is rising in the northeast. And the constellation Hercules, with its globular cluster, is rising in the east in the evening at this time of year.

The Summer Sky

Summer, of course, is a comfortable time to watch the stars because of the warm weather. Spica is over toward the southwest in the evening. A bright reddish star, Antares, is in the constellation Scorpius, the Scorpion, to the south. ("Antares" means "compared with Ares," another name for Mars, because Antares is also reddish.)

Hercules and Cygnus are high overhead, and the star Vega is prominent near the zenith. Cassiopeia is in the northeast. The center of our galaxy is in the direction of the dense part of the Milky Way that we see in the constellation Sagittarius, the Archer, low in the south.

Around August 12 every summer is a wonderful time to observe the sky, because that is when the Perseid meteor shower occurs. One bright meteor a minute may be visible at maximum. Just lie back on the grass and watch the sky in general—don't look in any direction specifically. (An outdoor concert is a good place to do this, if you can find a bit of grass away from spotlights.) Although the Perseids are the most observed meteor shower, partly because it occurs at a time of warm weather in the northern part of the country, many other meteor showers occur during the year.

The summer is a good time of year for observing a variable star, Delta Cephei; it appears in the constellation Cepheus, which is midway between Cassiopeia and Cygnus. Delta Cephei varies in brightness with a 4.3-day period. As we shall see later, studies of its variations have led to our being able to tell the distances to galaxies. And this takes us back to the real importance of studying the sky—not in just learning *where* things are but *what* they are and *how* they work. The study of the sky has led us to understand the universe, and this is the real importance and excitement of astronomy.

Appendix 12

Interstellar Molecules

Name of Molecule	Chemical Symbol	Year of Discovery	Part of Spectrum	First Wavelength Observed	Telescope Used for Discovery	
methylidyne	CH	1937	visible	4300 Å	2.5-m	Mt. Wilson
cyanogen radical	CN	1940	visible	3875 Å	2.5-m	Mt. Wilson
methylidyne ion	CH^+	1941	visible	4232 Å	2.5-m	Mt. Wilson
hydroxyl radical	OH	1963	radio	18 cm	26-m	Lincoln Lab
ammonia	NH_3	1968	radio	1.3 cm	6-m	Hat Creek
water	H_2O	1968	radio	1.4 cm	6-m	Hat Creek
formaldehyde	H_2CO	1969	radio	6.2 cm	43-m	NRAO/Green Bank
carbon monoxide	CO	1970	radio	2.6 mm	11-m	NRAO/Kitt Peak
hydrogen cyanide	HCN	1970	radio	3.4 mm	11-m	NRAO/Kitt Peak
cyanoacetylene	HC_3N	1970	radio	3.3 cm	43-m	NRAO/Green Bank
hydrogen	H_2	1970	ultraviolet	1013-1108 Å		NRL rocket
methyl alcohol	CH_3OH	1970	radio	36 cm	43-m	NRAO/Green Bank
formic acid	HCOOH	1970	radio	18 cm	43-m	NRAO/Green Bank
"X-ogen"	HCO^+	1970	radio	3.4 mm	11-m	NRAO/Kitt Peak
formamide	$HCONH_2$	1971	radio	6.5 cm	43-m	NRAO/Green Bank
carbon monosulfide	CS	1971	radio	2.0 mm	11-m	NRAO/Kitt Peak
silicon monoxide	SiO	1971	radio	2.3 mm	11-m	NRAO/Kitt Peak
carbonyl sulfide	OCS	1971	radio	2.7 mm	11-m	NRAO/Kitt Peak
methyl cyanide, acetonitrile	CH_3CN	1971	radio	2.7 mm	11-m	NRAO/Kitt Peak
isocyanic acid	HNCO	1971	radio	3.4 mm	11-m	NRAO/Kitt Peak
methylacetylene	CH_3C_2H	1971	radio	3.5 mm	11-m	NRAO/Kitt Peak
acetaldehyde	CH_3CHO	1971	radio	28 cm	43-m	NRAO/Green Bank
thioformaldehyde	H_2CS	1971	radio	9.5 cm	64-m	Parkes
hydrogen isocyanide	HNC	1971	radio	3.3 mm	11-m	NRAO/Kitt Peak
hydrogen sulfide	H_2S	1972	radio	1.8 mm	11-m	NRAO/Kitt Peak
methanimine	H_2CNH	1972	radio	5.7 cm	64-m	Parkes
sulfur monoxide	SO	1973	radio	3.0 mm	11-m	NRAO/Kitt Peak
(no name)	N_2H^+	1974	radio	3.2 mm	11-m	NRAO/Kitt Peak
ethynyl radical	C_2H	1974	radio	3.4 mm	11-m	NRAO/Kitt Peak
methylamine	CH_3NH_2	1974	radio	3.5, 4.1 mm	11-m, 6-m	NRAO and Tokyo
dimethyl ether	$(CH_3)_2O$	1974	radio	9.6 mm	11-m	NRAO/Kitt Peak
ethyl alcohol	CH_3CH_2OH	1974	radio	2.9-3.5 mm	11-m	NRAO/Kitt Peak
sulfur dioxide	SO_2	1975	radio	3.6 mm	11-m	NRAO/Kitt Peak
silicon sulfide	SiS	1975	radio	2.8, 3.3 mm	11-m	NRAO/Kitt Peak
acrylonitrile	H_2CCHCN	1975	radio	22 cm	64-m	Parkes
methyl formate	$HCOOCH_3$	1975	radio	18 cm	64-m	Parkes
nitrogen sulfide radical	NS	1975	radio	2.6 mm	5-m	Texas
cyanamide	NH_2CN	1975	radio	3.7 mm	11-m	NRAO/Kitt Peak
cyanodiacetylene	HC_5N	1976	radio	3.0 cm	46-m	Algonquin
formyl radical	HCO	1976	radio	3.5 mm	11-m	NRAO/Kitt Peak
acetylene	C_2H_2	1976	infrared	2.4 μ	4-m	KPNO
cyanohexatriyne	HC_7N	1977	radio	3 cm	46-m	Algonquin
cyanoethynyl radical	C_3N	1977	radio	3.4 mm	11-m	NRAO/Kitt Peak
ketene	H_2C_2O	1977	radio	3.0 mm	11-m	NRAO/Kitt Peak
nitroxyl	HNO	1977	radio	3.7 mm	11-m	NRAO/Kitt Peak
ethyl cyanide	CH_3CH_2CN	1977	radio	2.6-3.4 mm	11-m	NRAO/Kitt Peak
methane	CH_4	1977	radio	3.9 mm	11-m	NRAO/Kitt Peak
diatomic carbon	C_2	1977	infrared	1 μ	1.5-m	Mt. Hopkins

Appendix 12 continued on the opposite page

Appendix 12 *Continued*

Interstellar Molecules

Name of Molecule	Chemical Symbol	Year of Discovery	Part of Spectrum	First Wavelength Observed	Telescope	Used for Discovery
butadiynl radical	C_4H	1978	radio	3 mm	11-m	NRAO/Kitt Peak
cyanoethylyne	C_3N	1978	radio	3 mm	11-m	NRAO/Kitt Peak
cyano-octatetra-yne	HC_9N	1978	radio	3 cm	46-m	Algonquin
nitric oxide	NO	1978	radio	2 mm	11-m	NRAO/Kitt Peak
methyl mercaptan	CH_3SH	1979	radio	4 mm	7-m	Bell Labs
isothiocyanic acid	HNCS	1979	radio	3 mm	7-m	Bell Labs
ozone	O_3	1980	radio	1 mm	10.4-m	Owens Valley

Notes: Only the first wavelength or wavelengths observed are listed. Discoveries of forms including isotopes not included. Mt. Wilson is one of the Hale Observatories. Hat Creek is the site of the University of California's radio observatory. NRAO has most of its telescopes at Green Bank, West Virginia, its millimeter-wave telescope on Kitt Peak in Arizona, and the VLA in New Mexico. The Naval Research Laboratory (NRL) is in Washington, D.C. The Australian National Radio Astronomy Observatory is at Parkes, N.S.W. The millimeter telescope at Fort Davis, Texas, is operated by the University of Texas. The Herzberg Institute's radio telescope is at Algonquin Park, Canada. The Kitt Peak National Observatory (KPNO) and Mt. Hopkins are in Arizona. The Lincoln Lab of MIT is in Lexington, Massachusetts. Bell Labs is in Holmdel, N.J. Caltech's Owens Valley Radio Observatory is in California.

Appendix 13

Color Index

A measure of temperature often used for the horizontal axis in Hertzsprung-Russell diagrams is the difference between the apparent magnitudes measured in two spectral regions, for example B-V, blue minus visual. This standard system of filters is defined in Figure A13–1.

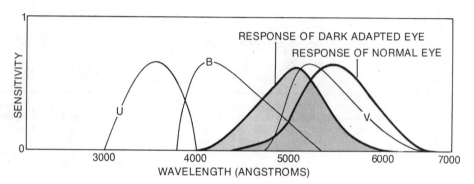

Figure A13–1 The U (ultraviolet). B (blue), and V (visual, which is more-or-less yellow) curves represent the standard set of filters used by many astronomers. The response of the eye under normal conditions and the response of the dark-adapted eye are also shown.

The difference B-V is called the *color index* (Fig. A13–2). The blue magnitude, B, measures bluer radiation than the visual magnitude, V. A very hot star is brighter in the blue than in the visible; thus V is fainter, that is, a higher number, than B. Therefore B-V is negative. Do not be confused by the fact that the magnitude of a brighter star is a lower number (more negative) than that of a fainter star.

Thus labeling the horizontal axis with the color index is equivalent to labeling it with temperature in kelvins. Since the color index can be measured easily with a telescope, and does not require theoretical interpretation or analysis of spectra before graphing, it is the measure of tempera-

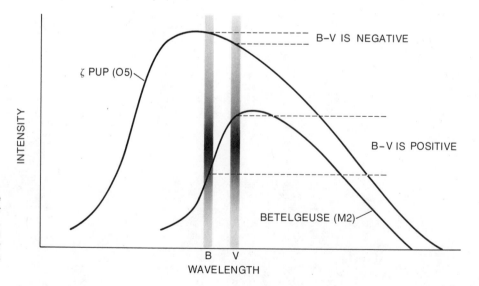

Figure A13–2 Hotter stars have negative color indices while cooler stars have positive color indices. (The star is brighter in the blue than in the visible, that is, B is a lower magnitude than V, so B-V is less than 0.)

ture ordinarily plotted. These plots, a type of H-R diagram, are often called *color-magnitude diagrams.*

The color index is zero for an A star of about 10,000 K. It falls in the range from −0.3 for the hottest stars to about +2.0 for the coolest. One can also compute a color index for the U and B (ultraviolet and blue) magnitudes, U-B, a color index for the m_{pg} and m_v (photographic and visual) magnitudes, or indeed a color index for magnitudes measured in any two spectral regions. These other color indices in the visual part of the spectrum have the same sense as the color index for B-V, that is, negative for hot stars and positive for cool ones. We must merely be certain that we are subtracting the longer wavelength measurement from the shorter wavelength measurement. Infrared astronomers use other standard systems of filters in their region of the spectrum.

Examples:
ζ Pup (blue-white) O5
B = +1.96 V = +2.25
C.I. = B-V = −0.29
Betelgeuse (red) M2
B = +2.55 V = +0.69
C.I. = B-V = +1.86

Appendix 14

Opacity and Limb Darkening

The Solar Opacity and the Limb

When we look at the sun (which we must do through special filters, for the sun is bright enough to burn our retinas if we look at the photosphere directly) it appears as a disk. But why, since the sun is just a ball of gas, does it appear to have a sharp edge? Understanding the answer to this question leads us to a fuller understanding of the properties of the photosphere and of radiating gases in general.

A gas may be transparent, partially transparent, or opaque. On a very clear day, we can see a long distance through the air. If the day is hazy, we can no longer see distant objects. The air has become opaque. Actually, it is not completely opaque, but only partially opaque or partially transparent (it is all in the point of view, like calling the weather either partly cloudy or partly sunny), because we can still see objects at intermediate distances. If the air is very foggy, then we cannot see very far at all.

The *opacity* (Fig. A14–1) is a measure of how opaque a gas is per unit of

A   B

Figure A14–1   On a day when the opacity of the atmosphere is high, we don't see very far *(A)*. When the opacity of the atmosphere is lower, we can see much farther *(B)*. In either case, the optical depth does not amount to much by the time our line of sight reaches nearby objects such as the foreground buildings. But in case *A* it becomes large at a distance closer to us than the Eiffel Tower. The view is from the Paris Observatory.

Optical depth *is a measure of how far one can see through a gas, and depends on both the opacity (per cm) and the distance.*

The sun's radius is so large that any small region of the surface seems pretty flat. Thus we can speak of levels, of heights above those levels, and of up and down as though we were speaking of a flat surface. Up and higher correspond to increasing distance from the center of the sun.

length. If the opacity of a gas is large, then we cannot see very far through it. If the opacity of a gas is very small, then the gas is fairly transparent and we can see a long way through it. However, even for a gas of low opacity (which is measured per cm), if we look through a lot if it, the opacity adds up. When the gas becomes completely opaque, we say that the *optical depth* is great. On the other extreme, when the optical depth is small, a gas is partially transparent; in other words, we can see things through it.

When we look at the sun, we see through the solar gas until the optical depth becomes so great that we cannot see any farther. We define the level to which we see when we look directly at the center of the solar disk as the *base of the photosphere.*

But though we can consider the solar gas in terms of its optical depth, which is related to how far we can see through it, each bit of the gas is not only absorbing radiation but also simultaneously emitting radiation. In fact, for a given temperature, the amount of emission is proportional to the amount of absorption, a basic fact about radiation that is known as Kirchhoff's law. The brightness that we see is a result of the emission summed up from all the gas we can see.

When we look from the earth at the center of the visible disk that is facing us, the solar gases obviously become opaque at some level; after all, we cannot see stars shining through the sun from behind it. As we look farther and farther away from the center of the disk, for a while the gas remains sufficiently opaque that we cannot see through it. Eventually, however, we reach a point (Fig. A14–2) where the gas is pretty transparent. We are still seeing through some gas, but its optical depth isn't very high.

As we look farther and farther away from the center of the solar disk, the smallest angle between a line of sight along which the solar gas is pretty opaque and a line of sight along which the solar gas is pretty transparent is only about 1 second of arc (which, let us recall, is only 1/60 of a minute of arc and, therefore, only 1/3600 of a degree of arc). There is no jump from being transparent to being opaque; the optical depth decreases continuously over this 1 arc sec as our line of sight changes. If this angle were a large angle, we could see the solar opaqueness gradually diminish. In that case, the sun's edge would look fuzzy. But the human eye cannot resolve angles smaller than 1 arc min. Compared to 1 arc min, 1 arc sec is a small angle. Thus to the human eye, the change from complete opacity to complete transparency appears sudden, and the solar edge (which is called the *limb*) appears sharp.

Limb Darkening

When we take a photograph of the sun in white light, we find that the regions near the limb are noticeably darker than the regions near the disk center. This phenomenon, which shows clearly on the photograph opening Chapter 7, is called *limb darkening.*

The fact that the sun looks darker near the limb, which is an **observational** result, has been **interpreted** by astronomers in terms of variation of

Figure A14–2 The angle over which the solar gas changes from transparent to opaque is smaller than we can resolve with our eyes, so the solar limb appears sharp.

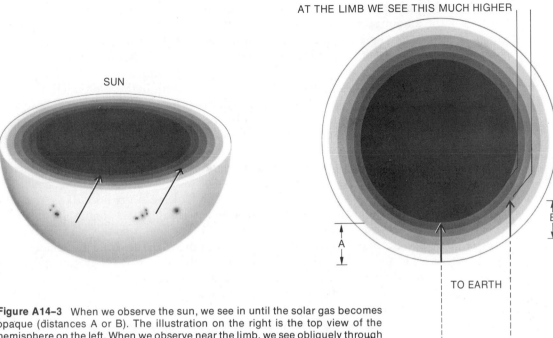

AT THE LIMB WE SEE THIS MUCH HIGHER

SUN

A

B

TO EARTH

Figure A14–3 When we observe the sun, we see in until the solar gas becomes opaque (distances A or B). The illustration on the right is the top view of the hemisphere on the left. When we observe near the limb, we see obliquely through the gas. At the limb, the atmosphere becomes opaque to us at a higher, cooler level in the solar atmosphere than it does when we look at the center of the solar disk, even though the optical depth along arrow A equals that along arrow B. The cooler level radiates less than hotter levels (according to the Stefan-Boltzmann law), so we observe limb darkening. There is no significance to the shades of gray shown in the layers; the tones merely differentiate layers.

the temperature with height through the sun's outer layers. Thus our knowledge of the temperature of the solar atmosphere is derived, in large part, from studying the solar limb darkening, using the reasoning that we shall now describe. The knowledge we gain in this manner, and the methods that have been developed through study of the sun, also apply to other stars.

When we look at any opaque mass of gas, we see in through the gas for a certain distance depending on the opacity of the gas. For example, when we look at the center of the sun's disk, we see to a level that we have just defined as the base of the photosphere. Looking through the solar atmosphere can be thought of as looking through concentric spherical shells of gas. When we look at the center of the disk, we are looking perpendicularly to the surfaces of these shells (Fig. A14–3). The gas is getting denser and denser, and thus each bit of the same size is more and more opaque (has more opacity) as we go into the sun.

We must now realize that we essentially see only the gas at the level of the base of the photosphere (this "level" may be 200 km or so thick). Why is this? First, the gas above this level is fairly transparent and so we see through it. Second, since the gas becomes opaque by the level of the base of the photosphere, we obviously can't see any radiation from gas still farther down. This leaves essentially only the radiation from in between, that is, the level where the gas becomes opaque, which we have defined as the base of

When we speak of the center of the sun, we usually mean the center of the 3-dimensional object we call the sun. When we speak of the center of the disk of the sun, we mean the center of the apparent disk that we observe.

the photosphere. The amount of radiation we receive at each wavelength is the amount that corresponds to the Planck curve for the temperature of the gas at that level. This gas is at a temperature of about 5800 K. If it were hotter, we would get more radiation (that is, it would be brighter), and if it were cooler, we would get less (that is, it would be fainter).

There is no way that we can see any farther into the sun than the base of the photosphere, so our knowledge of the interior of the sun is all based on very indirect studies. But for levels in the few thousand kilometers above the base of the photosphere—which corresponds to the upper part of the photosphere and the lower part of the chromosphere—we can in fact directly deduce what the temperature is from studying limb darkening.

We can do this because when we look at any point on the sun other than the center of its disk, we are no longer looking perpendicularly through the shells of gas but are rather looking through the gas at an oblique angle (as shown on Fig. A14–3). The gas still becomes opaque when we have looked through a certain amount of it, but the place at which this happens is now farther out from the center of the sun than it was when we looked at the center of the solar disk. Thus the gas becomes opaque at a level that is above the base of the photosphere. We now see gas the intensity of whose radiation is determined by the temperature at this new level. Just as we discussed above, the gas above this new level is too transparent to see, and the gas below the new level is hidden.

Note that in studying limb darkening, we are still looking on the disk of the sun. Our line of sight terminates at the level at which the gas becomes opaque. When we look off the edge, on the other hand, we are seeing different regions of the solar gas.

The gas at this higher level that we see when we look near the limb emits less radiation than the gas at the center of the disk. This is just the observational result that we call limb darkening. From the fact that it emits less, we deduce by Planck's law that it is cooler. Since the sun appears darker and darker as we look closer and closer to the limb, we deduce that the temperature of the photosphere declines with height.

In this manner, by studying the solar limb darkening we can determine how the temperature of the sun changes with height in the solar atmosphere. We are thus getting information about the third, radial dimension of the sun to add to the information about the two dimensions on the solar surface of which we can take pictures. The temperature actually declines to about 4150 K at about 500 km above the base of the photosphere. This height corresponds to the lower part of the chromosphere.

Appendix 15

The Equation of Time

Standard time is based on the hour angle of the sun, which is the angle in the sky that the sun has traveled since it last passed through the meridian. (The hour angle is also equal to the sidereal time minus the right ascension of an object.) When the sun is at its highest point in the sky, its hour angle is zero. In reckoning civil time from the hour angle of the sun on a 24-hour day, we add 12 hours to the hour angle of the sun. Thus, on a 24-hour clock, the time at noon (when the sun's hour angle is zero) is 12 hours, and when the hour angle of the sun is 3 hours, the time is 15 o'clock. This gives the convenience of having the date change in the middle of the night rather than in the middle of the day.

Each day, the sun moves ahead of the stars. Its R.A. increases by 4. However, this increase is not a regular one. The variations arise because the earth does not move at a constant rate around the sun, and because the earth's axis of rotation is inclined with respect to the plane of its orbit. Since we wish civil time to increase at a constant rate we define a fictitious object called the *mean sun*. The mean sun moves along the ecliptic such that its right ascension changes at a constant rate, equal to the average rate of increase in right ascension for the true sun. Our civil timekeeping is then kept by the mean sun.

Sometimes the mean sun is ahead of the true sun and sometimes it is the other way around. We define the *equation of time* as the difference between the right ascension of the mean sun and that of the true sun:

$$E = R.A. \text{ (mean sun)} - R.A. \text{ (true sun)}.$$

Since the right ascension of an object is equal to the sidereal time minus the hour angle of the object, this becomes

$$E = H.A. \text{ (true sun)} - H.A. \text{ (mean sun)},$$

so, to tell time by the sun, you must know the value of E for the day on which you are observing. You then observe the hour angle of the sun (as with a sundial) and subtract E. That is,

$$\text{mean solar time} = \text{true solar time} - E.$$

Of course, to find your standard time, you must then know how far you are from the center of the time zone.

We can get some idea of how E varies throughout the year by separately considering the two major effects that cause its variation. First, since the earth's orbit is not a perfect circle, it does not move at a constant rate. The earth is moving fastest on about January 2, and slowest about 6 months later. Let us consider the effect of this non-uniform revolution of the earth on E. On January 2, the mean sun and true sun would be in the same place, so E would

Figure A15–1 The equation of time, or the difference between true solar time and mean solar time. Curve A shows the effect of the fact that the earth's orbit is elliptical, and the earth does not move at a constant speed. Curve B shows the effect of the changing declination of the sun, since the earth's axis is tilted by 23½°. Curve C is the sum of curves A and B. To use this curve, and the sun, to tell time, first determine the hour angle of the sun (as with a sundial). Then use this graph to find the value of curve C for the day you are observing, and subtract that value from the hour angle of the sun. You then have the hour angle of the mean sun. To convert to standard time, you must add 12 hours and subtract the difference between your longitude and the longitude of the center of your time zone.

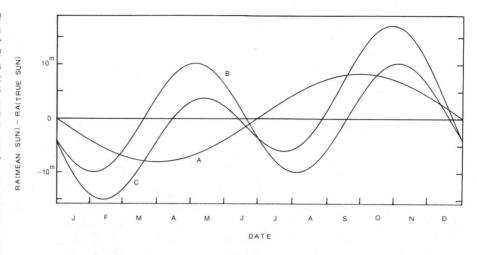

be 0. However, since the earth is moving faster than average, the true sun will get ahead of the mean sun and continue to increase its lead for about three months. Then the earth begins to move more slowly than average. For the next three months the mean sun will catch up; then it, in turn, moves ahead, extending its lead for three months. For the last three months of the year the earth will again be moving faster than average, and the true sun will finally catch up to the mean sun on January 2. The maximum difference between the mean and true suns resulting from this effect is 7.7 minutes. The behavior we have just described is graphed as curve A in Figure A15–1.

Now we consider the effect of the tilt of the earth's axis on E. This effect is zero when the sun is at the vernal equinox. As the sun moves along the ecliptic, some of its change in position shows up as a change in right ascension and some as a change in declination. When the change in declination is greatest, the change in right ascension is smallest. As the sun is crossing the equator, its change in declination is greatest, so its change in right ascension is slowest. Thus, after the sun passes through the vernal equinox, the true sun falls behind the mean sun, and falls farther behind for 1½ months. For the next 1½ months it will catch up, moving fastest with respect to the average sun when it is at the summer solstice, so it will move ahead for the next month and a half. The cycle will continue until the effect is zero when the sun is in the vernal equinox. In the next half year the cycle will again repeat. The largest contribution to E due to this effect is 10 minutes. This behavior is graphed as curve B in Figure A15–1.

Finally, we find E by adding the curves A and B in Figure A15–1, to get curve C. Thus, for any day of the year we can read the difference between the right ascension of the mean sun and the true sun, and do our solar time-keeping.

GLOSSARY

The first phrase following the word or term is the *definition* or *defining property*. Further explanation or examples may follow in a separate sentence.

absolute magnitude–The magnitude that a star would appear to have if it were at a distance of ten parsecs from us.

absolute zero–The lowest point on a temperature scale; in particular, for kinetic temperature, the point at which all classical motion of particles stops. In the Celsius scale it is about $-273°C$.

absorption line–Wavelengths (or frequencies) at which the intensity of radiation is less than it is at neighboring wavelengths (or frequencies).

accretion–The addition of matter by a body. For example, the accretion of mass by a star is the addition of mass to a star, perhaps of interstellar matter or of matter drawn off a companion star.

achromatic–Non-color-dependent; a more limited technical definition is an optical system that is constructed so that two wavelengths are focused at the same point, as distinguished from an apochromat, in which three colors are focused at the same point.

active sun–The group of solar phenomena that vary with time, or in a regular pattern over the solar surface. Examples are sunspots, flares, and prominences.

albedo–The fraction of light reflected by a body.

alpha particle–A helium nucleus; consists of two protons and two neutrons.

alt-azimuth–A two-axis telescope mounting in which motion around one of the axes, which is vertical, provides motion in azimuth, and motion around the perpendicular axis provides up and down (altitude) motion. Continuous motion along both axes is necessary to track the stars, but with modern minicomputers, this is no longer a serious limitation for large telescopes.

altitude–(a) Height above the surface of a planet, or (b) for a telescope mounting, elevation in angular measure above the horizon.

amino acid–A type of molecule containing the group NH_2 (the amino group). Amino acids are fundamental building blocks of life.

angstrom–A unit of length equal to 10^{-8} cm.

angular momentum–An intrinsic property of a system corresponding to the amount of its revolution or spin. The amount of angular momentum of a body orbiting around a point is the mass of the orbiting body times its (linear) velocity of revolution times its distance from the point. The amount of angular momentum of a spinning sphere is the moment of inertia, an intrinsic property of the distribution of mass, times the angular velocity of spin.

angular velocity–The rate at which a body rotates or revolves expressed as the angle covered in a given time (for example, in degrees per hour).

annular eclipse–A type of solar eclipse in which a ring (annulus) of solar photosphere remains visible around the moon.

anorthosite–A type of rock, common in the lunar highlands though rare on earth.

antimatter–A type of matter in which each particle (antiproton, antineutron, etc.) is opposite in charge and certain other properties to a corresponding particle (proton, neutron, etc.) of the same mass of the ordinary type of matter from which the solar system is made. Particles of antimatter are known to exist, but it is not known why matter is dominant in this region of the universe or whether regions exist in which antimatter is common.

apastron–For an orbit around a star, the farthest point from that star.

aperture–The diameter of the lens or mirror that defines the amount of light focused by an optical system.

aphelion–For an orbit around the sun, the farthest point from the sun.

apparent magnitude–The brightness of a star as seen by an observer, given in a specific system in which five magnitudes corresponds to a brightness ratio of one hundred times and the scale is fixed by correspondence with a historical background.

*apsides (*sing. *apsis)*–The points at the ends of the major axis of an elliptical orbit. The *line of apsides* is the line that coincides with the major axis of the orbit.

association–A physical grouping of stars; in particular, we talk of O and B associations or T associations.

asteroid–A "minor planet"; a non-luminous chunk of rock smaller than planet-size but larger than a meteoroid, in orbit around a star.

asteroid belt–A region of the solar system, between the orbits of Mars and Jupiter, in which most of the asteroids orbit.

astrology–A non-scientific system based on superstition, that purports to explain or predict human actions by study of celestial positions.

astrometric binary–A system of two stars in which the existence of one star can be deduced by study of its gravitational effect on the proper motion of the other star.

astrometry–The branch of astronomy that involves the detailed measurement of the positions of stars and other celestial bodies.

astronomical unit–The average distance from the earth to the sun.

atom–The smallest possible unit of a chemical element. When an atom is subdivided the parts no longer have properties of any chemical element. An atom consists of a nucleus with orbiting electrons.

atomic clock–A system that uses atomic properties to provide a measure of time.

atomic number–The number of protons in an atom.

atomic weight–The number of protons and neutrons in an atom, averaged over the abundances of the different isotopes.

aurora–Glowing lights visible in the sky, resulting from processes in the earth's upper atmosphere.

autumnal equinox–Of the two locations in the sky where the ecliptic crosses the celestial equator, the one that the sun passes each year when moving from northern to southern declinations.

azimuth–The angular distance, around the horizon from the northern direction, usually expressed in angular measure from 0° for an object in the northern direction, to 180° for an object in the southern direction, around to 360°. Together, the altitude and the azimuth define the direction to an object.

background radiation–See *primordial background radiation.*

Baily's beads–Beads of light visible around the rim of the moon at the beginning and end of a total solar eclipse. They result from the solar photosphere shining through valleys at the edge of the moon.

Balmer lines–Spectral absorption or emission lines resulting from transitions to or from the second energy level (first excited level) of hydrogen.

barred spiral galaxy–A type of spiral galaxy in which the arms unwind from a straight structure in the plane of the galaxy called a "bar."

basalt–A type of rock resulting from the cooling of lava.

beta particle–An electron or positron outside an atom.

big bang theory–A cosmological model, based on Einstein's general theory of relativity, in which the universe was once compressed to infinite density and has been expanding ever since.

binary star–Two stars revolving around each other.

black body–A hypothetical object that, if it existed, would absorb all radiation that hit it and would emit radiation that exactly followed Planck's law.

black dwarf–A non-radiating ball of gas that results either when a white dwarf radiates all its energy or when gas contracts gravitationally but contains too little mass to begin nuclear fusion.

black hole–A region of space from which, according to the general theory of relativity, neither radiation nor matter can escape.

Bode's law–A numerical scheme, known for 200 years, that gives the radii of the orbits of the seven innermost planets and the radius of the asteroid belt.

Bohr atom–Niels Bohr's model of the hydrogen atom, in which the energy levels are depicted as concentric circles of radii that increase as (level number)^2.

bolometer–A device for measuring the total amount of radiation from an object.

bolometric magnitude–The magnitude of a celestial object corrected to take account of the radiation in parts of the spectrum other than the visible.

bound-free transition–An atomic transition in which an electron starts bound to the atom and winds up free from it.

breccia–A type of rock made up of fragments of several types of rocks. Breccias are common on the moon.

brightness–The amount of energy received per unit of time in a given spectral region.

burster–One of the sources of bursts of x-rays.

calorie–A measure of energy in the form of heat, originally corresponding to the amount of heat required to raise the temperature of 1 gram of water by 1° C.

carbon cycle–A chain of nuclear reactions, involving carbon at its intermediate stages, that transforms four hydrogen atoms into one helium atom with a resulting release in energy. The carbon cycle is only important in stars hotter than the sun.

Cassegrainian (or Cassegrain)–A type of reflecting telescope in which the light focused by the primary mirror is intercepted short of its focal point and refocused and reflected by a secondary mirror through a hole in the center of the primary mirror.

Cassini's division–The major division in the rings of Saturn.

catalyst–A substance that participates in a reaction but that is left over in its original form at the end of the reaction.

celestial equator–The intersection of the celestial sphere with the plane that passes through the earth's equator.

celestial poles–The intersection of the celestial sphere with the axis of rotation of the earth.

celestial sphere–The hypothetical sphere centered at the center of the earth to which it appears that the stars are affixed.

center of mass–The "average" location of mass; the point in a body or system of bodies at which we may consider all the mass to be located for the purpose of calculating the gravitational effect of that mass or its mean motion when a force is applied.

Cepheid variable–A type of supergiant star that oscillates in brightness in a manner similar to the star δ Cephei. The periods of Cepheid variables, which are between 1 and 100 days, are linked to the absolute magnitude of the stars by known relationships; this allows the distances to Cepheids to be found.

charm–An arbitrary name that corresponds to a property that distinguishes certain elementary particles, including types of quarks, from each other.

chemical element–See *element*.

chondrite–A type of stony meteorite that contains small crystalline spherical particles called chondrules.

chromatic aberration–A defect of lens systems in which different colors are focused at different points.

chromosphere–The part of the atmosphere of the sun (or another star) between the photosphere and the corona. It is probably entirely composed of spicules and probably roughly corresponds to the region in which mechanical energy is deposited.

circumpolar stars–For a given observing location, stars that are close enough to the celestial pole that they never set or close enough to the opposite celestial pole that they never rise.

classical–When discussing atoms, not taking account of quantum mechanical effects.

closed universe–A type of big bang universe in which the universe will eventually contract. A closed universe has finite volume.

cluster–(a) Of stars, a physical grouping of many stars in space; (b) of galaxies, a physical group of at least a few galaxies in space.

cluster variable–A star that varies in brightness with a period of 0.1 to one day, similar to the star RR Lyrae. They are found in globular clusters. All cluster variables have approximately the same brightness, which allows their distance to be readily found.

CNO cycle–Same as carbon cycle.

color–(a) Of an object, a visual property that depends on wavelength; (b) an arbitrary name assigned to a property that distinguishes three kinds of quarks.

color index–The difference, expressed in magnitudes, of the brightness of a star or other celestial object measured at two different wavelengths. The color index is a measure of temperature.

coherent radiation–Radiation in which the phases of waves at different locations in a cross-section of radiation have a definite relation to each other; in non-coherent radiation, the phases are random. Only coherent radiation shows interference.

color-magnitude diagram–A Hertzsprung-Russell diagram in which the temperature on the horizontal axis is expressed in terms of color index.

coma–(a) Of a comet, the region surrounding the head; (b) of an optical system, an off-axis aberration in which the images of points appear with comet-like asymmetries.

comet–A type of object in the solar system, in a very elongated orbit around the sun, that when relatively near to the sun shows a coma and a tail.

comparison spectrum–A spectrum of known elements on earth usually photographed on the same photographic plate as a stellar spectrum in order to provide a known set of wavelengths or zero Doppler shift.

conjunction–The closest apparent approach in the sky of two celestial objects. When only one body is named it is understood that the second body is the sun.

conservation law–A statement that the total amount of some property (angular momentum, energy, etc.) of a body or set of bodies does not change.

constellation–One of the 88 areas into which the sky has been divided for convenience in referring to the stars or other objects therein. Many of the constellations were delineated, not necessarily in their current shape, long ago, and names and pictures from mythology attached.

continental drift–The informal name for the theory of plate tectonics that describes the surface of the earth as a set of shifting regions called plates.

continuous spectrum–A spectrum with radiation at all wavelengths but with neither absorption nor emission lines.

continuum–The continuous spectrum that we would measure from a body if no spectral lines were present.

convection–The method of energy transport in which the rising motion of masses from below carries energy upward in a gravitational field. Boiling is an example.

convergent point–The point in the sky toward which the members of a star cluster appear to be converging or from which they appear to be diverging. Referred to in the moving cluster method of determining distances.

corona–The outermost region of the sun (or of other stars), characterized by high temperatures of millions of degrees.

coronagraph–A type of telescope with which the corona can be seen in visible light at times other than that of a total solar eclipse.

cosmic rays–Nuclear particles or nuclei travelling through space at high velocity.

cosmogony–The study of the origin of the universe, usually applied in particular to the origin of the solar system.

cosmological constant–A constant arbitrarily added by Einstein to an equation in his general theory of relativity in order to provide a solution in which the universe did not expand. It was only subsequently discovered that the universe did expand after all.

cosmological principle–The principle that on the whole the universe looks the same in all directions and in all regions.

cosmology–The study of the universe as a whole.

coudé focus–A focal point of large telescopes in which the light is reflected by a series of mirrors so that it comes to a point at the end of a polar axis. The image does not move even when the telescope moves, which permits the mounting of heavy equipment.

deceleration parameter (q_0)–A particular measure of the rate at which the expansion of the universe is slowing down.

declination–Celestial latitude, measured in degrees north or south of the celestial equator.

deferent–In the Ptolemaic system of the universe, the larger circle, centered at the earth, on which the centers of the epicycles revolve.

degenerate matter–Matter whose properties are controlled, and which is prevented from further contraction, by quantum mechanical laws as expressed by the Pauli exclusion principle.

density–Mass divided by volume.

density-wave theory–The explanation of spiral structure of galaxies as the effect of a wave of compression that rotates around the center of the galaxy and causes the formation of stars in the compressed region.

deuterium–An isotope of hydrogen that contains one proton and one neutron.

diamond ring effect–The last Baily's bead glowing brightly at the beginning of the total phase of a solar eclipse, or its counterpart at the end of totality.

differential rotation–Rotation of a body in which different parts have different angular velocities (and thus different periods of rotation).

differentiation–For a planet, the existence of layers of different structure or composition.

diffraction–A phenomenon affecting light as it passes any obstacle, spreading it out in a complicated fashion.

diffraction grating–A very closely ruled series of lines that, through their diffraction of light, provide a spectrum of radiation that falls on it.

disk–(a) Of a galaxy, the disk-like flat portion, as opposed to the nucleus or the halo; (b) of a star or planet, the two-dimensional projection of its surface.

dispersion–(a) Of light, the effect that different colors are bent by different amounts when passing from one substance to another; (b) of the pulses of a pulsar, the effect that a given pulse, which leaves the pulsar at one instant, arrives at the earth at different times depending on the different wavelength or frequency at which it is observed. Both of these effects arise because light of different wavelengths travels at different speeds except in a vacuum.

D lines–A pair of lines from sodium that appear in the yellow part of the spectrum.

DNA–Deoxyribonucleic acid, a long chain of molecules that contains the genetic information of life.

Doppler effect–A change in wavelength that results when a source of waves and the observer are moving relative to each other.

dwarf–A main-sequence star.

dynamo–A device that generates electricity through the effect of motion in the presence of a magnetic field. The solar dynamo explains sunspots and the solar activity cycle.

earthshine–Sunlight illuminating the moon after having been reflected by the earth.

eccentricity–A measure of the flatness of an ellipse, defined as half the distance between the foci divided by the semi-major axis.

eclipse–The passage of all or part of one astronomical body into the shadow of another.

eclipsing binary star–A binary star in which one member periodically hides the other.

ecliptic–The path followed by the sun across the celestial sphere in the course of a year.

ecliptic plane–The plane of the earth's orbit around the sun.

electromagnetic force–One of the four fundamental forces of nature, the force that gives rise to electromagnetic radiation.

electromagnetic radiation–Radiation resulting from changing electric and magnetic fields.

electric field–A force field set up by an electric charge.

electron–A particle of one negative charge, 1/1830 the mass of a proton, that is not affected by the strong force. In an atom, the electron is normally thought of as orbiting the nucleus.

electron temperature–The kinetic temperature of the electrons present.

electron volt (*eV*)–The energy necessary to raise an electron through a potential of one volt.

element–A kind of atom characterized by a certain number of protons in its nucleus. All atoms of a given element have similar chemical properties.

elementary particle–One of the constituents of an atomic nucleus freed from the nucleus.

ellipse–A curve with the property that the sum of the distances from any point on the curve to two given points, called the foci, is constant.

elliptical galaxy–A type of galaxy characterized by elliptical appearance. Elliptical galaxies have no interstellar matter.

emission line–Wavelengths (or frequencies) at which the intensity of radiation is greater than it is at neighboring wavelengths (or frequencies).

emission nebula–A glowing cloud of interstellar gas.

energy–A fundamental quantity usually defined in terms of the ability of a system to do something that is technically called "work," that is, the ability to move an object by application of force, where the work is the force times the displacement.

energy level–A state corresponding to an amount of energy that an atom is allowed to have by the laws of quantum mechanics.

ephemeris–A listing of astronomical positions and other data that change with time. From the same root as *ephemeral*.

epicycle–In the Ptolemaic theory, a small circle, riding on a larger circle called the deferent, on which a planet moves. The epicycle is used to account for retrograde motion.

equator–(a) Of the earth, a great circle on the earth, midway between the poles; (b) celestial, the projection of the earth's equator onto the celestial sphere.

equatorial mount–A type of telescope mounting in which one axis, called the polar axis, points toward the celestial pole and the other axis is perpendicular. Motion around only the polar axis is sufficient to completely counterbalance the effect of the earth's rotation.

equinox–An intersection of the ecliptic and the celestial equator. The center of the sun is geometrically above and below the horizon for equal lengths of time on the two days of the year when the sun passes the equinoxes; if the sun were a point and atmospheric refraction were absent, then day and night would be of equal length on those days.

erg–A unit of energy in the metric system, corresponding to the work done by a force of one dyne (the force that is required to accelerate one gram by one $cm\ s^{-2}$) producing a displacement of one centimeter.

ergosphere–A region surrounding a rotating black hole (or other system satisfying Kerr's solution) from which work can be extracted.

Euclidean space–A space with zero curvature; a space where the sum of the angles of a triangle is 180°.

event horizon–The sphere around a black hole from within which nothing can escape; the place at which the exit cones close.

excitation–The raising of atoms to higher energy states than the lowest possible.

exclusion principle–The quantum mechanical rule that certain types of elementary particles cannot exist in completely identical states.

exit cone–The cone that, for each point within the photon sphere of a black hole, defines the directions of rays of radiation that escape.

exobiology–The study of life located elsewhere than on earth.

extinction–The dimming of starlight by scattering and absorption as the light traverses interstellar space.

extragalactic–Exterior to the Milky Way Galaxy.

filament–A feature of the solar surface seen in Hα as a thin, dark wavy line. A filament is a prominence projected on the solar disk.

filtergram–A photograph taken through a filter that passes only a very narrow band of wavelengths; usually applied to solar photographs.

fireball–An exceptionally bright meteor.

fission, nuclear–The splitting of an atomic nucleus.

flare–An extremely rapid brightening of a small area of the surface of the sun, usually observed in hydrogen-alpha and other strong spectral lines and accompanied by x-ray and radio emission. The hottest part of the flare, from which the x-rays originate, reaches millions of kelvins. Flares can be detected on other stars as transient brightenings.

flash spectrum–The solar chromospheric spectrum seen in the few seconds before or after totality at a solar eclipse.

fluorescence–The transformation of photons of relatively high energy to photons of lower energy through interactions with atoms, and the resulting radiation.

flux–The amount of something (such as energy) passing through a surface per unit time.

focal length–The distance from a lens or mirror to the point to which rays from an object at infinity are focused.

focus–(a) A point to which radiation is made to converge; (b) of an ellipse, one of the two points the sum of the distances to which remains constant.

force–In physics, something that can or does cause a change of momentum, measured by the rate of change of momentum with time.

force field–A way of describing phenomena that result from action at a distance, that is, even though objects are not touching.

Fraunhofer lines–The absorption lines of a solar or other stellar spectrum.

frequency–The rate at which waves pass a given point.

full moon–The phase of the moon when the side facing the earth is fully illuminated by sunlight.

fusion–The amalgamation of nuclei into heavier nuclei.

galactic cluster–An asymmetric type of collection of stars that shared a common origin.

galaxy–A collection of stars, gas, and dust, typically containing millions of solar masses, that is, a basic building block of the universe. When we speak of *the galaxy* or *our galaxy*, we mean the Milky Way Galaxy.

Galilean satellites–The four brightest satellites of Jupiter.

gamma rays–Electromagnetic radiation with wavelengths shorter than approximately 1 angstrom.

giant planets–Jupiter, Saturn, Uranus, and Neptune.

giant star–A type of star brighter than main-sequence stars of the same spectral type.

gibbous moon–The phases between half moon and full moon.

globular cluster–A spherically symmetric type of collection of stars that shared a common origin.

gluon–The as yet undiscovered particle that would carry the color force (and thus the strong—nuclear—force).

granule–One of the bright features, approximately one arc sec in diameter, that cover the face of the quiet sun.

grating–A surface ruled with closely spaced lines that, through diffraction, breaks up light into its spectrum.

gravitation–One of the four fundamental forces of nature, the force by which two masses attract each other.

gravitational lens–In the gravitational lens phenomenon, a massive body changes the path of electromagnetic radiation passing near it so as to make more than one image of an object. The double quasar was the first example to be discovered.

gravitationally–Controlled by the force of gravity.

gravitational radius–The radius that, according to Schwarzschild's solutions to Einstein's equations of the general theory of relativity, corresponds to the event horizon of a black hole.

gravitational redshift–A redshift of light caused by the presence of mass according to the general theory of relativity.

gravitational waves–As yet undiscovered waves that many scientists consider to be a consequence according to the general theory of relativity of changing distributions of mass.

great circle–The intersection of a plane that passes through the center of a sphere with the surface of that sphere; the largest possible circle that can be drawn on the surface of a sphere.

greenhouse effect–The effect by which the atmosphere of a planet heats up above its equilibrium temperature because it is transparent to incoming visible radiation but opaque to the infrared radiation that is emitted by the surface of the planet.

Gregorian–A type of reflecting telescope in which the light focused by the primary mirror is reflected by a secondary mirror after it passes its prime focus through a hole in the center of the primary mirror.

Gregorian calendar–The calendar in current use, with normal years that are 365 days long, with leap years every fourth year except for years that are divisible by 100 but not by 400.

H_o–The Hubble constant.

H I region–An interstellar region of neutral hydrogen.

H II region–An interstellar region of ionized hydrogen.

H line–The spectral line of ionized calcium at 3968 angstroms.

Hα–The first line of the Balmer series of hydrogen, at 6563 angstroms.

half-life–The length of time for half a set of particles to decay through radioactivity or instability.

halo–Of a galaxy, the region of the galaxy that extends far above and below the plane of the galaxy, as opposed to the disk or the nucleus.

head–Of a comet, the nucleus and coma together.

heliacal rising–The first time in a year that an astronomical body rises sufficiently far ahead of the sun that it can be seen in the morning sky.

heliocentric–Sun-centered; using the sun rather than the earth as the point to which we refer. A heliocentric measurement, for example, omits the effect of the Doppler shift caused by the earth's orbital motion.

helium flash–The rapid onset of fusion of helium into carbon through the triple-alpha process that takes place in most red giant stars.

hertz–The measure of frequency, with units of s^{-1}; formerly called cycles per second.

Hertzsprung gap–A region above the main sequence in a Hertzsprung-Russell diagram through which stars evolve rapidly and thus in which few stars are found.

Hertzsprung-Russell diagram–A graph of temperature or equivalent vs. luminosity or equivalent for a group of stars.

horizontal branch–A part of the Hertzsprung-Russell diagram of a globular cluster, corresponding to stars that are all at approximately zero absolute magnitude and have evolved past the red giant branch and are moving leftward on the diagram.

hour angle–Of a celestial object as seen from a particular location, the difference between the local sidereal time and the right ascension (H.A. = L.S.T. — R.A.).

H–R diagram–Hertzsprung-Russell diagram.

Hubble constant (H_o)–The constant of proportionality in Hubble's law linking the velocity of recession of a distant object and its distance from us. The most widely accepted current value is 50 km s^{-1} Mpc^{-1}.

Hubble law–The linear relation between the velocity of recession of a distant object and its distance from us, $v = H_o d$.

hyperfine level–A subdivision of an energy level caused by such relatively minor effects as changes resulting from the interactions among spinning particles in an atom or molecule.

IC–Index Catalogue, one of the supplements to Dreyer's New General Catalogue.

image tube–An electronic device that receives incident radiation and intensifies it or converts it to a wavelength at which photographic plates are sensitive.

inclination–Of an orbit, the angle of the plane of the orbit with respect to the ecliptic plane.

Index Catalogue–See *IC*.

inferior conjunction–The closest alignment of an inferior planet as it passes nearly between the earth and the sun.

inferior planet–A planet whose orbit around the sun is within the earth's, namely, Mercury and Venus.

infrared–Radiation beyond the red, from about 7000 angstroms up to 1 mm.

interference–The property of radiation, explainable by the wave theory, in which waves that are in phase can add (constructive interference) and waves that are out of phase can subtract (destructive interference); for light, this gives alternate light and dark bands.

interferometer–A device that uses the property of interference to measure such properties of objects as their positions or structure.

International Date Line–A crooked imaginary line on the earth's surface, roughly corresponding to 180° longitude, at which, when crossed from east to west, time jumps by one day.

interplanetary medium–Gas and dust between the planets.

interstellar medium–Gas and dust between the stars.

interstellar reddening–The phenomenon by which the extinction of blue light by interstellar matter is greater than the extinction of red light.

inverse Compton scattering–A method of converting photons from lower to higher energy through interaction with electrons that are moving with velocities close to the speed of light.

ion–An atom that has gained or lost one or more electrons, usually referring to the case where one or more electrons are lost. See also *negative ion*.

ionization temperature–A temperature defined by the number and types of ions present.

irregular galaxy–A type of galaxy that lacks defined shape or symmetry.

isotope–A form of a chemical element with a different number of neutrons.

isotropic–Being the same in all directions.

Jeans length–The minimum diameter of gas that is unstable to gravitational collapse.

Jovian planet–Same as giant planet.

Julian calendar–The predecessor to the Gregorian calendar, the calendar with 365 day years and leap years every fourth year without exception.

Julian day–The number of days since noon on January 1, 4713 B.C. Used for keeping track of variable stars or other astronomical events.

K line–The spectral line of ionized calcium at 3933 angstroms.

Keplerian–Following Kepler's law.

kiloparsec–One thousand parsecs.

kinetic temperature–The temperature defined by the velocities of particles in a gas.

laser–An acronym for "light amplification by stimulated emission of radiation," a device by which certain energy levels are populated by more electrons than normal, resulting in an especially intense emission of light at a certain frequency when the electrons drop to a lower energy level.

latitude–Number of degrees north or south of the equator measured from the center of the earth.

leap year–A year in which a 366th day is added.

libration–The effect by which we can see slightly more than half the lunar surface even though the moon basically has one half that always faces us.

light–Electromagnetic radiation between about 3000 and 7000 angstroms in wavelength; sometimes used more generally for other wavelengths of radiation in the ultraviolet or infrared.

light curve–The graph of the magnitude of an object vs. time.

light year–The distance that light travels in a year.

limb–The edge of a star or planet.

limb darkening–The decreasing brightness of the disk of the sun or another star as one looks from the center of the disk closer and closer to the limb.

line profile–The graph of the intensity of radiation vs. wavelength for a spectral line.

Local Group–The two dozen or so galaxies, including the Milky Way Galaxy, that form a subcluster.

local standard of rest–The system in which the average velocity of nearby stars is zero. We usually refer measurements of velocity of distant objects to the local standard of rest.

logarithmic–Increasing by multiples of ten. This is as opposed to linear, in which increases are additive.

longitude–The angular distance around a body measured along the equator from some particular point; for a point not on the equator, it is the angular distance along the equator to a great circle that passes through the poles and through the point.

luminosity–The total amount of energy given off by an object per unit time.

luminosity class–Regions of the H–R diagram that define classes of luminosity of objects of the same spectral type.

lunar eclipse–The passage of the moon into the earth's shadow.

Lyman alpha–The spectral line that corresponds to a transition between the lowest two major energy levels of a hydrogen atom.

Lyman lines–The spectral lines that correspond to transitions to or from the lowest major energy level of a hydrogen atom.

Magellanic Clouds–Two small irregular galaxies, the closest neighbors to the Milky Way Galaxy, visible in the southern sky.

magnetic field–A force field set up of a type that affects magnets in a certain way, attracting one part of the magnet while repelling another part.

magnetohydrodynamics–The study of the interaction of moving matter and a magnetic field (magneto-, from the magnetic field, and hydro-, which means water, from the historical use of the term hydrodynamics to mean the study of motion of water).

magnetosphere–A region of magnetic field around a planet.

magnitude–A factor of $\sqrt[5]{100} = 2.511887\ldots$ in brightness. See *absolute magnitude* and *apparent magnitude*. An *order of magnitude* is a power of ten.

main sequence–A band on a Hertzsprung-Russell diagram in which stars fall during the main, hydrogen-burning, phase of their lifetimes. The main sequence runs diagonally from hot, bright stars at the upper left to cool, faint stars at the lower right.

major axis–The longest diameter of an ellipse; the line from one side of an ellipse to the other that passes through the foci. Also, the length of that line.

mantle–The shell of rock separating the core of a differentiated planet from its thin surface crust. The earth's mantle starts about 40 km below the surface and continues nearly half-way toward the center.

mare–One of the smooth areas on the moon or on some of the other planets.

mascon–A concentration of mass under the surface of the moon, discovered from its gravitational effect on spacecraft orbiting the moon.

maser–An acronym for "microwave amplification by stimulated emission of radiation," a device by which certain energy levels are more populated than normal, resulting in an especially dense emission of radio radiation at a certain frequency when the system drops to a lower energy level.

mass–A measure of the inherent amount of matter in a body.

mass function–A complicated term involving the masses of both components of a binary star system. When the spectrum of only one member of the pair is visible, one can derive only the mass function and not the individual masses.

mass-luminosity relation–A well-defined relation between the mass and luminosity for main-sequence stars.

megaparsec (Mpc)–A million parsecs.

meridian–The great circle on the celestial sphere that passes through the celestial poles and the observer's zenith.

Messier object–One of the objects in Messier's catalogue of nonstellar objects or in the additions to Messier's list.

metal–(a) For stellar abundances, any element higher in atomic number than 2, that is, heavier than helium. (b) In general, neutral matter that is a good conductor of electricity.

meteor–A track of light in the sky from rock or dust burning up in the earth's atmosphere.

meteorite–An interplanetary chunk of rock after it impacts on a planet or moon, especially on the earth.

meteoroid–An interplanetary chunk of rock smaller than an asteroid.

micrometeorite–A tiny meteorite. The micrometeorites that hit the earth's surface are often sufficiently slowed down that they can reach the ground without being vaporized.

micron–The former name, still in use by astronomers, for a micrometer (10^{-6} meter).

Milky Way–The band of light across the sky from the stars and gas in the plane of the Milky Way Galaxy.

minor axis–The shortest diameter of an ellipse; the line from one side of an ellipse to the other that passes midway between the foci and is perpendicular to the major axis. Also, the length of that line.

Mira variable–A long-period variable star similar to Mira (omicron Ceti).

molecule–A collection of atoms bound together that is the smallest collection that exhibits a certain set of chemical properties.

momentum–A measure of the tendency that a moving body has to keep moving. The momentum in a given direction (the "linear momentum") is equal to the mass of the body times its component of velocity in that direction. See also *angular momentum.*

nebular hypothesis–The particular nebular theory for the formation of the solar system advanced by Laplace.

nebular theories–The theories that the sun and the planets formed out of a cloud of gas and dust.

neutrino–Spinning, neutral elementary particle. A neutrino has been thought to have no rest mass and always travels at the speed of light, though some experimental evidence now exists that neutrinos may have some mass, which would make them travel slower than light, and may change from one type of neutrino to another.

neutron–A massive, neutral elementary particle, one of the fundamental constituents of an atom. Neutrons and protons are similar in mass, 1830 times more massive than electrons.

neutron star–A star that has collapsed to the point where it is supported against gravity by neutron degeneracy.

New General Catalogue–The common name for "A New General Catalogue of Nebulae and Clusters of Stars," put together by J. L. E. Dreyer in 1888.

new moon–The phase of the moon when the side of the moon facing the earth is the side that is not illuminated by sunlight.

Newtonian telescope–A type of reflecting telescope where the beam reflected by the primary mirror is reflected by a flat secondary mirror so that the focus falls to the side of the telescope tube.

N galaxy–A type of galaxy with a blue nucleus that dominates the galaxy in that it provides most of the radiation. The emission lines from N galaxies are generally broader than emission lines from Seyferts. N galaxies are probably a type of elliptical galaxy. Historically, N galaxies and Seyfert galaxies were defined by different astronomers on the basis of different information, and the difference is not always clearcut.

node–A point of intersection between two great circles. Eclipses of the sun and moon occur when these bodies are simultaneously near the nodes of their paths in the sky.

nonthermal radiation–Radiation that cannot be characterized by a single number (the temperature). Normally, we derive this number from Planck's law, so that radiation that does not follow Planck's law is called nonthermal.

nova–A star that suddenly increases in brightness. Most novae are thought to result in binary systems when matter from the giant component falls on the white dwarf component.

nucleosynthesis–The formation of the elements.

nucleus–(a) Of an atom, the core of an atom, which has a positive charge, contains most of the mass, and takes up only a small part of the volume; (b) of a comet, the chunks of matter, taking up a volume no more than a few kilometers across, at the center of the head of a comet; (c) of a galaxy, the innermost regions of a spiral galaxy; it does not show spiral structure and is visible from the sky as a bulge in the otherwise flat disk of the galaxy.

O and B association–A group of O and B stars close together in space. The members of an O and B association were formed at roughly the same time.

objective–The principal lens or mirror of an optical system.

oblate–Having an equatorial diameter greater than the polar diameter.

occultation–The hiding of one astronomical body by another, such as the occultation of a star by the moon.

Olbers' paradox–The observation that the sky is dark at night contrasted to a simple argument that shows that the sky should be uniformly bright.

Oort comet cloud–The hundreds of millions of incipient comets surrounding the solar system in a sphere of perhaps 50,000 A.U. in radius.

opacity–The lack of complete transparency of a gas.

open cluster–A galactic cluster, an asymmetric type of star cluster.

open universe–The version of big bang cosmology in which the universe will expand forever. An open universe has infinite volume.

opposition–The passage of a planet through the point most directly opposite the sun on the other side of the earth.

optical depth–The number of factors of the transcendental number e (2.71828 . . .) that radiation is dimmed in passage through a gas.

optical double–A pair of stars that appear extremely close together in the sky even though they are at different distances from us and are not physically linked.

organic–Containing carbon in its molecular structure.

Ozma–One of two projects involving the search for radio signals from extra-terrestrial civilizations on nearby stars.

parallax–(a) When used by itself, the word "parallax" refers to trigonometric parallax, half the angle through which a star appears to be displaced when the earth moves from one side of the sun to the other, that is, through 2 A.U. The parallax of a star is inversely proportional to the distance to the star from the sun. (b) Some of the other ways of measuring distances, usually in those cases referred to with an adjective, as in spectroscopic parallax.

parsec–The distance from which one astronomical unit would subtend one second of arc. A parsec works out to be approximately 3.26 light years.

Pauli Exclusion Principle–See *exclusion principle*.

peculiar velocity–The velocity of a star with respect to the local standard of rest.

penumbra–(a) For an eclipse, the part of the shadow from which the sun or other radiating body is only partially occulted; (b) of a sunspot, the outer region of the sunspot, not as dark as the central umbra.

perfect cosmological principle–The assumption that on a large scale the universe is homogeneous and isotropic in space and unchanging in time.

periastron–The near point of the orbit of a body to the star around which it is orbiting.

perihelion–The near point to the sun of the orbit of a body, orbiting the sun.

phase–(a) Of a planet, the varying shape of the lighted part of a planet or moon as seen from some vantage point; (b) the relation of the maxima or minima of a set of waves. For example, when light waves are in phase (coherent), all the waves have their maxima simultaneously cross a plane that is perpendicular to the direction in which the waves are travelling. Colloquially speaking, waves that are in phase go up and down together, that is, they are in step.

photometry–The measurement of the amount of light from an astronomical object.

photomultiplier–An electronic device that through a series of internal stages multiplies the small current that is given off when light is incident on the device so that a relatively large current results.

photon–A packet of energy that can be thought of as a particle of light travelling at the speed of light. A photon of energy E is equivalent to an electromagnetic wave of wavelength hc/E, where h is Planck's constant and c is the speed of light in a vacuum.

photosphere–The region of a star from which most of its light is radiated.

plage–The part of a solar active region that appears bright when viewed in $H\alpha$.

Planck's constant–The constant of proportionality between the frequency of an electromagnetic wave and the energy of an equivalent photon. $E = h\nu$, or, since $\lambda\nu = c$, $E = hc/\lambda$.

Planck's law–The formula that predicts, for gas at a certain temperature, how much radiation there would be at any given wavelength.

planet–A celestial body of substantial size (more than about 1000 km across), basically non-radiating and of insufficient mass for nuclear reactions ever to begin, ordinarily in orbit around a star.

planetesimal–One of the small bodies into which the primeval solar nebula condensed and from which the planets formed.

plasma–An electrically neutral gas composed of approximately equal numbers of ions and electrons.

plate tectonics–See *continental drift*.

polar axis–The axis of an equatorial telescope mounting that is parallel to the earth's axis of rotation.

polarization–The arrangement of electromagnetic waves so that all the planes in which the waves are oscillating are parallel to each other.

Population I–The class of stars set up by Walter Baade to describe the younger stars typical of the spiral arms. These stars have relatively high abundances of metals.

Population II–The class of stars set up by Walter Baade to describe the older stars typical of the galactic halo. These stars have very low abundances of metals.

positron–The antiparticle corresponding to an electron. Its charge is +1.

precession–The slowly changing position of stars in the sky with respect to earth-based coordinates, resulting from the slowly varying orientation (the precession) of the earth's axis. The apparent precession of the stars is the result of the actual precession of the earth's axis.

primary cosmic rays–The cosmic rays arriving at the top of the earth's atmosphere.

prime focus–The location at which the main lens or mirror of a telescope focuses an image without being reflected or refocused by another mirror or other optical element.

primordial background radiation–Radiation detected in millimeter and submillimeter wavelength regions that is coming from all directions in space and interpreted to be the remnant of the big bang. Also known as 3° background radiation, background radiation, or the remnant of the primeval fireball.

prolate–Having the diameter along the axis of rotation longer than the equatorial diameter.

prominence–Solar gas protruding over the limb, visible to the naked eye only at eclipses but also observed outside of eclipses by its emission line spectrum. Prominences are at approximately the same temperature as the chromosphere.

proper motion–Motion across the sky with respect to a framework of galaxies or fixed stars, usually measured in seconds of arc per century.

proto-–A prefix from the Greek for "before." When used in conjunction with the name of a celestial body, means the state of the body just before it is considered to have formed.

proton–Massive elementary particle with positive charge 1, one of the fundamental constituents of an atom. Protons and neutrons are similar in mass, 1830 times more massive than electrons.

proton-proton chain–A set of nuclear reactions by which four hydrogen nuclei combine one after the other to form one helium nucleus with a resulting release of energy.

pulsar–A celestial object that gives off pulses of radio waves. A very few objects are known to pulse gamma rays, x-rays, or visible light.

q_0–The deceleration parameter, a cosmological parameter that describes atoms and at which the expansion of the universe is slowing up.

quantum mechanics–The branch of 20th century physics that describes atoms and radiation; the theory involves bundles of energy known as quanta.

quark–One of the subatomic particles from which many modern theoreticians believe such elementary particles as protons and neutrons are composed. The various kinds of quarks have positive or negative charges of 1/3 or 2/3.

quasar–One of the very large redshift objects that is almost stellar (point-like) in appearance. The word quasar derives from QSR, which stands for quasi-stellar radio source, although radio-quiet quasars are now also known.

quiet sun–The collection of solar phenomena that do not vary with the solar activity cycle.

radar–The acronym for **r**adio **d**etection **a**nd **r**anging, and active rather than passive radio technique in which radio signals are transmitted and their reflections received and studied.

radial velocity–The velocity of an object along a line (the radius) joining the object and the observer; the component of velocity toward or away from the observer.

radian–The unit of angular measure, defined as the ratio of a length of arc intercepted by two radii to the length of the radius. π radians = 180°.

radiant–The point in the sky from which all the meteors in a meteor shower appear to be coming.

radiation–Electromagnetic radiation.

radioactive–Having the property of spontaneously changing into another isotope or element.

radio galaxy–A galaxy that emits radio radiation many orders of magnitude stronger than radio radiation from normal galaxies.

radio telescope–An antenna or set of antennas, often together with a focusing reflecting dish, that is used to detect radio radiation from space.

radio waves–Electromagnetic radiation with wavelengths longer than about one millimeter.

ray–(a) A light ray, a wave of electromagnetic radiation; (b) on the surface of a moon or planet, a streak of material that is relatively light in shade, presumably representing material ejected when a crater was formed.

recombination–The addition of an electron to an ion, usually resulting in radiation (recombination lines) when the electron subsequently jumps down to lower energy states.

recurrent nova–An object that undergoes the nova phenomenon more than once.

red giant–A post-main-sequence stage of the lifetime of a star; the star becomes relatively bright and relatively cool.

reddening–The phenomenon by which the extinction of blue light by interstellar matter is greater than the extinction of red light so that the redder part of the continuous spectrum is relatively enhanced.

redshift–The shift of a spectrum, usually of spectral lines in particular, to longer wavelengths.

reflecting telescope–A type of telescope that uses a mirror or mirrors to form the primary image.

reflection nebula–Interstellar gas and dust that we see because it is reflecting light from a nearby star or stars.

refracting telescope–A type of telescope in which the image is formed by a lens or lenses.

refraction–The bending of electromagnetic radiation as it passes from one medium to another or between parts of a medium that has varying properties. The *index of refraction* of a substance is the ratio of the speed of light in a vacuum to that in the substance.

relativistic–Having a velocity that is such a large fraction of the speed of light that the special theory of relativity must be applied.

relativity–Either of the theories of relativity worked out by Albert Einstein. The special theory of relativity (1905) is a theory of relative motion. The general theory of relativity (1916) is a theory of gravitation.

resolution–The ability of an optical system to distinguish fine detail.

retrograde motion–The apparent motion of the planets when they appear to move backwards with respect to the stars from the direction that they move ordinarily.

retrograde rotation–The rotation of a moon or planet opposite to the dominant direction of orbiting and rotating of the sun and planets.

revolution–The orbiting of one body around another.

right ascension–Celestial longitude, measured eastward along the celestial equator in hours of time from the vernal equinox.

Roche's limit–The sphere for each mass inside of which blobs of gas cannot agglomerate by gravitational interaction without being torn apart by tidal forces.

rotation–Spin on an axis.

Russell-Vogt theorem–The theorem that the evolution of a star is completely determined by its mass and chemical composition.

RR Lyrae star–A short period "cluster" variable. All RR Lyrae stars have approximately equal absolute magnitude and so are used to determine the distances to globular clusters.

saros–The 18 year 11⅓ day (sometimes 18 year 10⅓ day or 18 year 12⅓ day, depending on the number of intervening leap years) period at which similar eclipses repeat.

Schmidt camera–A type of telescope that uses a spherical mirror and a thin lens to provide photographs of a wide field.

Schwarzschild radius–The radius that, according to Schwarzschild's solutions to Einstein's equations of the general theory of relativity, corresponds to the event horizon of a black hole.

secondary cosmic rays–High energy particles that are generated in the earth's atmosphere by primary cosmic rays.

seeing–The steadiness of the earth's atmosphere as it affects the resolution that can be obtained in astronomical observations. Good seeing corresponds to a steady atmosphere, and bad seeing corresponds to an unsteady atmosphere.

seismology–The study of waves propagating through a body and the resulting deduction of the internal properties of the body. The prefix "seismo-" comes from the Greek word for earthquake.

semimajor axis–Half the major axis, that is, for an ellipse, half the longest diameter.

Seyfert galaxy–A type of spiral galaxy that has a bright nucleus and whose spectrum shows emission lines. Historically, N galaxies and Seyfert galaxies were defined by different astronomers on the basis of different information, and the difference between them is not always clearcut.

shock wave–A front marked by an abrupt change in pressure caused by the motion of an object faster than the speed of sound in the medium through which the object is travelling.

sidereal–With respect to the stars.

sidereal time–The hour angle of the vernal equinox. Equivalent to the right ascension of objects on the observer's meridian.

singularity–A point in space where quantities become exactly zero or become infinitely large; a singularity is present in a black hole.

solar activity cycle–The 11- or 22-year cycle with which such solar activity as sunspots, flares, and prominences varies.

solar constant–The total amount of energy that would hit each square centimeter of the top of the earth's atmosphere at the earth's average distance from the sun.

solar time–A system of time-keeping with respect to the sun such that the sun is overhead of a given location at noon.

solar wind–An outflow of particles from the sun representing the expansion of the corona.

solid angle–A three-dimensional angle.

solstice–The point on the celestial sphere of northernmost or southernmost declination of the sun in the course of a year; colloquially, the time when the sun reaches that point.

space velocity–The velocity of a star with respect to the sun.

spectral type–One of the categories O, B, A, F, G, K, M, C, into which stars can be classified from study of their spectral lines, or extensions of this system. The sequence of spectral types corresponds to a sequence of temperature.

spectrograph–A device used to make and photograph a spectrum.

spectroheliograph–A device that makes an image of the sun at a particular wavelength by scanning a spectrograph slit across a solar image and simultaneously scanning a second slit that defines the wavelength being observed.

spectroscope–A device used to make and look at a spectrum.

spectroscopic binary–A type of binary star that is known to have more than one component because of the changing Doppler shifts of the spectral lines that are observed.

spectroscopic parallax–The distance to a star derived from comparison of its apparent magnitude with its absolute magnitude deduced from study of its position on an H–R diagram as determined by observation of its spectrum (spectral type and luminosity class).

spectrum–A display of electromagnetic radiation spread out by wavelength or frequency.

spicule–A small jet of gas at the edge of the quiet sun, approximately 1000 km in diameter and 10,000 km high with a lifetime of about 15 minutes.

spiral galaxy–A class of galaxy characterized by arms that appear as though they were unwinding like a pinwheel.

sporadic meteor–A meteor that is not associated with a shower.

star–A self-luminous ball of gas that shines or has shone because of nuclear reactions in its interior.

steady state theory–The cosmological theory based on the perfect cosmological principle.

Stefan-Boltzmann Law–The radiation law that states that the energy emitted by a black body varies with the fourth power of the temperature.

stones–A stony type of meteorite, including the chondrites.

stratosphere–One of the upper layers of the atmosphere of a planet, above the weather. The earth's stratosphere ranges from about 20 to 50 km in altitude.

steradian–The unit of solid angular measure, defined as the ratio of the surface area of that section of a sphere intercepted by a solid angle to the square of the radius. A full sphere subtends 4π steradians from its center.

strong force–The nuclear force, the strongest of the four fundamental forces of nature.

subtend–The angle that an object appears to take up in your field of view; actually, the angle between lines drawn from opposite sides of the object to your eye. For example, the full moon subtends ½°.

sunspot–A relatively dark area of the solar surface. Sunspots appear dark because they are relatively cool; they represent regions of extremely high magnetic field.

sunspot cycle–The 11-year cycle of variation of the number of sunspots visible on the sun.

supergiant–A post-main-sequence phase of evolution of stars of more than about 4 solar masses. Supergiants fall in the extreme upper right of the H–R diagram.

supernova–The explosion of a star with the resulting release of tremendous amounts of radiation and presumably high-energy particles as well.

synchrotron radiation–A type of non-thermal radiation emitted by electrons spiralling at relativistic velocities in a magnetic field.

synodic–Measured with respect to an alignment of astronomical bodies other than or in addition to the sun or the stars (usually the moon or a planet). For example, a synodic month depends on the positions of the sun, earth, and moon. The synodic period of Mars depends on the relative positions of the earth and Mars as they orbit the sun.

syzygy–An alignment of three celestial bodies. Sometimes applied more specifically to an alignment of the sun, earth, and moon.

T association–A grouping of several T Tauri stars, presumably formed out of the same cloud of interstellar dust and gas.

tektites–Small glassy objects found scattered around the southern part of the southern hemisphere of the earth.

terminator–The line between nighttime and daytime on a moon or planet; the edge of the part of a moon or planet that is lighted by the sun.

terrestrial planets–Mercury, Venus, Earth, and Mars.

thermal radiation–Radiation whose distribution of intensity over wavelength can be characterized by a single number (the temperature). Black-body radiation, which follows Planck's law, is an example of thermal radiation.

tidal force–A force caused by the differential effect of the gravity from one body being greater on the near side of a second body than on the far side.

transit–The passage of one celestial body in front of another celestial body. When a planet is *in transit*, we understand that it is passing in front of the sun. Also, *transit* is the moment when a celestial body crosses an observer's meridian, or the special type of telescope used to study such events.

trigonometric parallax–See *parallax*.

triple-alpha process–A chain of fusion processes by which three helium nuclei (alpha particles) combine to form a carbon nucleus.

Trojan asteroids–A group of asteroids that precede or follow Jupiter in its orbit by 60°.

tropical year–The length of time between two successive vernal equinoxes.

T Tauri star–A type of irregularly varying star, like T Tauri, whose spectrum shows broad and very intense emission lines. T Tauri stars have presumably not yet reached the main sequence and are thus very young.

UBV system–A system of photometry that uses three standard filters to define wavelength regions in the ultraviolet, blue, and green-yellow (visual) regions of the spectrum.

U.L.E.–Ultra-Low Expansion, Corning Glass' successor to Pyrex as a material out of which telescope mirrors are made. U.L.E. expands or contracts very little when the temperature changes, and thus a mirror made out of it holds its shape as the temperature varies without distorting the image.

ultraviolet–The region of the spectrum between about 100 and 4000 angstroms; also used in the restricted sense of ultraviolet radiation that reaches the ground, namely, that between about 3000 and 4000 angstroms.

umbra–(a) Of a sunspot, the dark central region; (b) of an eclipse shadow, the part from which the sun cannot be seen at all.

uvby–A system of photometry that uses four standard filters to define wavelength regions in the ultraviolet, violet, blue, and yellow regions of the spectrum.

Van Allen belts–Regions of high energy particles trapped by the magnetic field of the earth.

variable star–A star whose brightness changes over time.

velocity–The measure of the speed and direction of an object; the distance travelled by an object per unit of time.

vernal equinox–The intersection of the ecliptic and the celestial equator that the sun passes each year when moving from southern to northern declinations.

visual binary–A binary star that can be seen through a telescope to be double.

VLA–The Very Large Array, a set of radio telescopes of the National Radio Astronomy Observatory in New Mexico, used to make aperture synthesis measurements.

VLBI–Very-long-baseline interferometry, the technique of using simultaneous measurements made with radio telescopes at widely separated locations to obtain extremely high resolution.

wavelength–The distance over which a wave goes through a complete oscillation.

weak force–One of the four fundamental forces of nature, weaker than the strong force and the electromagnetic force. It is important only in the decay of certain elementary particles.

weight–The force determined by the gravitational pull on a mass.

white dwarf–The final stage of the evolution of a star of between 0.07 and 1.4 solar masses; a star supported by electron degeneracy. White dwarfs are found to the lower left of the main sequence of the H–R diagram.

Wien's displacement law–The expression of the inverse relationship of the temperature of a black body and the wavelength of the peak of its emission.

Wolf-Rayet star–A type of O star whose spectrum shows very broad emission lines.

W Virginis star–A Type II Cepheid, one of the fainter class of Cepheid variable stars characteristic of Cepheids in globular clusters.

x-rays–Electromagnetic radiation between about 1 and 100 angstroms.

year–The period of revolution of a planet around its central star; more particularly, the earth's period of revolution around the sun.

ZAMS–See *zero-age main sequence.*

Zeeman effect–The splitting of certain spectral lines in the presence of a magnetic field.

zenith–The point in the sky directly overhead an observer.

zero-age main sequence–The curve on an H–R diagram determined by the location of stars at the time they begin nuclear fusion.

zodiac–The band of constellations through which the sun, moon, and planets move in the course of the year.

zodiacal light–A glow in the nighttime sky near the ecliptic representing sunlight reflected by interplanetary dust.

SELECTED READINGS

Monthly Non-Technical Magazines on Astronomy

Sky and Telescope, 49 Bay State Road, Cambridge, MA 02238.
Astronomy, 411 East Mason Street, P.O. Box 92788, Milwaukee, WI 53202.
Mercury, Astronomical Society of the Pacific, 1290 24th Avenue, San Francisco, CA 94122.
The Griffith Observer, Griffith Observatory, 2800 East Observatory Road, Los Angeles, CA 90027.

Magazines and Annuals Carrying Articles on Astronomy

Science News, 1719 N Street, N.W., Washington, DC 20036. Published weekly.
Scientific American, 415 Madison Avenue, New York, NY 10017.
National Geographic, Washington, DC 20036.
Natural History, Membership Services, Box 6000, Des Moines, IA 50340.
Physics Today, American Institute of Physics, 335 East 45 Street, New York, NY 10017.
Science Year (Chicago, IL: Field Enterprises Educational Corp.). The World Book Science Annual.
Smithsonian, 900 Jefferson Drive, Washington, DC 20560.
Yearbook of Science and the Future (Chicago, IL: Encyclopaedia Britannica).

Observing Reference Books

Donald H. Menzel, *A Field Guide to the Stars and Planets* (Boston: Houghton Mifflin Co., 1975).
Charles A. Whitney, *Whitney's Star Finder,* Second Edition (New York: Alfred A. Knopf, Inc., 1977).
Arthur P. Norton, *Norton's Star Atlas and Reference Handbook,* regularly revised by the successors of the late Mr. Norton (Cambridge, Mass.: Sky Publishing Corp.).
The Observer's Handbook (yearly), Royal Astronomical Society of Canada, 252 College Street, Toronto M5T IR7, Canada.
Guy Ottewell, *Astronomical Calendar* (yearly), sponsored by the Department of Physics, Furman University, Greenville, SC 29613, in cooperation with the Astronomical League.
The American Ephemeris and Nautical Almanac (yearly), U.S. Government Printing Office, Washington, DC 20402.
Hans Vehrenberg, *Atlas of Deep Sky Splendors* (Cambridge, Mass.: Sky Publishing Corp., 3rd edition, 1978). Photographs, descriptions, and finding charts for hundreds of beautiful objects.

General Reference Books

C. W. Allen, *Astrophysical Quantities,* 3rd ed. (London: The Athlone Press of the University of London, 1973). Tables and lists of almost every conceivable kind.
Kenneth R. Lang, *Astrophysical Formulae* (New York: Springer-Verlag New York, Inc., 1980). Formulas of all kinds, plus many tables.
Jeanne Hopkins, *Glossary of Astronomy and Astrophysics,* Revised Edition (Chicago: University of Chicago Press, 1980). A technical glossary.

For Information about Amateur Societies

American Association of Variable Star Observers (AAVSO), 187 Concord Avenue, Cambridge, MA 02138.
Association of Lunar and Planetary Observers (ALPO), 8930 Raven Drive, Waco, TX 76710.

Careers in Astronomy

Education Officer, American Astronomical Society, Sharp Laboratory, University of Delaware, Newark, DE 19711. A free booklet, *Careers in Astronomy,* is available on request.

Space for Women, derived from *The Earth in the Cosmos: Space for Women*, a Symposium for Women on Careers in Astronomy, Astrophysics, and Earth and Planetary Sciences. Sponsored by Center for Astrophysics, Harvard University, Radcliffe College, and Smithsonian Institution. For free copies, write to: Center for Astrophysics, 60 Garden Street, Cambridge, Mass. 02138.

General Reading

Herbert Friedman, *The Amazing Universe* (Washington: National Geographic Society, 1975). National Geographic's survey of modern astronomy, written by a pioneer in space science and profusely illustrated in color.

Kenneth R. Lang and Owen Gingerich, eds., *A Source Book in Astronomy and Astrophysics, 1900–1975* (Cambridge, MA: Harvard University Press, 1979). Reprints of fundamental articles.

Donald H. Menzel, *Astronomy* (New York: Random House, 1970). A well-illustrated large-format popular description of the universe by the former director of the Harvard College Observatory.

Simon Mitton, ed., *The Cambridge Encyclopedia of Astronomy* (New York: Crown Publishers, 1977).

Jay M. Pasachoff and Marc L. Kutner, *University Astronomy* (Philadelphia: Saunders College/HRW, 1978). A deeper treatment of the topics covered in this book.

Harlow Shapley and H. E. Howarth, eds., *A Source Book in Astronomy* (New York: McGraw-Hill, 1929). Reprints of fundamental articles.

Harlow Shapley, ed., *Source Book in Astronomy 1900–1950* (Cambridge, Mass.: Harvard University Press, 1960). Reprints of fundamental articles.

Otto Struve and Velta Zebergs, *Astronomy of the Twentieth Century* (New York: Macmillan, 1962). A historical view.

Collections of Articles

Owen Gingerich, ed., *New Frontiers in Astronomy* (San Francisco: W. H. Freeman & Co., 1975). A collection of reprints from *Scientific American*.

John C. Brandt and Stephen P. Maran, *The New Astronomy and Space Science Reader* (San Francisco: W. H. Freeman & Co., 1977). A varied collection of articles.

The Solar System (San Francisco: W. H. Freeman & Co., 1975). A reprint of the special *Scientific American* issue of September 1975.

Owen Gingerich, ed., *Cosmology + 1* (San Francisco: W. H. Freeman & Co., 1977). Ten *Scientific American* reprints on cosmology, quasars, and black holes, plus one on the search for extraterrestrial intelligence.

Michael A. Seeds, ed., *Astronomy: Selected Readings* (Menlo Park, CA: Benjamin/Cummings, 1980). Articles from *Astronomy* magazine, including all their color illustrations.

Advanced Books

Eugene H. Avrett, ed., *Frontiers in Astrophysics* (Cambridge, Mass.: Harvard University Press, 1976). A series of chapters, each written by an expert, on contemporary research.

Martin Harwit, *Astrophysical Concepts* (New York: John Wiley & Sons, 1973). For mathematical upper-division courses.

Elske v. P. Smith and Kenneth C. Jacobs, *Introductory Astronomy and Astrophysics* (Philadelphia: Saunders College/HRW, 1973). Uses calculus.

SOME ADDITIONAL BOOKS

Observatories and Observing

H. T. Kirby-Smith, *U.S. Observatories: A Directory and Travel Guide* (New York: Van Nostrand Reinhold, 1976).

David O. Woodbury, *The Glass Giant of Palomar* (New York: Dodd Mead, 1970). The story of the construction of the 5-m telescope.

Helen Wright, Joan N. Warnow, and Charles Weiner, eds., *The Legacy of George Ellery Hale* (Cambridge, MA: MIT Press, 1972). A beautifully illustrated historical treatment.

W. M. Smart, *Text-Book on Spherical Astronomy* (Cambridge: Cambridge University Press, 1977). The old standard, revised from the first (1931) edition and the later (1944) edition.

D. McNally, *Positional Astronomy* (New York: John Wiley & Sons, 1975). A more modern treatment than Smart.

Stars

Lawrence H. Aller, *Atoms, Stars, and Nebulae*, Revised Edition (Cambridge, MA: Harvard University Press, 1971). One of the Harvard Books on Astronomy series of popular works.

Bart J. Bok and Priscilla F. Bok, *The Milky Way*, 5th edition (Cambridge, MA: Harvard University Press, 1981). A readable and well-illustrated survey from the Harvard Books on Astronomy series; includes good discussions of star clusters and H-R diagrams.

John A. Eddy, with Rein Ise, ed., *A New Sun: The Solar Results from Skylab* (NASA SP-402, 1979, GPO 033-000-00742-6).

Donald H. Menzel, *Our Sun*, Revised Edition (Cambridge, MA: Harvard University Press, 1959). A delightful general survey in the Harvard Books on Astronomy series.

Robert Jastrow, *Red Giants and White Dwarfs*, 2nd ed., (New York: W. W. Norton Co., 1979).

Frederick Golden, *Quasars, Pulsars, and Black Holes* (New York: Charles Scribner's Sons, 1976).

Walter Sullivan, *Black Holes* (New York: Anchor Press/Doubleday, 1979).

Henry L. Shipman, *Black Holes, Quasars, and the Universe*, Second Edition (Boston: Houghton Mifflin, 1980). A careful and thorough discussion of several topics of great current interest.

Cecilia Payne-Gaposchkin; *Stars and Clusters* (Cambridge, MA: Harvard University Press, 1979).

Solar System

Harold Masursky, G. W. Colton, and Farouk El-Baz, eds., *Apollo Over the Moon: A View from Orbit* (NASA SP 362, 1978, GPO 033-000-00708-6).

Edgar M. Cortright, ed., *Apollo Expeditions to the Moon* (NASA SP 350, 1975, GPO 033-000-00630-6).

Clark R. Chapman, *The Inner Planets: New Light on the Rocky Worlds of Mercury, Venus, Earth, the Moon, Mars, and the Asteroids* (New York: Charles Scribner's Sons, 1977).

Fred L. Whipple, *Orbiting the Sun: Planets and Satellites of the Solar System*, enlarged edition of *Earth, Moon, and Planets* (Cambridge, MA: Harvard University Press, 1981). In the Harvard series.

Brain O'Leary and J. Kelly Beatty, *The New Solar System* (Cambridge, MA: Sky Publishing Co., 1981). Each chapter written by a different expert on a level meant for general audiences. Fully illustrated.

David D. Morrison, *Voyager to Jupiter* (NASA SP 439, 1980). The story of the Voyager missions and what we learned, masterfully told and profusely illustrated.

Bevan M. French, *The Moon Book* (New York: Penguin, 1977). A clear, authoritative, thorough, and interesting to read report on the lunar program and its results.

Bruce Murray and Eric Burgess, *Flight to Mercury* (New York: Columbia University Press, 1977).

James A. Dunne and Eric Burgess, *The Voyage of Mariner 10* (NASA SP 424, 1978).

Viking Lander Imaging Team, *The Martian Landscape* (NASA SP 425, 1978, GPO 033-000-00716-7).

Donald Goldsmith and Tobias Owen, *The Search for Life in the Universe* (Menlo Park, Benjamin/Cummings, 1980).

Galaxies and the Outer Universe

Richard Berendzen, Richard Hart, and Daniel Seeley, *Man Discovers the Galaxies* (New York: Neale Watson Academic Publications, 1976). A historical review.

Charles A. Whitney, *The Discovery of Our Galaxy* (New York: Alfred A. Knopf, 1971). A historical discussion on a more popular level.

J. S. Hey, *The Evolution of Radio Astronomy* (New York: Neale Watson Academic Publications, 1973).

Gerrit L. Verschuur, *The Invisible Universe* (New York: Springer-Verlag, 1974). A nontechnical treatment of radio astronomy.

David A. Allen, *Infrared, the New Astronomy* (New York: John Wiley & Sons, 1975). Includes a personal narrative.

Halton C. Arp, *Atlas of Peculiar Galaxies* (Pasadena, CA: California Institute of Technology, 1966). Worth poring over.

Allan Sandage, *The Hubble Atlas of Galaxies* (Washington, DC: Carnegie Institution of Washington, 1961), Publication No. 618. Beautiful photographs of galaxies and thorough descriptions. Everyone should examine this carefully.

Harlow Shapley, *Galaxies*, Third Edition, revised by Paul W. Hodge (Cambridge, MA: Harvard University Press, 1972). A non-technical study of galaxies, written by the master. In the Harvard series.

Timothy Ferris, *The Red Limit* (New York: William Morrow & Co., 1977). Written for the general reader.

Nigel Calder, *Einstein's Universe* (New York: Viking, 1979). Mostly relativity.

William J. Kaufmann, *The Cosmic Frontiers of General Relativity* (Boston: Little, Brown, 1977).

William A. Fowler, *Nuclear Astrophysics* (Philadelphia: American Philosophical Society, 1967). A thin volume of popular lectures on the origin of the elements.

George Gamow, *One, Two, Three . . . Infinity* (New York: Bantam Books, 1971). A reprinting of a wonderful description of the structure of space that has introduced at an early age many a contemporary astronomer to his or her profession.

Joseph Silk, *The Big Bang* (San Francisco: W. H. Freeman & Co., 1979). A wide-ranging discussion on an elementary level.

Steven Weinberg, *The First Three Minutes* (New York: Basic Books, 1977). A readable discussion of the first minutes after the big bang, including a discussion of the background radiation.

ILLUSTRATION ACKNOWLEDGMENTS

COLOR PLATES

Plates 1, 64, and 67   Copyright by The Association of Universities for Research in Astronomy, Inc. The Kitt Peak National Observatory;

Plate 2   Courtesy of I. M. Kopylov, Special Astrophysical Observatory, U.S.S.R.;

Plates 3, 4, 10, 11, 12, and 25   Jay M. Pasachoff;

Plate 5   Deutsches Museum;

Plate 6   NASA and TRW Systems Group;

Plate 7   National Radio Astronomy Observatory;

Plate 8   Mario Grassi;

Plate 9   © 1972 Gary Ladd;

Plate 13   Dennis Di Cicco photograph; Williams College expedition;

Plates 14 and 15   Naval Research Laboratory/NASA;

Plates 16 and 17   High Altitude Observatory/NASA;

Plates 18 and 19   Lewis House, Ernest Hildner, William Wagner, and Constance Sawyer/ High Altitude Observatory, National Center for Atmospheric Research, National Science Foundation, and NASA;

Plates 20 and 21   Courtesy of Bruce E. Woodgate, Einar Tandberg-Hanssen, and colleagues at NASA's Marshall Space Flight Center;

Plate 22   Charles Eames;

Plate 23   Martin Grossmann;

Plates 24, 26, 27, 28   NASA;

Plate 29   International Planetary Patrol;

Plates 30-32, 37-44, 46, front and back covers   Jet Propulsion Laboratory/NASA; Saturn photos obtained with the assistance of Jurrie van der Woude and Steven Edberg;

Plates 33 and 34   NASA; color-corrected version courtesy of Friedrich O. Huck;

Plate 35   Experiment and data — Massachusetts Institute of Technology; maps — U.S. Geological Survey; NASA/Ames spacecraft; courtesy of Gordon H. Pettengill;

Plate 36   Gustav Lamprecht;

Plate 45   Lunar and Planetary Laboratory, University of Arizona;

Plates 47, 57, and 70   Copyright by The Association of Universities for Research in Astronomy, Inc. The Cerro Tololo Inter-American Observatory;

Plates 48, 49, 50, 51, 54, 55, 56, 59, 60, 61, 62, 63, 68, 69, and 71   © by the California Institute of Technology and the Carnegie Institution of Washington. Palomar Observatory photographs. 5-m telescope — 49, 50, 54, 56, 61, 63, 69, 71; 1.2-m Schmidt — 48, 51, 55, 59, 60, 62, 68;

Plate 52   Courtesy of Philip E. Angerhofer, Richard A. Perley, Bruce Balick, and Douglas Milne with the VLA of NRAO;

Plate 53   Courtesy of S. S. Murray and colleagues at the Harvard-Smithsonian Center for Astrophysics;

Plate 58   Copyright by The Association of Universities for Research in Astronomy, Inc. The Cerro Tololo Inter-American Observatory. Courtesy of Nolan R. Walborn and Thomas E. Ingerson;

Plate 65   Hans Vehrenberg;

Plate 66   U.S. Naval Observatory;

Plate 72   Westerbork data from Arnold H. Rots and William W. Shane; imaged at National Radio Astronomy Observatory; photograph by The Kitt Peak National Observatory;

Plate 73   Alan Stockton, Institute for Astronomy, University of Hawaii.

BLACK-AND-WHITE ILLUSTRATIONS AND MARGINAL NOTES

Frontispiece Mt. Wilson and Las Campanas Observatories, Carnegie Institution of Washington;

CHAPTER 1 — Opener Lick Observatory Photograph;
Figs. 1-2 and 1-3 Jay M. Pasachoff;
Fig. 1-4 Skyviews Survey, Inc.;
Figs. 1-5, 1-6, 1-7, and 1-9 NASA;
Fig. 1-12 Harvard College Observatory;
Figs. 1-13 and 1-14 Palomar Observatory, California Institute of Technology.

CHAPTER 2 — Opener The Kitt Peak National Observatory;
Fig. 2-3 From Leo Goldberg, "Ultraviolet Astronomy," *Scientific American,* June 1969, and *Frontiers in Astronomy,* reprinted courtesy of *Scientific American;*
Fig. 2-7 Chris Jones, Union College;
Fig. 2-8B Jay M. Pasachoff;
Fig. 2-11 Yerkes Observatory;
Fig. 2-12 American Institute of Physics, Niels Bohr Library;
Fig. 2-18 Copyright Royal Greenwich Observatory;
Figs. 2-20 and 2-24 Palomar Observatory, California Institute of Technology;
Fig. 2-21 From the historical collection of Mt. Wilson and Las Campanas Observatories, Carnegie Institution of Washington;
Fig. 2-23 Palomar Observatory — National Geographic Society Sky Survey. Reproduced by permission from the California Institute of Technology;
Fig. 2-25 Courtesy of I. M. Kopylov, Special Astrophysical Observatory, U.S.S.R.;
Fig. 2-26 The Kitt Peak National Observatory;
Fig. 2-27 The Kitt Peak National Observatory and The Cerro Tololo Inter-American Observatory/Arthur A. Hoag, Lowell Observatory;
Fig. 2-28 Institute for Astronomy, University of Hawaii, photo by Duncan Chesley;
Figs. 2-29 and 2-30 Jay M. Pasachoff;
Fig. 2-31 NASA/Marshall Space Flight Center;
Figs. 2-32, 2-39B and 2-40A Perkin-Elmer;
Fig. 2-35 Palomar Observatory, California Institute of Technology;
Fig. 2-36 (left) Jay M. Pasachoff, *(middle)* Harvard College Observatory, *(right)* Palomar Observatory, California Institute of Technology;
Figs. 2-37 and 2-46 Jay M. Pasachoff;
Fig. 2-40B Gordon P. Garmire, California Institute of Technology;
Fig. 2-41 Giovanni Fazio, Harvard-Smithsonian Center for Astrophysics;
Fig. 2-42 Wyoming Infrared Observatory;
Figs. 2-43 and 2-44 Courtesy of Bell Laboratories;
Fig. 2-49 Palomar Observatory, California Institute of Technology;
Fig. 2-50 A. G. Davis Philip, Dudley Observatory;
Fig. 2-51 Lick Observatory Photograph, courtesy of Sandra Faber;
Figs. 2-52, 2-53, and 2-54 Jay M. Pasachoff.

PART II — Opener The Cerro Tololo Inter-American Observatory.

CHAPTER 3 — Opener By permission of The Houghton Library, Harvard University;
Fig. 3-1 From *The Little Prince* by Antoine de Saint-Exupéry, copyright 1943, by Harcourt Brace Jovanovich, Inc. in America and William Heinemann Ltd. in England; copyright 1971, by Consuelo de Saint-Exupéry; reproduced by permission of the publishers;
Fig. 3-6 Palomar Observatory — National Geographic Society Sky Survey; reproduced by permission from the California Institute of Technology;
Fig. 3-7 Bundesarchiv Koblenz;
Fig. 3-16 American Institute of Physics, Niels Bohr Library, Margrethe Bohr Collection;
Fig. 3-18 Harvard College Observatory;
Fig. 3-19 The Kitt Peak National Observatory;
Fig. 3-20 Jay M. Pasachoff and Donald H. Menzel;
Fig. 3-21 *Harvard Circular 256,* 1925.

CHAPTER 4 — Opener © 1977 Anglo-Australian Observatory;
Fig. 4-2 By permission of The Houghton Library, Harvard University;

CHAPTER 4

Fig. 4-10 Richard E. Hill from the site of Case Western Reserve University's Burrell Schmidt telescope on Kitt Peak; 3 hr. exposure, f/5.6, Ektachrome 400 film;

Fig. 4-11 Yerkes Observatory;

Fig. 4-15 Emil Schulthess, Black Star;

Fig. 14-16A After a drawing by Michael M. Shurman, University of Wisconsin — Milwaukee;

Fig. 4-17 © 1970 United Features Syndicate, Inc.;

Fig. 4-20 Mount Vernon Ladies' Association of the Union;

Fig. 4-21 Drawing by Handelsman; © 1978 *The New Yorker Magazine, Inc.;*

Fig. 4-22 Chapin Library, Williams College;

Fig. 4-23 Courtesy of the Royal Library of Copenhagen;

Marginal Note, Section 4.8 Written by Larry Rhine and Mel Tolkin. © 1975 Tandem Productions, Inc. All rights reserved.

CHAPTER 5 — *Opener* The Kitt Peak National Observatory;

Fig. 5-1 Reproduced by special permission of *Playboy Magazine,* copyright © 1971 by *Playboy;*

Fig. 5-7 Dorrit Hoffleit — Yale University Observatory, courtesy of American Institute of Physics, Niels Bohr Library;

Fig. 5-8 American Institute of Physics, Niels Bohr Library, Margaret Russell Edmondson Collection;

Fig. 5-9 After Bok and Bok, *The Milky Way,* courtesy Harvard University Press;

Fig. 5-10 © 1974 Anglo-Australian Observatory;

Fig. 5-15 Peter van de Kamp, Sproul Observatory.

CHAPTER 6 — *Opener* Palomar Observatory, California Institute of Technology;

Fig. 6-1 Mt. Wilson and Las Campanas Observatories, Carnegie Institution of Washington;

Figs. 6-2, 6-20, and 6-22 Lick Observatory Photograph;

Fig. 6-6 Peter van de Kamp;

Fig. 6-7 From W. D. Heintz, *Double Stars* (1978), updated by W. D. Heintz;

Fig. 6-9 R. Hanbury Brown;

Fig. 6-10 Peter van de Kamp and Sarah Lee Lippincott, from *Vistas in Astronomy 19,* 231 (1975), courtesy Pergamon Press;

Fig. 6-11 Courtesy of Harold McAlister, Georgia State University;

Fig. 6-12 American Association of Variable Star Observers/Janet Mattei;

Figs. 6-13, 6-17, and 6-21 After Bok and Bok, *The Milky Way,* courtesy Harvard University Press;

Fig. 6-15 Harvard College Observatory;

Fig. 6-16 Shigetsugu Fujinami;

Fig. 6-18 Williams College — Hopkins Observatory;

Fig. 6-19 By permission of The Houghton Library, Harvard University;

Fig. 6-24 After Bok and Bok, *The Milky Way,* courtesy Harvard University Press, and a graph by Harold L. Johnson and Allan R. Sandage in the *Astrophysical Journal 124,* 379. Reprinted by permission of the University of Chicago Press, © 1956 by the American Astronomical Society;

Fig. 6-25 Photograph courtesy Harvard College Observatory. X-ray data courtesy George W. Clark, Massachusetts Institute of Technology;

Fig. 6-26 Astronomy Netherlands Satellite data/Space Research Laboratory, Utrecht;

Fig. 6-27 Jonathan E. Grindlay and colleagues, Harvard-Smithsonian Center for Astrophysics.

CHAPTER 7 — *Opener* William C. Livingston, The Kitt Peak National Observatory;

Figs. 7-2 and 7-27B Courtesy of Instituto de Astrofisica de Canarias, Tenerife, and Kiepenheuer-Institut für Sonnenphysik, Freiburg;

Figs. 7-3 and 7-7 Mt. Wilson and Las Campanas Observatories, Carnegie Institution of Washington;

Figs. 7-4, 7-5, 7-6, and 7-8 Jay M. Pasachoff;

Figs. 7-9 and 7-27A Sacramento Peak Observatory;

Fig. 7-10 The Aerospace Corporation/David K. Lynch;

Fig. 7-11 M. Kanno, Hida Observatory, University of Kyoto;

Fig. 7-12 High Altitude Observatory and Southwestern at Memphis, photo by John L. Streete and Leon B. Lacey;

CHAPTER 7

Figs. 7-14 and 7-38 Haleakala Observatory, Institute for Astronomy, University of Hawaii;

Fig. 7-15 Naval Research Laboratory *(interior)* and High Altitude Observatory *(exterior)*/NASA;

Fig. 7-16 American Science and Engineering, Inc./NASA;

Fig. 7-17 Naval Research Laboratory/NASA;

Figs. 7-20, 7-21A, 7-23, 7-24, 7-25, and 7-26 Jay M. Pasachoff;

Figs. 7-21B Charles F. Keller, Jr., William H. Regan, and Maxwell T. Sanford, II, Los Alamos Scientific Laboratory/University of California;

Fig. 7-22 Bryan Brewer, Earth View, Inc.;

Fig. 7-28B William C. Livingston, The Kitt Peak National Observatory;

Fig. 7-29 A plot of Zurich sunspot numbers; updated from data provided by M. Waldmeier, Swiss Federal Observatory;

Fig. 7-30A NASA Goddard Space Flight Center;

Fig. 7-30B Lewis House, Ernest Hildner, William Wagner, and Constance Sawyer, HAO/NCAR/NSF and NASA;

Fig. 7-31 Robert F. Howard and John M. Adkins, Mt. Wilson Observatory, Carnegie Institution of Washington;

Figs. 7-34 and 7-35 Courtesy of Harold Zirin, California Institute of Technology, Big Bear Solar Observatory photographs;

Fig. 7-36 Gustav Lamprecht;

Fig. 7-37 Haleakala Observatory, Institute for Astronomy, University of Hawaii;

Fig. 7-39 Harvard College Observatory/NASA;

Fig. 7-40 Updated from Stephen H. Schneider and Clifford Mass, *Science 190,* 741, copyright 1975 by The American Association for the Advancement of Science;

Fig. 7-41 Palomar Observatory, California Institute of Technology;

Fig. 7-42A Courtesy of the Director of the Mount Wilson and Las Campanas Observatories and of Otto Nathan;

Fig. 7-42B Lick Observatory Photograph;

Fig. 7-43 Courtesy of The Archives, California Institute of Technology;

Fig. 7-46 Einstein Archives/American Institute of Physics;

Fig. 7-47 Data provided by Harriet H. Malitson/NASA Goddard Space Flight Center, mostly from data included in *The Physical Output of the Sun,* O. R. White, editor (Boulder: University of Colorado Press, 1976).

PART III — Opener Palomar Observatory, California Institute of Technology;

CHAPTER 8 — Opener Theodore R. Gull/The Kitt Peak National Observatory;

Fig. 8-2 George H. Herbig, Lick Observatory;

Fig. 8-6 American Institute of Physics, Niels Bohr Library, Segre Collection;

Figs. 8-11 and 8-14 Raymond Davis, Jr., Brookhaven National Laboratory.

CHAPTER 9 — Opener Lick Observatory Photograph;

Fig. 9-1 After Richard L. Sears, *Journal of the Royal Astronomical Society of Canada,* No. 1, Feb. 1974; originally from B. E. Paczyński, *Acta Astron. 20,* 47, 1970;

Figs. 9-3, 9-5, and 9-12 Palomar Observatory, California Institute of Technology;

Figs. 9-4 and 9-10 Lick Observatory Photograph;

Fig. 9-7 From C. R. O'Dell, in IAU Symposium No. 34, "Planetary Nebulae," eds. D. E. Osterbrock and C. R. O'Dell. Reprinted by permission of the International Astronomical Union;

Fig. 9-9 Irving Lindenblad, U. S. Naval Observatory;

Fig. 9-11 © Ben Mayer, Los Angeles.

CHAPTER 10 — Opener Lick Observatory Photograph;

Fig. 10-1 After Richard L. Sears, *Journal of the Royal Astronomical Society of Canada,* No. 1, Feb. 1974; originally from B. E. Paczyński, *Acta Astron. 20,* 47, 1970;

Fig. 10-2 By permission of The Houghton Library, Harvard University;

Fig. 10-3 Palomar Observatory, California Institute of Technology;

Fig. 10-4 Paul Griboval, McDonald Observatory, University of Texas;

Fig. 10-5 Palomar Observatory, California Institute of Technology; courtesy of Sidney van den Bergh; photograph by R. Minkowski;

Fig. 10-6A Stephen S. Murray and colleagues, Harvard-Smithsonian Center for Astrophysics;

Fig. 10-6B A. R. Thompson and colleagues, VLA/NRAO, courtesy of Robert M. Hjellming;

CHAPTER 10
Fig. 10-7 William Miller and Museum of Northern Arizona;
Fig. 10-8 NASA;
Fig. 10-10 Jocelyn Bell Burnell;
Fig. 10-11 Joseph H. Taylor, Marc Damashek, and Peter Backus, University of Massachusetts — Amherst at the NRAO;
Fig. 10-13 Joseph H. Taylor;
Fig. 10-16 After Paul H. Serson;
Fig. 10-17 Lick Observatory Photograph;
Fig. 10-18A H. Y. Chiu, R. Lynds, and S. P. Maran, photographed at The Kitt Peak National Observatory;
Fig. 10-18B F. R. Harnden, Jr., and colleagues, Harvard-Smithsonian Center for Astrophysics;
Fig. 10-21A Joseph Weber, University of Maryland;
Fig. 10-21B Peter Kramer, Williams College;
Fig. 10-22 Palomar Observatory — National Geographic Society Sky Survey; reproduced by permission from the California Institute of Technology;
Figs. 10-23 and 10-24 Bruce Margon, University of Washington;
Fig. 10-25 Jonathan E. Grindlay and Frederick D. Seward, Harvard-Smithsonian Center for Astrophysics, and Ernest R. Seaquist and William Gilmore, University of Toronto;
Fig. 10-26 Ernest R. Seaquist and William Gilmore, University of Toronto, and John T. Stocke, University of Arizona.

CHAPTER 11 — *Opener and 11-8* Palomar Observatory, California Institute of Technology/ Jerome Kristian;
Fig. 11-5 Drawing by Charles Addams; reprinted with permission of *The New Yorker Magazine, Inc.* © 1974;
Fig. 11-7 Chapin Library, Williams College;
Fig. 11-9 Lois Cohen — Griffith Observatory;
Fig. 11-10 Riccardo Giacconi and colleagues, Harvard-Smithsonian Center for Astrophysics.

PART IV

CHAPTER 12
Fig. 12-2 Lick Observatory Photograph;
Fig. 12-4 NASA;
Fig. 12-8 George Lovi, Vanderbilt Planetarium;
Figs. 12-9, 12-15, 12-21A&B, and 12-30 Chapin Library, Williams College;
Fig. 12-10 Burndy Library, photographed by Owen Gingerich;
Fig. 12-12 Courtesy of Owen Gingerich;
Fig. 12-13 Charles Eames;
Fig. 12-14 Charles Eames, courtesy of Owen Gingerich;
Fig. 12-17 University of Michigan Library, Dept. of Rare Books and Special Collections, translation by Stillman Drake, reprinted courtesy of *Scientific American;*
Fig. 12-18 By permission of The Houghton Library, Harvard University;
Fig. 12-22 Jay M. Pasachoff;
Fig. 12-30 Owen Gingerich;
Fig. 12-31 B.C. by permission of John Hart and Field Enterprises, Inc.;
Fig. 12-34 United Press International photograph;
Fig. 12-37 Harold E. Edgerton, MIT.

CHAPTER 13 — *Opener* NASA;
Figs. 13-2, 13-3, and 13-11 Lick Observatory Photographs;
Figs. 13-5, 13-6A, 13-10, 13-13, 13-14A, 13-15, and 13-16 NASA;
Fig. 13-6B Drawing by Alan Dunn; reprinted with permission of *The New Yorker Magazine, Inc.* © 1971;
Figs. 13-7 and 13-8 Lunar Receiving Laboratory, Johnson Space Center, NASA;
Fig. 13-9 General Electric Research and Development Center;
Fig. 13-12 Drawing by Donald E. Davis under the guidance of Don E. Wilhelms of the U. S. Geological Survey;
Fig. 13-14B Harold E. Edgerton, MIT;

CHAPTER 13
Fig. 13-18 Judith W. Frondel, Harvard University;
Fig. 13-19 Data from Gerald J. Wasserburg and D. A. Papanastassiou, courtesy of Gerald J. Wasserburg, California Institute of Technology.

CHAPTER 14 — *Opener* NASA;
Fig. 14-1 New Mexico State University Observatory;
Figs. 14-5 to 14-12 NASA.

CHAPTER 15 — *Opener* NASA/Ames;
Fig. 15-1 Palomar Observatory, California Institute of Technology;
Fig. 15-2 Lick Observatory Photograph;
Figs. 15-8 and 15-9 Courtesy of D. B. Campbell, Arecibo Observatory;
Figs. 15-11 and 15-15 to 15-22 NASA/Ames; 15-22 courtesy of David J. Diner;
Fig. 15-12 Tass from Sovfoto;
Figs. 15-13 and 15-14 Science Service.

CHAPTER 16 — *Opener* NASA;
Fig. 16-1 U. S. Geological Survey, EROS Data Center;
Figs. 16-2, 16-3, 16-15A&B and 16-16 NASA;
Figs. 16-5 to 16-8, 16-11 and 16-18 Jay M. Pasachoff;
Fig. 16-9 From W. J. Webster, T. C. Chang, L. T. Darby, and H. M. Finkelstein, *Icarus 24,* 144, 1975, copyright © 1975 Academic Press, Inc.;
Fig. 16-10 From Raymond Siever, "The Earth," *Scientific American,* September 1975, and *The Solar System* (San Francisco: W. H. Freeman and Co., 1975), reprinted courtesy of *Scientific American;*
Fig. 16-12 Courtesy of Wilbur Rinehart, National Geophysical and Solar-Terrestrial Data Section, Environmental Data Service, National Oceanic and Atmospheric Administration;
Fig. 16-13 © 1973 National Geographic Society;
Fig. 16-22 National Environmental Satellite Service, National Oceanic and Atmospheric Administration;
Fig. 16-23 J. G. Zeikus, University of Wisconsin — Madison;
Quotation, Section 16.1 From Paul A. Weiss (abridged from *Rockefeller Institute Review 2,* pp. 8-14) with permission of Dr. Weiss and the Rockefeller University Press. Reprinted in *A Random Walk in Science,* R. W. Weber, ed. (London: The Institute of Physics, 1973).

CHAPTER 17 — *Opener* NASA/JPL;
Figs. 17-1 and 17-2 Lick Observatory Photographs;
Figs. 17-4 to 17-9, 17-11 to 17-19 NASA/JPL;
Fig. 17-10 Jay M. Pasachoff.

CHAPTER 18 — *Opener* NASA/JPL;
Fig. 18-2 Lunar and Planetary Laboratory, University of Arizona;
Fig. 18-5 *Sky and Telescope;*
Fig. 18-6 Yerkes Observatory photograph;
Fig. 18-7 Charles T. Kowal, California Institute of Technology; Palomar Observatory — National Geographic Society Sky Survey. Reproduced by permission from the California Institute of Technology;
Fig. 18-8A Jay M. Pasachoff;
Fig. 18-11 NASA;
Figs. 18-9, 18-12 to 18-15, 18-17, 18-30 to 18-34 NASA/JPL;
Fig. 18-16 NASA/Lunar and Planetary Laboratory, University of Arizona.

CHAPTER 19 — *Opener* NASA/JPL;
Fig. 19-2 New Mexico State University Observatory;
Figs. 19-3 and 19-23 Lowell Observatory;
Figs. 19-5 and 19-21 Harold J. Reitsema, Lunar and Planetary Laboratory, University of Arizona;
Fig. 19-6 NASA/Ames;
Figs. 19-7 to 19-13 JPL/NASA; obtained with the assistance of Jurrie van der Woude, Steven Edberg, and Corona Publishing Co., Inc.;
Fig. 19-14 Lunar and Planetary Laboratory, University of Arizona;

CHAPTER 19

Fig. 19-16 Observations by J. L. Elliot, E. Dunham, and D. Mink;

Figs. 19-17 and 19-18 Data from J. L. Elliot, E. Dunham, L. H. Wasserman, R. L. Millis, and J. Churms;

Fig. 19-19 William Liller, with the 4-m telescope at The Cerro Tololo Inter-American Observatory;

Fig. 19-20 Master and Fellows of St. John's College of Cambridge;

Fig. 19-22 Lick Observatory Photograph;

Fig. 19-24 Palomar Observatory, California Institute of Technology;

Fig. 19-25 U. S. Naval Observatory/James W. Christy, U. S. Navy Photo.

CHAPTER 20 — *Opener* Institut d'Astrophysique de Liège, Belgium with the Schmidt camera at the Observatoire de Haute Provence, France;

Fig. 20-1 William Liller, Harvard College Observatory;

Figs. 20-2 and 20-9 Mt. Wilson and Las Campanas Observatories, Carnegie Institution of Washington;

Figs. 20-5 and 20-12 Palomar Observatory, California Institute of Technology;

Fig. 20-3 European Southern Observatory/Jean Pierre Swings;

Figs. 20-4, 20-6 and 20-11 NASA;

Fig. 20-7 Updated with rings on Jupiter and Uranus from a drawing by Leon Tadrick in *The Sciences 15*, No. 8, Nov. 1975. © 1975 The New York Academy of Sciences;

Fig. 20-8 National Portrait Gallery, London, painted by R. Phillips prior to 1721;

Fig. 20-10 *Sky and Telescope* by Roger Sinnott from orbital elements by Joseph Brady and Edna Carpenter, Lawrence Radiation Laboratory, University of California;

Fig. 20-13 Smithsonian Astrophysical Observatory;

Fig. 20-14 John Pazmino, Amateur Astronomers Association, Inc., New York;

Fig. 20-15 Peter Bloomer, *Horizons West*, by permission of Meteor Crater Enterprises, Inc.;

Fig. 20-16A Arthur A. Griffin;

Fig. 20-16B Wide World Photos;

Fig. 20-17 Palomar Observatory, California Institute of Technology/John Huchra;

Fig. 20-18A&B Harvard-Smithsonian Center for Astrophysics;

Fig. 20-21 Charles T. Kowal, California Institute of Technology; Palomar Observatory — National Geographic Society Sky Survey. Reproduced by permission from the California Institute of Technology.

CHAPTER 21 — *Opener* © Lucasfilm Ltd. (LFL) 1980. All rights reserved. From the motion picture: *The Empire Strikes Back*, courtesy of Lucasfilm, Ltd.;

Fig. 21-1 © 1974 United Features Syndicate, Inc.;

Fig. 21-2 Sproul Observatory;

Fig. 21-3 Peter van de Kamp, Sproul Observatory;

Fig. 21-5 J. G. Zeikus, University of Wisconsin — Madison;

Fig. 21-6A From *The Day the Earth Stood Still*. (© 1951 Twentieth Century-Fox Film Corporation. All rights reserved.);

Figs. 21-6B, 21-13, and 21-16 © Lucasfilm, Ltd. (LFL) 1980. All rights reserved. From the motion picture: *The Empire Strikes Back*, courtesy of Lucasfilm, Ltd.;

Fig. 21-7A NASA;

Fig. 21-7B Jay M. Pasachoff;

Figs. 21-8 and 21-9 Woodruff T. Sullivan, III;

Figs. 21-10 and 21-11 Cornell University photographs;

Fig. 21-12 NASA/Ames Research Center;

Figs. 21-14 and 21-15 Robert M. Sheaffer.

PART V — *Opener* Reproduced by courtesy of The Trustees, The National Gallery, London;

CHAPTER 22 — *Opener* Palomar Observatory, California Institute of Technology;

Fig. 22-2 Harvard College Observatory;

Figs. 22-3, 22-6, and 22-7 Palomar Observatory, California Institute of Technology;

Figs. 22-4, 22-5, 22-8, and 22-9 Lick Observatory Photographs;

Fig. 22-10 Sgr. A West marked — William Liller, Harvard-Smithsonian Center for Astrophysics, with the 4-m telescope of the Cerro Tololo Inter-American Observatory;

Fig. 22-11 Dennis Downes, Max Planck Institut für Radioastronomie, Bonn, G.F.R.;

CHAPTER 22

Fig. 22-12A Eric E. Becklin, Institute for Astronomy, University of Hawaii, and Gerry Neugebauer, Palomar Observatory, California Institute of Technology;

Fig. 22-12B Walter Baade, Palomar Observatory, California Institute of Technology, with the 1.2-m Schmidt telescope;

Fig. 22-13 Lund Observatory, Sweden;

Fig. 22-14 Riccardo Giacconi and Herbert Gursky/Harvard-Smithsonian Center for Astrophysics, with permission of D. Reidel Publishing Co.;

Fig. 22-15 Riccardo Giacconi and Herbert Gursky/Harvard-Smithsonian Center for Astrophysics;

Fig. 22-16 Leon P. Van Speybroeck and colleagues, Harvard-Smithsonian Center for Astrophysics;

Fig. 22-17 Frederick D. Seward, Harvard-Smithsonian Center for Astrophysics; ultraviolet photo (3300-3900 A) by William Liller, with the CTIO Curtis Schmidt;

Fig. 22-18 H. A. Mayer-Hasselwander, K. Bennett, G. F. Bignami, R. Buccheri, N. D'Amico, W. Hermsen, G. Kanbach, F. Lebrun, G. G. Lichti, J. L. Masnou, J. A. Paul, K. Pinkau, L. Scarsi, B. N. Swanenburg, and R. D. Willis; COS-B Observation of the Milky Way in High-Energy Gamma Rays, Ninth Texas Symposium on Relativistic Astrophysics, Eds. J. Ehlers, J. J. Perry, M. Walker, *Annals of the New York Academy of Sciences 336*, 211, 1980; courtesy of K. Pinkau;

Fig. 22-19 Data from W. Becker, photo from Palomar Observatory, California Institute of Technology;

Fig. 22-20 Agris Kalnajs, Mount Stromlo Observatory;

Figs. 22-21 and 22-22 Courtesy of Humberto Gerola and Philip E. Seiden, IBM Watson Research Center.

CHAPTER 23 — *Opener* Jay M. Pasachoff;

Fig. 23-4 Palomar Observatory, California Institute of Technology;

Fig. 23-9 Harvard University/E. M. Purcell;

Fig. 23-12A Gart Westerhout, U. S. Naval Observatory;

Fig. 23-12B Gerrit Verschuur;

Fig. 23-13 Leo Blitz, University of California at Berkeley.

CHAPTER 24 — *Opener* Courtesy of Robert D. Gehrz, J. A. Hackwell, and Gary Grasdalen, Wyoming Infrared Observatory;

Fig. 24-2 Palomar Observatory, California Institute of Technology;

Fig. 24-3 Marc L. Kutner;

Figs. 24-4 and 24-13A National Radio Astronomy Observatory;

Fig. 24-6 Graphic Films, Inc., Hollywood;

Fig. 24-7 Joseph H. Taylor, Five College Radio Observatory;

Fig. 24-8 Richard S. Cohen, NASA Goddard Institute for Space Studies;

Fig. 24-10 Bart J. Bok, Steward Observatory;

Fig. 24-11 Data from Marc L. Kutner, Rensselaer Polytechnic Institute; Lick Observatory Photograph;

Fig. 24-13B National Radio Astronomy Observatory and E-Systems, Inc.;

Figs. 24-14 and 24-15 Jay M. Pasachoff;

Marginal note, Section 24.1 Reproduced, with permission, from *Annual Review of Astronomy and Astrophysics*, Vol. 12, copyright 1974 by Annual Reviews Inc.

PART VI — *Opener* Lick Observatory Photograph;

CHAPTER 25 — *Opener* Palomar Observatory, California Institute of Technology;

Fig. 25-1 Harvard College Observatory;

Figs. 25-2, 25-3, 25-7, 25-10, 25-14A, 25-16A, and 25-18 Palomar Observatory, California Institute of Technology;

Figs. 25-4 and 25-5 Mt. Wilson and Las Campanas Observatories, Carnegie Institution of Washington;

Fig. 25-6 Courtesy of the U. K. Schmidt Telescope Unit, Royal Observatory, Edinburgh;

Figs. 25-8A and 25-12 I. M. Kopylov, Special Astrophysical Observatory, U.S.S.R.;

Fig. 25-8B From the 1966 *Atlas of Peculiar Galaxies* by Halton C. Arp, Mount Wilson and Las Campanas Observatories, Carnegie Institution of Washington;

CHAPTER 25

Fig. 25-11 NASA Langley Research Center/Frank Hohl;

Figs. 25-13 and 25-14B Alar Toomre and Juri Toomre;

Fig. 25-15 Hyron Spinrad, University of California at Berkeley;

Fig. 25-16B Diagram by John Tonry, Marc Davis, and John Huchra, Harvard-Smithsonian Center for Astrophysics;

Fig. 25-17A National Academy of Science;

Fig. 25-17B From E. Hubble and M. L. Humason, *Astrophysical Journal 74*, 77, 1931, courtesy of University of Chicago Press;

Fig. 25-18 Palomar Observatory, California Institute of Technology;

Fig. 25-21 Vera C. Rubin, Department of Terrestrial Magnetism, Carnegie Institution of Washington;

Fig. 25-22 The Kitt Peak National Observatory/Hyron Spinrad, University of California at Berkeley;

Fig. 25-23 Middle photograph, Lick Observatory Photograph; top left photograph from Palomar Observatory, California Institute of Technology; radio map based on Phillip J. Hargrave and Martin Ryle, *Monthly Notices of the Royal Astronomical Society 166*, 305, copyright 1974 by the Royal Astronomical Society;

Fig. 25-24 Radio map based on B. F. C. Cooper, R. M. Price, and D. J. Cole, *Australian Journal of Physics 18*, 589, copyright 1965; photograph from Palomar Observatory, California Institute of Technology;

Fig. 25-25 Leftmost photograph from Palomar Observatory, California Institute of Technology;

Fig. 25-30A The Kitt Peak National Observatory;

Fig. 25-30B Halton C. Arp, Palomar Observatory, California Institute of Technology;

Fig. 25-33 National Radio Astronomy Observatory and E-Systems, Inc.;

Fig. 25-34 National Radio Astronomy Observatory, courtesy of John Lancaster and Robert M. Hjellming;

Fig. 25-35 Westerbork Synthesis Telescope, A. G. Willis, R. G. Strom, and A. S. Wilson, *Nature 250*, 625, copyright 1974 by Macmillan Journals Limited, London;

Fig. 25-36 Courtesy of Richard G. Strom, George K. Miley, Jan Oort, and *Scientific American;*

Fig. 25-37 Westerbork Synthesis Telescope, George K. Miley, G. C. Perola, P. C. van der Kruit, and H. van der Laan, *Nature 237*, 269, copyright 1972 by Macmillan Journals Limited, London;

Fig. 25-38 Courtesy of F. N. Owen, J. O. Burns, and L. Rudnick, National Radio Astronomy Observatory;

Fig. 25-39 Lick Observatory Photograph.

CHAPTER 26 — *Opener and 26-2* Maarten Schmidt/Palomar Observatory, California Institute of Technology;

Figs. 26-1 and 26-3 Palomar Observatory, California Institute of Technology;

Fig. 26-8 Bruce Margon and E. A. Harlan at the Lick Observatory;

Fig. 26-9 Halton C. Arp/Palomar Observatory, California Institute of Technology;

Fig. 26-10 The Cerro Tololo Inter-American Observatory;

Figs. 26-11 and 26-17 Alan Stockton, Institute for Astronomy, University of Hawaii;

Figs. 26-12 and 26-13 Harvey Tananbaum and colleagues, Harvard-Smithsonian Center for Astrophysics;

Fig. 26-15 B. F. Burke, P. E. Greenfield, D. H. Roberts, with the VLA of the National Radio Astronomy Observatory;

Fig. 26-16 Jerome Kristian, James A. Westphal, and Peter Young/Palomar Observatory, California Institute of Technology;

Fig. 26-18 Palomar Observatory, California Institute of Technology/montage by W. W. Morgan, Yerkes Observatory;

Fig. 26-19 R. M. West, A. C. Danks, and G. Alcaino, European Southern Observatory, *Astronomy and Astrophysics 62*, L113, 1978.

CHAPTER 27 — *Opener* Lotte Jacobi;

Fig. 27-1B From the collection of Mr. and Mrs. Paul Mellon;

Fig. 27-2 Jay M. Pasachoff;

CHAPTER 27

Fig. 27-3 Edward J. Groth, P. James E. Peebles, and Michael Soneira, based on a survey by C. Donald Shane and Carl A. Wirtanan of the Lick Observatory, *Scientific American* 237, No. 5, p. 84, November, 1977; courtesy of *Scientific American;*

Fig. 27-6 Jerome Kristian, Allan Sandage, and James Gunn;

Fig. 27-7 Wide World Photos;

Fig. 27-9 Bell Laboratories;

Fig. 27-12 Based on Patrick Thaddeus, *Annual Review of Astronomy and Astrophysics 10,* 305, copyright 1972 by Annual Reviews Inc.;

Fig. 27-13 D. Woody and P. L. Richards;

Fig. 27-14 Courtesy of George F. Smoot;

Fig. 27-15 NASA;

Figs. 27-16 and 27-17 Robert V. Wagoner, Stanford University;

Fig. 27-18 John B. Rogerson, Jr., and Donald H. York, Princeton University Observatory. Reprinted from the *Astrophysical Journal 186,* L95, with permission of the University of Chicago Press, © 1973 by the American Astronomical Society;

Fig. 27-19 Amos Yahil, Allan R. Sandage, and Gustav Tammann;

Fig. 27-20 X-ray image from Riccardo Giacconi and colleagues at the Harvard-Smithsonian Center for Astrophysics; optical image from Wallace Sargent and Charles Kowal, Palomar Observatory, California Institute of Technology.

CHAPTER 28 — *Opener* JPL/NASA;

Fig. 28-1 The Kitt Peak National Observatory;

Figs. 28-2 and 28-3 NASA;

Figs. 28-4, 28-5, 28-6, 28-8, and 28-9 Jay M. Pasachoff;

Fig. 28-10 JPL/NASA;

Fig. 28-11 Bruce E. Woodgate and Einar Tandberg-Hanssen, NASA Marshall Space Flight Center;

Marginal Note, Section 28.1 From *The Lives of a Cell* by Lewis Thomas; copyright © 1973 by The Massachusetts Medical Society; reprinted by permission of Viking Penguin Inc.

APPENDIX 11

Box A11.1 From Homer, "The Odyssey," Richmond Lattimore, trans., copyright© 1965, 1967 by Richard Lattimore, reprinted by permission of Harper & Row.

INDEX

References to illustrations, either photographs or drawings, are followed by i. References to marginal notes are followed by m. References to tables are followed by t. References to Color Plates are prefaced by CP. References to Appendices are prefaced by A.

Significant initial numbers are alphabetized under their spellings; for example, 21 cm is alphabetized as *twenty-one*. Less important initial numbers and subscripts are ignored in alphabetizing; for example, 3C 273 is at the beginning of the C's, M1 appears at the beginning of the M's, and CO_2 appears after CO. Greek letters are alphabetized under their English spellings.